6th International Symposium on Applied Chemistry (ISAC 2020)

IOP Conference Series: Materials Science and Engineering
Volume 1011

Tangerang, Indonesia
18 – 20 November 2020

Editors:

Abdi Wira Septama
Ahmad Randy
Lucia Dwi Antika
Aprilia Nur Tasfiyati

ISBN: 978-1-7138-4546-1
ISSN: 1757-8981

Printed from e-media with permission by:

Curran Associates, Inc.
57 Morehouse Lane
Red Hook, NY 12571

Some format issues inherent in the e-media version may also appear in this print version.

Institute of Physics
Dirac House, Temple Back
Bristol BS1 6BE UK

Phone: 44 1 17 929 7481
Fax: 44 1 17 920 0979

techtracking@iop.org

Additional copies of this publication are available from:

Curran Associates, Inc.
57 Morehouse Lane
Red Hook, NY 12571 USA
Phone: 845-758-0400
Fax: 845-758-2633
Email: curran@proceedings.com
Web: www.proceedings.com

TABLE OF CONTENTS

IOP Conf. Series: Materials Science and Engineering 1011 (2021) 011001 doi:10.1088/1757-899X/1011/1/011001

PREFACE

This present volume contains the *Proceeding of the 6th International Symposium on Applied Chemistry (ISAC) 2020*. In this year, ISAC 2020 is co-organized by Research Center for Chemistry, Research center of Biotechnology, and Research Unit for Natural Product Technology Indonesian Institute of Science (LIPI) which virtually delivered through online event on November 18-20, 2020.

By carrying the theme **"Natural Medicine and Nutraceutical"**, the 6th ISAC 2020 brought together more than 70 presenters and participants from several background such as university lecturer, researcher, and student. With this theme, the symposium brought the latest trends of natural product research in various applications ranging from traditional medicine and pharmaceutical chemistry, pharmaceutical technology and drug delivery system, food technology and nutraceutical, as well as analytical chemistry and chemical engineering. This meeting provides a forum for exchange of information and share of experiences on all aspects of natural medicine and nutraceutical related researches as well as aims to initiate scientific cooperation and international collaboration among the research institutions.

We really appreciated all the attention and support given for the success of the 6th ISAC 2020 and would like to express our deep gratitude to everyone who has taken part in this event:

- The scientific reviewers for delivering the manuscript into very good quality,
- The distinguished keynote speakers: **Prof. Dr. Muhammad Iqbal Choudhary** (International Center for Chemical and Biological Sciences, University of Karachi, Pakistan), **Prof. Dr. rer. nat. Gunawan Indrayanto** (Faculty of Pharmacy, Universitas Surabaya, Indonesia), **Dr. Ambara Rachmat Pradipta** (Department of Chemical Science and Engineering, Tokyo Institute of technology, Japan), **Dr. Nilesh Nirmal** (Institute of Nutrition, Mahidol University, Thailand), **Assoc. Prof. Nurul Huda** (Faculty of Food Science and Nutrition, University Malaysia Sabah, Malaysia), and **Prof. Dr. Nina Artanti** (Research Center for Chemistry, Indonesian Institute of Sciences, Indonesia),
- All the participants for their contributions on this event, and
- The committee members for the tremendous efforts, commitment, and hard work in accomplishing the symposium and proceeding.

Their active assistance and participations, as already demonstrated, have greatly contributed to the success of this conference.

We wish all attendees of ISAC 2020 an enjoyable scientific gathering and a technically rewarding experience. We look forward to seeing all of you at the conference.

Sincerely,
Abdi Wira Septama, Ph.D
The 6th ISAC 2020 Chairman

IOP Conf. Series: Materials Science and Engineering 1011 (2021) 011001 doi:10.1088/1757-899X/1011/1/011001

COMMITTEES

Advisory Board:
Dr. Laksana Tri Handoko (Head of Indonesian Institute of Sciences)
Dr. Eng. Agus Haryono (Deputy for Engineering Sciences, Indonesian Institute of Sciences)

Organizing Committee:
Chairman: Abdi Wira Septama, Ph.D.
Vice Chairman: Khoirun Nisa, Ph.D.

Technical Program Committee:
Chairman: Dr. Marissa Angelina, M.Farm, Apt
Vice Chairman: Lucia Dwi Antika, M.S.
Secretariat: Ahmad Iskandar Setyawan, M.Sc.
Ahmad Randy, Ph.D.
Aprilia Nur Tasfiyati, M.Si.
Ayu Septi Anggraeni, M.Sc.
Dr. Hafiizh Prasetia
Hikmat, M.Si.
Iskandar Azmi Harahap, M.S.
Muhammad Safaat, M.T.
Setyani Budiari, M.Si.

Scientific Board:
Prof. Dr. Nina Artanti (Indonesian Institute of Sciences, Indonesia)
Prof. Dr. Silvester Tursilo Adi (Indonesian Institute of Sciences, Indonesia)
Mohd. Fadhlizil Fasihi Mohd Aluwi, Ph.D (Universiti Malaysia Pahang, Malaysia)
Md Ifthekar Ahmad, PhD (Shri Gopichan College of Pharmacy, India)
Aida Maryam Basri, PhD (Universiti Teknologi Brunei, Brunei Darussalam)
Nik Nurul Najihah, PhD (Universiti Sultan Zainal Abidin, Malaysia)
Nurul Hikmah Harun, Ph.D. (Universiti Sultan Zainal Abidin, Malaysia)
Ahmad Randy, PhD (Indonesian Institute of Sciences, Indonesia)
Dr. Hafiizh Prasetia (Indonesian Institute of Sciences, Indonesia)
Abdi Wira Septama, PhD (Indonesian Institute of Sciences, Indonesia)
Dr. Osi Arutanti (Indonesian Institute of Sciences, Indonesia)
Ajeng Arum Sari, PhD (Indonesian Institute of Sciences, Indonesia)
Anastasia Fitria Devi, PhD (Indonesian Institute of Sciences, Indonesia)
Dr. Dewi Sondari (Indonesian Institute of Sciences, Indonesia)
Dr. Galuh Widiyarti (Indonesian Institute of Sciences, Indonesia)
Dr. Marissa Angelina (Indonesian Institute of Sciences, Indonesia)
Dr. Sofa Fajriah (Indonesian Institute of Sciences, Indonesia)
Hakikie Melanie, PhD (Indonesian Institute of Sciences, Indonesia)
Khoirun Nisa, PhD (Indonesian Institute of Sciences, Indonesia)
Rizna Triana Dewi, PhD (Indonesian Institute of Sciences, Indonesia)
Roni Maryana, PhD (Indonesian Institute of Sciences, Indonesia)
Gagus Ketut Sunnardianto, PhD (Indonesian Institute of Sciences, Indonesia)
Siti Nurul Aisyiyah Jenie, PhD (Indonesian Institute of Sciences, Indonesia)
Dr. Badrut Tamam (Polytechnic of Health Denpasar, Indonesia)
Dr. Djabal Nur Basir (Hassanudin University, Indonesia)
Vilia Darma Paramita, PhD (State Polytechnic of Ujung Pandang)
Alfian Noviyanto, PhD (Nano Center Indonesia Research Institute)
Helen Tuhumury, PhD (Pattimura University, Indonesia)
Dr. Handajaya Rusli, S.Si, M.Si (Bandung Institute of Technology, Indonesia)
Dr. Oman Zuas (National Standardization Agency of Indonesia, Indonesia)
Dr. Untung Triadhi (Bandung Institute of Technology, Indonesia)
Dr. Andri Hardiansyah (Indonesian Institute of Sciences, Indonesia)
Dr. Edi Suprayoga (Indonesian Institute of Sciences, Indonesia)
Dr. Fitri Dara (Indonesian Institute of Sciences, Indonesia)

IOP Conf. Series: Materials Science and Engineering 1011 (2021) 011001 doi:10.1088/1757-899X/1011/1/011001

IOP Conf. Series: Materials Science and Engineering 1011 (2021) 011002 doi:10.1088/1757-899X/1011/1/011002

Peer review declaration

All papers published in this volume of *IOP Conference Series: Materials Science and Engineering* have been peer reviewed through processes administered by the Editors. Reviews were conducted by expert referees to the professional and scientific standards expected of a proceedings journal published by IOP Publishing.

- **Type of peer review:**
  Single-blind. Reviewers' identities are kept hidden from authors.

- **Conference submission management system:**
  EDAS Conference and Journal Management System (https://www.edas.info).

- **Number of submissions received:**
  84

- **Number of submissions sent for review:**
  73

- **Number of submissions accepted:**
  71

- **Acceptance Rate (Number of Submissions Accepted / Number of Submissions Received X 100):**
  84.5%

- **Average number of reviews per paper:**
  1-3

- **Total number of reviewers involved:**
  47

- **Contact person for queries:**
  Abdi Wira Septama, Chair of ISAC 2020 (email: abdiwiraseptama@gmail.com)

IOP Conf. Series: Materials Science and Engineering 1011 (2021) 012001 doi:10.1088/1757-899X/1011/1/012001

Study of antioxidant activities, acceptability, and shelf life prediction of Ciplukan (*Physalis angulata* L.) juice drinks

R Luthfiyanti[1+], A C Iwansyah[1+*], Y Rahayu[2] and N S Achyadi[2]

[1]Research Center for Appropriate Technology, Indonesia Institute of Sciences. Jl. Aipda K.S. Tubun No 5, Subang 41213, West Java, Indonesia
[2]Department of Food Technology, Faculty of Engineering, Pasundan University, Bandung. Jl. Setiabudi No.9, Bandung, West Java 40153 Indonesia
[+]These authors contributed equally

*Corresponding author email: chandra.iwansyah@gmail.com

Abstract. Antioxidants are free radical scavenger compounds that can reduce or prevent the adverse effects of reactive oxygen species and reactive nitrogen species (ROS / RNS). This study aims to study the antioxidant activity, acceptability, and shelf life of fruit extract drinks and ciplukan buds (*Physalis angulata* L.). The experimental design used factorial completely randomized design (FCRD), with two factors, namely the ratio of juice extracts and ciplukan buds (a) and stabilizer concentrations (b), three replications. The organoleptic analyzed by hedonic test, antioxidant activity by DPPH method, and total phenolic contents, total flavonoid contents and alkaloid using each of the Folin-Ciocalteau, aluminum chloride and gravimetric methods. The estimated shelf life is calculated using the Arrhenius method at 28°C for 28 days. The results showed that the ratio of fruit extracts and buds (a) affected the preference for taste, aroma, and color attributes. The interaction between juice extract ratio with a bud (a) and stabilizer concentration, (b) influences antioxidant activity, color attributes, and viscosity. The best ciplukan drink samples based on organoleptic calculations are juice with fruit extracts with buds 15:1 and stabilizer concentration of 0.1% (a3b1) and fruit extracts with buds 20:1 and stabilizer concentration of 0.1% (a4b1). The highest total phenolic and flavonoid contents obtained sample code a3b1, while the highest alkaloid content from code sample a4b1. Based on the estimated shelf life of total microbes, ciplukan bud extract has a shelf life of 9 months at 28°C.

1. Introduction

Ciplukan (*Physalis angulata* L.) has been used for decades as alternatives to traditional medicine due to its functional properties, including antihyperglycemic, antibacterial, antiviral, immunomodulatory, anti-inflammatory, antioxidant and anti-analgesic [1]. Ciplukan contains of various active compounds such as saponins, flavonoids, polyphenols, alkaloids, physalin B, physalin D, physalin F, protein, and chlorogenic citric acid, withangulatin A, palmitic acid, acetate acid, protein, vitamin C, tannins, malic acid. Some of these contents are considered as antioxidant compounds [2].

Antioxidants are compounds that are able to ward off free radicals, with mechanisms as electron-giving compounds or reductants. In the food sector, antioxidants are compounds contained in food and when present in low concentrations compared to oxidized substrates significantly reduce or prevent the adverse effects of oxidative stress on normal physiological functions of humans [3]. Oxidative stress is a condition where reactive oxygen species (ROS), reactive nitrogen species (RNS) and endogenous antioxidants found in the human body are not in a balanced state. This imbalance can cause damage to cells and various degenerative diseases, one of which is diabetes mellitus. Components that are present in the human body when interacting with ROS or RNS will form free radical compounds or compounds capable of producing free radicals [4].

Secondary metabolites of flavonoids, polyphenols, and alkaloids contained in ciplukan are natural antioxidant compounds. Flavonoids and polyphenols are antioxidant compounds that are included in the phenolic group, the mechanism of action of phenolic antioxidants, namely potential radical scavengers. The phenolic compound of a biologically active component is a substance that can contribute hydrogen to free radicals. It will break the lipid oxidation reaction chain at the initial initiation stage [5]. Alkaloids are compounds that have a basic hydrogen atom and are part of a cyclic system that can function as an antioxidant compound [6]. With the presence of antioxidants in ciplukan plants, it is expected to be able to prevent and reduce non-communicable diseases such as diabetes mellitus, coronary heart disease, and so on, by inhibiting damage to body cells due to the process of free radical oxidation.

Diabetes mellitus is caused by a disruption in the β cells of the islets of Langerhans, which causes reduced insulin secretion and the development of hyperglycemia [7]. Hyperglycemia is a condition where fasting glucose levels in the blood are higher than 110 mg/dL due to interference with insulin secretion, insulin action or both [8]. The content of alkaloids, polyphenols, and flavonoids in ciplukan leaf and fruit extracts tested was instrumental in the anti-hyperglycemia activity in pancreatic β cells [9-10]. Our previous study, regarding the characteristics of each part of the ciplukan plant, showed the highest antioxidant activity in leaves, fruit and buds (fruit peels) [11]. Utilizing the antioxidant content in ciplukan carried out the development of functional food-based products, namely fruit juice drinks from fruit extracts and ciplukan buds. Therefore, this study aimed to study the antioxidant activity and acceptability of fruit juice drinks and ciplukan buds. In this study estimating the shelf life at room temperature (28°C) was also carried out. This is because the antioxidant content is susceptible to microbiological, chemical and physical damage, causing a decrease in the quality of beverage products that affect the shelf life.

2. Materials and Methods

2.1. Materials

The materials used for making ciplukan drinks were fresh fruit and bud ciplukan (*Physalis angulata* L.) obtained from the village area of Dawuan, Dawuan District, Subang Regency. Ciplukan plants were collected from January until June 2018. The botanical identification of these ciplukan plants has been carried out at Herbarium Bogoriense, Center for Biological Research-LIPI (No. 886/IPH.1.01/If.07/IV/2018). Citric acid, non-calorie sugar (stevia), and CMC were obtained from Setia Guna Chemical Store, Bogor; while honey is bought at CV. Nutrima, Bogor. The materials used for the analysis were methanol, DPPH, distilled water, quercetin, 10% $AlCl_3$, 1M sodium acetate, gallic acid, Folin-Ciocalteu, 7.5% Na_2CO_3, and ammonium hydroxide were purchased from Sigma-Aldrich, Singapore.

2.2. Preparation samples

The Fresh ciplukan fruits and buds were sorted, weighed, and then washed with running water. The washed ciplukan fruits and buds, were blended at 70°C for two minutes. The extraction process of ciplukan juice and buds was carried out by weighing the ingredients and extracting them using a water solvent, each ratio of 1:2 and 1:20 (w/v). Each extract was filtered using filter paper, then collected for the next process (11). The production of ciplukan juice drinks was processed by mixing ciplukan juice extract, and bud ciplukan bud extracts with a ratio, namely: 5:1, 10:1, 15:1, and 20:1 (v/v). Then, the extract was mixed and added with honey, non-calorie sweetener, and stabilizer. Further, pasteurization process was done at the temperature of 70°C for 15 minutes. After that, citric acid was added, homogenized, and packaged in the sterilized bottles [12].

2.3. Experimental design
This current study uses a factorial completely randomized design (FCRD), with 2 factors, namely (1) the ratio of juice extracts and buds of ciplukan plants (a), including: ratio of 5: 1 (a1), 10: 1 (a2), 15: 1 (a3), and 20: 1 (a4), and (2) stabilizing concentrations (b), including: 0.1% (b1), 0.2% (b2), and 0.3% (b3), with three replications, 36 trial units were obtained. The parameters tested were: sensory quality (hedonic test) and antioxidant activity. Analysis of total phenols, total flavonoids, and alkaloids was carried out on the two best samples based on sensory test results and antioxidant activity. Ciplukan juice drinks with the best formula were analyzed for estimating shelf life by Arrhenius for 28 days, every seven days observing parameters of vitamin C levels, pH, and total microbial count (total plate count).

2.4. Procedur analysis

2.4.1 Organoleptic test. The sensory quality of ciplukan juice drinks was analyzed using the preferred method (hedonic) with the parameters: color, aroma, taste and viscosity [13]. The preferred test was conducted at the Organoleptic Laboratory, Center for Appropriate Technology Research - LIPI, using 30 semi-trained panelists, with a rating scale of 1 (very dislike) - 6 (very like).

2.4.2 Total phenolic, flavonoids, alkaloids contents and antioxidant activity. Total phenolic content of ciplukan juice drinks was determined by Folin-Cioceltaeu method [14]. The total flavonoids content of ciplukan juice drinks were analyzed using the method [15]. Total alkaloids from ciplukan drinks were analyzed using gravimetric methods [16]. Analysis of antioxidant activity of ciplukan juice drinks was measured by using the 1,1-diphenyl-2-picrylhydrazyl (DPPH) method [17].

2.4.3 Vitamin C. Vitamin C was analyzed using the titration method [18]. A total of ten mL of the ciplukan juice drinks sample was pipetted into a clonical measuring flask and KI (5 mL; 10%) added with H_2SO_4 (1 mL; 0.3 M) into the measuring flask. After that, KIO_3 (10 mL; 0.01 M) was added to the measuring flask, and homogenized. The excess iodine was titrated using $Na_2S_2O_3$ (0.01 M) with a blank solution in the form of titration of distilled water (10 mL). The vitamin C content can be calculated by the following equation (1):

$$Vitamin\ C = \ meq\ KIO_3 - \ meq\ N_2S_2O_3 \tag{1}$$

2.4.4 pH value. The pH value was measured by using a pH meter (Metler Toledo Pro2Go)[19]. pH meter before use, first turned on and stabilized for 15-30 minutes. Calibration is carried out using a buffer solution at pH 4 and pH 7. Samples of ciplukan drinks are put into a beaker glass, then the pH meter electrode is dipped into the drink sample and left to show a stable number. Samples were measured triplicates.

2.4.5 Total microbes (total plate count). The total microbes in ciplukan juice drinks were counted using the total plate number (TPC) method [20]. A serial ten-fold dilution was performed until 10^-4. Furthermore, 1 mL of each dilution was taken and put into a petri dish which contained 15 mL of PCA media in duplicate, then homogenized and covered, then incubated at 37°C for 48 hours. The calculation was done for dilution with the number of colonies of 30-300. The total plate number (TPC) is calculated by the following equation (2):

$$Number\ of\ colonies\ x\ \frac{1}{dilution\ factor} \tag{2}$$

2.5. Statistical analysis.
The data are presented in mean ± standard deviation (sd). Data were analyzed by analysis of variance (ANOVA), with a confidence interval of 95% and tested for normality of the data first. Duncan's comparison test was used to determine the average difference between each treatment. The data processing program used is Microsoft Excel 2013 and SPSS Statistics Software 24.00 for Windows.

Linear regression analysis is used to determine the relationship between the variables measured by storage time. The k value obtained from the regression equation is applied to the Arrhenius equation to estimate the shelf life of ciplukan juice drinks [21].

3. Results and Discussion

3.1. Sensory parameter

The sensory quality of fruit juice extracts and ciplukan buds was analyzed using the preferred method (hedonic) with the test parameters: color, aroma, taste, and viscosity attributes [13]. Taste is one of the characteristics that can determine the level of consumer preference in response to the sense of taste and smell. Winarno [22] states that the factors that can affect the taste are interactions between taste components in food ingredients or products, concentration, temperature, and chemical compounds. Comparison of fruit juice extracts and ciplukan buds significantly affected the flavor attribute of ciplukan juice drinks ($P<0.05$). Duncan's multiple comparison tests showed that the extract ratio of 20:1 between juice and bud (a4) had the highest average value of preference (Figure.1). The more ciplukan juice extracts are added, the tendency for panelists to increase. This tendency is because, in ciplukan fruit, there is the sugar content that can add a sweet taste in drinks. In contrast, in ciplukan buds, there are alkaloids, tannin, and saponin compounds that have bitter taste characteristics [23].

Aroma determines whether or not consumers accept a product. Comparison of fruit juice extracts and buds significantly affected the aroma attributes in fruit juice drinks and ciplukan buds ($P<0.05$). At the same time, the concentration of stabilizers and their interactions were not significantly different ($P>0.05$). Duncan's multiple comparison test showed that fruit extract drinks and constipation buds of 10:1 (a2) had the highest average value of flavor attributes (Figure 1). The aroma associated with volatile compounds in food, the more volatile compounds, and the form of aroma will be sharper [24]. The aroma of functional drinks of fruit juice and ciplukan buds is estimated to be influenced by additional ingredients such as honey. Honey has high volatility and hygroscopic properties, hydrocolloid can reduce volatile compounds [25].

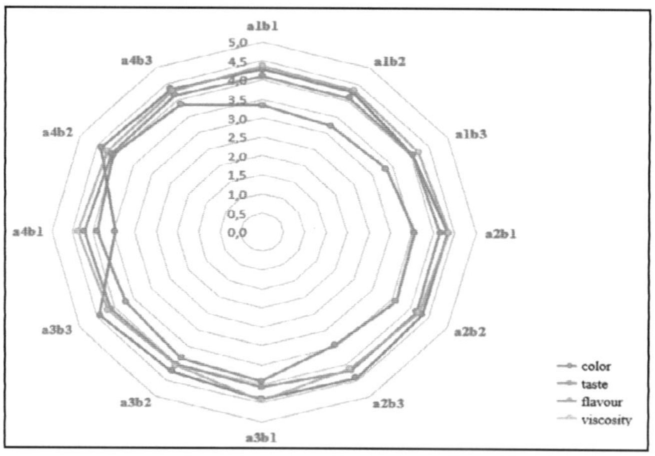

Figure 1. Effect of fruit extract ratio + bud and stabilizer concentration on taste, color, aroma and viscosity attributes of ciplukan functional drinks (n=3).

Color has produced the stimulation of a beam of radiation energy that falls into the retina of the eye. It is associated as a factor that describes the level of maturity, freshness, purchasing power, and safety of a product or foodstuff [26]. The interaction between the juice extract ratio with the ciplukan buds and stabilizer concentration significantly affected the color attributes ($P<0.05$). Duncan's multiple comparison test showed that extracts of fruit juice and buds with a ratio of 15: 1 and 20: 1 with a stabilizer concentration of 0.1%, namely a3b1 and a4b1, had the highest preference for the color attribute of the ciplukan bud extract juice (Figure 1). This result is thought to be due to differences in the ratio of fruit extract and ciplukan bud extract. The higher the percentage of fruit juice extracts added to the buds, the brighter the colors produced. The brownish-yellow color of this functional drink is derived from the antoxantin flavonoid pigment, but it also comes from tannins and alkaloids, owned by ciplukan fruits and buds [24].

Viscosity is very influential on the level of consumer preference on the product. Factors that affect viscosity include temperature, the concentration of the solution, molecular weight of solution, presence of other substances, and pressure of the material used. The interaction between the juice extract ratio and the ciplukan buds and stabilizer concentration significantly affected the viscosity attribute ($P<0.05$). Duncan's multiple comparison test showed that the treatment with a ratio of 15: 1 with a stabilizer concentration of 0.1% (a3b1) had the highest preference value on the ciplukan bud extract's color attribute. According to Kumalasari et al. [27] the higher the stabilizer concentration, the higher the viscosity that the drink will be thicker. The viscosity of the fruit juice and ciplukan buds is influenced by the amount of liquid added, the more liquid added, the lower the viscosity, and conversely, the more solids added, the higher the viscosity. The use of stabilizers will increase the continuous phase's viscosity into suspended particles so that they do not settle quickly [25].

3.2. Antioxidant activity

Antioxidant activity was tested by the DPPH method, which was classified according to IC_{50} values. The IC_{50} value is a large amount of concentration that can inhibit DPPH by as much as 50%. The smaller the IC_{50} value, the greater the antioxidant activity. The presence of antioxidant activity in the sample is marked by the fading of the color which results in a decrease in the absorbance value of the visible light from the spectrophotometer, so the fade the color the lower the absorbance value which shows, the more free radicals that react with antioxidants contained in functional drinks [28]. Comparison of fruit extracts with buds, stabilizer concentrations, and their interactions significantly affected antioxidant activity ($P<0.05$) (Figure 2).

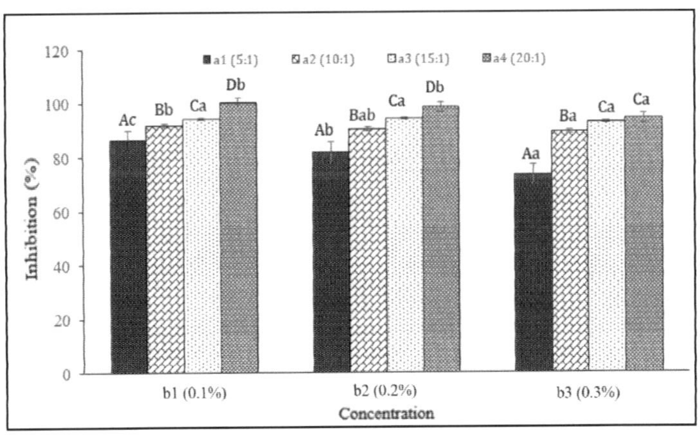

Figure 2. Effect of extract ratio of fruit juice + bud and stabilizer concentration on antioxidant activity of ciplukan functional drinks. Data presented as mean ± standard error (SE, n=3). The numbers followed by different letters on the graph show significant differences ($P<0.05$). Uppercase (extract ratio) and lowercase (stabilizer).

Figure 2 shows a better treatment of antioxidant activity (IC_{50}) demonstrated by a sample with a 5:1 juice and bud ratio and a CMC concentration of 0.3% (a1b3) 73.45 μg/mL. According to Iwansyah et al. [11], reported that ethanol extracts from ciplukan fruits and buds (fruit peels) had antioxidant activity, which was 133.76 μg/mL and 134.53 μg/mL, respectively. The results of this study indicate that the greater extract of fruit juice added, and the higher the concentration of the stabilizer mixed, the higher the antioxidant activity (Figure 2). These results agreement with Farikha et al. [29], who reported that the higher the stabilizer concentration, the higher the antioxidant activity of the dragon fruit juice drink.

A significant decrease in antioxidant activity in ciplukan juice drink can be caused by processing. According to Burda and Oleszek [30] states that materials that have the potential for an antioxidant activity that is processed with heat and are directly exposed to air will be easily oxidized and degraded so that it will affect its antioxidant activity. Antioxidants in food can eliminate ROS/RNS to stop radical chain reactions, which are called free radical scavengers (FRS) antioxidants. In contrast, those that can inhibit reactive oxidants from their formation are called preventive or secondary antioxidants. Several mechanisms of antioxidant reactions can function as: (1) protecting the body by preventing the formation of ROS and its responses in important parts of the body; (2) capturing or eliminating chemical compounds that absorb energy and electrons; (3) with catalytic systems that neutralize or reduce ROS, for example, the antioxidant enzyme SOD (superoxide dismutase), catalase, and glutathione peroxidase; (4) binding or inactivation of metal ions to prevent the formation of ROS, for example, ferritin, ceruloplasmin, catechins; and (5) antioxidants as chain breakers and damage ROS bonds, for example, ascorbic acid (vitamin C), tocopherol (vitamin E), glutathione, flavonoids [3].

3.3. Total phenolic contents, total flavonoids and alkaloids

Determination of the best sample is chosen based on the highest number of organoleptic test results from all of the attributes assessed and antioxidant activity. Based on this, two samples of the best ciplukan drinks were obtained, namely: (1) the ratio of juice extract with buds 15: 1 and the stabilizer concentration of 0.1% (a3b1), and (2) the ratio of juice extracts with buds 20:1 and stabilizer concentration of 0.1% (a4b1). Subsequent samples were tested for total phenolic, flavonoid, and alkaloid. The results of the analysis of the best samples of functional fruit drinks and ciplukan buds can be seen in Table 1.

Table 1. Total phenolic, flavonoid and alkaloid from the best formulations of Ciplukan juice drinks.

Sample	Parameter		
	Total flavonoid (mg QE/L)	Total fenolik (mg GAE/L)	Alkaloid (%)
a3b1	0.25±0.04[a]	18.74±0.67[a]	5.80±0.51[a]
a4b1	0.19±0.05[b]	15.57±0.71[b]	6.61±0.28[b]

Data are presented as means ± standard deviation (sd) (n = 3). QE: Equivalent to quercetin. GAE: Gallic acid equivalent. a3b1: extract ratio of 15:1 with a CMC concentration of 0.1%. A4B1: Extract ratio of 20:1 with CMC concentration of 0.1%. Independent t-test, different alphabetic in same colomn was significantly, $\alpha = 5\%$.

Table 1 shows that the highest total phenolic and flavonoids were obtained from beverage samples with a juice extract ratio and a bud of 15: 1 and a CMC concentration of 0.1% (a3b1), i.e., 18.74 mg GAE per liter and 0.25 mg QE per liter, respectively. Beverage samples obtained the highest levels of alkaloids with an extract ratio of juice and buds 20: 1 and CMC concentrations of 0.1% (a4b1), amounting to 6.61% ($P < 0.05$). These results are in line with Julianti et al. [23], which reported higher flavonoid and phenolic contents in the ciplukan buds while higher alkaloids contained in the ciplukan fruit.

Flavonoids play a role in giving flavor and color to various fruits and vegetables. The high content of flavonoids is also related to the high phenolic content, where flavonoids are a subset of phenolic compounds. Therefore, the high phenolic content in a material indicates the high content of flavonoids in the material [31]. As a natural antioxidant, flavonoids can capture free radicals. These antioxidant activities allow flavonoids to capture free radicals (such as ROS or RNS) associated with phenolic -OH groups to improve the condition of damaged tissue [32]. Flavonoids are reported to have an antidiabetic activity that can regenerate cells in the islets of Langerhans by clearing excessive free radicals, binding metal ions (chelating), breaking the chain of free radical reactions and blocking the polyol pathway by inhibiting the enzyme aldose reductase [33-34].

Polyphenols are bioactive and can reduce free radicals or antioxidants [35]. The antioxidant role of polyphenols is thought to protect pancreatic β cells from the toxic effects of free radicals produced under chronic hyperglycemia [36]. According to Barbosa [37], antioxidants of green tea polyphenols can reduce oxidative stress by preventing a chain reaction of converting superoxide to hydrogen superoxide by donating hydrogen atoms from the aromatic hydroxyl (-OH) polyphenol group to bind free radicals and remove them from the body through excretion system.

Alkaloids include alkaline compounds that contain one or more nitrogen atoms, usually in combination as part of a cyclic system. Alkaloids are often toxic to humans, and many have prominent physiological activities, so they are widely used in processing [38]. Alkaloids are antioxidant compounds that can regenerate damaged pancreatic β cells [39]. Cell damage caused by free radicals can be overcome by the presence of antioxidants that function as a lowering agent and decrease oxidizing agents before damaging cells so that cell damage can be reduced [40].

3.4. The shelf life of ciplukan juice drinks

The shelf life is determined based on the parameters of vitamin C test, pH, and total microbes. One selected sample is the sample in a4b1 treatment (comparison of fruit extracts with buds 20:1 and stabilizer concentration of 0.1%). Determination of the shelf life of fruit drinks and ciplukan buds packed in plastic bottles is carried out at room temperature (28 °C), and observations are carried out regularly every seven days starting from day 0 to 28.

Vitamin C is a non-enzymatic antioxidant that can act as a reducing agent. A donor of hydrogen atoms because it effectively captures free radicals, especially ROS or reactive oxygen species, prevents the occurrence of oxidized Low Density Lipoprotein (LDL), absorbs metals in the digestive tract, and has biological activity so that it can help maintain metabolism in the body [41]. In counteracting free radicals, vitamin C can directly react with superoxide anions, singlet oxygen, lipid peroxides, and hydroxyl radicals. Ascorbic acid, which acts as a reducing agent, will donate one electron semi-dehydroascorbic, to form dehydroascorbate, and degraded to form oxalic acid and treonic acid. Thus, vitamin C's ability as an inhibitor of free radicals is essential in maintaining the integrity of cell membranes [42]. The graph of decreased vitamin C, pH and increased total plate count levels during storage can be seen in Figure 3, 4 and 5.

Figure 3 shows that during storage at room temperature, there is a decrease in vitamin C levels. Regression equation for vitamin C levels in ciplukan functional drink is $Y = -0.4257x + 25.331$ ($R^2 = 0.973$) which means a constant decrease in quality (k) 0.4257. The activation energy (Ea) changes in Vitamins C levels of 0.8454 cal/mol, which means to initiate changes in vitamin C levels in functional drinks and require that much energy. Decreased levels of vitamin C during storage occur due to degradation can be caused by temperature, sugar concentration, pH, oxygen, and light or direct light. With the degradation of vitamin C, it will form compounds that are brown and not have vitamin C activity again. Temperature can cause loss of vitamin C, even in anaerobic conditions [43]. Ascorbic acid experiences quality degradation under anaerobic conditions to 2,3-dicogulonic acid, followed by lactone hydrolytic cleavage, and it produces deoxypentose, so it does not show vitamin C activity [44].

pH is a value that indicates the level of acidity in a solution. The size of the pH value is determined by the H + ions in the solution. Figure 4 shows that during storage, the room temperature decreased in pH. The regression equation for the pH value of functional drinks of fruit juice and ciplukan buds is $Y = -0.0229x + 6.09$ ($R^2 = 0.9531$), which means a constant decrease in quality (k) 0.0229. The activation energy (Ea) changes in pH value of 0.0455 cal/mol, which mean to initiate changes in pH in functional drinks requires that amount of energy. The decrease in pH value in functional drinks possibly due to the ability of microbial activity that can break down organic acids in fruit juices and ciplukan buds into carbon dioxide by oxidative yeasts. The decrease caused by temperature causes an increase in total acid. Low pH will result in sucrose being converted into sugar. In a solution in an acidic atmosphere or the presence of the enzyme hydrolysis, invertase can occur to produce sugar invert from sucrose, namely α-D-Glucose and β-D-Fructose. Glucose that is broken down will produce pyruvic acid, which in anaerobic conditions, pyruvic acid will be converted into acetic acid and alcohol, causing acid levels to increase, and pH values tend to decrease [45].

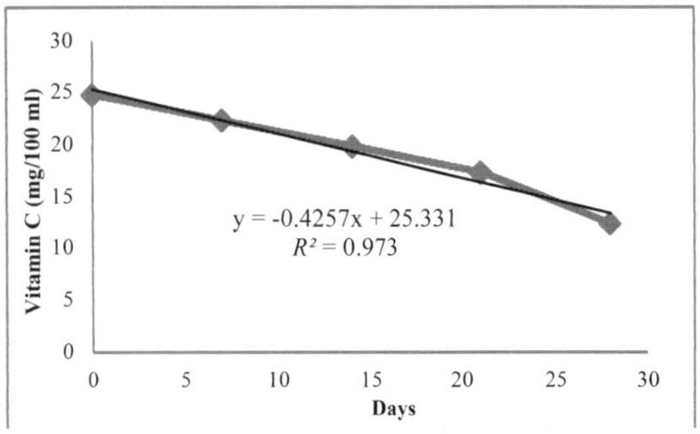

Figure 3. Linear regression graph of decreased vitamin C quality in functional Ciplukan juice drinks during storage (n = 3).

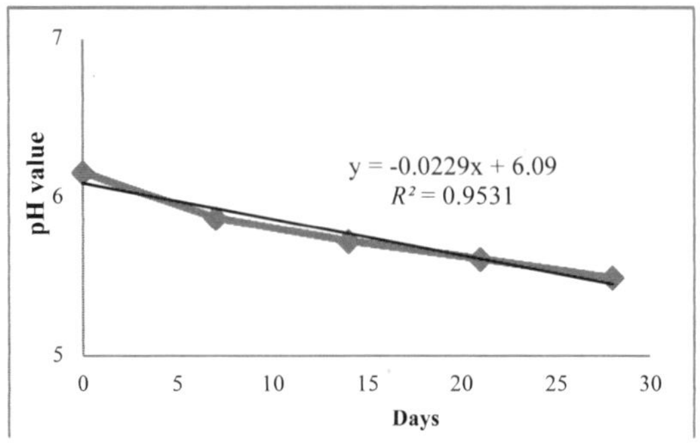

Figure 4. Linear regression graph of decreased pH value in functional Ciplukan juice drinks during storage (n = 3).

The total microbes contained in an ingredient or food product is one indicator of the level of safety and damage to the product. Increased growth of undesirable microbes can indicate that inside the work, there has been contamination from outside or insufficient processing so that it can reduce product quality. Total microbes are tested by the TPC method (total plate count). This method is to determine the number of microbes in a product by counting the bacterial colonies that grow on agar media. According to the Indonesian National Standard [46], it states that the maximum total amount of microbes that are still permitted to be contained in fruit drinks is $1 x 10^4$ colonies/mL. The increase in total microbes in fruit juice drinks and ciplukan buds until the end of storage is still below the maximum limit determined by SNI (Figure 5). it is estimated because the acidity or pH level decreases during the storage period. Acidic pH will cause the bacterial cell's internal pH to decrease so that it can interfere with bacterial cell activity, and bacterial growth is inhibited [47]. The pH value of the polluted environment also influences microorganisms' ability to carry out cellular functions, transport membrane cells, and balance reactions carried out by microorganisms. The pH value is inversely proportional to the acidity value, if the pH decreases, the acidity increases where the number of microorganisms during storage also affects the pH value [48].

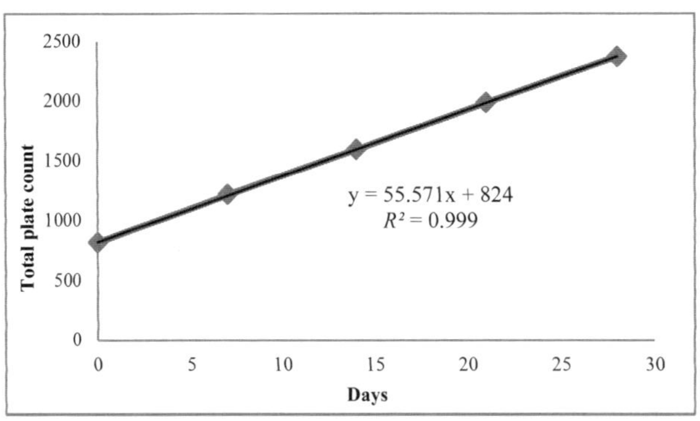

Figure 5. Linear regression graph of increased total plate count in functional ciplukan juice drinks during storage (n = 3).

Figure 5 shows that during room temperature storage, an increase in TPC value causes a decrease in product quality. The regression equation for the TPC value of the ciplukan functional drink is Y = 55.571x + 824 (R^2 = 0.999), which means a constant decrease in quality (k) 55.571. The k and time (t) values of the TPC ciplukan functional drink at room temperature (28 °C) are entered into the Arrhenius equation, which is $k = ko.e - E/RT$ atau $\ln k = \ln ko - E/RT$. Because ln ko and -E/ R are constants, then the equation can be written $\ln k = A + B.(\frac{1}{T})$, so that the equation $\ln k = -824 - 55.571(\frac{1}{T})$ is obtained. With this equation, the activation energy of 110.363 cal/mol will be obtained. So the value of the constant (k) rate of decline in the quality of the TPC parameters obtained is 5.5829/day. By using the equation At = A0 - K.t, the storability of ciplukan juice drink based on the TPC value at room temperature (28 °C), ie, for nine months (Table 2).

Table 2. The shelf life of ciplukan juice drinks (*P. angulata*).

Parameter	***Arrhenius* Equation**	R^2	**Energy Activation (kal/mol °K)**	**k value**	**Temp.**	**Shelf Life (months)**
TPC	$\ln k = -824 - 55.572(\frac{1}{T})$	0.999	110.363	5.5829	28° C	9 months

TPC: total plate count

4. Conclusion

Based on the results, it can be concluded that the extract ratio of juice and bud (a) influences the liking of flavor, aroma, and color attributes. The interaction between juice extract ratio with a bud (a) and stabilizer concentration, (b) influences antioxidant activity, color attributes, and viscosity. The best Ciplukan drink sample is based on organoleptic calculation results and antioxidant activity, namely a3b1 and a4b1. The highest total phenolic and flavonoid obtained sample code a3b1, while the highest alkaloid content from code sample a4b1. The selected sample is suspected to be of shelf life at room temperature (28oC). In the estimation of shelf life, a decrease in vitamin C levels and pH values, as well as an increase in total microbes during storage. Based on the calculation of the Arrhenius equation for total microbes, the shelf life of the Ciplukan bud extract is nine months.

Acknowledgment

We are grateful to the Ministry of Research, Technology and Higher Education, Republic of Indonesia for funding the Insinas Riset Pratama Individu (No.20/P/RPL-LIPI/INSINAS-/III/2019), as well as the Indonesian Institute of Sciences for supporting facilities and technical support (ELSA).

References

[1] Rohyani I S, Aryanti E and Suripto 2015 Phytochemical content of some of local plant species frequently used as raw materials for traditional medicine in Lombok Island *Prosiding Seminas Nasional Masyarakat Biodiversitas Indonesia* 1(2) 388-91

[2] Departemen Kesehatan Republik Indonesia 1995 Materia medika indonesia jilid VI. Jakarta: Departemen Kesehatan Republik Indonesia:Jakarta p 308

[3] Karadag A, Ozcelik B and Saner S 2009 Review of methods to determine antioxidant capacities *Food Anal. Methods* 2(1) 41–60

[4] Rohman A, Widodo H, Lukitaningsih E, Rafi M, Nurrulhidayang A F and Windarsih A 2019 Review on *in vitro* antioxidant activities of Curcuma species commonly used as herbal components in Indonesia. *Food Res.* 3(3) 1–8

[5] Isnindar S W, Wahyuono S and Setyowati E P 2011 Isolasi dan identifikasi senyawa antioksidan daun Kesemek *(Diospyros kaki* Thunb.*)* dengan metode DPPH (2,2-difenil-1- pikrilhidrazil). *Majalah Obat Tradisional* 16(3) 161–9

[6] Hanani E, Mun'im A and Sekarini R 2005 Identifikasi senyawa antioksidan dalam spons *Callyspongia* Sp. dari Kepulauan Seribu *Majalah Ilmu Kefarmasian* 2(3) 127-33

[7] Reddy P A, Vijay, Kumar R, Reddy G V, Reddy M K and Reddy Y N 2014 Anti-diabetic and hypolipidemic effect of aqueous and methanolic toot extracts of *Physalis angulata* in streptozotocin (STZ) -induced diabetic rats *Int. J. Pharm. Res. Sch.* 3(1) 402–9
[8] Price S A 2005 *Patofisilogi: konsep klinis proses-proses penyakit* (Jakarta: EGC)
[9] Sediarso, Sunaryo H and Amalia N 2008 Efek antidiabetes dan identifikasi senyawa dominan dalam fraksi kloroform herba Ciplukan *(Physalis angulata* L.*) J. Farm. Indones.* 8(1) 14–24
[10] Djajanegara I and Wahyudi P 2010 Uji sitotoksisitas ekstrak etanol herba Ceplukan *(Physalis angulata* Linn.*)* terhadap sel T47D secara *in vitro J Ilmu Kefarmasian Indones* 8(1) 41–7
[11] Iwansyah A C, Julianti W P and Luthfiyanti R 2019 Characterization of Nutrition, Antioxidant Properties, and Toxicity of Physalis Angulata L. Plant Extract. *Asian J. Pharm. Clin. Res.* 12(11) 95–9.
[12] Luthfiyanti R, Iwansyah A C, Herminiati A, Hidayat D and Rahman N 2019 Proses dan Formulasi Pembuatan Minuman Sari Buah dari Ekstrak Buah dan Kuncup Ciplukan *(Physalis angulata* Linn.*)*. [Internet]. Indonesia; P00201910060
[13] Soekarto B 1985 *Penelitian Organoleptik untuk Pangan dan Hasil Pertanian* (Jakarta: Bhrata Karya Aksara)
[14] Singleton V, Rosa R and Lamuela-Raventós M 1999 Analysis of total phenols and other oxidation substrates and antioxidants by means of folin-ciocalteu reagent. *Methods Enzymol.* 299 152–78
[15] Al-temimi A and Choudhary R 2013 Determination of antioxidant activity in different kinds of plants *in vivo* and *in vitro* by using diverse technical methods *J. Nutr. Food Sci.* 3(1) 1–6
[16] Mbagwu F N, Unamba C I N, Nwosu I C 2010 Phytochemical screening on the seeds of *Treculia aricana* and *Artocarpus atilis*. *New York Sci. J.* 3(12) 51–5
[17] Iwansyah A C, Damanik R M, Kustiyah L and Hanafi M 2016 Relationship between antioxidant properties and nutritional composition of some galactopoietics herbs used in indonesia: a comparative study. *Int. J. Pharm. Pharm. Sci.* 8(12) 236–43
[18] AOAC 2005 *Official Methods of Analysis of AOAC INTERNATIONAL* 18th Edition, 2005 Current through Revision 1, 2006
[19] Apriyantono A D, Fardiaz N L, Puspitasari S, Budiyanto S 1989 *Analisis Pangan* (Bogor: IPB Press)
[20] Fardiaz S 1992 *Mikrobiologi Pangan I* (Jakarta: PT Gramedia Pustaka Utama)
[21] Haryati, Estiasih T and Heppy F 2015 Shelf-life prediction of sterilized Mojokerto's fermented Black Glutinous Rice using accelerated shelf-life testing (ASLT) method by Arrhenius Equation Approach *Jurnal Pangan dan Agroindustri* 3(1)156–65
[22] Winarno F G 2008 *Kimia Pangan dan Gizi* (Jakarta: PT Gramedia)
[23] Julianti W P, Ikrawan Y, Iwansyah A C 2019 Effect of solvent on total phenolics content, antioxidant activity and toxicity of Ciplukan Fruit *(Physalis angulata* L.*) J. Ris. Teknol. Ind.* 50 70–9
[24] Arifin B and Ibrahim S 2018 Structure, bioactivity and antioxidant of flavonoid. *J. Zarah* 6(1) 21–9
[25] Christiana M A, Radiati L E and Purwadi 2017 Effect of Gum Arabic on organoleptic, color , pH, viscosity, and turbidity of apple concentrated honey drink *Jurnal Ilmu dan Teknologi Hasil Ternak* 10(2) 46–53
[26] Hatcher D W, Simons S J and Andeson M J 2000 Assesment of oriental noodle appearance as a function of flour refinement and noodle type by images analysis *J. Cereal Chem.* 77(2) 181–6
[27] Kumalasari R, Ekafitri R and Desnilasari D 2015 Pengaruh bahan penstabil dan perbandingan bubur buah terhadap mutu sari buah campuran pepaya-nanas *J. Hortikulura* 25(3) 266–76
[28] Molyneux P 2004 The use of the stable free radical diphenylpicryl-hydrazyl (DPPH) for estimating antioxidant activity *J. Sci. Technol.* 26(2) 211–9

[29] Farikha I N, Anam C and Widowati E 2013 Pengaruh jenis dan konsentrasi penstabil alami Terhadap karakteristik fisikokimia sari buah naga merah *(Hylocereus polyrhizus)* selama penyimpanan *J. Teknosains Pangan* 2(1) 30–8
[30] Burda and Oleszek W 2011 Antioxidant and antiradical activities of flavonoid *J. Agric. food Chem.* 49(6) 2774–9
[31] Maisuthisakul P, Pasuk S and Ritthiruangdej P 2008 Relationship between antioxidant properties and chemical composition of some Thai plants. *J. Food Compos. Anal.* 21 229–40
[32] Lugasi A J, Hovari K V, Sagi and Biro L 2003 The role of antioxidant phytonutrients in the prevention of disease *Acta Biol. Szeged.* 47(1) 119–25
[33] Mills S and Bone K 2002 *Principles and Practice of Phytotherapy: Modern Herbal Medicine.* (Edinburgh, Scotland: Churral Livingstone)
[34] Sandhar H K, Kumar B, Prashes S, Tiwari P, Salhan M and Sharma P 2011 A review of phytochemistry and pharmacology of flavonoids *Int. Pharm. Sci.* 1(1) 25–41
[35] Wojdyło A, Oszmian'ski J and Czemerys R 2007 Antioxidant activity and phenolic compounds in 32 selected herbs *Food Chem.* 105 940–9
[36] Prameswari O M and Widjanarko S B 2014 The effect of water extract of Pandan Wangi leaf to decrease blood glucose levels and pancreas histopathology at diabetes mellitus rats *Jurnal Pangan dan Agroindustri* 2(2) 16–27
[37] Barbosa D S 2007 Green tea polyphenolic compounds and human health *J. Consum .Prot. Food Saf.* 2 407–41
[38] Harborne J B 1973 Phytochemical methods (London: Chapman and Hall, L) p 49–188
[39] Arjadi F and Mustofa 2017 Ektrak daging buah mahkota dewa meregenerasi Sel Pulau Langerhans Pada tikus putih diabetes *Biogenesis* 5(1)27–33
[40] Robinovitch A, Pinzon W L S, Sorensen O, Bleackley RC and Power R F 1995 IFN-Gamma gene expression in Pancreatis Islet infiltrating mononuklear cells correlates with autoimmune diabetes in non obese diabetic mice *J. Immunol.* 154(9) 4874–82
[41] Levine M, Dhariwal K R, Welch R W, Wang Y and Park JB 1995 Determination of optimal vitamin C requirements in humans *Am. J. Clin. Nutr.* 62(6) 1347S-56S
[42] Suhartono E, Fachir H and Setiawan B 2007 *Kapita Sketsa Biokimia Stres Oksidatif Dasar dan Penyakit* (Banjarmasin: Pustaka Benua)
[43] Andarwulan N and Koswara S 1992 *Kimia vitamin* (Jakarta: Rajawali Pers)
[44] Ahmad I, Qadeer K, Zahid S, Sheraz M A, Ismail T, Hussain W, *et al.* 2014 Effect of ascorbic acid on the Degradation of Cyanocobalamin and Hydroxocobalamin in aqueous solution : a kinetic study *AAPS PharmSciTech.* 15(5) 1324–33.
[45] Sahambangun M J, Lengkey L C and Rumambi D 2016 Penentuan umur simpan sirup pala berdasarkan perubahan derajat keasaman pH *Jurnal Ilmu Fakultas Pertanian Universitas Sam Ratulangi* 7(3) 1–5.
[46] Standar Nasional Indonesia 2014 *Minuman sari buah SNI 3719:2014* (Jakarta: Badan Standarisasi Nasional)
[47] Molina E G, Perles R D, Moreno D A and Viguera C G 2009 Natural bioactive compounds of *Citrus limon* for food and health *J. Pharm. Biomed. Anal.* 51(2) 327–45
[48] Yuniani DR, Anam C, Khusniati T and Utami R 2017 Peranan jus buah mengkudu *(Morinda citrifolia)* dalam mempertahankan kualitas susu pasteurisasi kambing "peranakan etawa" selama penyimpanan *Jurnal Teknologi Hasil Pertanian* 10(1) 67–78

IOP Conf. Series: Materials Science and Engineering 1011 (2021) 012002 doi:10.1088/1757-899X/1011/1/012002

Baeckea frutescen L. essential oil was tested with *Aedes aegypti* larvae at various concentration levels

Renhart Jemi[1]*, Sumiati Simanullang[2], Sarinah[1], Nuwa[1], Gimson Luhan[1] and Desy Natalia Koroh[1]

[1]Departemen Forestry Faculty of Agriculture, University of Palangka Raya
[2]University of Palangka Raya Forestry students

*Corresponding author email: jemi@for.upr.ac.id

Abstract. Reducing the negative impact of using synthetic insecticides in a friendly manner. The alternative is to use a bioinsecticide. Ujung Atap (*Baeckea frutescens* L.) essential oil is one of them that has the potential as a bioinsecticide. Distillation of fresh leaves and dry leaves. *Aedes aegypti* larvicide mortality test on the resulting essential oil. *A.* aegypti mosquitoes that cause dengue fever. Analyze the essential oil content of *B. frutescens* L. by GC-MS. The Dried Leaf Distillation of *B. Frutescens* L. produces more and better essential oil than fresh leaves. Dried leaf essential oil is more deadly against larvicides than fresh leaf essential oil of *B. frutescens* L. Larvacide activity is classified as toxic. There are 25 compounds found in essential oils of fresh leaves and 23 compounds in dry leaves. The compound which is a gastric poison for insects is Limonene.

1. Introduction

Dengue Hemorrhagic Fever (DHF) is one of the main health problems of the people of Indonesia. This disease is caused by a dengue virus carried by *Aedes aegypti, which* attacks human blood, reduced blood platelet count. If the attack rate is high and the handling time is slow then it will result in death [1]. Hemorrhagic fever is a health problem in Indonesia, in 2015 there were 129,650 cases and the number of deaths was 1,070 people [2].

The use of synthetic mosquito repellent tends to be used by the community to prevent DHF [3,4]. The content is very dangerous because it can cause toxic effects both locally and systemically on humans [5,6]. Reducing this negative effect, prevention is needed by utilizing plants that are identified as having ingredients that can help repel and prevent mosquitoes that cause dengue fever, known as plant-based insecticides. Plant-based insecticides use biodegradable materials, do not pollute the environment, and are relatively safe for health.

One of the essential oil-producing plants is the *Ujung Atap* (*Baeckea frutescens* L.) [7,8]. *B frutescens* L grows in a tropical peat forest in the *Kerangas* forest area in Central Kalimantan. Central Kalimantan Dayak tribe people use this plant as a mosquito repellent by drying the leaves and then burning them. Several studies of *B. frutescens* L have been carried out to the leaves, stems, and roots, the results of which show potential as anti-inflammatory, anti-oxidant, anti-fungal, anti-bacterial and anti-microbial [9-13]. *B. frutescens* L. contains various bioactive compounds. However, study on the activity of essential oil larvicides, from fresh and dried leaves of *B. frutescens* L. has not been yet. Therefore, this study aimed to identify the activity of *B. frutescens* L. leaves in inhibition of *A. aegypti* larvae

2. Materials and methods

2.1. Place and material

Research was carried out at the Forest Products Laboratory of the Department of Forestry, Faculty of Agriculture, University of Palangka Raya, to carry out distillation and testing of the properties of essential oils. P2B2 Entomology Laboratory, Tanah Bumbu South Kalimantan, a place for raising and testing *A. aegypti* larvae. FMIPA Organic Chemistry Laboratory, Gadjah Mada University, Yogyakarta, a place to identify components of essential oil compounds using GCMS.

2.2. Preparation of leaf samples

The test sample was taken from the leaves of *B. frutescens* L. from the peat swamp forest in Karen Bangkirai Palangka Raya, Central Kalimantan, Indonesia. Cleaned leaves were then divided into two treatments which were fresh leaves and dry leaves. Fresh leaves were stored in an airtight plastic bag. While, another group were dried at a temperature of 28 °C, until the moisture content reached 12% - 13%.

2.3. Distillation B. frutescens L.

The Fresh and dry leaves were immersed in distilled water for 1 hour prior distillation process [14], then were sliced. The leaves that have been cut into small pieces were distilled as much as 1,000 g. Steam water distillation process, with a ratio of 1:2 (leaves:water) at the temperature of 70 °C for 3 hours [15]. The distillation product was added 0.5 grams of Na_2SO_4 and then separated to obtain essential oil [16]. The process of distillation of dry leaves was the same method as a distillation of fresh leaves.

2.4. Essential oil properties testing

Testing the properties of *B. frutescens* L essential oil includes yield, specific gravity, acid number, ester number and soluble in methanol

2.4.1. Rendement. Distillation essential oil was weighed to know its weight. By first known weight vial bottles where essential oil. Record the weight of essential oil and leaf weight of the B. frutescens L. The test procedure refers to SNI 80281-1:2014 [17]. Calculate the yield using the following formula:

$$Rendement\ (\%) = \frac{Essential\ oil\ Weight\ (g)}{Material\ weight\ (g)}\ x\ 100\% \qquad (1)$$

2.4.2. Specific gravity. Weigh an empty pycnometer. Fill a pycnometer with distilled water and weigh it to find its weight. Wash the pycnometer with ethanol thoroughly and dry it. Refill the pycnometer with essential oil and then put it on the scale. Take temperature measurements for each weight measurement. The measurement process refers to SNI 3954-2014 [18]. Record the results of weight measurements and calculate specific gravity using the following formula:

$$d_{T1}^{T1} = \frac{m2-m}{m1-m} \qquad (2)$$

$$d_T^T = d_{T1}^{T1} + 0{,}0007\ (T1 - T) \qquad (3)$$

Formula description : d_{T1}^{T1} is specific gravity at working temperature; d_T^T is specific gravity at standard temperature; mis empty pycnometer weight (g); m1 is weight pycnometer filled with water (g); 0,0007 is correction factor; T1 is working temperature (°C) and T is a reference temperature (20°C)

2.4.3. Acid number. Weigh 1 gram of essential oil, put it in 100 ml Erlenmeyer flask. Next, add 5 ml of ethanol and 5 drops of phenolptein solution as an indicator. Titrate with 0.1 N KOH and stop until the pink color dissolves. Record the amount of KOH used during the titration. The calculation of the acid number refers to the Kateren formula [19], as follows:

$$Acid\ number = \frac{V\ KOH\ x\ N\ KOH\ x\ 56{,}1}{Sample\ weight\ (g)} \quad (4)$$

Formula description: V KOH is volume titration (ml); N KOH is the normality of KOH (N); 56,1 is the KOH equivalent weight

2.4.4. Ester number. The test for the ester number is a continuation of the acid number. Add 25 ml of KOH 0.5 N into the Erlenmeyer flask, then reflux for 1.5 hours. Cool and titrate with 0.5 HCL and add 5 drops of phenolptein indicator until it turns pink. The calculation of ester numbers refers to the Sumangat and Risfaheri formula [20], as follows:

$$Ester\ number = \frac{V\ blanko - V\ sampel\ x\ N\ HCL\ x\ 56{,}1}{Sampel\ weight\ (g)} \quad (5)$$

Formula description: V blanko is volume titration (ml); V sample is the Volume of the acid number titration (ml) , N HCL is normality of HCL (N); 56,1 is the KOH equivalent weight; 56,1 is the KOH equivalent weight.

2.4.5. Solubility in methanol. The essential oil was mixed with 80% ethanol with volume ratio of 1:1 in a closed measuring cup. Then it is shaken until homogeneous. The testing process refers to the SNI 3954-2014 [21] procedure.

2.5. Larvicide activity test

2.5.1. Maintenance of A. aegypti mosquito larvae. A. aegypti larvae were obtained from the Center for Research and Development of Animal Resources Tanah Bumbu, South Kalimantan. The need for A. aegypti larvae is 260 instar III larvae. The maintenance and testing of larvae were carried out by the P2B2 Entomology Laboratory, Tanah Bumbu, South Kalimantan.

2.5.2. Supply of stock solutions. A total of 100 mg of essential oil is dissolved in 1 mL of DMSO and distilled water is added until the volume reaches the reserve fraction that has been made, diluted to essential oils, namely: 0; 10; 50; 100 and 150 μg / mL Solution stock was made and stored in a cupboard ice.

2.5.3. Procurement of larvae. Testing of the activity of essential oil larvicides using instar III larvae in a petri dish. Each sample used 10 larvae of three replications [22]. Distilled water was used as negative control, whole Abate 50 μg / mL was served as positive control.

2.6. Testing of larval activity

The Larvicidal activity of essential oils was tested with larval mortality parameters at various levels of the concentration of essential oils. Observation of larvae death after being exposed to essential oils for 24 hours at room temperature. Data on larval mortality are calculated using the formula of the percentage of mortality (mortality) of larvae agreed to how to calculate as follows [23-25]:

$$Mortality\ (\%) = \frac{Number\ of\ dead\ larvae}{The\ number\ of\ test\ larvae}\ x\ 100\% \quad (6)$$

2.7. GCMS test

The essential oil obtained was tested for Gas Chromatography-Mass Spectrometry (GC-MS) to find out the composition contained in essential oils. The mass spectrum obtained is then compared by comparing with the data centers of Wiley 7 and National Institute of Standards and Technology (NIST). Analysis of the results component structure, GCMS sponsored by the ChemSpider website (www.chemspider.com).

2.8. Statistical analysis

Mortality data were analyzed using probed to determine LC $_{(50)}$. A determination based on logarithmic equations between the concentration of essential oils (x-axis) and the percentage of mortality (y-axis). Calculate using the Microsoft Excel 2007 program.

3. Results and Discussions

3.1. The nature of essential oils from B.frutescens L.

The results of *B. frutescens* L. essential oil from fresh and dried leaves from distillation. Treat immersion treatment before distillation helps the distillation process and produces essential oils. It allows steam to penetrate into the leaf tissue, thereby dissolving essential oils, resins, waxes, and dyes [26]. The yield of essential oil from the dried leaves is 0.29% more yield than fresh leaves is only 0.12%. Because dry leaves are easily penetrated by steam and open oil cells, making it easier for steam to carry essential oils [27,28]. The acid number of essential oils from fresh leaves is 12.90, greater than the dry leaves, only 11.22. High acid number is unacceptable because the results of Harimutrti's study [29] showed that essential oils with high acid numbers cause irritation. The fresh leaf ester number is 11.22 and dried leaf is 14.03, the ester number is good [30]. The specific gravity of fresh leaves is 0.871 and dry leaves are 0.849. The specific gravity of essential oils is related to the mass fraction of the compositions therein [30]. Because mass fractions contain many long-chain molecules and relatively many bonds, unsaturated or oxygenated as they begin to react with oxidation. The results of testing fresh and dried essential oils of leaves in 80% ethanol showed not clear. Unclearness in 80% ethanol solution due to the presence of non-oxygenated terpene [31]. GCMS analysis results of *B. Frutescens* L. essential oil of terpene composition transferred by $C_{10}H_{16}$ and $C_{15}H_{24}$. Complete according to Table 1. Besides that, the oil color.

3.2. Larvicide Activity essential oil of B.frutescens L.

A mortality test was conducted to determine the mortality rate of *A. aegypti* larvae in each of *B. frutescens* L. essential oils. Percentage mortality each time an essential oil is presented in Figure 1, which displays fresh leaf essential oil from a concentration of 0-150 µg/mL with a trend of the increasing mortality percentage with a value of 20% and continuing to fall 16.67% at the concentration of 150 µg/mL. Essential oils from dried leaves have a mortality of 6.67% at the concentrations of 50 and 100 µg/mL and then increase at the concentration of 150 µg/mL by 40%. The difference in the percentage of mortality in each concentration of essential oil from fresh and dried leaves influences the toxicity of *B.frutescens* L. essential oil [32-34]. Insects have a sensitivity to bioactive compounds so that it affects their metabolic abilities [35]. Comparison of essential oils and abates at a ratio of 50 µg/mL with 80% mortality. High abate mortality because it contains active ingredients from temephos such as Tetramethyl Thiodi Phenylene, Phosphorothioate 1%, and 99% inert ingredients [36,37]. Temephos is toxic, so it affects the neurotransmitter system through inhibition of cholinesterase so that the larvae die before reaching adulthood [35,36]. Essential oils can kill *A. aegypti* larvae. But not very toxic compared to abate. These results are the same as the study of Vivekanandhan [38] comparing *Eucalyptus globulus* essential oil with abate. How to insecticide in fighting insects is very dependent on the shape and size of larvae, how to get into the body of insects, chemicals, concentration, and the number of doses [39,40]. The ability of *B.frutescens* L. essential oil from dried leaves and fresh leaves can kill larvacide *A, aegypti* at 40%, and 20% mortality, respectively. Based on WHO classification includes deadly effects on a scale of 10-96% [41].

Homogeneity mortality data analysis of *A. aegypti* larvicides showed the results of Fcount <Ftable (2.41 <10.97) on fresh leaves and Fcount <Ftable (2.45 <10.97) on dry leaves at a significant level of

1%. So that both A. Aegypti larvicide mortality data have the same variance, and probit analysis can proceed. The results of probit analysis on *B. frutescens* L. essential oil from fresh leaves showed LC_{50} = 177.83 μg/mL and dried leaves $LC_{(50)}$ = 156.20 μg/mL. Both essential oils are declared toxic because LC_{50} is included in the classification 31 <$LC_{(50)} \leq$ 1,000 μg / mL [42]. The toxic power of essential oils of *B. frutescens* L. dried leaves is greater than that of fresh. Because based on the results of the analysis, GCMS essential oils from dried leaves containing *limonene* composition, as presented in Table 1. Limonene compounds are monoterpenes that enter the digestive organs of A. Aegypti larvae and are absorbed by the midget dining then enter the blood so that it interferes with the metabolism of larval bodies. The larvae lack the energy to carry out their activities, eventually, the larvae cramp and die [43,44]. Wikandari and Surati [45] research stated that *Limonene* is a stomach poison that can kill *A. aegypti* larvacide [46].

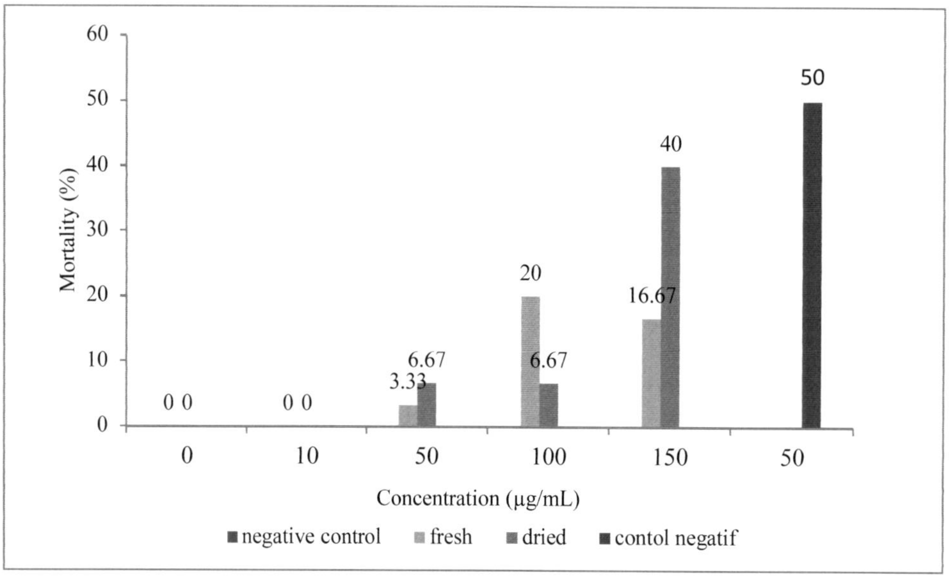

Figure 1. *A.aegypti* larvae mortality at several concentrations of *B. frutescens* L. essential oils from fresh leaves and dried leaves.

3.3. Essential Oil Compounds of of B.frutescens L.

GC-MS test results of *B. frutescens* L. essential oil from fresh and dried leaves displayed by their chromatograph in Figure 2 and Figure 3. A chromatogram showed 25 peaks in fresh leaves and 23 peaks in dry leaves. The compounds identified from the essential oil of *B. frutescens* L. fresh leaves and dried leaves are shown in Table 1.

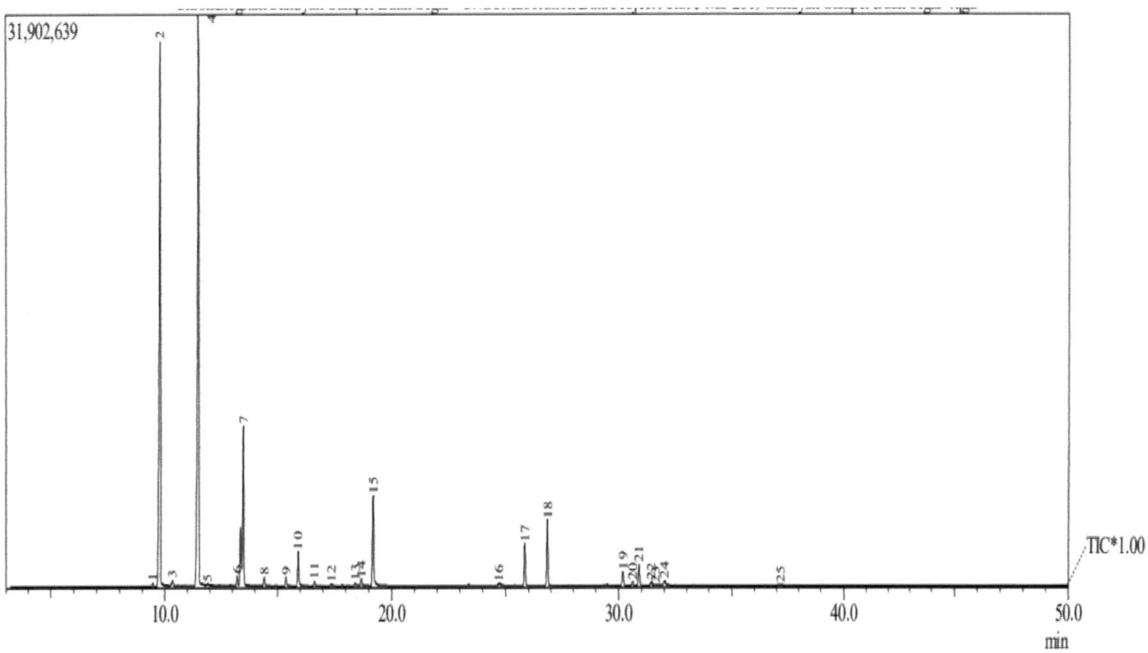

Figure 2. GCMS chromatogram of essential oils *B.frutescens* L. fresh leaves.

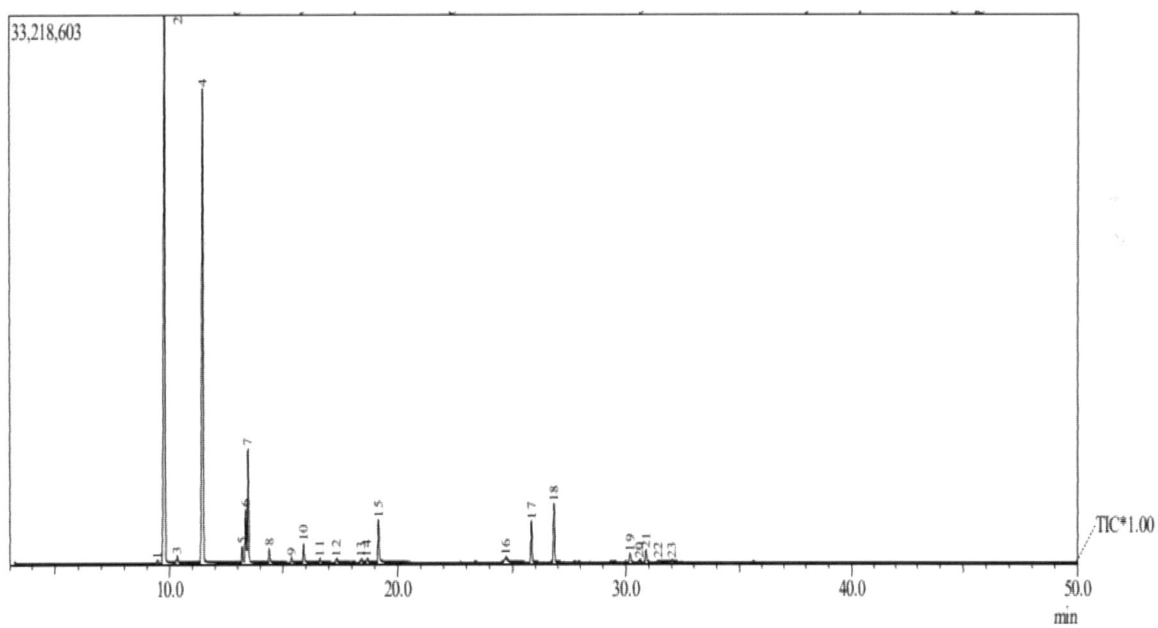

Figure 3. GCMS chromatogram of essential oils *B.frutescens* L. dry leaves.

Table 1. The compounds in the essential oil of *B. frutescens* L. from fresh and dried leaves.

No	Compound	Molecular Formula	BM	Content (%)	
				fresh	dried
1.	*α - Phellandrene*	$C_{10}H_{16}$	136	0.18	0.21
2.	*α – Pinene*	$C_{10}H_{16}$	136	32.52	41.24
3.	*Camphene*	$C_{10}H_{16}$	136	0.49	0.61
4.	*β – Pinene*	$C_{10}H_{16}$	136	36.51	33.31
5.	*Myrcene*	$C_{10}H_{16}$	136	0.16	-
6.	*Benzene 1-Methyl-2- (1-Methylethyl)*	$C_{10}H_{14}$	134	0.50	0.90
7.	*Limonene*	$C_{10}H_{16}$	136	-	3.02
8.	*Eucalyptol*	$C_{10}H_{18}O$	154	10.56	5.76
9.	*γ - Terpinene*	$C_{10}H_{16}$	136	0.48	0.81
10.	*α - Terpinolele*	$C_{10}H_{16}$	136	0.59	0.38
11.	*Linalool*	$C_{10}H_{18}O$	154	1.94	1.11
12.	*D - Fenchyl Alcohol*	$C_{10}H_{18}O$	154	0.33	0.28
13.	*Trans - Pinocarveol*	$C_{10}H_{18}O$	152	0.19	0.33
14.	*Isoborneol*	$C_{10}H_{18}O$	154	0.28	0.25
15.	*Terpineol-4*	$C_{10}H_{18}O$	154	0.52	0.27
16.	*α - Terpineol*	$C_{10}H_{18}O$	154	5.30	2.78
17.	*-(-) Methol*	$C_{10}H_{20}O$	156	0.38	-
18.	*Cyclohexane, 2 chloro-4methyl-1-(1-methylethyl)*	$C_{10}H_{19}Cl$	174	-	0.78
19.	*Trans-Caryophyllene*	$C_{15}H_{24}$	204	2.27	2.51
20.	*α - Humulene*	$C_{15}H_{24}$	204	3.56	3.60
21.	*(-)- Caryophyllene Oxide*	$C_{15}H_{24}O$	220	0.84	0.60
22.	*Trans-Ocimene*	$C_{10}H_{16}$	136	0.24	0.14
23	*(-)- CaryophylleneOxide*	$C_{15}H_{24}O$	220	1.16	0.80
24.	*Cyclohexana*	$C_{15}H_{24}$	204	0.27	0.13
25.	*Bicyclo*	$C_{15}H_{24}$	204	0.14	-
26	*Elemol*	$C_{15}H_{26}O$	222	0.41	0,16
27.	*1,2-Benzenedicarboxylic acid dioctyl ester (CAS) Dioctyl phthalase*	$C_{24}H_{38}O_4$	390	0.18	-
	Total			100	

In Table 1 the GCMS test results are dominated by monoterpene and sesciterpene compounds. There are differences in the content of compounds in the essential oil of *B. frutescens* L. from fresh and dried leaves. Minor compounds, namely, in fresh leaves there are components of the compound myrcene, (-) Metho, Bicyclo, *1,2*-Benzenedicarboxylic acid dioctyl ester *(CAS)* Dioctyl phthalase. While the dried leaves are components of the compositionof limonene and Cyclohexane. The main chemical composition (major) essential oil of *B. frutescens* L consists of compounds α - Pinene, β - Pinene, and Eucalyptol on fresh leaves and dry leaves. The major component of the compound is thought to kill *A. aegypti* larvae. Components of a mixture of α-Pinene and β-Pinene are organic compounds of the monoterpenoid group. According to Watanabe [47], components of monoterpenoid compounds have the potential as insecticides because they have higher toxicity and repellent activity against insects. Also, α-Pinene can be applied as an intermediate product used in the pharmaceutical, chemical fragrance, anti-bacterial, skin disease, disinfectant, and perfume industries. Eucalyptus is a cyclic ether compound and terpenoids. According to Klock, [48] eucalyptus can be used as an insecticide and insect repellent. The composition of the main components and limonene, shown in Figure 4.

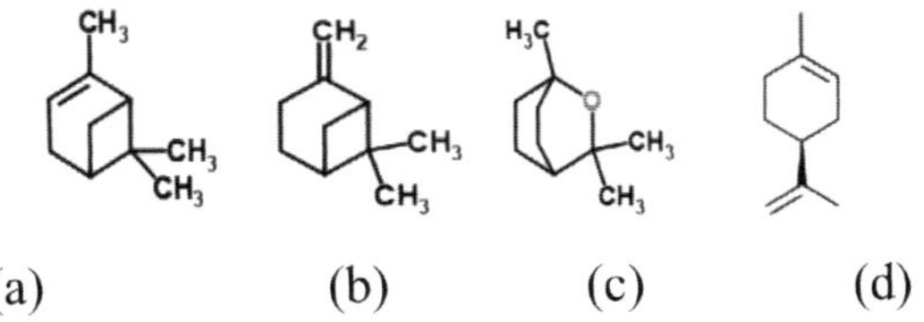

Figure 4. Compound (a) α – *Pinene* (b) β – *Pinene* (c) *Eucalyptol* and (d) *Limonene* (www. Chemspider.com).

4. Conclusion

Based on the results and discussion, it can be argued that *B. frutescens* L. essential oil, the result of the distillation of dry leaves, produces a lot of essential oil compared to wet leaves. And it can kill *A. aegypti* larvae with $LC_{(50)}$ = 177.83 μg/mL for wet leaves, and dry leaves LC_{50} = 156.20 μg/mL. Wet and dry leaf essential oils with $LC_{(50)}$ <1000 μg / mL were declared toxic to larvicides. As a gastric poison that can kill *A. aegypti* larvae. Chemical components of B. *frutescens* L. essential oil fresh leaves contain 25 compounds. . The compounds contained in the essential oil of fresh leaves are thought to kill larvae, namely Myrcene Methol, Bicyclo, and 1,2-Benzenedicarboxylic acid dioctyl ester (CAS) Dioctyl phthalates. While the dry leaves contain 21 compounds. Limonene compounds are thought to kill *A. aegypti* larvae. Other minor compound components present in fresh and dry leaves consist of, α-Phellandrene, Camphene, Benzene 1-Methyl-2- (1-Methylethyl), γ-Terpinene, α-Terpinolele, Linalool, D-Fenchyl Alcohol, Trans- Pinocarveol, α-Humulene, (-) - Caryophyllene Oxide, Cyclohexana, and Elemol.

References

[1] Boekoesoe L 2013 Kajian Faktor Lingkungan Terhadap Kasus Demam Berdarah Dengue (DBD) Studi Kasus di Kota Gorontalo Provinsi Gorontalo Disertasi Fakutas Kesehatan Masyarakat Universitas Gorontalo

[2] Suryani E T 2018 Gambaran kasus demam berdarah dengue di Kota Blitar tahun 2015-2017. *J Berkala Epidemiologi* **6** 260-7

[3] Pennetier C, Corbel V and Hougard J M 2005 Combination of a non-pyrethroid insecticide and a repellent: a new approach to controlling knockdown-resistant mosquitoes *Am. J. Trop. Med. Hyg.* **72** 739-44

[4] Raini M 2009 Toksitologi insektisida rumah tangga dan pencegahan keracunan *Media Penelitian dan Pengembangan Kesehatan* **19** 27-33

[5] Kim K, Kabir E and Jahan S A 2007 Exposure to pesticides and the associated human health effects *Sci. Total Environ.* **575** 525-35

[6] Dai D N, Thang T D, Olayiwola T O and Ogunwande I A 2015 Chemical composition of essential oil of *Baeckea frutescens* L. *Int. Res. J. Pure Appl. Chem.* **8** 26-32

[7] Tam N T, Thuam D T, Bighelli A, Castola V, Muselli A, Richomme P *et al.* 2004 *Baeckea frutescens* leaf oil from Vietnam: composition and chemical variability. *Flavour Frag. J.* **19** 217-20

[8] Zhaobin P, Feichao L, Yuee L and Xiasong L 2012 Antioxidant activity and anti inflammatory effect of total flavonoids from *Baeckea frutescens China Pharm.* **15** 477-8

[9] Jia B X, Zeng X L, Ren F X, Jia L, Chen X Q, Yang J *et al.* 2014 Baeckeins F–I, four novel C-methylated biflavonoids from the roots of *Baeckea frutescens* and their anti-inflammatory activities *Food Chem.* **155** 31-7

[10] Navanesan S, Wahab N A, Manickam S and Sim K S 2015 Evaluation of selected biological capacities of *Baeckea frutescens BMC Compl. Altern Medi.* **15** 186

[11] Razmavar S, Abdulla M A, Ismail S B and Hassandarvish P 2014 Antibacterial activity of leaf extracts of *Baeckea frutescens* against methicillin-resistant *Staphylococcus aureus Biomed Res. Int.* **2014** 521287
[12] An N T G, Huong L T, Satyal P, Tai T A, Dai D N, Hung N H *et al.* 2020 Mosquito larvicidal activity, antimicrobial activity, and chemical compositions of essential oils from four species of *Myrtaceae* from central Vietnam *Plants* **9** 544
[13] Rusli M S 2020 *Sukses Memproduksi Minyak Atsiri* (Jakarta: PT Agro Media Pustaka)
[14] Guenther E 2006 *Minyak Atsiri Jilid I* (Jakarta: UI Press)
[15] Badan Standardisasi Nasional 2014 *SNI 80281-1: 2014 Alat penyuling minyak atsiri - Bagian 1 : Sistem kukus – Syarat mutu dan metode uji* (Badan Standardisadi Nasional Jakarta Indonesia) p 6
[16] Schaduw J, Pojoh J A and Djabar T O 2012 Isolasi dan identifikasi minyak atsiri daun nilam (*Pogostemon cablin* Benth). *Jurnal Ilmiah Farmasi* **3** 61-3
[17] Badan Standardisasi Nasional 201 *SNI 3954-2014 Kayu Putih* (Badan Standardisadi Nasional Jakarta Indonesia) p 8
[18] Ketaren S 1986 *Minyak dan Lemak Pangan* (Jakarta: Penerbit Universitas Indonesia) pp. 6-11, 24-25, 139-151
[19] Sumangat D dan Risfaheri 2010 *Standar dan Masalah Mutu Minyak Nilam Indonesia. Monografi Nilam* (Bogor: Balai Penelitian Tanaman Rempah dan Obat) pp. 108-115.
[20] Rumengan A P 2010 Uji larvasida nyamuk (*Aedes aegypti*) dari ascidian (*Didemnum molle*). *Jurnal Perikanan dan Kelautan Tropis* **6** 83-6
[21] Ariyanto M W 1999 Pengujian Efikasi Kanidiospora Cendawan *Metharizium anisopliae* yang Dikultur pada Media Jagung sebagai Bioinsektisida Terhadap Larva Nyamuk *Aedes aegypti* Fakultas Kedokteran UGM Yogyakarta
[22] Florence A R and Solomon J 2016 Larvicidal activity of *Gmelina asiatica* L. leaf extracts against *Aedes aegypti* and *Culex quinquefasciatus Ann. Biol. Res.* **7** 12-20
[23] Amir H, Butt B Z and Vehra S E 2017 Evaluation of larvicidal activity of *Parthenium hysterophorus* against *Aedes aegypti*. *Int. J. Mosq. Res.* **4** 01-04
[24] Syahbana MS 2010 *Sukses Memproduksi minyak Atsiri* (Jakarta: PT. Agro Media Pustaka)
[25] Suryanto S, Sulaeman R and Budiani E S 2017 Pengaruh pola pengeringan terhadap rendemen dan kualitas minyak atsiri daun pucuk merah (*Syzygium oleana*) *Jurnal Online Mahasiswa Fakultas Pertanian Universitas Riau* **4** 1-8
[26] Khasanah L U, Kawiji K, Utami R and Aji Y M 2015 Pengaruh perlakuan Pendahuluan terhadap karakteristik mutu minyak atsiri daun jeruk purut (*Citrus hystrix* DC) *Jurnal Aplikasi Teknologi Pangan* **4** 48-55
[27] Harimurti N, Soerawidjaja T H, Sumangat D and Risfaheri R 2013 Ekstraksi Minyak Nilam (*Pogostemon cablin* Benth) dengan Teknik Hidrodifusi Pada Tekanan 1–3 Bar *Jurnal Pascapanen Pertanian* **9** 1-10
[28] Rohayati N 1997 Penggunaan Bentonit, Arang Aktif dan Asam Sitrat Untuk Meningkatkan Mutu Minyak Akar Wangi Skripsi (Institut Pertanian Bogor Indonesia)
[29] Idris A, Jura M R and Said I 2014 Analisis kualitas minyak nilam (*Pogostemon cablin* Benth) produksi kabupaten Buol *Jurnal Akademika Kimia* **3** 79-85
[30] Haditomo I 2010 Efek larvasida ekstrak daun cengkeh (*Syzgium Aromaticum* L) terhadap *Aedes Aegypti* L Skripsi (Fakultas Kedokteran Universitas Sebelas Maret, Surakarta Indonesaia)
[31] Kumar S, Mishra M, Wahab N and Warikoo R 2014 Larvicidal, repellent, and irritant potential of the seed-derived essential oil of Apium graveolens against dengue vector, *Aedes aegypti* L. (Diptera: *Culicidae*) *Front. Public Health* **2** 147
[32] Prijono D 2003 Teknik Ekstraksi, Uji Hayati, dan Aplikasi Senyawa Bioaktif Tumbuhan. Panduan bagi Pelaksana PHT Perkebunan Rakyat (Departemen Hama dan Penyakit Tumbuhan. Institut Pertanian Bogor Indonesia)

[33] Sulistiyani A 2015 Effectiveness of essential oil as larvacide on *Aedes aegypti Medical Journal of Lampung University* **4** 23-8
[34] de Oliveura Araujo A F, Ribeiro-Paes J T, de Deus J T, de Holanda Cavalcanti S C, de Souza Nunes R, Alves PB *et al.* 2016 Larvicidal activity of *Syzygium aromaticum* (L.) Merr and *Citrus sinensis* (L.) Osbeck essential oils and their antagonistic effects with temephos in resistant populations of *Aedes aegypti*. *Mem. Inst. Oswaldo Cruz* **111** 443-9
[35] Lawrens F I J, Wahoyongan G J, Bernaad J B 2014 Pengaruh dosis abate terhadap jumlah populasi jentik nyamuk *Aedes aegypti* di Kecamatan Malayang Kota Manado Skripsi (Fakultas Pertanian Universitas Sam Ratulangi Manado Indonesia)
[36] Sulistiyaningrum A 2015 Efectiveness of esential oil as larvacida an *Aedes aegypti*. *Medical Jurnal of Lampung University* **4** 23-8
[37] Vivekanandhan P, Usha-Raja-Nanthini A, Valli G and Shivakuma M S 2018 Comparative efficacy of *Eucalyptus globulus* (Labill) hydrodistilled essential oil and temephos as mosquito larvicide *J. Nat. Prod. Res.* **9** 1-4
[38] Hoedoj R dan Zulhasril 2004 *Insektisida dan Resistensi* (Parasitologi Kedokteran Edisi Ke-3. Fakultas Kedokteran Universitas Indonesia. Jakarta)
[39] Prijono D 2003 Teknik Ekstraksi, Uji Hayati, dan Aplikasi Senyawa Bioaktif Tumbuhan. Panduan bagi Pelaksana PHT Perkebunan Rakyat (Departemen Hama dan Penyakit Tumbuhan. Institut Pertanian Bogor Indonesia)
[40] WHO. 2005. *Guidelines for laboratory and field testing of mosquito larvicides*. (WHO Geneva)
[41] Rochmat A, Bahiyah Z and Adiati M F 2017 Pengembangan biolarvasida jentik nyamuk *Aedes aegypti* berbahan aktif ekstrak beluntas (*Pluchea indica* Less) *Reaktor* **16** 103-8
[42] Nirma N, Ausilawaty A, Hasbi I and Munawir A 2015 Efektivitas larvasida ekstrak kulit buah jeruk nipis (*Citrus aurantifolia*) dalam membunuh jentik nyamuk *Aedes* sp (studi di daerah epidemi DBD di wilayah kerja Puskesmas Antang Kecamatan Manggala). *Higiene* **3** 87-96
[43] Rocha D K, Matosc O, Novoa M T, Figueiredo A C, Delgado M and Moiteiro C 2015 Larvicidal activity against *Aedes aegypti* of *Foeniculum vulgare* essential oils from Portugal and Cape Verde. *Nat. Prod. Commun.* **10** 677-82
[44] Wikandari R J and Surati 2018 Effect of *Citrus hystrix* DC peels extract against morphology and histology of *Aedes aegypti* larvae *Aspirator* **10** 119-26
[45] Park H M, Kim J, Chang KS, Kim BS, Yang YJ, Kim GH *et al.* 2011 Larvicidal activity of Myrtaceae essential oils and their components against *Aedes aegypti*, acute toxicity on *Daphnia magna*, and aqueous residue *J. Med. Entomol.* **48** 405-10
[46] Watanabe K, Shono Y, Kakimizu A, Okada A, Matsuo N, Satoh A *et al.* 1993 New mosquito eepellent from *Eucalyptus camaldulensis J. Agric. and Food Chem.* **41** 2164-6
[47] Ringgani R, Budhijanto B and Budiman A 2016 Kinetika reaksi isomerisasi α-pinene. *Eksergi* **13** 7
[48] Klocke J A, Darlington M D and Balandrin M F 1987 1,8-Cineole (*eucalyptol*), a mosquito feeding and ovipositional repellent from volatile oil of *Hemizonia fitchii* (*Asteraceae*). *J. Chem. Ecol.*, **13** 2131-41

IOP Conf. Series: Materials Science and Engineering 1011 (2021) 012003 doi:10.1088/1757-899X/1011/1/012003

Optimization process for ethylene glycol production using the Pareto solution

M S Sulaiman, F S Rohman S and N Aziz*

School of Chemical Engineering, Engineering Campus, Universiti Sains Malaysia, Seri Ampangan, 14300 Nibong Tebal, Seberang Perai Selatan, Penang, Malaysia.

*Corresponding author: chnaziz@usm.my

Abstract. In this work, ε-constraint approach within ASPEN Plus environment is used to generate the Pareto solutions for two objectives: maximize productivity and minimize energy consumption, for the ethylene glycol production. Each point of Pareto solutions consists of different optimal temperature and hydrogen flowrate values, which lead to distinct amount of energy consumption and ethylene glycol product rate. These solutions give flexibility in evaluating the trade-offs and selecting the most suitable operating policy. The most satisfactory optimum and trade-off of both objectives is determined. The optimum value for ethylene glycol production and energy consumption are 109.92 kmol/hr and 11227860.10 kJ/hr, respectively, while the trade-off values are 18.02 kmol/hr and 10903846.2 kJ/hr, respectively. The comparison between multi-objective optimization and single-objective optimization concludes that a high benefit against disadvantages is achieved by implementing multi-objective optimization. The difference between multi-objective optimization and single-objective optimization for ethylene glycol production and energy consumption are –5.65 kmol/hr and 17526931.8 kJ/hr, respectively.

1. Introduction

Production of EG is rapidly increasing and is predicted that this increment will continue until 2030 [1]. EG is used as raw materials in the production of a wide range of products including polyester fibers, fiberglass and polyethylene terephthalate [2]. Industries involved in the production of EG have their own methods and approaches in attaining their target (high production rate or cost saving). Most of these techniques involved optimization of certain objectives.

Optimization is to make the best or most effective use of a situation or resource. Referring to this statement, the production of EG requires optimization for; energy consumption reduction, maximizing productivity, reduction of side products, minimizing feed requirements etc. These objectives usually conflict with one another and thus most industries prioritized only one resulting in inefficient plant operations. In the EG process, there are conflicting objective functions, i.e. maximum production and minimum energy consumption which cannot be solved by single objective optimization technique. Simultaneous optimization of these objectives yields in a MOO problem, which is characterized by a set of multiple solutions, known as Pareto solutions.

MOO is defined as an area of multiple criteria decision making that is concerned with mathematical optimization problems involving more than one objective function to be optimized simultaneously where optimal decisions need to be taken in the presence of trade-offs between two or more conflicting objectives. For a nontrivial MOO problem, no single solution exists that simultaneously optimizes each objective. In that case, the objective functions are said to be conflicting, and there exists a (possibly infinite) number of Pareto optimal solutions. A solution is called Pareto optimal, if none of the objective functions can be improved in value without degrading some of the other objective values. Different solution philosophies and goals are developed when setting and solving MOO problems due to the difference in points of views [3].

Pareto (optimal) solutions are range of solutions (objectives) used in acquiring the Pareto optimal set. Pareto optimal set is a set of all the solutions located in the feasible decision space. The boundary defined by the set of all point mapped from the Pareto optimal set is called the Pareto optimal front. Pareto optimal solutions are obtained through various applicable methods such as ε-Constraint Method - Keep just one of the objective and restricting the rest of the objectives within user-specific values, Weighted Sum Method - Scalarize a set of objectives into a single objective by adding each objective pre-multiplied by a user-supplied weight and Weighted Matrix Method - Combine multiple objectives using the weighted distance metric of any solution from the ideal solution [4].

MOO has been increasingly employed in chemical engineering and manufacturing. An optimization of the pressure swing adsorption process (cyclic separation process) involving dual maximization of nitrogen recovery and nitrogen purity using the multi-objective genetic algorithm (MOGA) was developed by Fiandaca and Fraga in 2009. A fine approximation of the Pareto frontier with acceptable trade-offs between the objectives results was produced [5]. In 2010, a multi-objective problem for the thermal processing of food involving two case studies (bi-objective and triple objective problems) with nonlinear dynamic models were solved by Sendín *et al.* using a hybrid approach consisting of the weighted Tchebycheff and the Normal Boundary Intersection approach [6].

A MOO of the combined carbon dioxide reforming and partial-oxidation of methane involving the objective functions of methane conversion, carbon monoxide selectivity and hydrogen to carbon monoxide ratio were performed by Ganesan *et al.* using the Normal Boundary Intersection (NBI) method in conjunction with two swarm-based techniques (Gravitational Search Algorithm (GSA) and Particle Swarm Optimization (PSO)) in 2013[7]. In the same year, an alternative technique to solve multi-objective optimization problems using the Aggregating Functions Approach, the Adaptive Random Search Algorithm, and the Penalty Functions Approach were used to compute the initial set of the non-dominated or Pareto-optimal solutions were proposed by Abakarov *et al.* [8]. In 2018, an approach using a Mixed-Integer Linear Program to solve the optimization problem for a weighted sum of the two objectives to calculate a set of Pareto optimal solutions were formulated by Pearce *et al.* The application of the approach showed improvements in at least one objective in most tasks and in both objectives in some of the processes [9].

Based on previous researches, various MOO techniques have been implemented to different kinds of processes. However, it is comprehended that the Single Objective Optimization (SOO) is used in the production of EG. This hinders the process to operate at its best optimal condition as SOO can mitigate only one objective. Thus, this research fills the gap by implementing MOO for the production of EG to maximize productivity and minimizing energy consumption simultaneously.

2. Methodology

2.1. Modelling and Initial Optimization

A reactor model is taken from Yu and Chien (2017) and simulated using ASPEN Plus V11 [10]. The model is constructed as in figure 1. It is a Plug Flow Reactor (PFR) model which comprises the characteristics stated in table 1.

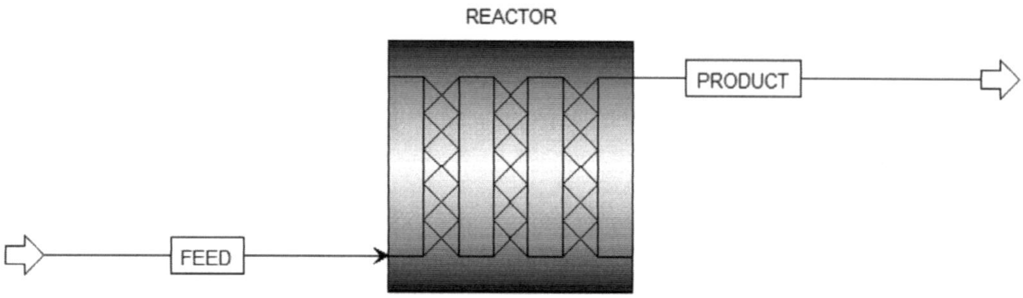

Figure 1. ASPEN Plus Plug Flow Reactor model.

Table 1. Specification of ASPEN Plus PFR model.

Feed Parameter	Quantity	Reactor Parameter	Quantity
Temperature	160 °C	Temperature	210 °C
Pressure	25 bar	Pressure	25 bar
Mole Flow	7059.97 kmol/hr	Length	5 m
DMO	0.0179072 %	Diameter	1 m
MEOH	0.264306 %	Bed Voidage	0.5
MG	0.0005002 %	Particle Density	980 kg/cum
H_2	0.717187 %		
CO	0.00010004 %		

The model is then optimized to increase the production of EG by obtaining the highest possible yield. This optimization refers to SOO. The purpose of this optimization is compared (benchmarking) with MOO. First, all related inputs and outputs are defined. The objective is set to attain maximum yield. There are no constraints applied to the optimization process. Feed hydrogen and reactor temperature is set as the manipulated variable (MV). The executable statement for the process is set as shown in Equation 1.

2.2. Correlation of Objectives
The objectives of MOO are determined which are EG production and energy consumption. The corresponding equations relating both objectives are constructed as shown equation 1 to 6.

$$a = (DMO\ IN \cdot 65.57 + MEOH\ IN \cdot 31.63 + MG\ IN \cdot 129.32 + H2\ IN \cdot 28.24 + CO\ IN \cdot 29.15)(TOUT - T\ IN) \quad (1)$$

$$b = (DMO\ IN \cdot 65.57 + MEOH\ IN \cdot 31.63 + MG\ IN \cdot 129.32 + H2\ IN \cdot 28.24 + CO\ IN \cdot 29.15)(25 - T\ IN) \quad (2)$$

$$c = (EG\ OUT \cdot 162.5 + ETOH\ OUT \cdot 209.9 - DMO\ IN \cdot 372.8) \quad (3)$$

$$d = (DMO\ OUT \cdot 65.57 + MEOH\ OUT \cdot 31.63 + MG\ OUT \cdot 129.32 + H2\ OUT \cdot 28.24 + CO\ OUT \cdot 29.15 + EG\ OUT \cdot 77.99 + ETOH\ OUT \cdot 65.21 + H20\ OUT \cdot 33.59)(T\ OUT - 25) \quad (4)$$

$$e = (111.7 + 77.99(T\ OUT - 25) \quad (5)$$

$$EG\ OUT = (a + b + c + d) \cdot e^{-1} \quad (6)$$

2.3. Optimization
For the ASPEN Plus optimization process, three separated simulation is carried out relating EG production to energy consumption which are Low Energy (LE) optimization, High Product (HP) optimization and MOO. LE and HP Optimization is to determine the boundary of the Pareto optimal front.

2.3.1. LE and HP Optimization. For HP optimization, the objective is set to gain maximum product. All related inputs and outputs required for the process are defined. There are no constraints applied to the optimization process. Feed hydrogen and reactor temperature is set as the manipulated variable (MV). The executable statement for the process is set as shown in Equation 1. As for LE optimization, the objective is set to gain minimum energy consumption. All related inputs and outputs required, constraints and MV are the same as HP optimization. The executable statement for the process is set as shown in equation 1.

2.3.2. MOO. For MOO, the steps are the same as HP optimization. However, there are constraints added to the optimization process. These constraints (EGOUT and Q) are obtained and defined based on the ε-constraint method. The ε-constraint method is stated in equation 7.

$$minimize\ f_{\mu}(x),$$

$$subject\ to\ f_m(x) \leq \varepsilon_m, m = 1, 2, \ldots, M\ and\ m \neq \mu$$

$$g_j(x) \geq 0, j = 1, 2, \ldots, J$$

$$h_k(x) = 0, k = 1, 2, \ldots, K$$

$$x_i^{(L)} \leq x_i \leq x_i^{(U)}, i = 1, 2, \ldots, n \qquad (7)$$

2.4. Pareto Chart
The Pareto front is plotted using the result obtained from the MOO section. The relationship between objectives, between MVs and between objectives and MVs are studied and discussed. The optimum values for EG production and energy consumption are determined. Comparison of MOO and SOO are made.

3. Result and Discussion
The reactor model was validated by experimental data from Li *et al.* (2015) [11] and Yu and Chien (2017) [10].

3.1. Objective Optimization
In order to optimize the objectives, it is first required to understand the characteristic of the respective objectives with changes in MVs. Figure 2, 3, 4 and 5 displays the characteristic of individual objectives against MV. Objective 1 represents product flow (EG), objective 2 represents energy consumption (Q), MV1 represents hydrogen flowrate and MV2 represents temperature. Hydrogen flowrate and temperature is chosen as the MVs because they are the factors directly affecting the rate of reaction [12].

From figure 2, it shows a nonlinear curve of which objective 1 increase as MV1 increase. The increment reduces as objective 1 reaches its maximum point. The slow increase is caused by maximum reaction rate limit. In other words, the production of EG is at its highest due to the rate of reaction reaching its maximum point. This is in line with the reaction rate and kinetic [10]. Thus, the EG production rate cannot be further increased.

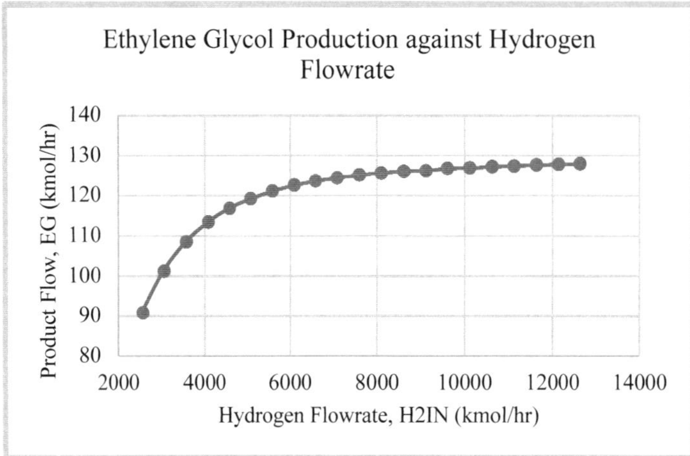

Figure 2. Relationship of MV1 and Objective 1.

Figure 3 shows that objective 1 increase exponentially as MV2 increase. However, objective 1 suffers steady fall once MV2 rises beyond 194 °C. This is due to more side products are formed as the reactor temperature increases. A chain reaction causing the rise in formation of side product due to more main product being consumed explains the fall of objective 1. This follows the Le Chatelier's principle [13].

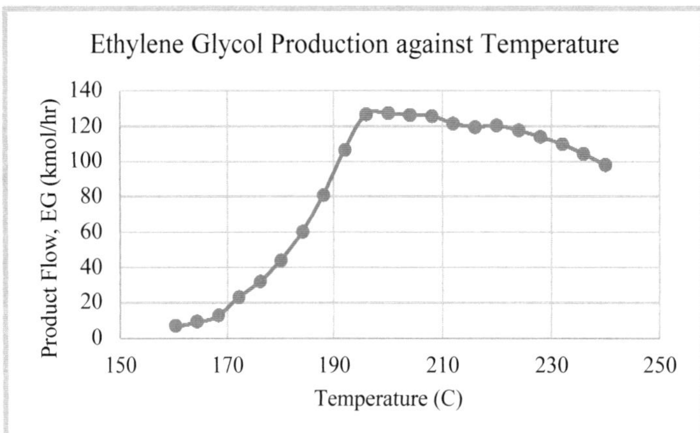

Figure 3. Relationship of MV2 and Objective 1.

From figure 4, it is seen that the graph has an almost linear increment. This shows that as MV1 rise, objective 2 increases. Since the reaction is exothermic, this leads to more energy released with increasing reaction thus, obeys the law of thermodynamics [14]. With higher hydrogen flowrate, lower energy consumption is achieved.

Figure 5 shows that as MV2 rise, objective 2 increases. This increment however continues with a fall. This is due to at first low temperature is used, so the energy released from the exothermic reaction helps in providing additional energy [14]. But as the temperature increases, more energy is required to heat up the reactor causing the rise in energy consumption.

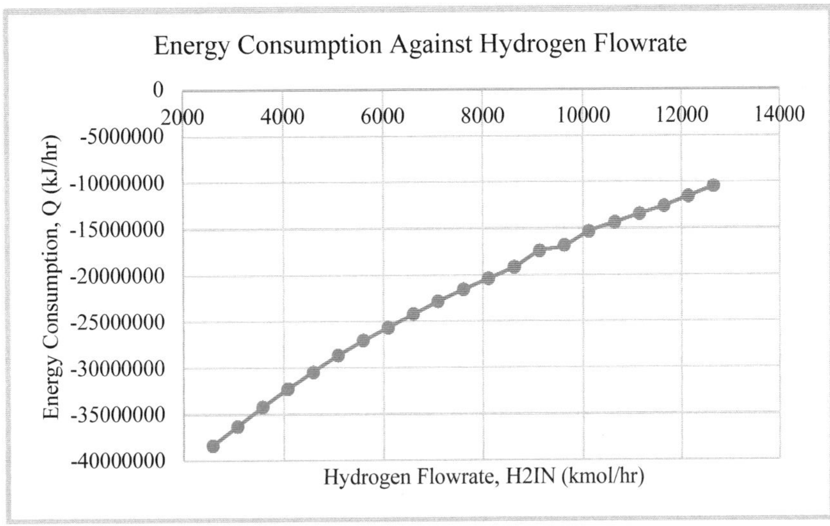

Figure 4. Relationship of MV1 and Objective 2.

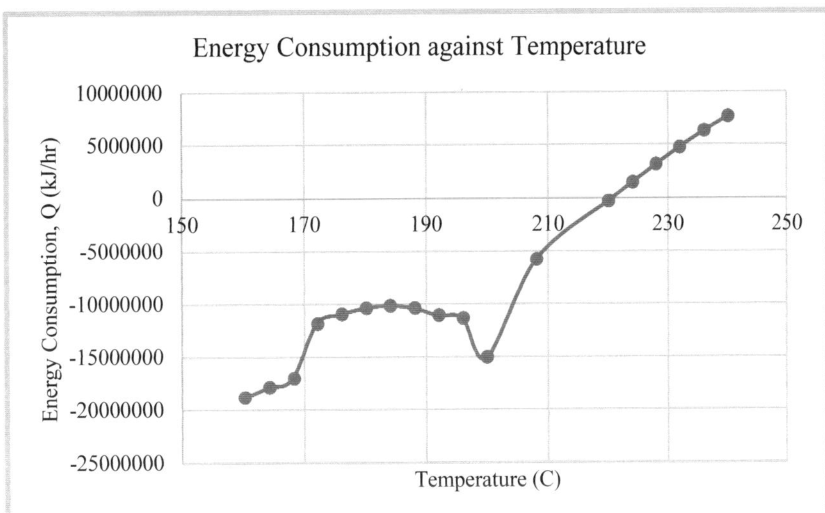

Figure 5. Relationship of MV2 and Objective 2.

To achieve a fine MOO, the hydrogen flowrate must be high for better EG production and lower energy consumption. The temperature must be strictly controlled in order to determine the best EG production while maintaining low energy consumption.

3.2. Multi-Objective Optimization

ε-constraint method is selected because it is applicable to either convex or non-convex problems [15]. That makes it perfect for this case since no solid grasp on the Pareto foundation can be made at this point. In addition, the method is thoroughly simple except in determining the epsilon (multi-objective) range. The range must fulfilled the requirements in which the objectives have to be within the range. This is the reason for the LE and HP optimizations. With the optimizations data, a range (constraints) for the objectives is developed as shown in equation 8.

$$max\ f_{EG}(x);\ f_Q = -24900468.30 \ll \varepsilon$$

IOP Conf. Series: Materials Science and Engineering 1011 (2021) 012003 doi:10.1088/1757-899X/1011/1/012003

$$f_{EG} = 127.93 \gg \varepsilon$$
$$\min f_Q(x);\ f_{EG} = 7.81 \gg \varepsilon$$
$$f_Q = -324013.90 \ll \varepsilon$$
$$-24900468.30 \ll \varepsilon \ll 127.93 \tag{8}$$

The constraints are manipulated within the range and a set of data are obtained. A Pareto front is plotted for production of EG against energy consumption. The purposed of the plot is to identify the optimum point which satisfied both objectives accordingly.

From figure 6, an inverse s-curved plot is obtained. A point named Utopia point is marked on the plot. The Utopia point is the ideal point in which is the ideal values for each objective. This point serves as a reference point in regards of finding the actual optimum point [16]. Thus, point B which is the true optimum point (value) is acquired. Point B is selected as the optimum point because it has the shortest distance with the Utopia point which justified that point B is the nearest to ideal-optimum value. From the plot, the trade-off between both objectives is also known. This shows that indeed both the objectives are in conflict without dominating each other.

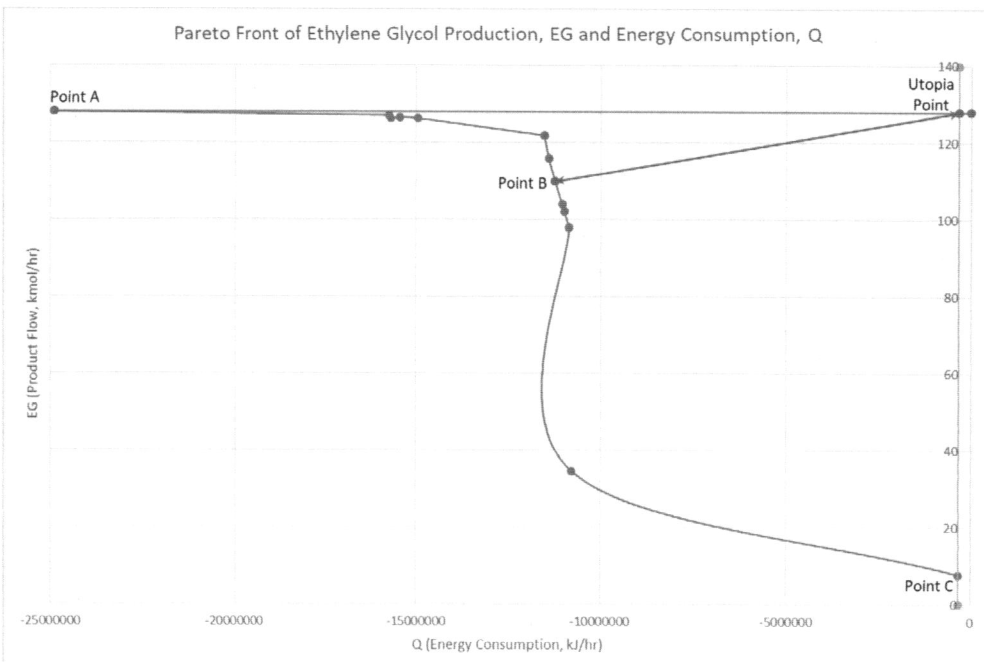

Figure 6. Pareto Optimal Front.

The optimum value for EG production and energy consumption are 109.92 kmol/hr and 11227860.10 kJ/hr, respectively. On the other hand, the trade-off value for EG production and energy consumption are 18.02 kmol/hr and 10903846.2 kJ/hr, respectively.

To further proved that the Pareto front in figure 6 is acceptable, figure 7 is plotted. From the plot, it is observed that MV2 rises with constant MV1. MV1 then decrease drastically once MV2 is beyond 194°C. The MVs for optimum point (Utopia point) is located before 194°C. This corresponds with the objective and MV relationship in figure 2, 3, 4 and 5. Thus, the validity of the Pareto optimal solution is confirmed.

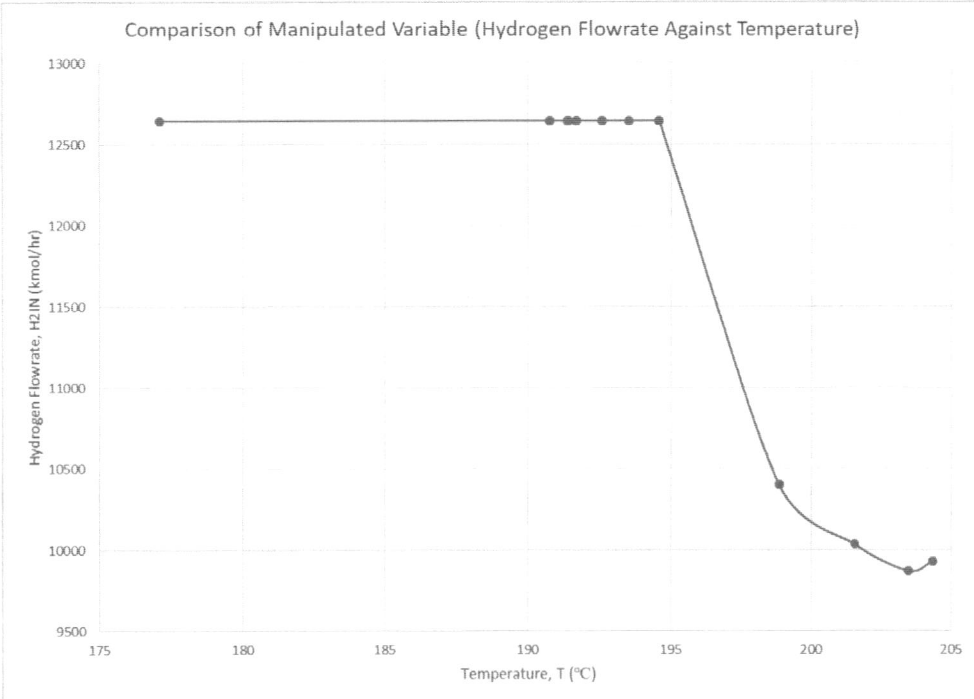

Figure 7. Pareto Optimal Solution.

3.3. Comparison of MOO and SOO

From table 2, it is seen that the production of EG for SOO is higher and with much lower feed hydrogen than MOO. However, the reactor temperature and energy consumption for MOO is lower than SOO. As observed in figure 2, 3, 4 and 5, energy consumption can be reduced by increasing hydrogen flowrate. The difference between MOO and SOO for ethylene glycol production and energy consumption are -5.65 kmol/hr and 17526931.8 kJ/hr, respectively. The difference in energy consumption is large with MOO having lower value and the difference in EG production is relatively small with SOO having higher value.

Table 2. Results for SOO and MOO.

	SOO	MOO	Unit
H2 IN	5063.32	12642.43	kmol/hr
H2 OUT	4574.38	12192.00	kmol/hr
EG OUT	115.57	109.92	kmol/hr
T OUT	210.00	192.56	°C
Q	-28754791.9	-11227860.1	kJ/hr

The decrease in EG production is acceptable because of the high drop in energy consumption which leads to a lower operating temperature creating a safer environment. The increase in hydrogen required is exceptionally high but can be countered by proper recycling of feed. Thus, the overall comparison showed high benefits with slight trade-off.

4. Conclusion

The overall optimization process is a success. A well-developed PFR model served excellently in error prevention. The interaction between MV1 and MV2 with objective 1 and objective 2 is highly dependent. Objective 2 is much more sensitive towards the change in MV1 and MV2 compared to objective 1. The interaction between MV1 and MV2 is unique in which it cannot be seen until 194°C.

The optimum value for EG production and energy consumption are 109.92 kmol/hr and 11227860.10 kJ/hr, respectively while the trade-off values are 18.02 kmol/hr and 10903846.2 kJ/hr, respectively. The difference between MOO and SOO for EG production and energy consumption are -5.65 kmol/hr and 17526931.8 kJ/hr, respectively. This shows that the comparison between MOO and SOO results in objective 1 for SOO is slightly higher than MOO but for objective 2, MOO is significantly higher than SOO. This concludes that a high benefit with slight trade-off (high energy consumption reduction and slight decrease in EG production) is achieved by implementing MOO.

Acknowledgement

The financial support from Universiti Sains Malaysia through Research University Grant (RUI) 203.PJKIMIA.8014146 is greatly acknowledged.

References

[1] American Chemistry Council (ACC) Inc 2020 What is Ethylene Glycol?

[2] Garside M 2019 Global production capacity of ethylene 2018 & 2030

[3] Fiandaca G, Fraga E S and Brandani S 2009 A multi-objective genetic algorithm for the design of pressure swing adsorption *Eng. Optimiz.* **41(9)** 833–54

[4] Deb K 2001 *Multi-Objective Optimization using Evolutionary Algorithms* John Wiley & Sons, Inc 13-80

[5] Rohman F S and Aziz N 2020 Multi-objective optimization of batch electrodialysis for minimizing energy consumption by using non-dominated sorting genetic algorithm (NSGA-II) *IOP Conference Series: Materials Science and Engineering* 736

[6] Sendín J O H, Alonso A A and Banga J R 2010 Efficient and robust multi-objective optimization of food processing: A novel approach with application to thermal sterilization *J. Food Eng.* **98(3)** 317–24

[7] Ganesan T, Elamvazuthi I, Ku Shaari K Z and Vasant P 2013 Swarm intelligence and gravitational search algorithm for multi-objective optimization of synthesis gas production *Applied Energy* **103** 368–74.

[8] Abakarov A, Sushkov Yu and Mascheroni R H 2012 Multi-criteria optimization and decision-making approach for improving of food engineering processes *Int. J. Food Stud.* **2** 1–21

[9] Pearce M, Mutlu B, Shah J and Radwin R 2018 Optimizing Makespan and Ergonomics in Integrating Collaborative Robots into Manufacturing Processes *IEEE T. Autom. Scie. Eng.* **15(4)** 1772–84

[10] Yu B-Y and Chien I L 2017 Design and optimization of dimethyl oxalate (DMO) hydrogenation process to produce ethylene glycol (EG) *Chem. Eng. Res. Des.* **121** 173-90

[11] Li S, Wang Y, Zhang J, Wang S, Xu Y, Zhao Y and Ma X 2015 Kinetics study of hydrogenation of dimethyl oxalate over cu/sio2 catalyst Industrial & Engineering Chemistry Research **54** 1243-50

[12] Tro N J *Chemistry: Structure and Properties, 2nd Edition* Pearson Chapter 14

[13] Knox K 1985 Le Châtelier's Principle *Journal of Chemical Education* **62** 863

[14] Stepanov I A 2000 The 1st Law of Thermodynamics in Chemical Reactions *Chemical Physics*

[15] Rohman F S and Aziz N 2019 Performance metrics analysis of dynamic multi-objective optimization for energy consumption and productivity improvement in batch electrodialysis *Chem. Eng. Commun.* 1-13

[16] Qi Y, Zhang Q, Ma X, Quan Y and Miao Q 2017 Utopian point-based decomposition for multi-objective optimization problems with complicated Pareto fronts *Appl. Soft Comput.* **61** 844-59

IOP Conf. Series: Materials Science and Engineering 1011 (2021) 012004 doi:10.1088/1757-899X/1011/1/012004

Fermentation of amylolytic yeast and lactic acid bacteria to improve the quality of modified cassava

Ema Damayanti[1*], Meylida Ichsyani[2], Lusty Istiqomah[1], Ayu Septi Anggraeni[1] and Muhamad Kurniadi[1]

[1]Research Unit for Natural Product Technology, Indonesian Institute of Sciences (BPTBA-LIPI)
Jl. Jogja Wonosari KM 31.5, Gunungkidul, D.I.Yogyakarta, 55861, Indonesia
[2]Faculty of Medicine, Universitas Jenderal Soedirman, Purwokerto, Indonesia

*Corresponding author: ema.damayanti@lipi.go.id

Abstract. The amylolytic activity has been studied in yeast and lactic acid bacteria (LAB) which are isolated from various types of traditional fermented cassava. The study aimed to obtain yeast and LAB isolates which high amylolytic activity for inoculum in cassava fermentation. A total of 23 yeasts and 22 LAB isolates were isolated from Gatot, Tiwul, Tapai, and Cassava fermented using chloramphenicol yeast glucose (CYG) media for yeast and de Man Rogosa Sharpe agar (MRSA) + $CaCO_3$ for LAB. Yeast amylolytic activity was tested using yeast extract peptone starch agar medium, while the amylolytic LAB used MRS + starch media with Lugol's iodine indicator. Colonies showing clear zone areas were measured as amylase activity on the iodine complex method. Isolates that have the highest amylolytic activity were used as inoculum in cassava fermentation. Yeast Tr7 and LAB G6 isolates showed the highest amylolytic activity (0.0624 U/mL and 0.0627 U/mL). Cassava fermentation products using yeast Tr7 and BAL G6 resulted 52.06% of hydrogen cyanide (HCN) reduction higher than without inoculum (32.32%). According to biochemical identification using API 50 CHL kit, LAB G6 isolate was 99.6% identical with *Pediococcys pentosaceus* whereas biochemical identification using API 20 C AUX showed that yeast Tr7 was 95.8% identical with *Candida tropicalis*. The biochemical identification result was similar to the molecular identification result. In molecular identification using 28S rRNA sequence analysis revealed that yeast Tr7 was 99% similar with *Candida tropicalis* NITCSK13 strain. LAB G6 was 100% similar with *Pediococcus pentosaceus* BSS1375 strain using 16S rRNA sequence analysis. *C. tropicalis* Tr7 and *P. pentosaceus* G6 potentially used for inoculum in cassava fermentation process to improve cassava quality as a functional food.

1. Introduction

Cassava (*Manihot esculenta* Crantz) is an important source of calories in tropical countries [1]. Cassava is a food crop commodity as a source of carbohydrates and raw material for food and animal feed. Fresh cassava is easily damage, contains high water and contains antinutrient, hydrogen cyanide (HCN) [2]. Various techniques were performed to make cassava safe for consumption, one of which is through the fermentation. Fermentation process was known to increase protein levels and reduce the HCN content of cassava products [3]. The fermentation involves specific microbes especially lactic acid bacteria (LAB), cellulolytic bacteria, and amylolytic yeast [4]. Yeast and lactic acid bacteria (LAB) are the microbial strains that have been frequently associated with the production of an enzyme for HCN reduction during fermentation of cassava and the development of flavor [5]

Indonesia is the fourth-largest world cassava supply country in the world at 5.8% [6]. In Indonesia, especially the Special Region of Yogyakarta, a variety of traditional processed products from cassava, such as Growol, Tapai, Tiwul, and Gatot, are processed using the natural fermentation without the addition of microbes. The traditional cassava fermentation process is dependent on the presence of natural microorganisms from raw materials and microorganisms from the environment [7]. Naturally in Gatot and Growol (fermented raw cassava) there are strains of LAB which are included in the genus Lactobacillus especially species of *L. plantarum*, *L. pentosus*, *Pediococcus pentosaceus* and *Streptococcus thermophilus* [8]. In tapai which is made traditionally was known contain a microbial consortium consisting of molds, yeasts, and bacteria [7]. In Gatot, homofermentative LAB of *L. manihotivorans*, heterofermentative LAB of *L. fermentum* [9], *L. manihotivorans*, *Bacillus licheniformis*, *Brevibacillus brevis*, and *L. fermentum* [10] were found.

Cassava is an energy source food which main components are starch and fibre [11]. Microbial and biochemical composition changes occur traditionally during the fermentation process of cassava [12]. Fermentation process of cassava using microbes has a very important role [13]. Polysaccharide starches are amylose polymers and amylopectin (branched polymers of glucose molecules that are linked by 1,6-glycoside bonds). Microbes are known to have the ability to break down starch in cassava through the production of amylase enzymes, including molds [14], yeast [15,16] and bacteria [17]. Modified cassava fermentation (Mocaf) now is one of the fermented cassava products that is being developed in Indonesia. The use of an inoculum for Mocaf fermentation is important to ensure the quality products according to the standard. This study aims to select microbes with amylolytic activity as an inoculum in the fermentation process of cassava and potentially reduce the content of cyanide acid in the final product.

2. Material and Method

2.1. Materials

The traditional fermented cassava used for samples in yeast isolation. Tapai, Tiwul, Gatot, and Ragi tape obtained from Demangan market and Wonosari market, Yogyakarta, Indonesia. Twenty-one of LAB isolates which used in this research were microbiology laboratory collection in BPTBA LIPI. LAB isolates were isolated from fermented cassava such as Gatot, Tiwul, Gaplek and modified cassava fermentation (Mocaf) products. LAB isolates were stored in 30% glycerol at -20 °C of temperature and were re-cultured on MRSA medium. Local variety of cassava root obtained from the Wonosari market, Gunungkidul, Yogyakarta, Indonesia.

2.2. Yeast isolation

A total of 5 grams of cassava fermented sample was mixed in 45 mL of chloramphenicol yeast glucose broth (CYG) and incubated for 72 hours at 24 °C of temperature. Samples were homogenized in sterile distilled water and serial dilutions were made from 10^1 – 10^7. The last three series dilutions were spread on chloramphenicol yeast glucose agar (CYGA) surface. Incubation was carried out aerobically at 24 °C of temperature for 48 hours. The yeast colonies that were grown were calculated, selected, and observed for morphological and biochemical characteristics.

2.3. Qualitative amylolytic assay

Twenty-two yeast isolates and 21 LAB isolates were inoculated in a medium containing 2% starch substrate. Yeasts were cultured on yeast extract peptone starch agar medium while LABs were cultured on modified MRS agar by replacing the carbon source into starch [18]. Yeast culture was incubated for 3 days at 24 °C and LAB culture were incubated for 2 days at 37 °C. Two 2 mL of 0.3% Lugol's iodine solution was dropped on the media surface and left for 10 minutes and then it was washed using 3 mL of physiological saline (0.85% NaCl) for10 minutes. The clear zone that appears around the colonies indicates the presence of amylolytic activity from the microbes [19].

2.4. Quantitative amylolytic assay

Quantitative amylolytic activity assay was carried out to determine degrading starch with α-amylase enzyme using iodine complex method [18]. One hundred μL of 24-hour-old LABs and yeasts were cultured on liquid media containing 2% starch (w/v) and 0.5 g/L CaCl2, 6.0 of. The experiment was performed in 3 replicate samples. Yeasts were incubated for 24 hours in a shaker incubator (Stuart Orbital Incubator, S1500), 120 of agitation, 24 °C of temperature. LABs were incubated for 20 hours, 120 rpm of agitation, 37 °C of temperature. The culture was centrifuged (Selecta BL II) at 7000 rpm for 5 minutes at 4 °C. The crude enzyme was obtained from the supernatant. One hundred μL of crude enzymes dissolved in starch and 0.1 mol/L citrate-phosphate buffer pH 5.5 and incubated at 55 °C for 30 minutes. One hundred μL of 1 mol/L H_2SO_4 was used to stop the reaction by the homogenizing process for 10 seconds. The absorbance of 100 μL of the solution was measured using a spectrophotometer (Dynamica Holo RB-10) at λ=620 nm with 2.4 mL of Iodine 4% reagent (g/L) and KI : I (30:3 ratio) addition. The standard curve was determined by using 1.2% (w/v) starch solution at 0.0 – 0.3% of concentration. One enzyme unit is equivalent to the amount of enzyme produced from the hydrolysis of 10 mg of starch for 30 minutes at 55 °C. Yeast and LABs isolates with high amylolytic activity were selected, purified, and stored in 30% CYG-glycerol medium for yeast and 30% MRS-glycerol for LAB at -20 °C.

2.5. Cassava fermentation

The modified cassava (Mocaf) fermentation process is carried out by the submerged fermentation method. Cassava was prepared, washed, and made chips using a slicer machine. Almost 1000 g of cassava chips were added with 1500 mL of distilled water in a plastic container. The fermentation process was carried out with 2 treatments, with inoculum addition (A) and without inoculum (B), 3 replicates of each. The inoculum consisted of 25 mL CYG broth containing 5% yeast (10^5 CFU/mL) and 25 mL MRS broth containing 5% LAB (10^7 CFU/mL). Fermentation was carried out for 48 hours with observed parameters including pH, temperature, total LAB, and total coliform at 0, 12, 24, 36 and, 48 of fermentation time. The pH value of 15 mL of fermentation liquid was measured using a pH meter. The LAB and coliform total of fermentation liquid was calculated using the total plate count (TPC) method MRS agar + $CaCO_3$ medium for LAB and EMB medium for coliform. The media was incubated at 37 °C for 24 – 48 hours. At the final fermentation process, cassava chips were separated from water and then it washed two times. Cassava chips were dried using a drying oven at 50 – 55 °C. Mocaf flour was obtained by grinding the chip using a grinder machine and was sieved to result in 120 mesh of particle size. HCN levels were determined in fresh cassava, before and after fermentation. HCN analysis was conducted in the Laboratory of Agriculture Technology and Food Product, Faculty of Agriculture Technology, Universitas Gadjah Mada, Yogyakarta. The picrate paper kit method was used for the semiquantitative determination of HCN cassava [20].

2.6. Molecular and biochemical identification

Molecular identification of lactic acid bacteria was performed using 16S rRNA sequences analysis. PCR amplification on 16 S rDNA using Primer 27 F: 5`-AGA GTT TGA TCC TGG CTC AG-3` and Primer

1492 R: 5`- GGT TAC CTT GTT ACG ACT T-3` GTT. Isolation of bacterial genomes, amplification, and sequencing of 16S rRNA gene sequences was carried out at InaCC, Research Center for Biology, Indonesian Institute of Sciences, Cibinong, Indonesia. Whereas molecular identification of yeast was performed using 28S rRNA sequence analysis. Isolation of yeast genomes, amplification, and sequencing of 28S rRNA gene sequences was carried out at Biotechnology Laboratory, Agency for Assessment and Application of Technology, Serpong, Indonesia. The sequenced data were trimmed and assembled with the BioEdit program and then converted in FASTA format. The DNA sequencing results in the following FASTA format at Basic Local Alignment Search Tool – Nucleotide (BLASTN) to locate the homology online at the DNA database center at National Center for Biotechnology Information (NCBI) (https://www.ncbi.nlm.nih.gov/). Biochemical identification procedures were carried out using the API 50 CHL kit (BioMerieux) for LAB isolates and API 20 C AUX kit (BioMerieux) according to manual instructions. Analysis of test results is carried out using the Web API.

2.7. Morphological analysis
Morphological characteristics of LAB and yeast were observed using a scanning electron microscope (SEM) after incubation at 37°C for 18 h on MRSA medium for LAB and 30 °C for 24 h on CYGA medium for yeast. After incubation, one loop of a single colony was added evenly on the surface of the carbon-tape covered stub. The sample was air-dried and coated with Au (using Au ion sputter-Hitachi MC1000) with a setting of 10 mA for 60 s and observed using SEM Hitachi SU3500. The SEM setting of high vacuum mode, accelerating voltage of 5 kV, spot intensity of 30%, and magnification of 10,000x.

2.8. Data Analysis
Quantitative data were analyzed statistically to find out the significance of the mean obtained. The analysis used was a one-way analysis of variance (ANOVA) followed by Duncan Multiple Range Test (DMRT). Statistical analysis was performed with SPSS Statistics 16.0 software.

3. Result and Discussion
Twenty yeast isolates were obtained from Tapai, Tiwul, Gatot and Ragi Tapai samples. Qualitative assay of amylolytic activities on yeast extract peptone – starch medium selected 13 yeast isolates. Whereas 21 LAB isolates were selected on MRS – starch medium. The quantitative assay with iodine complex method of selected yeast and LAB isolates resulted several yeast and LABs isolates with amylolytic activities (table 1). Amylolytic activity of yeast were higher than amylolytic activity of LAB. Yeast Tr7 which was isolated from Gatot dan LAB G6 which was isolated from Ragi Tapai resulted the highest amylolytic enzyme (0.0624 dan 0.0267 U/ml). Both of selected isolates used for inoculum in cassava fermentation.

The observed parameters of cassava fermentation using yeast Tr7 and LAB G6 as inoculum shown in table 2. Cassava fermentation process by inoculum resulted in higher on the LAB population, lower on coliform population, lower on pH value, and higher on temperature than without inoculum. The population of LAB at the beginning of fermentation, was 6.9 $\log_{10}$ CFU/mL in fermentation with inoculum while in fermentation without inoculum has not been detected even though at 12 hours after fermentation the population reached 7.89 $\log_{10}$ CFU/mL. On fermentation cassava using the inoculum, the LAB population increased until the end of fermentation at 48 hours. While on the fermentation of cassava without inoculum, the amount of LAB decreased at the end of fermentation. The coliform population in cassava fermentation by inoculum was less than the cassava without inoculum. A high of total LAB in the fermentation medium indicated that LAB was able to use cassava as a substrate for growth. During the fermentation process, the LAB converted the organic acids into lactic acid and resulted in a decrease in the pH of media. Acidic pH conditions made a selective environment for microbes that are intolerant of acidic conditions such as bacterial coliform pathogens [21].

Table 1. Amylolytic activity of lactic acid bacteria and yeast isolates.

Lactic acid bacteria			Yeast		
Isolates	Source	Amylolytic activity (U/mL)	Isolates	Source	Amylolytic activity (U/mL)
T1	Tiwul	0.0210±0.0016[bcd]	Ta1	Tapai	0.0507±0.0056[b]
T2	Tiwul	0.0178±0.0041[d]	Ta3	Tapai	0.0561±0.0072[ab]
T6	Tiwul	0.0188±0.0014[bcd]	Ta4	Tapai	0.0499±0.0081[b]
T10	Tiwul	0.0237±0.0023[ab]	Tc3	Gatot	0.0531±0.0024[ab]
G6	Gatot	0.0267±0.0016[a]	Ti2	Tiwul	0.0562±0.0013[ab]
G7	Gatot	0.0212±0.0030[bcd]	Ti3	Tiwul	0.0536±0.0098[ab]
G8	Gatot	0.0167±0.0007[d]	Ti4	Tiwul	0.0494±0.0065[b]
G9	Gatot	0.0202±0.0046[bcd]	Ti5	Tiwul	0.0571±0.0052[ab]
G10	Gatot	0.0182±0.0027[cd]	Ti7	Tiwul	0.0586±0.0050[ab]
M1	Mocaf	0.0206±0.0024[bcd]	Tr2	Ragi tapai	0.0569±0.0021[ab]
M6	Mocaf	0.0231±0.0023[abc]	Tr4	Ragi tapai	0.0571±0.0053[ab]
			Tr5	Ragi tapai	0.0580±0.0017[ab]
			Tr7	Ragi tapai	0.0624±0.0024[a]

Means in the same column with different superscript differ significantly P<0.05)

However, at the end of fermentation, the coliform did not completely disappear due to coliform still grow at pH = 4. In this study, the pH of media at the end of fermentation in all treatments were 4.25 and 4.30 (table 2). The best condition for the inactivation of coliform was pH 3.7 [22]. In another study showed that *Escherichia coli* which is one type of coliform bacteria were still detected in populations of 10^4 CFU/mL at pH 3 for 24 hours incubation time [23].

Table 2. The parameters of cassava fermentation using yeast Tr7 and lactic acid bacteria G6 inoculum.

Fermentation time (hours)	Total of lactic acid bacteria ($\log_{10}$ CFU/mL)		Total of coliform ($\log_{10}$ CFU/mL)		pH		Temperature (°C)	
	A	B	A	B	A	B	A	B
0	6.90±0.02[b]	0.00±0.00[a]	4.75±0.05[a]	4.84±0.06[a]	6.49	7.02	28.00	28.00
12	8.34±0.02[d]	7.87±0.56[c]	6.53±0.14[d]	6.80±0.05[e]	4.42	4.45	28.33	28.00
24	8.79±0.12[ef]	8.53±0.06[de]	7.19±0.05[f]	8.16±0.05[g]	4.36	4.39	29.00	28.33
36	8.97±0.06[f]	8.25±0.17[d]	6.93±0.03[e]	7.18±0.22[f]	4.28	4.34	29.33	29.00
48	8.55±0.39[de]	6.93±0.24[b]	5.76±0.11[b]	6.28±0.05[c]	4.25	4.30	29.67	29.00

Means in the same column with different superscript differ significantly P<0.05). Fermentation experiments using 1000 g fresh cassava with 2.5% inoculum (v/w). Inoculum composition was 1:1 (yeast : LAB) A: with inoculum and B: without inoculum.

Cassava used in this study was fresh cassava contained HCN content of 289.57 ppm. HCN content in this study was measured before and after fermentation. HCN content of cassava produced from fermentation with the addition of inoculum and without inoculum significantly were lower (p<0.05) than fresh cassava (table 3). In other study reported that the free cyanide (FC) reduction was recorded in cassava flour fermentation by *L. plantarum, L. mesenteroides*, and *Saccharomyces cerevisiae* and the FC content of all fermented sample at three fermentation time (24,36 and 48 h) and inoculums level (0,25 ml, 0.50 ml, and 0.75 ml) was lower than the unfermented samples [24].

Table 3. Cyanide content (HCN) in fermented cassava process.

Fermentation type	HCN content (ppm)		Average of HCN content (ppm)	HCN reduction (%)
	Before fermentation	After fermentation		
A	118.07±16.63	75.57±4.17^{a}	116.82±44.60^{a}	52.19
B	115.49±28.57	78.09±7.39^{a}	96.79±27.88^{b}	32.38
Average of HCN content (ppm)	136.78±31.48^{a}	76.83±5.87^{b}		

Means with different superscript differ significantly P<0.01). Fresh cassava contains 289.57 ppm of hydrogen cyanide (HCN) content. A: with inoculum and B: without inoculum

In this study also showed that the use of inoculums produced higher HCN formation than without inoculums. Table 3 showed that HCN residue and HCN reduction on cassava with inoculum addition was 75.57 ppm and 52.06%, these results were lower (P<0.05) than without inoculum addition which was 78.16 ppm and 32.32% (table 3). These differences occurred due to an increase in the population of bacteria, which accelerate cyanogenic glycoside damage [25]. Total HCN content in cassava chips can be substantially reduced by drying at temperature of 70 °C and air velocity of 2.75 m/s for TMS30572. The reduction in total HCN content of the cassava chips by this treatment might be due to enhanced hydrolysis process of cyanogenic glucosides by the enzyme linamarase. The higher the moisture content of the cassava the greater the loss in the cyanide content during drying [26]. HCN content in cassava could be removed using natural submerged fermentation methods. HCN is soluble in water and has a boiling point of 29 °C.

Table 4. Biochemical identification of lactic acid bacteria G6 using API 50 CHL kit.

No.	Substrate	Result	No.	Substrate	Result
0	Control	-	25	Esculin	+
1	Glycerol	-	26	Salicin	+
2	Erythritol	-	27	D-Cellobiose	+
3	D-Arabinose	-	28	D-Maltose	+
4	L-Arabinose	+	29	D-Lactose (bovine origin)	-
5	D-Ribose	+	30	D-Melibiose	-
6	D-Xylose	+	31	D-Saccharose (sucrose)	-
7	L-Xylose	-	32	D-Trehalose	+
8	D-Adonotil	-	33	Inulin	-
9	Methyl-βD-Xylopyranoside	-	34	D-Melezitrose	-
10	D-Galactose	+	35	D-Raffinose	-
11	D-Glucose	+	36	Amidon (starch)	-
12	D-Fructose	+	37	Glycogen	-
13	D-Mannose	+	38	Xylitol	-
14	L-Sorbose	-	39	Gentiobiose	-
15	L-Rhamnose	-	40	D-Turanose	-
16	Dulcitol	-	41	D-Lyxose	-
17	Inositol	-	42	D-Tagatose	+
18	D-Mannitol	-	43	D-Fucose	-
19	D-Sorbitol	-	44	L-Fucose	-
20	Methyl-αD-Mannopyranoside	-	45	D-Arabitol	-
21	Methyl-αD-Glucopyranoside	-	46	L-Arabitol	-
22	N-Acetylglucosamine	+	47	Potassium Gluconate	-
23	Amygdalin	+	48	Potassium 2-Ketogluconate	-
24	Arbutin	+	49	Potassium 5- ketogluconate	-

- = negative, + = positive

Statistical analysis results showed that fermentation process and the type of fermentation could significantly reduce the HCN levels. Additionally, submerged fermentation can dissolve linamarin and lotaustralin compounds, and stimulate the growth of microorganisms that can break down toxins into organic acids. The cassava fermentation method also inactivated the linamarase enzyme and catalyzed the formation of HCN [2].

The biochemical identification of LAB isolates using the API 50 CHL kit based on the ability of bacteria to hydrolyze carbohydrates. The results of identification using the API 50 CHL kit showed that the G6 isolate had 99.6% of similarity with *Pediococcus pentosaceus*. Positive results detected on several monosaccharide, disaccharide, and oligosaccharide (table 4). This result similar with other study that show that *Pediococcus pentosaceus* JCM 5890^T was able to ferment arabinose, ribose, xylose, glucose, fructose, galactose, mannose, cellobiose and maltose [27].

Biochemical identification of Tr7 yeast isolates using the API 20 C AUX kit (table 5). Test results using API 20 C AUX kit showed that the Tr7 yeast isolate had a 95.8% similarity with *Candida tropicalis*. The positive results were obtained in glucose, gluconate, xylose, adonitol, galactose, sorbitol, glucopyranoside, glucosamine, maltose, saccharose, trehalose, and melezitose. This result was similar with other study that *Candida tropicalis* has positive result in fermentation of glucose, maltose, galactose and xylose [28].

Table 5. Biochemical identification of yeast Tr7 using API 20 C AUX kit.

Tests	Substrate	Results	Tests	Substrate	Results
0	Aucun	-	SOR	D-Sorbitol	+
GLU	D-Glucose	+	MDG	Méthyl-DD-Glucopyranoside	+
GLY	Glycérol	-	NAG	N-Acétyl-glucosamine	+
2KG	Calcium 2-céto-Gluconate	+	CEL	D-Celobiose	-
ARA	L-Arabinose	-	LAC	D-Lactose	-
XYL	D-Xylose	+	MAL	D-Maltose	+
ADO	Adonitol	+	SAC	D-Saccharose	+
XLT	Xylitol	-	TRE	D-Trehalose	+
GAL	D-Galactose	+	MLZ	D-Mélézitose	+
INO	Inositol	-	RAF	D-Rafinose	-

- = negative, + = positive

Confirmation of identification of the two selected isolates was carried out by molecular methods using rRNA sequence analysis. The 16S rRNA amplification results of LAB G6 isolates obtained sample PCR products visualized using a gel documentation system with 1500 bp of gene size. G6 isolate was identified as *P. pentosaceus*. Homology BLAST nucleotide of 16S rDNA sequence with 1420 bp of length in the DNA bank database (http://www.ncbi.nlm.nlh.gov/) accessed January 17, 2016) has a 100% max identity against the nearest bacterial taxon *P. pentosaceus* strain BSS1375 (KT351726). While the PCR results from the 28S rRNA sequence after assembled and BLAST nucleotide in DNA bank database (http://www.ncbi.nlm.nlh.gov/) accessed May 15, 2017) showed that Tr7 yeast isolate was *Candida tropicalis* strain NITCSK1 with 99% of identity (table 6).

Table 6. Molecular identification of lactic acid bacteria G6 and yeast Tr7.

	Molecular identification	Reference organism	% similarity
TR7	28S ribosomal DNA	*Candida tropicalis* strain NITCSK1	99
G6	16S ribosomal DNA	*Pediococcus pentosaceus* strain BSS1375	100
	Biochemical identification		% identity
TR7	API 20 C AUX	*Candida tropicalis*	95.8
G6	API 50 CHL	*Pediococcus pentosaceus*	99.6

The morphological features of the two species are shown in figure 1. *Pediococcus pentosaceus* G6 isolate is a Gram-positive bacterium that has a cell size of 0.5 - 1.2 x 1.0 - 10 μm, coccus cells, and appears in pairs (diplococcus). Whereas yeast *C. tropicalis* Tr7 has ovoid cell and branching like the genus *Candida* in general. *Pediococcus pentosaceus* is commonly found in fermented food products such as rice fermentation, tempe, cassava fermentation and vegetable fermentation [29], corn, millet and sorghum based fermentation products [30], African pearl millet slurries [31], alkaline fermentations of cassava (Manihot esculenta Crantz) leaves, roselle (*Hibiscus sabdariffa*) and African locust bean (*Parkia biglobosa*) seeds [32], Tanzanian fermented food from sorghum, maize, millet and maize-sorghum [34], traditional sorghum fermentation [34]. According to Chahrour [35] study, the genera *Lactobacillus*, *Pediococcus* and *Enterococcus* were the amylolytic LABs that have founded on sorghum fermentation.

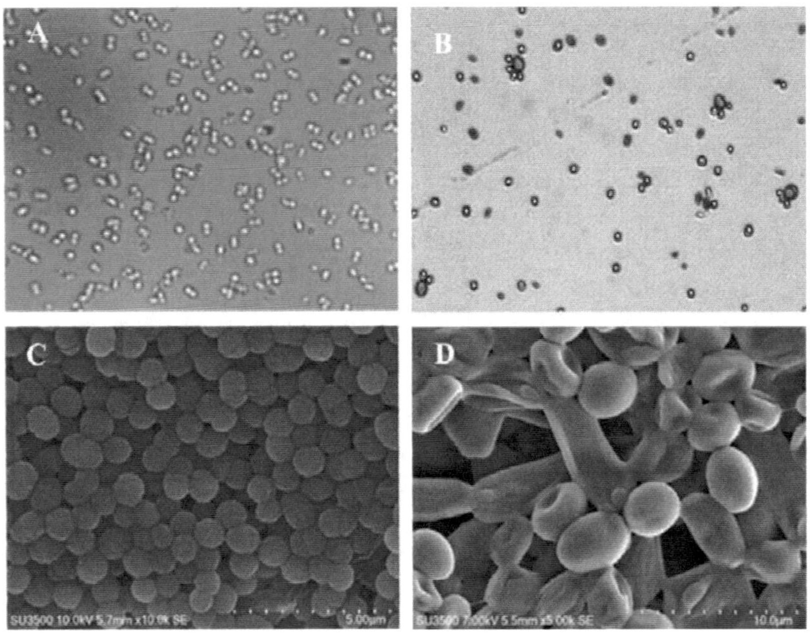

Figure 1. A. Cell morphology of *Pediococcus pentosaceus* G6 isolated from gatot (400x of magnification); B. Cell morphology of *Candida albicans* Tr7 isolated from ragi tapai 1000x of magnification) with methylene blue staining; C. Scanning electron micrograph of G6 (10,000x magnification) and D. Scanning electron micrograph of Tr7 (5,000x of magnification).

Candida is one of the amylolytic yeast that was founded in various types of fermented products. Oyewole [36] revealed that amylolytic microbial including *Candida krusei*, *Candida tropicalis*, *Pichia saitoi*, *Saccharomyces cerevisiae*, and *Zygosaccharomyces balii* were found in the fermentation of 'fufu'. Some studies reported that *Candida tropicalis* is commonly found and used in the fermentation process of various types of food because it has certain abilities. *C. tropicalis* BCC7755 is thermotolerant and is used in the saccharification process of ethanol fermentation from cassava pulp [37]. *C. tropicalis* is one of the dominant yeasts found in Brazilian fermented foods' cauim 'based on several substrates are used in its production, including cassava, rice, corn, maize and peanuts [38], Tanzanian fermented food' togwa 'from sorghum, maize, millet and maize – sorghum [33]. *C. tropicalis* with amylolytic activity was found in cassava-fermenting water in the early part of the fermentation [36]. Other studies related to the ability of yeast *C. tropicalis* in the cassava process. *C. tropicalis* was isolated from Agbelima, fermented cassava meal from Ghana, Togo and Benin having cellulase, polygalacturonase and linamarase processes that can break down cyanogenic glucoside in cassava [39]. *C. tropicalis* grows on soluble starch, corn, and cassava powders. That property has been used to develop a fermentation process whereby *C. tropicalis* can be grown directly on corn or cassava powders. The biomass and residual corn or cassava contains about 20% protein, which represents a balanced diet for either animal fodder or human food. The fact that no extra enzymes are required to hydrolyze starch results in a particularly efficient step of improving the nutritional value of products, through a single-step fermentation process [40].

4. Conclusion

Yeast Tr7 and LAB G6 isolates performed the highest amylolytic activity (0.0624 U/mL and 0.0627 U/mL). Cassava fermentation products using these microbes resulted in higher hydrogen cyanide (HCN) reduction (52.06%) than without inoculum (32.32%). Biochemical and molecular identification result showed that lactic acid bacteria G6 has closest similarity with *P. pentosaceus* and yeast Tr7 has closest identity with *Candida tropicalis*. *C. tropicalis* Tr7 and *P. pentosaceus* G6 potentially used for inoculum in cassava fermentation process to improve cassava quality as functional food.

Acknowledgments

This research was funded by Riset Tematik BPTBA LIPI 2014. The part of this work was funded by Insinas Ristek Pusat Inovasi LIPI 2013. Authors thank Prof. Langkah Sembiring as supervisor for MI in undergraduate program at Faculty of Biology, Universitas Gadjah Mada, Yogyakarta, Indonesia, 2014. Author also thank Rahmi Lestari Helmi for supporting this work.

References

[1] Nwokoro O and Anya FO 2011 *J. Mex. Chem. Soc.* **55** 246–50
[2] Adamafio N A, Sakyiamah M and Tettey J 2010 *Afr. J. Biochem. Res.* **4** 51–6
[3] Akindahunsi A A, Oboh G and Oshodi A A 1999 *Riv. Ital. Sostanze Gr.* **76** 437–40
[4] Achi O K and Akomas N S 2006 *Pakistan J. Nutr.* **5(3)** 224–9
[5] Behera S S, Ray R C, and Zdolec, N 2018 *Biomed. Res. Int.* **2018** 9361614
[6] Otekunrin O A and Sawicka B 2019 *Acta Sci. Agric.* **3** 194-202
[7] Aryanta W R 2000 *Japanese J. Lactic Acid Bacteria* **10** 90-102
[8] Rahayu E S, Djaafar T F, Wibowo D and Sudarmadji S 1996 *Indonesian Food Nutr. Prog.* **3** 21-8
[9] Nurhayati J, Subagio A and Widyatmoko H 2016 *Prosiding Seminar Hasil Penelitian Tanaman Aneka Kacang dan Umbi,* 517–22
[10] Astriani 2015 *Karakterisasi gatot terfermentasi oleh isolat indigenus gatot singkong (Rhizopus oligosporus dan Lactobacillus manihotivorans)*. Skripsi, Jurusan Teknologi Hasil Pertanian, Fakultas Teknologi Pertanian, Universitas Jember
[11] Meryandini A, Melani V and Sunarti T C 2011 *Afr. J. Food Sci. Technol.* **2** 030-5
[12] Tivana L D, Bvochora J, Mutukumira A N, Owens J D and Zvauya R 2007 *Proceedings of the 13th ISTRC Symposium*, 445–450Lambri M, Fumi M D, Roda A, and De Faveri D M 2013 *Afr. J. Biotechnol.* **12** 2685–91
[13] Pervez S, Aman A, Iqbal S, Siddiqui N N and Qader S A U 2014 *BMC biotechnology* **14** 49
[14] González C F, Fariña J I and de Figueroa L I C 2008 *Enzyme Microb. Technol.* **42** (3), 272–7.
[15] Carrasco M, Villarreal P, Barahona S, Alcaíno J, Cifuentes V and Baeza M 2016 *BMC Microbiology* **16** 21
[16] Barros F F C, Simiqueli A P R, De Andrade C J and Pastore G M 2013 *Biotechnol. Res. Inter.*

[17] Giraud E, Gosselin L, Marin B, Parada J L and Raimbault M 1993 *J. Appl. Bacteriol.* **75** 276-82
[18] Sinaga, D M 2011 *Screening aktivitas selulolitik, amilolitik dan xylanolitik strain bakteri dan khamir serta potensinya dalam akumulasi lipid dan penghasil etanol.* Skripsi. Fakultas Biologi. UGM. Yogyakarta.
[19] Egan S V, Yeoh H H and Bradbury J H 1998 *J. Sci. Food Agric.* **76** 39-48Thu T V, Foo H L, Loh T C, and Bejo M H 2011 *Afr. J. Biotechnol.* **10** 1362
[20] Castro-Rios K, Corpas E J, Cardenas V and Taborda G 2017 *J. Mater. Environ. Sci* **8** 364-9
[21] Jordan K N, Oxford L and O'Byrne C P 1999 *Appl. Environ. Microbiol.* **65** 3048-55
[22] Tefera T, Ameha K and Biruhtesfa A 2014 *Int. Food Res. J.* **21** 1751-56
[23] Obilie E M, Tano-Debrah K and Amoa-Awua W K. 2004 *Int. J. Food Microbiol.* **93** 115-121
[24] Famurewa J A V and Emuekele P O 2014 *Afr. J. Food Sci. Technol.* **5** 75-80
[25] Cai Y, Kumai S, Ogawa M, Benno Y and Nakase T 1999 *Appl. Environ. Microbiol.* **65** 2901-6
[26] Lachance M A, Boekhout T, Schorzetti G, Fell J W and Kurtzman C P. 2011 *The Yeast (Fifth Edition)* (Elsevier B.V) p 987-1278
[27] Suhartatik N, Cahyanto M N, Rahardjo S, Miyashita M and Rahayu E S 2014 *Inter. Food Res. J.* **21** 973–8
[28] Adimpong D B, Nielsen D S, Sørensen K I, Derkx P M and Jespersen L 2012 *BMC Microbiol.* **12** 75
[29] Turpin W, Humblot C, and Guyot J P 2011 *Appl. Environ. Microbiol.* **77** 8722–34
[30] Ouoba L I I, Nyanga-Koumou C A G, Parkouda C, Sawadogo H, Kobawila S C, Keleke S, and Sutherland J P 2010 *J. Appl. Microbiol.* **108** 2019–29
[31] Mugula J K, Nnko S A M, Narvhus J A and Sørhaug T 2003 *Inter. J. Food Microbiol.* **80** 187–99
[32] Mohammed S I, Steenson L R and Kirleis A W 1991 *Appl. Environ. Microbiol.* **57** 2529–33
[33] Chahrour W, Merzouk Y, Henni J E, Haddaji M and Kihal M 2013 *Afr. J. Biotechnol.* **12** 1703-9
[34] Oyewole O B 2001 *Inter. J. Food Microbiol.* **65** 213–8
[35] Rattanachomsri U, Tanapongpipat S, Eurwilaichitr L and Champreda V **2009** *J. Bioscie. Bioengineer.* **107** 488–93
[36] Schwan R F, Almeida E G, Souza-Dias M A G and Jespersen L 2007 *FEMS Yeast Res.* **7** 966–72
[37] Amoa-Awua W K, Frisvad J C, Sefa-Dedeh S and Jakobsen M 1997 *J. Appl. Microbiol.* **83** 288–96
[38] Azoulay E, Jouanneau F, Bertrand J C, Raphael A, Janssens J and Lebeault J M 1980 *Appl. Environ. Microbiol.* **39** 41

IOP Conf. Series: Materials Science and Engineering 1011 (2021) 012005 doi:10.1088/1757-899X/1011/1/012005

Application of feed forward neural network model to predict the limiting current of tin magneto electrodeposition

Sudibyo[1*], N Aziz[2], A Wijaya[3] and R S Fathona[3]

[1] Mineral Technology Research Unit, Indonesian Institute of Sciences, Lampung Indonesia
[2] School of chemical engineering, University of Science Malaysia, Penang, Malaysia
[3] Informatics Engineering, Sumatra Institute of Technology, Lampung, Indonesia

* E-mail: sudibyo@lipi.go.id

Abstract. Predicting the value of Tin Magneto electrodeposition (MED) is very important since the optimum mass transport occurred at the limiting current. The MED limiting current able to detect using electroanalytical chemistry, but this method is expensive; it needs some method, which able to predict the limiting current of tin MED. However, predicting the limiting current under magnetic field effect is more complicated due to the highly nonlinear characteristic and complicated of its multiple inputs single-output (MISO) system. The nonlinear model that able to predict the limiting current of tin MED is Artificial Neural Networks (ANNs). One of the ANNs which able to simulate the Multiple-Input-Single-Output (MISO) model is the Feed Forward Neural Network (FFNN). In this work, MISO FFNN will model a matrix data set with six variable inputs and one output. The data was obtained from the results of the experiments using electroanalytical chemistry. The output of this model is the limiting current of tin MED, meanwhile, the inputs are by the concentration of tin (Sn^{2+}) in the electrolyte (C), viscosity(v), diffusion coefficient (D), area of the electrode (A), the number of electroactive species (n) and magnetic field strength (B). To get the best model, the performance of FFNN was tested with three variations of the algorithm (Levenberg-Marquardt, Bayesian Regularization, and Scaled Conjugate Gradient) and ten variations of the number of neurons (10, 15, 20, 25, 30, 35, 40, 45 and 50). The best model obtained for this MISO FFNN model is which uses the Levenberg-Marquardt algorithm and the highest number of neurons (50 neurons).

1. Introduction

Tin electrodeposition has been applied to microelectronics as an alternative for tin/lead finishes [1]. This process has a problem to obtain a uniform, dense, and compact deposition. Numerous studies have been addressed to reduce it. One of the methods available to solve is by adding a magnetic field in the electrodeposition process, this process known as magneto electrodeposition (MED) [2–4]. This MED technology has a great impact on the electrodeposition process to synthesize metal alloy, thin-film, multilayer, nanowires, multilayer nanowires, dot array, and nanocontacts which are the technology of the future to build the next generation of microelectronics devices [2].

The magnetic field effect caused the increase of mass transport on an electrochemical reaction in MED which indicated by the increase of limiting current. The optimum mass transport occurs at the limiting current, if the process conducted over the limiting current, this will cause the excess of hydrogen evolution reaction which will damage the surface quality of electroplating. Hence, knowing the limiting current (ib) is very important both in conventional electrodeposition processes or MED. Many researchers have studied that the limiting current in MED is affected by many parameters such as the magnetic field strength(B), the number of electrons of the redox process (n), the electrode area (A), the diffusion coefficient of the electroactive species (D), the concentration of the electroactive species (C_{bulk}); and the kinematic viscosity of the electrolyte (v) [2].

The MED limiting current able to detect using electroanalytical chemistry. Since the electroanalytical chemistry is expensive and only in small scale (lab. Scale); it needs some method, which able to predict the limiting current of tin MED in the large-scale process. However, the limiting current under magnetic field (ib) is a very highly nonlinear process and complicated process which has multiple inputs with a single output. The nonlinear model that able to predict the limiting current of tin MED is Artificial Neural Networks (ANNs). One of the ANNs which able to simulate the Multiple-Input-Single-Output (MISO) problem is the Feed Forward Neural Network (FFNN). FFNN model also reported has been successfully simulated many nonlinear models such as temperature and pressure in the distillation and reactor process [5–7].

In this work, MISO FFNN will model a matrix data set with six variable inputs and one output. The data were obtained from the results of the experiments using linear sweep voltammetry (LSV) technique. LSV technique is one of the electroanalytical chemistry methods, which use potentiostat-galvanostat and three-electrode electrochemical reactor as shown in Figure 1. Platinum electrodes were used as a working electrode (WE) and counter electrode (CE), meanwhile, Ag/AgCl was used as a reference electrode (RE) in the three-electrode electrochemical reactor as shown in Figure 1.

The output of MISO - FFNN model is the limiting current of tin MED, meanwhile, the inputs are by the concentration of tin (Sn^{2+}) in the electrolyte (C), viscosity(v), diffusion coefficient (D), area of the electrode (A) and magnetic field strength (B). To get the best model, the performance of FFNN was tested with three variations of the algorithm (Levenberg-Marquardt, Bayesian Regularization, and Scaled Conjugate Gradient) and ten variations of the number of neurons. The FFNN model was generated using MATLAB software. The best model of MISO FFNN will be developed in the Simulink model which easy and simple to use.

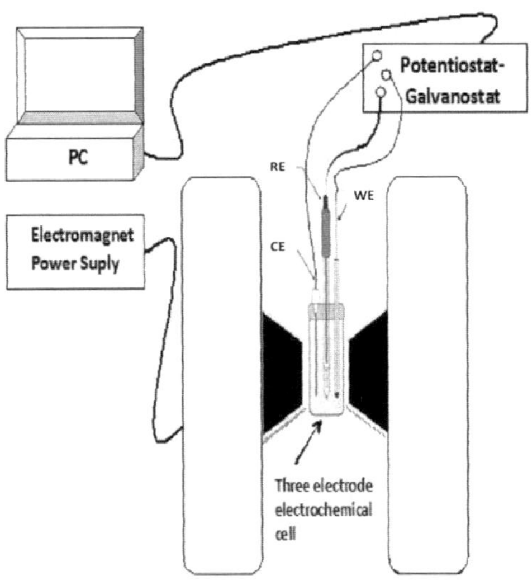

Figure 1. Schematic illustration of an electrochemical cell.

2. Research methodology

2.1 Experimental work of Magneto electrodeposition (MED) Process in Tin

Tin (II) sulfate ($SnSO_4$) (≥99%, from Merck Co. Ltd.), Natrium sulfate (Na_2SO_4) (≥99% from Merck Co. Ltd), and Sodium Gluconate (≥99.8%, from R & M Co. Ltd.) are ingredients used in previous studies. All ingredients are prepared with the concentration determined using distilled water. The magnet used in this research is a water-cooled superconductor electromagnet (Lake Shore EM 4, USA) with a 5 cm diameter pole and 5 cm distance. The intensity of the magnetic field is adjusted from 0 to 0.3 T. The analytical electrochemical techniques used are linear-sweep voltammetry (LSV) and Choroamperometry (CA) using Potentiostat/ Galvanostat. In this work, LSV was used to measure the limiting current, meanwhile, CA was used to measure the diffusion coefficient of electroactive. The LSV results are then used as output in the FF NN modeling.

2.2 Development of the FFNN MISO Model

Feedforward neural networks (FFNN) develop from the approach of input and output data relationships from the existing experimental results. FFNN in this study will use 6 inputs, namely Tin concentration (M), electrode area (A), electroactive diffusion coefficient (D), kinematic electrolyte viscosity (v), magnetic field strength (B) and some electrons involved in redox process (n), from the 6 inputs will produce 1 output that is the limiting current value. To see the real results of each influence given this study uses 6 inputs on FFNN of Tin MED. Then, the entire input data is formed 6 x 1 matrix, while the output is in the form of limiting current (ib) in the form of a 1x1 matrix in Matlab. In this work, 120 data are divided into three parts, namely train (40%), validation (30%), and test (30%).

The structure of the neural network model consists of the input layer, hidden layer, and output layer network. The hidden layer functions, which consist of many neurons as a controller of complex estimations performed by non-linear transfer functions. The identification activity is ended until the best model is obtained from the simulation[5–7]. In this study, the identification command is typed in the m-file in Matlab. Simulations with MISO FFNN have varied the neurons from 10, 15, 20, 25, 30, 35, 40, 45, and 50 neurons and with 3 variations of the algorithm used namely, Levenberg-Marquardt, Bayesian Regularization, and Scaled Conjugate Gradient. The performance of the FFNN model was analyzed using mean square error (MSE). The best model is then imported into Simulink on Matlab.

3. Results and Discussion

The performance of MISO FFNN using various neurons and three variations of the algorithm used, namely, Levenberg-Marquardt, Bayesian Regularization, and Scaled Conjugate Gradient were listed in Table 1. Table 1 shows that the best performance of MISO FFNN using 50 of neurons and Levenberg-Marquardt of algorithm training with the smallest of MSE value. Figure 2 shows the validation performance of FFNN using the algorithm of Levenberg-Marquardt and 50 neurons. The figure shows that the model has MSE of 5.13 for training, 7.78 of MSE for validation, and 2.85 of MSE for the test of the neural network model. The suitable neuron number will give the best performance of the neural network. In this case, the highest number of neurons gives the best performance to simulate the limiting current of tin MED. The increasing of neurons numbers will increase the accuracy, but it will increase the time consuming for calculation. In this work, the accuracy is more important than the higher calculation time, hence this result is applicable for the limiting current prediction of tin MED.

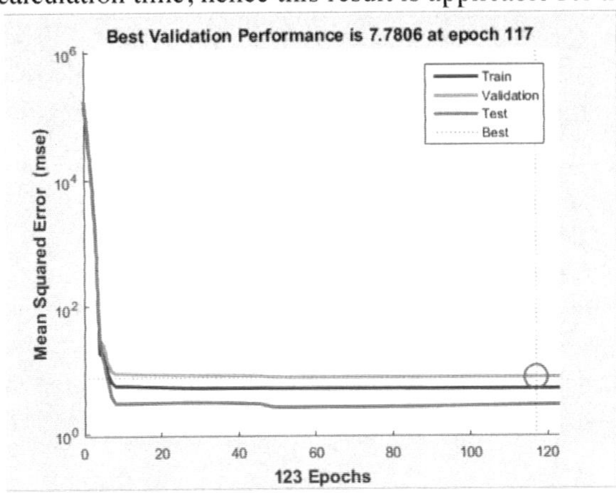

Figure 2. Validation performance of FFNN using the algorithm of Levenberg-Marquardt and 50 neurons.

Table 1. Model identification results with 3 variations of training algorithms and neurons variations.

Algorithm Training	Neurons	TRAINING (MSE) (40 % of Data)	VALIDATION (MSE) (30% of Data)	TESTING (MSE) (30% of Data)
Levenberg - Marquardt	10	3.05	5.84	7.74
	15	3.82	5.95	5.95
	20	6.79	3.40	6.42
	25	3.06	8.98	5.86
	30	3.41	7.60	6.87
	35	5.65	4.10	5.53
	40	3.82	11.50	4.67
	45	5.53	4.10	5.53
	50	5.13	7.78	2.85
Bayesian Regularization	10	5.15	0.00	5.94
	15	4.40	0.00	7.55
	20	6.28	0.00	2.95
	25	5.55	0.00	4.64
	30	5.60	0.00	4.44
	35	4.86	0.00	7.41
	40	5.58	0.00	4.60
	45	6.20	0.00	3.44
	50	4.86	0.00	6.16
Scaled Conjugate Gradient	10	0.157	13.90	22.60
	15	0.150	16.80	25.50
	20	0.163	16.80	9.23
	25	0.169	18.00	19.70
	30	10.30	17.60	14.50
	35	9.84	10.10	8.00
	40	8.06	12.20	16.30
	45	9.06	12.20	9.85
	50	10.90	11.80	12.20

3.1 The Development of MISO FFNN in the Simulink Model.

The performance test of the model shows that MISO FFNN using the *Levenberg-Marquardt* algorithm with 50 neurons provides the best performance in modeling the Tin MED process. This model is written in the form of m-files which are then exported into Simulink in Matlab. The Simulink model was developed to simplify the use of MISO FFNN to predict the limiting current of tin MED as shown in Figure 3 [8]. This Simulink model has six inputs, namely Tin concentration (M), electrode area (A), the diffusion coefficient of electroactive (D), kinematic electrolyte viscosity (v), magnetic field strength (B) and the number of electrons involved in the redox process (B) n). The six inputs will be processed by FFNN and will produce an output in the form of a limiting current value (*ib*).

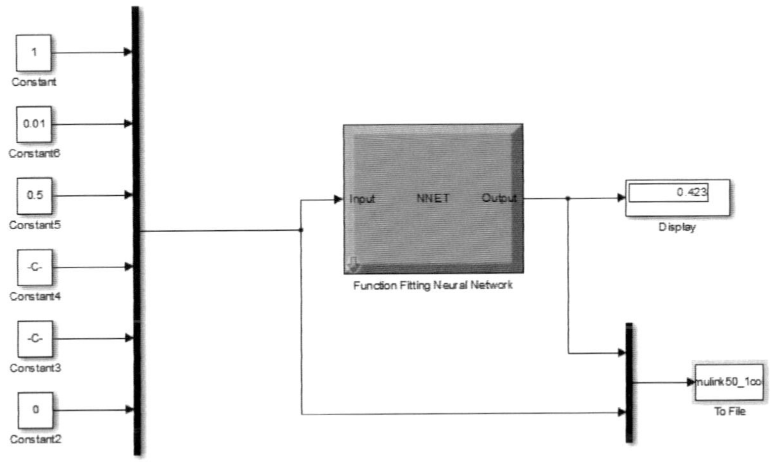

Figure 3. Model FFNN MISO with Simulink.

3.2 The Effect of Tin Concentration towards the Limiting Current of Tin Magneto Electrodeposition

The second experiment was carried out by predicting the value of the limiting current (ib) by the various Tin concentration (M) using the FFNN model. The concentration of tin given was varied from 0.01 to 0.1 M and obtained the MSE of the limiting current value ratio (i_b) data origin and value of the limiting current (ib) the data FFNN of 3.6302 x 10^{-6} as shown in Figure 4.

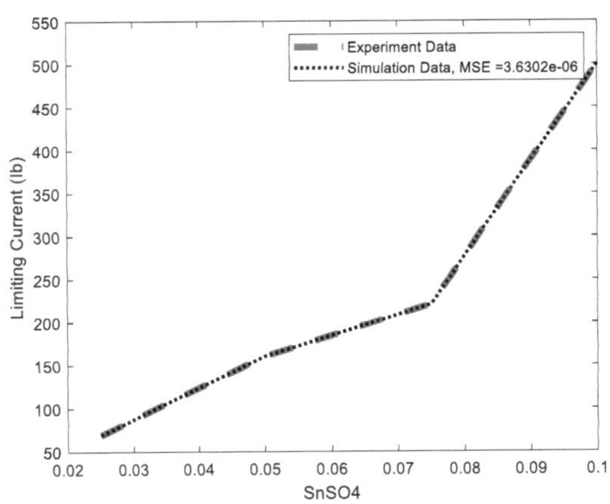

Figure 4. Comparison between simulation results using the MISO FFNN model (identification using the Levenberg-Marquardt algorithm model with 50 of neurons) with original data (variation of tin concentration).

3.3 The Effect of Electrode Area towards the Limiting Current of Tin Magneto Electrodeposition

The third experiment was carried out by predicting the value of the limiting current (ib) by electrode area variation (A) using the FFNN model that had been obtained. Given the wide variation electrode is from 0.5 to 0.953 cm^2 and obtained the MSE of the limiting current value ratio (ib) data origin and value of the limiting current (ib) the data FFNN of 29.3542, shown in Figure 5. The error value for the small area of the electrode is big, hence this model is not good enough to simulate for a small area.

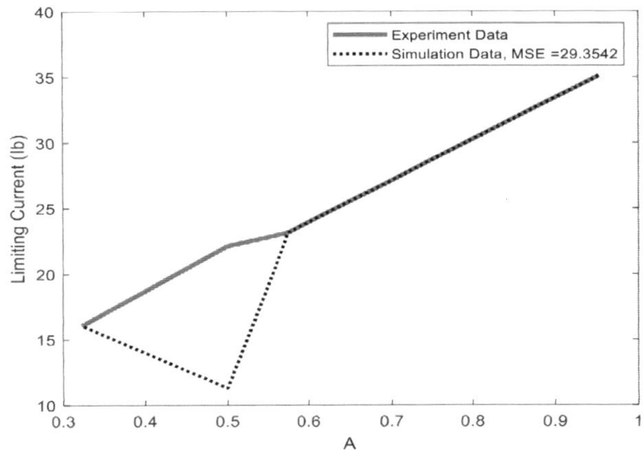

Figure 5. Comparison between simulation results using the MISO FFNN model (identification using the Levenberg-Marquardt algorithm model with 50 of neurons) with original data (variation of electrode area).

3.4 The Effect of Supporting Electrolytes Concentration towards the Limiting Current of Tin Magneto Electrodeposition

To find out the effect of supporting electrolytes on current limiting, the concentration of supporting electrolytes (Na_2SO_4) which varies from 0.075 to 1 M. The increase of supporting electrolyte concentration leads to the decreasing of diffusion coefficient. This phenomenon caused by the increase of electrolyte kinematic viscosity as the increment of supporting electrolyte concentration. Comparison i_b between simulation results using the MISO FFNN model (identification using the Levenberg-Marquardt algorithm model with 50 of neurons) with original data (variation of coefficient diffusion as a function of increasing of supporting electrolyte concentration) gives the MSE value of 0.95113 as shown in Figure 6. The result shows that the error value is not suitable for the highest value of supporting electrolyte concentration.

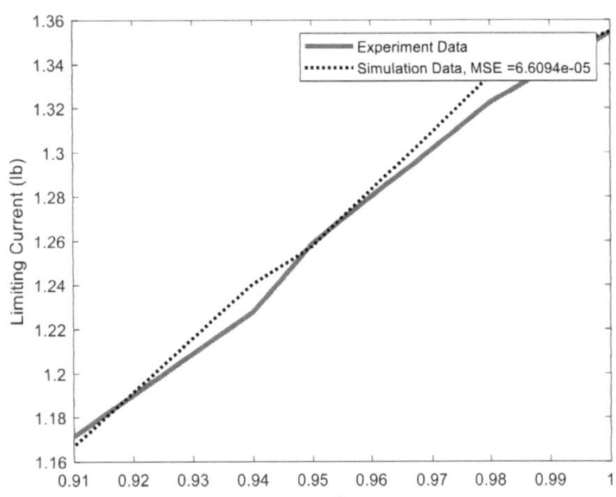

Figure 6. Comparison between simulation results using the MISO FFNN model (identification using the Levenberg-Marquardt algorithm model with 50 of neurons) with original data (variation of coefficient diffusion as a function of increasing of supporting electrolyte concentration).

3.5 The Effect of Additive Electrolytes Concentration towards the Limiting Current of Tin Magneto Electrodeposition

The effects of the current limiting electrolyte additive (i_b) for tin MED are studied by varying the gluconate shown in Figure 7. The increase of additive electrolytes will be caused by the decrease of limiting current (i_b). The decrease in the limiting current leads to increased friction as the viscosity electrolyte increases which reduces the magnetohydrodynamic (MHD) convection of the solution [2].

Although the presence of an additive will decrease the limiting current, the additive is useful for produce better electrodeposits such as to avoid a roughness on the metal surface, to reduce corrosion [9], to increase the purity of electrodeposits. Figure 7 also shows the MSE value of the experimental i_b towards the i_b value of FFNN gives very small of MSE value (0.00023015). This proves that this model has a good capability to predict i_b in different kinematic viscosity.

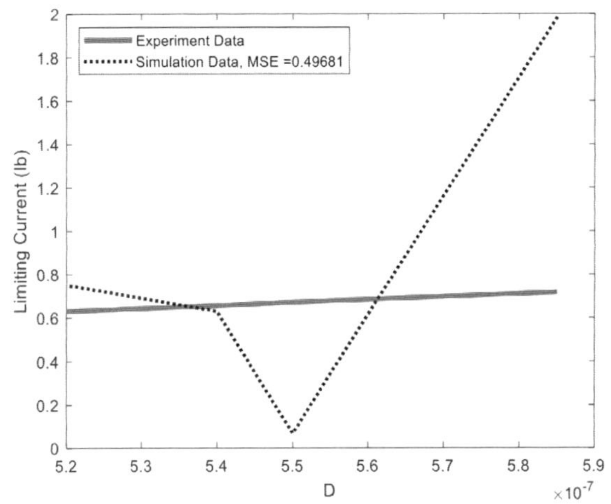

Figure 7. Comparison between simulation results using the MISO FFNN model (identification using the Levenberg-Marquardt algorithm model with 50 of neurons) with original data (variation of kinematic viscosity as a function of increasing of gluconate concentration).

3.6 The Effect of Magnetic Strenght (B) towards the Limiting Current of Tin Magneto Electrodeposition

Figure 8 shows the effect of a magnetic field variation (B) towards the value of the limiting current (*ib*), in which the FFNN model able to predict the limiting current from the real experiment. The variation of the magnetic field (B) given is from 0.05 T - 0.3 T and the MSE value is obtained from the comparison of the limiting current value (*ib*) of the experimental data with the FFNN result is 3.2877 x 10^{-5}. The result shows that the FFNN model is has a good performance to predict the i_b at the varying magnetic field.

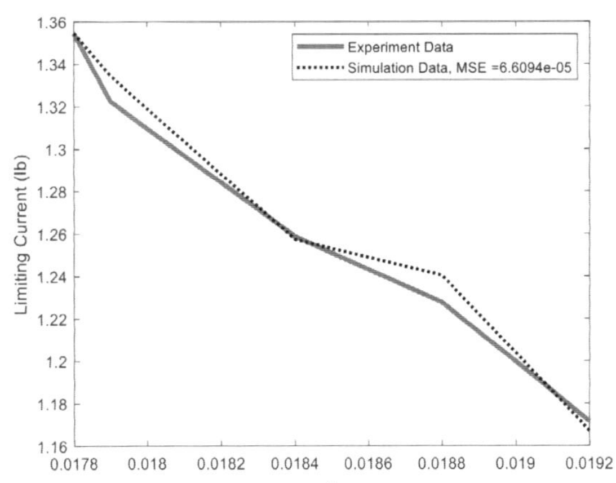

Figure 8. The comparison of the limiting current model FFNN using (the model algorithm Levenberg - Marquardt and 50 of neurons) in predicting the limiting current toward the variations of magnetic fields.

3.7 The Effect of the number of electrons involved in the Redox Process (n) towards the Limiting Current of Tin Magneto Electrodeposition

The effect of the number of electrons involved in the redox (n) process on limiting the current of tin MED was studied using various supporting electrolytes which has different electron transfers such as Na_2SO_4 (Na^{2+} has two-electron transfer), HCL (H^+ has one electron transfer). The effect of supporting electrolyte that has a single electron transfer was studied by a varied of HCl concentration from 1 – 3 M as shown in Figure 9. The figure shows the comparison of the limiting current model FFNN using (the model algorithm Levenberg - Marquardt and 50 of neurons) in predicting the limiting current toward the variations of HCl concentration gives MSE value of 0.00040895. This also proves that FFNN also suitable to simulate ib of tin MED using a variation of HCl.

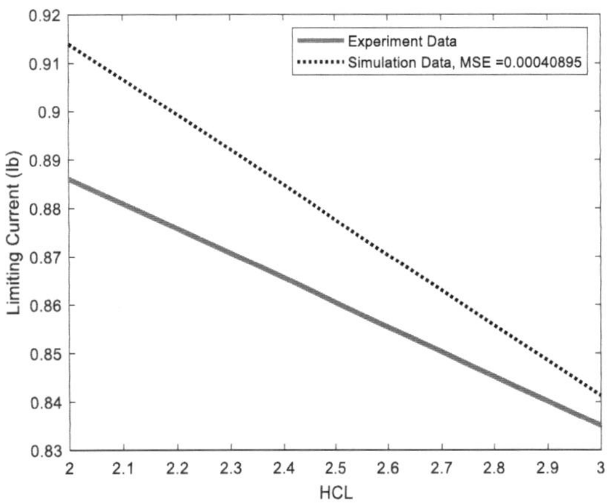

Figure 9. The comparison of the limiting current model FFNN using (the model algorithm Levenberg-Marquardt and 50 of neutrons) in predicting the limiting current towards the variation of HCl concentration.

4. Conclusion

The MISO FFNN model has successfully developed using Simulink on Matlab to predict the limiting current value of Tin MED and user friendly. The results show that the best performance of the FFNN model was achieved when the model generated using 50 neurons and the Levenberg-Marquardt algorithm. Several tests with a different value of input parameters (C, A, D, v, B, and n) shows that the FFNN model can predict the limiting current of tin MED.

Acknowledgments

This work was supported by the Ministry of Research and Technology - The Republic of Indonesia and the Indonesian Institute of sciences through Mandiri research grants are greatly acknowledged.

References

[1] Uzir M H, Sudibyo, Aziz N and Othman M R 2015 Magneto-electro deposition of tin dendrites *Surf. Coatings Technol.* **264** 66–71

[2] Legeai S, Chatelut M, Vittori O, Chopart J P and Aaboubi O 2004 Magnetic field influence on mass transport phenomena *Electrochim. Acta* **50**(1) 51–7

[3] Santos J S, Matos R, Trivinho-Strixino F and Pereira E C 2007 Effect of temperature on Co electrodeposition in the presence of boric acid *Electrochim. Acta* **53**(2) 644–9

[4] Ebadi M, Basirun W J and Alias Y 2010 Influence of magnetic field on the electrodeposition of Ni-Co alloy *J. Chem. Sci.* **122**(2) 279–85

[5] Mishra S, Prusty R and Hota P K 2016 Analysis of Levenberg-Marquardt and Scaled Conjugate gradient training algorithms for artificial neural network based LS and MMSE estimated channel equalizers *Proc. - 2015 Int. Conf. Man Mach. Interfacing, MAMI 2015*, **no. April, 2016**

[6] Parmar R, Shah M and Shah M G 2017 A Comparative study on different ANN techniques in wind speed forecasting for generation of electricity *IOSR J. Electr. Electron. Eng.* **12**(01) 19–26

[7] Baghirli O 2015 Comparison of Lavenberg-Marquardt, Scaled Conjugate Gradient and Bayesian

Regularization Backpropagation Algorithms for multistep ahead wind speed forecasting using multilayer perceptron feedforward neural network *Dissertation* **June, p. Uppsala University**

[8] Bashah N A A, Othman M R, and Aziz N 2015 Feed forward neural network model for production of isopropyl myristate in a semibatch reactive distillation: An evaluation of extrapolation prediction capability *Chem. Eng. Trans.* **45** 1831–6

[9] Uzir M H, Sudibyo, Aziz N and Othman M R 2015 Magneto-electro deposition of tin dendrites *Surf. Coatings Technol.* **264** 66–71

The influence of coffee cherry fermentation on the properties of *Cascara arabica* from Subang, West Java

E Sholichah[1*], D Desnilasari[1], R J Subekti[2], M A Karim[1] and B Purwono[2]

[1]Research Center for Appropriate Technology, Indonesian Institute of Sciences, JL KS Tubun No.5, Subang, West Java, Indonesia
[2]Department of Chemistry, Faculty of Mathematic and Natural Science, Universitas Gadjah Mada, Jl Sekip Utara, Bulaksumur, Yogyakarta, Indonesia

*Corresponding author and main contributor: enny.sh002@gmail.com

Abstract. Cascara is an infusion beverage of dried coffee cherry pulp prepared from a by-product of coffee production. Cherry of arabica coffee was fermented with spontaneous fermentation, *Lactobacillus plantarum* and yeast (*Saccharomyces cerevisiae*) at room temperature for 24 hrs. The properties of cascara including total phenolic, flavonoid, and caffeine contents, antioxidant activity, colour, and sensory acceptance were evaluated. Results showed that cascara from *L. plantarum* fermentation (Y) exhibited the highest polyphenols and caffeine contents of 46.78 ppm and 52.8 ppm, respectively. Cascara from spontaneous fermentation (X) resulted the highest flavonoids content and antioxidant activity of 0.28 ppm and 89.52%, respectively. The result of sensory evaluation showed that cascara from yeast fermented cherry obtained the highest overall acceptance score of 4 out of 6.

1. Introduction

Cascara also is known as cherry tea coffee is defined as dried coffee cherry pulp which is a by-product of coffee bean production [1]. Cascara beverage is usually prepared by infusing dried coffee cherry porridge with hot water to extract active compounds that provide health benefits. Polyphenol compounds such as chlorogenic acid, protocatechuic acid, gallic acid, and are present in cascara and result in high antioxidant activity [2]. Mullen, *et al.* (2013), reported that cascara from China, India, and Mexico contain active compound that are chlorogenic acids (CGA), caffeine, and polyphenolic content [3]. The utilization of cascara as a derivative product from coffee production can increase the economic value of coffee cherry pulp and its nutritional benefits [4].

The processing of coffee production was developed continuously to improve the coffee bean quality (flavor and unique taste) and fulfil the costumer or market demand. Coffee cherry fermentation was used to reduce the caffeine content and obtained the best flavor quality [5–7]. Caffeine is the major compound of whole coffee cherry, coffee bean and coffee husk [3]. Caffeine consumption can enhance mood, exercise performance and reduce the symptoms of Parkinson's disease and tremors. It has been reported that high caffeine consumption also stimulates the central nervous system, increases the blood circulation and respiration [8]. Coffee cherry fermentation is intended to reduce the caffeine content in coffee beverage.

Luwak coffee is a very popular fermented coffee because of the good flavor, taste, aroma and lower acidity and caffeine [5]. Luwak (*Paradoxurus hermaphroditus*) eats coffee cherry in the coffee plantation, then it defecates the remaining seeds as feces. Coffee cherry was fermented by the digestion system in luwak body. Usman *et al* (2015) reported that the predominant bacteria in the luwak feces was lactic acid bacteria. Hatiningsih *et al* (2018) reported that lactic acid bacteria and yeast were the main microbes in the spontaneous fermentation for Green Coffee (*Coffea arabica*) [9]. In vitro fermentation method using Lactobacillus bacteria and yeast has been developed to improve the quality of coffee. Fermentation of green coffee resulted green coffee bean with higher acid content, lower caffeine and polyphenol while the antioxidant activity was stable [9]. Lactic acid bacteria act in solubilization digests the complexed pectin producing more soluble pectin, while *Saccharomyces cerevisiae* hydrolyses cellulose to become glucose, reduced crude fiber and tannin effecting the increase of amino acids [10]. Coffee cherry fermentation influences the physicochemical characteristic of coffee bean, moreover it also affects the properties particularly the chemical content such as phenolic, alkaloid compounds and others soluble components of cherry pulp as the raw material of cascara. The aim of this study was to investigate the effect of in-vitro fermentation condition on cascara infused water includes colour, total phenolic, flavonoid, antioxidant activity and sensory acceptability.

2. Materials and methods

2.1. Materials

Arabica coffee cherry was obtained from coffee plantation of BUMDES Mukti Harja Subang-West Java. The chemicals for analysis were sodium carbonate (Na_2CO_3), Folin-Ciocalteu reagent, gallic acid, DPPH (2,2-diphenyl-1-picrylhydrazyl), ethanol absolute, quercetin, aluminum chloride ($AlCl_3$), sodium acetate (CH_3COONa), methanol, caffeine, and distilled water. Microbes used in the fermentation process were *Lactobacillus plantarum* (isolated from coffee cherry and identified based 16S rRNA) and commercially available yeast of *Saccharomyces cerevisiae* (fermipan) which were in the dried powder form.

2.2. Preparation of fermented cascara

Arabica coffee cherry was fermented in 3 fermentation methods (Table 1) including spontaneous fermentation as the control (X), fermentation by *L. plantarum* starter (Y), and *S. cerevisiae* starter (Z) in room temperature for 24 h. The medium for fermentation was prepared by dissolving 2g of starter powder in 2 l of distilled water and mixing well. The fermentation process was done by soaking 1 kg of arabica coffee cherry to the fermentation liquid in triplicate. Fermented coffee cherry was pulped using coffee pulper to separate the pulp and coffee bean. The coffee pulp was dried using oven dryer in temperature 50°C for 12 h. The dried of coffee pulp was ground and packed in the tea bag which was 3 g per pack, the secondary packaging was aluminum foil (10x10cm). Cascara samples for analysis were prepared by pouring 150 ml of the boiled water into a tea bag of cascara and infusing for 15 minutes and then removing the cascara. The cascara infusion water was the sample for analysis steps.

Table 1. Arabica Coffee Cherry fermentation methods.

Sample	Fermentation Methods
X	spontaneous fermentation
Y	*Lactobacillus plantarum* starter
Z	*Saccharomyces cerevisiae* starter

2.3. Colour analysis

The colour of cascara arabica infused water was measured by colourimeter (3NH-Taiwan). The sample was placed in a cuvette then colour reading was carried out. The obtained value was 3 colour co-ordinates namely, L* (represents lightness index), a* (represents the tons from red to green colour) and b* (represents the tons from yellow to blue colour) [11].

2.4. Total phenolic content

Total phenolic content was determined according to the method of the Indonesian National Standard 3753: 2014 for black tea bag with some modifications [12]. Gallic acid was used as the standard which the concentrations were of 10, 20, 30, 40, and 50 ppm. An amount of 1 ml of standard solution or sample was added by 5.0 ml of Folin-ciocalteu 10% and allowed for 3-8 minutes. Four ml of 7.5% Na_2CO_3 solution was added and mixed homogenously. The mixture then placed for 50 minutes at room temperature. All steps done in the dark bottles. Distilled water was used as a blank solution. The samples solution was treated as the standards solution. The next step was measuring the absorbance of blank, standards and samples solutions using the Shimadzu 1900 UV-Vis spectrophotometer at 765 nm. The total phenolic content was calculated based on the standard curve of gallic acid (y=0.0147x + 0.0074; R^2=0.99).

2.5. Flavonoid content

Determination of flavonoid content of samples used the aluminium chloride colourimetric method [13] with slightly modification. Quercetin was used as the standard calibration curve with serial concentration is 0.2; 0.4; 0.6; 0.8 and 1 ppm, ethanol was as a blank. An amount of 1.5ml of standard, samples and blank were added by 1.5 ml of 1.2% AlCl3 solution and 1.5 ml of CH3COONa. Then all dark bottles cover tightly and mix using vortex, then were placed in a dark room for 30 minutes. The absorbance was measured by Spectrophotometer UV-Vis Shimadzu 1900 in wavelength of 415 nm. (y= 0.7843x-0.0055; R^2= 0.9958).

2.6. Caffeine content

Determination of caffeine in the cascara infused water was use HPLC Infinity II 1260-AGILENT according to the method of Shrestha et al [14]. A C-18 column was used for HPLC analysis (ODS 5 µm, internal diameter 4.6 nm and length 150 mm) with isocratic mobile phase methanol : water (40:60) at flow rate of 1 ml/min and column temperature at 40°C. The peak of caffeine was observed at 2.56 minutes using UV detector set at 275 nm. The concentration of caffeine standard was 2, 4, 6, 8, and 10 ppm. Distilled water was used for blank. Caffeine content of samples was determined based on caffeine standard curve (y= 25.156x+4.3998; R^2=1).

2.7. Antioxidant Activity

2,2-Diphenyl-1-picrylhydrazyl (DPPH) assay was used for antioxidant activity measurement [15] . Quercetin was used as positive control in range concentrations were 5, 10, 15, 20, 25 and 30 ppm. DPPH solution was made in 0,3mM and ethanol was the solvent. An amount of 2 ml standard and samples were placed in dark bottles, then added by 2 ml of DPPH 0.3 mM and 1 mL of ethanol. The mixed solution was homogenized and incubated in the dark room for 30 minutes. The absorbance of standard and samples were measured using Spectrophotometer UV-Vis Shimadzu 1900 at 517 nm. Antioxidant capacity reported as percent radical scavenging activity (% RSA) and calculated based on equation 1 [16].

$$(\%RSA) = \frac{A_b - A_s}{A_b} \text{ x } 100\ \% \qquad \text{(Eq. 1)}$$

Where:
Ab : absorbance of blank
As : absorbance of sample

2.8. Sensory analysis

Sensory test of the cascara infused water was assessed by hedonic test with 30 panelists, in which the samples consisted 3 cascara infused water (X, Y, Z). The panelist evaluated the samples through hedonic parameters including colour, aroma, taste, after taste and overall acceptability with the hedonic scale from 1-6 in which 1 represents "extremely dislike," 6 indicates "extremely like,". The data was calculated the average value of every parameter for quantification the cascara acceptability. The mean value of each sensory parameter was reported.

2.9. Statistical analysis

Statistical analysis for all parameters determination including colour, total phenolic, flavonoid, caffeine content, and antioxidant activity was conducted by Analysis of variance (ANOVA) and post-hoc ($P<0.05$) test was performed to find out the difference between the variables. Microsoft Excel version 2016 program was employed for Statistical analysis.

3. Results and discussions

3.1. Colour Analysis

Colour of the cascara infused water reported in Table 1, expressed in L, a and b. L* was indicate the lightness, a* represent the difference in the colours red (+) and green (-), and b* was the indicator coordinates yellow (+) and blue (-) [17]. Cascara infused water were resulted from L. plantarum and S. cerevisiae fermentation methods had lower value of lightness, redness and yellowness than spontaneous fermentation, although insignificantly different ($p<0.05$). Yellowish colour of cascara infused water can be produced by epicatechin compound, and the redness was resulted by rutin, beta carotene and anthocyanin [18,19]. Colour of unfermented cascara infused water (3:100) had value of L was 32.27, a* was 7 and b* 17 [20]. The sample X showed higher lightness, redness and yellowness than sample Y and Z. Organic compounds such as phenolic, flavonoid and alkaloid can dissolve in the water during fermentation thereby reducing the lightness, redness and yellowness. Microbes digest the matrix of coffee cherry which consists of macromolecules such as complexed carbohydrate, lipid and protein due to may increase solubility of flavonoid, alkaloid and phenolic [10].

Table 2. Colour of fermented Cascara.

	L	a	b
X	28.33 ± 0.29^{n}	2.62 ± 0.14^{m}	3.94 ± 0.19^{o}
Y	28.15 ± 0.44^{n}	2.58 ± 0.22^{m}	3.69 ± 0.27^{o}
Z	28.18 ± 0.4^{n}	2.57 ± 0.17^{m}	3.53 ± 0.25^{o}

3.2. Total phenolic content

The major component of polyphenols in the whole coffee fruit were flavan-3-ols and flavonols, further the arabica coffee contain greater diversity of polyphenol compound then robusta [3]. The result showed that yeast fermentation (Z) produced cascara with the lowest value of total phenolic (Figure 1), meanwhile the *L. plantarum* fermentation displayed no effect on total phenolic compound. *S. cerevisiae* produced extracellular enzyme including amylase which can hydrolyze the α(1,6)- bonding of amylopectin and change sugar become alcohol [21]. In the coffee cherry, phenolic compounds conjugated with amino acids and glycosides [3,22], so it was possible the hydrolysis reaction can increase the solubility of glycosides and phenolic compound in the water. Lactobacillus act to convert carbohydrates of substrate into lactic acid. Carbohydrate in coffee cherry is a complex carbohydrate such as cellulose and pectin, so fermentation by *L. plantarum* failed to decrease the content of phenolic compounds in cascara.

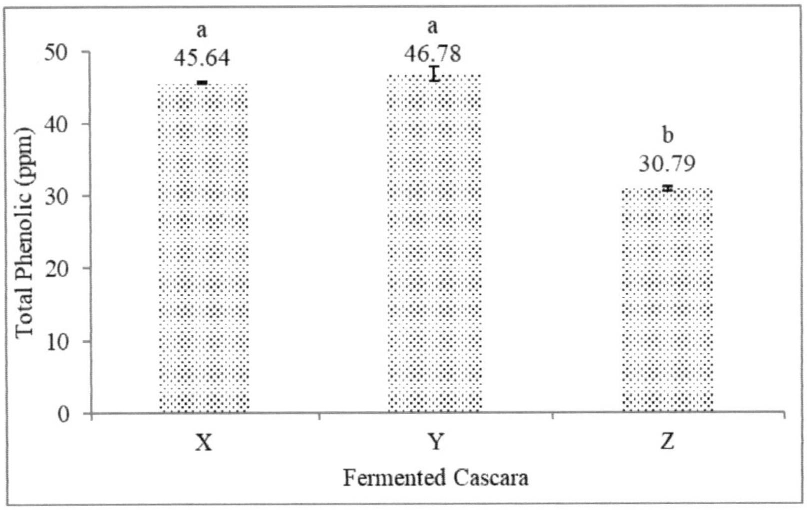

Figure 1. Total Phenolic Content of Fermented Cascara.

3.3. Flavonoid content

Colourimetric method for flavonoid determination based on the formation of complexed between $AlCl_3$ and keto groups of C-4 and hydroxyl groups of C-3 or C-5 in the flavon and flavonols [23]. Quercetin is a flavonol compound used as standard in flavonoid determination. Figure 2 shows that the highest total flavonoid content was obtained of sample X that was 0.278 ppm. Fermentation by *L. plantarum* and *S. cerevisiae* decreased the total flavonoid content even though it was not significantly. *L. plantarum* and *S. cerevisiae* cause amount of the flavonoid release into water or fermentation medium.

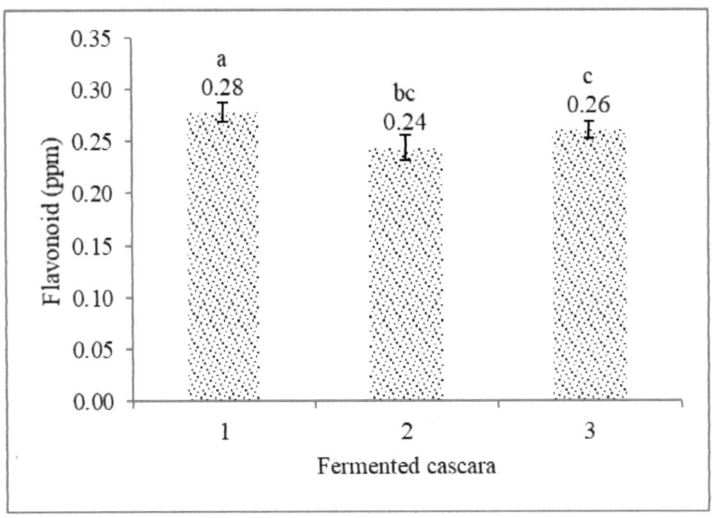

Figure 2. Flavonoid content of fermented cascara.

3.4. Caffeine content

The amount of caffeine is influenced by many factors that are the processing method, type of coffee cherry, plantation [2,3,9]. Whole arabica coffee cherry contain 1.3-5.2 mg/g of caffeine, coffee bean 1.3-5.4 mg/g, coffee husk are between 1-1.3 mg/ g [24]. The concentration of caffeine in the arabica coffee cherry is lower than robusta coffee [3]. Caffeine is alkaloid compound contained mostly in the coffee, tea and cocoa. Caffeine consumption recommended in moderate level to get mood booster effect and health benefits [24].

IOP Conf. Series: Materials Science and Engineering 1011 (2021) 012006 doi:10.1088/1757-899X/1011/1/012006

Determination of caffeine using HPLC detected the peak of caffeine in the retention time was 2.5 minute. The obtained caffeine concentration in the cascara infused water showed in Figure 3. The result express that spontaneous fermentation produced cascara with the lowest caffeine significantly ($p<0.05$). In the spontaneous fermentation may consist many kind of microbes but the dominants were Lactic acid bacteria and yeast which could produce pectinolytic and proteolytic enzymes [9]. Furthermore, the microbes have abilities to hydrolyze and degrade the caffeine content in the cherry coffee during the fermentation. Caffeine can be derived become chlorogenate acid and continuing degradation to form smaller compound that can release or soluble in the water [7].

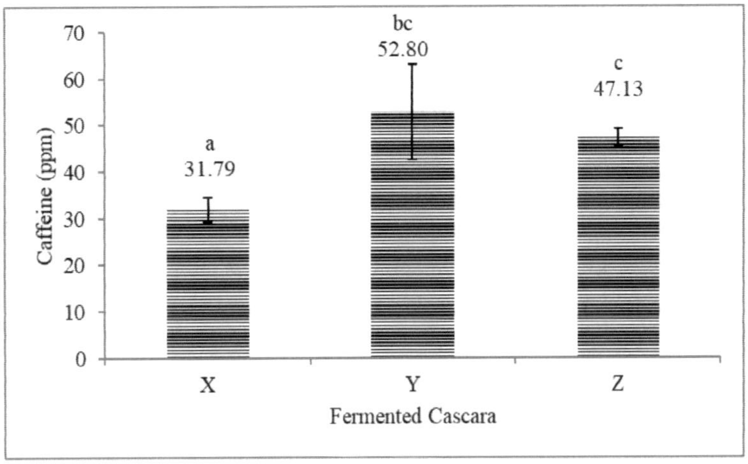

Figure 3. Caffeine content of fermented cascara.

3.5. Antioxidant activity
Coffee cherry was reported as a potential source of antioxidant because it was confirmed containing rich of the active compound such as phenolic, flavonoid, alkaloid and organic acids. Antioxidant activity indicate the capability to inhibit the oxidation process or radical reaction [16]. Sources of antioxidant compound are required to improve the immunity and maintain the healthy especially in this pandemic Covid-19 [25]. Mullen *et al* [3], reported that the antioxidant activity of coffee cherry related to the phenolic content but it did not related to flavonoid content .

Fermentation of coffee cherry in 3 fermentation methods resulted cascara which were stable in its antioxidant activity. Spontaneous fermentation produced higher antioxidant activity that was 89.52 % followed by *L. plantarum* (87.63%) and *S. cerevisiae* (84.76%) (Figure 4).

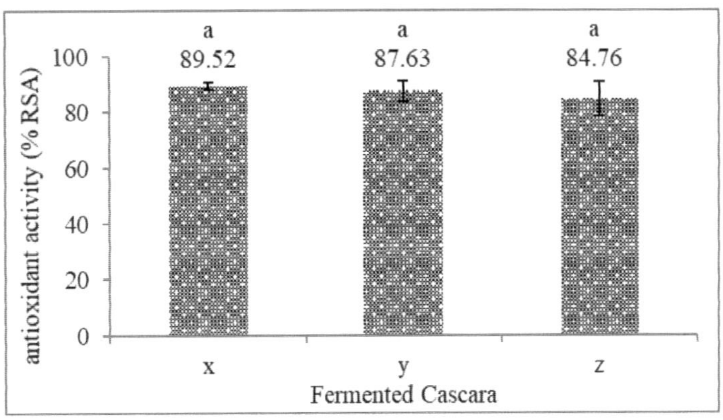

Figure 4. Antioxidant activity of fermented cascara using DPPH method.

3.6. Sensory analysis

Figure 5 shows that sample Z possessed the highest value of overall obtained acceptability with the score was 4 from 6. While the values for samples X and Y were 3. These results indicated that the fermentation process by *S. cerevisiae* can improve the taste quality of cascara beverage better than with *L. plantarum* and spontaneous fermentation. This was in line with research by Neto *el al* [26] which was fermented coffee cherry by yeast produced cascara with better overall acceptance than spontaneous fermentation. Cascara fermentation by *L. plantarum* was not showed significantly different from spontaneous fermentation in the overall acceptance. The results of sensory test of fermented coffee beverages using yeast get score 91 points [27], while fermented coffee beverage *L. plantarum* showed 88.3 points, based on the results of the Specialty Coffee Association of America's Cupping (https://sca.coffee/ research / coffee-standards). This was due to the fermentation process using *S. cerevisiae* in coffee cherries, the mucilage of coffee cherry was acted as a source of carbon and nitrogen which could produce ethanol, lactic acid, and other metabolites that produce better taste, aroma, flavor and acceptance. The fermentation process can produce volatile compounds that are diffused in the beans and cascara that have an impact on the final quality of the product. Yeast, in this case S. *cerevisiae* can produce compounds, such as esters, alcohols, aldehydes, ketones, and terpenoids, that could affect the product's aroma through carbon metabolism so that a distinctive aroma is produced and differentiates it from other drinks

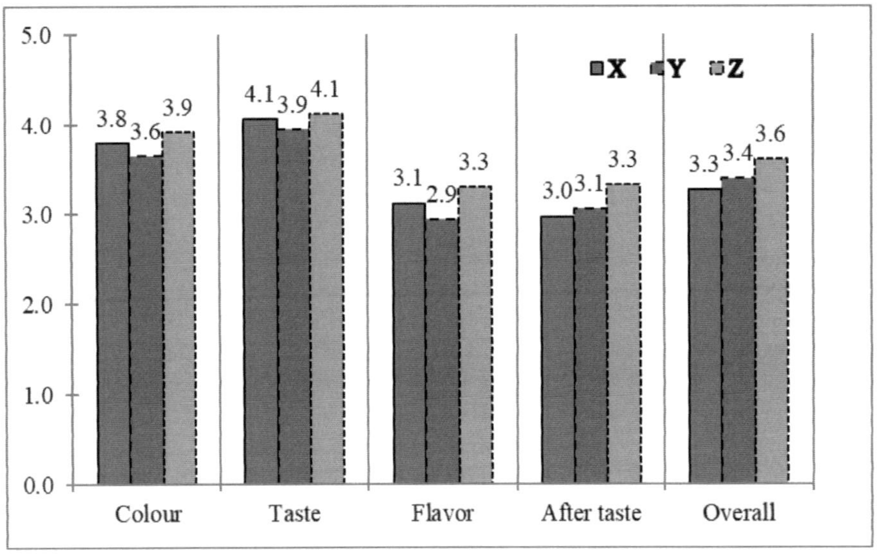

Figure 5. Result of organoleptic analysis.

4. Conclusions

Fermentation processing method of arabica coffee cherry influenced the total phenolic, flavonoid, caffeine content and sensory acceptability. *Lactobacillus plantarum* fermentation (Y) showed the highest polyphenols and caffeine contents of 46.78 ppm and 52.8 ppm, respectively. Cascara from spontaneous fermentation (X) resulted the highest flavonoids content and antioxidant activity of 0.28 ppm and 89.52%, respectively. The result of sensory evaluation showed that cascara from yeast fermented cherry obtained the highest overall acceptance score of 4 out of 6. Fermentation method of coffee cherry suggest to be continuously developed to optimize the fermentation condition to improve the coffee bean and cascara quality.

Conflict of interest

The authors have no conflict of interest to state regarding this manuscript.

Acknowledgement
We thank to BAPPENAS who funded this research in the 2018-2019 in National Priority Program. Thanks to BUMDES Mukti Harja Subang-West Java for the raw materials supporting and all colleagues in Research Center for Appropriate Technology- Indonesian Institute of Sciences LIPI who supported this research.

References

[1] Garis P, Romalasari A, Purwasih R and Kunci K 2019 Pemanfaatan Limbah Kulit Kopi Cascara Menjadi Teh Celup *Industrial Research Workshop and National Seminar* (Bandung) pp 279–85
[2] Heeger A, Kosińska-Cagnazzo A, Cantergiani E and Andlauer W 2017 Bioactives of coffee cherry pulp and its utilisation for production of Cascara beverage *Food Chem.* **221** 969–75
[3] Mullen W, Nemzer B, Stalmach A, Ali S and Combet E 2013 Polyphenolic and hydroxycinnamate contents of whole coffee fruits from China, India, and Mexico *J. Agric. Food Chem.* **61** 5298–309
[4] Widyotomo S 2013 Potency and technology of coffee trash diversification product to increase good quality and added value *Rev. Penelit. Kopi dan Kakao* **1** 63–80
[5] Usman D, Suprihadi A and Kusdiyantini E 2015 Fermentasi kopi robusta *(Coffea canephora)* menggunakan isolat bakteri asam laktat dari feces luwak dengan perlakuan lama waktu inkubasi *J. Akad. Biol.* **4** 31–40
[6] Towaha J and Rubiyo R 2013 The effect of fermentation on flavor quality of probiotic Civet Coffee *Bul. RISTRI* **4** 175–82
[7] Tawali A B, Abdullah N and Wiranata B S 2018 The influence of fermentation using bacteria lactic acid yoghurt to the flavor of coffe robusta *(Coffea robusta) Canrea J. Food Technol. Nutr. Culin. J.* **1** 90–7
[8] Reich M S, Dietrich M S, Finlayson A J R, Fischer E F and Martin P R 2008 Coffee and cigarette consumption and perceived effects in recovering alcoholics participating in Alcoholics Anonymous in Nashville, Tennessee, USA. *Alcohol. Clin. Exp. Res.* **32** 1799–806
[9] Hatiningsih S, Antara N S and Gunam I B W 2018 Microbiological and physicochemical changes of Green Coffee *(Coffea arabica)* fermentation in Kintamani, Bangli, Bali *Sci. J. Food Technol.* **5** 123–38
[10] Riki Saumi Nuryana, Rachmat Wiradimadja D R 2016 Pengaruh dosis Dan waktu fermentasi Kulit Kopi *(Coffea arabica)* menggunakan *Rhizopus oryzae* dan *Saccharomyces cerevisiae* terhadap kandungan protein kasar dan serat kasar *Student e-Journal* **5**
[11] Sholichah E, Deswina P, Sarifudin A, Andriansyah C E and Rahman N 2019 Physicochemical, structural and morphological properties of some arrowroot *(Maranta arundinacea)* accessions growth in Indonesia *AIP Conf. Proc.* **2175** (020008-1)-(020008-9)
[12] Nasional B S 2014 *SNI 3753:2014 (Teh hitam celup)* (Indonesia: Badan Standarisasi Nasional)
[13] Chandra S, Khan S, Avula B, Lata H, Yang M H, Elsohly M A *et al.* 2014 Assessment of total phenolic and flavonoid content, antioxidant properties, and yield of aeroponically and conventionally grown leafy vegetables and fruit crops: a comparative study *Evidence-based Complement. Altern. Med.* **2014**
[14] Pokhrel P, Shrestha S, Rijal S K and Rai K P 2016 A simple HPLC method for the determination of caffeine content in tea and coffee *J. Food Sci. Technol. Nepal* **9** 74–8
[15] Molyneux P 2004 The use of the stable free radical diphenylpicryl-hydrazyl (DPPH) for estimating antioxidant activity *Songklanakarin J. Sci. Technol.* **26** 211–9
[16] Santoso U 2016 Antioksidan pangan *Penerbit Gadjah Mada Univ. Press. Yogyakarta*
[17] Pathare P B, Opara U L and Al-Said F A J 2013 Colour measurement and analysis in fresh and processed foods: a review *Food Bioprocess Technol.* **6** 36–60
[18] Ramirez-Martinez J R 1988 Phenolic compounds in coffee pulp: quantitative determination by HPLC *J. Sci. Food Agric.* **43** 135–44
[19] Ariadi H P and Windrati W S 2015 Compound extraction of coffee fruit cod : study of species

and maceration duration of Coffee Berk. *Ilm. Pertan.* **x** 1–5

[20] Nafisah D and Widyaningsih T D 2018 Kajian metode pengeringan dan rasio penyeduhan pada proses pembuatan teh cascara kopi arabika *(Coffea arabika L .) J. Pangan dan Agroindustri* **6** 37–47

[21] Kustyawati M E, Sari M and Haryati T 2013 Effect of fermentation using *Saccharomyces cerevisiae* on the biochemical properties tapioca *Agritech* **33** 281–7

[22] Esquivel P and Jiménez V M 2012 Functional properties of coffee and coffee by-products *Food Res. Int.* **46** 488–95

[23] Azizah D N, Kumolowati E and Faramayuda F 2014 Determination of flavonoid level method of $AlCl_3$ in the extract of Cocoa Fruit *(Theobroma cacao L.) Kartika J. Ilm. Farm.* **2** 45–9

[24] Bondesson E 2015 A nutritional analysis on the by- product coffee husk and its potential utilization in food production 1–25

[25] Yousfi N, Bragazzi N L, Briki W, Zmijewski P and Chamari K 2020 The COVID-19 pandemic: how to maintain a healthy immune system during the quarantine - a multidisciplinary approach with special focus on athletes. *Biol. Sport* **37(3)** 211-6

[26] Neto D P D C, Pereira G V D M, Finco A M O, Rodrigues C, Carvalho J C De and Soccol C R 2020 Microbiological, physicochemical and sensory studies of coffee beans fermentation conducted in a yeast bioreactor model *Food Biotechnol.* **34** 172–92

[27] Pereira G V de M, de Carvalho Neto D P, Medeiros A B P, Soccol V T, Neto E, Woiciechowski A L and Soccol C R 2016 Potential of lactic acid bacteria to improve the fermentation and quality of coffee during on-farm processing *Int. J. Food Sci. Technol.* **51** 1689–95

IOP Conf. Series: Materials Science and Engineering 1011 (2021) 012007 doi:10.1088/1757-899X/1011/1/012007

Pumpkin flour effects on antioxidant activity, texture, and sensory attributes of flat tubers noodle

N Novita Indrianti*, E Sholichah and N Afifah

Research Center for Appropriate Technology, Indonesian Institute of Sciences, JL KS Tubun No.5, Subang, West Java, Indonesia

*Corresponding author and main contributor: novitaindrianti@gmail.com

Abstract. Canna and arrowroot are local tubers which have high amount of amylose content and suitable for noodle. Flat noodles are one type of noodles that are characterized with a thin flat shape and is generally made from starch. Pumpkin flour is one of the processed products of pumpkin fruit, which can be used in manufacturing of formulated foods as functional ingredient. The aim of this research was to study the effect of pumpkin flour on texture, antioxidant activity, and sensory attributes of flat tubers noodle. In this study, flat noodle was made from canna or arrowroot starch, which is added with pumpkin flour. We added the pumpkin flour (0%, 5%, 10%, and 15% w/w, canna or arrowroot starch basis) in the flat noodle formulation. Result showed that the addition of pumpkin flour increased antioxidant activity of flat tubers noodle. The stickiness of flat tubers noodle increased with the increasing of pumpkin flour level. Flat noodle made form arrowroot starch was stickier than flat canna noodle. Sensory analysis was carried out on flat canna noodle with hedonic scoring method to identify which product is the most preferred by the panelist. Generally the aroma and taste of flat canna noodle did not vary with increasing of pumpkin flours level. Flat noodle made from canna starch with 10% of pumpkin flour obtained the highest score for color, texture (stickiness), and overall acceptability assessed. Antioxidant activity and stickiness measurement of flat canna noodle with 10% of pumpkin flour were 39.80% and 3350 gf respectively.

1. Introduction

Noodles are one the most popular Indonesian foods owing to its taste and ease of preparation [1]. According to data of World Instant Noodles Association (WINA), instant noodle consumption in Indonesia reached 12.52 million servings in 2019 [2]. In wheat noodle making the presence of gluten protein produce an elastic and compact structure of noodle so it can avoid noodle stickiness and cooking losses. The absence of gluten in non-wheat noodle compositions affects noodle quality. Therefore, in non-wheat noodles, the gluten protein network has to be replaced by using a gelling agent the gelling properties of starch [3]. Flat noodles are one type of noodles that are characterized with a thin flat shape and is generally made from starch [4].

Canna (*Canna edulis Kerr*) and arrowroot (*Maranta arundinaceae*) are local tubers which can be used as sourches of carbohydrate content and have nutritional value similar to that of wheat. Canna and arrowroot starch contained amounts of amylose (35% and 21.9%, respectively) [5]. According to Chansri et al. [6], high amylose starch is an ideal starch as raw material for noodles. Flour and starch from tubers can be used to substitute wheat flour in noodle preparation. As stated by previous researchers, canna flour and arrowroot flour can be used as raw materials in the making of noodles [7,8].

The characteristics of noodle, such as texture and sensory attributes, are important factors affecting product quality [9]. The high stickiness of noodles is an undesirable texture parameter for noodle quality. Sensory evaluation has been carried out to evaluate sensory attributes and consumer acceptance of the product [9]. Color is one of the critical parameter that indicated visual appearance and overall acceptability of noodles and can influences purchasing decisions [10]. The color of noodle can be improved by adding some natural ingredients to the noodle. Pumpkin (*Cucurbita moschata*) is one of the most desirable vegetables and abundantly available in Indonesia [11]. Dabash et al. [12] stated that pumpkin can be processed into flour which has a longer shelf-life and can be used in the manufacture of formulated foods. Pumpkin flour is popular due to its highly-desirable flavour, sweetness and deep yellow-orange color. The yellow-orange characteristic color of pumpkin is due to the presence of caroteniods that can act as antioxidants in the human body [13]. A previous study has shown that the replacement of corn flour with pumpkin flour affected texture and sensory parameters in gluten-free pasta [14]. The purpose of this study was to study the effect of pumpkin flour on texture, antioxidant activity, and sensory attributes of flat tubers noodle.

2. Materials and methods

2.1. Materials

The materials used in this study were canna starch, arrowroot starch and yellow pumpkin flour (purchased from local producer 'Kusuka Ubi' in Yogyakarta), salt and water. While the instruments used in this research were digital scales, a mixer, an aluminum tray, a steamer, an infrared dryer, scissors, a texture analyzer, a spectrophotometer, and glassware.

2.2. Flat Noodle Preparation

The flat tubers noodle were prepared by mixing 100 g canna starch/arrowroot starch with pumpkin flour (PF). As a control was used flat noodle made from 100% tubers starch without the addition of pumpkin flour. The treatments which were carried out in this research were the pumpkin flour concentrations (5%; 10%; and 15% pumpkin flour/100 g tubers starch). The additional ingredients were salt (1%) and water (1:1) based on tubers starch. All ingredients mixed and stirred using a mixer to form a slurry. The slurry (80 g) poured into an aluminum tray (25cm x 28cm x 2cm), steamed at 100 °C for 3 minutes using a steamer, and then cooled at a room temperature. The flat tubers noodle sheet dried in an infrared dryer at 45-50 °C to half-dry. Half-dry flat noodle sheets cut off into small sheets and dried again at 45-50 °C for ± 1 hour. Flat noodle packed and stored for analysis.

2.3. Antioxidant Activity

Antioxidant activity of flat tubers noodle were determined by using 2.2′-difenil-1- picrylhydrazyl radical (DPPH) method [15]. Amount of 0.2 g of flat tubers noodle was extracted with 5 mL methanol using vortex for 1 minute and filtered to obtain the extract. Two hundred and eighty microliters of 0.1 mM DPPH solution was pipetted into each well plate followed by 20 μL of sample, or solvent for the blank. The mixture was incubated at 30°C for 1h, and the absorbance at 515 nm was measured with a microplate reader. The inhibition percentage of the radical scavenging activity was calculated using the equation:

$$\% DPPH = \frac{1 - sample\ absorbance}{blank\ absorbance} x\ 100\% \qquad (1)$$

Where A0 is absorbance of the blank and As is absorbance of the sample at 515 nm

2.4. Texture

Flat noodle texture, expressed as stickiness, was measured using a texture analyzer (TAXT-Plus, Stable Micro Systems, Surrey, UK) and determined using a Heavy Duty Platform (HDP/90) probe. Flat noodles were boiled in 150 ml of water until completely cooked. The test was performed by placing sample (cooked flat noodles) under the retaining plate in as flat a position as possible (and centrally aligned if testing stranded products) at mode : pre-test Speed : 1.0 mm/s, test speed: 0.5 mm/s, post-test speed: 10.0 mm/s, compression force : 1000g, compression time: 2 sec, withdrawal distance: 10.0mm, trigger type: Auto - 20g, tare mode: Auto.

2.5. Sensory Attributes
Three flat tubers noodle with the lowest stickiness were evaluated by 30 untrained panelist. The panelist evaluated the flat tubers noodle based on how much they liked the color, aroma, taste, texture (stickiness) and overall acceptability. Sensory evaluation of cooked flat noodle was carried out using 7-point hedonic scale (1 representing dislike extremely and 7 representing like extremely). Panelist was provided with a cup of water to refresh their palate after testing each sample.

2.6. Statistical analysis
All measurement was carried out in triplicate and results were expressed as mean ± standard deviation. Analysis of variance (ANOVA) was used to determine the significant differences between the samples. Means were compared according to Fisher's least significant difference (LSD) procedure at a significance level of $P < 0.05$. All statistical analyses were performed using Microsoft Excel 2007.

3. Results and discussions

3.1. Antioxidant Activity
The antioxidant activity of flat tubers noodle was evaluated using DPPH assay.

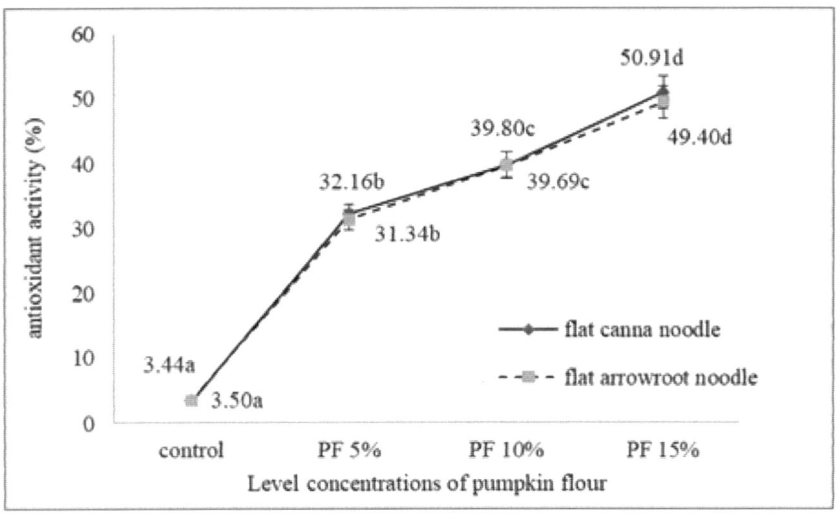

Figure 1. Antioxidant activity of flat tubers noodle.

Figure 1 demonstrates that the scavenging activity of flat tubers noodle with pumpkin flour addition ranged from 32.16 to 50.91% on flat canna noodle and 31.34 to 49.40% on flat arrowroot noodle, respectively. The addition of pumpkin flour increased the antioxidant activity of flat tubers noodle compared to control. This result can be due to the fact that pumpkins are rich in β-carotene, which functions as an antioxidant [16]. Malkanthi *et al.* [17] reported that antioxidant activity was increased in 5% pumpkin seed powder incorporated biscuits compared to control. Nyam *et al.* [18] found that bread supplemented with 5% pumpkin had a 37.99% increase in DPPH radical scavenging activity in pumpkin seed bread as compared to control bread. Similar results also reported that fortified cupcakes with 10% pumpkin powder had the highest level of antioxidant activity [19].

3.2. Texture

Stickiness is defined as the maximum peak force to separate the probe from the sample's surface upon probe retraction (the higher the force value, the stickier is the sample). The texture of flat noodle with respect to the stickiness is shown in Figure 2.

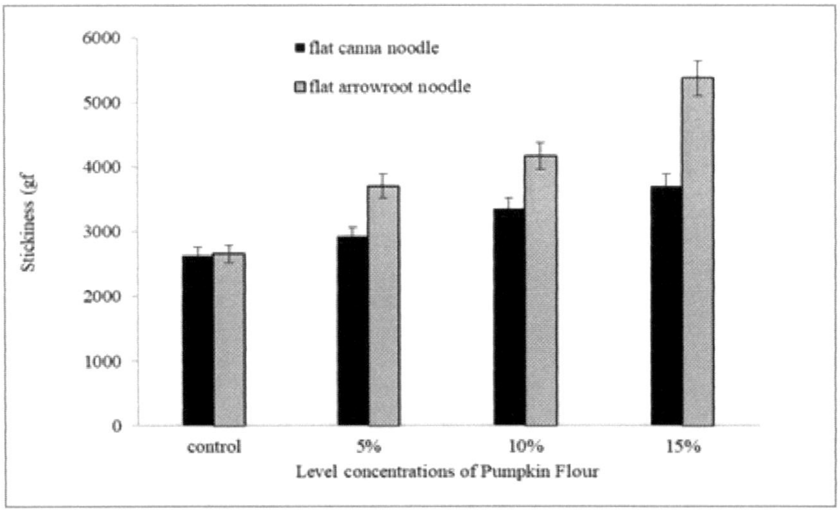

Figure 2. Stickiness of flat tubers noodle.

Figure 2 showed that the addition of pumpkin flour determined an increase in the stickiness of flat tubers noodle. The stickiness of flat canna noodle with addition of pumpkin flour ranged from 2920 to 3704 g was lower than the stickiness of flat arrowroot noodle (3701 to 5369 g). The stickiness depends on the quantity and quality of starch and its gelatinization condition [20]. Canna starch has a high amylose content resulting noodle with low stickiness and more compact. Amylose have inter molecular bonding force stronger than the amylopectin thereby inhibiting leaching In addition, starches with high setback viscosity were retrogradated quickly so that noodles texture becomes more cohesive and non-sticky because in the surface of noodles formed a layer of starch which has been retrogradated [21]. Furthermore, the addition of pumpkin flour increased the water absorption of composite flour (tubers starch and pumpkin flour) which resulted in noodle more sticky. Li et al. [22] stated that a higher water absorption of Djulis flour produces very soft and sticky noodles. Our result was similar to the adhesiveness of pasta which stated that the adhesiveness was increased by increasing pumpkin flour content in the pasta formulation [14][23].

3.3. Sensory Attributes

Three flat tubers noodle with the lowest stickiness i.e flat canna noodle with 5% of pumpkin flour, flat canna noodle with 10% of pumpkin flour and flat canna noodle with 15% of pumpkin flour were evaluated by 30 untrained panelist. Results from the sensory attributes of flat canna noodles are shown in Figure 3. Generally the aroma and taste of flat canna noodles did not vary with increasing of pumpkin flours level. Color indicated visual appearance that is a key quality index in noodles which influences purchasing decisions [10]. Panelists rated the color of flat canna noodles with 10% of pumpkin flour highest (mean score of 5.5) than other flat canna noodles. The observed trend for color may be attributed to the fact that flat canna noodles with 10% of pumpkin flour appeared brighter compared to others. Panelist were more liked towards the texture (stickiness) of flat canna noodle with 10% of pumpkin flour, and this obtained a mean score of 4.5 ("rather like"). Flat noodle made from canna starch with 5% and 10% of pumpkin flour obtained the same mean score for overall acceptability assessed.

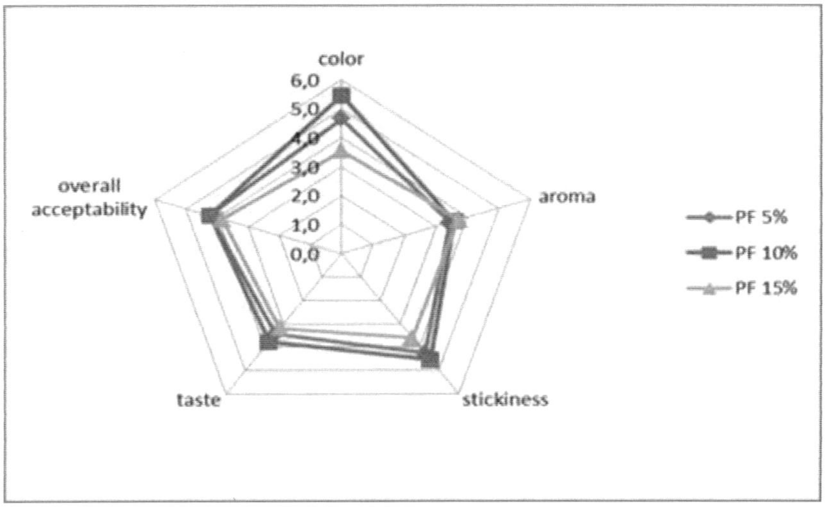

Figure 3. sensory attributes of flat canna noodle with different levels of pumpkin flour.

4. Conclusions

The addition of pumpkin flour increased antioxidant activity of flat tubers noodle. The stickiness of flat tubers noodle increased with the increasing of pumpkin flour level. Generally the aroma and taste of flat canna noodle did not vary with increasing of pumpkin flours level. Flat noodle made from canna starch with 10% of pumpkin flour obtained the highest score for color, texture (stickiness), and overall acceptability assessed. Antioxidant activity and stickiness measurement of flat canna noodle with 10% of pumpkin flour were 39.80% and 3350 gf respectively.

Conflict of interest

The authors have no conflict of interest to state regarding this manuscript.

Acknowledgement(s)

We thank the Ministry of Research, Technology and Higher Education of the Republic of Indonesia by Insentif Riset Sistem Inovasi Nasional (INSINAS) program for in this research.

References

[1] Abidin A Z and Devi C 2013 Development of wet noodles based on cassava flour *J. Eng. Technol. Sci.* **45** 97–111
[2] World Instant Noodles Association 2020 Global Demand for Instant Noodles https://instantnoodles.org/en/noodles/market.html. 1–3
[3] Santacruz S, Pennanen M and Ruales J 2009 Protein enrichment of oriental noodles based on Canna edulis starch *Asian J. Food Agro-Industry* **2** 521–38
[4] Klinmalai P, Hagiwara T and Sakiyama T 2017 Chitosan effects on physical properties, texture , and microstructure of fl at rice noodles *LWT - Food Sci. Technol.* **76** 117–23
[5] Aprianita A, Vasiljevic T, Bannikova A and Kasapis S 2014 Physicochemical properties of flours and starches derived from traditional Indonesian tubers and roots *J. Food Sci. Technol.* **51** 3669–79
[6] Chansri R, Puttanlek C, Rungsadthong V and Uttapap D 2005 Characteristics of clear noodles prepared from edible canna starches *J. Food Sci.* **70** 337–42
[7] Miftakhussolikhah, Ariani D, Ervika R N H, Angwar M, Wardah, *et al.* 2016 Cooking characterization Of Arrowroot *(Maranta arundinaceae)* noodle in various arenga starch substitution Ber. *Biol.* **15** 141–8

[8] Handajani S and Pangesthi L T 2019 Optimization of ganyong starch *(Canna edulis)* on Making of Dry and Instant Noodles *IOP Conf. Series: Earth Environ. Sci.* **347** 012082

[9] Astuti R D, David W and Ardiansyah 2020 Sensory evaluation of noodles substituted by sweet potato flour and rice bran *Curr. Res. Nutr. Food Sci.* **08** 144–54

[10] Akonor P T, Tortoe C, Buckman E S and Hagan L 2017 Proximate composition and sensory evaluation of root and tuber composite flour noodles *Cogent Food Agric.* **45** 1–7

[11] Wahyono A, Dewi A, Oktavia S, S J and Kang W 2020 Antioxidant activity and total phenolic contents of bread enriched with pumpkin flour *IOP Conference Series: Earth and Environ. Sci.* **411** 012049

[12] Dabash V, Buresova I, Tokar M, Zacharova M and Gal R 2017 The effect of added pumpkin flour on sensory and textural quality of rice bread *J. Microbiol. Biotechnol. Food Sci.* **6** 1269–72

[13] Satihya Mala K, Aathira P, Anjali E, Srinivasulu K and Sulochanamma G 2018 Effect of pumpkin powder incorporation on the physico-chemical, sensory and nutritional characteristics of wheat flour muffins *Int. Food Res. J.* **25** 1081–7

[14] Mirhosseini H, Farhana N, Rashid A, Tabatabaee B, Whye K and Kazemi M 2015 Effect of partial replacement of corn fl our with durian seed fl our and pumpkin fl our on cooking yield, texture properties, and sensory attributes of gluten free pasta *LWT - Food Sci. Technol.* **63** 184–90

[15] Payet B, Sing A S C and Smadja J 2005 Assessment of antioxidant activity of Cane Brown Sugars by ABTS and DPPH radical scavenging assays: determination of their polyphenolic and volatile constituents *J. Agric. Food Chem.* **53** 10074–9

[16] Koh S and Loh S 2018 In vitro bioaccessibility of β-carotene in pumpkin and butternut squash subjected to different cooking methods Int. Food Res. J. 25 188–95

[17] Malkanthi H, Umadewi S and Jamuna K 2018 Glycemic response and antioxidant activity of pumpkin seed powder *(Cucurbita maxima)* blended biscuits *J. Pharmacogn. Phytochem.* **7** 1877–82

[18] Nyam K, Lau M and Tan C 2013 Fibre from Pumpkin *(Cucurbita pepo* L.*)* seeds and rinds: physico-chemical properties, antioxidant capacity and application as bakery product ingredients *Malay. J. Nutr.* **19** 99–109

[19] Sello A A and Mostafa M Y A 2017 Enhancing antioxidant activities of cupcakes by using pumpkin powder during storage *J. Food Dairy Sci.* **8** 103–10

[20] Sozer N, Dalgıc A and Kaya A 2007 Thermal, textural and cooking properties of spaghetti enriched with resistant starch *J. Food Eng.* **81** 476–84

[21] Indrianti N, Afifah N and Sholichah E 2019 The production of composite flour from Canna/ Arrowroot starch and pumpkin flour as flat noodle ingredient *Biopropal Ind.* **10** 49–63

[22] Li P, Lu W, Hsieh C, Li T and Huang D 2015 Rheological properties of dough and quality of salted noodles supplemented with Djulis (*Chenopodium formosanum* Koidz.) flour *J. Agric. Sci.* **7** 84–92

[23] Purwandari U, Khoiri A, Muchlis M, Noriandita B, Zeni N, Lisdayana N et al. 2014 Textural, cooking quality, and sensory evaluation of gluten-free noodle made from breadfruit, konjac, or pumpkin flour *Int. Food Res. J.* **21** 1623–7

IOP Conf. Series: Materials Science and Engineering 1011 (2021) 012008 doi:10.1088/1757-899X/1011/1/012008

Fe-Cr pillared clay as catalysts for the ethanol to gasoline conversion

Robert R Widjaya[1]*, Nanda Saridewi[2], Alda A Putri[2], Nino Rinaldi[1] and Adid A Dwiatmoko[1]

[1]Research Center for Chemistry Indonesian Sciences, Kawasan PUSPIPTEK Serpong, Tangerang Selatan 15314, Indonesia
[2]Department of Chemistry, UIN Syarif Hidayatullah Jakarta, Indonesia

*Corresponding author: robert.ronal.widjaya@lipi.go.id

Abstract. Bentonite is a clay mineral type of montmorillonite that can be used as a catalyst. The use of bentonite as a catalyst has weaknesses, namely small surface area, pore volume and low thermal stability, so it is necessary to modify the structure. Modification of bentonite structure carried out in this research used the pillarization method (Pillared Clay or PILC) with the addition of Fe and Cr metals to form Cr/PILC, Fe/PILC, and Fe-Cr/PILC catalysts. The PILC catalysts can be used as a catalyst in the ethanol to gasoline conversion processess for alternative fuels. The XRD result shows that the pillarization process with the addition of Fe and Cr metals was succeeded so it can increase the interlayer spacing by 22,03 Å, 15,79 Å, and 15,85 Å on the Cr/PILC, Fe/PILC, and Fe-Cr/PILC catalysts. The SAA result shows an increase in surface area and pore volume in the Cr/PILC, Fe/PILC, and Fe-Cr/PILC catalysts structure. The presence of Brønsted acid type is indicated through the absorption peaks at 1515-1640 cm^{-1} and Lewis acid type at 1435-1470 cm^{-1} from the FTIR result. The TGA result shows an increase in thermal stability on the Cr/PILC, Fe/PILC, and Fe-Cr/PILC catalysts. The Fe-Cr/PILC catalyst shows the best activity results on the ethanol to gasoline conversion process with the acquisition of percent conversion and selectivity by 43.30% and 87.15% respectively.

1. Introduction

Clays are rocks that contain minerals and are composed of alumina-silica. Minerals in clay are a group of mineral constituents of sedimentary rocks and also as the main element of soil [1]. Clay or bentonite is generally used as an ion-exchange, adsorbent, and catalyst in the paper, paint, and rubber industries, so clay is often referred to as a multipurpose material. The type of clay that has become a concern in recent years is bentonite. Bentonite can be used as a catalyst buffer because it has advantages, namely having a pore volume that is easy to modify and a flexible structure [2]. The use of bentonite as a buffer on the catalyst has disadvantages, its thermal and hydrothermal stability is low, which is below 400°C [3]. The catalytic activity of bentonite as a catalyst buffer can be increased by modifying the bentonite structure.

Bentonite engineering methods can be done by pillarization (Pillared Clay or PILC). The pillarization method is carried out by an intercalation process using a pillaring agent in the form of a metal hydroxide cation into a bentonite alumina-silica interlayer that is useful as a pillar or pillar between its layers [4]. The bentonite structure will dehydrate and dehydroxylate during the intercalation process, resulting in metal oxides between the alumina-silica layers. Bentonite which has been successfully polarized will have high thermal stability and produce interlayer space in molecular dimensions [3]. The pillarization method was chosen in this study because the pillarization method has the advantages of not changing the basic structure of the bentonite after an intercalation process and is easy to do. Metal polyhydroxy which is commonly used in the modification of pillared bentonite is Fe, Cr, Ti, Zr, and Al. Some previous studies that have used metals in the manufacture of PILC catalysts include Fe metals [5-7].

Research that uses variations of two metal oxides (bi-oxide metals) on the PILC catalyst is still rarely done. The use of two different metal oxides will complement each other's shortcomings. An alloy of two metal oxides used in the synthesis of the PILC catalyst can affect the acidity of the catalyst surface which will be used for the reaction in the conversion of ethanol to gasoline. Previous research has been carried out by Widjaya *et al.* (2012) using Cr metal in the production of Cr/PILC catalysts. The presence of Cr metal in the production of Cr/PILC catalysts can increase the interlayer distance of bentonite up to 19.25 Å, besides there is also an increase in specific surface area and pore volume after the pillarization process using Cr polycation. The increase that occurs in the Cr/PILC catalyst makes the catalyst not easily deactivated when tested for catalytic activity.

PILC catalyst using Fe metal in a study that there is an increase in acidity on bentonite surfaces due to the influence of the presence of Fe metal combined with Zr metal in the manufacture of Zr-Fe/PILC catalysts [8]. Based on the results of research on the synthesis of Cr/PILC and Fe/PILC catalysts, bentonite catalysts polarized by Fe and Cr metals are needed in conducting this research. The use of Fe and Cr metal alloys in the synthesis of Fe-Cr/PILC catalysts is expected to increase surface area, pore-volume, and acidity on the catalyst surface.

In this research, catalyst Fe/PILC, Cr/PILC, and Fe-Cr/PILC catalysts are made at a ratio of 1: 1, then the catalytic activity is tested through the process of converting ethanol to gasoline. The process of converting ethanol to gasoline using an Sn-Cr/PILC catalyst at a ratio of 1: 1 and obtained ethanol products which have been converted to benzene, toluene, octane, and naphthalene similar to commercial gasoline [6]. The catalytic process used in the conversion of ethanol to gasoline requires an acidic catalyst. The process of pillarization in bentonite will increase the comparison of Brønsted and Lewis acid sites where the formation of metal oxide pillars will dominate the surface properties of bentonites [9]. The synthesized Fe/PILC, Cr/PILC, and Fe-Cr/PILC catalysts were characterized by physical and chemical characteristics including increased interlayer distances, elemental distribution, acidity, specific surface area, and pore volume, thermal stability, and clusters functional contained in the catalyst. The products produced from the catalytic activity test used Fe/PILC, Cr/PILC, and Fe-Cr/PILC catalysts and then analyzed using GC/FID.

2. Experimental

2.1. Preparation Materials

Bentonite was obtained from commercial product Bentonite (From Sigma Aldrich 285234) and $CrCl_3.6H_2O$, $FeCl_3.6H_2O$, NaOH, ($AgNO_3$) 0.1 M from Merck. Ethanol with a concentration of 99,8% from Smart Lab A-1035 and distilled water. The nano-sized Sn-Bentonite was prepared by a pillarization method.

2.2. Single metal pillared clay synthesis (Fe/PILC and Cr/PILC)

One a half litter of distilled water was put into a 5L beaker glass, then 81.099 gr of $FeCl_3.6H_2O$ powder was added slowly while stirring with a magnetic stirrer for 2 hours. Next, 3 L of distilled water was put into a 5L beaker glass then added 24 gr of NaOH pellets slowly while stirring with a magnetic stirrer for 2 hours. Fe polycation solution was prepared by adding 3L of 0.2 M NaOH solution slowly into 1.5L of FeCl3.6H2O 0.2 M solution using a peristaltic pump at a speed of 0.25 mL/min while stirring rapidly for ± 12 hours at room temperature.

Distilled water (2L) was put into a 5L beaker glass, then 53.29 grams of $CrCl_3.6H_2O$ powder was added slowly while stirring with a magnetic stirrer for 2 hours. Next, 2L of distilled water was put into a 5L beaker glass then added 16 gr of NaOH pellets slowly while stirring with a magnetic stirrer for 2 hours. Cr polycation solution was prepared by adding 2L of 0.2 M NaOH solution slowly to 2L of 0.1

IOP Conf. Series: Materials Science and Engineering 1011 (2021) 012008 doi:10.1088/1757-899X/1011/1/012008

M $CrCl_3.6H_2O$ solution using a peristaltic pump at a speed of 0.25 mL/min while stirring rapidly for ± 12 hours at temperature room.

Five grams of bentonite were dissolved into 500 mL of distilled water, then heated to a temperature of 60 °C while stirring using a magnetic stirrer for 2 hours to obtain a 1% bentonite suspension. The preparation results of each Fe polycation solution was 4.5L and further 4L polycation solution was added little by little to a 1% bentonite suspension in a stirred state. Furthermore, after the addition of the polycation solution, the solution was mixed for 24 hours at the room temperature. The Fe / PILC and Cr / PILC suspensions were centrifuged at 7000 rpm for 5 minutes at 20 °C. The precipitate was washed using distilled water at 60 °C until the solution showed a negative reaction with no white precipitate formed when a 0.1 M $AgNO_3$ solution was dropped. Fe / PILC suspension and Cr/PILC were re-centrifuged, then the precipitate was dried in an oven at 100 °C for 100 °C 12 hours later weighed and put in a sample bottle.

2.3. Bimetallic pillared clay synthesis (Fe-Cr/PILC)

Nine hundred mL of distilled water was put into a 5L beaker glass, then added 48.6594 grams of $FeCl_3.6H_2O$ powder slowly while stirring with a magnetic stirrer for 2 hours. 1.8L of distilled water was put into a 5L beaker glass, then added 47.961 gr of $CrCl_3.6H_2O$ powder slowly while stirring with a magnetic stirrer for 2 hours.

Nine hundred mL of Fe polycation solution was put into the beaker glass and 1.8L of polycation solution was added then stirred with a magnetic stirrer for 2 hours. Next, 1.8 L of distilled water was put into a 5L beaker glass and then added 14.4 gr NaOH pellets slowly while stirring with a magnetic stirrer for 2 hours. Fe/Cr polycation solution was prepared by adding 1.8L of 0.2 M NaOH solution slowly to 2.7L of Fe/Cr metal solution using a peristaltic pump at a speed of 0.25 mL/min while stirring rapidly for ± 12 clock at room temperature.

Three grams of bentonite are dissolved into 300 mL distilled water, then heated to 60 °C while stirring using a magnetic stirrer for 2 hours to obtain a 1% bentonite suspension. The results of the preparation of 4.5L Fe/Cr polycation solution were added little by little to the 1% bentonite suspension in a stirred state. Then, after the addition of the polycation solution, the solution remains stirred for 24 hours at room temperature. The Fe-Cr/PILC suspension was then centrifuged at 7000 rpm for 5 minutes at 20 °C. The precipitate was washed using distilled water at 60 °C until the solution showed a negative reaction with no white precipitate formed when a 0.1 M $AgNO_3$ solution was dropped. Fe-Cr/PILC suspension was re-centrifuged, then the precipitate was dried in an oven at 100 °C for 12 hours then weighed and put in a sample bottle.

2.4. Characterization Materials

The catalysts were analyzed by X-Ray Diffraction (XRD) from Panalytical Varian AERIS using Cu-Kα, scan speed 2 deg/min, λ Cu 1.54Å, and step width 0.01 deg, radiation on generated 40 kV, 15 mA, it aims to determine the interlayer distance in the reflection field (d001) and crystallinity so that the composition of the composition is known in the catalyst matches the typical crystal diffraction pattern. Fe and Cr content in PILC is measured by X-Ray Fluorescence (XRF) from Brucker, under a flowing Helium gas. Surface Area Analyzer based on the Brunauer-Emmet-Teller (BET) method from TriStar II 3020 Version 2.00 manufactured by N_2 analysis adsorptive and equilibration interval 10 s, sample mass 0.35 gr, cold free space 32.81 cm^3, sample density 1 gr/cm3, Barret-Joyner-Halenda (BJH) method. Fourier Transform Infra-Red (FTIR) from shimadzu and used KBr phase method, prestige-21 with ATR 4000 attached to an automatic data acquisition center. The Thermogravimetric analysis (TGA) was performed by the LINSEIS STA Platinum Series apparatus, the temperature was increased at a rate of 10 °C/min from 10 to 1000 °C underflow of the nitrogen. Product analysis with GC-FID Carbowax/20M columns (30 m x 320 μm, 25 μm).

IOP Conf. Series: Materials Science and Engineering 1011 (2021) 012008 doi:10.1088/1757-899X/1011/1/012008

3. Results and Discussion

3.1. Comparison of XRD results for Bentonite and Sn-Bentonite

X-ray diffraction analysis using XRD (X-Ray Diffraction) on bentonite, Cr/PILC, Fe/PILC, and Fe-Cr/PILC measurement of the distance between layers (interlayer) was carried out at a diffraction angle of 2θ low angle between 2-10°. The results of XRD diffraction characterization of 2θ low angle can be seen in figure 1.

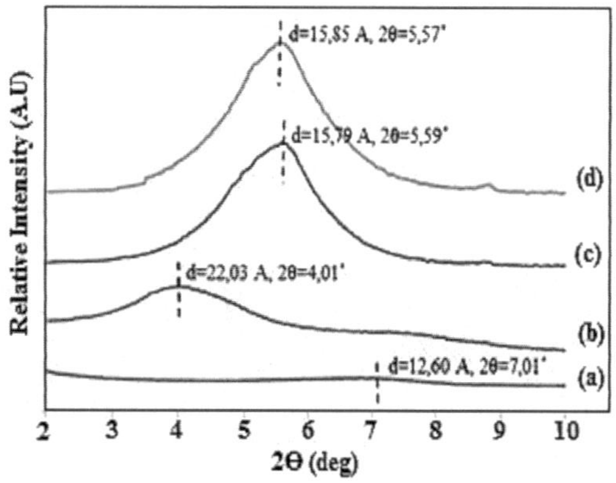

Figure 1. XRD low angle patterns of (a) bentonite, (b) Cr/PILC, (c) Fe/PILC, and (d) Fe-Cr/PILC.

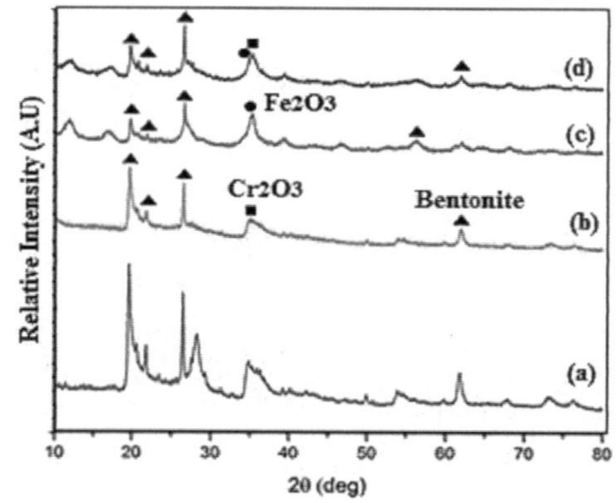

Figure 2. XRD high angle patterns of (a) bentonite, (b) Cr/PILC, (c) Fe/PILC, and (d) Fe-Cr/PILC.

The low angle X-ray diffraction pattern in Figure 1 shows that there has been a 2θ shift of the reflection peak d_{001} to the left (in the smaller direction) of the bentonite sample after the pillarization process. The peak of reflection d_{001} in bentonite before pillarization shows 2θ at 7.01°. This is consistent with the low angle X-ray diffraction pattern of bentonite which has a main peak at an angle of $2\theta = 7°$ [5]. The shift of reflection peak d_{001} that occurs in bentonite after pillarization with Cr metal on the Cr/PILC catalyst shows an angle of 2θ of 4.01°. Based on research conducted by Widjaya *et al.*, the Cr/PILC catalyst has a reflection peak d_{001} at $2\theta = 4.76°$. The same thing happened to bentonite after pillarization with Fe metal on Fe/PILC catalyst which showed an angle of 2θ at 5.59°. Sekewael (2008) has made montmorillonite iron oxide composites with the results of low angle X-ray diffraction patterns at $2\theta = 5.24°$. Bentonite which has been pillarized with Fe and Cr metals at a ratio of 1:1 in Fe-Cr/PILC also shows a shift of angle of 2θ to 5.57°. The low angle X-ray diffraction pattern on the Fe-Cr/PILC catalyst has an angle of 2θ which is similar to the Fe/PILC catalyst, this can be indicated by the presence of a more dominant Fe metal than the Cr metal on the Fe-Cr/PILC catalyst.

Shifting peak 2θ reflection d_{001} shows an increase in the distance between layers (interlayer) on the bentonite that has been carried out the pillarization process. The increase that occurred due to the exchange of cations that exist in the interlayer of bentonite with polycation of Fe and Cr, thus causing the expansion of the distance between layers in bentonite. The interlayer distance of the bentonite before the pillarization process is 12.60 Å. The bentonite has a distance between layers between 12.0-15.0 Å. Bentonite which has been carried out by the pillarization process using Fe and Cr metal has increased the interlayer distance to 22.03 Å for Cr/PILC, 15.79 Å for Fe/PILC, and 15.85 Å for Fe-Cr/PILC. Cr/PILC catalyst has the highest interlayer distance compared to Fe/PILC and Fe-Cr/PILC catalysts, this is because in the pillarization process using Cr metal can make the molecular size larger [10]. The Fe-Cr/PILC catalyst has decreased interlayer when compared to the Cr/PILC catalyst. The decrease occurred due to diffusion barriers by the number of molecules of one of the most intercalated metal pillars, namely Fe metal so that the Fe-Cr/PILC catalyst has a similarity between the interlayer and the Fe/PILC catalyst [11].

The increased interlayer distance is caused by the exchange of hydrated cations on the surface between bentonite layers and larger cations, namely Cr^{3+} and Fe^{3+}. An increase in the distance between layers is one of the success factors in the process of pillarization due to the intercalation of the pillaring agent in the form of Fe and Cr metal into the interlayer of the bentonite. The increase in interlayer distance can be concluded that the pillarization process using Fe and Cr metals in the synthesis of Cr/PILC, Fe/PILC, and Fe-Cr/PILC catalysts has been successfully carried out. The composition of the catalyst can be determined through crystallinity with a 2-high angle X-ray diffraction pattern between 10-80°. The results of the XRD analysis showed typical peaks which were then compared with data from the JCPDS (Joint Committee of Powder Diffraction Standards).

The high angle X-ray diffraction pattern of 2θ can be seen in Figure 2. High angle X-ray diffraction patterns in bentonite produce peak angles of 2θ = 19.48°; 21,65°; 26.3°; 35,14°; 54,11°; and 61.48 °. These results are following JCPDS No. 29-1499 which shows that these peaks are typical peaks for montmorillonite type. Research conducted by Dewi *et al.* states that montmorillonite X-ray diffraction patterns are at 2θ = 19.58°; 20,64°; and 26.33°. Based on these results, the bentonite structure is a montmorillonite type [12]. Bentonite with montmorillonite type is distinguished through alumina-silica composition which has a 2:1 configuration ratio with the characteristic of swelling in the bentonite framework. The high angle X-ray diffraction pattern on the Cr/PILC catalyst results in a peak angle of 2θ = 19.72°; 20,71°; 26.52°; 34,61°; and 61.79°. Peak 2θ = 34.61° indicates that the Cr_2O_3 compound that was successfully interpreted had a rhombohedral crystal system. These results are following JCPDS No. 84-1616 which shows typical Cr_2O_3 peaks at 2θ = 24.52°; 34.6°; and 36.5°. Based on research by Rahmani *et al.*, typical peaks of the rhombohedral Cr_2O_3 phase appear at 2θ = 24.5°; 34.6°; 36.2°; 41.5°; and 50.2°. Figure 2 shows the peak shift 2 peaks and the change in intensity which proves that the addition of Cr_2O_3 into the interlayer of the bentonite was successfully carried out and formed a crystalline phase [13].

The high angle X-ray diffraction pattern on the Fe/PILC catalyst produces a peak angle of 2θ = 19.94°; 26,75°; 35.3°; and 55.73°. The peak of X-ray diffraction that appears in the region 2θ = 35.3° indicates that the Fe_2O_3 compound with the hematite phase was successfully synthesized. These results are following JCPDS No. 13-5340 which shows the typical peaks of Fe_2O_3 at 2θ = 24.19°; 35.3°; and 62.47°. The peak indicating the presence of Fe_2O_3 was at 2θ = 24.02°; 33,36°; 35.3°; and 54.02 [14]. Based on these results, Fe_2O_3 which was intercalated into the interlayer of the bentonite was successfully carried out. The high angle X-ray diffraction pattern on the Fe-Cr/PILC catalyst produces an angle peak of 2θ = 19.77°; 22.47°; 26,65°; 35.3°; 36.5°; and 61.93°. The typical peaks that appear in the region 2θ = 35.3° and 36.5° indicate that there are Fe_2O_3 compounds with hematite phase and Cr_2O_3 compounds with the rhombohedral phase. These results are following JCPDS No. 13-5340 which shows the typical peaks of Fe_2O_3 at 2θ = 24.19°; 35.3°; and 62.47° and JCPDS No. 84-1616 which shows typical Cr_2O_3 peaks at 2θ = 24.52°; 34.6°; and 36.5°.

3.2. Measurement with X-Ray Fluorescence (XRF)

The results of XRF characterization on bentonite, Cr/PILC, Fe/PILC, and Fe-Cr/PILC can be seen in Table 1. The XRF analysis results in Table 3 show that the main components of bentonite are SiO_2 and Al_2O_3. This is because SiO_2 and Al_2O_3 in bentonite have the largest percentage composition compared to other metal oxide components, such as Fe_2O_3, Cr_2O_3, MgO, CaO, and K_2O. The content of SiO_2 and Al_2O_3 in bentonite before the pillarization process was 60.51% and 19.12%, respectively.

Table 1. Results of XRF analysis for catalysts.

Catalysts	Elements (% wt)						
	SiO_2	Al_2O_3	Fe_2O_3	Cr_2O_3	MgO	CaO	K_2O
Bentonite	60.51	19.12	4.84	0.01	12.2	1.59	0.41
Cr/PILC	46.78	13.7	3.01	17.67	14.51	2.96	0.25

Fe/PILC	32.93	6.74	44.25	0.03	13.25	1.26	0.21
Fe-Cr/PILC	39.22	9.1	31.89	5.42	10.68	1.86	0.33

Modification of the structure of bentonite by performing the pillarization process using metal oxides Fe and Cr showed an increase in the percentage of composition in Fe_2O_3 and Cr_2O_3 on each catalyst Cr/PILC, Fe/PILC, and Fe-Cr/PILC. Cr/PILC catalyst showed an increase in Cr_2O_3 composition which was originally in bentonite before the pillarization process was 0.01% to 17.67%, whereas in Fe/PILC catalyst there was an increase in Fe_2O_3 composition from 4.84% to 44.5%. Alloys of bimetallic oxides Fe and Cr which were interpreted in the Fe-Cr/PILC catalyst showed the presence of Fe_2O_3 and Cr_2O_3 content of 31.89% and 5.42%, respectively. Based on the results of the XRF analysis, before the pillarization process using two metal oxides of Fe and Cr, the composition of Fe_2O_3 on the Cr/PILC catalyst was 3.01% and the composition of Cr_2O_3 on the Fe/PILC catalyst was 0.03%. The composition of the element content when the process of pillarization using bimetallic Fe and Cr on the Fe-Cr/PILC catalyst indicates that the elements of Fe and Cr are bonded to each other in interlayer bentonite [6]. A comparison of the composition of Fe_2O_3 is greater when compared to the composition of Cr_2O_3 on the Fe-Cr/PILC catalyst. This confirms the results of the low angle XRD analysis of diffusion barriers by the number of molecules of one of the most intercalated pillars of the metal oxide Fe, so the XRF analysis results show that the composition of Fe_2O_3 has a greater percentage content than the composition of Cr_2O_3 (Gil et al., 2000).

3.3. Measurement with Surface Area Analyzer (SAA)

Table 2 showed the results of surface area and pore volume analysis of bentonite, Cr/PILC, Fe/PILC, and Fe-Cr/PILC using SAA. Bentonite before being modified through the pillarization process showed a surface area of 26.48 m^2/g and a pore volume of 0.08 cm^3/g. After the pillarization process, there was an increase in surface area and pore volume respectively to 158.79 m^2/g and 0.15 cm^3/g for the Cr/PILC catalyst and 184.24 m^2/g and 0.31 cm^3/g for the Fe/PILC. The increase in surface area that occurs indicates that the insertion of the Fe^{3+} and Cr^{3+} cations has been stably bound in the interlayer of the bentonite in the Fe/PILC and Cr/PILC catalysts. The addition of transition metals, Fe and Cr can cause an increase in the surface area of the catalyst, this occurs because the Fe and Cr metals that are embodied through the pillarization process may have been distributed to the interlayer of the bentonite, thereby increasing the surface area and volume of the portion.

Table 2. Results of SAA analysis.

Material	SSA (m^2/g)	Pore Volume (cm^3/g)
Bentonite	26.48	0.08
Cr/PILC	158.79	0.15
Fe/PILC	184.24	0.31
Fe-Cr/PILC	174.11	0.26

The SAA analysis graph in Table 2 showed that the Fe-Cr / PILC catalyst has a surface area and pore volume of 174.11 m^2/g and 0.26 cm^3/g. These results show an increase in surface area and pore volume in the interlayer of the bentonite after the pillarization process using Fe and Cr metal on the Fe-Cr/PILC catalyst, but when compared with the Fe/PILC catalyst, the Fe-Cr/PILC catalyst has decreased surface area and volume of the portion. The presence of Cr metal affects the decrease in surface area and pore volume in the Fe-Cr/PILC catalyst. The Cr metal cannot increase surface area and pore volume that is too large in bentonite, this is in accordance with the results of surface area and pore volume analysis in Cr/PILC catalysts which are smaller when compared to Fe/PILC. The decrease occurred due to Cr metal trapped interlayer (interlayer) and on the surface of pillared bentonite, thus covering the pores of bentonite. The surface area is one of the factors that determine how the catalyst activity takes place in the catalysis reaction process, so with an increased surface area from 26.48 m^2/g to 174.11 m^2/g making the site of the reaction bigger [15]. A large surface area will provide a large contact area between the reactants and the catalyst, so the size of the contact will affect the overall catalysis process. The reactant molecule will move freely before undergoing adsorption on the surface of the catalyst which will then be activated and react to produce a product.

3.4. Measurement with Thermogravimetric Analysis (TGA)

Figure 3 shows a decrease in the relative mass percent curve in bentonite, Cr/PILC, Fe/PILC, and Fe-Cr/PILC as the heating temperature increases to 1000°C. Decrease in mass percent of total thermal decomposition of bentonite, Cr/PILC, Fe/PILC, and Fe-Cr/PILC catalysts respectively by 19.07%; 33.13%; 31.07%; and 32.04%. The results of measurements by TGA in the first stage show the release of physically bound water molecules (dehydration) on the surface of the catalyst structure. Bentonite shows a mass change of 7.65% at temperatures between 76 °C to 140 °C. The bentonite material is very stable to heat-stable to temperatures below 73°C [6]. The process of pillarization using metals and Fe and Cr on Cr/PILC, Fe/PILC, and Fe-Cr/PILC catalysts affects the release of water molecules in the temperature range 74-225°C with changes in mass with a range of 17.19-18.58%. This finding indicated by changes in molecular structure due to the addition of Fe and Cr metals to the bentonite so that the decomposition temperature increases.

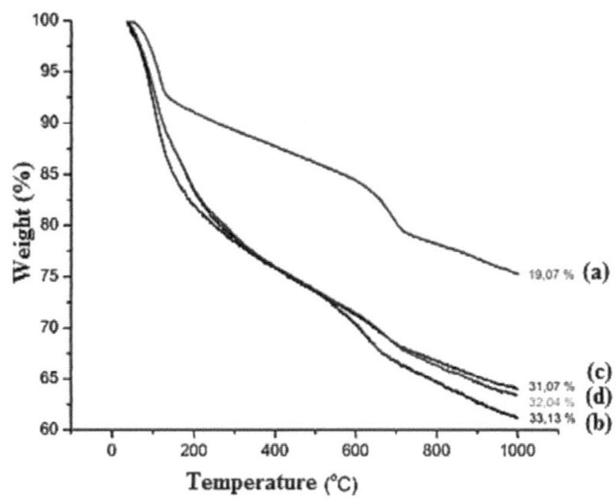

Figure 3. TGA Curve of (a) Bentonite, (b) Cr/PILC, (c) Fe/PILC, and (d) Fe-Cr/PILC.

Bentonite in the second stage showed a change in mass at a temperature of 140-200°C of 1.35%, while bentonite pillarized metals Fe and Cr on catalyst Cr/PILC, Fe/PILC, and Fe-Cr/PILC experienced mass changes with a range of 8.28-8.58% at temperature 210-520°C. The temperature there is a release of water molecules (dehydration) and changes in the structure of salt (Mg, Na, Ca, and K) in the cation oligomer in the interlayer (interlayer) bentonite [15]. The third step shows the chemical decomposition. Third stage mass changes in bentonite occur at temperatures of 200-686°C of 10.05%. The results of the third stage TGA measurements, catalyst Cr/PILC, Fe/PILC, and Fe-Cr/PILC showed a change in mass percent with a range of 5.01-6.25% at temperatures between 500-710°C. The temperature range of 500-680°C occurs maximum dehydroxylation of pillared metals in interlayer bentonite which results in damage to the proton structure on bentonite alumina-silica sheets [16].

The results of measurements with TGA at temperatures above 710°C indicate the maximum temperature limit that can be accepted by bentonite material before the total damage caused by the process in the first to third stages. The addition of metals, namely Fe and Cr metals to Cr/PILC, Fe/PILC, and Fe-Cr/PILC catalysts, provides physical property resistance to the bentonite structure thereby resulting in increased thermal stability compared to bentonite which has not been carried out by the pillarization process [17].

IOP Conf. Series: Materials Science and Engineering 1011 (2021) 012008 doi:10.1088/1757-899X/1011/1/012008

3.5. Measurement with Fourier Transform Infrared Spectroscopy (FTIR)

The FTIR spectrum results in Figure 4 showed that bentonite has 6 characteristic absorption peaks detected at a wavelength of 453.7 cm^{-1}; 524.38 cm^{-1}; 1081.41 cm^{-1}; 1635.64 cm^{-1}; 3433.29 cm^{-1}; and 3630.03 cm^{-1}. The absorption peak at the wave number 453.7 cm^{-1} is the bending vibration of Si-O-Si in the bentonite silicate aluminum framework, while the absorption peak at the wavenumber area 1081.41 cm^{-1} shows the existence of stretch vibration (stretching) Si-O-Si. The bending vibrations and stretching of Si-O-Si occur in each of the absorption bands 470 cm^{-1} and 1008.7 cm^{-1}. The existence of Si-O-Si buckling vibrations and shows the formation of an aluminosilicate framework on the catalyst [18].

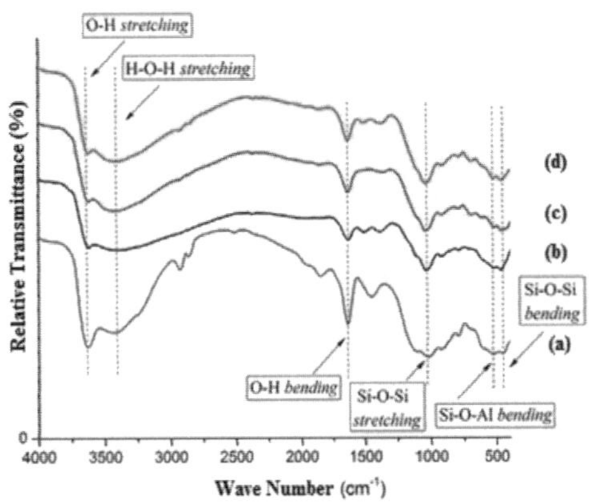

Figure 4. FTIR Results analysis of (a) bentonite, (b) Cr/PILC, (c) Fe/PILC, and (d) Fe-Cr/PILC.

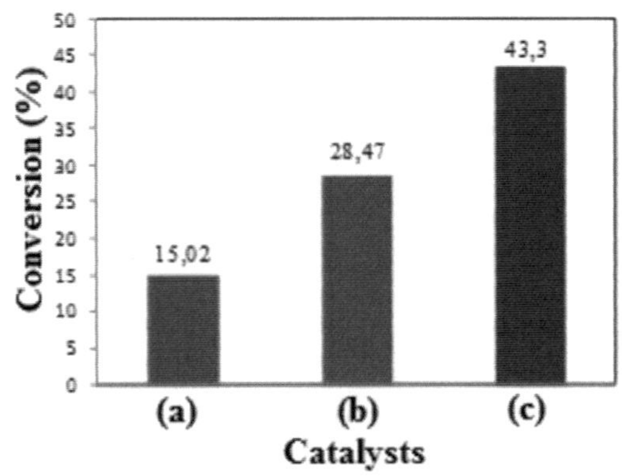

Figure 5. GC-FID Results analysis of (a) Cr/PILC, (b) Fe/PILC, and (c) Fe-Cr/PILC.

FTIR analysis showed that the success of the pillarization process using Fe and Cr metals with a shift in absorption peaks which indicates the addition of compounds derived from metals that are intercalated into the interlayer of bentonite. Figure 4 shows the shifts in 6 typical bentonite absorption peaks after pillarization of the Cr/PILC, Fe/PILC, and Fe-Cr/PILC catalysts. Also, there is a new absorption peak in the Cr/PILC catalyst in the wavelength region of 920.05 cm^{-1} which shows the bending vibration of -OH with the Cr^{3+} cation. The bend -OH vibrations with cations occur in the region of 918.88 cm^{-1}. New absorption peaks on the Fe/PILC catalyst also appear at wavelengths of 794.67 cm^{-1} due to the presence of a stretching -OH vibration which indicates the binding of the Fe^{3+} cation to the interlayer of the bentonite. The Fe-Cr/PILC catalyst showed new absorption peaks in the region of 794.67 cm^{-1} and 921.97 cm^{-1} indicating the presence of both Cr^{3+} cations from Cr and Fe^{3+} metals from Fe.

FTIR analysis results were useful to find out the type of acidity the catalyst has. There are two types of acidity that catalysts have, namely Brønsted and Lewis acids. Brønsted acid type is at a wavelength of 1515-1640 cm-1 and Lewis acid type is at a wavelength of 1435-1470 cm-1 [19]. The Brønsted acid site tends to donate protons or H+ ions, while the Lewis acid site tends to capture or accept electrons. The peak yield of FTIR uptake in bentonite before and after pillarization the Cr/PILC, Fe/PILC, and Fe-Cr/PILC catalysts have both acidity types, both Brønsted and Lewis acid types.

3.6. Measurement with Gas Chromatography- flame ionization detector (GC-FID)

The success of ethanol compounds that have been converted to gasoline compounds in Figure 5 showed that the Fe-Cr/PILC catalyst is able to produce the highest percent conversion, which is 43.3% compared to both Cr/PILC and Fe/PILC catalysts. The results of GC/FID analysis of gasoline compound products formed from the results of ethanol conversion determined each percent of selectivity which can be seen in Table 3.

Table 3 shows that the gasoline compound resulting from the conversion of ethanol produces product components with a range of C6 to C12 hydrocarbon bonds in the form of paraffin and aromatic compounds. Product compounds produced from the conversion reaction of ethanol into gasoline are the main components contained in gasoline or commercial gasoline. The total selectivity of gasoline in Table 3 shows the highest selectivity results obtained from the use of Fe-Cr/PILC catalysts with a percent selectivity of 87.15% in which the highest percentage of product selectivity is produced in benzene compounds (C6H6), which is 48.52 %. The benzene is one of the most important compounds in the gasoline component. The use of the Fe-Cr/PILC catalyst in the activity of the catalyst to the conversion process of ethanol to gasoline produces the highest percent conversion and selectivity values [6].

Table 3. Product selectivity results from GC-FID ethanol analysis to gasoline.

No.	Product	Selectivity (%)		
		Cr/PILC	Fe/PILC	Fe-Cr/PILC
1.	n-Heptane (C_7H_{16})	4.97	2.29	1.21
2.	Cyclohexane (C_6H_{12})	3.07	2.29	31.87
3.	Toluene (C_7H_8)	1.30	1.19	1.52
4.	Benzene (C_6H_6)	15.02	28.18	48.52
5.	n-Dodecane ($C_{12}H_{26}$)	-	1.92	-
6.	Fenol (C_6H_6O)	3.79	1.54	0.42
7.	n-Hexane (C_6H_{14})	19.35	15.66	3.61
Total Selectivity Product		47.48	53.07	87.15

The Fe-Cr/PILC catalyst has a good surface area and pore volume and acidity of the catalyst although the results are lower than the Fe/PILC catalyst for surface area and pore volume and with the Cr/PILC catalyst as seen from the measurement results of SAA. Fe metal alloy as an active component with Cr metal as a promoter on Fe-Cr/PILC catalyst can influence the catalytic reaction so that it can produce the best percent conversion and selectivity of the catalyst in the catalyst activity test of the conversion process of ethanol to gasoline.

4. Conclusion

Based on the results of the characterization, modification of the bentonite structure using the pillarization method in the synthesis of Cr/PILC, Fe/PILC, and Fe-Cr/PILC catalysts increased physical and chemical properties possessed by the catalyst, namely increased surface area, pore-volume, thermal stability and acidity in the catalyst to increase the activity of the catalyst in the process of converting ethanol to gasoline. The best results on the catalyst activity test on the process of converting ethanol to gasoline are obtained from the use of the Fe-Cr/PILC catalyst which results in percent conversion, and selectivity of 43.30%, 87.15% respectively.

Acknowledgments

The authors gratefully acknowledge to Research Center for Chemistry, and Physics - Indonesian Institute of Sciences for research funding and facility, as well as Department of Chemistry, UIN Syarif Hidayatullah Jakarta, Indonesia for research and study facility.

References

[1] Utami D N 2018 Kajian jenis mineralogi lempung dan implikasinya dengan gerakan tanah *Jurnal Alami : Jurnal Teknologi Reduksi Risiko Bencana* **2**(2) 89
[2] Widjaya R R, Soegijono B and Rinaldi N 2012 Characterization of Cr/Bentonite and HZSM-5 Zeolite as catalysts for ethanol conversion to biogasoline *Makara J. Sci.* **16**(1) 65–70

[3] Kloprogge J T, Duong L V and Frost R L 2005 A review of the synthesis and characterization of pillared clays and related porous materials for cracking of vegetable oil to produce biofuels *J. Environ. Geol.* **47** 1–36
[4] Darmawan A, Suseno A and Purnomo S A 2005 Sintesis lempung terpilar titania *Jurnal Kimia Sains Dan Aplikasi* **8**(3) 61–8
[5] Widjaya R R, Juwono A L and Rinaldi N 2019 Chromium pillared montmorillonite as catalyst for liquid biofuel conversion. *Asian J. App. Sci.* **07**(05) 587–96
[6] Widjaya R R, Juwono A L and Rinaldi N 2019 Development yin-chromium pillared in bentonite as catalyst for ethanol to gasoline conversion *Asian J. App. Sci.* **07**(04) 425–34
[7] Okoye I P and Obi C 2011 Synthesis and characterization of titanium pillared bentonite clay Mineral *Research J. App. Sci.* **6**(7) 443–6
[8] Molina C B, Casas J A, Zazo J A and Rodríguez J J 2006 A comparison of Al-Fe and Zr-Fe pillared clays for catalytic wet peroxide oxidation *Chem. Eng. J.* **118**(1–2) 29–35
[9] Fatimah I and Wijaya K 2011 Effect of aluminum content in aluminum pillared montmorillonite on its surface acidity properties *ITB Journal Science* **43**(2) 123–38
[10] Ding M, Zuo S and Qi C 2015 Preparation and characterization of novel composite AlCr-pillared clays and preliminary investigation for benzene adsorption *App. Clay Sci.* **115** 9–16
[11] Gil A, Gandía L M and Vicente M A 2000 Recent advances in the synthesis and catalytic applications of pillared clays *Catal. Rev. Sci. Eng.* **42**(1–2) 145–212
[12] Dewi D A D N, Simpen I N and Suarsa I W 2020 Synthesis and characterization of photocatalyst Fe_2O_3 pillared montmorillonite doped TiO_2 and its application for rhodamine B photodegradation using visible light irradiation *J. Chem.* **14**(1) 82–8
[13] Rahmani F, Haghighi M and Amini M 2015 The Beneficial utilization of natural zeolite in preparation of Cr/Clinoptilolite nanocatalyst used in CO_2-oxidative dehydrogenation of ethane to ethylene. *J. Ind. Eng. Chem.* **2563** 1–14
[14] Dewi M A K, Suprihatin I E and Sibarani J 2017 Fotodegradasi zat warna remazol brilliant blue dengan bentonit terimpregnasi Fe_2O_3 *Jurnal Kimia* **11**(1) 82–7
[15] Rinaldi N and Kristiani A 2017 Physicochemical of pillared clays prepared by several metal oxides *AIP Conference Proceedings* **1823** 1
[16] Kar P and Mishra B G 2013 Silicotungstic acid nanoparticles dispersed in the micropores of Cr-pillared clay as efficient heterogeneous catalyst for the solvent fee synthesis of 1,4-dihydropyridines *Chem. Eng. J.* **223** 647–56
[17] Liang H, Long Z, Yang S and Dai L 2015 Organic modification of bentonite and its effect on rheological properties of paper coating *App. Clay Sci.* **104** 106–9
[18] Saraswati I 2016 Zeolite-A synthesis from glass *Jurnal Sains Dan Matematika* **23**(4) 112-5
[19] Rinaldi N and Dwiatmoko A A 2011 studi awal pada Preparasi katalis berbasis lempung berpilar untuk reaksi etanol Menjadi gasoline (ETG) *Jurnal Kimia Terapan Indonesia* **13**(2) 78–85

IOP Conf. Series: Materials Science and Engineering 1011 (2021) 012009 doi:10.1088/1757-899X/1011/1/012009

Using kefir glucomannan to modify the diversity and composition of cecum bacterial in Sprague Dawley rats with metabolic syndrome

D Desnilasari[1*], E Harmayani[2], J Widada[3] and Nurliyani[4]

[1]Research Center for Appropriate Technology, Indonesian Institute of Sciences, JL KS Tubun No.5, Subang, West Java, Indonesia
[2]Department of Food and Agricultural Technology, Faculty of Agricultural Technology, Universitas Gadjah Mada,Jl Flora No.1, Bulaksumur, Yogyakarta, Indonesia
[3]Faculty of Agriculture, Gadjah Mada University, JL Flora, Bulaksumur, Yogyakarta, Indonesia
[4]Department of Animal Product Technology, Faculty of Animal Science of Gadjah Mada University, JL Fauna No.3, Bulaksumur, Yogyakarta, Indonesia

*Corresponding author and main contributor: dewi.desnilasari@gmail.com

Abstract. This study aimed to determine the effect of kefir and kefir glucomannan (Kefir-GM) to the diversity and composition cecum bacteria and bodyweight and feed intake in rat metabolic syndrome. Research carried out used four group's treatment, namely (1) control, normal rat (2) rats with metabolic syndrome (3) diet kefir in rats with metabolic syndrome, and (4) diet kefir-GM in rats with metabolic syndrome. After 4 weeks, cecum was taken for analysis of the diversity and composition of bacterial. Kefir and Kefir-GM in rats metabolic syndrome did not significantly affect the diversity at $p<0.05$, but increased Lactobacillus 14.61% and Bifidobacterium 2.2%, and decreased Clostridium 38.15% and Bacteroides 22.51%. Bodyweight and feed intake did not significantly different at $p<0.05$. Based on this research, kefir and Kefir-GM did not affect diversity but it is able to modify composition bacteria on cecum and also can keep the bodyweight and feed intake in rats.

1. Introduction

Diet and lifestyle nowadays are the important factor that leading to the rapid occurrence of metabolic syndrome. Former research was finding that this regarding with the function of gut microflora [1]. The changes composition gut microflora because bad diet effect to increasing gain energy in gastrointestinal tract and become easy gain weight [2]. Moreover, gut microflora has function related to the health of the host.

Research on obese and non-obese patients showed that those who were obese have a poor diversity microbiome compare with non-obese [3]. In the contrary, research by Nakayama et al [4] showed that the high Shannon Weiner's diversity index in 7-10 years old school-age children in Asia does not related to the good bacterial community. The results showed that Japanese children with the smallest microbial diversity index actually have greater abundance of Bifidobacterium compare with Indonesian children who have higher diversity index. This showed that food influences the diversity of gut microbiome.

Therefore, many researchers assume that the administration of probiotics, prebiotics and symbiotic as dietary supplements can increase the microflora which is beneficial to the health of the host.

Kefir is the traditional fermented milk using kefir grain that is consists of a polysaccharide composed of a complex microbial association among bacteria and yeasts. Kefir is widely known as an excellent source of probiotics with potential health benefits [5] and is produced based on cow milk or goat milk. Satir and Guzel-seydim [6] reported that kefir based goat milk consist of abundant microbe (*Lactobacillus* spp., *Lactococcus* spp., *Lactobacillus acidophilus*, *Bifidobacterium* and Yeast) and antioxidant compound (phenolic) higher compare with kefir based cow milk. Furthermore, this research showed that goat milk kefir contain bioactive peptide that act as immunomodulatory, antioxidant, and hypercholesteremic substance [7]. Moreover, Likotrafitri *et al.* [8] research showed that *Lactobacillus kefiri* isolated from kefir grain promoted the stabilization of gut microbiome.

To enhance the efficacy of probiotics and promote the growth of autochthonous beneficial bacteria, prebiotics were introduced as potential food ingredients. Glucomannan has potential as prebiotic that promoted the growth of good microflora. Research by Chen *et al.* [9] showed that mice feeding with glucomannan increased fecal *Bifidobacteria* count and suppressed Clostridium perfringens counts. Moreover Harmayani *et al.* [10] research showed that diet supplemented with porang glucomannan enhanced the production of total SCFA that promoted decreased cholesterol content in blood, inhibited the growth of *Escherichia coli*, and was not significantly supported growth of the *Bifidobacteria* and *Lactobacilli*.

In this current research, we produced Kefir-GM product based on goat milk that supplementation glucomannan from porang (*Amorphopallus oncophylus*) using kefir grain as starter. This Kefir-GM to be expected has good effect to the metabolic syndrome especially the effect to the gut microbiome because of the product contains of prebiotic and probiotic that is called symbiotic product. Therefore, the aim of this research is to evaluate the effect of Kefir-GM supplementation on the diversity and composition bacteria in the rats *Sprague Dawley* with the metabolic syndrome.

2. Materials and methods

2.1. Kefir and kefir-GM preparation

Kefir in this research was made with two different formulations, without glucomannan and with the addition of glucomannan 0.3% b/v. Kefir was made based goat milk and was added with whey 0,1% (v/v). The solution was pasteurized in 75°C for 15 minutes, and cooled at room temperature. The mixture was added with 2% (b/v) kefir grain, and then was incubated in 28 °C for 20 hours. After incubation, the kefir grain was separated with the kefir used sterilized filter [11].

2.2. Animal study

The group of male Sprague Dawley rats age of 8-12 weeks old were divided into 4 groups (each group used 6 rats): 1) Normal control (negative control rats) that received standard diet only, 2) rat with metabolic syndrome (fed high-fat/high-fructose (HFHF)) (positive control), 3) rat with metabolic syndrome fed HFHF supplemented with kefir, 4) rat with syndrome metabolic fed HFHF supplemented with Kefir-GM. The dose of kefir was 3.6 mL/200 g body weight/ day, for 4 weeks. Before treatment, the rats were acclimated with standard diet AIN-93 for 1 week, and then fed high fat and high fructose for 2 weeks. The dose of kefir was 3.6 mL/200 g body weight/ day, for 4 weeks. Before treatment, the rats were acclimated with standard diet AIN-93 for 1 week, and then fed high fat and high fructose for 2 weeks. The rats were then divided into 4 groups as above. High fat and high fructose diet were administered until the end of experiment (4 weeks). The composition of standard diet and high-fat/high-fructose diet were formulated according to Reeves *et al.* [12] and de Castro et al [13]. Every week the body weight of the rat was weighted and every day the leftover feed not consumed by rat was weighted to. The sample from digesta was collected from cecum in the end period of the experiment. All procedures related to animal experiment in this study were approved by Medical and Health Research

Ethics Committee (MHREC), Faculty of Medicine Universitas Gadjah Mada, Indonesia (Approval Number: KE/FK/95/EC/2015).

2.3. Analyze the Diversity and Composition Cecum Bacteria Using TRFLP (Terminal Restriction Fragment Length Polymorphism)

2.3.1. DNA isolation, PCR amplification, and restriction enzyme digestion. Isolation DNA bacteria was used Favor Prep™ Stool DNA Isolation Mini Kit. After that, the isolate DNA was amplification gen 16s rRNA using universal primer 27F (5'-AGAGTTTGATCCTGGCTCAG-3') and 1492R (5'-GGTTACCTTGTTACGACTT-3') that has been added with FAM dye at the forward primer [14]. Amplification 16S rRNA was used Gotaq Green Mixture with total reaction 25µl (12.5µl Gotaq Green Mix, 1µl forward primer (27FFAM), 1µl reverse primer (1492R), 1µl sample DNA (250ng), and 9.5µl nuclease free water). PCR program cycle condition: an initial denaturing step at 95°C for 2 min, 30 cycles of denaturing step 95°C for 1 min, annealing at 56°C for 30 s and extension at 72°C for 1 min. The final extension was performed at 72°C for 5 min. The PCR product was digested with MspI (Thermo Scientific) and HhaI (GeneMark enzyme) [14]. Formulation for digested product PCR was used 1.5µl buffer tango, 12 µl DNA, 1 µl enzyme and 0.5µl nuclease free water. This was incubated in 37°C for 4 hours.

2.3.2. Identification product restriction enzyme. The fragment that was produced from digested by restriction enzyme was identification used Capillary Electrophoresis Applied Biosystems 3130 series Genetic. This was performed by 1st Base. The profile of the fragment was obtained in this process was analyzed next.

2.3.3. Analyzed the TRFLP data. The profile of the fragment from the identification process was analyzed using Software Peak Scanner™ v1.0 (Thermo scientific). The length of fragment T-RFs was selected between 50bp until <500bp. To reduce the noise data, the sample was selected again. The T-RFs must have value of peak area was >1% (by dividing the value of the peak area with the total peak area of the T-RFs) [15]. The selected T-RFs analyzed using T-align program in http://inismor.ucd.ie/~talign/. The data was obtained was binomial data for the clustered analysis. Analysis clustered was performed using NTSYSPc version 2.1. Dendrogram was performed using underweight pair group with arithme average (UPGMA). The diversity and distribution frequency was performed using Shannon Weinner and Shimpson index [16]. Prediction bacteria was performed using web-based program MICA III (mica.ibest.uidaho.edu), RDP 16S bacterial rRNA database, and also database NCBI 16S rRNA that has been simulation cutting in silico using Bio-edit.

2.4. Statistical analysis.
The body weight rat, daily feed intake, and diversity index data was analyzed using ANOVA and if there was significant different data was analysis next using Duncan.

3. Results and discussions

3.1. Diversity and prediction composition bacteria in cecum
Following this a dendrogram of the results of running TRFs data through the Group Average clustering algorithm was showed in Figure 1.

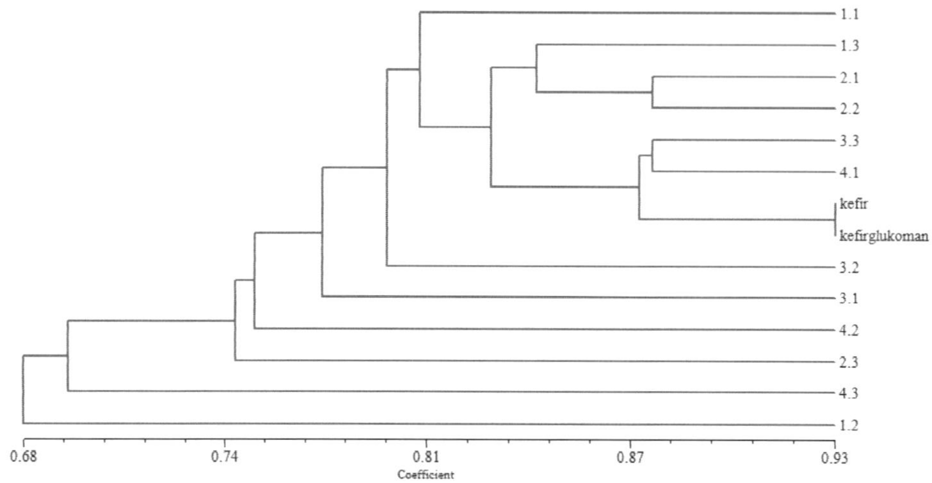

Figure 1. Similarity dendrogram composition bacteria according TRFs that digested using *Hha*I and *Msp*I. In which: Normal rat (1.1-1.3), metabolic syndrome rat (2.1-2.3), diet kefir in metabolic syndrome rat (3.1-3.3), diet kefir-gm in metabolic syndrome rat (4.1-4.3).

Figure 1 shows that the similarity TRFs of Kefir and Kefir-GM were the highest compare with TRFs bacteria in gut rat treatment was 0.93. Composition bacteria in 3.3 group (rat with metabolic syndrome diet kefir) and 4.1 group (rat with metabolic syndrome diet kefir-gm) has 0.86 similarity with kefir and kefir-gm. Normal rat (1.1 and 1.3) group has similarity score 0.79 with kefir and kefir-gm. Group 3.2 and 3.3 (rat with metabolic syndrome diet kefir) showed similarity score 0.78 and 0.76 with kefir and kefir-gm. Group 4.2 (rat with metabolic syndrome diet kefir-gm) showed similarity score 0.74. It is all indicated that composition bacteria in kefir and kefir-gm similar with the composition bacteria in cecum of rat with metabolic syndrome that was supplemented with kefir and kefir-gm. This similarity indicated that kefir and kefir-gm could modify microflora of cecum in rat with metabolic syndrome.

Number of species, diversity index Shannon Weiner, and distribution frequency bacteria in cecum rat was showed in table 1. Statistical analysis diversity index bacteria in all group treatment rats showed was not significantly different.

Table 1. Diversity index bacteria cecum based on TRFs digested with *Hha*I and *Msp*I enzymes.

	Community TRFs											
	Normal rat			Metabolic syndrome rat			Metabolic syndrome + Kefir			Metabolic syndrome + Kefir-GM		
Parameter	1.1	1.2	1.3	2.1	2.2	2.3	3.1	3.2	3.3	4.1	4.2	4.3
S^a	20	32	14	17	11	24	26	21	22	18	19	38
H^b	1.86	2.39	0.97	1.71	0.97	2.40	2.43	2.20	2.07	1.78	1.36	2.13
$H/Hmax^c$	0.62	0.69	0.37	0.61	1.00	0.77	0.75	0.68	0.58	0.62	0.46	0.59

a : number of species, it was count based on number different TRFs in the group
b: diversity index, Shannon Wiener
c : frequency distribution that was count from Shannon Wiener
Normal rat (1.1-1.3), metabolic syndrome rat (2.1-2.3), diet kefir in metabolic syndrome rat (3.1-3.3), diet kefir-gm in metabolic syndrome rat (4.1-4.3)

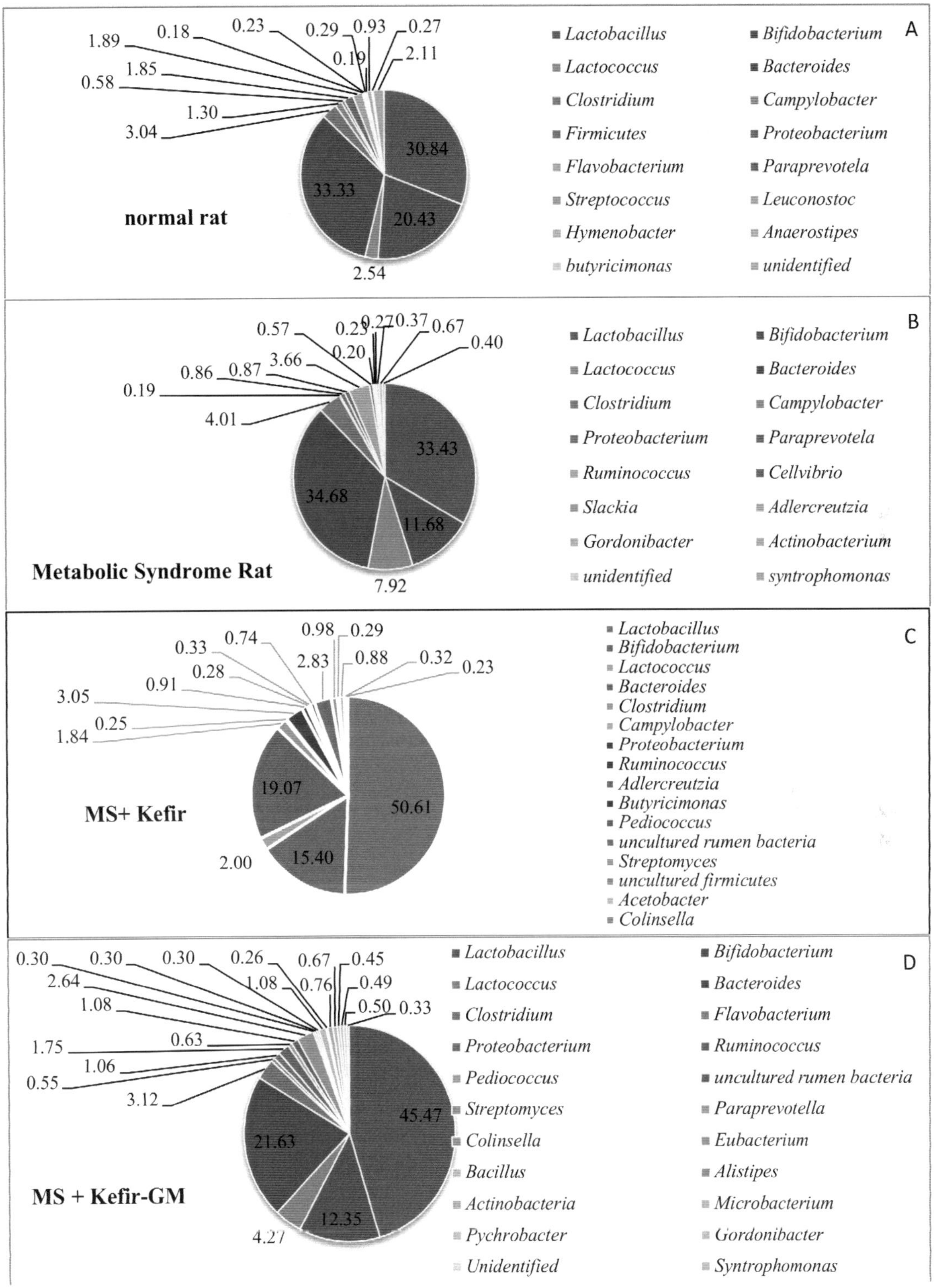

Figure 2. Composition bacteria in cecum normal rat (a), metabolic syndrome rate (b), metabolic syndrome rat +diet kefir (c), and metabolic syndrome rat + diet kefir-gm (c) based on TRFs.

Figure 2 displays the bacterial composition in cecum sample based on peak area of electrogram. Result showed that *Lactobacillus*, *Bifidobacterium*, *Bacteroides*, *Lactococcus*, *Clostridium*, and *Proteobacterium* were found in the cecum rat. Previously, Kovatcheva-datchary and Arora [2] reported that microflora *Lactobacillus*, *Bifidobacterium*, *Bacteroides* and *Clostridium* were found in human gut. Diverse microbiota composition in human gut is strongly related with human diet. *Campylobacter* was found in normal rat, metabolic syndrome rat, and metabolic syndrome rat that was supplemented with kefir. *Flavobacterium* was found in cecum normal rat, metabolic syndrome supplemented with kefir and kefir-gm.

Average percentage of *Lactobacillus* in cecum rat with metabolic syndrome that supplemented kefir or kefir-gm was increase until 14.61% compare with metabolic syndrome rat. Percentage *Lactobacillus* in cecum rat with metabolic syndrome that was supplemented with kefir was higher than that was supplemented with kefir. Average percentage of *Bifidobacterium* in rat with metabolic syndrome that was supplemented kefir and kefir-gm also was increase 3.7%. These finding was indicated that supplementation kefir and kefir-gm could increase composition the beneficial bacteria in gut. *Lactobacillus* and *Bifidobacterium* according Hemaiswarya *et al.* [17] were genera that recognize as probiotic to help and to build up the beneficial bacterial flora in the intestine and completely exclude the pathogenic bacteria. Lactobacillus as probiotic has been shown to exhibit a mannose specific adhesion by which it can adhere to human colonic cells. Once the probiotic adheres to the cell, various biological activities take place, which primarily include the release of cytokines and chemokines. These then exert their secondary activity such as stimulation of mucosal and systemic host immunity.

Supplementation kefir and kefir-gm to the rat with the metabolic syndrome could decrease percentage *Clostridium* and *Bacteroides* compare with the metabolic syndrome rat was not supplemented with kefir. The average percentage *Clostridium* was decrease 38.15%, for the diet kefir 54.11% and 22.19% for the diet kefir-gm, respectively. The average percentage *Bacterioides* in metabolic syndrome rat with diet kefir and kefir-gm was decrease 22.51%, for diet with kefir 45.01% and diet with kefir-gm 37.63%, respectively. According to Gibson and Roberfroid [18], *Bacteroides* and *Clostridium* dominated microflora in gastrointestinal in rat and human. These bacteria were commensal that some was beneficial to support the health of the host and much become pathogen.

The low increase of *Bifidobacterium* in cecum rat with metabolic syndrome with diet kefir-gm was associated with the decrease of the *Clostridium*. *Clostridium* was needed to degrade the glucomannan by the enzyme that they produced. Enzyme glucanase and 1,4-D mannose was performed to degrade glucomannan become oligosaccharide and monosaccharide that could be utilized by *Lactobacillus* and *Bifidobacterium*. This process used Embden Meyerhoff Parnas pathway. In the other hand, *Bifidobacterium* produced glycoside hydrolase that actively degrades polysaccharide in plant [10]

Den-Besten *et al.* [19] described the degradation process of glucomannan in which the microbiota hydrolyze non digestible carbohydrates into oligosaccharides and monosaccharides with fermentation in the anaerobic environment inside the gut. Major bacterial metabolic routes are the Embden-Meyerhof-Parnas pathway (glycolysis, for six-carbon sugars) and the pentose-phosphate pathway (for five-carbon sugars), which convert monosaccharides into phosphoenolpyruvate (PEP). Subsequently, PEP is converted into fermentation products such as organic acids or alcohols. At the level of glyceraldehyde-3-phosphate dehydrogenase (GAPDH) the electron carrier NADH is formed. An-aerobically, there are three types of pathways to get rid of excess reducing equivalents. Major end products of the described fermentation pathways are the SCFAs (Acetate, Propionate, and Butyrate) in abundant number. Prediction bacteria in cecum rat with metabolic syndrome that was supplemented kefir and kefir-gm showed existence of *Pediococcus acidilacti*. *P. acidilactici* is lactic acid bacteria that acid and bile salt resistant. This bacteria also has a role to fermented carbohydrate [20]. *P. acidilactici* is a type of probiotics typically found as a member of normal flora of alimentary tract, oral cavity gastrointestinal tract and has been commonly used in the food industry [21]. *P. acidilactici* was produced pediocin that including bacteriocin that has a role as antimicrobial [22]. Furthermore *P.acidilactii* could stand in the gastrointestinal tract temporary. Moreover supplementation *P. acidilactici* with FOS in rat did not significantly showed differences in metabolic parameter [20]. Previous study showed that supplementation *P. acidilactici* strain M76 in the hyper cholesterol rat could decrease cholesterol total in blood serum and liver through the mechanism forming exopolysaccharide that effected reducing cholesterol contain in the body [23]. Moreover, other study also showed that administration of *P. acidilactici* R037 in mice and rat lowered serum triglycerides, possibly through suppressing intestinal absorption and potentiating lipolytic pathways [24], indicating suppression of intestinal digestion and absorption of food, and suppression of appetite.

3.2. Effect diet kefir and kefir-GM on body weight and feed consumed in metabolic syndrome (MS) rats

The body weight of the rat was shown in figure 3. Analysis body weight was performed using the difference of body weight before and after diet with kefir and kefir-GM. Statistical analysis showed did not significantly different between group of treatment in p>0.05. It was indicated that diet kefir or kefir-gm could stabilize body weight. This result same with the research result by Judiono *et al.* [25] that was supplemented clear kefir in the hyperglycemic rat.

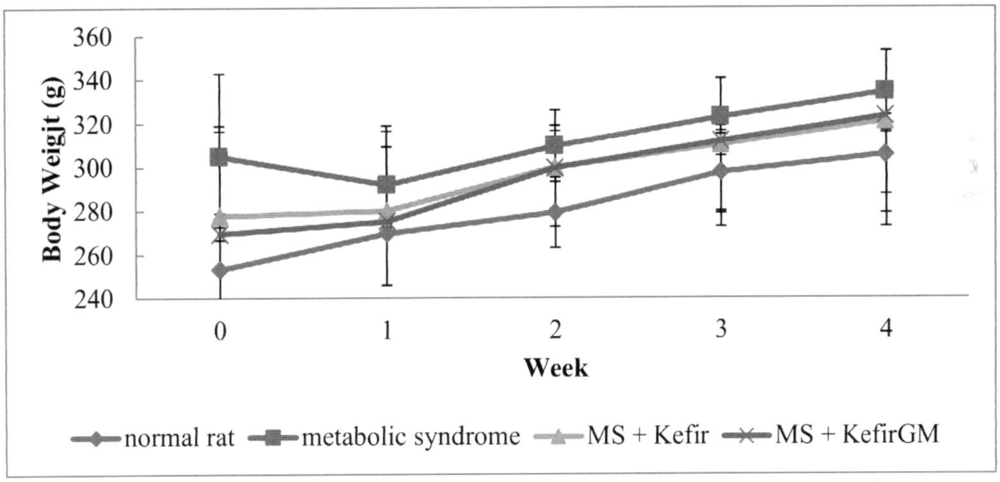

Figure 3. The body weight of the rat during the treatments.

The body weight gain of the rat in contrast with the average feed intake every day (Figure 4). It was shown that there were no significant differences between groups. Supplementation of konjac glucomannan, porang, inulin, and normal feed in the group treatment of the rat in the research by Harmayani *et al.* [10] was shown the same trend that the difference body weight and the average feed intake of the rat was not significantly different.

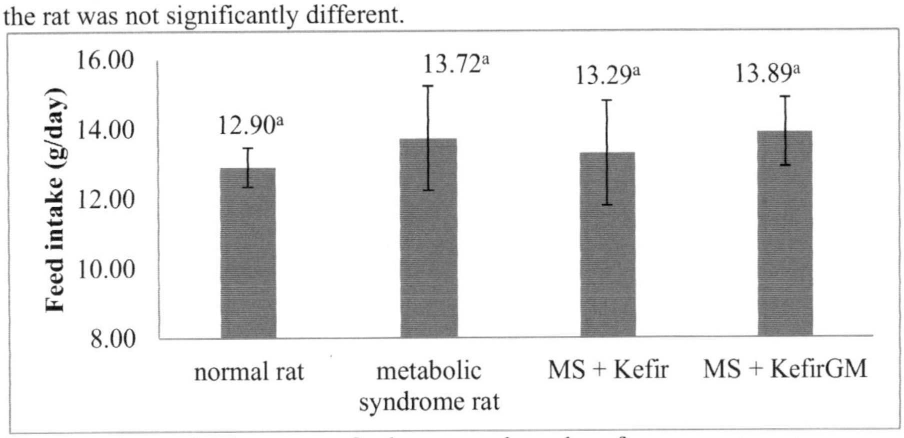

Figure 4. The average feed consumed per day of group treatment.

The difference body weight and the average food intake did not significantly different could be indicated that supplementation kefir and kefir-gm could stabilize body weight gain with inhibited the rat to increase the feed consume. Bogsan *et al.* [26] declared that supplementation probiotic such as *L.plantarum*, *L.rhamnosus*, and *Propionibacterium* L66-5 in human and rat could promote satiety effect. Flint et al [27] assumed that complex carbohydrate has the low energy density. Then the low energy density affected the rate of stomach emptying so then effect the feeling of satiety. In the same satiety, consumed complex carbohydrate affected low energy intake to the body. This is because of diet complex carbohydrate could increase the energy expenditure by feces. Protein, lipid or carbohydrate from food that we consumed could be trap by complex carbohydrate that it could not degraded in intestine. Kaats *et al.* [28] assumed that glucomannan has ability to quickly and profoundly absorb water suggests that it may lead to a swelling in the stomach, resulting in consequent feelings of fullness and decreased appetite.

St-Onge *et al.* [29] was recommendation minimal number microbe in kefir was 10^{10}CFU/ml. This number of microbes could support the health of the hyperlipidemia patient. Furthermore in 2010 European Food Safety Authority (EFSA) declared that glucomannan could decrease the body weight in overweight adult if it was consumed >3g per day. The doses was 1g for 3 times and addition with consumed 1-2 glass of water before eat [30].

4. Conclusions

Kefir and Kefir-GM did not affect diversity but it is able to modify microbiota composition on digesta by increasing *Lactobacillus* and *Bifidobacterium* and suppressing *Clostridium* and *Bacteroides*. Kefir and Kefir-GM can keep the weight body and feed intake in rats with metabolic syndrome.

Conflict of interest

The authors have no conflict of interest to state regarding this manuscript.

Acknowledgement(s)

We thank for Directorate General of Higher Education, The Ministry of Research, Technology and Higher Education of the Republic of Indonesia for research support, through "Penelitian Unggulan Perguruan Tinggi" 2015 (No. 165/LPPM/2015)

References

[1] He M and Shi B 2017 Gut microbiota as a potential target of metabolic syndrome : the role of probiotics and prebiotics *Cell Biosci.* **7** 1–14

[2] Kovatcheva-datchary P and Arora T 2013 Best practice & research clinical gastroenterology nutrition , the gut microbiome and the metabolic syndrome *Best Pract. Res. Clin. Gastroenterol.* **27** 59–72

[3] Olden C Van, Groen A K and Nieuwdorp M 2015 Role of intestinal microbiome in lipid and glucose metabolism in diabetes mellitus *Clin. Ther.* 1–6

[4] Nakayama J, Watanabe K, Jiang J, Matsuda K, Chao S, Haryono P *et al.* 2015 Diversity in gut bacterial community of school-age children in Asia *Sci. Rep.* **5** 1–11

[5] Fiorda F A, Pereira G V de M, Thomaz-Soccol V, Rakshit S K, Pagnoncelli M G, Vandenberghe L P de S *et al.* 2017 Microbiological , biochemical , and functional aspects of sugary ke fi r fermentation - A review *Food Microbiol.* **66** 86–95

[6] Satir G and Guzel-seydim Z B 2015 LWT - Food science and technology influence of kefir fermentation on the bioactive substances of different breed goat milks *LWT - Food Sci. Technol.* **63** 852–8

[7] Ebner J, Arslan A A, Fedorova M, Hoffmann R, Kucukcetin A and Pischetsrieder M 2015 Peptide profiling of bovine kefir reveals 236 unique peptides released from caseins during its production by starter culture or kefir grains *J. Proteomics* **117** 41–57

[8] Likotrafiti E, Valavani P, Argiriou A and Rhoades J 2015 In vitro evaluation of potential

antimicrobial synbiotics using Lactobacillus kefiri isolated from kefir grains *Int. Dairy J.* **45** 23–30

[9] Chen H, Fan Y, Chen M and Chan Y 2005 Unhydrolyzed and hydrolyzed konjac glucomannans modulated cecal and fecal microflora in Balb / c mice *Nutrition* **21** 1059–64

[10] Harmayani E, Aprilia V and Marsono Y 2014 Characterization of glucomannan from Amorphophallus oncophyllus and its prebiotic activity in vivo *Carbohydr. Polym.* **112** 475–9

[11] Nurliyani, Harmayani E and Sunarti 2018 Goat Milk Kefir Supplemented with Porang Glucomannan Improves Lipid Profile and Haemotological Parameter in Rat Fed High Fat and High Fructose Diet *Rom. J. Diabetes Nutr. Metab. Dis.* **25** 11–21

[12] Reeves P G, Nielsen F H and Fahey G C 1993 AIN-93 Purified Diets for Laboratory Rodents : Final Report of the American Institute of Nutrition AdHoc Writing Committee on the Reformulation of the AIN-76A Rodent Diet *J. Nutr.* **123** 1939–51

[13] de Castro U G, dos Santos R A, Silva M E, de Lima W G, Campagnole-Santos M J and Alzamora A C 2013 Age-dependent effect of high-fructose and high-fat diets on lipid metabolism and lipid accumulation in liver and kidney of rats *Lipids Health Dis.* **12** 1–11

[14] Claassen S, Kaba M, Moodley C, Zar H J and Nicol M P 2013 A comparison of the ef fi ciency of fi ve different commercial DNA extraction kits for extraction of DNA from faecal samples *J. Microbiol. Methods* **94** 103–10

[15] Chen L, Teasdale M T, Kaczmarczyk M M, Freund G G and Miller M J 2012 Development of a Lactobacillus speci fi c T-RFLP method to determine lactobacilli diversity in complex samples *J. Microbiol. Methods* **91** 262–8

[16] Sari P M, Harmayani E and Widada J 2014 *Effect of Lactobacillus plantarum Mut7 and high Fiber of Sweet Potato Powder Dietto University of Microbial Communities of Digesta in Sprague Dawley Rats* (Universitas Gadjah Mada)

[17] Hemaiswarya S, Raja R, Ravikumar R and Carvalho I S 2013 Mechanism of action of probiotics *Brazilian Arch. Biol. Technol.* **56** 113–9

[18] Gibson G and Roberfroid M B 1995 Critical review dietary modulation of the human colonie microbiota : introducing the concept of prebiotics *J. Nutr.* **125** 1401–12

[19] den Besten G, van Eunen K, Groen A K, Venema K, Reijmgound D-J and Bakker B M 2013 The role of short-chain fatty acids in the interplay between diet , gut microbiota , and host energy metabolism *J. Lipid Res.* **54** 2325–40

[20] Juśkiewicz J, Semaskaite A, Zduńczyk Z, Wróblewska M, Gružauskas R and Juśkiewicz M 2007 Minor effect of the dietary combination of probiotic Pediococcus acidilactici with fructooligosaccharides or polysaccharidases on beneficial changes in the cecum of rats *Nutr. Res.* **27** 133–9

[21] Halim M, Anizawanie N, Mustafa M, Othman M, Wasoh H, Rizal M *et al.* 2017 LWT - food science and technology rffect of encapsulant and cryoprotectant on the viability of probiotic Pediococcus acidilactici ATCC 8042 during freeze-drying and exposure to high acidity , bile salts and heat *LWT - Food Sci. Technol.* **81** 210–6

[22] Papagianni M and Anastasiadou S 2009 Pediocins: The bacteriocins of *Pediococci.* sources, production, properties and applications *Microb. Cell Fact.* **8** 1–16

[23] Moon Y J, Baik S H and Cha Y S 2014 Lipid-lowering effects of pediococcus acidilactici M76 Isolated from Korean traditional makgeolli in high fat diet-induced obese mice *Nutrients* **6** 1016–28

[24] Takahiro U, Airo T, Kazuya H, Hideyuki K, Kazunori H, Hideyuki M *et al.* 2018 Effects of *Pediococcus acidilactici* R037 on serum triglyceride levels in mice and rats after oral administration *J. Nutr. Sci. Vitaminol.* **64** 41–7

[25] Judiono, Djokomoeljanto and Hadisaputro S 2011 Effect of oral clear kefir probiotic on glycemi status, lipid peroxidation, antioxidative properties of streptozotocin induces hyperglycemia Wistar rats *Gizi Indones.* **34** 1–6

[26] Bogsan C S B, Florence A C R, Perina N, Barbuti R C, Navarro-Rodriguez T, Eisig J N *et al.*

2011 Probiotics intake and metabolic syndrome: a proposal *Trends Food Sci. Technol.* **22** 457–64

[27] Flint H J, Scott K P, Duncan S H, Louis P, Forano E, Flint H J *et al.* 2012 Microbial degradation of complex carbohydrates in the gut *Gut Microbes* **3** 289–306

[28] Kaats G R, Bagchi D and Preuss H G 2015 Konjac glucomannan dietary supplementation causes significant fat loss in compliant overweight adults *J. Am. Coll. Nutr.* **0** 1–7

[29] St-Onge M-P, Farnworth E R, Savard T, Chabot D, Mafu A and Jones P J 2002 Effect of skim milk supplementation on blood cholesterol concentration, blood pressure, and triglycerides in a free-living human population *J. Am. Coll. Nutr.* **11** 56–67

[30] Zalewski B M and Szajewska H 2015 Effect of glucomannan supplementation on body weight in overweight and obese children: Protocol of a randomised controlled trial *BMJ Open* **5** 1–5

IOP Conf. Series: Materials Science and Engineering 1011 (2021) 012010 doi:10.1088/1757-899X/1011/1/012010

Bio-modified soybean meal as a new protein source for food

L Istiqomah[1*], E Damayanti[1], A E Suryani[1], I O Susilawati[2] and Lutfiani[2]

[1]Research Unit for Natural Products Technology, Indonesian Institute of Sciences, (BPTBA-LIPI). Jl. Jogja-Wonosari Km. 31.5, Gading, Playen, Gunungkidul, 55861 D. I. Yogyakarta, Indonesia

[2]Pharmacy and Food Analyst, Faculty of Mathematics and Sciences, Lambung Mangkurat University, Banjar Baru, Kalimantan, Indonesia

*Corresponding author: lusty.istiqomah@gmail.com

Abstract. Soybean meal has long been considered as a plant protein source particularly for animal feedstuff. Nowadays, there is a promising prospect of this material for food processing. However, its protein utilization has drawbacks, such as poor protein solubility and anti-nutritional factors called phytic acid. The phytic acid could be reduced by the addition of phytase, whereas nutrient content enhanced by microbial fermentation. In this study, we analyzed the bio-modified of soybean meal (SBM) through fermentation by phytase producing *Lactobacillus plantarum* A1-E to evaluate its effect on nutritional quality and anti-nutrient factor called phytic acid and aflatoxin. The study was designated into two treatments consisted of unfermented soybean meal (USBM) and fermented soybean meal (FSBM) by *L. plantarum* A1-E. Parameters observed were microbial population, phytic acid content, aflatoxin content, and nutrient composition (moisture, ash, crude protein, crude fiber, and lipid). Data were analyzed using analysis of variance (ANOVA) and a T-test to compare treatment means. FSBM by *L. plantarum* A1-E increased population (6.8×10^7 CFU/ml) during 72 h incubation ($P<0.05$) than 0 h (2.1×10^6 CFU/ml). FSBM reduced phytic acid content (34.13%) and aflatoxin (6.12%) compared to control (USBM). Fermentation during 72 h had on average more crude protein (1.33-fold), crude fiber (2.09-fold), ash (1.13-fold), and less crude lipid (0.71-fold) than SBM. Thus, it can be concluded that FSBM by phytase producing *L. plantarum* A1-E could be applied for degradation phytic acid during food processing and represents an improvement in nutritional quality thereby becomes potential raw material as a new protein source for food.

1. Introduction

Soybean meal (SBM) provide useful sources of nutrients such as complex carbohydrates, proteins (44.40%), and fiber sources (6.75%) [1]. Furthermore, SBM is known for its high protein content, balanced amino acid composition, and high level of lysine in comparison with other plant protein sources thereby it is considered as a protein source for food and animal feedstuff. However, its protein utilization has several limitations, such as undesirable anti-nutritional components caused poor protein solubility and low digestibility [2]. Among all the anti-nutritional components, phytic acid is of prime concern for human nutrition. Phytic acid is the primary storage form of both phosphate and inositol in plant seeds and presence in the form of complexes with metal cations viz. iron, zinc, and calcium cause a mineral-related deficiency in humans and also negatively impact protein and lipid utilization [3,4]. By binding plant protein, phytic acid decreased protein solubility and digestibility and the formation of phytate-carbohydrate complexes making carbohydrate less degradable thereby reduced the nutritional value of feed. Moreover, the human small intestine is a lack of a phytate-degrading enzyme in the upper part of the digestive tract. Therefore, the phytate-mineral complex remains partially hydrolyzed in the human gut [3]. Another anti-nutrient component in SBM was aflatoxin [5-7]. Consumption of aflatoxin-contaminated food can cause a severe health risk to humans (carcinogenic) [8].

In order to decrease its anti-nutrient factor and increase nutritional content, several approaches including physical treatment (over-heating), chemical treatment, and alcohol extraction were used. However, these methods were not fully eliminating them. For this purpose, the addition of exogenous phytase is needed. So far, no phytase product for human food applications has found its way to the market. Several researchers have reported a convincing improvement of food products by adding microbial-based phytase during food processing [9] or certain microbes for food fermentation [10,11]. Microbial-producing phytase is becoming promising approaches as inoculum for the fermentation process of SBM. *Lactobacillus plantarum* A1-E has been reported to produce extracellular phytase (41.9 U/mL) *in vitro* [12].

Solid-state fermentation (SSF) is a process in which a solid substrate is deposited in a fermenter, cultivated with single or multiple strains, and incubated in a bioreactor for a few days [13]. As a processing technology, SSF of SBM improves several physiological functions and the nutritional value, particularly protein. Fermented soybean meal (FSBM) by *Rhizopus oligosporus* removed ANF (trypsin inhibitor, phytic acid, and tannins) [14]. FSBM by microorganisms enhanced the bioavailability of nutritious components [10,11,15].

However, the application of phytase producing microbial in soybean fermentation and its effect on phytic acid content has not been reported yet. Therefore, in this study, we investigated the effect of solid-state fermentation by phytase producing *L. plantarum* A1-E on soybean meal (SBM) chemical composition and anti-nutrient factor with emphasis on phytic acid.

2. Materials and methods

2.1. Materials
The lactic acid bacteria (LAB) strain used was *L. plantarum* A1-E [12]. SBM sample was acquired from the commercial feed mill in Yogyakarta. Reagents used for microbial growth were purchased from Oxoid and chemicals for analysis were obtained from Merck.

2.2. Microbial strain and growth condition
Lactobacillus plantarum A1-E was grown on *de Man Rogosa* (MRS) agar media and maintained by periodic transfer. *Lactobacillus plantarum* was propagated in MRS broth media at 37 °C for 24 hours and stored at 4 °C.

2.3. Soybean meal processing
The sample of USBM was prepared by a scoop in plastic bag then grinded and homogenized to obtain particle size 1.0 mm. The sample of this SBM was subjected to fermentation process (FSBM) performed in triplicate according to Mandviwala and Khire [16] with slight modifications. Erlenmeyer flasks (250 ml) containing 10 g of SBM moistened with 10 ml distilled water were sterilized in an autoclave at 121°C for 40 min, cooled, inoculated with 5% of *L. plantarum* A1-E suspension containing 10^6 CFU/ml from a 1-day-old culture grown on MRS broth at 37 °C. The contents of each flask were mixed thoroughly with a sterile inoculating needle for uniform distribution then incubated at 37 °C for 72 h. The flask was harvested then the sample was collected for further analysis.

2.4. Microbial population and growth curve
The colony population of *L. plantarum* A1-E during fermentation was counted every 24 h using the Total Plate Count (TPC) method according to Seeley et al. [17]. One gram of FSBM was put into the falcon tube filled with 9 mL of 0.85% NaCl and homogenized then serial dilution (up to 10^{-5}) was carried out. A total of 100 μL culture in media was taken using a micropipette and put into 900 μL of 0.85% NaCl in microtube (10^{-1} dilution). Dilution then performed again to obtain 10^{-2} dilution and so on. A

total of 100 μL suspension from two of the last dilution was plated into a petri dish containing MRS agar media using the spread plate method then incubated at 37 °C for 48 h. After 48 h incubation, the colony numbers that grow counted by colony counter. The colony population was calculated by the Standard Plate Count (SPC) and performed in $\log_{10}$ CFU/ml, then plotted in the growth curve.

2.5. *Phytic acid determination*

Phytic acid from sample was measured by the K-PHYT assay kit developed by Megazyme. The extraction process of phytic acid was carried out by the addition of one gram of sample (USBM and FSBM) in a glass beaker with 20 ml of HCl solution (0.66 M) then covered with foil and stirred vigorously overnight at room temperature. One milliliter of the extract was transferred to a microfuge tube at 13,000 rpm for 10 min. After that, immediately 0.5 ml of the resulting extract supernatant to a fresh microfuge tube and neutralized by the addition of 0.5 ml of NaOH solution (0.75 M). This neutral sample extract was used for the next step.

Enzymatic dephosphorylation reaction was subjected to two steps: free phosphorus reaction and total phosphorus reaction. Free phosphorus reaction was performed with reaction mixture consist of 0.62 ml of distilled water, 0.2 ml of solution I (buffer), 0.05 ml of sample extract) then mixed by vortex and incubated in water bath set at 40 °C for 10 min. After 10 min, the next reaction by addition of 0.02 ml of distilled water and 0.2 ml of solution 3 (ALP) then mixed by vortex and incubated in water bath set at 40 °C for 15 min. After 15 min, the reaction was stopped by addition of 0.3 ml of trichloroacetic acid (50% w/v) then centrifuged the terminated reaction at 13,000 rpm for 10 min. The supernatant was carefully transferred into fresh microfuge tube for colorimetric determination of phosphorus.

Colorimetric determination of phosphorus was performed according to Fiske and Subbarow [18] and Lowry and Lopez [19]. One milliliter of the supernatant sample was added with 0.5 ml color reagent. Color reagent consisted of 1 part of solution B (5% of ammonium molybdate in 25 ml of distilled water) and 5 part of solution A (10% of ascorbic acid in 100 ml sulphuric acid 1 M). The mixture was mixed by vortex and incubated in a water bath set at 40 °C for 1 h. After 1 h, mixed by vortex and transferred 1 ml to a semi-micro cuvette and read the absorbance at 655 nm.

The calibration curve was prepared by making the various standard concentration of phosphorus solutions (0, 0.5, 2.5, 5, and 7.5 μg) by the addition of distilled water. One milliliter of the standard solution was added with 0.5 ml color reagent as described before then mixed by vortex and incubated in a water bath set at 40 °C for 1 h. After 1 h, mixed by vortex and transferred 1 ml to a semi-micro cuvette and the absorbance was read at 655 nm.

2.6. *Aflatoxin determination*

Aflatoxin content was measured by the AgraQuant Total Aflatoxin Assay kit (Romer Lab) using Enzyme-linked Immunosorbent Assay (ELISA) technique. Initially, aflatoxin from USBM and FSBM samples were extracted by ground each sample so that 75% passed through a 20-mesh screen, then thoroughly mixed the subsample portion. A total of 20 g sample was added with 100 ml of 70/30 (v/v) methanol/water extraction solution then shake or blend for 3 min. The sample was kept to settle, then filtered the top layer of extract through a Whatman #1 filter, and the filtrate was collected.

Aflatoxin assay was carried out by the addition of 100 μL of each standard or sample into the appropriate dilution well contained 200 μL of conjugate using a single channel pipettor. Each dilution-well was mixed by carefully pipetting it up using an 8-channel pipettor with fresh tips. A total of 100 μL of the contents from each dilution well was immediately transferred into a corresponding antibody-coated microwell then incubated at room temperature for 15 min. After incubation, the contents of microwell strips were empty into a waste container. Each microwell washed with distilled or deionized water then dumped the water from the microwell strips and this step was repeated 4 times for a total of 5 washes. Several layers of tissue were put on a flat surface, and microwell strips were tapped on towels to expel as much residual water as possible after the fifth wash. The bottom of the microwells was dried with a tissue. A total of 100 μl of the substrate was added into each microwell strip using an 8-channel pipettor then incubated at room temperature for 5 min followed by the addition of 100 μl of stop solution into each microwell strip using an 8-channel pipettor the color change from blue to yellow. The absorbance was read at 450 and 630 nm.

2.7. Nutrient composition
Nutrient content of USBM and FSBM, including moisture content, ash, crude fiber, crude protein, and lipid were performed in triplicate. Dry matter was measured at 105 °C until constant weight. Crude protein content was carried out by the Kjeldahl method based on AOAC [20] (Helrich, 1990). Crude fiber was determined as AOAC [20] (Helrich, 1990). Crude lipid was obtained by Soxhlet extraction, as described by Novakofski *et al.* [21].

2.8. Data analysis
One-way ANOVA followed by the Duncan Multiple Range Test (DMRT) comparison test was performed using Costat [22] to analyze the LAB population. Phytic acid, aflatoxin, and nutrient content were analyzed using a T-test to compare treatment means [23].

3. Result and discussion

3.1. Microbial population and growth curve
Lactobacillus plantarum is considered as a GRAS (Generally Regarded as Safe) and a genus Lactobacillus is a heterogeneous group of LAB with important implications in food fermentation. Lactobacilli are currently used as probiotics, silage inoculants, and as starters in fermented food [24]. Table 1 showed that the population of *L. plantarum* A1-E increased during fermentation time until 48 h ($P<0.05$) and stable between 48 and 72 h ($P>0.05$). Charalampopoulos *et al.* [25] reported that *L. plantarum* (0.20-0.41 $\log_{10}$ CFU/ml) during 48 h fermentation attained the highest cell populations growing in malt medium (10.11 $\log_{10}$ CFU/ml) compared to barley medium (9.43 $\log_{10}$ CFU/ml) and wheat medium (9.29 $\log_{10}$ CFU/ml).

Table 1. Population of *L. plantarum* A1-E in FSBM.

Fermentation time	Population	
	CFU/ml	$\log_{10}$ CFU/ml
0	$2.1 \times 10^{6\,c}$	6.33 ± 0.05^{c}
24	$6.9 \times 10^{6\,b}$	6.80 ± 0.22^{b}
48	$6.4 \times 10^{7\,a}$	7.79 ± 0.12^{a}
72	$6.8 \times 10^{7\,a}$	7.82 ± 0.11^{a}

SBM contained complex carbohydrates, proteins, vitamins, minerals, and fiber sources. Carbohydrates were converted by *L. plantarum* A1-E into simple sugars particularly glucose as the fermentation product. According to a prior study, natural fermentation increases glucose in soybean [26], thereby the growth of LAB was faster.

Log of number of viable bacteria in Table 1 then plotted in a graph against time by subtracting each data from 24, 48 and 72 h fermentation period with data from the time of inoculation of the bacteria (0-time) to obtain the growth curve (Figure 1). Growth of *L. plantarum* begins with an initial phase (lag) which is an adaptation phase, then followed by an exponential phase until 48 h incubation. Growth enters steady growth or a stationary phase at 48 to 72 h. Istiqomah [11] reported that the lag phase of *L. plantarum* A1-E on the modified MRS broth media at 37 °C incubation without agitation occurred in the first two hours and showed slow growth whereas the exponential phase was reached between the 2 – 10 h incubation and the stationary phase started at the 10 up to 24 h incubation, and afterward enters the death phase. Smetanková *et al.* [27] revealed that at 37 °C and 45 °C, the exponential phase of *L. plantarum* 2L2 was observed between 2-8 h during the transition to the stationary phase, there is a stimulation of the formation of some proteins and in the stationary phase, there is a lack of essential nutrients.

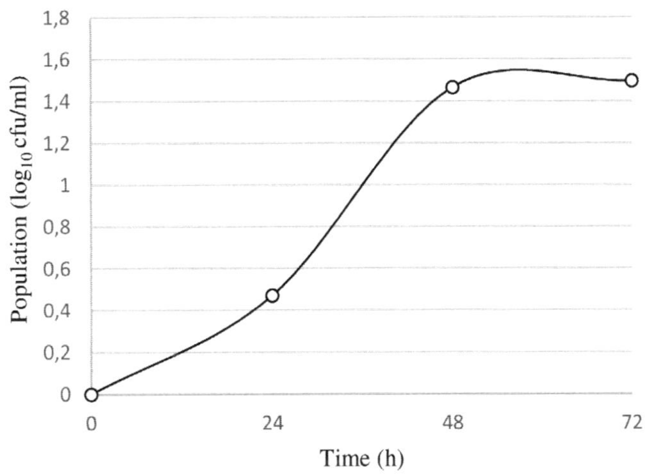

Figure 1. Growth curve of *L. plantarum* A1-E in SBM during fermentation.

The different exponential phase tends in this study compared to previous studies probably due to differences in a growth medium. SBM medium contains a greater amount of nutrients than MRS broth media thus *L. plantarum* A1-E cleaved carbohydrates in the media to be converted into simple sugars used for cell propagation so that cell growth continues to increase up to 48 h incubation. Charalampopoulos et al. [25] stated that *L. plantarum* grew exponentially until 10–12 h of fermentation and the viable cell densities declined slightly during the stationary phase (12–48 h).

Generally, LAB produced primary metabolites such as lactic acid, acetic acid, and hydrogen peroxide in the logarithmic phase and the stationary phase. In this phase, the synthesis of enzymes by bacterial cells is used for the metabolism of metabolites. After certain substrates or compounds needed for bacterial growth in the culture medium are nearing exhausted and there is a built-up of inhibitory products, there is a decrease in the rate of growth of LAB [28].

3.2. Phytic acid and aflatoxin content

Table 2 revealed that fermentation by *L. plantarum* A1-E (FSBM) during 72 h reduced 34.13% of phytic acid content ($P<0.05$) compared to control (USBM). *Lactobacillus plantarum* A1-E produced phytase with 41,9 U/ml of activity and 5.5 U/mg of specific activity [11]. Sofyan et al. [29] reported that fermentation of rice bran by *Rhizopus* sp. combined with deproteinated chitin waste (DCW) decreased phytic acid (35,1%). It has been reported that fermented millet grain reduced food inhibitors, phytic acid, and tannins [30]. Germinated pearl millet sprouts that fermented by mix pure cultures of *L. brevis*, *L. fermentum*, *Saccharomyces diasticus*, and *S. cerevisiae* at 30 °C for 72 h resulted in 88.3 % reduction in phytate content [31]. Natural fermentation can achieve a large reduction in anti-nutritional factors (phytic acid) in substrates by the action of microbial accompanied by a significant improvement in the protein digestibility. This producing phytase microbes were able to hydrolyze phytic acid in SBM into Myo-inositol (phosphate). This Myo-inositol (phosphate) then hydrolyzed by alkali phosphate in producing Myo-inositol and inorganic phosphorus (Pi). Hydrolysis of phytate during food processing and then preparation, for example by fermentation is a result of the phytate-degrading activity of phytase, which is naturally present in plants and microorganisms [3].

Table 2. Phytic acid and aflatoxin content of USBM and FSBM.

Material	Phytic acid (g/100 g)	Aflatoxin (µg/kg)
USBM	1.38±0.11^{a}	3.08±0.11^{a}
FSBM	0.91±0.07^{b}	2.89±0.17^{a}

Reduction (%)	34.13	6.12

Differences superscript on the same column shown a real difference (P<0.05)

Aflatoxin content in SBM was 3.08 µg/kg. Al-Seeni [5] reported that the level of aflatoxin detected in feeding soybean was ranged from 1 to 14.8 ppb. Daga et al. [6] revealed that the aflatoxin level was low in natural SBM (0.5µg/kg). Nomura et al. [7] (2018) reported that the mean concentration of AFB1 in SBM was 1.1 µg/kg. FSBM during 72 h incubation reduced 6.12% aflatoxin content in SBM. Two *L. plantarum* strains Lb7 and Lb9 isolated from sourdough performed poorly AFB_1 binding abilities only 2.1% and 5.2%, respectively [32]. Istiqomah et al. [33] reported that *L. plantarum* G1 addition in the aflatoxin-contaminated diet decreased the aflatoxin content (2.71 ppb) compared to control (9.01 ppb). Mechanism of aflatoxin binding ability in LAB due to the cluster on the LAB cell wall [34]. Moreover, the binding of aflatoxin by LAB appears extremely strain-specific and different binding compounds. The Lactobacillus genus (*L. helveticus, L. brevis, L. acidophilus*, and *L. crispatus*) has been found to have an S-layer made of protein subunits, while *L. rhamnosus* has polysaccharides (cell wall polysaccharide, peptidoglycan, and teichoic or lipoteichoic acids) where it binds aflatoxins [8].

3.3. Nutrient composition

The nutrient content of SBM (Table 3) reflected the differences in composition between SBM and FSBM. SBM contained ash 6.10%, crude protein 43.14%, crude fiber 4.56%, and crude lipid 5.43%. Banaszkiewicz [1] (2014) reported that SBM provides crude ash 6.65%, crude protein 44.40%, crude fiber 6.75%, and crude fat 2.18%. FSBM after 72 h fermentation had on average more protein (1.33-fold), total fiber (2.09-fold), and less lipid (0.71-fold) than USBM (P<0.05) whereas the crude fat content was kept at a stable level (P>0.05).

Table 3. Nutrient composition of USBM and FSBM (in % dry matter basis).

Material	Moisture	Ash	Crude protein	Crude fiber	Crude lipid
USBM	9.18 ± 0.30^b	6.10 ± 0.09^b	43.14 ± 0.42^b	4.56 ± 0.07^b	5.43 ± 0.64^a
FSBM	32.72 ± 0.27^a	6.91 ± 0.21^a	57.48 ± 0.43^a	9.53 ± 0.06^a	3.87 ± 0.93^a
Enhancement (fold)	3.56	1.13	1.33	2.09	(0.71)

Differences superscript on the same column shown a real difference (P<0.05)

Chi and Cho [9] reported that FSBM by *Bacillus amyloliquefaciens* U304 could substantially improve both the nutritional quality and bioactivity of SBM. FSBM by *S. cerevisiae* improved nutritional quality (50% increase in fiber and 32%, 83% and 69% decreases in carbohydrate, lipid, and trypsin inhibitors, respectively) and increased antioxidant activity (up to 206%) [10]. Protein content in FSBM by *B. subtilis* or *Aspergillus oryzae* microbes was increased (8.37% and 0.34%, respectively) due to the degradation of allergens into peptides thereby peptides can be easily absorbed by an animal and transported within an organism [14]. Fermented rice bran by *Rhizopus* sp. combined with deproteinated chitin waste (DCW) decreased crude fiber contents (38.4%) [16]. Hydrolysis in SBM as a substrate during fermentation by *L. plantarum* A1-E probably due to its stronger proteolysis at 37 °C. Moreover, *L. plantarum* A1-E produced phytase that breaks the formation of phytate-protein and phytate-carbohydrate complexes making protein and carbohydrate more available and degradable thereby increased the nutritional value of SBM.

4. Conclusion

Producing phytase *L. plantarum* A1-E had an important application in human nutrition during fermentation of soybean meal (SBM). FSBM increased *L. plantarum* A1-E population (6.8×10^7 CFU/ml) during 48 h incubation, reduced 34.13% of phytic acid and 6.12 of aflatoxin content, and also resulted more crude protein (1.33-fold), crude fiber (2.09-fold), ash (1.13-fold), and less crude lipid

(0.71-fold) than USBM. Therefore, this bio-modified SBM represents an improvement in nutritional quality and becomes potential raw material as a new protein source for food.

Acknowledgement
The authors gratefully thank the DIPA Indonesian Institute of Sciences (LIPI) for financial support and Mr. I. Nyoman Gunadarma for technical assistance.

References
[1] Banaszkiewicz T 2014 Nuritional value of soybean meal in *Soybean Nutrition* editor H A El-Shemy (Croatia: Intech)
[2] Yin H, Jia F and Huang J 2019 The variation of two extracellular enzymes and soybean meal bitterness during solid-state fermentation of *Bacillus subtilis Grain Oil Sci Technol* **2** 39-43
[3] Kumar V, Sinha A K, Makkar H P S, and Becker K 2010 Dietary roles of phytate and phytase in human nutrition: A review *Food Chem* **120** 945-59
[4] Gupta R K, Gangoliya S S and Singh N K 2015 Reduction of phytic acid and enhancement of bioavailable micronutrients in food grains *J Food Sci Technol* **52** 676-84
[5] Al-Seeni 2012 Natural occurrence of heavy metal, fungi and mycotoxins in soybean meal samples used in animal feeding in Saudi Arabia *African J Biotecnol* **11** 9288-94
[6] Daga A, Horn M B, Kottwitz L B M and Fariña 2015 Bromatological and mycotoxin analysis on soybean meal before and after the industrial process of micronization *Cienc. Rural Santa Maria* **45** 1336-41
[7] Nomura M, Aoyama K and Ishibashi T 2017 Sterigmatocystin and aflatoxin B1 contamination of corn, soybean meal, and formula feed in Japan *Mycotoxin Res* **34** 21-7
[8] Ahlberg S H, Joutsjoki V and Korhonen H J 2015 Potential of lactic acid bacteria in aflatoxin risk mitigation: a review *Int J Food Microb* **207** 87-102
[9] Haros M, Rosell C M, and Benedito C 2001 Fungal phytase as a potential bread-making additive. European Food Research and Technology **213** 317-22
[10] Chi C H and Cho S J 2016 Improvement of bioactivity of soybean meal by solid-state fermentation with *Bacillus amyloliquefaciens* versus *Lactobacillus* spp. and Saccharomyces cerevisiae *LWT-Food Sci Technol* **68** 619-25
[11] Silva F de O, Miranda T G, Justo T, Frasão, Conte-Junior C A, Monteiro M *et al.* 2018 Soybean meal and fermented soybean meal as functional ingredients for the production of low-carb, high-protein, high-fiber and high isoflavones biscuits *LWT-Food Sci Technol* **90** 224-31
[12] Istiqomah L 2015 *Isolation and characterisation of lactic acid bacteria as phytase producer from digestive tract of poultry and the phytase characterisation* (Indonesia: Gadjah Mada University, Thesis)
[13] Qin H B, Yang H J, Qiao Z J, Gao S S and Liu Z 2013 Identification and characterization of a *Bacillus subtilis* strain HB-1 isolated from Yandou, a fermented soybean food in China *Food Control* **31** 22-7
[14] Egounlety M and Aworh O C 2003 Effect of soaking, dehulling, cooking and fermentation with Rhizopus oligosporus on the oligosaccharides, trypsin inhibitor, phytic acid and tannins of soybean (*Glycine max* Merr.), cowpea (*Vigna unguiculata* L. Walp) and groundbean (*Macrotyloma geocarpa* Ha.) *J Food Engineer* **56** 249-54
[15] Teng D, Gao M, Yang Y, Liu B, Tian Z and Wang J 2012 Bio-modification of soybean meal with *Bacillus subtilis* or *Aspergillus oryzae Biocatalysis Agric Biotechnol* **1** 32-8
[16] Mandviwala T N and Khire J M 2000 Production of high activity thermostable phytase from thermotolerant *Aspergillus niger* in solid state fermentation *J Industrial Microbiol. Biotechnol* **24** 237-43
[17] Seeley H W, VanDemark P J and Lee J J 2001 *Microbes in Action* 8th ed. (Newyork: W.H. Freeman and Company)
[18] Fiske C H and Subbarow Y 1925 The colorimetric determination of phosphurus *J Biol Chem* **66**

375-400
[19] Lowry O H and Lopez J A 1946 The determination ofinorganic phosphate in the presence of labile phosphate esters *J Biol Chem* **162** 421-28
[20] Helrich K 1990. *Official Methods of Analysis of the Association of Official Analytical Chemists* 15th ed. (Airlington: Association of Official Analytical Chemists Inc.) p. 807.
[21] Novakofski, J, Park S, Bechtel P J, Mckeith F K 1989 Composition of cooked pork chops: effect of removing subcutaneous fat before cooking *J Food Sci* **54** 15-7
[22] Cohort 2008 *CoSTAT Version 6.400* Cohort Software 798 (Monterey)
[23] Gomez K A and Gomez A A 2007 *Statistical Procedures for Agricultural Research* 2nd ed (Jakarta: UI Press)
[24] Giraffa G, Chanishvili N, Widyastuti 2010 Importance of Lactobacilli in food and feed biotechnology *Res in Microbiol* **161** 480-7
[25] Charalampopoulos D, Pandiella S S, and Web C 2002 Growth studies of potentially probiotic lactic acid bacteria in cereal-based substrates *J Appl Microbiol* **92** 851-59
[26] Kaczmarska K T and Chandra-Hioe M V 2017 Effect of germination and fermentation on carbohydrate composition of Australian sweet lupin and soybean seeds and flours *J Agric Food Chem* 1-34 DOI: 10.1021/acs.jafc.7b02986
[27] Smetanková J, Hladíková Z, Valach F, Zimanová M, Kohajdová Z, Greif G *et al.* 2012 Influence of aerobic and anaerobic conditions on the growth and metabolism of selected strains of *Lactobacillus plantarum Acta Chimica Slovaca* **5** 204-10
[28] Ayuti S R, Nurliana, Yurliasni, Sugito and Darmawi 2016 The dynamics of *Lactobacillus casei* and the characteristics of fermented milk based on temperature and storage time *Agripet* **16** 20-37
[29] Sofyan A, Damayanti E and Julendra H 2010 Reduction of phytic acid and aflatoxin content to rice bran through fermentation by Rhizopus sp. combined with deproteinated chitin waste addition *Proceed. Int. Seminar on Tropical Animal Production Community Empowerment and Tropical Animal Industry* (Yogyakarta) p 104-8
[30] Coulibaly A, Kouakou B and Chen J 2011 Phytic acid in cereal grains: Healthy or harmful ways to reduce phytic acid in cereal grains and their effects on nutritional quality *Am J plant Nutr Fert Technol* **1** 1–22
[31] Kaur K D, Jha A, Sabikhi L, and Singh A K 2014 Significance of coarse cereals in health and nutrition: a review *J Food Sci Tech* **51** 1429-41
[32] Zinedine A, Faid M, and Benlemlih M 2005 In vitro reduction of aflatoxin B1 by strains of lactic acid bacteria isolated from Moroccan sourdough bread *Int J Agric Biol* **7** 67-70
[33] Istiqomah L, Damayanti E, Julendra H, Suryani A E, Sakti A A, and Anggraeni A S 2017 Effect of methionine and lactic acid bacteria as aflatoxin binder on broiler performance *Int. Biology Conference* (Surabaya) *AIP Conference Proceedings* **1854** p 020017 DOI: 10.1063/1.4985408
[34] Hernandez-Mendoza A, Rivas-Jimenez L and Garcia H S 2011 Assessment of aflatoxin B_1 binding to *Lactobacillus reuteri* by microscopy and fluorescence techniques *J. Food Biotechnology* **25** 140-50

IOP Conf. Series: Materials Science and Engineering 1011 (2021) 012011 doi:10.1088/1757-899X/1011/1/012011

The dispersion pattern of PM_{10} and SO_2 on Highway Kuin Utara and Kuin Selatan Banjarmasin City based on GIS spatial model

H Prasetia[1*], N Annisa[2], R Riduan[2], E R Setyowati[2], A N Tasfiyati[1] and R Maryana[1]

[1] Research Center for Chemistry, Indonesian Institute of Sciences (LIPI), Indonesia, Kawasan PUSPIPTEK Serpong, 15314, Tangerang Selatan, Banten
[2] Department of Environmental Engineering, Faculty of Engineering, Lambung Mangkurat University, Indonesia, Jalan A. Yani Km 37.5, 70714, Banjarbaru, Kalimantan Selatan

*Corresponding author: hafi002@lipi.go.id

Abstract. The transportation sector primarily causes urban air pollution in Indonesia and has a significant role in environmental emissions. Emissions resulting from vehicles are particulate matter, SO_2, CO_2, CO, HC, and NO_x. The concentration and dispersion of PM_{10} and SO_2 from the activity of the traffic on the roads Kuin Utara and Kuin Selatan, Banjarmasin City, observed in this study. The dispersion model of the concentration of PM_{10} and SO_2 from traffic activity using GIS Spatial Models has been conducted. Research has been carried out for two days as representing weekend and weekdays at the three stations. After modeling and mapping the highest value for the concentration of PM_{10} on the weekend, it was obtained at station three, which was 167.3 $\mu g.m^{-3}$. The highest level of PM_{10} on weekdays was achieved at station 3 with an amount of 236.3 $\mu g.m^{-3}$. Based on modeling, the highest level SO_2 on weekdays reached in station 3, which was 423 $\mu g.m^{-3}$.

1. Introduction

At this time, the high level of air pollution in the urban environment is a concern of society, especially its negative influence on public health [1-6]. Developing countries have problems that are more complex if compared with developed countries, and this is due to factors of urbanization and the quantity of the dense population [7-9]. Urbanization leads to rapid development in the industrial sector, commercial business activities, transportation, which leads to significant energy consumption [10]. Level PM_{10} and $PM_{2.5}$ have been reported frequently above the established quality threshold [11-12]. Furthermore, the health burden associated with air pollution is the highest predicted in LMICs (low- and middle-income countries) in the Western Pacific Region and Southeast Asia.

Rapid technological development in urban areas is closely related to environmental problems caused, one of which is air pollution [13]. The occurrence of air pollution comes from the industrial sector, transportation, and household activities. The transportation sector is the dominant source of pollution in big cities. The transportation sector is proven to contribute to the highest air pollution in Indonesia, around 85% [14]. The transportation sector is a source of moving pollution, including cars, trucks, buses, motorcycles, planes, and ships. The use of fuel oil by the transportation sector is known to have a significant impact on the environment. According to the Ministry of Environment in 2012, air pollution from motor vehicles contributed 70.5% Carbon Monoxide, 18.34% Hydrocarbons, 8.89% Nitrogen Oxides, 1.33% Particulates, and 0.88% Sulfide Oxides.

The WHO reported that 92% of the world's population had been exposed to air pollution are concentrated, where 90% of the cause of death in LMICs [2] as many as seven out of ten countries in Asia have been reported to have a concentration of PM_{10} that exceeds the WHO average annual standard [15]. Air pollution outside of the room can result in public health is disturbed, and contributes to the risk of premature death in Asia and that about 2.1 million deaths annually [16]. Air pollution outside of space has been known as the ten most significant health risks in the world and among the top five risks in developing countries in Asia [11]. The impact of PM_{10} can reduce lung and heart function, nervous system, and blood vessel disorders that come in contact for several days, weeks, months, even years [17]. Emissions SO_2 can cause an irritating effect on the airways causing symptoms of coughing, shortness of breath, and increasing asthma [18]. Based on data from the S. Parman Community Health Center, Upper Respiratory Tract Infection is the number 2 disease most complained by the community and obtained 2410 people in 2017.

South Kalimantan is a province that has many rivers. River Kuin that crosses The Kuin Utara, Kuin Selatan, and Kuin Cerucuk villages are still living on the banks of the river or above the river. Besides, increasing population growth has led to increased housing and road construction [19]. Based on observations, The Kuin Utara and Kuin Selatan roads are roads leading to tourism, industrial, and residential areas. This road section is crossed by motorized vehicles, cars, and trucks. Increased transportation through highway affects air quality. Also, The Kuin River is the primary access crossed by tourist or commercial motorized boats. Transport of water that separates The road Kuin Utara and Kuin Selatan is also expected to contribute from ship emissions motor. Based on this background and the data centers that show the number of respiratory diseases hence the importance of this study due to PM_{10} and SO_2 in the air can cause disturbances in health, especially breathing. The concentration and dispersion of PM_{10} and SO_2 from the activity of the traffic on the roads Kuin Utara and Kuin Selatan, Banjarmasin City, observed in this study.

2. Materials and methods

2.1. Study area

This research was conducted at Road Kuin Utara, and Kuin Selatan, Banjarmasin City. Sampling was carried out at three stations representing each region. The coordinates of the research location can be seen from Table 1, and visually it can be seen in **Figure 1**.

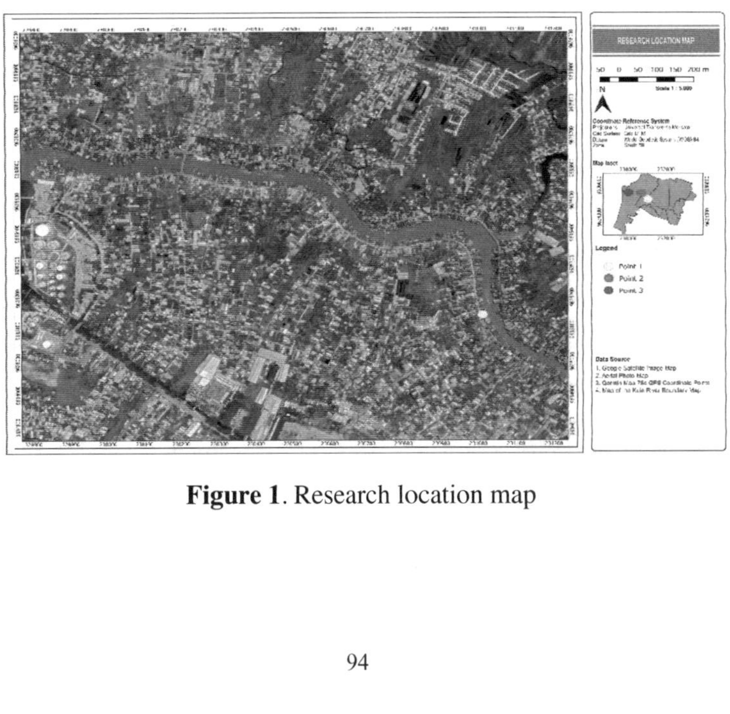

Figure 1. Research location map

Table 1. Research location coordinates

	Location	Coordinates	
		South Latitude	East Longitude
Station I	Road Kuin Selatan Neighbourhood 16	03° 17.879'	114° 34.763'
Station II	Yard of Sultan Suriansah Mosque	03° 17.406'	114° 34.342'
Station III	Industrial Area Pertamina	03° 17.649'	114° 34.216'

2.2. Research design

This research was conducted for two days. This measurement has been carried out for two days, where one day on weekdays representing Monday to Friday, and weekend s serving Saturday and Sunday. PM_{10} concentration measurements were carried out for 24 hours based on SNI 7119.15: 2016 and South Kalimantan Governor Regulation Number 53 (2007). While SO_2 measurements were carried out for 1 hour based on SNI 7119.7: 2005 and South Kalimantan Governor Regulation number 53 the Year 2007. Analysis of SO_2 will be carried out on peak hours morning, afternoon, and evening. This research was conducted at three observation stations.

The calculation of the amount of land transportation that passes is calculated for 24 hours with a span of 1 hour ago converted to get the value of the volume of traffic in passenger car units (PCU). Then the emission rate of PM_{10} concentration is calculated. The amount of water transportation that passes for 24 hours is then calculated the emission rate of the PM_{10} level. As for SO_2, land and water transport will be calculated for 1 hour during the measurement. Determination of peak hours is done in the morning, afternoon, and evening based on the results of counting obtained. Then the results are converted into units of passenger cars that will get traffic volume [20].

2.3. Data analysis

The wind direction obtained from the laboratory results is then plotted into the WRPlot program. The result of WRPlot used is the dominant wind direction that will be inputted on the Caline4 program. Then the results from Caline4 produce a prediction of the concentration at the receptor station to be patterned on the Quantum GIS program. But there are differences for SO_2, after getting the WRPlot results and then entering it into the AERMOD View program. The dispersion pattern directly obtains the results of the AERMOD View program. So the dispersion patterns can be mapped in QGIS [21]. Using QGIS has an advantage that makes it easier to get information that has been processed and stored as attributes of a location or object. The main feature of data that can be utilized in geographic information systems is data that has been tied to the area and is primary data that has not been specified. The geographic Information system is a reliable tool for handling spatial data [22-24].

3. Results and Discussion

3.1. Measurement results concentration of PM_{10} and SO_2

Particulate matter (PM_{10}) is particles of air in the form of solid with a diameter of fewer than 10 micrometers. The existence of those particles in a relatively long time will hover and enter into the human body through the respiratory tract, causing health disorders [14]. PM_{10} can be either non-carcinogenic or carcinogenic. The nature of the carcinogenic PM_{10} generated from the content contained in the $PM_{10,}$ such as PAHS and lead. Carcinogenic properties can be seen from the acute effects and the effects of chronic caused by exposure to PM_{10}. The source of the main pollutant PM_{10} derived from transportation activities [17]. Particulate matter is mostly generated from the presence of residue in the fuel. Such residues are not burnt and wasted through the exhaust pipes because of the processing of the fuel not better [25].

The measurement of the concentration of PM_{10} in this study using the method of gravimetric. The results of the analysis of PM_{10} on the weekend obtained at all points do not exceed the quality standard.

However, on weekdays only in station three, that exceeds the quality standard that is 205.0 μg.m^{-3}. Based on The Governor Decree of South Kalimantan No. 53 (2007) for Maximum Limit Air Quality and Noise, the value PM_{10} of 150 μg.m^{-3}. The results of the measurement of PM_{10} can be seen in Table 2.

Table 2. Research measurement results from concentration of PM_{10}

	Time	Unit	Concentration of PM_{10} Measurement results	Concentration of PM_{10} Maximum Limit
Station I	Weekend	μg.m^{-3}	65.8	150
	Weekdays	μg.m^{-3}	56.4	150
Station II	Weekend	μg.m^{-3}	131.4	150
	Weekdays	μg.m^{-3}	116.2	150
Station III	Weekend	μg.m^{-3}	147.9	150
	Weekdays	μg.m^{-3}	205.0	150
	Average		120.45	150

PM_{10} concentration values in Banjarmasin are lower than in several cities in other countries. Pakistan has the most pollution levels. The PM_{10} concentration value for each country is that Pakistan (2010) has average PM_{10} pollution of 282 μg.m^{-3}, Qatar (2012) has average PM_{10} pollution of 165 μg.m^{-3}, and Bangladesh (2013) has average PM_{10} pollution of 163 μg.m^{-3}[26].

SO_2 is a component of pollutants in the atmosphere obtained from the burning of fossil fuels and other processes that have sulfate content. SO_2 can cause acid rain, which is the accumulation of acidic substances in the air so that it is dangerous for the health of living things. SO_2 has a characteristic odor that is sharp and does not burn. SO_2 can cause an irritating effect on the airways causing symptoms of coughing, shortness of breath, and increasing asthma. SO_2 is a gas that is very soluble in water. SO_2 in the air can dissolve in water vapor, then form acid and fall as acid rain. The impact of acid rain can occur in areas far from SO_2 polluting sources due to the influence of wind [18].

The measurement of the concentration of SO_2 in this study using the method of pararosaniline. The results of the analysis of SO_2 on weekends and weekdays are obtained at all points and do not exceed the quality standard. Based on the Governor decree of South Kalimantan No 53 (2007) for Maximum Limit Air Quality and Noise, the value SO_2 of 900 μg.m^{-3}. The results of the measurement of SO_2 can be seen in Table 3.

Table 3. Research measurement results from concentration of SO_2

	Time	Unit	Concentration of SO_2 Measurement results	Concentration of SO_2 Maximum Limit
Station I	Weekend	μg.m^{-3}	5.4	900
	Weekdays	μg.m^{-3}	5.4	900
Station II	Weekend	μg.m^{-3}	5.4	900
	Weekdays	μg.m^{-3}	5.4	900
Station III	Weekend	μg.m^{-3}	154.1	900
	Weekdays	μg.m^{-3}	5.4	900

Based on the highest SO_2 concentration values from the measurement results, it is known that SO_2 pollution in Banjarmasin City is lower than in several cities in other countries. China has the highest pollution level. The SO_2 concentration value of each country is that China has average SO_2 pollution of 17300 μg.m^{-3}, Malaysia has a common SO_2 infection of 8300 μg.m^{-3}, and South Korea has average SO_2 pollution of 8600 μg.m^{-3}[27].

3.2. The dispersion pattern of the concentration of PM_{10}

Map the dispersion pattern of PM_{10} is obtained from the program Caline4. The visualization of the dispersion pattern of concentrations of PM_{10} using the QGIS program. The QGIS Program used for making contour maps by doing plotting of the data of XYZ to be the grid. The concentration of PM_{10} produced from the pollutants generated by the activity of traffic on a toll road. Figure 2 showed the dispersion pattern of the level PM_{10} on the weekend at station 1 to a maximum was 90.6 $\mu g.m^{-3,}$ and the level of the minimum amount of 63.3 $\mu g.m^{-3}$. At station 1, the direction of the wind towards the south at an average speed of 0,8025 $m.s^{-1}$. Figure 3 shows the dispersion pattern of PM_{10} concentration on weekend s at the maximum station 2 of 146.8 $\mu g.m^{-3}$ and the minimum level of 132.7 $\mu g.m^{-3}$.

At station 2, the wind direction goes east with an average speed of 2.387 $m.s^{-1}$. Figure 4 shows the dispersion pattern of PM_{10} concentration on weekend s at the maximum station 3 of 167.3 $\mu g.m^{-3}$ and the minimum level of 146.7 $\mu g.m^{-3}$. At station 3, the wind direction goes east with an average speed of 0.756 $m.s^{-1}$. The high degree of pollutants is influenced by the direction and speed of the wind, and the volume of traffic. According to Huda, the direction and speed of the wind that blows will affect the spread of concentration. The farther away from the source of pollutants from the receptor station, the level of contaminants will decrease [28].

Figure 2. PM_{10} dispersion patterns station I (weekend)

Figure 3. PM_{10} dispersion patterns station II (weekend)

Figure 4. PM_{10} dispersion patterns station III (weekend)

Figure 5 showed the dispersion pattern of PM_{10} concentration on weekdays at station one maximum of 132.5 $\mu g.m^{-3}$ and a minimum level of 32.5 $\mu g.m^{-3}$. At station 1, the direction of the wind is going west with an average speed of 0.37 $m.s^{-1}$. Figure 6 shows the dispersion pattern of PM_{10} concentration on weekdays at the maximum station 2 of 164.8 $\mu g.m^{-3}$ and the minimum level of 119.5 $\mu g.m^{-3}$. At Station 2, the direction of the wind is heading east with an average speed of 0.707 $m.s^{-1}$. Figure 7 shows the dispersion pattern of PM_{10} concentration on weekdays at the maximum station 3 of 236.3 $\mu g.m^{-3}$ and the minimum level of 203.1 $\mu g.m^{-3}$. At Station 3, the wind direction is heading south with an average speed of 0.292 $m.s^{-1}$. The high concentration of PM_{10} is influenced by the direction and speed of the wind, and the volume of traffic. According to Huda, the direction and speed of the wind that blows will affect the spread of concentration. The farther away from the source of pollutants from the receptor station, the level of PM_{10} will decrease [28]. Based on Appendix E, the partial t-test shows that the wind speed variable has a probability value <0.05. So, it can be concluded that the wind speed variable has a significant effect on PM_{10} concentration. PM_{10} is very dangerous for human health. The existence of PM_{10} in the air is relatively long will enter the human body through breathing, and into the lungs. PB is one of the ingredients of PM_{10}. The presence of Pb in ambient air can cause anemia, increased blood pressure, kidney damage, nervous system disorders, damage the brain, and decrease IQ [14].

Figure 5. PM_{10} dispersion patterns station I (weekdays)

Figure 6. PM_{10} dispersion patterns station II (weekdays)

Figure 7. PM_{10} dispersion patterns station III (weekdays)

3.3. The dispersion pattern of the concentration of SO_2

This SO_2 dispersion pattern map was obtained from the AERMOD View program. Visualization of SO_2 concentration dispersion patterns using the AERMOD View program. The AERMOD View program is used for air quality simulations and can directly create contour maps based on a grid. SO_2 concentrations generated from pollutants generated by traffic activities on a road section. Figure 8 shows the dispersion pattern of the SO_2 level of weekend s at peak daylight hours at the maximum station of 67.5 $\mu g.m^{-3}$ and the minimum concentration of 0.7 $\mu g.m^{-3}$. At station 1, the wind direction goes east with an average speed of 1.98 $m.s^{-1}$. Figure 9 shows the dispersion pattern of the SO_2 concentration of weekend s at peak daylight hours at a maximum station of 38.7 $\mu g.m^{-3}$ and a minimum level of 0.4 $\mu g.m^{-3}$. At station 2, the wind direction goes east with an average speed of 3.49 $m.s^{-1}$. Figure 10 shows the dispersion pattern of the SO_2 concentration of weekend s at peak afternoon hours at the maximum station of 423 $\mu g.m^{-3}$ and the minimum level of 4 $\mu g.m^{-3}$. At station 3, the direction of the wind is heading south with an average speed of 0.81 $m.s^{-1}$.

Figure 8. SO_2 dispersion patterns station I (weekend)

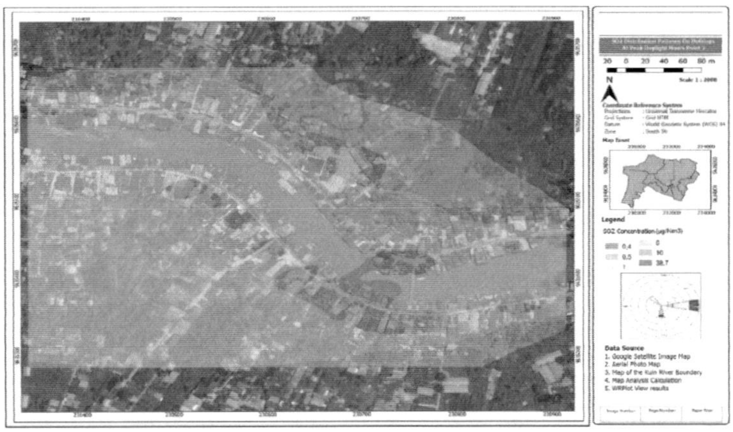

Figure 9. SO_2 dispersion patterns station II (Weekend)

Figure 10. SO_2 dispersion patterns station III (weekend)

The high concentration of pollutants is influenced by the direction, speed wind, and the volume of traffic. According to Huda, the direction and speed of the wind that blows will affect the spread of concentration. The farther away from the source of pollutants from the receptor station, the concentration of pollutants will decrease [28]. Based on Appendix E, the partial t-test shows that the wind speed variable has a probability value <0.05. So, it can be concluded that the wind speed variable has a significant influence on SO_2 concentration. The impact of SO_2 in the air can cause respiratory tract irritation and an increase in mucosal secretions. SO_2 emissions are known as substances that are harmful to health, especially in patients with chronic respiratory diseases and the elderly. At very high concentrations can cause death [29-30].

4. Conclusion

The results of direct measurements on the Kuin Utara and Kuin Selatan highway sections obtained only PM_{10} concentrations that exceeded the quality standard station three, which was 167.3 $\mu g.m^{-3}$. The highest level of PM_{10} on weekdays was achieved at station 3 with an amount of 236.3 $\mu g.m^{-3}$. The results of the mapping dispersion pattern of concentration of SO_2 obtained in a station three at the peak hours of the day to a maximum was 423 $\mu g.m^{-3}$. Whereas on immediate measures of SO_2 level, nothing exceeded the quality standards based on South Kalimantan Governor Regulation No. 53 (2007). Although the SO_2 level is small, however, if exposed for long periods will cause changes in lung function.

References

[1] Biswas P *et al.* 2006 Health effects of fine particulate air pollution: Lines that connect *J. Air Waste Manag. Assoc.* **56**(6) pp. 707–8
[2] World Health Organization 2014 Methods for the burden of disease attributable to 795 ambient air pollution for the year 2012 Acessed online from: http://www.who.int/phe/health_topics /outdoorair/databases/AAP_BoD_methods_March2014.pdf?ua=1
[3] Hankey S and Marshall J 2017 Urban Form, Air Pollution, and Health *Curr. Environ. Heal. Reports* **4** (4) 491–503
[4] Wang Q 2018 Urbanization and global health: The role of air pollution *Iran. J. Public Health,* **47**(11) 1644–52
[5] Schiavon M *et al.* 2015 Assessing the air quality impact of nitrogen oxides and benzene from road traffic and domestic heating and the associated cancer risk in an urban area of Verona (Italy)," *Atmos. Environ.* **120** 234–43
[6] Lancet Planet Health 2019 The impact of air pollution on deaths, disease burden, and life expectancy across the states of India: the Global Burden of Disease Study 2017 *Lancet* **18** 30261–4
[7] Baldasano J M, Valera E and Jimenez P 2003 Air quality data from large cities *Sci. Total Environ.* **307** 141–65
[8] UN 2007 World Urbanization Prospects: The 2007 Revision Population Database New York
[9] Mannucci P M and Franchini M 2017 Health effects of ambient air pollution in developing countries *Int. J. Environ. Res. Public Health* **14**(9) 1–8
[10] Gulia S., Nagendra S M S , Khare M and Khanna I 2015 Urban air quality management–A review *Atmos. Pollut. Res.* **6** 286–304
[11] ADB 2014 Improving Air Quality Monitoring in Asia: A Good Practice Guidance, Final report, Asian Development Bank, Mandaluyong City, Philippines: Asian Development Bank, 2013 Accesed online from: http://cleanairasia.org/wp-content/uploads/portal/files/improving_ aqmt_in_asia.pdf.
[12] Worobiec A., Vermaak S S P, Berghmans P, Winkler H, Burger R and Van Grieken R 2011 Air Particulate Emissions in Developing Countries: A Case Study in South Africa *Anal. Lett.* **44**(11) 1907–24
[13] Bachtiar V S 2013 Studi Paparan Konsentrasi Gas Karbonmonoksida (CO) di Lingkungan Kerja Petugas Parkir Dan Polisi Lalu Lintas Di Kota Padang *J. Dampak* **10**(1) 60–72
[14] Roza V, Ilza M dan Anita S 2015 Korelasi Konsentrasi Particulate Matter (PM10) di Udara dan

Kandungan Timbal (Pb) dalam Rambut Petugas SPBU di Kota Pekanbaru *Din. Lingkung.* **2**(1) 52–60

[15] CAA (Clean Air Asia) 2016 Guidance Framework for Better Air Quality in Asian Cities. Pasig City, Philippines Accesed online from: http://cleanairasia.org/ibaq/wp-content/themes/ibaq/pdf/Ambient Air Quality Standards and Monitor.pdf.

[16] Lim S S *et al.* 2012 A comparative risk assessment of burden of disease and injury attributable to 67 risk factors and risk factor clusters in 21 regions, 1990-2010: A systematic analysis for the Global Burden of Disease Study 2010 *Lancet* **380**(9859) 2224–60

[17] Wulandari A, Darundiati Y H and Raharjo M 2016 Analisis Risiko Kesehatan Lingkungan Pajanan Particulate Matter (PM10) pada Pedagang Kaki Lima Akibat Aktivitas Transportasi (Studi Kasus: Jalan Kaligawe Kota Semarang) *J. Kesehat. Masy.* **4**(3) 677–91

[18] Muziansyah D, Sulistyorini R and Sebayang S 2015 Model Emisi Gas Buangan Kendaraan Bermotor Akibat Aktivitas Transportasi (Studi Kasus: Terminal Pasar Bawah Ramayana Koita Bandar *J. Rekayasa Sipil Dan Desain* **3**(1) 57–70

[19] Rochgiyanti 2011 Fungsi Sungai Bagi Masyarakat di Tepian Sungai Kuin Kota Banjarmasin *Komunitas* **3**(1) 51–9

[20] Arisnawati R S, Riduan R and Annisa N 2016 Pemetaan Distribusi Konsentrasi Karbon Monoksida (CO) Dihubungkan Dengan Aktivitas Kendaraan Bermotor di Kampus Universitas Lambung Mangkurat Banjarbaru *Univ. Lambung Mangkurat. Banjarbaru*

[21] Prahasta E 2002 *Konsep-konsep Dasar Sistem Informasi Geografis* (Bandung: Informatika)

[22] Dulbahri 1993 *Sistem Informasi Geografis* (Jakarta: Gramedia)

[23] Riduan R, Sasmalini, Prasetia H and Annisa N 2019 Evaluation of Tidal Swampland Suitability Based on GIS Spatial Model on Barambai Reclamation Unit, South Kalimantan *MATEC Web Conf.* **280** 05020

[24] Annisa N, Prasetia H and Riduan R 2019 Green Configuration-Based GIS Spatial Model in Riparian Area of the River Kuin Banjarmasin, Indonesia *MATEC Web Conf.* **280** 04006

[25] Purwanto C, Arthana I W and Suarna I W 2015 Inventarisasi Emisi Sumber Bergerak di Jalan (On Road) Kota Denpasar *ECOTROPHIC J. Ilmu Lingkung. (Journal Environ. Sci.*, **9**(1) 1–9

[26] Rodriguez-Urrego D and Rodriguez-Urrego L 2020 Air quality during the COVID-19: $PM_{2.5}$ analysis in the 50 most polluted capital cities in the world *Environ. Poll.* **266**(1) 115042

[27] Susanto J P 2005 Kualitas udara beberapa kota di asia *J.Tek.Ling* **6**(1) 324–9

[28] Huda S N 2017 *Pengaruh Lalu Lintas Terhadap Sebaran PM10 di SDN 1 Syamsudin Noor Kota Banjarbaru* (Banjarbaru: Universitas Lambung Mangkurat)

[29] Damri, Ilza M and Afandi D 2016 Analisis Paparan CO Dan SO2 Pada Petugas Parkir di Basement Mall Ska di Kota Pekanbaru *Din. Lingkung.* **3**(1) 42–7

[30] Turyanti A 2016 *Pemodelan Dispersi PM10 Dan SO2 Dengan Pendekatan Dinamika Stabilitas Atmosfer Di Lapisan Perbatas Pada Kawasan Industri.* (Bogor: Institut Pertanian Bogor)

The stability study of electrolytic conductivity synthetic wastewater

Fransiska Sri Herwahyu Krismastuti

Research Centre for Chemistry – Indonesian Institute of Sciences, Building 452, Kawasan PUSPIPTEK, Serpong, Tangerang Selatan, Banten 15314, Indonesia

fransiska.krismastuti@gmail.com, fran008@lipi.go.id

Abstract. The accurate and traceable measurement of electrochemical parameters, such as electrolytic conductivity, in water environment is crucial for water quality monitoring. This can be achieved by the use of metrological traceable certified reference material produced according to ISO 17034. It requires stability test of the reference material candidate as part of the certification for a new reference material. This study is focused on monitoring of the stability of electrolytic conductivity synthetic wastewater, as an example of complex solution containing high concentration of ions. The stability study is focused on the material of the packaging bottle, the shelf-life duration and the high temperature representing a shipping condition. The results demonstrate that the electrolytic conductivity synthetic wastewater is stable for all the conditions tested, but the synthetic wastewater packaged in glass bottle has longer shelf-life. The results confirmed that the synthetic wastewater has potential to be developed as a certified reference material.

1. Introduction

Electrolytic conductivity is one of water quality standards that has been widely used in various field including monitoring of the aquatic environment. It measures of how well a solution conducting electricity or estimates the total concentration of ionized substance in an electrolyte solution [1, 2]. It has a symbol of $\varkappa$ with the unit of S m^{-1}. The value of $\varkappa$ is obtained from a measurement of solution resistance (R_S) of a homogenous, isotropic electrolyte placed between electrodes in a conductivity cell under specified conditions [1]. The electrolytic conductivity value has a relationship with the total dissolved solid (TDS), which has a symbol of γ [3, 4]. The ratio between TDS and electrolytic conductivity, namely conversion factor (K), defines the concentration of ions indicating roughly the concentration of major ions and dissolved solids (non-conductive materials) in the water [3-5]. Therefore, the value of conversion factor can be different depending on the different waters.

For the routine measurement for electrolytic conductivity and/or TDS, a conductivity meter is an easy and simple instrument but it can produce an accurate and reliable result. This meter is equipped with a probe, usually handheld, having two electrodes inside. After the probe is placed in the liquid to be measured, the meter applies voltage between two electrodes inside the probe. Electrical resistance from the solution causes a drop-in voltage, which is read by the meter. The meter converts this reading to milli- or microSiemens per centimetre.

In order to get accurate and reliable results, it is necessary to calibrate the conductivity meter regularly using certified reference materials (CRM) with established traceability to the International System of Unit (SI). Metrological traceability of the measurement results is crucial to fulfil the legal requirement and to ensure the required measurement correctness with consideration to the expanded measurement uncertainty [6, 7]. Besides using traceable CRM, metrological traceability can also be ensured by taking part in an interlaboratory comparison, such as bilateral comparison or key comparison [6].

The needs for new CRMs are growing and become more specific as laboratories raise their awareness. This emphasises on how representative the reference materials are. ISO 17034:2016 describes the requirements for CRMs production including the assessments required in order to certify a reference material into a CRM, such as assessment of homogeneity, stability and property values [8]. The homogeneity assessment includes the homogeneity in the final packaging and assessment of its contribution to the uncertainty. The stability assessment monitors the stability and its uncertainty due to the effect of storage and transportation, while the property value includes identification of contributions to uncertainty [8]. The specific guidance on technical issue of the assessment of homogeneity, stability and characterization for the certification of reference materials is explained in ISO Guide 35:2017 [9].

In the water environment, monitoring of industrial wastewater plays an essential role for environmental protection to ensure process efficiency and to verify final quality of the effluent before discharging the wastewater to the environment. Reliable results for determination of electrolytic conductivity in wastewater are important for checking the performance of industrial wastewater treatment and detecting possible sources of industrial contamination. It needs accurate and reliable analytical results as well as the use of CRMs in order to draw valid conclusions. The CRMs are used to achieve traceability in chemical measurement, to calibrate the instruments, to validate the methods, to monitor laboratory performance and to permit comparison of methods [10].

This study is focus on investigating the stability of synthetic wastewater, as part of the assessment of a reference materials candidate. Until now, there is lack of CRMs for electrolytic conductivity measurement in complex or matrix solution. The only CRMs matrices for electrolytic conductivity measurement is the one in bioethanol matrices [11]. Thus, this study is important as preliminary research in developing CRMs in synthetic wastewater matrix. For this purpose, the synthetic wastewater was designed mimicking the real wastewater but without any biological compound. The feasibility of this synthetic wastewater has been studied before confirming that this composition had an electrolytic conductivity value of 7.430 mS cm^{-1} and a conversion factor of 0.58 which was relevant to the real wastewater [12]. In this study, the stability of the synthetic wastewater is further assessed by considering the effect of the container used to package the synthetic wastewater, the shelf-life or storage duration and the transport condition.

2. Experimental section

2.1. Materials

Sodium hydrogen carbonate ($NaHCO_3$), potassium chloride (KCl), sodium chloride (NaCl) and hydrogen chloride (HCl) were purchased from Merck. All the chemicals are reagent grade.

2.2. Methods

The synthetic wastewater was prepared from 60 mg of $NaHCO_3$, 14 mg of KCl, 92 mg of NaCl and 2 mL of HCl 3 % for 1000 mL solution, as studied before [12]. The solution was stirred overnight to homogenize the mixture before it was transferred into the clean and dry container. There are two types container used; clear glass screw-cap bottles and high-density poly ethylene (HDPE) bottles. All containers were thoroughly cleaned prior using. Upon received, the bottles were rinsed using deionized water, filled with the deionized water and capped. These filled bottles were kept overnight before being emptied and dried at the drying oven for overnight. The synthetic wastewater was filled into the clean and dry bottles and then stored away from daylight at a room temperature. The stability of electrolytic conductivity wastewater was checked in both types of bottle used to storage the solution. The shelf-life was determined by measuring the electrolytic conductivity synthetic wastewater at 25 °C for the storage duration of 0, 1, 3, 6, 9, 12, 18 and 24 months after the production date. The stability of electrolytic conductivity synthetic wastewater was also tested at higher temperature (40 °C) representing the stability of the solution during shipping or other transportation condition.

The electrolytic conductivity was measured using conductivity meter LAQUA F-74BW (Horiba) which has been calibrated using certified reference materials (IDNRM-MC-2-002) produced by Chemical Metrology Laboratory with the conductivity value of 6.67 mS cm^{-1} and uncertainty of 0.03 mS cm^{-1} at 25 °C. This reference materials were traceable to International Standard through CRM 1603 produced by Danish National Metrology Institute (DFM).

3. Results and discussion

The synthetic wastewater used in this study would be developed for a reference material candidate thus the stability assessment was conducted according to ISO 17034:2016 by considering the effect of the container used to package the synthetic wastewater, the shelf-life and the transport condition. The synthetic wastewater contained conductive ions and it was prepared from sodium hydrogen carbonate, potassium chloride, sodium chloride and hydrogen chloride [12]. The hydrogen chloride was added into the mixture because it could suppress carbon dioxide dissociation which is crucial especially for low electrolytic conductivity [13]. This synthetic wastewater was designed mimicking the real industrial wastewater in the absence of microbial, pathogenic compound and inorganic toxic metal for the safety reason. The synthetic wastewater was also free form any organic and volatile substances to maintain the solution stability during certain period of time and prevent the unnecessary biofouling [14].

The synthetic wastewater, as reference material candidate, were designed and produced with the electrolytic conductivity value of 7.430 mS cm^{-1}. Two types of container, clear glass and HDPE bottles, were prepared to package the synthetic wastewater. The reference value for electrolytic conductivity synthetic wastewater stored in each type of container was then assigned, as presented in Table 1. It shows the reference value and the expanded uncertainty for electrolytic conductivity synthetic wastewater in each type of container was still in agreement with the designed electrolytic conductivity value. The expanded uncertainty was calculated by considering the sources of uncertainty.

Table 1. The reference electrolytic conductivity value of synthetic wastewater packaged at different container and measured at 25 °C.

Container type	Reference value (mS cm^{-1}) ± expanded uncertainty (*k*=2)
Glass bottle	7.431 ± 0.0037
HDPE bottle	7.430 ± 0.0039

All bottles containing synthetic wastewater were stored at room temperature and protected from direct sunlight. The electrolytic conductivity synthetic wastewater in each type bottle was measured at 25 °C as a function of storage time (0, 1, 3, 6, 9, 12, 18 and 24 months), as presented in Figure 1. This figure includes horizontal dotted lines (blue lines in Figure 1) showing the expanded uncertainty (U_{RM}, k=2) assigned to the reference value of the electrolytic conductivity synthetic wastewater filled in the glass and HDPE bottle. Figure 1 shows that the electrolytic conductivity value for the synthetic wastewater packaged in glass and HDPE bottles is in between the reference value with its expanded uncertainty (inside the area of horizontal dotted lines). After 12 months of storage, the electrolytic conductivity value synthetic wastewater packaged in both types of bottle starts to increase closer the upper limit of expanded uncertainty assigned to its reference value. It could be affected by the instability of the electrolytic conductivity synthetic wastewater filled in the bottles. Therefore, based on the type of container used to store the synthetic wastewater, it confirms that both types of container are suitable for packaging of the synthetic wastewater because the electrolytic conductivity values are adequately stable during the testing period.

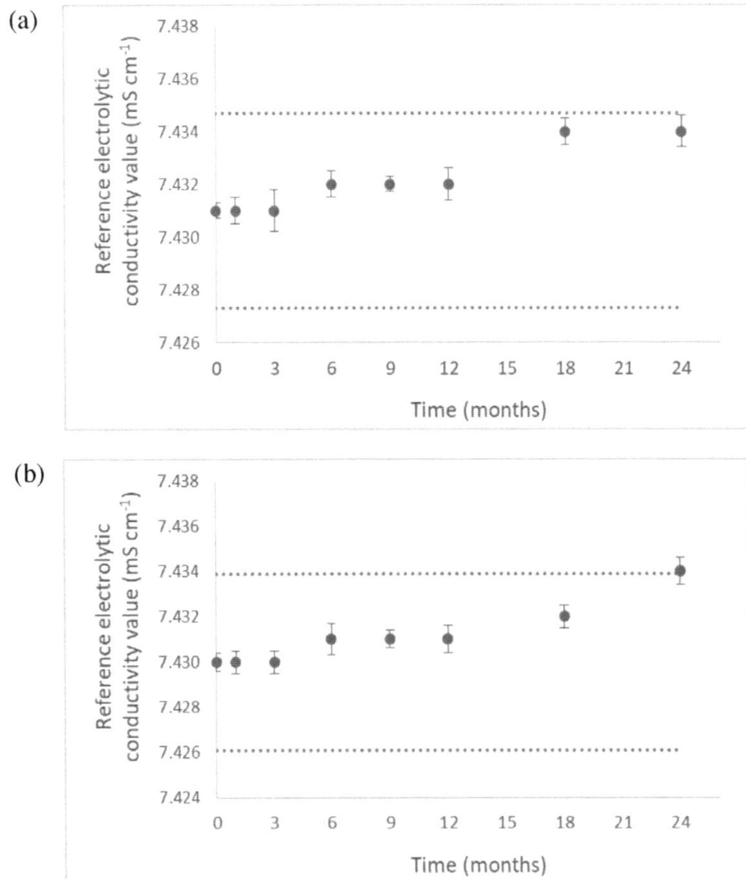

Figure 1. The electrolytic conductivity synthetic wastewater packaged in glass bottle (a) and HDPE bottle (b) in a function of time (0, 1, 3, 6, 9, 12, 18 and 24 months) measured at 25 °C. The red dots are the electrolytic conductivity value with the standard deviation as the error bars and the blue dotted lines are the reference values with their expanded uncertainties (k=2).

The stability of electrolytic conductivity synthetic wastewater packaged in the glass and HDPE bottles as a function of time was also tested at the temperature of 40 °C. This temperature is considered representing the elevated temperature that potentially encountered during transportation. The stability during transportation can be influenced by some factors, such as temperature and type of transportation used. However, this study is only focused on the temperature during transportation. This result can be used as a recommendation for transporting the reference material packaged in certain type of container. The stability at the elevated temperature (40 °C) for both glass and HDPE bottles as a function of time is presented in Figure 2. Similar to Figure 1, this figure also includes horizontal dotted lines (blue lines in Figure 2) showing the expanded uncertainty (U_{RM}, k=2) assigned to the reference value of the electrolytic conductivity synthetic wastewater filled in the glass and HDPE bottles.

(a)

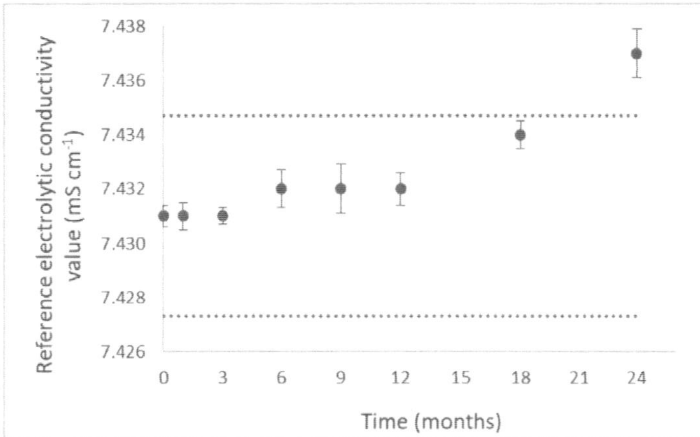

(b)

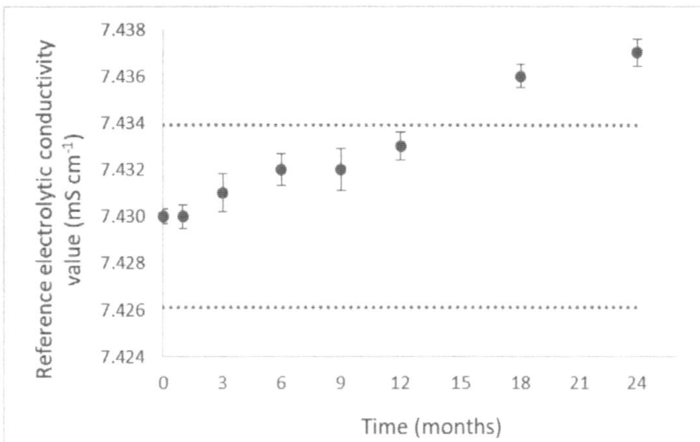

Figure 2. The electrolytic conductivity synthetic wastewater stored in glass bottle (a) and HDPE bottle (b) in a function of time (0, 1, 3, 6, 9, 12, 18 and 24 months) measured at 40 °C. The red dots are the electrolytic conductivity value with the standard deviation as the error bars and the blue dotted lines are the reference values with their expanded uncertainties (k=2).

The electrolytic conductivity value for the synthetic wastewater packaged in both type of bottles and measured at elevated temperature has a different trend with the electrolytic conductivity synthetic wastewater measured at 25 °C. As shown in Figure 2, not all of the reference values of electrolytic conductivity (red dots) are in between the area of the reference value and its expanded uncertainty (blue dotted lines). When measured at 40 °C, the synthetic wastewater packaged in the glass and HDPE bottle have the electrolytic conductivity value above the reference value and its expanded uncertainty after 24 months and 18 months of storage, respectively. It shows that the synthetic wastewater at the elevated temperature would have longer shelf-life if packaged in the glass bottle than the one packaged in the HDPE bottle confirming the glass bottle is a more ideal container for packaging the reference material of the electrolytic conductivity.

Figure 3 summarises the percentage of the relative change in electrolytic conductivity synthetic wastewater over the time at different temperature and packaging container.

(a)

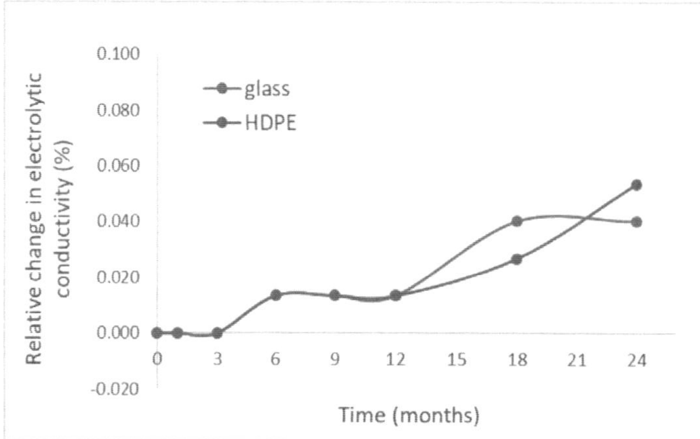

(b)

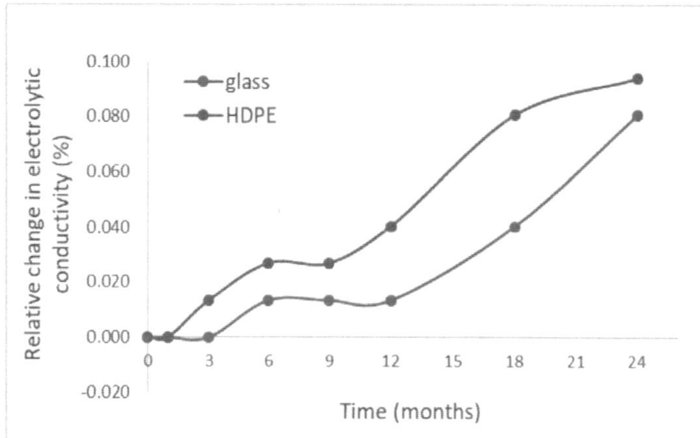

Figure 3. Relative change in electrolytic conductivity synthetic wastewater as a function of time packaged in different container, glass bottle (blue traces) and HDPE bottle (red traces), and measured at different temperature, 25 °C (a) and 40 °C (b).

Figure 3 shows the effect of temperature in the electrolytic conductivity of the synthetic wastewater. In general, the relative change in the electrolytic conductivity measured at elevated temperature (Figure 3(b)) is higher compared to the electrolytic conductivity measured at room temperature (Figure 3(a)) confirming that temperature is a sensitive parameter in electrolytic conductivity measurement. According to the previous reported study, increasing one degree Celsius increases the conductivity value about 1.5 % up to 5 % [13, 15]. When the electrolytic conductivity sample measured at 25 °C (Figure 3(a)), the relative change in electrolytic conductivity for the wastewater packaged in the glass and HDPE container is almost the same, until about 12 months shelf-life. However, at 40 °C (Figure 3(b)), the relative change in electrolytic conductivity for the wastewater stored in HDPE bottle is higher than the electrolytic conductivity of the wastewater packaged in glass bottle. It confirms that the glass container is a more ideal container for packaging the synthetic wastewater as a reference material candidate.

4. Conclusion

The stability study of electrolytic conductivity synthetic wastewater as a reference material candidate has successfully conducted according to ISO 17034:2016. The stability study was conducted in two types of container used as packaging of the synthesis wastewater; the clear glass screw-cap bottles and the HDPE bottles. The stability of electrolytic conductivity synthetic wastewater in both containers was monitored for 24 months with the time interval of 0, 1, 3, 6, 9, 12, 18 and 24 months at two different temperature; 25 °C and 40 °C representing the storage and transport temperature, respectively. The results confirm that both types of container are suitable for using as reference material packaging but having different shelf-life. The reference material candidate stored in glass bottle has better stability in higher temperature compared to the reference material candidate stored in HDPE bottle. Therefore, it is recommended to use the glass bottle as the packaging container for the reference material of electrolytic conductivity.

References

[1] Máriássy M, Pratt K W and Spitzer P 2009 *Metrologia* **46**: 199-213.
[2] Orrù E. 2014 Traceability of Electrolytic Conductivity Measurements for Ultra Pure Water. Giugno: Politecnico Di Torino.
[3] Ali N S, Mo K and Kim M 2012 *KSCE Journal of Civil Engineering* 16: 708-713.
[4] Walton NRG 1989 *Desalination* 72: 275-292.
[5] Hubert E and Wolkersdorfer C 2015 *African Journals Online* 41: 490-500.
[6] Breuel U, Werner B and Jehnert D 2009 *Chimia* 63: 643-646.
[7] Asakai T, Maksimov I, Onuma S, Suzuki T, Miura T and Hioki A 2017 *Accreditation and Quality Assurance* **22**.
[8] ISO 2016 ISO 17034:2016 General requirements for the competence of reference material producers. Genewa: International Organization for Standardization.
[9] ISO 2017 ISO Guide 35:2017 Reference Materials - Guidance for Characterization and Assessment of Homogeneity and Stability. International Organization for Standardization. Genewa.
[10] Camara C, Madrid Y, Rebollo C, Azcarate J, Kramer G N, Gawlik B M, Lamberty A and Quevauviller P 2004 *Trends in Analytical Chemistry* **23**: 194-202.
[11] Fraga I C S, Ribeiro C M, Sobral S P, Dias J C, Goncalves M A, Borges P P and Gonzaga F B 2012 *Talanta* **99**: 99-103.
[12] Krismastuti F S H and Hamim N 2019 *Accreditation and Quality Assurance* **24**: 437-441.
[13] Shreiner R H, Pratt K W 2004 *National Institute of Standards and Technology Special Publication* **260-142**: 1-31.
[14] Mortula M and Shabani S, editors. 2012 Removal of TDS and BOD from Synthetic Industrial Wastewater via Adsorption. 2012 International Conference on Environmental, Biomedical and Biotechnology; Singapore: IACSIT Press.
[15] Jameel R H, Wu Y C and Pratt K W. 2000 *Primary Standards and Standard Reference Materials for Electrolytic Conductivity*. Ch. Chapter (US Government Printing Office: Washington).

Potential of carbonized rice husk as a filter media rain garden to decrease the turbidity of water and Coli bacteria in the Stormwater Runoff. a review of current research

N Annisa[1*], H Prasetia[2] and Q Sholihah[3]

[1]Department of Environmental Engineering, Faculty of Engineering, Lambung Mangkurat University, Jalan A. Yani Km 37.5, Banjarbaru 70714, Kalimantan Selatan, Indonesia
[2]Research Center for Chemistry, Indonesian Institute of Sciences (LIPI), Kawasan PUSPIPTEK Serpong, 15314, Tangerang Selatan, Banten
[3]Department of Industrial Engineering, Faculty of Engineering, Brawijaya University, Jalan. MT Haryono 167, Kec Lowokwaru 65145, Malang, Indonesia

*Corresponding author: aiyuvasha@ulm.ac.id

Abstract. The rapid development of physical infrastructure and infrastructure activities in several metropolitans in Indonesia, followed by industrial, transportation, construction, trade, and office canters, have had an impact on the environment. Such facilities cause a reduction in a green open space. The existence of high rainfall causes uncontrolled stormwater run-off. As a result of the massive floods that frequently happen. The problem that often found that stormwater run-off tends to result in high turbidity and pollution of Coli bacteria, so processing of the treatment is necessary. One of the methods of maintaining the quality of water can use rain gardens that can be made in a run-off water source, by using carbonized rice husk filters. This research method was carried out with the technique of literature review, such as national and international journals, books, websites, and other documents related to the topic of study. Based on the results of the study, carbonized rice husks were able to decrease turbidity and Coli bacteria with decent removal efficiency approaching 90%. The carbonized rice husk becomes an alternative for the media filter rain garden to decrease the value of turbidity and the number of Coli bacteria, as it has large pores and has a content of high Silica (Si).

1. Introduction

The rapid development of physical infrastructure and infrastructure activities in several metropolitan areas in Indonesia, followed by industrial, transportation, construction, trade, and office centers, has had an impact on the environment. The installation also resulted in a decreased number of GOS (green open space) cities. The lack of GOS has led to a decline in the quality of the environment marked by increased air temperature, decreased soil water, flooding or inundation, dirt surface degradation, the intrusion of seawater, coastal abrasion, and so on (1)(2).

The unpleasant urban atmosphere makes it uncomfortable for urban communities living in the cities, so it decreases productivity and the welfare of human populations living in the cities (3)(4). The conditions of the environment of the urban areas become increasingly severe when the rain fell. The average rainfall passed down by the tropical region is very dense and, in high intensity, were more than 2,000 mm.year^{-1} (5). Only two hours of rain can cause a prevent puddle, even flooding. The reason is that people are only able to rely on water canals as stormwater run-off. The problem is, most of the drains didn't function properly because they're clogged with waste or because sediments never dredged. Recently, communities have been introduced with stormwater absorption techniques to ease the water canal. Such absorption technique included the production of bio pores, infiltration wells, conblock, or grass block. These techniques can be applied to individual courtyards. But the problem is, these techniques still have flaws, Bio pores, infiltration wells, conblock, and grass block can only help absorb stormwater, but not for dangerous substances brought by rain. Hazardous materials like this, if incorporated into the soil, can interfere with the fertility of the ground. If carried away by the channels of water heading into the canal, the river and then the ocean can also harm the ecosystem.

Water is an essential environmental component of life. Communities also use water functions besides as constituent to live creatures for household needs such as bathing, washing, cooking food, and other household needs. Water is an essential resource for human life, both for consumption and other purposes. However, there is very little water supply due to many polluted water resources (6). The daily needs of Indonesian people generally use water resources in the environment, such as rivers, wells, and swamps. They do this due to limited associated water support facilities. The limited supply of qualified freshwater needs to be appropriate technologies adapted following the environment for manufacturing the water system (7).

Often the problems are found with stormwater run-off received that they carry pollutants or even chemicals that are dangerous for people such as copper, cadmium, chrome, tin, and zinc, which are dissolved in stormwater. Besides, the number of Coli bacteria tends to be high (8). In other studies, the concentration of pollutants on rainfall in urban areas holds up to 35-1500 mg. L^{-1} total suspended (TSS), 1-6 mg. L^{-1} total nitrogen, 0.6-3 mg. L^{-1} the total phosphorus, and 50-1000 μg. L^{-1} heavy metal (such as Cu, Pb, and Zn) (9)(10). The presence of herbicides, such as glyphosate (N-phosphonomethyl glycine) in stormwater run-off urban, also many have already reported (11)(12)(13)(14). The rain garden is one of the methods to manage the run-off of the rain that can be applied near the drainage of stormwater. The technology of rain gardens is also referred to as a biofilter or which has multifunction in maintaining the environment (15). This paper will be discussed how to manage the run-off of the stormwater using the rain garden method and the effectiveness of the carbonized rice husk as the alternative filter media for a rain garden.

2. Materials and methods

2.1. Rain Garden Concept

The rain garden is a natural expanse like a garden comprising a combination of soil and leaves and plants. A rain garden is also called a bioretention area, in which it's designed to hold stormwater filters temporarily, and it helps the infiltration and evaporation process. A rain garden is designed to allocate a quick infiltration using an accelerated penetration. Rain gardened on the scale of households, no natural engineering, and used the original land (4).

Rain is designed to stream fast run-off by infiltrating the accelerated infiltration. Rain gardens are made on the household scale, with no natural engineering and use of native land. Using original native species plants in the region that contain aesthetics, and assist in the process with an inclusion zone as macropore. Rain garden not only beautifies a space but has great benefits for the environment. To create a rain garden involves several things, hydrological cycle, layout, conservation of resources, creation of artificial habitat, psychologist, soil chemical, horticulture, surface design, and ecology.

The process that occurs in the rain garden system should be understood so that this technology can be used optimally. The primary method in rain garden for local rain began with the intersection, namely the process of trapped the stormwater by the plant leaf and the closing layer, thus slowing or decreasing the flow of the surface. The second process is infiltration. This is the primary process in the rain garden, both those with underdrains and those with no channels. The sediment will occur as a result of the slow stream in bioretention so that particles in the water will be left at the surface of the rain garden. Absorption is the process of water containment in the space between the soil particles that will then be absorbed by the roots of the plants. The next process is evapotranspiration occurring in a rain garden and will turn some of the run-offs into water vapors (16). This development causes changes to the landscape from natural to more rigid and hard. It causes stormwater that drops in this area to be unabsorbed and to become run-off water that will disturb the surrounding environment—one way to be able to manage the stormwater by developing a garden in the location. The Rain garden is made commonly and does not require special preparation or treatment (4).

Besides, according to the previous studies, the effects of the better plant election are expected to improve the effectiveness of the rain garden. The election of the plants is based on the soil conditions, sunlight exposures, and other conditions around the rain garden. Some plants can grow wholly with full sunlight, some sun lighting, or even just being able to live in the shade. Other than that, we should also consider plant capacity in tolerating droughts. Selected plants should be able to survive in dry circumstances or drought season. In that season, the plant needs a livelihood to be able to live, growth and crop size. Consider the height of the plants, the spread, and the root system. The layout of the plant should not disrupt any other facility, such as a buried pipe, cable, or disposal tank. If placed near the road or car park, consider the rooting problem, do not case the problem resulting from the root that appears to surface (17)(18).

2.2. Hydrological Cycles and Run-Off

The hydrological cycle is the circulation of water from the ocean to the atmosphere and then back to the Earth and forth. The water from the sea level evaporates into the air, moves and upwards to the atmosphere. Then condenses and turned into a cloud-shaped water point and subsequently plunged into the Earth and the seas as rain. The rain that fell to Earth was partly held up by the plants, and some others soaked into the ground. If the land had become saturated, then the water would flow on the soil surface, which filled with basins, lakes, rivers, and back to the sea (19).

The flow of water can enter its soil via soil pores. This process is called infiltration. In the land media, the direction of water flows laterally as an interflow, which leads to the spring, lake, and river, or vertically, which is known as percolation to the ground. The gestures on the soil media through the pores are affected by the gravitational force and capillaries. The gravitational force causes the flow always to go lower places, while the capillary style causes the water to move in all directions. If the intensity of the rain falling on a DAS exceeds the infiltration capacity, after the infiltration rate was filled, the water would fill up gaps on the ground's surface. After the basins are full, further water will flow (abundant) on the surface of the land (20).

Concentration time is the time required for water to flow from the farthest point in the catchment are to the end of a view (control point). By the time of this concentration, all of the catchment areas contributed to the stream at the control point. Concentration time depends on catch area characteristics, land use, and water track distance from the farthest point to the end of being reviewed. Stormwater, which falls across the ocean areas, will be concentrated (flowing into) a control point. Stormwater, which hit the entire captured area, was running as a run-off of the surface, which then went into smaller channels and then joined the larger chains and then concentrated at the control point (20).

When the rain falls in the area that nature is of land covered in trees, reeds, and grass on the territory of forest or plantation then the storm is at first relative to just wet the foliage (canopy storage) and after that twigs of trees that began soaked then the branches and the tree trunk (streamflow). The water will flow continues to moisten the dry leaves that have fallen, then moisten the grass. After all, the relatively new wet the surface of the soil begins to moisten. Once the land starts to wet, then seep into the ground until it becomes saturated groundwater flow or surface water run-off starts to occur (4). Surface run-off is stormwater that drains through thin layers above the surface of the land, going into trenches and sewers that eventually join into river water and eventually become a stream. In the mountainous area (the upstream part of the watershed), The surface run-off sprawl could sink into the river quickly, which could cause the river to increase. If the stream is larger than the river's capacity to flow into the flows, there will be a spillover on the river's banks and cause flooding (15).

2.3. The Requirement of the Water Quality

The requirement of the water quality is right according to the Regulation of the Minister of Health of the Republic of Indonesia Number 32 the Year 2017 About the Standard of the Health of the Environment and Health Requirements of Water For purposes of Hygiene and Sanitation, Swimming Pool, Solus Per Aqua, and the Public Baths are aware that for the media the water used for Hygiene and Sanitation includes the parameters of the physical, biological, and chemical which can be in the form a mandatory parameter and additional parameters (Table 1). A mandatory Parameter is a parameter that should be checked periodically following the provisions of the legislation. In contrast, other parameters are only required to be checked if the conditions of geohydrology indicate the presence of potential contamination associated with the additional parameters. Water for Hygiene, Sanitation, are used for the maintenance of personal hygiene such as a bath and brush your teeth, as well as for purposes of washing food, eating utensils, and clothing. Table 2 contains a list of mandatory parameters for biological parameters, which should be examined for sanitary hygiene purposes, which include total coliform and Coli bacteria (colony-forming units in 100 ml of water samples).

Table 1. The physical parameters in the quality standard of the health environment.

No.	Required Parameter	Unit	Quality Standards (Maximum Level)
1.	Turbidity	NTU	25
2.	Colour	TCU	50
3.	Total dissolved solids	mg. L^{-1}	1000
4.	Temperature	^{o}C	Air temperature ± 3
5.	Taste	-	No Taste
6.	Odour	-	No odour

Table 2. The biology parameters in the quality standard of the health environment.

No.	Required Parameter	Unit	Quality standards (maximum level)
1.	Total coliform	CFU.$100ml^{-1}$	50
2.	Coli bacteria	CFU.$100ml^{-1}$	0

2.4. Data Analysis

The method of this research was conducted with the technique of Literature reviews such as national and international journals, books, websites, and other documents related to the topic of study. This search focused on the keywords provided by Coli bacteria, turbidity, rain garden, carbonized rice husk, stormwater run-off. The problem of research is studied by describing the study subject circumstances based on facts and research of various sectors. Analyses the data have done descriptively by displaying the data in table and image shapes.

3. Results and Discussion

3.1. The Effectiveness of Rain Garden

Rain garden developed by Dietz and Clausen, according to specifications in the recent manual design of Prince George County, MD (21). At the same time, the sighting of the model in **Figure 1**. In that model, the land located in a rain garden tub coated with a seam coming from a wooden piece, which is 5 cm. There are two waste pipes, which serve as the water sample site. The tube is located at the top and on the base of the model. The diameter pipe is 10.2 cm. The pipe is located at the bottom of the rain garden, covered by rocks with a height of 2.5 cm. Media rain garden has the characteristics of sand amounting to 83.0-84.0 %, Silt 7-10%, and clay 6.0-8.0% (21).

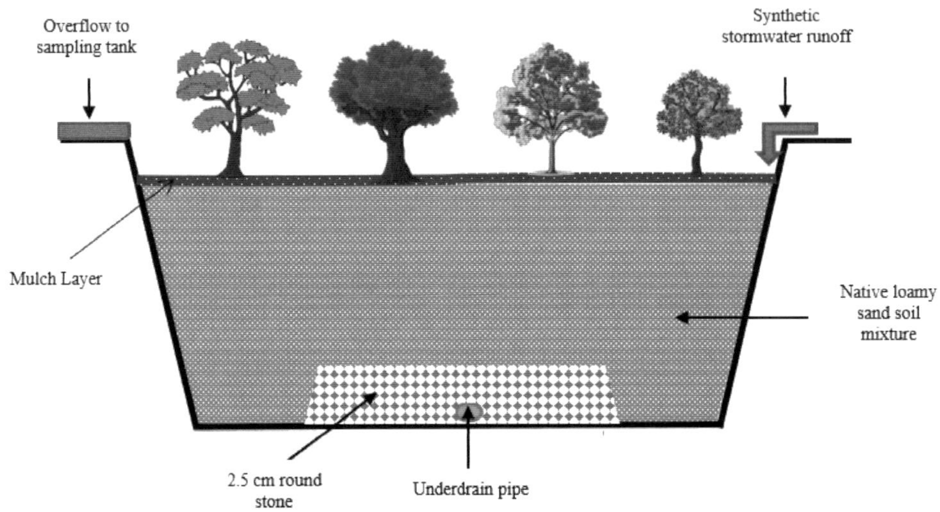

Figure 1. Layout rain garden.

Based on the results of the study Dietz and Clausen, by making two pieces of a rain garden as a test, showed that only about 0.8% of the flow of the incoming water leaving the rain garden as water run-off. Most of the 99.2% of entry as the flow under the surface. Based on the results of the same study showed that the value of ammonia - N (NH_3 - N) and total N is very significant absorbed by the rain garden. Dietz and Clausen added that there was no significant difference in the concentration of NO_3 - N, TKN (Total Kjeldahl Nitrogen), or ON (organic nitrogen) found in between the rainfall, the inlet or the intensity of the underdrain. However, the concentration of NH_3 – N from the underdrains significantly ($p = 0.001$) lower than NH_3 – N concentration in precipitation and the inlet. This means that the concentration of NH_3 - N approach the limit of detection for rainfall and the inlet, and at the threshold of detection for underdrains (21).

The concentration of TN (total nitrogen) of the underdrain of rain garden 1 is significantly ($p = 0.05$) lower than the rainfall and the concentration of TN inlet; TP (total phosphorus) concentration in the inlet was significantly ($p= 0.001$) greater than the precipitation. Interestingly, the concentration of TP in the effluent from the second underdrains has a value substantially higher than both the rainfall and the concentration of the inlet. However, TP concentration higher from the underdrains than the inlet of the after was analyzed using ANOVA. The average concentration geometric mean TP at the inlet (regression not significant) is 0.019 mg. L^{-1}, while the average of the two underdrains, shows the trend of exponential decay (R2 = 0.6). The amount of concentration of phosphorus at the inlet and underdrains towards the end of the study period is similar. Based on these results, the researchers suspect that the disorders of the land at the beginning of the study period is the cause of the increase in the concentration of phosphorus at the outlet. Rain garden application can be said as one form of implementation of biofilter that can remove hazardous materials from stormwater. Based on research, the performance of a rain garden can significantly decrease the number of Coli bacteria (0.6 logs) and protozoa (2 logs) (21). A Rain garden not only serves as a water-absorbent but can also purify it. Even in some studies, the existence of a rain garden can suppress the pollution in the river up to 30 percent (22). Many studies have also reported benefits from the presence of a rain garden (Table 3).

Table 3. Summary of rain garden effectiveness in removal pollutants.

Location	Removal Pollutant		Source
Study of two rain garden constructions, Haddam. CT	– NH_3-N	: 84.6%	(21)
	– NO_3	: 35.4%	
	– TKN	: 31.2%	
	– TN	: 32.0%	
	– ON	: 21.3%	
Studies in the box to the laboratory for six hours, the average flow and the time of 4.1 $cm.h^{-1}$	– Phosphorus ($\mu g. L^{-1}$)	: 99 ± 0	(23)
	– Nitrate nitrogen ($\mu g. L^{-1}$)	: 97 ± 3	
	– TKN ($\mu g. L^{-1}$)	: 97 ± 2	
Studies in boxes laboratory: double flow with a speed of 8.1 $cm.h^{-1}$	– Phosphorus ($\mu g. L^{-1}$)	: 73 ± 13	(23)
	– Nitrate nitrogen ($\mu g. L^{-1}$)	: 70 ± 14	
	– TKN ($\mu g. L^{-1}$)	: 73 ± 14	
Largo, MD	– Copper ($\mu g. L^{-1}$)	: 43% ± 11	(22)
	– Lead ($\mu g. L^{-1}$)	: 70% ± 23	
	– Zink ($\mu g. L^{-1}$)	: 64% ± 42	
	– Calcium ($\mu g. L^{-1}$)	: 27% ± 14	
	– Phosphorus ($\mu g. L^{-1}$)	: 87% ± 2	
	– Nitrate nitrogen ($\mu g. L^{-1}$)	: 15% ± 12	
	– TKN ($\mu g.L^{-1}$)	: 67% ± 9	

3.2. Factors that affect filtration with carbonized

According to Ronny and Saleh, several factors affect the filtration process in the carbonized filter during the process of adsorption. The first is the discharge that is too large will lead to the not proper functioning of the filters efficiently so that the filtration process cannot occur with the perfect, due to the water flow too fast through the cavity between the granules of the sand media. This results in a decreased time of contact between the surface of the granular filter media with the water to be filtered. The flow velocity is too high while passing through the cavity between the granules to cause the particles are also finely filtered will pass. The thick-thin medium will determine the duration of the jetting and filter power. The Media, which is too thick, has very high filtered power, but it takes time jetting long. The larger the surface area of the adsorbent, the more adsorbate that can be absorbed so that the process of adsorption can be more effective, the smaller the particle diameter than the surface area of the adsorbent (24).

The contact time is a very determining factor in the adsorption process, and longer contact time allows for diffusion and paste of adsorbate molecules to work better. The concentration of organic matter and later solutes will be well adsorbed if the contact time is longer with slow water flow. Besides, the media that is used must be dry, so this can improve the process of oxidation, as water will permeate into the filter media for longer than water which is permeating into the wet media filter, this will extend the time of the contact so that the process of oxidation will run more efficiently.

3.3. The effectiveness of carbonized rice husks to removal turbidity and Coli bacteria

Turbidity in water rain comes from the mud that dissolves with stormwater consists of suspended material carried by the flow of the surface run-off from high to low portions (25). Besides that, also household wastes that are disposed of haphazardly and include decomposition processes in the soil dissolve with stormwater. Turbidity is also influenced by the presence of the objects mixed or objects colloid in the water. Turbidity is caused by materials or suspended particles that range in size from small colloidal to materials coarse discontinued that range from the scope of small colloidal to rough stuff. Turbidity can be generated from a physical process of mechanical and chemical such as sedimentation and decomposition. Turbidity also makes a real difference in terms of aesthetics and in terms of the quality of the water itself (7). The results of the research of Syarifudin and Santoso in lowering the value of turbidity is known that the water used has a value of turbidity without treatment within 35,7 NTU, after the managed method sieve the rice husk value is decreased until it reaches the 0.95 NTU, from research it is known that factors affecting the decline are the level of thickness of the filter material of carbonized rice husk (7).

Coli bacteria in stormwater run-off were originating from manure and organic waste contained in the soil that can enter into the flow of rainwater. The human factor and the animals influence the presence of the Coli bacteria. Coli bacteria generated from biological processes resulting from the process of provision of fertilizer and animal manure. The water containing the Coli bacteria is thought to have been polluted by dung. The Coli bacteria is used as an indicator of determining whether water has become tainted by feces because it is more resistant to life on the condition of the environment compared to other pathogenic bacteria (6).

Research result from Syarifudin in managing Coli bacteria in water with a carbonized rice husk filter is capable of dropping compared with the water before processing, and this can be seen from the findings of Coli bacteria before the treatment of 21,333 colonies, after being managed, the best-known results for various variations in the thickness of the carbonized rice husk filters can lower up to 110 colonies, in standard quality still inappropriate because only up to 50 colonies of 100 ml of the bacteria can do so they can decrease the thickness Coli bacteria to reach the quality required for human beings (6). In addition to what Syarifudin has reported, the effectiveness of rice husk charcoal in reducing turbidity, microorganisms, and other pollutants has been widely reported. The summary can be seen in the following Table 4.

Table 4. The effectiveness of carbonized rice husks to removal turbidity and Coli bacteria.

Method	Removal Pollutant	Source
Two-stage purification	– Turbidity : 90-95% – Total Coliform : 90-95% (Coli Bacteria)	(26)
The laboratory-scale filtration tests	– Turbidity : 95%	(27)
Natural coagulants	– Turbidity : 96%	(28)
Gravity sand filtration with carbonized rice husk	– Turbidity : >65%	(29)
Filtration, combined sand (0.7-1.3mm) and carbonized rice husks in 1:1 proportion	– Turbidity : 98.97% – Hardness : 12.19% – Color : 90.63%	(30)
Used as adsorbent	– Turbidity : 98,2% – COD : 88% – TSS : 80% – Colour : 95%	(31)
Gravity sand filtration with carbonized rice husk	– Turbidity : >65%	(32)
Ceramic membrane with 20% rice husk combination	– Total Coliform :93-96% (Coli Bacteria)	(33)
Used as adsorbent	– Turbidity : 95%	(34)

3.4. SEM Analysis

According to the research that the carbonized rice husk activated at 700-degree Celcius indicates the carbonized rice husks have huge pores (Figure 2) (35). With this, it can be assumed that the carbonized rice husk absorbs turbidity and Coli bacteria. In conjunction with the research, Adediran et al. stated that most of the carbonized rice husks had the content of Silica (Si) (Figure 3) (36). These elements of Si existed significantly in the adsorbs of pollutants from stormwater.

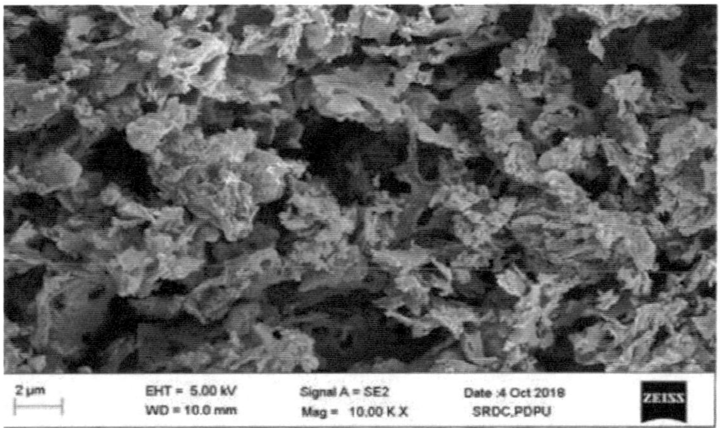

Figure 2. SEM (35).

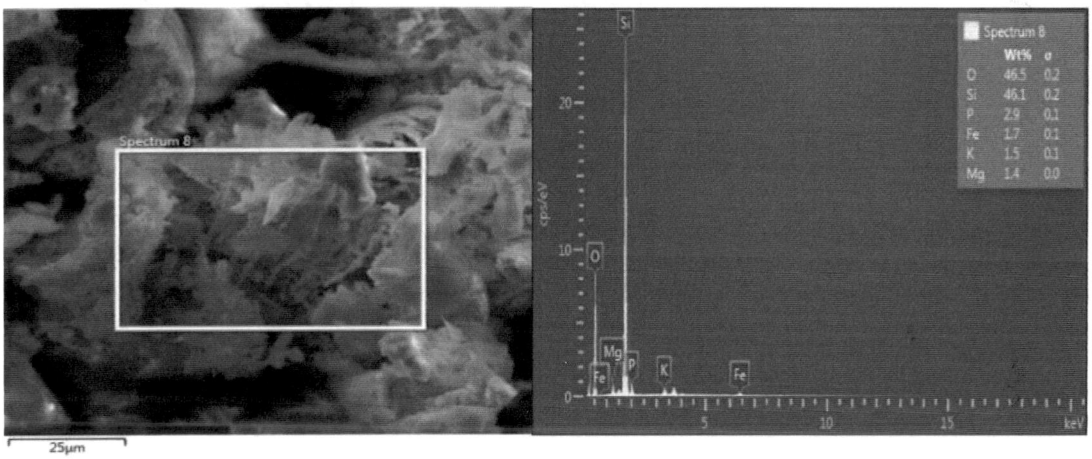

Figure 3. SEM image representative and EDX (36).

4. Conclusion

The rain garden is a natural expanse like a garden comprising a combination of soil and leaves and plants. A rain garden is also called a bio resistant area, in which it is designed to temporarily hold stormwater, filters and it helps the infiltration and evaporation process. The rice husk carbonized barriers were alternative media filter rain gardens to degrade the plot's value and the number of Coli bacteria, as it has large pores and contains the fertility of the superior. Based on the results of the study, carbonized rice husk can be removed from turbidity and Coli bacteria. The efficiency rate decreased turbidity between 65%-99%, while the decrease in Coli bacterial between 90-96%. The Rain Garden with filter rice husk carbonized is expected to be a new idea in the management of stormwater to decrease the value of turbidity and also Coli bacteria to be able to increase the effectiveness of its control to be researched further.

References

[1] Sukawi. Taman Kota dan Upaya Pengurangan Suhu Lingkungan Perkotaan (Studi Kasus Kota Semarang). Semarang: Universitas Diponegoro; 2008.

[2] Soemarno. Hutan Kota dan Pengelolaannya. Bahan Kajian. Malang: PPSUB; 2010.
[3] Annisa N, Riduan R, Prasetia H 2016 Aplikasi rain garden untuk memperindah dan meningkatkan kualitas lingkungan kota. Jukung J Tek Lingkung 2(2) 22–37
[4] Annisa N, Riduan R and Prasetia H 2016 Model rain garden untuk penanggulangan limpasan air 2(1) 78–92.
[5] Dibyo S. Persebaran Curah Hujan di Indonesia [Internet]. 2014. Available from: http://ssbelajar.blogspot.com/2013/07 /persebaran-curah-hujan-di-indonesia.html
[6] Syarifudin A 2017 Penurunan bakteri Coli pada air Sungai Martapura menggunakan saringan arang sekam padi Hilos Tensados 1 391–8
[7] Syarifudin A and Santoso I 2019 Efektivitas saringan abu sekam padi untuk menurunkan kekeruhan pada air Sungai Martapura Jurnal Kesehatan Lingkungan 15(2) 647–54
[8] Li Y L, Deletic A, Alcazar L, Bratieres K, Fletcher T D and McCarthy D T 2012 Removal of Clostridium perfringens, Escherichia coli, and F-RNA coliphages by stormwater biofilters. Ecol. Eng. 49 137–45
[9] Davis A P, Shokouhian M, Sharma H and Minami C 2001 Laboratory study of biological retention for urban stormwater management. Water Environ. Res. 73(1) 5–14
[10] Roy-Poirier A, P.Champagne and Filion Y 2010 Review of bioretention system research and design: past, present, and future J. Environ. Eng. -ASCE 9 878–89
[11] Carlisle S and Trevors J T 1988 Glyphosate in the environment Water Air Soil Pollut. 39:409–420.
[12] Bucheli T D, Müller S R, Heberle S and Schwarzenbach R P 1998 Occurrence of pesticides in rain water roof runoff, and artificial stormwater infiltration. Environ. Sci. Technol. 32 3457–64
[13] Huang X, Pedersen T, Fischer M, White R, Young T M 2004 Herbicide run-off along highways.1. field observations. Environ. Sci. Technol. 38 3263–71
[14] Wilson PC, Boman BJ. Characterization Of Selected Organo-Nitrogen Herbicides In South Florida Canals: Exposure And Risk Assessments. Sci Total Env. 2011;412:119–126.
[15] Jufianto I, Ikhsan J, Barid B. Analisis Pengaruh Metode Taman Hujan dalam Menurunkan Debit dan Kekeruhan Air Limpasan Permukaan. 2013;(2012):131–8.
[16] Halief K, Ningsih R and Nuryanto N 2011 Pengembangan teknik bioretention dalam mengatasi limpasan air hujan Proceeding PESAT (Psikologi, Ekon Sastra, Arsit &Sipil) 4(7) 51–6
[17] Annisa N and Prasetia H 2017 Manajemen limpasan air hujan di daerah perkotaan dengan rain garden dan menjaganya dari keberadaan nyamuk. Jukung (Jurnal Tek Lingkungan) 3(2) 47–54
[18] Prasetia H, Harijati N, Annisa N. Rain Garden. Malang: CV IRDH; 2018.
[19] Hidayat A K and Empung 2016 Analisis curah hujan efektif dan curah hujan dengan berbagai periode ulang untuk wilayah Kota Tasikmalaya dan Kabupaten Garut. J Siliwangi. 2(2):121–6
[20] Muharomah R 2014 Analisis run-Off sebagai dampak perubahan lahan sekitar pembangunan underpass Simpang Patal Palembang dengan memanfaatkan teknik GIS. Tek Sipil dan Lingkung. 2(3) 424–33
[21] Dietz M E, Clausen J C 2005 A field evaluation of rain garden flow and pollutant treatment. Water Air Soil Pollut. 167(1–4) 123–38
[22] Chisholm H 2008 An Analysis of The Efficacy of Rain Gardens for The Protection of Water Resources in Annapolis Royal, NS. The University of Guelph
[23] Davis A P, Shokouhian M, Sharma H and Minami C 2006 Water quality improvement through bioretention media: nitrogen and phosphorus removal. Water Environ. Res. 78(3) 284–93
[24] Ronny and Saleh M 2018 Penurunan kadar COD dengan metode filtrasi multimedia filter pada air limbah laundry J Penelit. 4 51
[25] Sholihah Q, Kuncoro W, Wahyuni S and Puni S 2020 The analysis of the causes of flood disasters and their impacts in the perspective of environmental law IOP Conference Series: Earth and

Environmental Science 437 012056
[26] Malhotra C, Patil R, Kausley S and Ahmad D 2013 Novel uses of rice-husk-ash (a natural silica-carbon matrix) in low-cost water purification applications AIP Conference Proceeding 1538 113–9
[27] Adams F V and Mulaba-Bafubiandi A F 2014 Application of rice hull ash for turbidity removal from water. Phys. Chem. Earth 72 73–6
[28] Adams F V, Ikotun B D, Patrick D O and Mulaba-Bafubiandi A F 2014 Characterization of rice hull ash and its performance in turbidity removal from water. Part. Sci. Technol. 32(4) 329–33
[29] Kainth G S 2015 Removal of turbidity from water using low-cost adsorbents BTech (thesis) 2015 1–33.
[30] Emanuelle C, Viana M, Wilmar J and Mourad K A 2016 Using rice husks in water purification in Brazil. Int. J. Environ. Plan. Manag. 2(3) 15–9
[31] Zahid I, Hussain S, Malghani N, Naeem Z, Amin M, Mushtaq F, et al. 2016 Municipal wastewater treatment using rice husk and kikar Proceeding International Research Symposium on Engineering Advancements (IRSEA) 2016(Irsea) 56–9
[32] Singh S 2017 Turbidity removal from water by use of different additive materials Int. J. Sci. Eng. Sci. 1(10) 55–7
[33] Sari S F and Sutrisno J 2018 Penurunan total Coliform pada air tanah menggunakan membran keramik. Waktu 16(1) 30–8
[34] Al-Dulaimi N S F and Al-Dulaimi S S F 2019 Experimental Measurement of Rice Husk Effectiveness as an Alternative Adsorbent for Turbidity Reduction in Synthetic Water. Al-Nahrain J. Eng. Sci. 22(4) 329–34
[35] Pillai P, Dharaskar S, Shah M and Sultania R 2020 Determination of fluoride removal using silica nano adsorbent modified by rice husk from water. Ground Sustain. Dev. 11 100423
[36] Adediran A A, Alaneme K K, Oladele I O and Akinlabi E T 2019 Structural characterization of silica-based carbothermal derivatives of rice husk. Procedia Manuf. 35 436–41

IOP Conf. Series: Materials Science and Engineering 1011 (2021) 012014 doi:10.1088/1757-899X/1011/1/012014

Quality of physicochemical and sensory of mushroom (*Pleurotus ostreatus*) chicken nuggets with carrageenan and konjac as hydrocolloids

W Setiaboma[a] and D Kristanti[b]

Research Center for Appropriate Technology, Indonesian Institute of Sciences, JL KS Tubun No.5, Subang, West Java, Indonesia

Corresponding author and main contributor: [a]wrboma@gmail.com/
[b]dita.kristanti@gmail.com

Abstract. Nuggets are mostly made of meat however reducing meat in meat product impact on characterization products. The present study was to investigate the effect of various hydrocolloids (carrageenan and konjac) on the physicochemical characteristic of chicken mushroom nuggets. Four formulations of mushroom chicken nuggets were N0 (non-hydrocolloid), N1 (sodium tripolyphosphate -STPP), N2 (konjac), and N3 (carrageenan). All nuggets were evaluated for moisture content, ash content, color, texture, pH, cooking loss, and sensory evaluation. The moisture content was less than 60 %. The cooking loss and texture of the nuggets was affected ($P<0.05$) additional of hydrocolloids. The N3 increased cooking loss (11.34) and hardness (1360.14 N). The sensory attributes were not influenced of the additional hydrocolloids. This current study indicated that N2 could be potential as thickener in mushroom chicken nuggets caused the pH value, cohesiveness, and chewiness of N2 were similar ($P>0.05$) to N1.

1. Introduction

Convenience foods impact on lifestyle at present hence frozen food product becomes a lifestyle choice. Chicken nuggets is one of popular frozen food, that is chicken product made from meat cut to shape and coated of bread flour [1]. Today's consumer preferences are healthy food. Consumptions of fiber have been important because it effects on our system body. Substitution of convenience product is one of ways to improve product being better. However, the desirable sensory of product is substantial factor. Decision of a consumer is not only healthiness food but also sensory expectation.

Mushrooms have been known as source fiber hence they have been used as meat like product. They have unique flavor and beneficial effect. They have rich in protein, high fiber, low cholesterol and rich in amino acid [2]. The nutrition of fresh oyster mushroom ware 88.75% of moisture content, 9.76% of ash, 28.85% of protein, 2.47% of fat, 48.16% of carbohydrate, and 12.87% of fiber [3]. In addition, the texture of mushrooms is similar to chicken meat. The mushrooms had been used to formulation in sausages [4], nuggets [5], patty [6], and meatballs [7]. Chicken nugget is made from meat and flours that is establish batter on emulsion system. The emulsion system will impact on quality of the nuggets. Additional thickening agent is one of methods to preserve the quality of nuggets. Sodium tripolyphosphate (STPP) is food additive that commonly used as a tenderizer [8]. However, the STPP is chemical and its additional is regulated [8]. The hydrocolloids function is food additive as a thickener in coated product like nuggets moreover It will absorb water easily and form a gel [9]. However, it is

IOP Conf. Series: Materials Science and Engineering 1011 (2021) 012014 doi:10.1088/1757-899X/1011/1/012014

important to understand substitute product and food additive prior to the quality of nuggets. The aim of this current study was to investigate the effect of various hydrocolloids (carrageenan and konjac flours) on the physicochemical characteristic of chicken mushroom nuggets.

2. Materials and Methods

2.1. Materials

The mushroom was purchased from *BumDes* Daya Maju Bersama Subang, West-Java, Indonesia. Chicken meat, seasoning (onion, garlic, shallots, pepper powder, nutmeg powder, salt, sugar) were obtained from local market. While STPP, konjac and carrageenan were obtained from Bandung, West-Java.

2.2. Preparation of nuggets

The mushroom was steamed for 15 minutes. The formulation of the nuggets is given in Table 1. All ingredients were mixed and chopped in chopper (Philips). The batter was placed on a baking pan and steamed for 30 minutes. The mixture was cooled at room temperature (±28 °C) and was cut to size 1x1x5 cm. The mixture was coating using flour, egg whites, and bread flour after that the nuggets were baked in oven at 105 °C for 10 minutes.

Table 1. Compositions of nuggets.

Materials	Formulas (%)			
	N0	N1	N2	N3
Oyester Mushroom	30.27	30.27	30.27	30.27
Chicken meat	15.13	15.13	15.13	15.13
STPP	-	0.34	-	-
Konjac flours	-	-	0.34	-
Carrageenan flours	-	-	-	0.34
Flours	15.89	15.89	15.89	15.89
Condiment	27.02	27.02	27.02	27.02
Ice water	4.54	4.54	4.54	4.54

2.3. The characteristics of physicochemical analysis

The moisture and ash were determined using the gravimetric method according to AOAC [10]. The protein content was determined using Dumas combustion method by DuMAster Buchi D-480. The cooking loss was assayed using method Sheard *et al.* [11] measured weight of cooked product and raw. The percentage of weight loss was calculated according to Equation 1:

$$\%\,cooking\ loss = (initial\ mass - final\ mass)x\frac{100}{initial}mass \quad (1)$$

The texture profile analysis was examined using texture analyzer TA-XT2 (Microsystems Ltd. Surrey, England, UK) using method of Kumar *et al.* [12]. All nuggets were cut into cube sized (1x1x2) cm. Setting the condition were pre-test speed 2mm/Sec, test speed 1 mm/Sec, post-test speed 10 mm/Sec, 30% strain mode, 5 seconds time and trigger force 5 g, and stainless-steel cylinder probe 5mm diameter (P35). pH value was analyzed using Kim *et al.* [13] with modified. Two grams of sample and 10 ml distillated water were homogenized using vortex. pH was determined using a pH meter analyzer (Mettler-Toledo). The color analysis was measured using Konica Minolta Chromameter and color score was performed as CIE Lab.

2.4. Sensory evaluation

The sensory evaluation was processed using hedonic testing. The test was conducted on 30 untrained panelists. The panelist assessed the attributes of color, flavor, texture and overall acceptability assessed through 5 levels (1 very dislike to 5 very like).

2.5. Statistical analysis

The statistical analysis was performed by ANOVA and Duncan's multiple range test at 5% signification. All measurements were replicated in triplicate and the experiment was done twice.

3. Result and Discussion

3.1. Characteristic of chemical properties

The chemical properties of all nuggets are shown in Table 2. The moisture content of N0, N1, N2, and N3 was 49.95%, 52.00%, 50.73% and 51.96%, respectively. The moisture content was significantly different ($P<0.05$). The moisture content of added hydrocolloid nuggets was higher than control nugget (N0). This result is similar with previous studies. The addition of carrageenan on the chicken patty affect the increasing of moisture content [14]. The research of Gadekar *et al.* [15] showed that carrageenan had effect of increasing the moisture content on the goat meat product but not significantly different with control. The use of konjac were increasing 11-12% the moisture content of the reduce reduced-fat frankfurters [16,17]. The result of Ruiz-capillas *et al.* [18] and Triki *et al.* [19] research presented that the moisture content of sausage added with konjac were significant higher than control. The moisture content of added carrageenan (N3) was higher than N0 and N2, its similar with the result study of Husain and Huda-Faujan [20]. The moisture content of all nuggets in this study were in accordance with Indonesian National Standard (SNI) for chicken nugget. Base on the SNI, the moisture content of chicken nugget must less than 60% [21]. The water content determine shelf life of product which increasing of water content results microorganisms to multiply [22]. According to additional of hydrocolloid, the moisture content of N2 was the best.

The ranged of ash content for all nuggets were between 1.94 % and 2.02 % and the ash content of N0, N1, N2 and N3 were not significantly different (Table 2). This result was corresponding with SNI that the ash content of chicken nugget must less than 2.5%. The result of previous studies showed that that ash content of chicken nugget of shiitake mushroom powder was 2.37% [23]. Formerly, Husain and Huda-Faujan [20] reported that the ash content of mushroom steams and chickpea flour ranged between 2.93%-3.03%. The ash content of chicken patty and goat meat product that addition of carrageenan was not significantly different when compared to control [14,15]. The addition of konjac on sausage and reduce reduced-fat frankfurter were not affect their ash content [17,19]. Others research showed that konjac was affect the increasing of ash content of sausage and reduce reduced-fat frankfurter [16,18].

The ranged of protein contents for all nuggets were between 19.77 % and 22.11 %. The protein contents of the nuggets were N0 (20.02 %), N1 (22.11 %), N2 (20.09 %), and N3 (19.77). The use of hydrocolloids affects the decrease of the protein content but there were not significantly different. This result is consistent with previous studies. The carrageenan was effect on the had effect on the decline of protein content of the chicken patty [14]. Furthermore, the konjac also affect the decrease of protein content in sausage and reduce reduced-fat frankfurter [16,19]. However, the results of this study are not in accordance with the research of Ruiz-capillas *et al.* [18] and Lin et. al. [17], where the protein content of sausage and reduce reduced-fat frankfurter were increase with the addition of konjac.

The pH value of all nuggets was 6.10-6.23. Similar study were reported by Husain and Huda-Faujan [20] that pH value of mushroom nugget had range 6.50-6.60. The pH value of konjac, xanthan and shiitake powder nugget formulations was 6.13-6.22 [23]. The pH value of N1 (6.26) and N2 (6.23) were significantly higher than N0 (6.10). Furthermore, the pH value of N3 (6.23) was also higher than N0 but not significantly different. The addition of various hydrocolloids has affected the increasing of pH value on reduce reduced-fat frankfurter and sausage [16,18]. Das *et al.* [14] and Gadekar *et al.* [15] described that the carrageenan affected the increasing of pH value on the chicken patty and goat meat product but not significantly different when compared to control.

Table 2. Chemical properties of various meat mushroom nuggets.

Chemical properties	Nuggets samples			
	N0	N1	N2	N3
Moisture (%)	49.95^{a}	52.00^{b}	50.73^{c}	51.96^{b}
Ash (db%)	1.94^{a}	2.02^{a}	1.99^{a}	1.96^{a}
pH values	6.10^{a}	6.26^{c}	6.23b^{c}	6.14^{ab}
Protein (db%)	20.02^{a}	22.11^{a}	20.09^{a}	19.77^{a}

Values are expressed as mean ± standard deviation. Means in the same row with different letters were significantly different at $p<0.05$

3.2. Characteristic of physical properties

The physical properties of the nuggets are presented in Table 3. Cooking loss of N1, N2, N3 were lower than N0. The previous research Luckose *et al.*[24], the cooking loss of chicken nuggets is 19.61% where is higher than N0. The cooking loss is character where indicated water holding capacity (WHC). The lower cooking loss, the better WHC[25]. Meat protein formed firm stable net of restructured product like nuggets where water and denaturation of protein effected restructured product [26]. Jeong and Han [27] showed that additional fruit with high fiber reduced the cooking loss of sausages. The fiber contributes to bonding water and to keeping fat [26]. The cooking loss of N1 and N2 were not significant different. The STPP is chemical salt while konjac is hydrocolloid. This result indicated that hydrogen bond of hydrocolloid was interacted with meat along mushrooms [13].

The lightness value of the nuggets was N0 (60.84), N1 (61.43), N2 (60.13), and N3 (60.85), respectively (Table 3). The lightness value of nuggets commercial is 64.38-68.41 [28]. The nuggets commercial is made from chicken while added mushrooms can change of the color. The mushrooms have greyish white color that additional mushrooms can decrease lightness [6]. The a-value of N0, N1, and N2 were not significant different, but significantly higher than N3. The b-value was significantly lower ($P<0.05$) in N1, N2 and N3 than N0. The b-value of mushroom chicken nuggets is lower than chicken nuggets or commercial chicken nuggets where additional mushroom is decreases the yellowness [6]. The lightness, a value, and b value of N2 ware significantly lower than control, its consistent with previous studies. The addition of konjac affect the decline of lightness, a value, and b value of sausage and reduced-fat frankfurter [16,19].

The hardness was affected by the additional of hydrocolloids. The hardness of N0, N2 and N3 were no significantly different than N1. According to Wan Rosli *et al.* [6], additional level of mushroom on patty decreased the hardness. The flour has function as fillers that adequate to bind and to hold water in system [25]. The additional of hydrocolloid may have caused the increased in the hardness. The hardness of chicken patty that added carrageenan was higher than control but not significantly different [14]. The carrageenan forms a rigid gel because of cross linking [29]. The additional of konjac caused the increasing of hardness on the reduced-fat frankfurter and sausage [16,18]. In this study additional of STPP was the highest of the hardness. The STPP increases ionic force that causing meat protein are more elastic[8].

The springiness of N0, N1 and N3 were not significantly different, while N3 was higher than other. It is indicated N0 had enough elasticity. The cohesiveness and the chewiness of N0 was not significant different from N2 and N3. The N1 was the best of chewiness and both of N0 and N2 were the best of springiness. The cohesiveness of N0 was not significantly different than all treatments. The cohesiveness impact on interaction of ingredients [30]. The texture properties of the nuggets in this study are similar with previous studies. The addition of konjac and carrageenan caused the increase of hardness and there were not affect the springiness, cohesiveness, and chewiness of sausage, reduced-fat frankfurter, and chicken patty [16,18]. The additional STPP increased springiness in ham that was highly bounded in STPP because of interaction between ionic strength of STPP and interference on structure of myofibril [31]. In this study, the nuggets were solid interaction with other ingredients. The N2 had better texture profile than N0 and N3.

Table 3. Physical properties of various meat mushroom nuggets.

	Nuggets			
	N0	N1	N2	N3
Cooking loss	14.72 ^{a}	12.65^{b}	11.34^{c}	12.17^{b}
L*	60.84^{ab}	61.43^{b}	60.13^{a}	60.85^{ab}
a*	2.65^{a}	2.42^{a}	2.25^{a}	1.77^{b}
b*	18.43^{a}	16.82^{b}	17.09^{b}	15.52^{c}
Hardness (N)	1254.10^{a}	1616.94^{b}	1360.14^{a}	1360.14^{a}
Springiness	0.89^{ac}	0.86^{ab}	0.94^{c}	0.82^{b}
Cohesiveness	0.75^{ab}	0.75^{ab}	0.79^{a}	0.72^{b}
Chewiness (N)	827.96^{a}	1038.73^{b}	898.09^{ab}	808.41^{a}

Means in the same row with different letters were significantly different at $p<0.05$

3.3. Sensory evaluation

The sensory evaluation of mushroom chicken nuggets with various hydrocolloids are shown in Fig. 1. The flavor, texture, and taste of N0, N1, N2, and N3 was not significantly different. The color and overalls of N1 was the highest. However, the color and overalls of others treatment (N0, N2, and N3) were accepted by panelist which based on the overalls was more than 3 (like). The additional hydrocolloids had positive result on sensory characteristics of the nuggets. Previously, Nayak *et al.* [1] and Nafiah *et al.* [32] disclosed that additional carrageenan was not affected preference panelist. The result of this study consistence with previous that the additional of konjac was not change sensory evaluation [19]. This result indicated that additional konjac (N2) or carragenan (N3) could be accepted as well as additional STPP (N1). In this study seem additional hydrocolloids (N2 and N3) were suitable alternative to STPP (N1) substitutes.

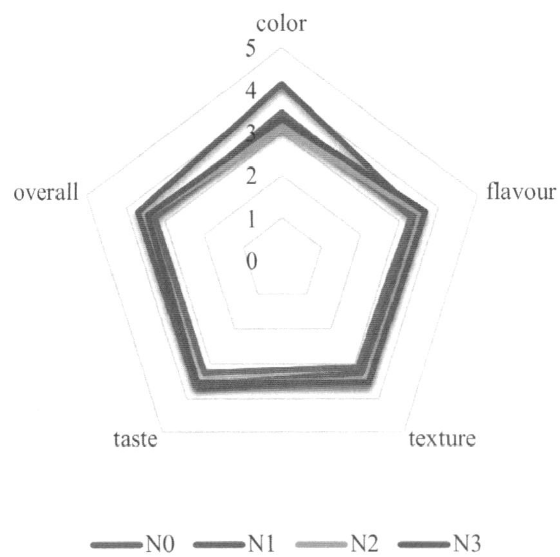

Figure 1. Sensory evaluation of mushroom chicken nuggets with various hydrocolloids.

4. Conclusion

The mushroom chicken nuggets expose possibilities of hydrocolloids to improve texture. The moisture content and pH value were affected by additional of hydrocolloid, whereas N2 was better than N3. Furthermore, the additional hydrocolloids were affected the cooking loss and textures. The cooking loss, lightness (L-value), and hardness of N2 were higher ($P<0.05$) than N0 and N3. Sensory attributes of the mushroom chicken nuggets were unaffected by additional of hydrocolloids. The study describes that konjac and carrageenan can be used to replace STPP without noticeable changes in sensory. Hence, the mushroom chicken nuggets can be made without added hydrocolloids or salt (STPP) that the sensory evaluation is not different from addition of hydrocolloids. The additional of konjac (N2) was better of physicochemical properties of mushroom chicken nuggets.

Acknowledgments

The authors are grateful to thank for Research Center for Appropriate Technology, Indonesian Institute of Sciences (LIPI) for providing access and technical support.

References

[1] Nayak N K, Pathak V, Singh V P, Goswami M and Bharti S K 2015 Quality of Carrageenan Incorporated Low Fat Chicken Nuggets during Refrigerated Storage at 4°C *Livestoks Research International* **3** 7–13
[2] Saragih R 2015 Nugget Jamur Tiram *(Pleurotus ostreatus)* Sebagai Alternatif Pangan Sehat Vegetarian *J. WIDYA Kesehat. dan Lingkung.* **1** 90–5
[3] Tolera K D and Abera S 2017 Nutritional quality of Oyster Mushroom (Pleurotus Ostreatus) as affected by osmotic pretreatments and drying methods *Food Sci. Nutr.* **5** 989–96
[4] Prisilia F H, Praptiningsih Y and Fauziah R R 2017 Karakteristik Sosis Berbahan Baku Campuran Jamur Tiram Putih (Pleurotus ostreatus) dan Otak Sapi *J. Agroteknologi* **11**
[5] Sumantri B, Ali A and Johan V S 2015 Pemanfaatan Tempe dengan Jamur Tiram (Pleurotus ostreatus) dalam Pembuatan Nuggets *Jom Faperta* **2** 1–12
[6] Wan Rosli W I, Solihah M A, Aishah M, Nik Fakurudin N A and Mohsin S S J 2011 Colour, textural properties, cooking characteristics and fibre content of chicken patty added with oyster mushroom (Pleurotus sajor-caju) *Int. Food Res. J.* **18** 621–7
[7] Komariah N U and Hendrarti E N 2005 Sifat Fisik Bakso Daging Sapi Dengan Jamur Tiram Putih (Pleurotus Ortreatus) Sebagai Campuran Bahan Dasar *J.Indon.Trop.Anim.Agric* **30** 34–41
[8] Fitrianingsih, Isnaeni P, Yaddi Y, Libriani R, Auza F A and Prasanjaya P N K 2020 Physical and organoleptic properties of chicken meatball prepared with varied gelling agents *IOP Conf. Ser. Earth Environ. Sci.* **465**
[9] Herawati H 2018 Potensi Hidrokoloid Sebagai Bahan Tambahan Pada Produk Pangan Dan Nonpangan Bermutu *J. Penelit. dan Pengemb. Pertan.* **37** 17
[10] AOAC 2000 Official Methods of Analysis. Method 930.15
[11] Sheard P R, Nutea G R and Chappellb A G 1998 The Effect of Cooking on the Chemical Composition of Meat Products with Special Reference to Fat Loss **49** 175–91
[12] Kumar V, Biswas A K, Sahoo J, Chatli M K and Sivakumar S 2013 Quality and storability of chicken nuggets formulated with green banana and soybean hulls flours *J. Food Sci. Technol.* **50** 1058–68
[13] Kim T, Shim J, Hwang K, Kim Y, Sung J, Paik H and Choi Y 2009 Effect of hydrocolloids on the quality of restructured hams with duck skin Restructured Duck Ham Preparation and *Poult. Sci.* 1–8
[14] Das S K, Prabhakaran P, Tanwar V K and Biswas S 2015 Effect of some plant starches and carrageenan as fat substitutes in chicken patties *J. Anim. Sci.* **93** 3704–12
[15] Gadekar Y, Sharma B D, Shinde A K, Das A and Mendiratta S K 2016 Effect of incorporation of functional ingredients on quality of low fat restructured goat meat product *Nutr. Food Sci.*
[16] Jiménez-Colmenero F, Cofrades S, López-López I, Ruiz-Capillas C, Pintado T and Solas M T

2010 Technological and sensory characteristics of reduced/low-fat, low-salt frankfurters as affected by the addition of konjac and seaweed *Meat Sci.* **84** 356–63
[17] Lin K W and Huang H Y 2003 Konjac/gellan gum mixed gels improve the quality of reduced-fat frankfurters *Meat Sci.* **65** 749–55
[18] Ruiz-Capillas C, Triki M, Herrero A M, Rodriguez-Salas L and Jiménez-Colmenero F 2012 Konjac gel as pork backfat replacer in dry fermented sausages: Processing and quality characteristics *Meat Sci.* **92** 144–50
[19] Triki M, Herrero A M, Jiménez-Colmenero F and Ruiz-Capillas C 2013 Effect of preformed konjac gels, with and without olive oil, on the technological attributes and storage stability of merguez sausage *Meat Sci.* **93** 351–60
[20] Husain H and Huda-Faujan N 2020 Potential application of grey oyster mushroom stems as halal meat replacer in imitation chicken nuggets *Food Res.* **4** 179–86
[21] Badan Standardisasi Nasional 2002 *Naget ayam (Chicken nugget)* (Jakarta: Badan Standardisasi Nasional)
[22] Hasniar, Rais M and Fadilah R 2019 Analysis of Nutrition Content and Organoleptic Test in Tempe *J. Pendidik. Teknol. Pertan.* **5** 189–200
[23] Akesowan A 2016 Production and storage stability of formulated chicken nuggets using konjac flour and shiitake mushrooms *J. Food Sci. Technol.* **53** 3661–74
[24] Luckose F, Pandey M C and Radhakrishna K 2015 Development of low salt restructured chicken nugget by response surface methodology and its quality evaluation *Int. Food Res. J.* **22** 2403–12
[25] Ma'ruf W, Rosyidi D, Radiati L E and Purwadi P 2019 Physical and Organoleptic Properties of Chicken Nugget from Domestic Chicken (Gallus domesticus) Meat with Different Corn Flours as Filler *Res. J. Life Sci.* **6** 162–71
[26] Polizer Y J, Pompeu D, Hirano M H, Freire M T D A and Trindade M A 2015 Development and evaluation of chicken nuggets with partial replacement of meat and fat by pea fibre *Brazilian J. Food Technol.* **18** 62–9
[27] Jeong Y and Han Y 2019 Effect on the emulsification stability and quality of emulsified sausages added with Wanggasi-Chunnyuncho (Opuntia humifusa f. Jeollaensis) fruit powders *Food Sci. Anim. Resour.* **39** 953–65
[28] Lukman I, Huda N and Ismail N 2009 Physicochemical and Sensory Properties of Commercial Chicken Nuggets *J. Food Ag-Ind* **2** 171–80
[29] Sidi N C, Widowati E and Nursiwi A 2014 Pengaruh Penambahan Karagenan pada Karakteristik Fisikokimia dan Sensoris Fruit Leather Nanas (Ananas Comosus L. Merr.) dan Wortel (Daucus Carota) *J. Apl. Teknol. Pangan* **3** 122–7
[30] Sharima-Abdullah, N., Hassan, C.Z., Arifin, N. and Huda-Faujan N 2018 Physicochemical properties of imitation chicken nuggets produced from different ratio of chickpeas flour and textured vegetable protein *Int. Food Res. J.* **25** 1016–25
[31] Resconi V C, Keenan D F, García E, Allen P, Kerry J P and Hamill R M 2016 The effects of potato and rice starch as substitutes for phosphate in and degree of comminution on the technological, instrumental and sensory characteristics of restructured ham *Meat Sci.* **121** 127–34
[32] Nafiah H, Winarni and Susatyo E B 2012 Pemanfaatan karagenan dalam pembuatan nugget ikan cucut *IJCS - Indones. J. Chem. Sci.* **1** 27–31

IOP Conf. Series: Materials Science and Engineering 1011 (2021) 012015 doi:10.1088/1757-899X/1011/1/012015

Artocarpanone isolated from *Artocarpus heterophyllus* heartwoods enhances cytotoxic effect of cisplatin against H460 and MCF-7 cell lines

Abdi Wira Septama[1*], Nik Nurul Najihah Nik Mat Daud[2]

[1]Research Center for Chemistry, Indonesian Institute of Sciences, Kawasan PUSPIPTEK Serpong, Tangerang Selatan, Banten 15314, Indonesia.

[2]Faculty of Medicine, Universiti Sultan Zainal Abidin, Jalan Sultan Mahmud 20400, Kuala Terengganu, Terengganu, Malaysia.

*Corresponding author's email: abdiwiraseptama@gmail.com
abdi.wira.septama@lipi.go.id

Abstract. *Artocarpus heterophyllus* has been used as a folk medicine in Asian countries. Flavonoid possessed a wide-range of therapeutic compounds for the treatment of several diseases including cancer prevention. Artocarponanone is s flavonoid isolated from the heartwoods of *A. heterophyllus*. This study was undertaken to investigate the interaction of natural compound and commercial anticancer, cisplatin against two cancer cell lines. MTT assay was performed to determine the cytotoxic effect of the combination. Isobologram analysis was used to predict the interaction of combination against cancer cell lines. The effect of this combination was then confirmed using AO/PI staining method. It showed that artocarpanone demonstrated cytotoxic effect against H460 and MCF-7 cell lines with IC50 values of 16.88 and 23.34 μg/mL, respectively. While cisplatin exhibited the cytotoxic effect against H460 and MCF-7 cell lines with IC50 values of 12.95 and 10.29 μg/mL. The combination of these compounds produced the synergistic effect on MCF-7 cell lines with a combination index (CI) of 0.09 and 0.13, respectively. Fluorescence microscope using AO/PI staining showed that the combination caused the morphological change of cancer cell lines and lead to necrosis and apoptosis. This result indicated that artocarpanone may enhance the cytotoxic effect of cisplatin against H460 and MCF-7 cell lines.

1. Introduction

Cancer is still a major cause of death all around the world. Lung cancer is one of the most common cancer found particularly in men. There are 2 types of lung cancer cells which are non-small cell lung cancer and small cell lung cancer cells. About 85% of non-small cell lung cancer cells have occurred in women regardless of smokers or non-smokers [1]. It has been reported 1 million people death worldwide due to this cancer. In the case of breast cancer, it has been known that this cancer is the most common type of cancer found in women, and caused high rate of the mortality and morbidity [2]. Conventional cancer therapy methods including surgery, radiotherapy, and chemotherapy can offer major curative effects against the localized tumors [3,4]. Hence, the use of those therapies may cause the incidence of tumor recurrence and lead to the development of any malignancy as well as drug resistance. Therefore, it needs to find alternative approach to overcome the problem. One of appealing strategy is the use of combination drug with natural compound that possessed anticancer activity. The interaction of those combination may lead the synergistic effect of commercial drugs toward cancer cell development.

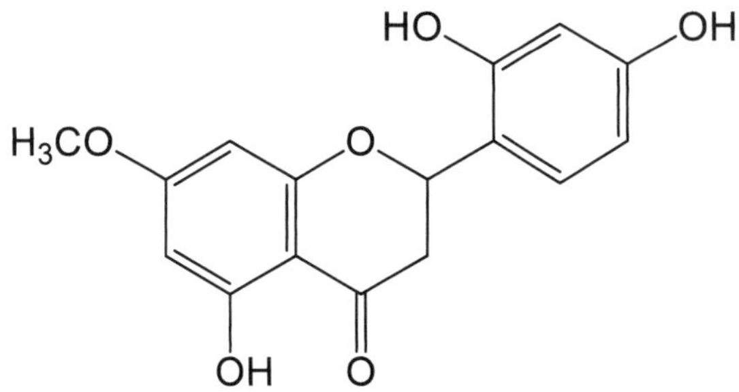

Figure 1. Chemical structure of artocarpanone

Natural product is a source of alternative therapeutic agents that can be used in treating of several diseases. *Artocarpus heterophyllus* has long been used as traditional medicine in many Asian countries. This plant is the main source of many bioactive compounds including flavonoid [5]. Artocarpanone (Figure 1) is flavonoid compound found in the heartwood of this plant. This compound possessed several pharmacological properties including for antibacterial, immunomodulator, and cancer prevention [6-8]. Although anticancer activity of this compound has been reported, however less information available about the effect of artocarpanone in combination with anticancer against cancer cell lines. This study was undertaken to investigate the interaction of artocarpanone in combination with cisplatin against two cancer cell lines including H460 and MCF-7 cell lines.

2. Materials and methods

2.1. Materials and Cell Culture
RPMI-1640, Fetal Bovine Serum, Penicillin-Streptomycin, 3-(4,5-dimetylthiazol-2-yl)-2,5-diphenyl tetrazolium bromide (MTT), and dimethylsulfoxide (DMSO), and cisplatin were from Sigma Aldrich, St. Louis, MO. Acridine orange (AO) and propopidium iodode (PI) were obtained from Merck. Artocarpanone was isolated from *A. heterophyllus* as previously described [5].

2.2. Cell Culture
In brief, MCF-7 and H460 cell lines (ATC.HTB-177) were purchased from the American Type Culture Collection (ATCC) and cultured in complete medium containing RPMI-1640 media supplemented with 10% v/v fetal bovine serum, and 1% v/v penicillin-streptomycin in 25 cm2 T-Flask (GIBCO, USA). The cells were humidified atmosphere with 5% CO2 at 37°C.

2.3. Cytotoxic assay
The assay was conducted as previously described a slightly modification [9]. Briefly, 5×10^3 cells/well of H460 and MCF7 were plated in a microplate, incubated for 24 h. Then, compounds (31.25, 15.63, 7.81, 3.91, 1.95 μg/mL) with five different concentrations were added into each well, and incubated for 72 h in 37°C at 5% CO2. Commercial anticancer (cisplatin) as positive control and untreated cell was used as a negative control. After incubation, 20 μL of thiazolyl blue tetrazolium bromide (MTT) reagent (5 mg/mL) was plated into the 96-wells plate, incubated for 4 h. Next, 150 μL of DMSO was added after aspiration of media. The cytotoxicity activity was counted using microplate reader at 570 nm. The study was performed in triplicate and the percentage of inhibition was counted using the following formula:

$$\text{Percentage inhibition (\%)} = ((\text{Abs control-Abs treated}))/((\text{Abs control})) \times 100\%$$

2.4. Synergistic study
Isobologram analysis was performed to assess the interaction of artocarpanone with cisplatin for its cytotoxic effect against H460 and MCF7 cell lines. The assay was conducted according to previous method with slight modification [9]. In brief, the dose-dependent effects of artocarpanone and fixed concentrations of cisplatin were evaluated using MTT assay. The tested cells were exposed with a mixture of cisplatin and artocarpanone. The combination index (CI) was counted to evaluate their interaction; synergy, antagonism, and additive by using the following formula:

$$CI=(D)1/((Da)1) + (D)2/(Da)2$$

Da1 = the concentration of artocarpanone alone required to cause certain percentage, D1 = the concentration of artocarpanone in equal percentage when combined with cisplatin. Da2 = the concentration of cisplatin alone required to cause certain percentage, D2 = the concentration of cisplatin in equal percentage when combined with artocarpanone. The CI values were defined as synergy if $CI < 1$, additive if $CI = 1$, and antagonist if $CI > 1$.

2.5. Fluorescence microscope
Morphological changes and apoptosis features of cells were evaluated using fluorescence microscope according to Goh et al., with slight modifications [10]. In brief, H460 and MCF-7 cell lines after treated with compounds were seeded and incubated for 72 h of incubation. Next, the medium was discarded. The treated cells were stained using combination dye containing 10 μL of 1 mg/mL acridine orange (AO) and 10 μL of 1 mg/mL of propopidium iodode (PI). The stained cells were observed under inverted fluorescence microscope.

2.6. Statistical analysis
The results were in triplicate and express as the means ± standard error of the mean (SEM). Any significance different was considered where a value of $p < 0.05$.

3. Results and Discussion

The cytotoxic effect of artocarpanone and cisplatin on H460 and MCF7 cell lines was evaluated using MTT assay. Artocarpanone exhibited strong activity with IC_{50} values of 16.88 and 23.34 μg/mL, respectively (Table 1). Meanwhile, cisplatin showed the cytotoxic effect against H460 and MCF7 cell lines with IC_{50} values of 12.96 and 10.29 μg/mL, respectively. These results were confirmed using fluorescence microscope, morphological changes of those tested cell line were observed after treatment with artocarpanone and cisplatin for 72 h. This result indicated that those compounds cause cell damage which in turn inducing cell death. It has been reported that natural products possess anticancer activity. Several reports have confirmed the potency of natural product, particularly flavonoid as anticancer as well as cancer prevention. Artocarpin, a flavonoid compound isolated from *Artocarpus* sp., demonstrated strong cytotoxic effect against a wide-range of cancer cell lines including T47D, MCF7 and H460 [9]. In addition, artocarpin also exhibited strong anticancer activity in non-small lung cancer cell line by inducing apoptosis through ROS mediated MAPKs and Akt activation [11].

Primary cancer treatments have incorporated both chemotherapeutic agents and ionizing radiation to eliminate the bulk of tumor mass. There is the evidence, however, that the combination of these therapies brings the incidence of tumor relapse that the results in development of drugs resistance in tumor cells. Therefore, alternative approach is needed to overcome this problem [12,13]. One of the strategies is combination two drugs all together to eliminate cancer cells. The use of drug combination may increase the biological activities of those compound. It may be due to different compounds have different target of action as well as the different compounds may have the same target of action that lead to agonistic activity [14]. In the present study, we combined artocarpanone and cisplatin against two tested cancer cell lines. The interaction of the mixture was then analysed using isobologram analysis. Combination index (CI) represents the interaction of two compounds in combination, CI less than 1 indicates the synergistic effect of compounds in the mixture, on the other hand, CI more than 1 indicates the antagonistic.

Table 1. IC_{50} values of artocarpanone and cisplatin on H460 and MCF7 cancer cell lines

No.	Compounds	IC_{50} (μg/mL)	
		H460	**MCF7**
1.	Artocarpanone	16.88 (55.89 μM)	23.34 (77.28 μM)
2.	Cisplatin	12.96 (43.18 μM)	10.29 (34.29 μM)

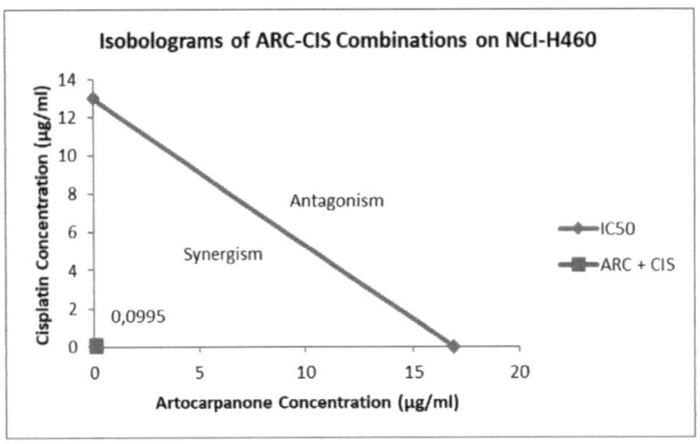

(a)

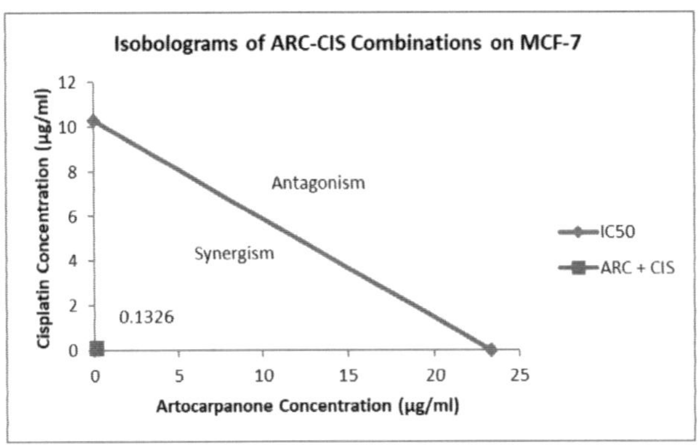

(b)

Figure 2. Isobologram analysis of H460 (a) and MCF-7 (b) when treated with combination of artocarpanone and cisplatin.

As shown in Figure 2 (a), IC_{50} value of the mixture artocarpanone and cisplatin against H460 cell lines was bellow than straight line. The combination index (CI) using isobolgram analysis was then used to confirm the interaction. It showed that CI of the mixture of artocarpanone and cisplatin was 0.09, and it represented the synergistic effect. This result indicated that artocarpanone may enhance the cytotoxic effect of cisplatin against H460 non-small lung cancer cell line. In the case of cytotoxic effect against MCF7 (Figure 2b), the combination of artocarpanone and cisplatin produced an IC_{50} which was bellow than straight line. Isobologram analysis was then calculated to confirm the interaction. It found that CI of the mixture was 0.13. This result indicated that a synergistic effect of those compounds in the combination by which artocarpanone may enhance the activity of cisplatin against breast cancer cell line.

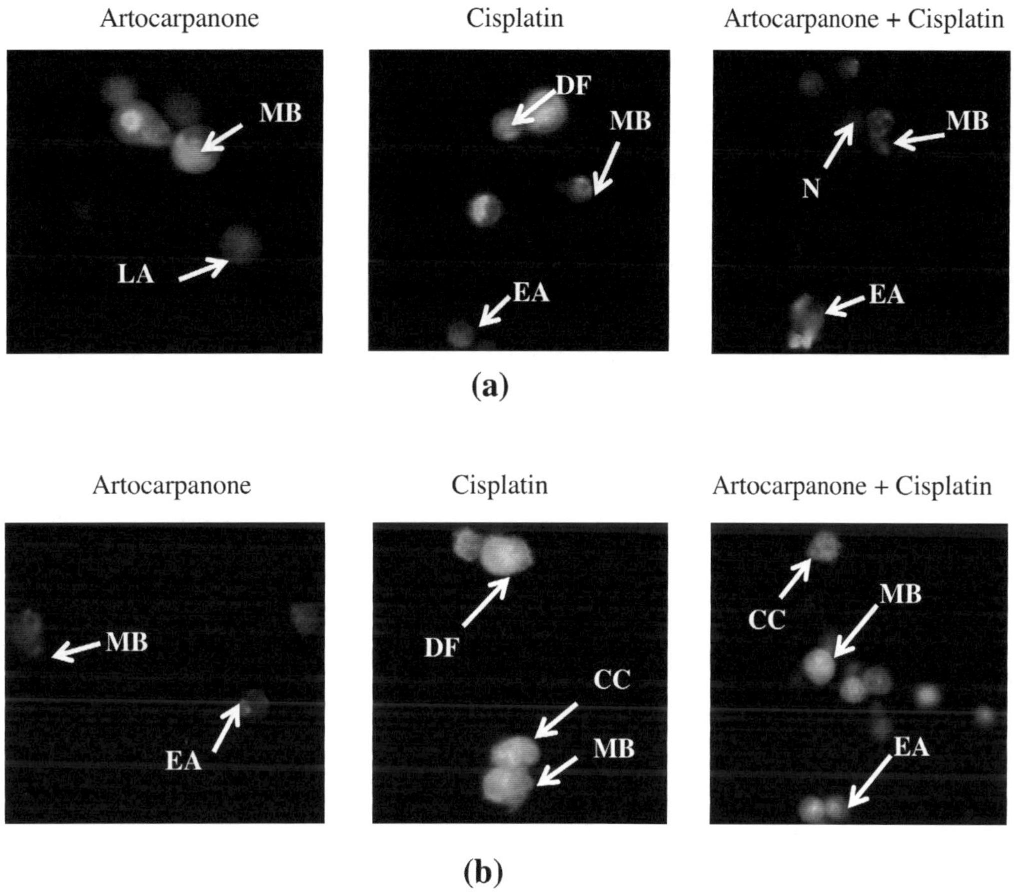

Figure 3. Morphological changes of H460 (a) and MCF-7 (b) induced by cisplatin, artocarpanone, and cisplatin-artocarpanone in combination. Arrows show (CI) Combination Index, (FL) Fluorescence, (AB) Apoptotic Body, (MB) Membrane Blebbing, (EA) Early Apoptosis, (LA) Late Apoptosis, (CC) Chromatin Condensation, (VI) Viable Cells, (DF) DNA Fragmentation.

In order to confirm any synergistic effect of the combination against H460 and MCF7 cell lines, fluorescence microscope analysis was then performed. Morphological change was observed using dual staining AO/PI method. This AO/PI staining method was performed to observe any morphological changes after treating with compounds such as viable, necrotic, or apoptosis cells [15]. It has been known that AO will bind to live cells via intercalation with DNA and RNA, while PI is able to bind into

DNA and RNA of dead cells. Dual staining AO/PI was relatively stable for image-based viability analysis [16]. The results exhibited that the treatment with combination of artocarpanone and cisplatin (Figure 3 a), the number of H460 cells was lower than treatment with single compound. The morphological observation after treatment with the mixture for 72 h demonstrated that the combination lead to significant changes in apoptotic cell and some membrane blebbing compared to treatment with single compound. On the other hand, treatment with the combination of artocarpanone and cisplatin against MCF7 cell line (Figure 3 b) also indicated significant changes in cell morphology compared to compound when used alone. It can be observed that the mixture of artocarpanone and cisplatin cause some morphological changes in MCF7 cell line including membrane blebbing (MB), early apoptosis (EA), and chromatin condensation (CC).

The result of this study proved the potency of artocarpanone in enhancing cytotoxic effect cisplatin against cancer cell lines. In this combination, low concentration of cisplatin as well as artocarpanone to achieve their therapeutic concentration on cancer cell lines. It is important in order to reduce and avoid drug resistance due to the use of drugs in high concentration. It has been known that cisplatin is platinum anticancer drug in which this compound works via DNA damage signal transduction in cancer cells, also induce apoptosis and necrosis. Meanwhile, the mechanism of action of artocarpanone on cancer cells is not fully understood. The result of this study was consistent with another previous work, it has been recorded that flavonoid deguelin enable to enhance cytotoxic activity of cisplatin on cancer cell lines [17]. In addition, a diterpene compound named triptolide also demonstrated a synergistic effect when combined with cisplatin in inducing cancer cells apoptosis [18]. The mechanism of synergistic interaction, however, needs to be explore through molecular study.

4. Conclusion

It can be concluded that artocarpanone may enhance cytotoxic effect of cisplatin in a synergy interaction. In the combination only required low concentration to afford their therapeutic doses against tested cancer cell lines including H460 non-small lung cancer cell lines and MCF7 cell lines. The results of this study propose the potency of artocarpanone as adjuvant of cisplatin *in vitro*. Nevertheless, further experiments which involved molecular study are needed to justify the mechanism of action in this combination.

Acknowledgements

We would thanks to Ministry of Education, Malaysia under the Fundamental Research Grant Scheme (FRGS 2017-2020) and Universiti Sultan Zainal Abidin (FRGS/1/2017/SKK06/UNISZA/03/1) for financial and technical supports.

References

[1] Torre LA, Bray F, Siegel RL, Ferlay J, Lortet-Tieulent J and Jemal A 2015 *CA Cancer J. Clin.* **65** 87-108
[2] Oser MG, Niederst MJ, Sequist LV and Engelman JA 2015 *The Lancet Oncol.* **16** e165-e172
[3] Tao Z, Shi A, Lu C, Song T, Zhang Z and Zhao J 2015 *Cell Biochem. Biophys.* **72** 333-8
[4] Borghei Y S, Hosseini M, Dadmehr M, Hosseinkhani S, Ganjali MR and Sheikhnejad R 2016 *Anal. Chim. Acta.* **904** 92-7
[5] Septama A W and Panichayupakaranant P 2015 *Pharm. Biol.* **53** 1608-13
[6] Septama A W and Panichayupakaranant P 2016 *Pharm. Biol.* **54** 686-91
[7] Sato M, Fujiwara S, Tsuchiya H, Fujii T, Iinuma M, Tosa H and Ohkawa Y 1996 *J. Ethnopharmacol.* **54** 171-6
[8] Septama A W, Jantan I and Panichayupakaranant P 2018 *J. Pharm. Pharmacol.* **70** 1242-52
[9] Daud NN, Septama AW, Simbak N, Bakar NH and Rahmi EP 2019. *Nat Prod Sci.* **25(4)** 311-6
[10] Goh SH, Mohamed Alitheen NB, Md Yussoff F, Yap SK and Loh SP 2014 *Pharmacogn. Mag.* **10** 1–8
[11] Tsai MH, Liu JF, Chiang YC, Hu SC, Hsu LF, Lin YC *et al.* 2017 *Oncotarget* **8** 28342–58
[12] Ahmad M, Khan MA, Marwat KS, Zafar M, Khan MA, Hassan TU *et al.* 2009 *American Eurasian*

J. Agric. Environ. Sci. **5** 126–40
[13] Cai Y, Luo Q, Sun M and Corke H 2004 *Life Sci.* **74** 2157–84
[14] Yang Y, Zhang Z, Li S, Ye X, Li X and He K 2014 *Fitoterapia* **92** 133–47
[15] Chan LL, Kuksin D, Laverty DJ, Saldi S and Qiu J 2015 *Cytotechnology* **67** 461–73
[16] Stankovic MS, Curcic MG, Zizic JB, Topuzovic MD, Solujic SR and Markovic SD 2011 *Int. J. Mol. Sci.* **12** 4190–05
[17] Alabsi AM, Ali R, Ali AM, Al-Dubai SA, Harun H, Abu Kasim NH *et al.* 2002. *Asian Pac. J. Cancer Prev.* **13** 5131–6
[18] Li CJ, Chu CY, Huang LH, Wang MH, Sheu LF, Yeh JI *et al.* 2012 *Cancer Lett.* **319** 203–13

IOP Conf. Series: Materials Science and Engineering 1011 (2021) 012016 doi:10.1088/1757-899X/1011/1/012016

Inhibition activity of α-glucosidase enzyme, phenolic content and toxicity of Mara (*Macaranga tanarius (L.) Mull. Arg*) Leaf

Megawati[1]*, Hariyanti[2], Resta Pebriani[2] and Sofa Fajriah[1]

[1]Research Center for Chemistry, Indonesian Institute of Sciences, Kawasan PUSPIPTEK Serpong, Tangerang Selatan, Banten 15314, Indonesia.
[2]Faculty of Pharmacy and Science, Universitas Muhammadiyah Prof. DR. HAMKA, Jakarta, Indonesia

*Corresponding author: megarafandi@gmail.com

Abstract. Previous studies have shown that mara leaves (*Macaranga tanarius* (L) Mull. Arg*)* can be used as medicinal plants because they have antibacterial and anti-inflammatory activity. Compounds containing flavonoids, tannins, saponins. But there is still little research on mara leaves as inhibition α-glucosidase. The purpose of this study was to determine the antidiabetic activity by the *α-glucosidase* method and its toxicity by the BSLT method and the results of fractionation. This test can be done by looking at the IC_{50} values obtained from linear regression analysis. Whereas the acute toxicity test uses BSLT (method of Shrimp Lethality Test). Using concentrations of 1000 ppm, 500 ppm, 100 ppm and 10 ppm. At each concentration using 10 larvae *Artemia salina* Leachand analyzed mortality. LC_{50} values were obtained from linear regression of mortality percentages and log doses. The results of this study prove the ethyl acetate fraction of *Macaranga tanarius* (L) Mull. Arg leaves has a high value inhibition activity that is 0.74 μg / mL and is non-toxic so it can be considered as inhibition α-glucosidase.

1. Introduction

The genus Macaranga is a useful species of 250 species in tropical forests where its distribution is in tropical forests of Asia and the Pacific. Investigation of the potential of Macaranga plants as medicinal plants and cosmetic ingredients is still not optimal. However, several natural chemical chemistry researchers from the Faculty of Forestry and FMIPA of Mulawarman University have conducted searches in the form of phytochemical screening and isolation of active compounds from Macaranga in East Kalimantan. Phytochemical and pharmacological studies of Macaranga plants that have been carried out generally show that there are variations in the content of secondary metabolites such as terpenoids, terpenylated flavonoids and stylbenoid, and some phenolic derivatives [1].

More than 26 species of *Macaranga* have been studied phytochemically with more than 190 secondary metabolite compounds have already been identified, with phenolic compound as a major compound group [2]. Our previous study was successfully identified some phenolic compounds such as macarangin, apigenin, apigenin glycoside, and scopoletin [3-5] from *M. Gigantifolia*, 5,7,3',4'-tetrahydroxy-3,6-diprenylflavone and kaempferol 7-O-β-glucose from *M. gigantifolia* and *M. hispida (Blume)* [3,6] and scopolotin compound from *M. hispida (Blume)* [7]. Previous studies reported that flavonoids and terpenoids are the main components of the genus *Macaranga* [8]. Apigenin and active compounds were isolated from *Portulaca oleracea* L. which has potential antibacterial activity and potentially used as an antibacterial for the treatment of diseases associated with pathogenic bacteria

[9]. Phenolic or polyphenols are one of the most widely distributed groups of natural products in the plant kingdom. More than 8,000 phenolic structures are currently known, and more than 4,000 flavonoids have been supported [10].

The α-glucosidase inhibitors are one of the therapeutic approaches in decreasing postprandial hyperglycemia by delaying the digestion of poly- and oligosaccharides to absorbable monosaccharides [11]. Furthermore, α-glucosidase inhibitors have been a huge concern to researchers working in the field of medicinal chemistry since their antidiabetic and anti-obesity activity, is also associated with their activity against human immunodeficiency virus (HIV) and hepatitis [12-14].

One of the most common diseases throughout the world is diabetes mellitus, a chronic metabolic disorder characterized by hyperglycemia accompanied by various chronic vascular complications. Therefore, control of postprandial blood glucose spikes is crucial for the treatment of diabetes. A number of treatments has been used by patients with type II diabetes are synthetic amylase inhibitors and glucosidases, such as, acarbose, however, they report side effects. As a result, safer natural amylase inhibitors and glucosidases and many compounds have been reported from plant sources [15-19].

Therefore, a further study of the phytochemical content of the *Macaranga* genus needs to be done especially from *M. tanarius (L.) Mull. Arg*. In this study, we evaluated the total phenol content from methanol extract, n-hexane, ethyl acetate and buthanol fractions, antidiabetic activities from *M. tanarius*.

2. Research Methods

2.1. General

Sample extraction performed using methanol (Merck), evaporated with Buchi R214-Switzerland rotary evaporator, biological activity reagent for antidiabetic enzyme using *α-glucosidase* enzyme, total phenolic content and toxicity test using *Artemia salina* Leach larvae.

2.2. Plant material

Macaranga tanarius leaves was collected in March 2012 from Mekongga forest, Southeast Sulawesi, Indonesia. The plant was identified by the staff of Herbarium Bogoriense, Research Center for Biology, LIPI and a voucher specimen had been deposited at the herbarium.

2.3. Extraction and isolation

The dried leaves (2 kg) of *M. tanarius* was macerated with methanol for 24 hours (3 times). The methanol extract (220 g) was partitioned with n-hexane, ethyl acetate and n-butanol, respectively, and all fractions were evaluated for their total phenolic content using Follin-Ciocalteu methods.

2.4. Inhibition assay for α-glucosidase activity

α-glucosidase inhibitory activity evaluation of the extracts was performed using an established procedure [13,20,21]. The *α-glucosidase* enzyme solution was dissolved in a phosphate-buffer solution (pH 7) containing 200 mg albumin serum. Prior to its application, enzyme solution was diluted 25 times in buffer solution. The mixture contained of 20 mM p-nitrophenyl α-D-glucopyranose as the substrate and 100 mM phosphate buffer (pH 7). Ten μL of the extract dissolved in DMSO was prepared. The reaction mixture was then water-bath incubated at 37°C for 5 min. The enzyme solution (250 μL) was added and the solution was incubated for 15 min. The enzyme reaction was stopped by the addition of 200 mM sodium carbonate solution. The resulted p-nitrophenol from the reaction was measured at λ 400 nm. Quercetin solution was measured. The commercial α-glucosidase anti-diabetic drug, glucobay, was available in the laboratory only in sustain release tablets. Sample concentrations for activity evaluation were 6.25, 12.5, 25, 50, 100 μg/mL. Inhibition was calculated using formula:

$$\text{Inhibition (\%)}: [(C - S) / C] \times 100 \qquad \text{(Eq. 1)}$$

In which: S = Sample absorbance; C = blank absorbance

2.5. Total Phenolic Content in Extract

The total phenolic content of the extract was determined by the Folin–Ciocalteu method [22]. Various concentration of gallic acid solution were pipetted and placed into test tubes. Each tube was added with 1 mL of the sample, and distilled water to 7.5 mL volume. Added with 0.5 mL of Folin–Ciocalteu reagent, and mixed thoroughly for 8 min, followed by the addition of 1.5 mL of 20% (w/v) of sodium carbonate. The mixture was allowed to stand for further 60 min in the dark, and absorbance was measured at 765 nm. The total phenolic content was calculated based on the calibration curve, and the results were expressed as mg of gallic acid equivalent per g of dry weight.

2.6. Toxicity test

Toxicity test conducted by *brine shrimp lethality* test method developed by Meyer [23]. *Artemia salina* Leach. The eggs incubated in sea water for 48 hours until hatch into larvae. About 10 *A. salina* larvae put in microplate and added with sample in various concentration of samples (10, 100, 500, and 1000 ppm), incubated for 24 hours at room temperature. Toxicity calculated based on the number of dead larvae after incubated. LC_{50} value indicated of sample concentration required to make 50% of the *A. salina* larvae dead.

3. Result and Discussion

In this study the simplisia used was leaves which had been pollinated. The mara leaves were then extracted using 96% Methanol. The extraction method used is maceration. Maceration has the advantage of simple procedures and equipment; the extraction method is not heated so that the natural ingredients do not decompose. Cold extraction allows many extracted compounds, although some compounds have limited solubility in extraction solvents at room temperature. The maceration process is carried out by soaking the simplicia with methanol 96% for 24 hours in a tightly closed container and protected from light. After 24 hours filtering is done and the pulp is macerated again with a new solvent. Maseration is done three times by soaking 2 kg of simplicia powder in a maceration bottle. After all the macerates were collected, solvent separation was carried out using vacuum rotary evaporator at a temperature of 48-50 °C until thick extract was obtained. Through a rotary vacuum evaporator, the solvent liquid can be separated without any overheating of the samples by lowering the pressure above the liquid and thus lowers the boiling point [24].

Table 1. Methanol extract of *M. tanarius* (L) Mull. Arg.

Extract	Weight	% Yield
Simplisia *M. Tanarius*	2 kg	-
Methanol Extract	220 g	10.23

After obtaining methanol extract of Mara leaves as much as 220 g with percent yield of 10.23 percent then partitioned (fractionation) using *n*-hexane-water (1: 1) obtained *n*-hexane fraction and water fraction, then the results were concentrated by using vacuum rotary evaporator and the water fraction was subsequently partitioned with ethyl acetate - water as much as (1:1), then the results of the ethyl-acetate fraction were concentrated using *vacuum rotary evaporator* and The water fraction was then partitioned again with butanol-water as much as (1:1), then the results of the butanol fraction were concentrated using *vacuum rotary evaporator*. The last result was the fraction of the methanol extract of Mara leaves as follows: hexane fraction (1.52 gram), ethyl acetate fraction (9.005 gram), butanol fraction (5.881 grams) and water fraction (3.219 gram). Further, each extract and fraction were tested for inhibition alpha glucosidase [7].

Table 2. Yield of Extract *M. tanarius* (L) Mull. Arg.

Samples	Extract Weight (gram)	% Yield
n-Hexane Fraction	1.522	7.60
Ethyl Acetate Fraction	9.005	45.22
Butanol Fraction	5.881	29.40
Water Fraction	3.219	16.09

Methanol extract and all fractions (hexane, ethyl acetate, butanol) were evaluated for their antidiabetic using *α-glucoside*, total phenolic content using Folin-Ciocalteu method.

Table 3. Total phenolics content of extract and fraction from *M. tanarius* (L) Mull. Arg.

Samples	% Total Phenolic Content (GAE (w/w))
Methanol extract	27.117 ± 1.06
Hexane Fraction	10.435 ± 0.45
Ethyl Acetate Fraction	37.518 ± 1.37
Butanol Fraction	34.646 ± 1.05
Water Fraction	6.775 ± 0.346

GAE= Gallic Acid Equivalent

In this study, total phenolic content of methanol extract was 27.117 ± 1.06 (w/w). It means that each 100 g dry weight of methanol extract contained total phenolics that was equivalent to gallic acid of (Table 2). Partition step of methanol extract with n-hexane, ethyl acetate and n-butanol yielded highest total phenolic content in the ethyl acetate fraction with the total phenolic content was 37.518 ± 1.37 (Table 2). Ethyl acetate fraction was further purified using various chromatography techniques.

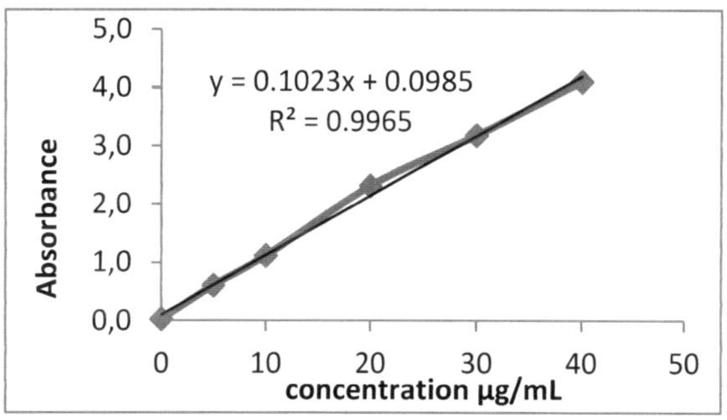

Figure 1. Comparative solution calibration curve (gallic acid), total phenolic content.

Supported by the data analysis of all biological activities test resulted, showed that ethyl acetate extract of *M. tanarius* leaves has strong activity as antidiabetic with IC_{50} 0.74 ppm compared with quercetin as positive control (IC_{50} 5.601 ppm) (Table. 4). IC_{50} value indicated the concentration of sample that can inhibit enzyme activity 50%.

Toxicity tests are used to determine the toxic effects of a compound that will occur in a short time exposure period or the administration of a single concentration of test compounds in test animals. The BSLT (*Brine Shrimp Lethality Test*) method is used for initial screening of active compounds contained in plant extracts since it is easy, inexpensive, fast, does not require special labs and the results are reliable [25].

Table 4. Inhibition activity of α-glucosidase from extract and fractionasi from *M. tanarius* leaves.

Samples	Concentration	% inhibition	IC_{50}
Quarcetin	10	60.207	5.601
	5	48.932	
	2.5	36.276	
	1.25	15.141	
	0.625	10.468	
Methanol extract	10	58.672	8.628
	5	30.272	
	2.5	29.508	
	1.25	28.464	
	0.625	22.150	
Hexane Fraction	10	40.312	12.46
	5	12.531	
	2.5	10.157	
	1.25	10.014	
	0.625	9.020	
Ethyl Acetate Fraction	10	96.445	0.74
	5	94.939	
	2.5	93.972	
	1.25	39.550	
	0.625	37.555	
Butanol Fraction	10	97.709	1.619
	5	90.785	
	2.5	90.800	
	1.25	37.437	
	0.625	36.935	

Table 5 shows that only the fraction *n*-hexane which is a toxic compound that has an IC value of $_{50}$ (301.99 µg / mL) compared to other extracts is not classified as a toxic compound because it has an LC_{50} of more than 1000 µg / mL, which according to the theory that an extract is declared toxic according to the BSLT method if it has an LC_{50} of less than 1000 µg / mL [26].

Table 5. Toxicity test results with BSLT method extract and fraction.

No	*Macaranga tanarius* (L) Mull.Arg.	LC_{50} (µg/mL)	Conclusion
1	Extract of Methanol	>1000	Non-toxic
2	Fraction n-Hexane	301,99	Toxic
3	Ethyl Acetate	>1000	Non-toxic
4	Butanol Fraction	>1000	Non-toxic
5	Water Fraction	>1000	Non-toxic

From the results obtained it turns out that the ethyl acetate fraction has a resistance value of *α-glucosidase* of 0.74 µg / mL more active and approaching the standard, quarcetin (5.601 µg / mL) when compared to other extracts, in the BSLT test the ethyl acetate extract did not show any salina artemia larvae death after adding this extract. Meyer [23] in Farihan [27] reported that an extract showed toxic activity in BSLT if the extract could cause the death of 50% of test animals at concentrations of less than 1000 ppm. The statement above shows that ethyl acetate extract is not toxic to Artemia because it has an LC50 value> 1000 ppm. The ethyl acetate extract in the phenolic content test has a higher content. *α-glucosidase* and safe to consume without showing mortality in this extract when tested for toxicity by the BSLT method.

Differences in antidiabetic activity were obtained from total phenol levels and total flavonoids. Phenol and flavonoid compounds have a linear increase in anti-diabetic activity, so the higher the level, the better the anti-diabetic [28]. Phenolics and flavonoids have an antidiabetic or antihyperglycemic effect which is possible in several ways by inhibiting absorption, stimulating insulin secretion or acting like insulin and can help the work of enzymes that play a role in carbohydrate metabolism. One of the flavonoid compounds is quercetin which is known as a strong anti-diabetes. Quercetin as part of a flavonoid which is considered as a glucose transport inhibitor in the small intestine which is responsible for the absorption of glucose in the small intestine so that it can reduce blood sugar levels [29,30].

4. Conclusion

This study displayed that ethyl acetate fraction of *Macaranga tanarius (L) Mull* leaf extract. Arg, consists of phenolic compounds that potentially inhibit α-glucosidase enzyme. There is forceful correlation between the ability to inhibit *α-glucosidase* and the safety of the compounds contained in the ethyl acetate extract, which does not show its toxicity. Taken together, ethyl acetate extract from *M. tanatius* leaves potentially developed as a candidate for the prevention and treatment of diabetes. Further pre-clinical study using animal model is necessary to figure out potent anti-diabetic properties of *M. tanatius* leaves extract as anti-diabetic.

References

[1] Ramadhan R and Kusuma I W 2017 Samarinda: Mulawarman University Press pp 47-63
[2] Magadula J J 2014 *J. Med. Plant Res.* **8**(12) 489-503
[3] Darmawan, A, Suwarso, W P, Kosela S and Kardono L B S 2015 *Ind. J. of Pharm.* **26** 52–6
[4] Primahana, G and Darmawan A 2017 *J. Pure App. Chem. Res.* **6**(1) 22-6
[5] Fajriah S, Megawati and Darmawan A 2016 *J. Life Sci.* **6**(1) 7-9
[6] Megawati, Saepudin E, Hanafi M, Darmawan A and Lotulung P D N 2015 *Makara J. Sci.* **19** 96–100
[7] Megawati, Hanafi M, Saepudin E and Fajriah S 2016 *J. Ilmu Kefarmasian Indonesia* **14**(1) 38-42
[8] Cai J, Zhao X L, Liu AW, Nian H and Zhang S H 2011 *Phytomedicine* **18** 366-73
[9] Nayaka H B, Ramesh L L, Madire K U and Asha T 2014 *Int. J. Bacteriol.* **2014** 175851
[10] Cheynier V 2005 *Am. J. Clin. Nutr.* **81** 223–9
[11] Takahashi T and Miyazawa M 2012 *Phytother. Res.* **26** 722-6
[12] Mehta A, Zitzmann N, Rudd P M, Block T M and Dwek R A 1998 *FEBS Lett.* **430** 17–22
[13] Du Z Y, *et al* 2006 *Eur. J. Med. Chem.* **41**(2) 213-8
[14] Zhu Y, Yin L, Cheng Y and Yamaki K 2008 *Food Chem.* **109** 737–42
[15] Yang Z, Wang Y C, Wang Y and Zhang Y 2012 Food Chem. **13** 617-25
[16] Zhang L, Bai B, Liu X, Wang Y, Li M. and Zhao D 2011 *Food Chem.* **126** 203-6
[17] Ye X P, Song C Q, Yuan P and Mao R G 2001 Chin. *J. Nat. Med.* **8**(5) 0349-52.
[18] Ieyama T, Puteri M D P T G and Kawabata J 2011 *Food Chem.* **128** 308-11
[19] Lee S S, Lin H C and Chen C K 2008 *Phytochem.* **69** 2347-53
[20] Lee D S and Lee S H 2001 *FEBS Lett.***501** 84-6
[21] Fajriah S, Darmawan A, Megawati and Hanafi M 2018 *Res. J. Chem. Environ.* **22** (Special Issue II) 114 -9
[22] Kaur C and Kapoor H C 2002 *J. Food Sci. Technol.* **37** 153-61
[23] Meyer B N, Ferrigni N R, Putnam J E, Jacobsen L B, Nichols D E and McLaughlin 1982 *J. Plant Med.* **45** 31-4
[24] Miller K A and Liu D 2017 *Rotary Evaporator Standard Operating Procedure* Department of Material Science ang Engineering. Becman Institute p 1
[25] I Made A G and Rasmaya N 2006 *General toxicology.* F. Pharmacy Pharmacy of Udayana University: Edition II. Bali.
[26] Jazilah N, Fasya N and Abtokhi 2014 *Alchemy* **3**(2) 118-24

[27] Farihah *2006 Uji Toksisitas Ekstrak Daun Ficus benjamina L terhadap Artemia salina Leach dan Profil Kromatografi Lapis Tipis*. Skripsi Diterbitkan. Surakarta: Fakultas Farmasi Universitas Muhammadiyah Surakarta.
[28] Ghasemzadeh A dan Ghasemzadeh N 2011 *Med. Plants Res.* **5**(31) 6697-703
[29] Ratya A 2014 *J. Agromed. Unila.* **1**(1) 61-6
[30] Sasmita F W, Eko S, Husamah dan Yuni P 2017 Biosfera. **34** 22-31

IOP Conf. Series: Materials Science and Engineering 1011 (2021) 012017 doi:10.1088/1757-899X/1011/1/012017

Investigation of the potential covid-19 inhibitor from *Cassia alata* and *Dendrophthoe petandra* using computational approach

Teni Ernawati*, Marissa Angelina, Rizna Triana Dewi, Sofa Fajriah, Rokip

Research Center for Chemistry - Indonesian Institute of Sciences (LIPI), Kawasan Puspiptek, 452 Building, Tangerang Selatan, Banten 15314, Indonesi

*Corresponding author's email: teni.ernawati001@gmail.com

Abstract. COVID-19 is a recent coronavirus outbreak caused by SARS-CoV-2 beginning in early December 2019 in Wuhan China raising global health problems. The SARS coronavirus main peptidase (SARS-CoV M(pro)) plays an essential role in the life-cycle of the virus and is a primary target for the development of anti-SARS agents. Also, the viral 3-chymotrypsin-like cysteine protease 3 (3CLpro) enzyme controls coronavirus replication and is essential for its life cycle. Therefore, herein, we performed in silico virtual screening of the compounds from *Cassia alata* and *Dendrophthoe petandra* such as; aloe-emodin, emodin, kaempferol, quercetin and quercitrin against SARS-CoV-2 M^{pro} and also $3CL^{pro}$ through molecular docking approach using Autodock 4.2, with the Lamarckian Genetic Algorithm. The binding energies obtained from the result of molecular docking of SARS-CoV-2 M^{pro} and also $3CL^{pro}$ quercetin, kaempferol and aloe-emodin appeared to have the best potential to act as COVID-19 inhibitors.

1. Introduction

In the past ten years, many new coronaviruses have been identified. They infect large numbers of hosts from mammals to birds and closely related coronaviruses have been identified [1-5]. Corona viruses (CoV) belong to a group of viruses that can infect humans and vertebrate animals. CoV infection affects the respiratory, digestive, liver, and central nervous systems of humans and animals [6-8]. This research focuses on proteases in CoV (Mpro/3CLpro), especially PDB ID 6LU7,2GTB, 3M3V and also through Homology modeling as a potential target protein for COVID-19 treatment [9-12]. Protein is taken from a protein data bank that can be accessed by the public since early February 2020 and there are also those who use homology modeling using Swiss-Model [13-17]. The use of proteins that have similarities with SARS-Cov is used to tether the active compounds contained in *Cassia alata* and *Dendrophthoe petandra* plants namely aloe-emodin, emodin, kaempferol, quercetin and quercitrin. Molecular docking was carried out in this study to predict the activity of the five compounds in the SARS-CoV active site protein. In addition, bioactivities of the five compounds and commercial antiviral compounds were compared to see the extent to which the antiviral activity of the compounds contained in *C. alata* and *D. petandra*. SARS-CoV protein is very important for viral proteolytic maturation and has been examined as a potential target protein to prevent the spread of infection by inhibiting the division of the polyprotein virus [13,18,19]. This molecular docking study provides a great opportunity to identify potential drug candidates for antiviral treatment.

IOP Conf. Series: Materials Science and Engineering 1011 (2021) 012017 doi:10.1088/1757-899X/1011/1/012017

2. Materials and Methods

The 3D structures of the tested compounds were built by Chem Draw 12.0. The crystal structures of main protease covid-19 (PDB ID: 6LU7, 2GTB, 3M3V, and homology docking) were used as enzyme models and processed by AutoDock 4.2. The hydrogen bond network was refined by means of the optimize option. The grid boxes were centered on x, y, and z coordinates -11.683, 14.686, 65.363 (6LU7); 18.216, -12.873, 11.035 (2GTB); -27.887, -20.584, 25.196 (3M3V) and also for homology modelling 14.313, 14.502, 26.864. The homology modeling template using SWISS-MODEL, application of molecular visualization using UCSF-Chimera 1.9, and Autodock 4.2 RAMPAGE for testing Ramachandran plot. In this paper, the active compound from *C. alata* and *D. petandra* such as aloe-emodin, emodin, kaempferol, quercetin and quercitrin were selected as ligand to dock into the main protease covid-19 (PDB ID: 6LU7, 2GTB, 3M3V, and homology docking). And also, the commercial antiviral medicines were used as ligand to dock into the same main protease covid-19 for comparative.

3. Results and Discussion

The five compounds from *C. alata* and *D. petandra* as well as five commercial antiviral compounds were subjected to docking studies using AutoDock 4.2. The proteins used in this docking simulation are the 2 main SAR-CoV-2 proteases obtained from protein bank data and also the $3CL^{pro}$ homology modeling using the Swiss Model. These four SAR-CoV-2 proteins are used to calculate the binding affinity and free energy of the SAR-CoV-2 receptor binding. Table 1 was shown the structures and amino acids found in the active site pockets of 2 main Protease M^{pro} (6LU7, 2GTB) and $3CL^{pro}$ (3M3V and its homology model). 6LU7 is the main protease (M^{pro}) found in COVID-19, which been structured and repositioned in PDB and can be accessed by the public, as of early February 2020. 2GTB is the main protease found in the CoV associated with the severe acute respiratory syndrome (SARS), which can be accessed in PDB and was suggested to be a potential drug target for 2019-nCov. Xu et al mentioned that the main protease in 2019-nCov shares 96% similarity with that in SARS [1]. Ligands and several drug candidates compounds have been previously selected, based on adherence to Lipinski's rule of five. The selected ligands that did not incur more than 2 violations of Lipinski's rule could be used in molecular docking experiments with the target protein.

The results of molecular docking of chemical substituents from *C. alata* and *D. petandra* on several SAR-CoV-2 proteins are shown in Table 1. The less free energy generated from the results of the inhibition of compounds with the protein SAR-CoV-2, the more active the prediction of its compounds according to the docking simulation theory. Likewise, the smaller the value of the inhibition, the faster the kinetics of the required inhibition of the compound against the SAR-CoV-2 protein. Figure 1 showed the docking score data for all active compounds in the sample and commercial antiviral compounds showed that among the 10 simulated chemical compound structures; nelfinavir as a commercial antiviral drug had the lowest docking score while for the sample compound (almost all the SAR-CoV proteins used, except 3M3V); quercetin which is an active compound contained in *D. petandra* had the lowest docking score (for SAR-CoV-2 6LU7 and 3M3V); kaempferol as active compound from *C. alata* had the lowest docking score (SAR-CoV-2 2GTB and $3CL^{pro}$ model Homology).

The binding of hydrogen with Cys-145 and His-163 plays an important role in the inhibition of the SAR-CoV-2 protein. The binding of hydrogen with Cys-145 and His-163 plays an important role in the inhibition of SAR-CoV because cysteine and histidine residues are the main constituents of the SAR-CoV-2 receptor. In addition to the main residues that play an important role, there are also related amino acid residues that are no less important in producing small inhibitory energy so that the SAR-CoV-2 molecules are inhibited, among others; Pro168 Glu166 Leu167 Met165 Gln189 Gln192 Met49 Cys44 Asp187 Arg188. From the results of docking simulations of 10 chemical structures, it can be concluded that those that have consistent active predictions on all the SAR-CoV-2 proteins tested are: quercetin, kaempferol and aloeemodin for chemical substituents from *C. alata* and *D. petandra* compounds as well as nelfinavir, hydroxychloroquine and remdesivir for commercial antiviral compounds.

Table 1. Result of docking score of chemical substituents from *C. alata* and *D. petandra*.

No	Protein (PDB ID)	Ligand	ΔG(kkal /mol)	Ki (uM)	Amino acid residue interactions
1	**M^{pro}** 6LU7	Aloeemodin	-7.84	1.78	Val3 Leu4 His163 Phe140 Leu141 Asn142 Ser144 Cys145 Leu27 Gly143 Thr26
		Emodin	-7.30	4.47	His172 Glu166 Val3 Leu4 His163 Phe140 Leu141 Asn142 Gly143 Cys145 Leu27
		Kaempherol	-7.26	4.80	Glu166 Val3 Leu4 Phe140 His163 Leu141 Asn142 Gly143 Ser144 Cys145 Thr26
		Quercetin	-8.18	1.01	Thr26 Gly143 Leu27 Cys145 Ser144 Leu141 Asn142 Phe140 His163 Val3 Glu166
		Quercitrin	-5.49	93.8	Thr26 Gly143 Leu27 Ser144 Asn142 Leu141 Leu4 Phe140 His163 Val3 Gln189
		Hidroxychloro quinin	-8,23	0.930	Leu4 Met49 Thr45 Ser144 Cys145 Leu27 Gly143 Thr25 Asn142 Leu141
		Lopinavir	-5.56	0.083	Gln189 Met165 Leu4 Met49 Val3 Glu166 Ser46 Thr45 His163 Asn142 Cys145 Leu27 Gly143 Thr26 Thr25 Thr24
		Nelfinavir	-8.79	0.360	Ser46 Thr45 Met49 Val3 Leu4 His41 His163 Cys145 Leu27 Thr25 Gly143 Thr26 Thr25 Asn142
		Remdesivir	-6.81	0.01	Gln189 Met49 Met165 His172 Glu166 Val3 Leu4 His163 Phe140 Leu141 Asn142 Gly143 Thr26 Thr25
		Ritonavir	-7.53	0.003	Asn142 Leu141 Gly143 Leu27 Cys145 His167 Met165 Leu4 Met49 Val3 Glu166 Gln189
2	**M^{pro}** 2GTB	Aloeemodin	-7.25	4.85	Phe185 Val186 Ala173 Met165 Gln189 Thr190 Gln192 Leu167 Pro188
		Emodin	-6.37	21.57	Met165 Gln189 Ala173 Thr190 Phe185 Gln192 Ala191 Leu167 Pro168
		Kaempherol	-7.87	1.69	Tyr54 His164 His41 Cys44 Cys145 Ser144 Asp187 Met165 Met49 Glu168 Gln189
		Quercetin	-7.59	2.72	Glu166 Gln189 Met165 Met49 Asp187 His164 His41 Tyr54 Ser144 Cys44 Cys145
		Quercitrin	-5.98	41.52	Pro168 Leu167 Ala191 Thr190 Glu166 Met165 Gln189 His164
		Hidroxychloro quinin	-9.45	0.117	Asn142 Ser144 Cys145 His163 Met165 Met49 His41 His164 Arg188
		Lopinavir	-10.7	0.014	Leu141 Asn142 Pro168 Glu166 His163 Met165 Gln189 Thr190 Met49 Cys44 His41 His164 Asp187 Arg188
		Nelfinavir	-11.64	0.002	Pro168 Glu166 Leu167 His163 Met165 Gln189 Cys145 Gln192 Met49 Cys44 Asp187 Arg188
		Remdesivir	-8.54	0.546	Cys145 Cys44 His41 His154 Thr54 Asp187 Met49 Met165 His163 Asn142 Glu166 Leu167 Arg188 Thr190 Gln192
		Ritonavir	-7.77	0.002	Thr25 Cys145 Ser144 Gly143 Asn142 Glu166 Met165 His41 Met49 Gln189
3	**3CL^{pro}** 3M3V	Aloeemodin	-4.66	386.9 1	Arg40 Glu55 Tyr54 Asn53 Pro52 Arg188 Asn51
		Emodin	-4.27	744.4 7	Gln192 Ala191 Thr190 Gln189 Leu167 Glu166 Pro268
		Kaempherol	-4.85	278.5	Ala191 Thr190 Pro168 Gln189 Glu166 Leu50 Glu47

		Quercetin	-5.45	100.3	Pro168 Ala191 Gln192 Thr190 Leu167 Arg188 Glu166 Gln189 Met165 Leu50
		Quercitrin	-3.58	2360	Ala191 Thr190 Arg188 Asn51 Leu50
		Hidroxychloro quinin	-5.38	114.64	Leu50 Gln189 Pro168 Thr190 Ala191 Gln192
		Lopinavir	-2.31	2000	Asn51 Arg188 Leu50 Gln189 Thr190 Ala191
		Nelfinavir	-3.1	500.3	Gln189 Leu50 Thr190 Pro168
		Remdesivir	-4.49	513.3	Pro168 Glu166 Gln189 Leu50 Thr190 Ala191 Gln192
		Ritonavir	3.26	0	Gln189 Leu50 Asn51 Thr190 Ala191
4	**3CL^{pro}** Homology modeling using Swiss-Model	Aloeemodin	-7.15	5.75	Asn3154 Gln3152 His3006 Asp3151 Pro3153 Tyr3018 Ile3129 Thr3012
		Emodin	-5.59	80.5	Asn3154 Gln152 Pro3153 Ile3129 Thr3012 Asp3151 Tyr3018
		Kaempherol	-7.91	1.6	Leu3155 Gln3156 Asn3154 Gln3152 Asp3151 Ile3016 Tyr3018 Leu3131 Ile3129 Glu3130
		Quercetin	-7.69	2.3	Leu3155 Asn3154 Gln3156 Gln3152 Asp3151 Tyr3018 Glu3130 Ile3129 His3006
		Quercitrin	-5.83	53.03	Leu3155 Gln3156 Asn3154 Gln3152 Leu3131 Pro3153 Thr3013 Asp3151 Ile3016 Glu3130 Thr3012 His3006
		Hidroxychloro quinin	-9.85	60.59	His3006 Thr3012 Gln3128 Ile3129 Ile3016 Asp3151 Pro3153 Glu3130 Leu3131 Gln3156
		Lopinavir	-7.77	2.01	Cys3109 Tyr3016 His3006 His3127 Gln3128 Asp3151 Ile3016 Ile3129 Thr3012 Glu3130 Gln3152 Pro3153
		Nelfinavir	-10.78	0.012	Ile3016 Gln3152 Asn3154 Pro3153 Ile3129 His3006 Leu2992 Cys3109 Gly3107 Glu3130
		Remdesivir	-6.89	8.85	Tyr3018 Ile3016 Asp3151 His3006 Thr2990 Leu2992 Cys3109 Val2991 Gly3105 Gln3152 Ile3129 Thr3012 Pro3153 Glu3130
		Ritonavir	-8.33	0.782	Ile3018 Gln3152 Asn3154 His3006 Thr2990 Leu2992 Gln3128 Thr3012 Ile3129 Pro3153 Leu3131 Glu3130 Asn3106

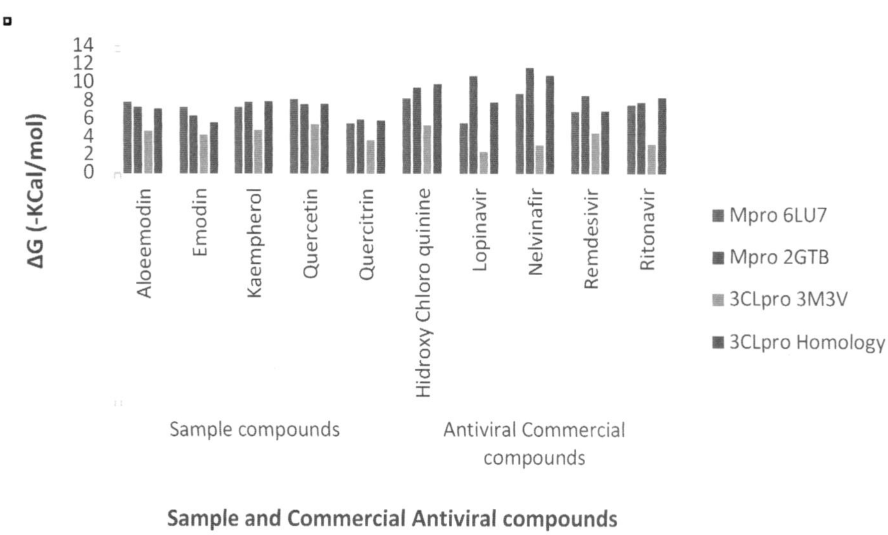

Figure 1. Histogram showing molecular docking results Sample compounds from *C. alata* and *D. petandra*

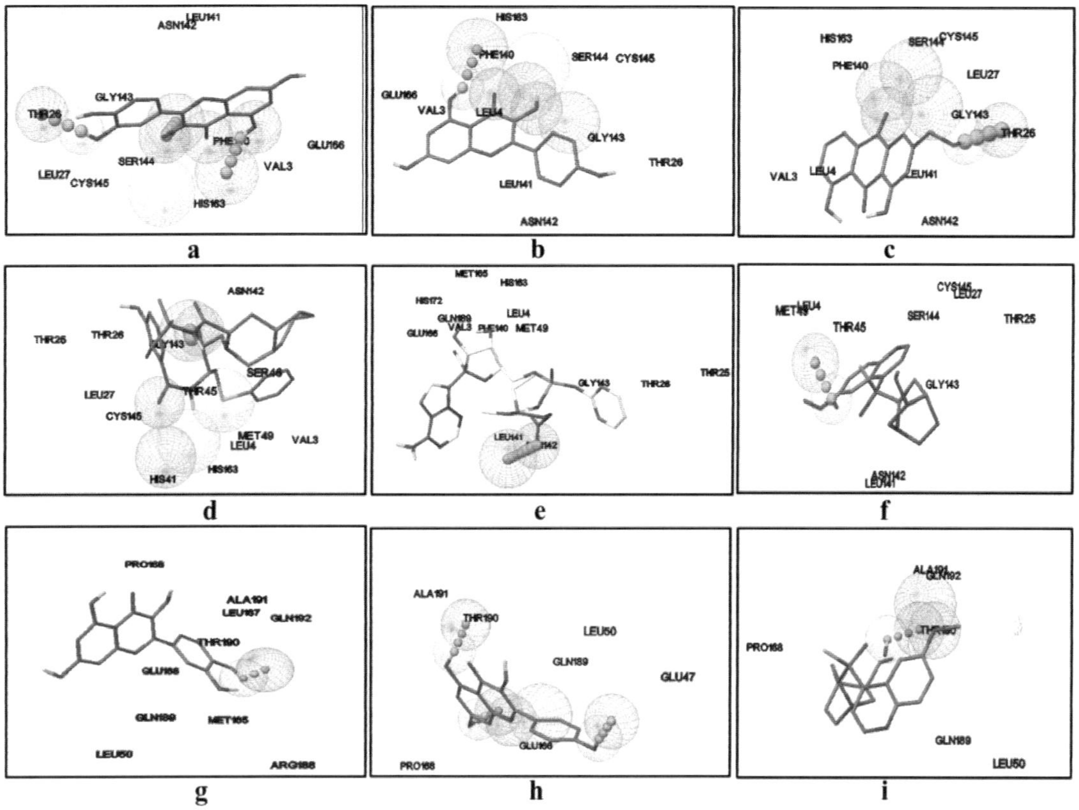

Figure 2. Interaction of ligand binding of 6LU7 SAR-CoV M^{pro} with; (**a**) quercetin, (**b**) kaempferol, (**c**) aloemodin, (**d**) nelfinavir, (**e**) remdesivir, (**f**) hidroxychloroquinine, 3M3V SAR-CoV 3CL^{pro} with; (**g**) quercetin (**h**) kaempferol (**i**) hidroxychloroquinine. The green dots show H-bonds.

Proteases represent potential targets for the inhibition of CoV replication, and the protein sequences of the SARS-CoV M^{pro}/3CL^{pro} and the 2019-nCoV M^{pro}/3CL^{pro} are 96% identical, and the active sites in both proteins remain free from mutations. The disruption of protease activity can lead to various diseases; thus, commonly, host proteases can be used as potential therapeutic targets. In many viruses, proteases play essential roles in viral replication; therefore, proteases are often used as protein targets during the development of antiviral therapeutics.

4. Conclusion

Virtual screening with docking simulation in this study was carried out on five chemical compounds contained in *C. alata* and *D. petandra* on the SAR-CoV-2 protein and also five commercial antiviral compounds. Molecular docking analysis produces ligand interactions with the binding site. The interaction of ligand binding with SAR-CoV-2 shows that the interaction of amino acid residues plays an important role in maintaining functional conformation. The mechanism of enzyme and ligand interaction in this study is useful for understanding the prospective mechanism of the interaction of SAR-CoV-2 and ligands. Two parameters of the virtual filtering simulation that can predict SAR-CoV-2 inhibitory activity are low free energy Gibs and low kinetic inhibition value. Gibs free energy (ΔG) resulting from docking simulation were shown quercetin, kaempferol and aloe-modin have lower free energy Gibs and low kinetic inhibition compared to other sample compounds. Nelfinavir, lopinavir and hydroxychloroquine as commercial antivirals show good results as antivirals compared to other commercial compounds.

Acknowledgments
We are grateful for the funding supported by The Deputy of Engineering Sciences-LIPI. We are also very thankful for all the support and help from Research Center for Chemistry – Indonesian Institute of Sciences in facilitating the research. The first author who is also the correspondent author is the main contribution to this work.

References

[1] Xu X, Chen P, Wang J, Feng J, Zhou H, Li X *et al* 2020 *Sci. China Life Sci.* **63** 457-60
[2] Ji W, Wang W, Zhao X, Zai J and Li X 2020 *J. Med. Virol.* **92** 433-40
[3] Zhu N, *et al.* 2020 *N. Engl. J. Med.* **382** 727-33
[4] Shu Y and McCauley J J E 2017 *Euro Surveill.* **22**(13) 30494
[5] Chen Y, Liu Q and Guo D 2020 *J. Med. Virol.* **92** 418-23
[6] Zhou P, *et al.* 2020 *Nature* **579** 270–73
[7] Wu F, *et al* 2020 Nature **579** 265-69
[8] Needle D, Lountos G T and Waugh D S 2015 *Acta Crystallogr. D Biol. Crystallogr.* **71** 1102-111
[9] Anand K, Ziebuhr J, Wadhwani P, Mesters J R and Hilgenfeld R 2003 *Science* **300** 1763-67
[10] Ghosh AK, *et al.* 2005 *J. Med. Chem.* **48** 6767-71
[11] Kumar V, Tan KP, Wang YM, Lin SW and Liang PH 2016 *Bioorg. Med. Chem.* **24** 3035-42
[12] Pillaiyar T, Manickam M, Namasivayam V, Hayashi Y and Jung SH 2016 *J. Med. Chem.* **59** 6595-628
[13] Qamar M T, *et al.* 2019 *Sci. Rep.* **9** 1-16
[14] Notredame C, Higgins DG, Heringa J and Coffee T 2000 *J. Mol. Biol.* **302** 205-17
[15] Gouet P, Courcelle E, Stuart DI and Metoz F 1999 *Bioinformatics* **15** 305-8
[16] Eswar N, Webb B, Marti-Renom MA, Madhusudhan MS, Eramian D, Shen MY, *et al.* 2006 *Bioinformatics* **5**(1) 5.6.1-5.5.30
[17] Johnson M, Zaretskaya I, Raytselis Y, Merezhuk Y, McGinis S and Madden TL 2008 *Nucleic Acids Res.* **36** 5-9
[18] Gasteiger E, Gattiker A, Hoogland C, Ivanyi I Appel RD and Bairoch A 2003 *Nucleic Acids Res.* **31** 3784-88
[19] Pettersen EF, Goddard TD, Huang CC, Couch GS, Greenblatt DM, Meng EC, *et al.* 2004 *J. Comput. Chem.* **25** 1605-12

IOP Conf. Series: Materials Science and Engineering 1011 (2021) 012018 doi:10.1088/1757-899X/1011/1/012018

Wastewater treatment based on remazol blue using DBD plasma reactor under acidic and basic conditions

Badzlina Khairunizzahrah[1], Setijo Bismo[1]*

Chemical Engineering Department, Engineering Faculty, Universitas Indonesia

Email corresponding author: setijo.bismo@ui.ac.id

Abstract. Synthetic dyes are one of the most dangerous components of textile wastewater. Waste that contains various organic compounds, especially synthetic dyes, causes a high Chemical Oxygen Demand (COD) values. Remazol blue is a reactive dye that is widely used for the textile dyeing process. Therefore, water pollution occurs when the dye waste produced from textile activities is discharged into water bodies without prior treatment. In this study, innovation in waste treatment technology was applied in the Dielectric Barrier Discharge (DBD) plasma reactor. DBD are non-thermal plasma reactors designed to degrade organic and inorganic compounds in wastewater through oxidation process. This is carried out by active species, such as ozone (O_3), hydrogen peroxide (H_2O_2) and hydroxyl radical (•OH), which are formed in the reactor. The study aims to evaluate the performance of non-thermal DBD plasma reactor in treating wastewater containing one of the textile dyes, namely remazol blue. The results showed that the percentage of degradation achieved was 98.43% within a process time of 120 minutes with a final concentration of 0.75mg/L. Furthermore, it showed that optimum conditions were obtained at pH value of 4, air flow rate of 2.5L/min, liquid flow rate of 50mL/min and a primary Neon sign transformer (NST) voltage of 19kV.

1. Introduction

The textile industry has developed rapidly in recent years and has become one of the most polluted industries in the world. The increase in textile production by this industry has resulted to a rise in wastewater. During its manufacturing processes, the textile industry consumes a substantial amount of water and produces wastewater which contains hazardous substances and chemical dyes. Therefore, the wastewater produced becomes very colourful and contains organic pollutants that are difficult to degrade. The wastewater from the textile industry has a concentrated colour generally derived from the dye remnants. They form complex aromatic compounds in the form of azo compounds and derivatives of benzene groups which are difficult to degrade [1].

Various conventional methods for dealing with textile waste such as biological oxidation, physical-chemical treatment and adsorption have been carried out. However, these methods are less effective because most dyes are soluble in water, have complex structures and are resistant to chemical and biological degradation [2]. In addition, they have a high operational cost and toxic by-products are formed during the process. Various modern methods have been carried out for the removal of dye pollutants from wastewater before its final disposal into water bodies. They include biodegradation, chlorination and ozonation methods. However, these methods do not provide satisfactory results and require a high operational cost which makes it less effective in Indonesia. Therefore, in this study the proposed waste treatment technology innovation was the use of a Dielectric Barrier Discharge (DBD) plasma reactor.

DBD are non-thermal plasma reactors designed to degrade organic and inorganic compounds in wastewater through oxidation process by active species, such as O_3, H_2O_2 and •OH, which are formed in the reactor. It also has advantages, which include low energy consumption to induce chemical reactions in gases at relatively low temperatures (T< 200°C), operates at room temperature and atmospheric pressure, the reactor design allows the process scale-up to operate with larger volumes of gas, exhibits high electron density and glow discharge extensions [3].

2. Material and Methods

The main tool used in this study was the Dielectric Barrier Discharge (DBD) plasma reactor with a spiral cylindrical coaxial pipe configuration. The unity series of the reactor system which show how the DBD plasma reactor is connected with other supporting components is illustrated below in figure 1.

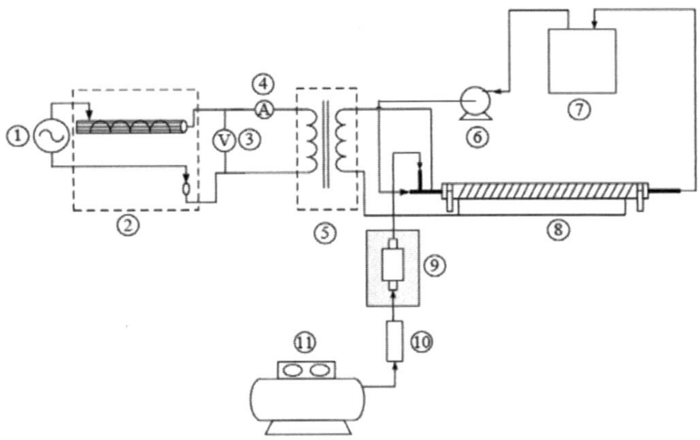

1	Electricity Source
2	Slide Regulator
3	Voltmeter
4	Amperemeter
5	Neon-Sign Transformer (NST)
6	Pump
7	Reservoir
8	DBD Plasma Reactor
9	Ozonator
10	Flowmeter
11	Compressor

Figure 1. The schematic diagram of experimental setup for wastewater degradation.

The synthetic wastewater which was degraded in this study was remazol blue. The initial concentration of remazol blue used was 50ppm. The total liquid volume used was 500mL. The test was carried out with an air flow rate of 2.5L/min, and an NST secondary voltage of 19kV with a processing time of 120 minutes by varying the pH of the solution i.e. pH 4 and 6 for acidic conditions pH 7 for neutral conditions and pH 10 for basic conditions.

3. Results and Discussion

The study began with a preparatory stage which included the reactor hydrodynamic and ozone solubility test. The number of hydroxyl radicals were quantified using two methods, iodometric and permanganometric titration, to find the optimum conditions. Furthermore, the degradation of remazol blue dye wastewater was carried out by varying the pH value at optimum conditions.

3.1. Hydrodynamic Testing

Hydrodynamic testing was carried out to determine the optimum residence time of the fluid in the DBD plasma reactor. With optimum residence time, the contact time between the feed gas fluid and plasma were optimal therefore, the reaction was effective. The calculation of the liquid phase residence time is shown in equation below.

$$\tau = \frac{V}{Q_L} \quad (1)$$

Where τ is the average residence time of the reactants in the reactor, V is the reactor volume, and Q_L is the flow rate of the liquid feed.

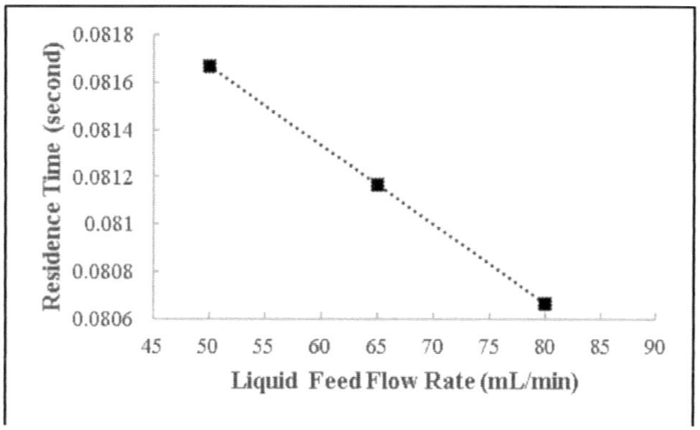

Figure 2. Effect of fluid flow rate on average residence time with a fixed air flow rate = 2.5 L/min.

Figure 2 shows that the higher the flow rate of liquid feed entering the DBD plasma reactor, the lower the residence time of the feed in the reactor. When the liquid flow rate was 50mL/min, residence time had its highest value. This was because at a low liquid flow rate not all parts of the space in the reactor were occupied by liquid therefore, the gas phase occupied the space that was not filled, which slightly compressed the liquid phase. Conversely, the liquid flow rate of 80mL/min had the smallest residence time value due to the flow being in the pulse regime (high interaction). Therefore, the liquid phase pressure was strong enough to withstand the gas phase pressure and the residence time was less influenced by the gas phase. Furthermore, based on the hydrodynamic test, it was predicted that the most optimum liquid feed flow rate to be operated in a non-thermal DBD plasma reactor was 50mL/min.

3.2. Hydroxyl Radical Quantification

The solution used in this study was aquades as a substitute for wastewater remazol blue compound with the aim that it does not affect the calculation of the quantity of •OH radicals formed. •OH radicals react with each other to form H_2O_2 (hydrogen peroxide) as shown in the equation below. H_2O_2 is a radical compound that is stable and has a long lifetime.

$$\bullet OH + \bullet OH \rightarrow H_2O_2 \quad (2)$$

To determine the production of hydroxyl radicals, the amount of H_2O_2 formed was measured.

3.2.1. Iodometric Titration Method. In the hydroxyl radical quantification of the iodometric titration method, the resulting •OH radical combined to produce H_2O_2, resulting in the addition of H_2O_2 to the KI used and iodine was produced as in the equation below [4].

$$H_2O_2 + 2KI + H_2SO_4 \leftrightarrow I_2 + K_2SO_4 + 2H_2O \quad (3)$$

$$I_2 + 2Na_2S_2O_3 \leftrightarrow Na_2S_4O_6 + 2NaI \quad (4)$$

From the above reaction, •OH radical calculation was compared with H_2O_2. KI solution produced other compounds besides iodine (I_2), which were basic in accordance with the oxidation reaction, as in the equation below:

$$2KI + 2\bullet OH \rightarrow I_2 + 2KOH \quad (5)$$

From this reaction, if the KI was oxidized by •OH radical therefore, the iodine formed will be half the number of moles of the •OH radical.

The test was carried out with an air flow rate of 2.5L/min, and an NST secondary voltage of 19kV with a processing time of 120 minutes. Samples were taken at 15, 30, 45, 60, 75, 90, 105 and

120minutes under optimum conditions i.e. when the air flow rate was 2.5L/min, liquid rate was 50mL/min and the NST secondary voltage was 19kV. This method was carried out by varying the pH of the solution i.e. pH 4 and 6 for acidic conditions pH 7 for neutral conditions and pH 9, 10 and 12 for basic conditions.

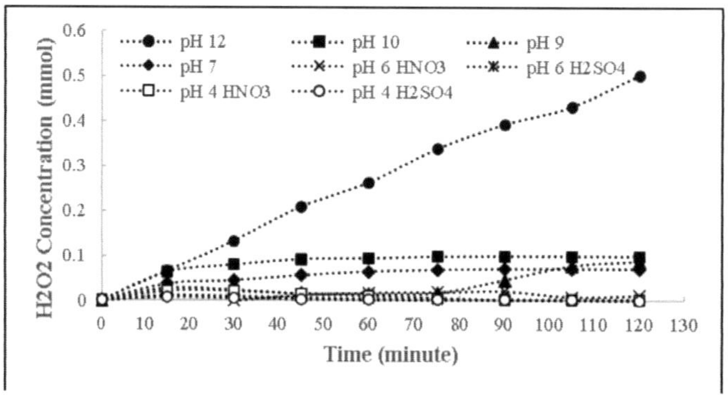

Figure 3. Effect of pH on the amount of H_2O_2 using iodometric titration method.

Figure 3 shows that the amount of H_2O_2 tends to rise with increasing pH. This was because the higher the pH, the faster the rate of ozone decomposition [5]. In basic conditions, ozone which is one of the active species produced in the discharge process was induced by the presence of hydroxide ions in solution. Therefore, it decomposed quickly into OH radicals [6].

The mechanism of ozone decomposition reaction into •OH radical is as follows [7; 8].

$$O_3 + OH^- \rightarrow HO_2^- + O_2 \quad (6)$$

$$HO_2^- + O_3 \rightarrow \bullet OH + \bullet HO_2^- + O_2 \quad (7)$$

3.2.2. Permanganometric Titration Method. The principle of permanganometric titration is based on the reduction-oxidation process. It involves the use of potassium permanganate ($KMnO_4$) as a strong oxidizer which also acts as a titrant. The H_2O_2 oxidation process by $KMnO_4$ follows the equation below:

$$5H_2O_2 + 2KMnO_4 + 3H_2SO_4 \rightarrow 2O_2 + 2MnSO_4 + K_2SO_4 + 8H_2O \quad (8)$$

The test was carried out with an optimized variable which was when the liquid flow rate was 50mL/min, air flow rate was 2.5L/min, and NST secondary voltage was 19kV with a processing time of 120minutes. Samples were taken at 15, 30, 45, 60, 75, 90, 105 and 120 minutes. Production of hydroxyl radicals in DBD plasma reactors in this method was analyzed from the influence of pH values.

From figure 4, it is seen that at higher pH, ozone showed more stable results, while at lower pH the amount of hydroxyl radicals fluctuated. Ozone stability was strongly influenced by the water matrix, specifically pH, Natural Organic Matter (NOM) and alkalinity. The pH value was used as an indication of the activity of hydrogen ions [9]. High pH made the formation of OH radicals easier due to the presence of OH ions, which initiated the decomposition of ozone as the first step in producing OH radicals.

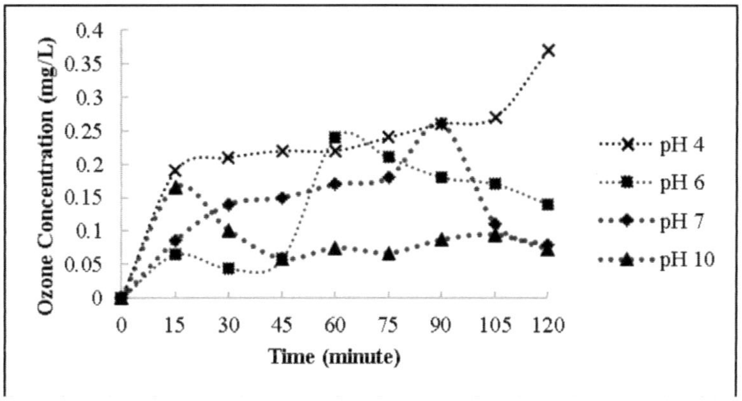

Figure 4. Effect of pH on the amount of H_2O_2 using permanganometric titration method.

3.3. Ozone Solubility Test

To carry out the ozone solubility test in water, aquades was used at an acidic, neutral and basic pH (pH of the solution = 4, 6, 7, and 10) under optimum conditions. It was used as a substitute for remazol blue to measure the solubility of ozone in water. The solution was treated for 120 minutes and the ozone levels were checked periodically at 15, 30, 45, 60, 75, 90, 105 and 120 minutes. The concentration of dissolved ozone in water was determined using a spectrophotometer. Furthermore, the sample was reacted with a reagent ozone kit, before it was measured with a spectrophotometer.

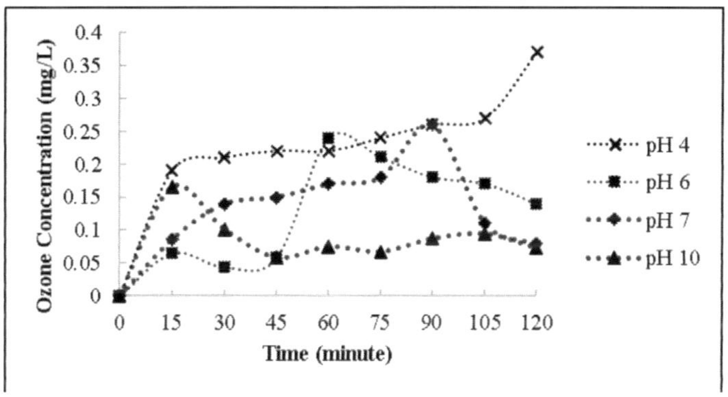

Figure 5. Effect of pH on ozone concentration in water.

From figure 5, it is seen that the concentration of dissolved ozone fluctuated due to the short lifetime it possesses when dissolved in water. Therefore, the dissolved ozone reacted with water molecules and decomposed easily into oxygen as in the equation below [10].

$$O_3 + H_2O + 2e \rightarrow O_2 + 2OH^- \quad (9)$$

In addition, the radicals formed in the DBD plasma reactor were reactive and interacted with ozone as shown by the following equations:

$$O_3 + \bullet HO_2 \rightarrow \bullet OH + 2O_2 \quad (10)$$

$$O_3 + \bullet O_2^- \rightarrow \bullet O_3^- + O_2 \quad (11)$$

Based on equation 10 and 11, it is seen that dissolved ozone returned to oxygen in the gas phase due to interactions with radicals. In addition, the fluctuation in the ozone solubility values was caused by unstable temperatures because, it was carried out at room temperature [11].

3.4. Degradation of Remazol Blue Compounds

In this study, the degradation of wastewater dyes was carried out using synthetic waste in the form of a remazol blue dye, which had an initial concentration of ≈ 50ppm. At this stage, the effect of the initial pH of wastewater on the percentage of degradation of remazol blue compounds was evaluated. This was carried out with variables that have been optimized at the quantification stage of hydroxyl radicals, namely at a liquid flow rate of 50mL/min, an air flow rate of 2.5L/min and a secondary NST voltage of 19kV. The results of the percentage degradation of remazol blue with variations in pH are shown in figure 6.

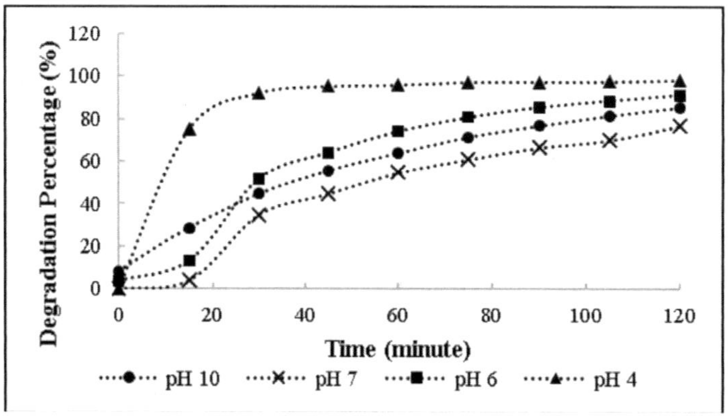

Figure 6. Effect of pH on percentage of remazol blue degradation.

Table 1. Remazol blue concentration in waste after degradation process.

pH	Final degradation (%)	Initial concentration (mg/L)	Final concentration (mg/L)
4	98.4397753	48.03	0.75

Figure 6 and table 1 show that the pH value affected the rate of degradation of the dye. The ability of ozone to degrade dyes was higher at low pH for certain dyes [12]. In remazol blue dye, its molecules oxidized specific chromophore molecules more effectively than hydroxyl radicals. In general, ozone oxidation pathways include direct oxidation by it or radical oxidation by OH. Direct oxidation by ozone molecules is more selective and dominates under acidic conditions, while radical oxidation by hydroxyl is less selective and dominates under basic conditions [13].

There are three possible reasons that may help to explain the above statement: 1) Anthraquinone dyes contain rigid structures that are not sensitive to oxidation reactions therefore, the ozonation mechanism may not be the dominant process for dye decomposition. 2) Anthraquinone may be transformed into leuko-forms (less colored) in acidic media, which reduces the colour content. 3) Amine group ($-NH_2$) in anthraquinone dye may be ionized to ($-NH_3^+$) at low pH, which increases the solubility of the dye [13].

4. Conclusion

Wastewater treatment on remazol blue compound using a non-thermal DBD plasma reactor was carried out with a degradation percentage of 98.43% and a final concentration of 0.75mg/L. The lower the pH, the higher the percentage of degradation. The optimum operating conditions for remazol blue compound degradation were obtained at pH 4, air flow rate 2.5L/min, liquid flow rate 50mL/min, secondary voltage NST 19kV, liquid volume 500mL, and waste concentration 50mg/L in 120 minutes processing time.

References

[1] Ghaisani A D, Mahayum A, Kusumandari K and Saraswati T E 2019 Degradation of dyes in textile industry wastewater using dielectric barrier discharge (DBD) plasma *AIP Conf. Proc.* **2097** 1 030097.

[2] He Z, Lin L, Song S, Xia M, Xu L, Ying H and Chen J 2008 Mineralization of C.I. reactive blue 19 by ozonation combined with sonolysis: peformance optimization and degradation mechanism *Sep. Purif. Technol.* **62**(2) 376-81.

[3] Rico V J, Hueso J L, Cotrino J and González-Elipe A R 2010 Evaluation of different dielectric barrier discharge plasma configurations as an alternative technology for green C1 chemistry in the carbon dioxide reforming of methane and the direct decomposition of methanol *J. Phys. Chem. A* **114(**11) 4009-16.

[4] Brandhuber P and Korshin G V 2009 Methods for the detection of residual concentations of hydrogen peroxide in advanced oxidation processes *WateReuse Foundation.*

[5] Ershov B G and Morozov P A 2009 The kinetics of ozone decomposition in water, the influence of pH and temperature *Russ. J. Phys. Chem.* A **83(**8) 1295-9.

[6] Beltran F J 2003 Ozone reaction kinetics for water and wastewater systems *crc Press*.

[7] Vanraes P, Ghodbane H, Davister D, Wardenier N, Nikiforov A, Verheust Y P, Van Hulle S W, Hamdaoui O, Vandamme J, Van Durme J and Surmont P 2017 Removal of several pesticides in a falling water film DBD reactor with activated carbon textile: energy efficiency *Water res.* **116** 1-2.

[8] Duan L, Shang K, Lu N, Li J and Wu Y 2013 March Study on the factors influencing phenol degradation in water by dielectric barrier discharge (DBD) In *J. Phys.: Conf. Ser.* **418** 012129.

[9] Sawyer C N, McCarty P L and Parkin G F 2003 *Chemistry for environmental engineering and science* New (York: McGraw-Hill)

[10] Nur M, Bonifaci N and Denat A 1998 Non-thermal electron mobility in high density gaseous nitrogen and argon in divergent electric field *TITLE AND SUBTITLE* **4** 12.

[11] Yao S, Wu Z, Han J, Tang X, Jiang B, Lu H, Yamamoto S and Kodama S 2015 Study of ozone generation in an atmospheric dielectric barrier discharge reactor *J Electrostat.* **75** 35-42.

[12] Liu B W, Chou M S, Kao C M and Huang B J 2004 Evaluation of selected operational parameters for the decolorization of dye-finishing wastewater using UV/Ozone *Ozone: Sci. Eng.* **26(**3) 239-45.

[13] Jdhfjksd Chu W and Ma C W 2000 Quantitative prediction of direct and indirect dye ozonation kinetics *Water Research* **34(**12) 3153-60.

Modified cassava flour (mocaf) as food additive on infant instant cream soup using fortificant of natural folic acid

Agustine Susilowati*, Yati Maryati, Hakiki Melanie, Puspa D. Lotulung, Aspiyanto, Nina Artanti and Setyani Budiari

Research Centre for Chemistry – Indonesian Institute of Sciences, PUSPIPTEK, Serpong, South Tangerang - 15314, Banten, Indonesia

*E-mail: agustine_1408@yahoo.co.id

Abstract. Modified cassava flour (Mocaf) can be used for food additive on the formulation of infant instant cream soup as thickener and binder of folic acid. The main objective of this study was to determine the optimum formulation condition of added mocaf and folic acid fortificant on infant cream soup for complementary feeding. Infant cream soup was prepared by adding mocaf at concentration 0, 6, 12, 18, 24, and 30% (w/w, base formula of cream soup), fortificant A (mixture of soy tempeh, nixtamalized corn, and fermented broccoli), and fortificant B (mixture of mung beans, nixtamalized corn and fermented broccoli). The optimum condition was achieved by adding mocaf at the concentration 30% and fortificant A, with composition of folic acid 191.14 µg/mL, dissolved protein 39.72 mg/mL, total solids 91.43 %, total sugars 926.73 mg/mL and reducing sugars 52.1 mg/mL. This formulation was able to increase folic acid by 10.27-folds in comparison with base formula of infant soup. LC-MS analysis showed folic acid monomer with molecular weight 442.75 m/z. Volatile compounds were dominated by acetic acid, furan and, fatty acid. Particle size distribution at quantities 10, 50 and 90% (v/w) had particle size of 10.58, 22.46, and 470.52 µm, respectively or average 141.26 µm.

1. Introduction

Instant cream soup is one of the Complementary Feedings (CF) apart from porridge and infant biscuit, which is introduced to infant with age of 6 - 24 months. Supplying infant cream soup is not only aimed for infant growth – development, but also used to adapt the digestive system since the taste, aroma, and other factors are related to the five senses of infant. Folic acid (Vitamin B9/Folacin) is one of the 13 essential vitamins required by women planning for a pregnancy, before and during early pregnancy, and the infant growth – development to prevent neural tube defects. Also, it has an important role as precursor in synthesis of nucleotide into re-methylation of homocysteine at period of fissing and growing cells, particularly to produce red blood cells and prevent the incidence of neural tube defects (NTDs) [1 , 2]. Instant cream soup for infants is prepared by adding fortificant of natural folic acid in order to support the utilization of safe and natural ingredient for infants. Fortificant of natural folic acid is produced from mixture of soy (*Glycine soya*) tempeh or mung bean (*Phaseolus radiatus* L.) fermented by *Rhizopus oligosporus* C1, combined with broccoli (*Brassica oleraceae* L.) or spinach (*Amaranthus* sp.) fermented by Kombucha culture, and nixtamalized yellow corn (*Zea mays* var. *indentata*) or white corn (*Zea mays* var. *indentata*) [3].

Furthermore, formulation of infant cream soup was performed by adding mocaf flour fermented by Lactic Acid Bacteria (LAB), enables to improve folic acid content in infant cream soup. Mocaf is modified cassava flour produced from fermentation with LAB microbes.These microbes induce characteristic alteration of cassava flour, including viscosity, gelation capability, power of rehydration, and easily dissolve. In overall, characteristic of mocaf is possibly affected by type and concentration of LAB culture, and fermentation time [4]. However, potential role of mocaf as food additive for thickener, binder, carrier, and emulsifier is less investigated.

The regulation concerning Food Additives is classified thickener or emulsifier or stabilizer agents as modified starch through acid or basic or oxidation or bleaching, and enzymatic processes as monostarch phosphate or hydroxy propyl distarch phosphate. However, it has yet specified the modified enzymatically process [5]. Mocaf possibly affects the existence of folic acid in instant cream soup, and it is closely impacted its composition and organoleptic quality. Mocaf used in the formulation of infant instant cream soup is expected to contribute in binding folic acid due to smaller size of granular and its functional group modification of cassava carbohydrate by LAB fermentation which is able to absorb and bind folic acid. Different mocaf concentration in formulation of infant instant cream soup may have different ability in binding folic acid and their physicochemical composition. Formulation of infant cream soup consist of mixture of flour and maize as source of soluble carbohydrate and other materials such as milk, dry vegetable, sugar and fortificant of folic acid from mixture of soy tempeh, fermented broccoli or spinach, and nixtamalized corn as main source of folic acid. Infant instant cream soup is formulated by equivalence to concentration of folic acid according to recommended dietary allowance (RDA) (80 µg/day or 30 – 50 µg per serving after pouring) [6].

The experimental study aims to evaluate the effect of mocaf concentration as thickener and carrier of folic acid on different type of fortificants in the formulation of infant instant cream soup and their physicochemical characteristics including folic acid, volatile compounds, and particle size.

2. Materials and Methods

2.1. Materials and equipment

Materials used in this experiment were soy beans, mung beans, broccoli obtained from a local market, dry yellow and white corn from type of horse dent procured from a local corn plantation center (South Tangerang), sucrose, inoculum of *R. oligosporus* strain C1 and Kombucha culture (Research Center for Chemistry – LIPI), Mocaf (Research Center for Biotechnology-LIPI), wheat flour (local, Bogasari), corn starch (Honig, Netherland), salt, powder full cream milk, dried carrot, dried onion, dried spring onion, white pepper, and garlic powder. All the chemical used in the process and analysis were analytical grade.

The equipment used in this experiment were analytical balance (Fujitsu, Japan), autoclave (CHENG YI, LS – 50 L, China), homogenizer (Ultra-Turrax, Ika Labortechnik, T50, Jane & Kunkel, Germany), blender (local, National), laminar flow chamber (local), incubator (local), cabinet dryer, and sieves of 60 and 80 mesh (Retsch, Germany). The analysis instrument used in this study were UV-Vis Spectrophotometer (Model RF-550, Shimadzu, Japan), Liquid Chromatography-tandem Mass Spectrometry (LC-MS) (Mariner Biospectrometry) with LC (Hitachi L 6200), Particle Size Analyzer (PSA) (Horiba Laser Scattering Particle Size Distribution Analyzer LA-960) and GC-MS (Shimadzu, Japan).

2.2. Experimental design

Experimental study was carried out by adding mocaf flour at concentration 0, 6, 12, 18, 24, and 30% (w/w, base formula of infant cream soup) to formula of infant cream soup consisting of base formula and fortificant A or fortificant B, equivalent to folic acid 1,000 µg, respectively. Fortificant A and fortificant B are mixture of soy tempeh, nixtamalized yellow corn and fermented broccoli, and mung beans, nixtamalized yellow corn and fermented broccoli in order to develop infant instant cream soup A and infant instant cream soup B, respectively. Analyses were carried out on initial fortificant materials and infant instant cream soup on dissolved protein (spectrophotometry) [7], total solids (Gravimetric method), total sugars (Phenol Sulphate method), reducing sugars (Somogyi Nelson method) [8] and folic acids (spectrophotometry) (Ruengsitagoon, & Hattanat, 2012) [9]. Identification of folic acid was conducted by LC-MS (Mariner Biospectrometry) equipped with LC (Hitachi L 6200) [10], distribution of particles size by Particle Size Analyser (Horiba LS-PSA LA-960) [11], and analysis of volatile compound by GC-MS [12]. Experimental data was determined by description method using average yield in triplicate analyses from duplicate process.

2.3. Procedures

2.3.1. Fermentation process of soy bean, mung bean tempeh and broccoli. Soy bean or mung bean were sorted, washed, blanched for 30 – 45 minutes, cooled, and steeped at pH 5 overnight. Next day, processed beans were hulled, washed and inoculated with tempeh inoculum of *R. oligosporus* strain-C1 at concentration 0.2% (w/w) uniformly, packaged in perforated plastic, and incubated at room temperature (28 – 30 °C) for 24 – 36 hours. Meanwhile, fermentation of broccoli was conducted by blanching broccoli at 80 °C for 5 minutes, and pulverized at the ratio of broccoli and water of 1:4 and resulted in broccoli suspension as substrate for folic acid. Broccoli suspension substrate was then poured into container, then added appropriate broccoli inoculum with broccoli concentration 15 % (v/w broccoli suspension) and sucrose 10 % (w/w, broccoli suspension). The suspension was stored in closed container with cheese cloth in dark room at room temperature for 3 days and then collected. All experimental processes were performed aseptically [3].

2.3.2. Nixtamalization process. A number of yellow corns were subsequently washed and steeped in water at the ratio of corn and water of 1:4. Then, $Ca(OH)_2$ were added with concentration of 20% (w/w dissolved protein of corn), and cooked at 90 °C for 60 minutes, cooled, rinsed to remove cooking water and excess lime, allowed, grounded and sieved through 80 mesh to produce nixtamalized yellow corn powder [3].

2.3.3. Preparation fortificant from mixture of tempeh paste, fermented broccoli and nixtamalized corn. Preparation process of mixture of tempeh paste and fermented broccoli was mixed at the ratio of tempeh paste and fermented broccoli of 1:1 [3], and homogenized at 8000 rpm for 30 minutes until it formed paste-like mixture. Nixtamalized corn was added into mixture of tempeh paste and fermented broccoli at the ratio of paste mixture and nixtamalized corn of 1 : 1, homogenized at 8000 rpm for 30 minutes and the mixture was obtained as fortificant paste. Next, the fortificant paste mixture was subsequently dried using cabinet dryer at 50 °C for 24 hours, and the sample size was reduced and sieved through 60 mesh until the desired powder was obtained, expressed as fortificant powder of A and B.

2.3.4. Formulation of instan infant cream soup using mocaf as binder of folic acid and thickener. The fortificant of natural folic acid being equivalent to folic acid 1,000 μg was subsequently added into base formula of infant cream soup consisting of mixture of wheat flour (26.7%), maize (53.3%), powder full cream milk (5%), salt (5%), sucrose (2%), dried carrot (3.4%), dried onion (2.4%), dried sliced spring onion (0.5%), white pepper (0.2%) and garlic powder (1.5%), and added mocaf flour at 0, 6, 12, 18, 24, and 30% (w/w, base formula of infant cream soup), mixed in dry condition, reduced size, sieved via 80 mesh, and packed. Infant instant cream soup was formulated according to recommended dietary allowance (RDA) on folic acid according to standard SNI No. 01-711.1, 2007 [13] for baby food (± 400 μg/day) so that it is predicted one (1) serving containing ± 75 – 150 μg) with weight 75 g powder cream soup and adding water 750 mL as intake for 1 (one) day.

3. Result and Discussion

3.1. Characteristic of materials

Preparation of instant cream soup by adding mocaf as carrier and thickener is related with composition of materials in its formulation. Initial composition of raw material prior to dry mixing process is shown in table 1. Based on the composition of each material, the highest folic acid (503.60 μg/mL) was obtained from soy tempeh, which is higher than mung bean tempeh (381.5 μg/mL), but had lower dissolved protein (1.15 mg/mL) than that of mung bean (1.80 mg/mL). Fermentation process of soy tempeh involves protease enzyme activity from *R. oligosporus* strain C_1 which causes protein degradation of soy tempeh to form amino acids, particularly glutamic acid as component in forming folic acid, in which expressed as dissolved protein [3]. As source of carbohydrate, modification of yellow corn through nixtamalization process [3] generates total sugars 146.84 mg/mL. Nixtamalized yellow corn also had the highest total solids (54.46%) which has main role on overall composition and fortificant. In addition, fermentation process of broccoli by Kombucha culture yields the highest concentration of reducing sugars (61.63 mg/mL) in comparison with soy tempeh and nixtamalized corn. This fermentation used sucrose as source of carbon to produce folic acid through synthesis *de novo* [14,15] by Kombucha culture. Remained reducing sugars from this synthesis are soluble in biomass and as a source of energy, and affect the *taste* of product.

Fortificant B has higher composition of folic acid, total solids, reducing sugars, and total sugars, but it has a lower composition of dissolved protein than fortificant A. This is possibly caused by difference in initial composition of material. Although, fortificant A used soy tempeh with higher concentration of folic acid (503.60 µg/mL) when compared with fortificant B using mung bean tempeh with folic acid 381.50 µg/mL. However, the presence of folic acid is also affected during processes such as homogenizing, drying, size reduction, and sieving. Folic acid is susceptible to mechanical treatment, light, and temperature [16]. Based on the composition of base formula of infant instant cream soup, it showed that folic acid is relatively high (109.09 µg/mL), however folic acid in mocaf is quite low (22.26 µg/mL). Table 1 summarizes material composition in formulation of infant instant cream soup, whereas figures 1a, 1b, 1c, and 1d shows subsequently fortificant A, fortificant B, base formula of infant instant cream soup and formula of mocaf flour.

Table 1. Composition of materials in formulating instant infant cream soup by using fortificant of natural folic acid and mocaf as carrier of folic acid.

Type of materials	Components				
	Folic acid (µg/mL)	Dissolved protein (mg/mL)	Reducing sugars (mg/mL)	Total sugars (mg/mL)	Total solids (%)
Mung bean tempeh	381.50	1.80	25.72	46.42	27.59
Soy tempeh	503.60	1.15	7.13	60.00	44.02
Fermented broccoli	113.26	1.25	61.63	65.03	8.08
Nixtamalized yellow corn	466.81	0.68	12.70	146.84	54.46
Fortificant A[a]	101.89	1.03	23.77	113.00	92.91
Fortificant B[b]	130.54	0.83	161.37	318.75	93.07
Base formula[c]	109.09	13.90	27.18	159.76	96.07
Mocaf flour	22.26	1.17	5.50	162.90	96.13

[a] Mixture of nixtamalized yellow corn, soy tempeh and fermented broccoli
[b] Mixture of nixtamalized yellow corn, mung bean tempeh and fermented broccoli
[c] Formula infant cream soup

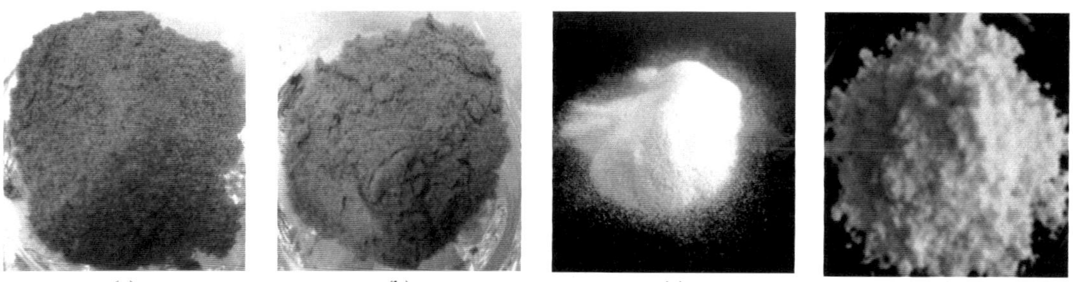
(a) (b) (c) (d)

Figure 1. (a) Fortificant A, (b) fortificant B, (c) base formula of instant porridge, (d) mocaf flour.

3.2. Effect of formulation on composition of infant cream soup

3.2.1. Folic acid (µg/mL) and dissolved protein (mg/mL). Preparing infant instant cream soup as source of folic acid is carried out through mixing the raw materials, fortificant A or fortificant B, base formula of cream soup and various mocaf concentration at rotation speed 4000 rpm for 15 – 30 minutes. In overall, it showed that the increase of mocaf concentration resulted in the increase of folic acid and dissolved protein in infant instant cream soup (Fig. 2). Both instant cream soup from fortificant A and fortificant B reached the highest folic acid (191.15 and 124.40 µg/mL) at mocaf concentration of 18% and 30%, respectively. These results were probably not only caused by dry mixing process, in which particle size from each formula and operation condition of dry mixing (4000 rpm, 15 minutes) affects on binding system of inter-components, but was also caused by difference in folic acid concentration of initial material. Folic acid in fortificant A is lower (101.89 µg/mL) than fortificant B (130.54 µg/mL) (Table 1). Thus, it required higher mocaf concentration to attain optimal folic acid (30%) in comparison with fortificant B (18%). In this optimum condition, the addition of mocaf with fortificants A and B in the formulation increased folic acid concentration by 1,027% (10-folds) and 129.05% (1.29-folds) when compared without adding mocaf, respectively.

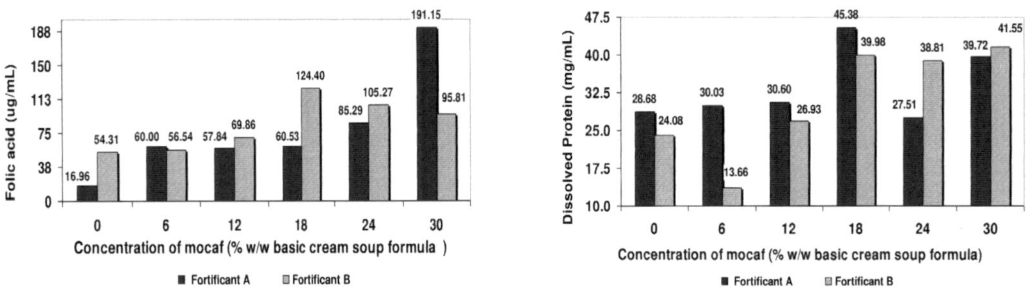

(a) (b)

Figure 2. Relationship of type of fortificant and mocaf concentration on recovery of folic acid (a) and dissolved protein (b) in infant instant cream soup.

Dissolved proteins were fluctuated and the optimum dissolved protein (45.38 mg/mL) were achieved at mocaf concentration of 18% by using fortificant A. This result is higher than that of fortificant B (41.55 mg/mL) at mocaf concentration of 30%, as shown in Figure 2b. The increment of dissolved protein concentration is not mainly due to the contribution of fortificant (fortificant A 5.8 and B 6.0 mg/mL), but also due to dissolved protein content in mocaf (1.17 mg/mL) (Table 1). Thus, the increasing of mocaf concentration resulted in the increase dissolved protein in infant instant cream soup, which has the binding ability in formulation. On the contrary, the reduction of dissolved protein is possibly caused by protein denaturation as a consequence of mechanical force during mixing process, in which lysis or partial dissociation as volatile compounds were occurred, that is unable to detect as dissolved protein according to Lowry method [8]. In optimum condition, the addition of mocaf by using fortificant A and fortificant B increased dissolved protein by 58.23% and 72.55%, respectively when compared without adding mocaf of 26.68 and 24.08 mg/mL, respectively.

3.2.2. Total sugar (mg/mL), reducing sugars (mg/mL) and total solids. Total sugars, reducing sugars, and total solids increased and then declined, as shown in Figure 3. Total sugar is the overall components in carbohydrate as source of energy. Optimization of formulation using both fortificants A and B is achieved at mocaf concentration of 30% (926.73 mg/mL) and 18% (623.75 mg/mL), as shown in Figure 3a. Total sugars were increased by 171% (1.7-folds) and 366.74% (3.67-folds) when compared without mocaf of 341.71 and 133.64 mg/mL, respectively. This increase is caused by mocaf concentration and composition of total sugars in initial material, although fortificant A has lower concentration of total sugars (78.28 mg/mL) in comparison with fortificant B (89.65 mg/mL) [8].

IOP Conf. Series: Materials Science and Engineering 1011 (2021) 012019 doi:10.1088/1757-899X/1011/1/012019

The reducing sugars presented a fluctuate trend. The cream soup with mocaf concentration of 30% and 12% had reducing sugars of 52.10 and 56.84 mg/mL, respectively, as demonstrated in Figure 3b. There was an increment of reducing sugars by 10.29% and 31% than cream soup with no added mocaf, which was 47.24 and 43.15 mg/mL, respectively. This is probably caused by mocaf contribution and interaction with other components during dry mixing process. Mocaf is soluble cassava starch as a result of fermentation by LAB. During fermentation process, carbohydrates are degraded by enzymes (invertase, amylase) of microbes to form monosaccharide. Meanwhile, the reduction of reducing sugar is possibly due to partial lysis during process [8].

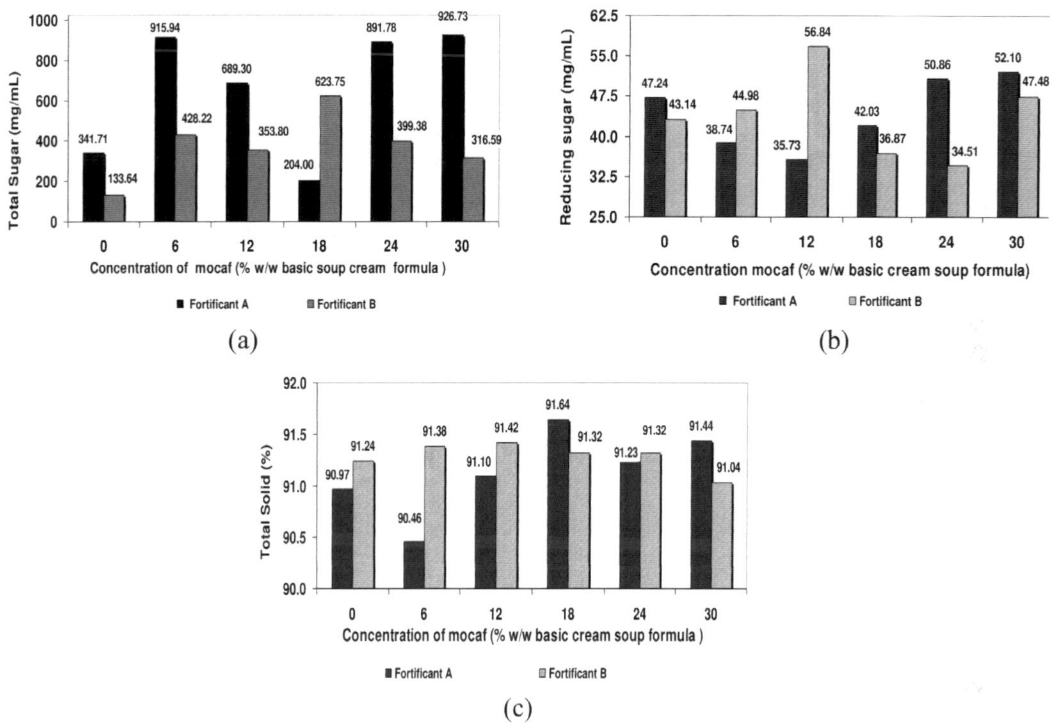

Figure 3. Relationship of type of fortificant and mocaf concentration on recovery of total sugars (a), reducing sugars (b), and total solids (c) in infant instant cream soup.

The infant cream soup with both fortificants A and B at mocaf concentration of 18% and 12% had total solids of 91.64% and 91.42%, respectively, as indicated in Figure 3c. Fortificant B had lower total solids since it required smaller mocaf concentration than fortificant A (12% > 18%). In optimum condition, adding mocaf at dry mixing process increased total solids of 0.74% and 0.2% in comparison with no added mocaf, which were 90.97% and 91.24%, respectively.

Based on the highest folic acid content, the optimum condition was achieved at adding mocaf concentration of 30% using fortificant A, which resulted in folic acid of 191.14 μg/mL. This indicates the ability of mocaf to bind folic acid in the formulation.

IOP Conf. Series: Materials Science and Engineering 1011 (2021) 012019 doi:10.1088/1757-899X/1011/1/012019

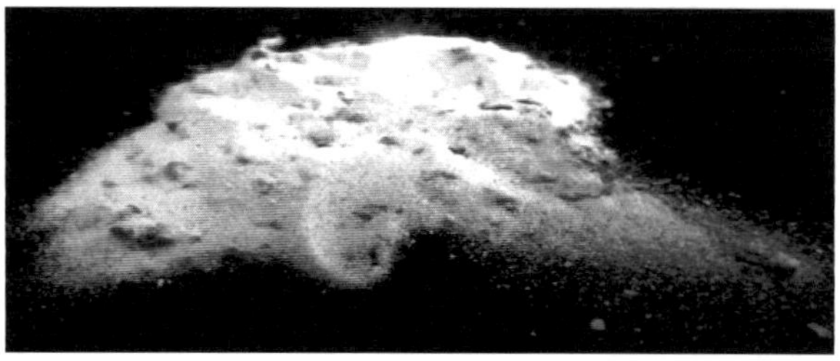

Figure 4. Infant instant cream soup by using fortificant powder and adding mocaf 30% (w/w, base formula of cream soup).

3.3. Characteristic of volatile compounds from infant instant cream soup.

Identification of volatile compounds in infant instant cream soup is performed by GC-MS on infant soup cream with 30% mocaf concentration using fortificant A since it has the highest folic acid content. Figure 5 showed chromatogram of volatile compounds with 40 peaks dominated by ethyl/methyl ester, acetic acid of 40.805%. This compound was possibly incurred from mocaf flour fermented by Lactobacillus sp., which is able to produce acetic acid, lactic acid, and their derivative compounds (ethyl/methyl ester). Other volatile compounds, such as benzaldehyde, 3-hydroxy-4-methoxy of 16.576% were probably derived from aromatic compounds from tempeh by its formation of amino acid (lisin, arginin, prolin, fenilalamin and valin) during fermentation. Bitter taste was resulted from degradation or oxidation of trigliseride of soy fat with lime taste caused by glycoside. Glycoside is isoflavon consisting of genistin and daidzin to contribute off-flavor on tempeh. Other compounds are furan and their derivatives, like furfural, 5-Hydroxymethylfurfural, D-Glucopyranose, 4H-Pyran-4-one of 13,945%. These compounds are produced from the whole flour in formulation (rice flour, wheat flour, mocaf), sucrose as raw materials and fortificant. Chemical reaction between sugars and amino acids and its interaction with dry mixing process probably formed these compound derivatives. Fatty acid formed in infant instant cream soup, such as tetradecanoic acid, hexadecanoic acid, octadecadienoic acid, linoelaidic acid and Octadecenoic acid are existed at 5,427%, and fatty acids derivatives (methyl ester, such as Glycerol, 3-Methoxy-2,2-dimethyloxirane, Mercaptoacetone, 2-Propenol, Cyclopropyl carbinol, 3-Hexene, Cyclopropanecarboxamide, Ribofuranosyl and 1,3-Propanediol of 29,588%). These compounds were possibly developed during formulation of cream soup consisting of full cream milk and fortificant of folic acid, particularly fat from soy tempeh. These fatty acids give specific aroma of cream soup.

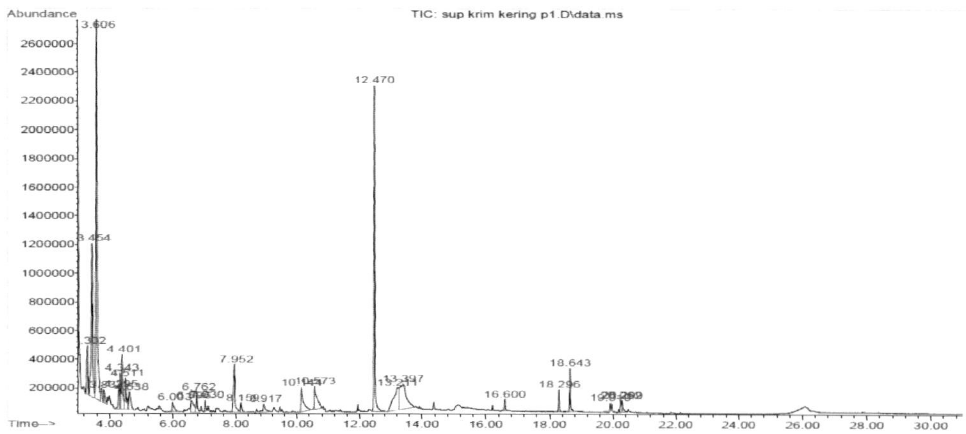

Figure 5. Volatile compounds in infant instant cream soup at mocaf concentration 30% (w/w, base formula of cream soup) by using fortificant A.

IOP Conf. Series: Materials Science and Engineering 1011 (2021) 012019 doi:10.1088/1757-899X/1011/1/012019

3.4. Identification of folic acid monomer

Folic acid has MW of 441 Da. Operation condition of LC-MS is at injection volume 5 μL, flow rate 0.2 mL/min. with eluent mixture of methanol and water at 80:20 ratio using column of C-8 (15 mm x 2 mm) [10]. Identification of standard folic acid standard by LCMS resulted in 1 peak at retention time (RT) at 3.3 minutes with relative intensity 100% and MW of 442.37 Da, as seen in Figure 6a and 6b.

Meanwhile, the chromatogram of infant soup cream at mocaf concentration of 30% using fortificant A showed one peak at retention time (RT) 1.6 minutes, dominated by 6 monomers of folic acid with MW ranged from 442.09 - 442.93 (Fig. 6c and 6d).

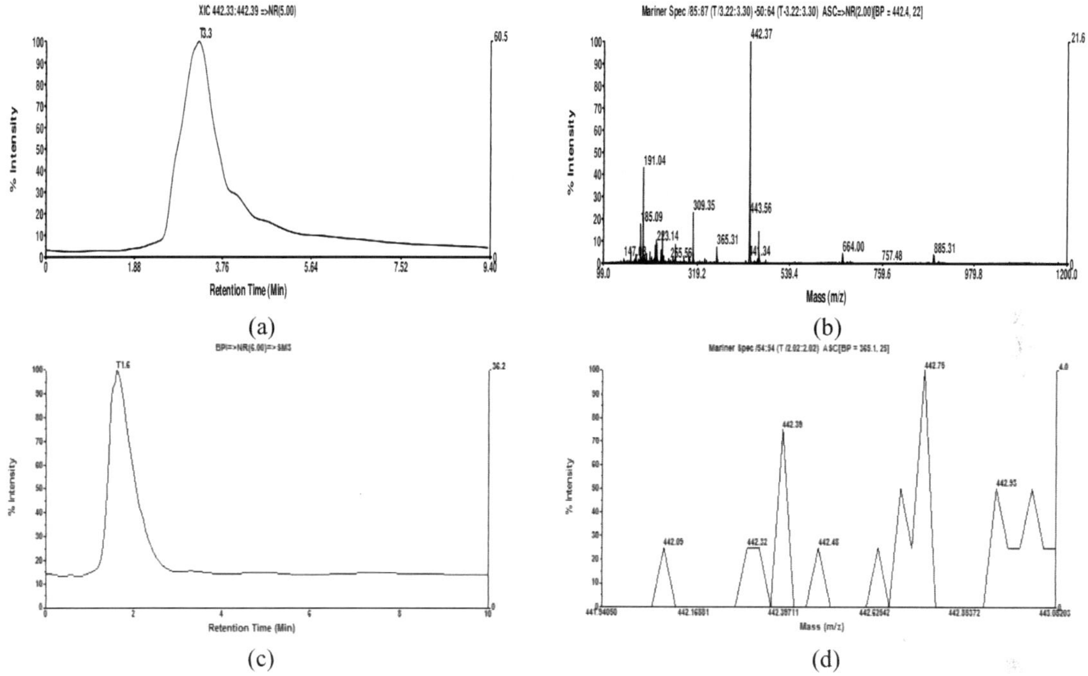

Figure 6. (a) Chromatogram and (b) mass spectra of standard folic acid, and (c) chromatogram and (d) mass spectra of infant instant cream soup at mocaf concentration 30% (w/w, base formula of cream soup).

3.5. Particle size and particle size distribution

The particle size distribution in infant instant cream soup showed the D10, D50 and D90 were 10.58 μm, 22.46 μm and 470.52 μm, respectively, with a mean diameter size 141.26 μm (Figure 7).

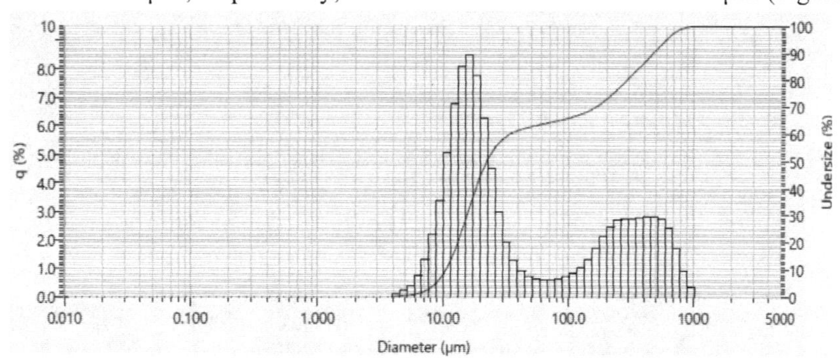

Figure 7. Particle size distribution in infant instant cream soup at mocaf concentration 30% (w/w, base formula of cream soup) by using fortificant A.

4. Conclusions

Mocaf flour has potential use as food additive, specifically as folic acid binder and thickener in preparation of infant instant cream soup using fortificant of natural folic acid. Based on folic acid recovery, the optimum condition was achieved at mocaf concentration 30% (w/w, base formula of cream soup) using fortificant A with compositions of folic acid 191.14 μg/mL, dissolved protein 39.72 mg/mL, total solids 91.43%, total sugars 926.73 mg/mL, and reducing sugars 52.1 mg/mL, respectively. The formulation with mocaf addition was able to recover folic acid by 1,027.06% (10.27-folds) than infant cream soup with no added mocaf. In addition, the LCMS analysis indicated the predominant monomer of folic acid with MW of 442.75 Da. Volatile compounds in infant cream soup were consisted of acetic acid compound and its derivatives 40.805%, sugar compounds and its derivatives 13,945%, fatty acid 5.427%, and fatty acid derivatives 29.588%. Particle size distribution in infant instant cream soup showed a mean diameter of 141.26 μm.

References

[1] Baley L B 2009 *Folate in Health and Disease* (Boca Raton : CRC Press).
[2] McKenzie S B 1999 *Megaloblastic and Nonmegaloblastic Macrocytic Anemias Hematology* (Baltimore: Williams adn Wilkins) pp 179-199
[3] Susilowati A, Iskandar Y M, Melanie H, Maryati Y, Lotulung P D and Aryani D G 2015 Pengembangan Konsentrat Sayuran Hijau dan Kacang-Kacangan Terfermentasi pada Jagung (Zea Mays L.) Pramasak sebagai Sumber Asam Folat untuk Formula Pangan Pintar Laporan Hasil Penelitian Program Tematik Kedeputian IPT Tahun Anggaran 2015 (Tangerang Selatan : Pusat Penelitian Kimia LIPI)
[4] Rahayu E S 2010 Lactic Acid Bacteria and Their Role in Food and Health: Current Research In Indonesia Skripsi Sarjana (Yogyakarta : UGM)
[5] Departemen Kesehatan Republik Indonesia 1992 Kumpulan Peraturan Perundang-undangan di bidang makanan Edisi II Jilid II Peraturan Menteri Kesehatan Republik Indonesia Nomor :722/Menkes/Per/IX/88 Tentang Bahan Tambahan Makanan (Jakarta : Ministry of Health)
[6] Anonymus 2012 Dietary Supplement Fact Sheet : Folate, Office of Dietary Supplements, National Institutes of Health http://ods.od.nih.gov/factsheets/folate.asp Accessed: Dec 8, 2014
[7] Lowry O H, Rosebrough N J, Farr A L and Randall R J 1951 Protein measurement with Folin phenol reagent *J Biol. Chem.* **193**(1) 265-76
[8] AOAC 1995 Official Methods of Analysis of Association of Official Analytical Chemists (Washington DC : AOAC)
[9] Ruengsitagoon W and Hattanat N 2012 Simple Spectrophotometric Method for Determination of Folic Acid Conf. The 4th Annual Northeast Pharmacy Research (Thailand : Khon Kaen University)
[10] Eichhorn P and Knepper T P 2001 Electrospray ionization mass spectrometric studies on the amphoteric surfactant cocamidopropylbetaine *J. Mass. Spectrom.* **36**(6) 677-84
[11] Hussain S Z and Maqbol K 2014 GC-MS: Principle, technique and its application in food science *Int. J. Curr. Sci.* **13** 116-26
[12] Backman 2018 *Laser diffraction measurement for non sperical particles Backman Coulter Life Sciences* (USA: Backman Coulter Inc)
[13] SNI 2007 Makanan Pendamping Air Susu Ibu (MP-ASI) Surat Keputusan Menteri Kesehatan Republik Indonesia No. 224/Menkes/SK II/2007 (Jakarta : Ministry of Health)
[14] De Crécy-Lagard V, El Yacoubi B, De la Garza R D, Noiriel A and Hanson A D 2007 Comparative genomics of bacterial and plant folate synthesis and salvage : predictions and validations *BMC Genomics* **8** 245.
[15] Rossi M, Amaretti A and Raimondi S 2011 Folate production by probiotic bacteria Folate Production by Probiotic *Bacteria Nutrients* **3**(1) 118 – 34
[16] Delchier N, Ringling C, Maingonnat J F, Rychlik M and Renard C M G C 2014 Mechanisms of folate losses during processing : diffusion vs. heat degradation *Food Chem.* **15**(157) 439– 47

Synthesis and characterization of magnetite (Fe_3O_4) nanoparticles from iron sand in Batanghari Beach

I Safitri[1*], Y G Wibowo[2], D Rosarina[3] and Sudibyo[4]

[1]Postgraduate Program of Physics Education, Universitas Ahmad Dahlan, Yogyakarta, Indonesia
[2]Postgraduate Program of Environmental Science, Universitas Jambi, Indonesia
[3]Faculty of Engineering, Universitas Muhammadiyah Tangerang, Indonesia
[4]Research Unit for mineral technology, Indonesian Institute of Science (LIPI), Lampung, Indonesia

*Email: Indrisafitri780@gmail.com

Abstract. The utilization of sand is still limited as a building material, so development is needed to increase the use-value of iron sand from the Batanghari River. No one study was found to improve iron sand in this field into magnetic material. This paper informs a case study of the synthesis and characterization of Fe_3O_4 nanoparticles from the Batanghari Jambi river iron sand using the coprecipitation method. Sintering temperature variations were carried out to see the effect of sintering temperature on crystal size, crystal structure, and optical properties of the sample. Sintering temperature variations used are 300°C, 400°C, and 500°C. The synthesis results showed that the sintering temperature of iron sand extract did not affect the crystal structure formed. However, the sintering temperature affects the size of the crystal; the higher the sintering temperature, the size of the crystal will be even greater. At a temperature of 300°C, a Fe_3O_4 crystal size was 6.94 nm, at a temperature of 400°C a Fe3O4 crystal size was 8.29 nm and at a temperature of 500°C a crystal size of 10.15 nm was obtained. The synthesis results also show that the higher the sintering temperature, the optical properties of the iron sand extract, the better. The decreasing energy gap value indicates this. At a temperature of 300°C, a gap energy value of 2.0 eV was obtained, at a temperature of 400°C a gap energy value of 1.9 eV was obtained and at a temperature of 500°C, the energy gap obtained was 1.8 eV.

1. Introduction

Sand is one of the abundant materials throughout the world. Indonesia, as an island, is one of the countries which has abundant sand reserves. Nowadays, sand is one of the materials that get special attention from many researchers in the world; this is caused by the unique characteristics possessed by sand in the nanoscale [1]. The unique characteristics of sand include high surface area, non-toxic, high surface energy, good biocompatibility, high absorption, superparamagnetic, and electron transfer [2]. The latest trend in utilizing sand is converting sand into metal oxide nanostructure [3]. This material is known to have a high surface area so it will be more sensitive in its use as nanosensors, gene delivery, cell therapy, magnetic scaffolds, magnetic resonance imaging, cell labeling, hyperthermia and drug delivery [4].

Various studies have reported the use of sand in the industrial world. One of the uses of iron sand is to produce Fe_3O_4 (Magnetite) nanoparticles. Fe_3O_4 is a form of iron oxide in nature besides maghemite (γ-Fe_2O_3) and hematite (α-Fe_2O_3) [5]. Fe_3O_4 is known as black iron oxide, which is the metal oxide with the most magnetic properties. A recent study also informs that sand can be used as a photocatalyst [6]. The photocatalyst is a material that can increase the rate of oxidation and reduction reactions induced by light [7]. The use of photocatalysts is considered an efficient method of separating pollutant compounds. The photocatalyst material is usually a semiconductor with a gap energy range of 1 to 3 eV. Synthesis of Fe_3O_4 with percolators $FeCl_3.6H_2O$ and $FeCl_2.4H_2O$ has also been carried out by observing the optical properties of the material [8]. Characterization using UV-Vis with a wavelength of 559 nm produces an energy-gap of 2.22 eV [9], it shows that Fe_3O_4 has the potential to be applied as a photocatalyst. Fe_3O_4 can be obtained from iron sand by the coprecipitation method. Coprecipitation is based on the deposition through organic compounds used so that this method becomes simpler, can be done at room temperature, and synthesis time is shorter than the sol-gel method, solvothermal method, hydrothermal method, sonochemical method, microemulsion method, and molten salts method.

Coprecipitation is one of the methods used to make nanoparticle material preparations. The working principle of this method is to convert metal salts into precipitates using a hydroxide or carbonate alkaline precipitant which is then converted to its oxidized form by heating. The coprecipitation method is the most effective because this method can be used in normal environmental conditions. In the synthesis, this method uses acid and base pairs. Acids function as solvents and bases carry solutes down to form the desired precipitate. The synthesis of magnetite nanoparticles by the coprecipitation method is expected to have monodispersive properties. Monodispersive properties mean that the magnetite particles are evenly or uniformly distributed.

The Batanghari river sand iron in Jambi Province, Indonesia is thought to contain Fe_3O_4. No studies were found explaining the characteristics of the iron sand deposited in this field. Thus, this material is could be improved into magnetic material. This paper will explain the synthesis and characterization of the Batanhgari River iron sand material using the coprecipitation method with different temperature variations. This study is important to improve the utilization of iron sand into nanomaterials that have better functions and selling points

2. Material and Method

2.1. Preparation sample

An intake of iron sand was carried out using permanent magnets. Sand retrieval utilizes the magnetic properties of magnetite minerals so that it could be separated from impurities (non-iron sand). Intake process was done by filtering dry sand to get smaller and uniform sized sand. Furthermore, filtered sand was placed near the permanent magnet so that the iron sand contained in the sand could be separated from non-iron sand. The process of drawing iron sand that has been obtained can be done ten times to obtain purer iron sand. Iron sand was filtered using 120 mesh sieves priot synthesize to obtain pure iron sand. Forty gram Batanghari Jambi iron sand was dissolved into 38 ml HCl (12 M) at 55 °C and stirred for about 60 minutes using a magnetic stirrer. hemical processes occured in the following reactions are:

$$3Fe_3O_4(s)+8HCl(l)\rightarrow 2FeCl_3(l)+FeCl_2(l)+3Fe_2O_3(s)+3H_2O(l)+H_2(g) \quad (1)$$

2.2. Coprecipitation method

After the solution was formed, it was filtered using filter papers. The solution added with 73 ml NH_4OH and placed for 1 hour to obtain a precipitate. Precipitated iron sand formed solid black separated from the solution, which is then washed using distilled water at seven times so that the results obtained are completely clean from impurities. This treatment fulfills the reaction equation:

$$2FeCl_3(l)+FeCl_2(l)+H_2O(l)+8NH_4OH(l)\rightarrow Fe_3O_4(s)+8NH_4Cl(l)+5H_2O(l) \quad (2)$$

2.3. Preparation of nanoparticle Fe_3O_4

The precipitate was dried in the oven at a temperature of 120 °C for 2 hoursto ovaporate its water content. Then sintering process was done with various temperatures, *i.e.* 300, 400 and 500 °C. The characterization of Fe_3O_4 nanoparticles was carried out using an X-ray Diffractometer (XRD), while UV-Vis spectrophotometer was used to determine the energy gap of the Fe_3O_4 nanoparticles.

3. Result and Discussion

3.1. Nanoparticle Fe_3O_4

One of the simplest and oldest methods of synthesis in solid-liquid reactions is the coprecipitation method. The solid produced by the coprecipitation method is stable and insoluble in solvents. The coprecipitation process involves metal cations from a certain medium that are deposited together in the form of hydroxide, carbonate, oxalate, or citrate. The precipitate is calcined at a certain temperature to give a product in powder form. The coprecipitation process involves controlling solvent and precipitating concentrations, temperature, and stirring speed in the manufacture of products. The coprecipitation method uses an acid solution to dissolve the main ingredient. Acid is used as a solvent because ferrous metal can react quickly with HCl to form iron (II) and H2 gas. The solution that has been stirred until it is homogeneous, it is necessary to add a settling solution in such a way that the precipitate has high homogeneity. NH4OH solution can be used as a precipitator. The use of NH4OH which contaminates the precipitate can be removed by heating.

The resulting precipitate is influenced by the concentration of solvent and precipitant, heating temperature, and the duration of stirring. The use of the coprecipitation method with various concentrations of iron salt solutions and settling solutions affects the amount of powder and the magnetic properties of the formed magnetite (Aji, 2008). In this study, the magnetite growth method was carried out by varying the temperature of the iron sand extract dissolving so that it could be seen its effect on the lattice parameters, crystal structure, crystal size, and the magnetic properties that appeared. Synthesis of Fe_3O_4 nanoparticles was carried out by the coprecipitation method. Fe_3O_4 nanoparticles were obtained by utilizing iron sand extract from the Batanghari river iron sand Jambi. The dissolution step of iron sand extract produces a black solution which indicates the presence of iron salt content. When Fe_3O_4 which is still in the form of the iron sand extract is mixed with high concentrated HCl where the iron metal contained in the iron sand extract can react quickly with HCl to form $FeCl_3$, $FeCl_2$, and Fe_2O_3 where all three act as precursors to produce Fe_3O_4. The iron salt solution formed is then precipitated using NH4OH solution. Precipitation of iron salts can occur due to the formation of mixed crystals or by the adsorption of ions during the precipitation process. The precipitate formed is the result of the oxidation process of ferrous metal material contained in the dissolved iron salt, according to the equation. Based on direct observations, it can be seen that the extract of iron sand and the results of the synthesis show a difference. The magnetite nanoparticles that were synthesized appeared to be more concentrated black compared to iron sand extract. The synthesized grains also look smoother than iron sand extracts.

3.2. XRD Characterization

Diffraction is an event that the wave is scattered by a disturbance (for example, a grid), followed by scattering in all directions, which causes strengthening or attenuation under certain conditions. Diffraction occurs when there is an equality of geometric order between the wavelength and the width of the lattice. The result of the amplification of the scattering describes the character of the scatter or the disturbance. If a beam with a wavelength corresponding to the distance between the crystal planes is shot into a crystalline material, there will be crystal diffraction. X-ray diffraction is used to identify the crystal structure of a solid by comparing the value of the d distance (crystal plane) and the intensity of the diffraction peaks with the standard JCPDS data issued by the ICDD (International Center of Diffraction Data).

XRD characterization was carried out to determine the crystal structure, lattice parameters, and crystal size of the synthesized Fe_3O_4 nanoparticles. The output of the characterization results was read using ORIGIN 9.0 software in the form of data 2θ and intensity. From the 2θ data and intensity, then plotted so that it forms a diffractogram showing the diffraction peaks that are formed. The results of the overall characterization of the sample data can be seen in figure 2. XRD characterization was carried out on all three samples, namely temperatures of 300, 400 and 500 °C. The control variable used in the characterization is the wavelength used for firing on the sample, Cu Kα wavelength 1.54059 Å, and the shooting angle from 20.01 ° to 69.99 °. The results of the characterization in the form of diffraction curves are then adjusted to JCPDS number 19-629 for magnetite and JCPDS number 39-1346 for maghemite.

Based on figure 2, it can be seen that temperatures of 300 °C, 400 °C, and 500 °C have peaks (2 2 0) and (3 1 1). The diffraction pattern that is formed for the maximum intensity representing Fe_3O_4 is at an angle of 2θ, which corresponds to the miller index (3 1 1). From the diffractogram, it can be stated that the iron sand used is dominated by Fe_3O_4 or the iron sand used is magnetite. The presence of these peaks indicates that the crystalline structure of Fe_3O_4, which is a spinel structure. The results of XRD characterization showed that the higher sintering temperature, the peak from the sample had become narrower. This study informed that the degree of crystallinity and the level of regularity of the arrangement of atoms in the sample increases with increasing sintering temperature. Due to the increase in sintering temperature, an increase in thermal energy that causes crystal growth so that the nucleus grows by attracting other atoms to fill the space in the lattice to be formed. Thus, the increasing temperature sintering crystal growth continues until the final transformation of the crystal. The highest intensity shows that the crystal has a good crystal order or many atoms that are arranged orderly and neat.

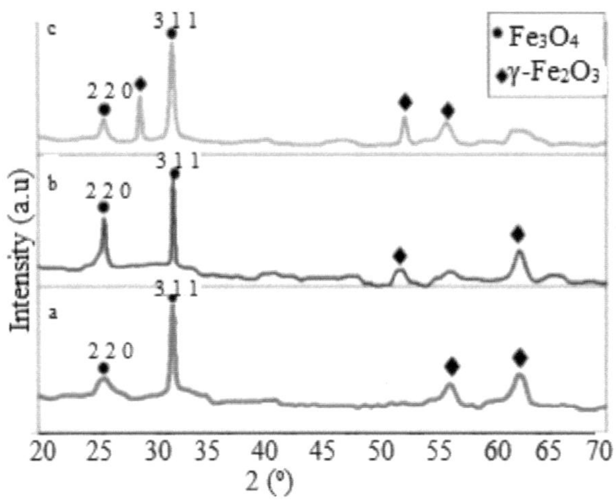

Figure 1. XRD pattern of Fe_3O_4 at 300 °C (a), 400 °C (b) and 500 °C (c).

In addition to the XRD results, there is another peak that is maghemite (it-Fe2O3). The presence of the γ-Fe2O3 phase is adjusted to JCPDS data no 39-1346. The appearance of γ-Fe2O3 caused by Fe_3O_4 material is very easy to change into other phases, wherein principle Fe_3O_4 undergoes oxidation when interacting with free air. Oxidation during the synthesis of Fe_3O_4 cannot be avoided so that Fe_3O_4 will always have oxides or sub-oxides on its surface.

Maghemite oxidation results from Fe_3O_4 occurs at a sintering temperature of 250 °C and will peak at a sintering temperature of 900 °C [10]. At the same time, the hematite phase will occur at sintering temperatures ranging from 550 to 700 °C [11]. The change in sintering temperature causes an increase in the heat energy that is given so that the atom vibrates and diffuses at the grain boundary, the process of phase transformation begins when the Fe^{2+} ion is oxidized to the Fe^{3+} ion [12]. This Fe^{3+} ion diffuses to form γ-Fe_2O_3. Also qualitatively, it can be predicted that the presence of γ-Fe2O3 phase in the sample is only in small levels because the sample obtained from this synthesis process is dominated by a deep black color that shows the characteristic Fe_3O_4. Based on the presence of γ-Fe_2O_3 phase, phase ratio calculation can be determined using the Hanawalt Method. Figure 1 shows that there are two different big groups of phases, namely the Fe_3O_4 and γ-Fe2O3 phases.

The Fe_3O_4 phase increases with increasing sintering temperature from 300 to 400 °C. However, it was seen that at a temperature of 500 °C, the phase ratio of Fe_3O_4 decreased. Due to Fe_3O_4 being very easily oxidized to γ- Fe_2O_3. The phase ratio of γ- Fe_2O_3 decreases from the sintering temperature of 300 °C to

400 °C due to the increased growth of Fe_3O_4 crystals, so that it can suppress the presence of γ- Fe_2O_3. Whereas at the sintering temperature of 500°C, an increase in the phase ratio γ- Fe_2O_3 is due to the angle of 2θ (32.23), a new phase appears as γ- Fe_2O_3 phase. The tendency of the Fe_3O_4 phase to change to the γ- Fe_2O_3 phase is quite high at a sintering temperature of 500°C. Besides, have the same structure, namely cubic, so they cannot be distinguished through simple structural analysis [13]. Characterization of Fe_3O_4 nanoparticles based on XRD results can not only be used to identify the crystal structure but also can be used to determine the lattice parameters and crystal size. From figure 1 at the maximum peak that is (3 1 1), we can know the value of parameters and crystal size. The material showed the width of the Fe_3O_4 lattice, which is 0.251 nm. Calculation of lattice parameters and crystallite size of Fe_3O_4 nanoparticles synthesized can use ORIGIN 9.0 software.

The results of quantitative data analysis using the ORIGIN 9.0 program indicate that the diffraction pattern formed is two phases, namely, phase 1 (magnetite) and phase 2 (maghemite). High peaks indicate that the sample has high crystallinity and a wide peak indicates that the sample is nanometer size. Furthermore, this can be justified from the results of the fitting using the Scherrer equation which is written in Equation 1, the approximate crystal size is obtained. It should be noted that as the crystal dimension enters the nanoscale, the peak widens with decreasing crystal size. It is known that the width of the diffraction peaks allows the determination of crystal size. In practical terms, the crystal size can be determined by a variant of the Schererr equation:

$$D = \frac{k\lambda}{B\cos\theta} \tag{3}$$

Where D is the thickness of the crystal, K is a constant that depends on the shape of the crystal, and B is the width of the peak at half the intensity in radians. If the Gaussian function is described to describe the peak of expansion, then the constant K is equal to 0.89, the Shcerrer equation is derived from Bragg's law and can be used to determine crystal size when the crystal is smaller than 1,000 Å. This fitting analysis was performed to determine the crystal size using the Schererr equation. It appears that magnetite has a cubic structure. The distance between atoms in 1 unit cell.

The synthesis results were characterized using XRD to determine the crystal structure, lattice parameters, and crystal size. Characterization using XRD tools. The data obtained from characterization using XRD is a diffractogram. The diffractogram shows the diffraction peaks at a certain angle. Based on the data obtained from the results of characterization using XRD then the data obtained is compared with the data stored in the JCPDS (19-629 as a reference for the correctness of the sample data made and to find out the lattice parameters of a crystal that has cubic symmetry, with a = b = c and α = β = γ = 90°, then the beam angle of the crystal plane (hkl) can be calculated from the relationship between the distances between the fields, you can use the equation:

$$a = d\sqrt{h^2 + k^2 + l^2} \tag{4}$$

The XRD phase ratio calculation can be calculated using the Hanawalt method:

$$\%Phase\ ratio = \frac{The\ desired\ phase\ intensity}{total\ of\ the\ intensities\ from\ all\ phases} \times 100\% \tag{5}$$

XRD is an effective method for determining the phase composition of unknown crystals. The analysis is performed by comparing the position of the diffraction peak and the intensity value against data already recorded in the JCPDS standard data (19-629). The output of characterization using XRD is in the form of 2θ data and intensity. From this data, it is plotted in a graphical form (diffractogram) so that the peaks of the diffraction resulting from the material crystals can be seen. By using ORIGIN 9.0 software, the peak of the diffraction peaks can be seen in which phase corresponds to it.

This research informed about the crystal size from each sample used. Through XRD diffractogram data, the crystal size can be determined by calculating FWHM (Full Width at Half Maximum) from the peak of the diffraction plane, which is the peak width of the diffraction peak of Fe_3O_4 samples. FWHM is used to determine crystal size using the Scherrer equation. From the calculation using the Scherrer equation the crystal size obtained at a sintering temperature of 300°C obtained a crystal size of 6.94 nm when the sintering temperature was enlarged to 400°C the crystal size was obtained which increased by 8.29 nm and the sintering temperature enlarged again to 500°C obtained the size crystal of 10.15 nm. Theoretically, the size of the crystal will increase along with the increase in sintering temperature, due to the crystal growth. At high temperatures, the constituent atoms have enough thermal energy for atomic diffusion so that the atoms move past the grain boundary causing an expansion of the area of the intersection between crystals that will increase the size of the grain. The larger grain size influences the shape and size of the crystals

Result of 2θ at the highest diffraction peak at the peak (311) for all three samples, were at a temperature of 300°C, an angle of 2θ = 35.73°, a temperature of 400°C at an angle of 2θ = 35.78°, and a temperature of 500°C at an angle of 2θ = 35, 72°. It can be seen that the three samples have an angle of 2θ, which did not change significantly. Following the Bragg equation, the width of the lattice is inversely proportional to the diffraction angle. The absence of angular shift means that there is no widening of the lattice distance in Fe_3O_4 with increasing sintering temperature [14]. Thus, in table 3, it can be seen that the lattice parameter values and lattice width are the same and do not change for the three samples. This phenomenon showed that the increase in sintering temperature in the sample does not change the crystal structure of Fe_3O_4. Based on the data, the three samples obtained indicate that magnetite has a cubic structure with a lattice parameter of 0.832 nm.

3.3. UV-VIS Characterization

When radiation interacts with a material, information about the characteristics of the material can be obtained. The interaction between radiation and matter can be in the form of scattering, absorption, and transmittance depending on the properties of the material. Characterization of optical properties was carried out using a UV-Vis spectrometer, where when light is passed through a material some of the light is reflected, scattered, absorbed, and partially transmitted. When light is passed from one medium to another, such as from air to solid material, several possibilities can occur. Some of the light radiation is transmitted across the medium, some are absorbed and some is reflected on the surface between the two mediums. The intensity of the incident rays towards the surface of the solid medium must be the same as the amount of light intensity that is transmitted, absorbed, and reflected.

Characterization using a UV-Vis Spectrophotometer to see the optical properties of Fe3O4 processed using the Touc Plot method. The determination of the energy gap is obtained by processing the transmission data, by plotting the correlation of the absorption coefficient to the photon energy given to the sample through the Touc Plot equation. The bandgap graph is a graph of the relationship between photon energy and the square of the transmission coefficient. The size of the bandgap can be generated by drawing a linear extrapolation line from the end of the absorption coefficient curve intersecting the x-axis.

$$ahv = A\,(hv - Eg)^n \qquad (6)$$

By being the absorption coefficient of a material, h is the Plank Kostanta (6.626x1034 Js), is the frequency, where c is the speed of light (3x108 m/s^2), Eg is the energy band gap width (energy gap), A is a constant depending on the material, and the values are n=2 for direct transitions and n=1/2 for indirect transitions. Determining the energy gap can be determined using the touc plot method by extrapolation of the linear graph of the Eg (eV) relationship as the abscissa on the x-axis and $((a\ h\ v)^2$ as the ordinate on the y-axis. Calculation using ORIGIN.9.0 software with extrapolation from the graph relationship Eg and $(a\ h\ v)^2$ to cut the energy axis to obtain the energy gap value. The result of the extrapolation is in the form of a graph where the intersection of abscissa and ordinate is obtained, which is the energy gap of Fe3O4. The optical properties of Fe_3O_4 are determined based on the measurement of the transmittance spectrum obtained from a UV-Vis spectrometer. Figure 3 is the result of transmittance to the wavelength of each sample. The transmittance graph shows the wavelength transmitted by the semiconductor material. The graph shows that the value of transmittance will continue to increase with increasing wavelength.

IOP Conf. Series: Materials Science and Engineering 1011 (2021) 012020 doi:10.1088/1757-899X/1011/1/012020

The graph shows that the photon energy that continues to occur in the range of ultraviolet light that is 200 to 400 nm and also occurs in the range of visible light that is 400 to 800 nm [15].

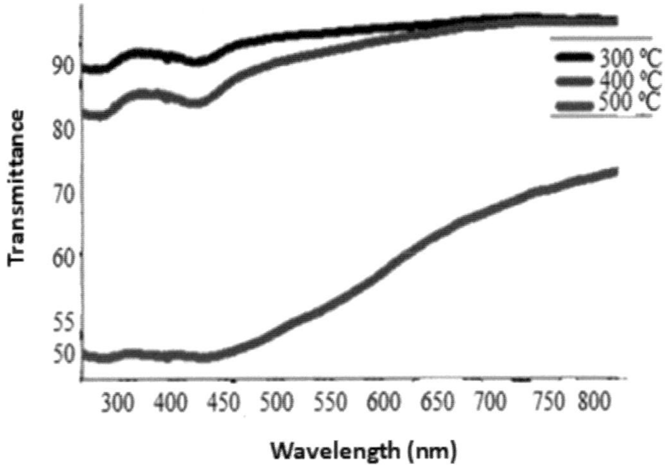

Figure 2. Wavelength relationship with transmittance.

The optical properties of Fe_3O_4 are determined based on the measurement of the transmittance spectrum obtained from a UV-Vis spectrometer. Figure 3 is the result of transmittance to the wavelength of each sample. The transmittance graph shows the wavelength transmitted by the semiconductor material. The graph shows that the value of transmittance will continue to increase with increasing wavelength. The graph shows that the photon energy that continues to occur in the range of ultraviolet light that is 200-400 nm and also occurs in the range of visible light that is 400 to 800 nm. The transmittance curve above also shows that the sample transmits visible light, which is larger than UV light. From several treatments with variations in sintering temperature, it can be seen the difference in transmission values. At a wavelength of 736 nm, the sample has a different transmittance value. At a temperature of 300 °C, the transmittance value obtained was 0.980%, 400 °C, the transmittance value was obtained at 0.976%, and at 500 °C, the transmittance value obtained was 0.777%.

Adding the sintering temperature to the sample will cause a decrease in the percentage of transmittance. According to the mechanism of sintering, where the higher the sintering temperature, there will be a process of compacting a group of powders, and there is a strong bond between the granules and pores contained between the grains so that the light of photons given is more absorbed [16]. Besides, a decrease in the energy gap can be related to figure 1, which shows a decrease in the percentage of transmittance in all three samples. The decrease in the percentage of transmittance shows that the photon light given to the sample is more absorbed. The lighter the photons absorbed means that the valence-banded electrons will have more additional energy. Thus, valence-banded electrons will more easily move to the conduction band.

The gap energy value is determined by the Touc's Plot method. Gap energy is the energy needed by electrons to break covalent bonds so that they can move from the valence band to the conduction band. In a semiconductor material, because the energy gap is narrow, then if the temperature rises, some electrons in the valence band go up to the conduction band by leaving a hole in the valence band. Electrons that are already in the conduction band or hole in the valence band will act as charge carriers for the occurrence of electric current. The determination of the gap energy starts from determining the transmittance value of Fe_3O_4. The energy gap can be determined by processing the transmittance data using the software. Transmittance measurements are carried out at wavelengths, ranging from ultraviolet light to visible light. So we get the wavelength value at each transmittance value, which is then used to get the gap energy value.

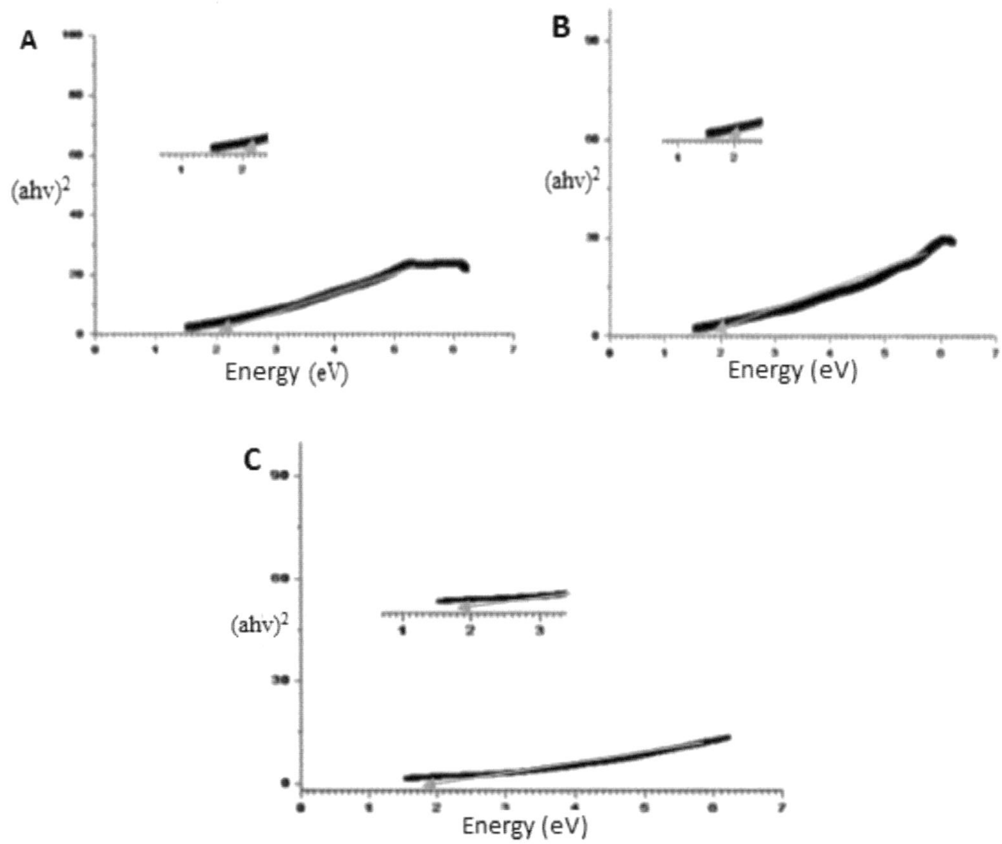

Figure 3. Gap energy of nanoparticle at 300 °C (a), 400 °C (b), 500 °C (c)

The energy gap can be seen in figure 3 a, b, and c by drawing a linear line on the curve so that it intersects with the x-axis (eV axis), then the gap energy value is obtained from the point of intersection. In figure 3(a) temperature of 300 °C has an energy gap value of 2.0 eV, then with an increase in figure 3(b) the temperature is 400 °C. The energy gap value becomes 1.9 eV, and figure 3(c) temperature of 500 °C, the gap energy needed is increasingly decreased by 1.8 eV. The following is the value of each sample, which can be seen in Table 1 below.

Table 1. Value of gap energy.

No	Temperature (°C)	GAP energy (eV)
1.	300	2.0 eV
2.	400	1.9 eV
3.	500	1.8 eV

Table 1 shows that the higher the sintering temperature, the gap energy value decreases. The size of the temperature affects the shape and size of the energy bandgap. The difference in the energy band gap produced in the sample is caused by differences in the size of the particles formed. The higher sintering temperature will cause a smaller pore size between particles and will make the crystal size bigger. The electron will make it easy to move from the valence band to the conduction band so that the energy gap obtained will be smaller where the optimum energy band gap is around 2.0 eV.

4. Conclusion

Based on the research that has been done, it can be concluded that the difference in sintering temperature in the Fe_3O_4 nanoparticles affects the diffraction peaks formed and the size of the crystals formed. Crystal size obtained for the temperature of 300, 400, and 500 °C were 6.94, 8.29 and 10.15 nm, respectively. The difference in sintering temperature in Fe_3O_4 nanoparticles also affects the gap energy obtained, where the greater the sintering temperature, the gap energy results obtained are also smaller. The gap energy obtained from the temperature of 300 °C is 2.0 eV, temperature of 400 °C 1.9 eV, and at the temperature of 500 °C 1.8 eV. Adding the sintering temperature to the sample will cause a decrease in the percentage of transmittance. Following the mechanism of sintering, where the higher the sintering temperature, there will be a process of compacting a group of powders, and there is a strong bond between the granules and pores contained between the granules so that the light of the photons given is more absorbed.

Reference

[1] Pickering K L, Khimi S R and Ilanko S 2015 The effect of silane coupling agent on iron sand for use in magnetorheological elastomers Part 1: Surface chemical modification and characterization *Compos. PART A* **68** 377–86
[2] Rahmawati R, Taufiq A, Sunaryono S, Fuad A, Yuliarto B, Suyatman S *et al.* 2018 Synthesis of Magnetite (Fe3O4) Nanoparticles from Iron sands by Coprecipitation-Ultrasonic Irradiation Methods *J. Mater. Environ. Sci.* **9** 155–60
[3] Taufiq A, Sunaryono, Hidayat N, Hidayat A, Putra E G R, Okazawa A *et al.* 2017 Studies on Nanostructure and Magnetic Behaviors of Mn-Doped Black Iron Oxide Magnetic Fluids Synthesized from Iron Sand *Nano* **12** 1–11
[4] Gupta A K and Vaidya V D 2007 Recent advances on surface engineering of magnetic iron oxide nanoparticles and their biomedical applications *Futur. Med.* **2** 23–39
[5] Jia C, Sun L, Luo F, Han X, Heyderman L J, Yan Z, Yan C, Zheng K, Ru U, Kasama T, Cervera-gontard L, Dunin-borkowski R E and Tzvetkov O G 2008 Large-scale synthesis of single-crystalline iron oxide magnetic nanorings *J. Am. Chem. Soc.* **130** 16968–77
[6] Eddy D R, Puri F N and Noviyanti A R 2015 Synthesis and photocatalytic activity of silica-based sand quartz as the supporting TiO_2 photocatalyst *Procedia Chem.* **17** 55–8
[7] Nayak S, Mohapatra L and Parida K 2015 Visible light-driven novel g-C_3N_4/NiFe-LDH composite photocatalyst with enhanced photocatalytic activity towards water oxidation and reduction reaction *J. Mater. Chem. A Mater. energy Sustain.* **3** 18622
[8] Hssan, Alshamsi and Hussein B S 2014 Synthesis, characterization and photocatalysis of γ-Fe2 O3 nanoparticles for degradation of Cibacron Brilliant Yellow 3G-P *Asian J. Chem.* **26** 70–3
[9] Kouotou M P, El Kasmi A, Wu LN, Waqas M and Tian ZY 2018 Particle size-band gap energy-catalytic properties relationship of PSE-CVD-derived Fe3O4 thin films *J. Taiwan Inst. Chem. Eng.* **93** 427-35
[10] Bora D K, Braun A, Erat S, Safonova O, Graule T and Constable E C 2012 Evolution of structural properties of iron oxide nano particles during temperature treatment from 250 C e 900 C : X-ray diffraction and Fe K-shell pre-edge X-ray absorption study *Curr. Appl. Phys.* **12** 817–25
[11] Pati S S and Philip J 2014 Effect of cation trapping on thermal stability of magnetite nanoparticles *J. Nanosci. Nanotechnol.* **14** 4114–23
[12] Alibeigi S and Vaezi M R 2008 Phase transformation of iron oxide nanoparticles by varying the molar ratio of Fe2+:Fe3+ *Chem. Eng. Technol.* **31** 1591–6
[13] Lu J, Jiao X, Chen D and Li W 2009 Solvothermal Synthesis and characterization of Fe_3O_4 and γ-Fe_2O_3 nanoplates *J. Phys. Chem.* **113** 4012–7
[14] Kazeminezhad I and Mosivand S 2014 Phase transition of electrooxidized Fe_3O_4 to γ and α-Fe_2O_3 nanoparticles using sintering treatment *Acta Phys. Pol. A* **125** 1210–4
[15] Zhu L, Kong X, Yang C, Ren B and Tang Q 2020 Fabrication and characterization of the magnetic

separation photocatalyst C-TiO_2/Fe_3O_4/ AC with enhanced photocatalytic performance under visible light irradiation *J. Hazard. Mater.* **381** 120910

[16] Cao C, Xiao L, Chen C and Cao Q 2015 Magnetically separable Cu2O/chitosan-Fe3O4 nanocomposites: Preparation, characterization and visible-light photocatalytic performance Chunhua *Appl. Surf. Sci.* **333** 110–8

Antifungal activities of lactic acid bacteria and yeast isolated from various types of Tempe

Ema Damayanti[1*], Nur Shabrina[2], Faisal Azis Prihantoro[2] and Maya Shovitri[2]

[1]Research Unit for Natural Product Technology, Indonesian Institute of Sciences (BPTBA-LIPI)
Jl. Jogja Wonosari KM 31.5, Gunungkidul, D.I.Yogyakarta, 55861, Indonesia
[2]Department of Biology, Faculty of Mathematic and Natural Science, Institute Teknologi Sepuluh Nopember, Surabaya, Indonesia

*Corresponding author: ema.damayanti@lipi.go.id

Abstract. Mycotoxin-producing fungi are harmful contaminants in food and feed products. Lactic acid bacteria (LAB) and yeast groups are known to produce antifungals. Indonesia is known as a Tempe-producing country, an important functional fermented food. Soybean Tempe has been commonly known and well-studied, while other types of Tempe such as Tempe Gembus, Tempe Koro, Tempe Lamtoro and Oncom have not been well studied. This study aimed to determine LAB and yeast population in various types of Tempe and to select antifungal activity against mycotoxin-producing fungi. Research methods included total aflatoxin analysis in Tempe products, enumeration and isolation of LAB and yeast, and antifungal assay. The results showed that the total aflatoxins in all Tempe products were 1.99 - 3.84 ppb and it was qualified as food for consumption. The total LABs were 7.16 - 8.25 $\log_{10}$ cfu/g while the total yeast was 4.48 - 7.38 $\log_{10}$ cfu/gram. The highest antifungal activities in mycotoxin-producing fungi such as *Aspergillus parasiticus* and *Penicillium citrinum*, shown by LAB G1 isolates which were identified as *Lactobacillus* sp. and yeast G6K1 and G6K2 which were identified as *Saccharomyces* spp. All the selected isolates were isolated from Tempe Gembus. The clear zone diameter of antifungals was 15-20 mm and 3.5 mm respectively for LAB G1 and G6K1-G6K2. These selected LAB and yeast had the potential to be used as natural bio-preservatives in functional food products to prevent the growth of mycotoxin-producing fungi.

1. Introduction

Mold is one of the main biological agents causing damage on food and feed products because it produces mycotoxins. Mycotoxins are produced by *Aspergillus*, *Fusarium*, and *Penicillium* spp. Mold is a natural contaminant that is generally present in grain-based foods (cereals) [1]. *Aspergillus* and *Penicillium* are two types of mold that dominant produce mycotoxins such as aflatoxin, ochratoxin, and patulin [2]. Aflatoxin is one of the important mycotoxins and it receives a lot of attention because it can cause health problems in humans and animals. In human and animal populations, aflatoxin is associated with both toxicity and carcinogenicity [2]. Health problems caused by the accumulation of aflatoxin in the body include carcinogens, mutagens, teratogens, immunosuppressants [3].

Lactic acid bacteria and yeasts are known to be potential natural agents that can inhibit the growth of mycotoxin-producing molds. Lactic acid bacteria especially *Lactobacillus* strains have the ability to

IOP Conf. Series: Materials Science and Engineering 1011 (2021) 012021 doi:10.1088/1757-899X/1011/1/012021

inhibit fungal growth and aflatoxin production [4]. Yeast is also reported to have potential antimicrobial activity against molds in food spoilage [5]. Some yeast species are also known to have enzymes that can degrade mycotoxins [1]. LAB and yeast are two groups of bacteria that are widely used in food processing because of their ability to improve food quality through metabolic processes such as fermentation.

Indonesia is a tropical country that has a large variety of traditional foods. Bean and legume-based foods are usually made by fermentation techniques using mold. Various types of *Rhizopus* mold are used in the fermentation process of soybean to make Tempe. Tempe that is commonly known is soybean (*Glicin max* L.) Tempe. The traditional process of making Tempe is by either steaming or boiling soybeans, dehulling the soybeans, mixing them with rice flour and fermenting them semi-aerobically using yeast. The most often used Tempe yeasts are *R. oligosporus* and *R. oryzae*, fermented for 1-2 days at room temperature. In addition to soybean Tempe, another type of fermented legume-based food commonly found in Java regions is Tempe Lamtoro (Klandingan or Manding) from *Leucaena leucocephala* bean. Tempe Koro made from tropical sword beans with white seeds (*Canavalia gladiata*) and jack beans with red seeds (*Canavalia ensiformis*) [6]. Tempe Gembus and also Oncom produced from by-products include peanut (*Arachis hypogaea*) press cake, solid waste from soybean curd production, solid waste from tapioca production (*onggok*) and dried coconut press cake [6,7] and fermented by *Neurospora sitophila* [8].

Lactic acid (LAB) bacteria and yeasts in Tempe have long been reported to be part of Tempe microbiota [9-11]. The types of yeast, *Candida* sp. and *Trichosporon* sp. isolated from Tempe are known to be able to inhibit the production of alfatoxin by *Aspergillus flavus* [12]. The information about LAB and yeast associated with various types of Tempe, apart from soybean Tempe, has not been widely reported. This study aimed to isolate lactic acid bacteria and yeast associated with various types of Tempe and to select LAB and yeast isolates which have antifungal activities against mycotoxin-producing molds.

2. Material and Method

2.1. Lactic acid bacteria and yeast enumeration

Samples of Tempe Koro, Tempe Lamtoro and Tempe Gembus were obtained from individual producers in Gading village, Gunungkidul, Yogyakarta, Indonesia, while Oncom sample was obtained from traditional markets in Bantul, Yogyakarta. Figure 1 shows the tempe sample used in this study. Five g of Tempe sample were mashed and dissolved in 45 ml of sterile NaCl 0.86% and then homogenized. Serial dilution was carried out up to 10^6. The total LAB and total yeast were enumerated by plating out 100 μL of sample using spread plate technique on agar media from three series of last dilution. The enumeration of LAB used de Man, Rogosa and Sharpe Agar (MRSA) + CaCO3 0.2%, incubated at 37 °C for 48 hours. The enumeration of yeast used Chloramphenicol Yeast Glucose Agar (CYGA), incubated at 28 °C for 48 hours, aerobics. The total colonies were counted using a colony counter.

2.2. Aflatoxin analysis

Supernatant samples (500 μL) were collected by extraction of Tempe using methanol (Merck, Germany) and analyzed by Elisa Reader (Multiscan, Thermo Scientific). The measurement of aflatoxin content was performed by ELISA method using AgraQuant Total Aflatoxin Assay procedure [13].

2.3. Lactic acid bacteria and yeast isolation

LAB and yeast colonies that grew as a single colony were isolated and refined on the same medium. LAB isolate was characterized through cell morphology observation including catalase test, motility test and Gram staining. Lactic acid bacteria have rod or cocci cell, negative catalase, non-motile and Gram positive. Yeast cells are usually larger than bacteria, vary in size with a width ranging from 1 to 5 μm and length ranging from 5 to 30 μm or larger. Yeast are usually egg-shaped, but some are elongated or ball-shaped and are positive catalase. Pure isolate was collected and stored in 20% glycerol at -20 ° C.

2.4. Antifungal assay

Antifungal assay was carried out on *Aspergillus parasiticus* FNCC 6033 and *Penicillium citrinum* FNCC 6111 obtained from the Food and Nutrition Culture Collection (FNCC), Gadjah Mada University Yogyakarta, Indonesia. *A. parasiticus* and *P. citrinum* were grown on Malt Extract Agar (MEA; Oxoid) medium at 30 °C for 5-7 days. The spores were then harvested from the plate by dripping 0.05% of Tween 80 solution on the surface of the mold spores and homogenized using sterile triangular stems. Dissolved spores were taken aseptically using a micropipette. The concentration of the mold spores was then calculated using a Haemocytometer (Marienfeld, Germany) and adjusted to 10^7/mL of the total spores [14]. LAB and yeast isolates were selected for qualitative antifungal assay using the overlay method [15]. LAB isolates were grown in MRSA medium using a streak plate method, incubated for 48 hours at 37 °C, while yeasts were grown on CYGA medium with a streak plate method and incubated for 48 hours at 30 °C. After growing, soft MEA (15%) contained mold spores was added to the plate, then incubated aerobically at 30 °C for 2-3 days. Antifungal activity was detected from the clear zone that appeared around the LAB or yeast colony. LAB and yeast isolates showing antifungal were selected for quantitative antifungal assay using the disc diffusion method. Secondary metabolites were obtained by separating the supernatant and biomass from the fermentation product in MRSB for LAB and YGB for yeast. A total of 50 µL supernatant containing secondary metabolites were dripped in a sterile disc paper and then placed on the surface of the MEA media containing mold spores. The inhibition zone diameter was measured after incubation at 30 ºC for 3 days.

3. Result and Discussion

Various types of Tempe samples used in this study and mold and yeast population found on CYEA media are shown in Figure 1. Various types of mold and yeast grew on the Tempe samples. There were differences in morphology, especially mold in various Tempe. Molds as a primary inoculum used to produce Tempe vary in types, depending on the raw materials. Tempe Gembus is Tempe derived from tofu pulp (soybeans) fermented with mold *Rhizopus* sp. [16], while Tempe Lamtoro (or referred to as Tempe Manding by the Javanese people) is made from lamtoro (*Leucaena leucocephala*) [6]. Tempe Koro is Tempe derived from velvet beans (*Mucuna pruriens* L) fermented with mold *Rhizopus* sp., while Oncom is a fermented food derived from groundnut or sometimes a mixture of groundnut cake and tofu waste, fermented by *Neurospora sitophila* [17]. The main genus for tempe production is *R. microsporus*, with varieties *microsporus*, *oligosporus*, *rhizopodiformis* and *chinensis*. In traditional process, cottonwood leaves (*Hibicus tiliaceus*) are used as a carrier for Tempe mould starter locally known as *usar* in Indonesia [18].

Some publication about indigenous fermented foods of Southeast Asia showed several processes to make various types of Tempe that exist in Indonesia, especially in Java regions. Tempe lamtoro is made by boiling the seeds with ash for about 2 hours. The cooked seeds are then dried overnight followed by dehulling process. The seeds are then soaked overnight, washed and soaked again until foam appears on the surface and boiled again for around 2.5 hours. After boiling, the seeds are dried, cooled and inoculated with mold wrapped in leaves and incubated for 36 - 48 hours [6].

Tempo Koro is also made in a similar way. *C. gladiata* beans are boiled in water containing wood ash. This method functions to absorb the odor and the bitter taste of the beans and prevent them from sticking back after dehulling. After boiling for about 1 hour, beans are dried and dehulled. The hull is soaked for 3 days by replacing the water daily to remove any poisonous content. The beans are then washed and cooked for about 30 minutes until the beans are soft. After cooking, the beans are dried again, ground and cut into small pieces and then inoculated with mold and wrapped in banana leaves. The fermentation process takes around 48 hours. Tempe Gembus is made from tofu waste. Waste from the process of making tofu is not only used for animal feed but also still consumed by humans. Tofu pulp is steamed for 1 hour, dried, cooled and inoculated with *R. oligosphorus* or mixed molds, wrapped in plastic or banana leaf and incubated at room temperature for around 24 hours. Tempe gembus is widely produced in both Central Java and Yogyakarta [6].

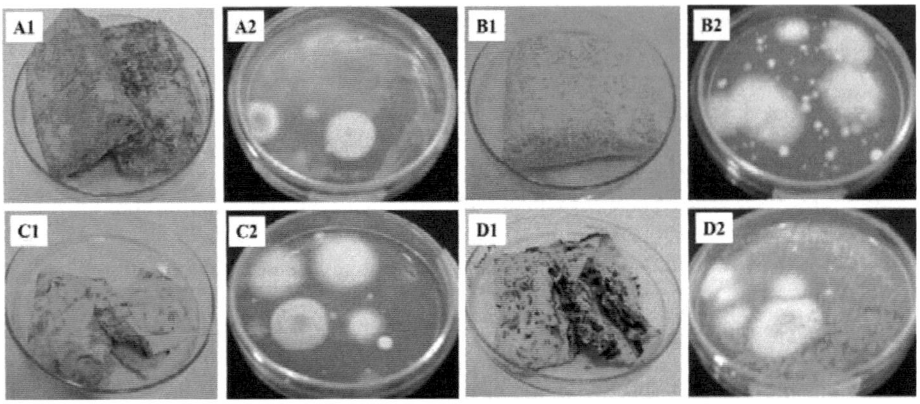

Figure 1. Oncom (A), Tempe Gembus (B), Tempe Koro (C), and Tempe Lamtoro (D). Number 1 shows the image of Tempe sample and number 2 shows various types of mold and yeast that grew on agar medium.

Oncom is a traditional food from West Java which is made from food waste such as peanut cake, and a mixture of tofu waste and sometimes added with tapioca waste (*onggok*) and dried coconut pulp. There are two types of oncom, namely black oncom which uses *R. oligosporus* and red oncom that uses *Neurospora intermedia* var. *oncomensis*. Oncom is made by soaking groundnut cake for 3-4 hours and soaking *onggok* for 1 hour, then mixing the two and boiling them for 1 hour, after which they are made into a thin layer. After cooling down, the mixture is inoculated with oncom yeast and placed in a bamboo container with a banana leaf. Incubation is carried out at room temperature 25-30 °C for 36-48 hours [7].

The nutrient value of various types of Tempe is shown in Table 1. Tempe Koro has the highest carbohydrate content and Oncom has the highest protein content. The aflatoxin content of all Tempe is less than 5 ppb. The Indonesian food and drug authority (BPOM RI) set a maximum limit of total aflatoxin in food products at 35 ppb [19]. Meanwhile the US FDA and EU's public health division limit the mycotoxin levels in food and animal feed at 20 ppb (US) and 4-15 ppb (EU) [20]. Based on these regulations, the total levels of aflatoxin in the four types of Tempe are within the safe limits for consumption. The low aflatoxin content in all the types of Tempe is possible due to the presence of lactic acid bacteria as shown in Figure 2. A study by Ghanbari shows that the inhibitory effect of LAB against *A. parasiticus* growth occurs in a population of 2×10^3 cfu/ mL of *L. plantarum* and *L. delbrueckii* subsp. Aflatoxin production is significantly inhibited, seen from the fact that the expression of the aflR gene greatly decreases after fungi is exposed to LAB [21]. Therefore, LAB acts as an antifungal against *A. parasiticus* and is effective in reducing aflatoxin production.

Table 1. Nutrition and aflatoxin content on various types of Tempe.

Parameter	Oncom	Tempe Gembus	Tempe Koro	Tempe Lamtoro
Carbohydrate (%)*	4	12	23	21
Crude protein (%)*	20 – 30	3,4	10	1
Fat (%)*	-	0.2	1.3	0.5
Crude fiber (%)*	2	3.9	-	-
Total aflatoxin (ppb)**	3.84 ± 1.00	2.15 ± 0.27	1.99 ±1.16	3.27 ± 1.62

*source: [22]
**source: this study

The number of LAB and yeast colonies from the samples of Tempe Gembus, Tempe Lamtoro, Tempe Koro and Oncom is shown in Figure 2. The highest number of LAB was obtained from Oncom with 8.25 $\log_{10}$ cfu/g while the lowest one was in Tempe Gembus (7.16 $\log_{10}$ cfu/g sample). The highest number of yeast was found in Tempe Gembus (7.38 log_{10} cfu/g) while the lowest number of yeast was found in Oncom i.e. 4.48 log_{10} cfu/g sample. The variation in the number of LAB and yeast populations in various types of Tempe is due to substrate, mold types and processes in each type of Tempe. Another study mentioned that LAB and yeast are part of a common microbial community during the production of soybean Tempe and remain in fresh Tempe products. Different Tempe production methods affect the presence of LAB and yeast. The highest population of LAB and yeast was found in fresh soybean Tempe, which was between 6.54 - 9.70 log_{10} cfu/g. Filotype identification found during the production of Tempe, there are several species and LAB strains that belong to the genus *Aerococcus*, *Enterococcus*, *Lactobacillus*, *Leuconostoc*, *Streptococcus*, and *Weissella* [9]. Another study stated that a high number of LAB at the initial step of Tempe processing is correlated with low pH at the initial step and this may also contribute to inhibiting the growth of *Enterobacteriaceae* and bacterial spores. The presence of yeasts and LAB up to 7 - 8 log_{10} cfu/g during fermentation indicates that those microorganisms contribute to fermentation process of Tempe [10]. In other studies, the LABs identified in the process of Tempe fermentation are the heterofermentative *Lactobacilli* group and *Streptococcus* non enterococci. *Lactobacillus* is the dominant LAB found in each stage of Tempe fermentation using *Rhizopus* mold [23].

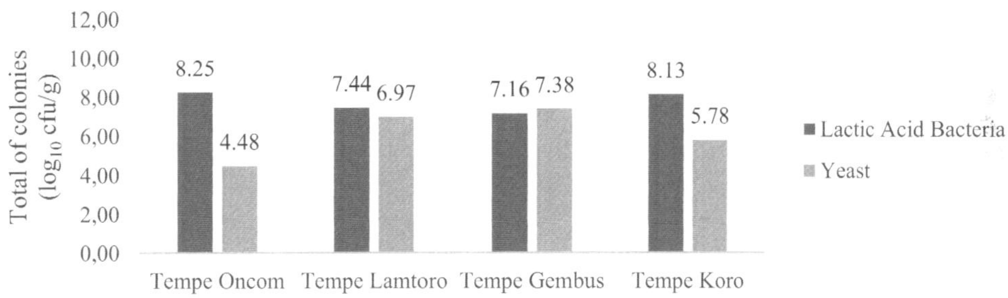

Figure 2. The number of lactic acid bacteria and yeast colonies in various types of Tempe samples.

The results of antifungal assay of LAB and yeast isolates obtained in this study are shown in Table 2. Lactic acid bacteria (LAB) isolates that showed qualitative and quantitative antifungal activity included O2 and O5 isolates from Oncom, KL2 and KL5 from Tempe Lamtoro and G1 from Tempe Gembus. The highest antifungal activity was produced by LAB G1, namely 15-20 mm. Yeast isolates showed higher antifungal activity. Antifungal activity produced by LAB in this study was relatively lower than the antifungal activity produced by other *Lactobacillus* sp. isolated from Kunu against *A. flavus* which was 5-20 mm [24]. Another study also revealed that *Lactobacillus* from some Nigerian fermented food resulted in 6 – 18 mm antifungal activity against *P. citrinum* [25]. In fact, this study showed that there were 8 yeast isolates which showed antifungal activity both in qualitative and quantitative method. Yeast G6K1 and G6K2 isolates from Tempe Gembus showed the highest antifungal activity (3.5 mm). The clear zone of LAB and yeast antifungals on overlay method is shown in Figure 3 (A1 and A2). Antifungals from lactic acid bacteria appear to be higher than those from yeast isolates.

Morphological analysis of LAB G1 isolates showed negative catalase non-motile gram-positive rod. Based on these characteristics, LAB G1 is the genus of *Lactobacillus* as in Figure 3 (B2). LAB is mostly used in the making of various fermented foods, where they contribute to improving their shelf life, organoleptic properties, and nutritional value. One of the most commonly used LABs in fermented foods is *Lactobacillus* [5]. LAB is widely used as a bio-preservative in food because it is known to produce bacteriocin compounds and antifungal compounds [26].

IOP Conf. Series: Materials Science and Engineering 1011 (2021) 012021 doi:10.1088/1757-899X/1011/1/012021

Table 2. Antifungal activities of lactic acid bacteria and yeast isolated from various types of Tempe.

Lactic acid bacteria	Clear zone*		Average of inhibition zone (mm)**		Yeast	Clear zone*		Average of inhibition zone (mm)**	
	Ap	Pc	Ap	Pc		Ap	Pc	Ap	Pc
O2	+	-	2	0	O5K1	-	-		
O4	-	-	0	0	G5K1	+	+	3	4
O5	+	-	10	0	G5K2	+	-	3	0
O6	-	-	0	0	G6K1	+	+	3.5	3.5
KL1	-	-	0	0	G6K2	+	+	3.5	3.5
KL2	-	+	0	10	G6K3	-	+	0	4
KL3	-	-	0	0	G6K4	+	-	0.3	0
KL4	-	-	0	0	KO5K1	-	-	0	0
KL5	++	-	15	0	KO5K2	-	-	0	0
KL6	-	-	0	0	KL15K1	+	+	0	4
G1	++	++	15	20	KL15K2	-	-	0	0
G3	-	-	0	0	KL15K3	+	-	2.5	0
G5	-	-	0	0	KL15K4	+	-	2.5	0
KO2	-	-	0	0	KL15K5	-	-	0	0
KO3	-	-	0	0					
KO4	-	-	0	0					
KO5	-	-	0	0					

* overlay method (qualitative)
** diffusion method (quantitative)
Ap = *Aspergillus parasiticus* FNCC 6033; Pc = *Penicilium citrinum* FNCC 6111
O = Oncom; G = tempe gembus; Ko = tempe koro; K; = tempe klandingan
–: 0 cm; +: 0,1-1 cm; ++: 1,1-2 cm; +++: 2,1-3 cm; dan ++++: ≥ 3,1 cm.

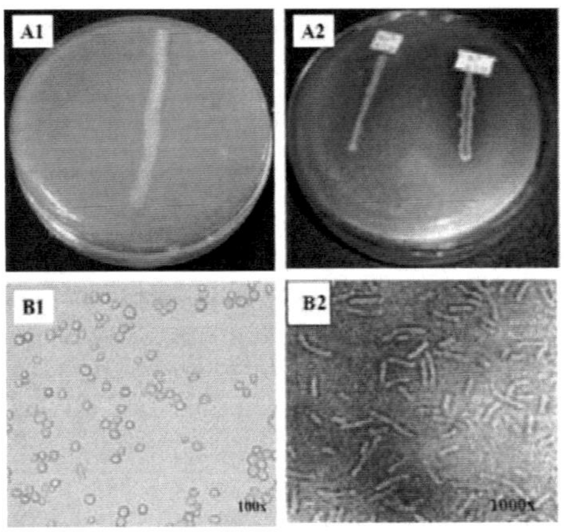

Figure 3. Clear zone on agar media indicated antifungal activities of yeast (A1) and lactic acid bacteria (A2). Morphological microscopic of yeast (B1) and lactic acid bacteria (B2).

Yeast G6K1 has characteristics categorized as the genus *Saccharomyces* as shown in Figure 3 (B1) which has budding and egg-shape and unicellular cell. *Saccharomyces* is known as a probiotic because it is non-toxic and able to survive in food products in the long-term condition [27]. Antagonistic ability of yeast has been attributed mainly to competition for nutrients, pH changes in the medium as a result of growth cells and organic acid production, tolerance to high concentrations of ethanol and the secretion and release of antimicrobial compounds, such as killer toxins or "mycocins" [28].

Thus, the presence of lactic acid bacteria and yeast in various types of Tempe products has a role as an inhibitor of the growth of mycotoxin-producing fungi. This is indicated by the low content of aflatoxin in Tempe products such as Tempo Koro, Tempe Lamtoro, Tempe Gembus or Oncom (Table 1). In addition, other studies mentioned that the total aflatoxin content in whole and fresh groundnuts was 42.3 ppb while those could reach 108 ppb in contaminated groundnuts [29,30]. After fermentation process to produce Oncom, the total aflatoxin decreased to 3.8 ppb as shown in this study. In addition to being able to inhibit the growth of aflatoxin-producing fungi, LAB and yeast are also known to have aflatoxin binding. Previous studies showed that *L. plantarum* G7 and *S. cerevisiae* B18 have aflatoxin binding ability [31,32].

4. Conclusion
The total aflatoxins in all the Tempe products were 1.99 - 3.84 ppb, thus qualified as food for consumption. The total LAB in Tempe product was 7.16 - 8.25 $\log_{10}$ cfu/g while the total yeast was 4.48 - 7.38 $\log_{10}$ cfu/gram. *Lactobacillus* sp. G1 and *Saccharomyces* spp. G6K1 and G6K2 which were isolated from Tempe Gembus showed the highest antifungal activities against mycotoxin-producing fungi *Aspergillus parasiticus* and *Penicillium citrinum.* These selected LAB and yeast were potential to be used as natural bio-preservatives in functional food products to prevent the growth of mycotoxin-producing fungi.

Acknowledgments
This research was funded by DIPA BPTBA LIPI 2014. Authors thank Sri Endartini and Madina Nurohmah for sample preparation, and Lusty Istiqomah for technical assistance in aflatoxin analysis.

References
[1] Reddy K R N, Salleh B, Saad B, Abbas H K, Abel C A and Shier W T 2010 *Toxin. Rev.* **29**
[2] Bennett J and Klich M 2003 *Clin. Microbiol. Rev.* **16**, 497–516
[3] Topcu A, Bulat T, Wishah R and Boyac I H 2010 *Int. J. Food. Microbiology* **139** 202–5
[4] Gomah N H, Ragab W S and Bullerman L B 2010 *Assiut. J. Agric. Sci.* **40** 27-36
[5] Salas M L, Mounier J, Valence F, Coton M, Thierry A and Coton E 2017 *Microorganism* **5**
[6] Astuti M 2015 Tempe from other Pulses in *Indigenous Fermented Foods of Southeast Asia*, edited by J. D. Owens. (Ner York: CRC press)
[7] Kuswanto K R 2015 Indonesian oncom (fermented food processing by product)" in *Indigenous Fermented Foods of Southeast Asia*, edited by J. D. Owens. (New York: CRC press)
[8] Nout M J R and Aidoo K E 2010 "Asian Fungal Fermented Food Industrial Applications", 2nd Edition *the Mycota X* edited by M. Hofrichter (Springer-Verlag Berlin Heidelberg)
[9] Efriwati, Suwanto A, Rahayu G and Nuraida L 2013. *Hayati J. Biosci.* **20** 57-64
[10] Nurdini A L, Nuraida L, Suwanti A and Suliantari 2005 *Int Food Res. J.* **22** 1668-74
[11] Pangastuti A, Alfisah R K, Istiana N I, Sari S L A, Setyaningsih R, Susilowati A, *et al.* 2019 *Biodiversitas* **20**
[12] Purwijantiningsih E, Hariyadi RD, Nurwitri C C and Istiania 2005 *Biota* **X** 146-53
[13] AgraQuant Total Aflatoxin. https://www.romerlabs.com/shop/inter_en/agraquant-r-total-aflatoxin-elisa-test/
[14] Yang E J and Chang H C 2010 *Int. Food Microbiol.* **139** 56-63

[15] Ström K 2005 "Fungal inhibitory lactic acid bacteria characterization and application of *Lactobacillus plantarum* MiLAB 393", Doctoral thesis, Swedish University of Agricultural Sciences Uppsala
[16] Sulchan M and Rukmi M G I 2007 *Med. J. Indones.* **16** 205
[17] Sastraatmadja D D and Saono S 1984 *Berita Biologi* **2** 9-10
[18] Nout M J R and Kier J L 2005 *J. Appl. Microbiol.* **98** 789–805
[19] BPOM RI (Badan Pengawas Obat dan Makanan RI), Keputusan Nomor HK.00.05.01.4057 Tentang Batas Maksimum Alfatoksin Dalam Produk Pangan 2004
[20] Alshannaq A and Yu J H 2017 *J. Environ. Res. Public.* Health. **14** 632
[21] Ghanbari R, Aghaee E M, Rezaie S, Khaniki G J, Alimohammadi M, Soleimani M *et al.* 2018 *J. Food Saf.* **38** e12413
[22] Owen J D 2015 "Indonesian oncom (fermented food processing by product)" in *Indigenous Fermented Foods of Southeast Asia*, edited by J. D. Owens. (New York: CRC Press)
[23] Pisol B, Nuraida L, Abdullah N, Suliantarari and Khalil K A 2013 *Proceeding 4th International Conference on Biology, Environ. Chem IPCBEE* **58**
[24] Olonisakin O O, Jeff-Agboola Y A, Ogidi O C and Akinyele B J 2017 *Prev. Nutr. Food Sci.* **22** 138-43
[25] Adebayo C O and Aderiye B I 2010 *Res. J. Microbiol.* **5** 1070–80
[26] T Bintsis 2018 *J. Bacteriol. Mycol.* **6** 89–94
[27] Lara-Hidalgo C E, Hernández-Sánchez H, Hernández-Rodríguez C and Dorantes-Álvarez L 2017 *J. Nutr. Metab.* **4** 1045
[28] Muccilli S and Restuccia C 2015 *Microorganisms* **3** 588-611
[29] Rubak Y T and Purawisastra S 2011 *PGM* **34** 21-8
[30] Windyarani A "Populasi *Aspergillus flavus* dan kandungan aflatoksin B1 pada biji kacang tanah mentah dan produk olahannya di Kecamatan Bogor Tengah, Kotamadya Bogor", Undergraduate Thesis, Department of Biology, Faculty of Mathematic and Natural Science, Institute Pertanian Bogor, 2009
[31] Damayanti E, Istiqomah L, Saragih J E, Purwoko T and Sardjono 2017 *IOP Conf. Series: Earth and Environmental Science* **101** 012030
[32] Istiqomah L, Damayanti E, Arisnandhy A, Setyabudi F M C S and Anwar M 2019 *AIP Conference Proceedings* **2099** 020009

IOP Conf. Series: Materials Science and Engineering 1011 (2021) 012022 doi:10.1088/1757-899X/1011/1/012022

Study of isothermic absorption model of moisture content of fibrous instant corn rice

D A Darmajana[1], A Indriati[1] and D Rahmawati[2]

[1]Research Center for Appropriate Technology-Indonesian Institute of Sciences, Jl. K.S. Tubun 5, Subang, Indonesia
[2]Food Technology Alumnus of Pasundan University, Bandung, Indonesia

*corresponding author: doddyandy@gmail.com

Abstract. In general, storage of food products made from grains is carried out in an environment with uncontrolled air temperature and humidity. In this situation, changes in temperature and humidity will continue according to daily weather patterns. Food products stored in this environment will experience adsorption and desorption processes alternately and continuously throughout the day. Dried food that has not reached its equilibrium moisture content tends to absorb moisture during storage so that it will trigger a decrease in quality through a series of physical, chemical, and microbiological processes. The relationship between water activity and equilibrium moisture content during storage can be approached with an isothermic sorption model. The purpose of this research is to determine the equilibrium moisture content of fibrous instant corn rice in various relative humidity (RH) values and to find a mathematical model from isothermic sorption curve of fibrous instant corn rice. The research methodologies include the production of fibrous instant corn rice, the determination of equilibrium moisture content at various RH (40%-90%), and mathematical modeling of isothermic sorption curves with standard model approaches (Caurie, Hasley, and Oswin). The results showed that the water vapor adsorption process occurred in the isothermic sorption experiments of fibrous instant corn rice at a temperature of 30°C which was marked by an increase in the moisture content of the instant fibrous instant corn rice at 40-90% RH. Hasley equation model: (log [ln (1/aw)] = -3.2810 - 3.0945 log Me), Caurrie: (ln Me = -3.3084 + 1.8163 aw) and Oswin: (ln Me = -2.3301 + 0.2805 ln [aw/(1-aw)]) can accurately describe the isothermic adsorption of fibrous instant corn rice at 30°C because it has an Mean Relative Determination (MRD) value of less than 5.

1. Introduction

Corn is a strategic commodity that can be developed into an alternative staple food as seen from its nutritional value which contains important protein in the Indonesian menu. The main (72-73%) nutrient content of corn is starch, with 25-30% of amylose and 70-75% of amylopectin. Simple sugar levels of corn are glucose, fructose and globulin, prolamin, glutelin and nonprotein nitrogen [1].

To support the development of corn as staple food, processing technology is needed to produce corn products that can be accepted and easy to prepare. One example of a product that can be developed from corn is instant corn rice [2]. Instant corn rice (ICR) is a type of food product that is easy to serve in a relatively short time.

ICR is stored in containers with a moisture content below 10%. The moisture content of ICR will change according to the environmental air conditions. Moisture content and water activity of foodstuff during storage can be studied using the isothermic water absorption model approach. The parameters of the relationship between equilibrium moisture content and water activity can describe sorption constants and material characteristics. The material absorption constant can inform the heat energy of the adsorption material and the amount of gas adsorbed. In addition, with the sorption isotherm model, it is possible to predict changes in temperature and partial pressure of air vapor against the amount of water absorbed by the material so that it can be used as data in the drying, pickling and packaging processes [3].

Indonesia's tropical climate with an average relative humidity of air above 60% increases the chance of moisture absorption and therefore the moisture content of dry food during storage also increases. High moisture content in food from agricultural products will degrade the quality through a series of physical, chemical and microbiological processes [4]. The greater the moisture content of food, the higher the water activity. Food with a relatively high moisture content tends to experience damage more quickly than food with lower moisture content.

Storage of dry food products in a humid environment lead to absorption of water vapor from the air until products reach the equilibrium moisture content with that of the surrounding air. The new moisture content that occurs in the stored products is called the equilibrium moisture content (EMC). This study aims to study or examine the equilibrium moisture content model of fibrous ICR. The resultant model will be useful for storing ICR in various circumstances (storage space humidity), so that steps can be taken to prepare the condition of storage space as well as for the drying operation of ICR.

2. Materials and methods

The main materials used in this study were fibrous ICR produced by the Research Center for Appropriate Technology – Indonesian Institute of Sciences, distilled water and chemical salts including Potassium Carbonate (K_2CO_3), Sodium Nitrite ($NaNO_2$), Sodium Chloride (NaCl), Ammonium Sulfate ($(NH_2)2SO_4$), and Potassium Sulfate (K_2SO_4).

The tools needed in this research were mercury thermometer, Trisense meter, analytical scale, slicer, oven, dryer, pyrex excavator, temperature incubator, cup clamp, stopwatch and a set of computers. Adsorption experiments were performed using static methods at a temperature of 30°C and five levels of relative humidity (40% - 90%) with two replications. Data fitting was done according to the Caurie, Hasley and Oswin equation models. Stages of the experiment consisted of sample preparation, determination of equilibrium moisture content, determination of absorption parameters for instant corn rice, comparison between experimental results and the data fitting to the equation model.

2.1. Determination of Equilibrium Moisture Content

Determination of equilibrium water content was carried out by making saturated salt solutions in glass jar by mixing chemical salt with distilled water using the Pyro Mag Stir tool (procedure for making saturated chemical salt solutions to produce relative humidity values based on ASTM 1991 vol. 14.02), then the sample container is weighed on a scale and then add 10 g of ICR. Weighing is done periodically every 24 hours. In addition to measuring weight, temperature and humidity are also observed using a thermometer and RH-meter and the observation is stopped after the equilibrium state indicated by a constant sample weight, e.g. Change in sample weight not more than 0.005 g.

After the sample weight was constant, the water content was calculated by the method of oven (gravimetry) based on SNI 01-2891-1992 and carried out with two replications. The water content obtained is the equilibrium water content (EMC). Determination of the isothermic adsorption curve is based on water activity and equilibrium water content data from the experimental results.

2.2. Isothermic Sorption Model for Fibrous Instant Corn Rice

Caurie Equation Model. This equation model is suitable for most foodstuffs in the α_w interval from 0.00 to 0.85 [5]. The form of the Caurie model in mathematical equations is:

$$\ln Me = \ln P1 - (P2\ Aw) \qquad \text{(Eq. 1)}$$

Equation 1 was transformed to linier equation with general form: y = a + b*x
In which:

P	=	Constant	Me	=	Equilibrium moisture content
y	=	ln Me	x	=	Aw
a	=	ln P1	b	=	-P2

Oswin equation model. This equation model is for food with α_w from 0.0-0.85 aw

$$Me = P1\,[\frac{aw}{1-aw}]^{P2} \qquad \text{(Eq. 2)}$$

Equation 2 was changed to straight line equation form with general form: y = a + b*x
In which:

P	=	Constant	Me	=	Equilibrium moisture content
y	=	ln Me	x	=	ln[aw/(1-aw)]
a	=	ln P1	b	=	P2

Hasley equation model. This equation model for foodstuffs at the interval aw 0.1 - 0.81.

$$Aw = \exp\,[\frac{-P1}{Me^{P2}}] \qquad \text{(Eq. 3)}$$

Equation 3 was changed to straight line equation form with general form: y = a + b*x

$$\log[\ln(1/aw)] = \log P1 - P2 \log Me \qquad \text{(Eq. 4)}$$

In which:

P	=	Constant	Me	=	Equilibrium moisture content
y	=	log [ln(1/aw)]	x	=	log Me
a	=	log P1	b	=	-P2

3. Results and Discussion

To obtain various constant values in the equilibrium moisture content equation, an experiment was carried out by storing samples of fibrous ICR in storage containers. The air in the storage package was adjusted for its RH by adding chemical salts. The results of setting the air RH in the storage are presented in Table 1.

Table 1. The comparison between the observed RH in the storage containers at 30°C saturated chemical salt and the theoretical RH.

Salt types	Theoretical RH (%)*	Observed RH (%)	Deviations
K_2CO_3	45.0	46.9	1.9
$NaNO_2$	65.0	66.6	1.6
NaCl	75.0	75.4	0.4
$(NH_2)2SO_4$	81.0	84.8	3.8
K_2SO_4	96.0	95.1	0.9

* Source: Bukle *et al.* 1987 [6]

Deviation values range from 0.4 to 3.8%. This deviation could occur due to several reasons, including the purity of the salt solution used or the fluctuations in temperature and humidity that occurred around the time of measurement. Based on ASTM 1967part 15 standard, the RH measurement deviation can reach 4.0 to 5.0%. According to Labuza (1968), in order to get good and accurate results, the purity of the saline solution used needs to be considered in determining the equilibrium moisture content [7].

The sample, in the form of ICR with a dry-based 7.90% moisture content, were stored in packages with a RH corresponding to a saturated chemical salt solution, at room temperature of 30°C. The sample was weighed every 24 hours until the sample weight reached constant value, meaning that an equilibrium of water vapor pressure between the stored material (sample) and the surrounding air (in the package) was reached. The moisture content measured in the material (sample) is expressed as the equilibrium moisture content. The results of the measurement of the equilibrium moisture content of ICR experiment results are presented in Table 2.

Table 2. Equilibrium Moisture Content (EMC) for Fibrous ICR and the Time needed to reach the EMC in several storage RH.

Salt type	RH (%)	Equilibrium Moisture Content (%)	Time needed (days)
K_2CO_3	47.4	8.90	4
$NaNO_2$	68.0	12.65	7
NaCl	75.9	14.00	9
$(NH_2)2SO_4$	86.1	16.09	11
K_2SO_4	95.7	22.65	16

Table 2 demonstrates that the higher the relative humidity of the surrounding air (in the packaging), the higher the equilibrium moisture content of the instant corn rice. The initial moisture content of instant corn rice was 7.90% dry basis and increased to 8.90% dry basis at 47.4% RH and reached 22.65% at 95.7%RH. The increase in moisture content will continue until equilibrium with the surrounding air was reached. The increase in the level of instant corn rice occurs because the partial pressure of environmental water vapor (in the experimental container) is higher than the partial pressure of water vapor in instant corn rice, so that the flow of water vapor into the instant corn rice or adsorption of water vapor occurs by instant corn rice. According to Hall (1957), migration or flow of water from places with high steam pressure to places with low pressure is proportional to the difference in partial pressure of water vapor [8].

The movement of water during the isothermal adsorption process can be caused by several factors including relative humidity (RH) of air, water activity (α_w), air flow velocity, diffusivity, capillarity, gravity, evaporation and condensation [9].

From Table 2, it can be explained that if instant corn rice is stored in an environment that has a temperature of 30°C with %RH as high as the %RH of air (second column), then the water content of ICR will go to the value as stated in column 3. From the results of the experiment, equilibrium moisture content of ICR was achieved in the range of 4 to 16 days. The higher RH storage value, the longer the time needed for fibrous ICR to reach its equilibrium point. In addition, the smaller the difference between the α_w of the product and its RH environment, the faster the time required for fibrous ICR to reach its equilibrium point. In addition, information on equilibrium moisture content at various air humidity values can be used as a factor in selecting packaging material. Each packaging material has a permeability value to water vapor or other gas present in the storage air.

3.1. Isothermic Adsorption Curves of Instant Corn Rice

3.1.1. Isothermic Sorption Model. Labuza (2002) states that the water activity in food can be calculated by comparing the water vapor pressure of the material (P) with the pure water vapor pressure (Po) under the same conditions, or by dividing the ERH of the environment by 100 [10].

$$Aw = \frac{P}{Po} = \frac{ERH}{100} \quad \text{(Eq. 5)}$$

In which:

Aw = water activity
P = partial pressure of water vapor material
Po = partial pressure of pure water vapor at the same temperature
ERH = relative humidity

The use of isothermic absorption curve equation models of equilibrium moisture content is intended to get a more reliable picture of the tendency of the relationship between water activity and equilibrium moisture content.

3.1.2. Hasley

The general form of Hasley equation is

$$a_w = e^{[\frac{-P1}{Me^{P2}}]} \quad \text{(Eq. 6)}$$

The equation is transformed into a linear regression equation of:

$$y = a + bx \quad \text{(Eq. 7)}$$

with x = log Me
y = log $[\ln\left(\frac{1}{aw}\right)]$
a = log P1 = - 3.2810
b = -P2 = - 3.0945

After the equation is obtained, the values of a and b can be calculated which will be used for the linear regression equation for the Hasley equation model. From the experimental data, it was found that the values of a = -3.2810 and b = -3.0945 so that the Hasley equation model becomes:

$$\log[\ln(1/aw)] = -3.2810 - 3.0945 \log Me$$

Next, the linear Hasley equation can be drawn into the isothermic absorption curve for the Hasley model. The curve illustrates the relationship between the equilibrium moisture content of the equation model in each salt RH.

3.1.3. Caurrie

The general form of the Caurrie equation is:

$$\ln Me = \ln P1 - (P2\ \alpha_w) \quad \text{(Eq. 8)}$$

The equation already linear with the form: y= a+ bx. From the calculation results, it was found that the values of a = -3.3084 and b =1,8163 so that the form of the Caurrie equation model becomes:

$$\ln Me = -3{,}3084 + 1{,}8163\ aw$$

Next, the linear Caurrie equation can be drawn into isothermic absorption curve for the Caurrie model. The curve illustrates the relationship between the equilibrium moisture content of the equation model in each salt RH.

3.1.4. Oswin

The general form of the Oswin equation is:

$$Me=P1\left[\frac{aw}{1-aw}\right]^{P2} \quad \text{(Eq. 9)}$$

The equation is transformed into a linear regression equation y = a + bx

with $x = \ln\left[\frac{aw}{1-aw}\right]$

$y = \ln Me$
$a = \ln P1 = -2.3301$
$b = P2 = 0.2805$

After the equation is obtained, the values of a and b can be calculated and the results will be used for the linear regression equation for the Oswin equation model. From the calculation results, it was found that the values of a = -2.3301 and b = 0.2805 so that the form of the Oswin equation model becomes:

$$\ln Me = -2.3301 + 0.2805 \ln[aw/(1-aw)] \quad \text{(Eq. 10)}$$

Then, the linear Oswin equation can be drawn into the isothermic absorption curve for the Oswin model. The curve illustrates the relationship between the equilibrium moisture content of the equation model in each salt RH.

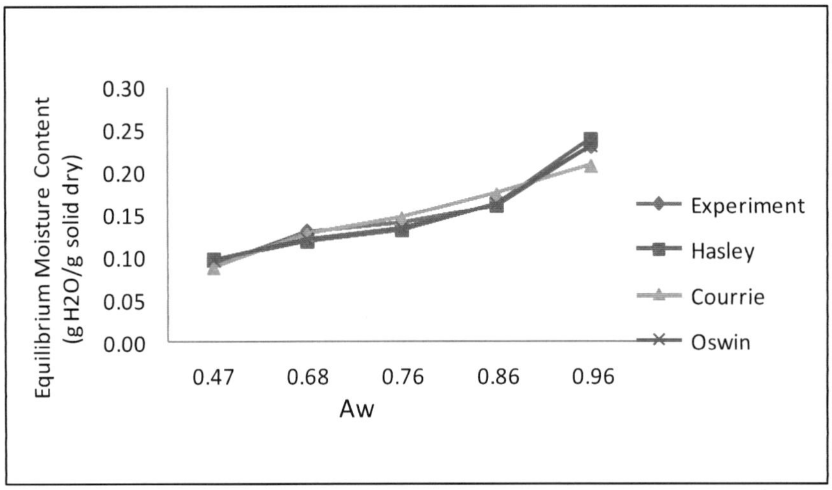

Figure 1. Model of isothermic sorption curve of fibrous instant corn rice.

3.1.5. Mean Relative Determination (MRD) Test

Comparison between the observed isothermic sorption curve with the predictive isothermic sorption models shows that some isothermic sorption models can describe the overall isothermic sorption curve at different levels of precision. This is reinforced by the MRD value which is a measure of accuracy between the predicted EMC and the observed EMC. MRD calculation uses the following formula:

$$MRD = \frac{100}{n}\sum_{i=1}^{n}\left|\frac{Mi-Mpi}{Mi}\right| \quad \text{(Eq. 11)}$$

in which,
a_w = water activity

Me = Equilibrium moisture content
Mi = Experimental moisture content
Mpi = Calculation result of moisture content
N = number of data

MRD values for each model can be seen in Table 3

Table 3. Results of MRD for the isothermic sorption curve equation model of fibrous ICR.

Model	Equation	MRD value
Hasley	$\log[\ln(1/aw)] = -3.2810 - 3.0945 \log Me$	5.0340
Caurrie	ln Me = -3,3084 + 1,8163 aw	4.7640
Oswin	ln Me = -2.3301 + 0.2805 ln[aw/(1-aw)]	3.7460

The three models could be used to predicted the isothermic absorption of ICR as indicated by MRD value <5. The model chosen was the Oswin model with the smallest MRD value of 3.7460. The Oswin model was chosen to describe the actual state of the phenomenon of isothermic sorption of fibrous ICR by the following formula:

$$\ln Me = -2.3301 + 0.2805 \ln[aw/(1-aw)] \quad \text{(Eq. 12)}$$

Best model (Oswin) for adsorption isotherm at different temperatures of lyophilized grugru palm powder with 8% maltodextrin [11]. Chen and Morey [12–13] evaluated four models: the modified Henderson model, modified Chung- Pfost model, the modified Halsey model and the modified Oswin model for fitting the sorption isotherms data from 18 grains and seed crops.

Aviara et al and Oyelade in Ndubisi [14] reported that the modified Oswin model gave the best fit to the EMC of soya bean and lafun, respectively. If the storage of fibrous ICR follows the chosen equation model, Oswin model, it must be stored in environmental conditions with relative humidity of 75.90% (RH) to reach a value of the equilibrium moisture content which is less than 14 percent.

4. Conclusion

Water vapor adsorption process occurred in isothermic sorption experiments of fibrous ICR at 30°C, which was marked by an increase in moisture content of fibrous ICR under the treatment of air relative humidity (RH) of 40-90%. Based on observations, the equilibrium water content of fibrous instant corn rice was achieved at intervals of between 4 and 16 days. The Caurrie and Oswin equation models could be used to predict the isothermic adsorption of fibrous ICR at 30°C because it had an MRD value below 5. Oswin equation was the best equation model with the highest accuracy because it had the smallest MRD value (3.7460). In contrast, the Hasley equation model was not suitable because it has an MRD than 5.

References

[1] Balai Besar Pengembangan Teknologi Tepat Guna Lembaga Ilmu Pengetahuan Indonesia 2010 *Perancangan Sistem Produksi Olahan Berbasis Jagung : Beras Jagung Instan Berserat* (Subang : Balai Besar Pengembangan Teknologi Tepat Guna Lembaga Ilmu Pengetahuan Indonesia)

[2] Supriadi A 2004 *Optimasi teknologi pengolahan beras jagung instan* (Bogor : Institut Pertanian Bogor)

[3] Alamri M S, Mohamed A A, Hussain S, Ibraheem M A and Qasem A A A 2018 Determination of moisture sorption isotherm of crosslinked millet flour and oxirane using GAB dan BET *J Chem* **2018** 2369762

[4] Bonazzi C. and Elisabeth Dumoulin 2014 *Quality Changes in Food Materials as Influenced by Drying Processes.* In book: Modern Drying Technology, pp.1-20. Accessed from: https://www.researchgate.net/publication/295091905_Quality_Changes_in_Food_Materials_as_Influenced_by_Drying_Processes.

[5] Kumendong J 1986 *Model Sorpsi Isotermik Jagung dan Gabah serta Penerapannya dalam Penyimpanannya* (Bogor :Institute Pertanian Bogor)

[6] Buckle K A 1987 *Ilmu Pangan* (Jakarta: Universitas Indonesia Press)

[7] Labuza T P 1968 *Sorption phenomena in foods: theoretical and practical aspects* Series in Food Material Science vol 1 (Dordrecht: Springer) pp 197-219

[8] Hall C W 1980 *Drying and Storage of Brain and Oil Seeds* (New York: Von Nostrand Reinhold)

[9] Srikiatden J and Roberts J S 2007 Moisture transfer in solid food materials: a review of mechanisms, models, and measurements *Int, J. Food Prop.* **10** 739–77

[10] Labuza T P 2002 *Determination of the shelf life of foods.* Accessed from: http://depa.pquim.unam.mx/amyd/archivero/ShelfLife1corto_8507.pdf.

[11] Oliveira D M, Clemente E, Afonso M R A and da Costa J M C 2013 Hygroscopic behavior of lyophilized powder of Grugru Palm *(Acrocomia aculeata) J. Anal. Chem.* **4** 10

[12] Chen C and Morey R V 1989 Comparison of four EMC/ERH equations *Trans ASAE* **32**(3) 0983–90

[13] Chen C and Morey R V 1989 Equilibrium relative humidity (ERH) relationships for yellow-dent corn. *Trans ASAE* **32**(3) 999–1006

[14] Aviara N A 2020 *Moisture Sorption Isotherms and Isotherm Model Performance Evaluation for Food and Agricultural Products, Sorption in 2020s*, George Kyzas and Nikolaos Lazaridis, Intech Open, Accessed from: DOI: 10.5772/intechopen.87996

IOP Conf. Series: Materials Science and Engineering 1011 (2021) 012023 doi:10.1088/1757-899X/1011/1/012023

Accelerated shelf life testing of mocatilla chip using critical moisture content approach and models of sorption Isotherms

D P Putri, L E Yulianti and N Afifah*

Research Center for Appropriate Technology, Indonesian Institute of Sciences, Indonesia

*Corresponding author: syabiljafa2008@gmail.com

Abstract. Mocatilla is a tortilla chip made from composite flour consists of modified cassava, corn, and rice flour which is enriched by tempeh. This study aimed to estimate the shelf life of mocatilla chips when packed in polyethylene terephthalate (PET), polypropylene (PP), and aluminum foil (alufo) using the Accelerated Shelf Life Testing (ASLT) method by the critical moisture content approach. The study was conducted by determining the initial, critical, and equilibrium moisture content of the sample by thermogravimetry. Six models of moisture sorption isotherm were fitted to determine the equilibrium moisture content using six salts with relative humidity (RH) values at a range of 0.076-0.903. The result showed that the initial and critical moisture content of mocatilla chips (g H_2O/g solid) was 0.0141 and 0.0532, respectively. Meanwhile the equilibrium moisture content at relative humidity 80% and temperature 30°C was 0.1471. The Guggenheim Anderson de Boer (GAB) equation was the most suitable model for describing the sorption isotherm phenomenon of mocatilla chips. Based on Labuza's equation, the shelf life prediction of mocatilla chips (80% RH, 30°C) packed in PET, PP, and alufo was 3, 8, and 62 months, respectively.

1. Introduction

Corn (*Zea mays* L.) is one of the numerous local commodities in Indonesia which has great potential to be developed into a variety of interesting and high-selling food products. Corn has a good nutritional content to be used as food. Corn flour contains starch 66.0-73.3%, protein 7.1-9.3%, fat 1.1-3.9%, fiber 2.5-6.4% and ash 0.3-1.5% [1]. Another local commodity that has long been known and consumed by the people of Indonesia is cassava. Currently, cassava has been modified by fermentation process using *Lactobacillus* microbial. This modified cassava (mocaf) has better characteristics in aroma and color, protein, fibre, and viscosity [2]. One of the food product innovations from corn-based combined with mocaf is mocatila chips. Mocatilla chip is a snack source of food fiber based on corn flour, corn starch, modified cassava flour, rice flour and enriched tempeh made through the extrusion process. Mocatilla chips have the characteristics of low water content (1.42%) with 1.42mm thickness and a crispy texture [3].

Shelf life is one of important attributes informed to consumers. Food industries need a relatively short time to determine the shelf life of their products. They usually use accelerated test that considerably shortens the process of obtaining the data [4]. Mocatilla chips are products that are easily damaged due to the process of absorption of water vapor, so estimating shelf life is done by using the Accelerated Shelf Life Testing (ASLT) method. ASLT is a method for estimating the shelf life of food products by storing products in the environment (temperature and RH) which can accelerate the decline in product quality. The decrease of mocatilla chips quality is seen from the value of water content and the level of crispness of the product. The approach used to determine the shelf life of mocatilla chips with the ASLT method is the use of critical water content as a product criterion has expired. The use of this method is suitable for products that are sensitive or easily damaged due to absorption of water vapor in food products [5]. Estimation of product shelf life using the critical water content approach can be done by the Labuza equation [6].

Parameters that need to be observed to be able to use the Labuza equation include initial water content, critical water content, isotherm sorption curves, and supporting parameters such as package permeability. Some mathematical models has been proposed to describe moisture sorption isotherm of food. The most common models which are applied for describing food products sorption are the Langmuir equation, the Guggenheim-Anderson-de Boer (GAB) model, the Brunauer-Emmett-Teller (BET) equation, the Smith model, the Oswin model, the Henderson model, the Halsey model, the Iglesias-Chirife equation, and the Peleg model [7]. Previous studies have estimated the shelf life by the ASLT method and the water sorption isotherm approach have been carried out on dry products such as tortillas [8], chocomix [6], mocaf cookies [9], anti-atherosclerosis herbs [10], and potato chips [11]. This study aimed to study the behaviour of moisture sorption isotherm of mocatilla chips and to estimate the shelf life of mocatilla chips in polyethylene terephthalate (PET), polypropylene (PP), and aluminum foil (alufo) packaging with ASLT method and critical water content approach.

2. Materials and Methods

2.1. Materials
The materials used in this study was mocaf (Mocaf Bandung, CV Karunia Maha Cipta), corn starch (Maizenaku, Egafood), rice flour (Rose Brand, BumiWaras Group), corn seed, *baking powder*, salt, cooking oil, and tempeh. The salts used for creating the specific RH were sodium hydroxide (NaOH), magnesium chloride ($MgCl_2$), potassium iodide (KI), sodium chloride (NaCl), potassium chloride (KCl), and Barium chloride ($BaCl_2$).

2.2. Preparation of mocatilla chips
Mocatilla chip was produced following procedure reported by Afifah *et al.* [3]. Modified cassava flour, corn starch and flour, rice flour, salt, baking powder, and steaming tempeh were mixed until homogen. The mixing flour was steamed, extruded, sheeted, cut, and dried until the moisture content reach below 10%. The chips were fried and drained to remove cooking oil.

2.3. Determination of initial, critical, and equilibrium moisture content
Moisture content was determined by drying the known weight of the sample at 105°C using Oven Memmert UFB until constant weight was achieved. Initial moisture content was immediately measured after the chips were drained. The critical moisture content of chips was the moisture content of product when the fracturability of chips was rejected by panelists through organoleptic testing.

The standard gravimetric method was applied to determine the equilibrium moisture content by exposing the chips to a constant RH environment. This condition is created by the saturated solution of a particular salt at room temperature (30°C). Six different salts e.g., NaOH, $MgCl_2$, KI, NaCl, KCl, $BaCl_2$ were used to maintain respective water activity (a_w) inside separate desiccators in the range of 0.076 to 0.903 [12]. The three samples were placed in porcelain dishes inside the desiccators containing those salt solutions. The samples were periodically weighed and the achievement of equilibrium moisture content was marked by a weighing difference of <2 mg for samples stored at RH <90% and <10 mg for samples stored at RH> 90% for three consecutive days [13]. After equilibration, the moisture content of the product was measured by drying at 105°C until a constant weight was achieved.

2.4. Determination of water vapor transmission Rate (WVTR)

The permeability of plastic packaging was measured at Balai Besar Kimia dan Kemasan (Jakarta, Indonesia) according to ASTM F 1249 that used Permatran Mocon W-3/33 at RH 90±3% and a temperature of 37.8°C. In this study, three plastic types were used namely polyethylene terephthalate (PET), polypropylene (PP), and aluminum foil (alufo).

2.5. Determination of sorption isotherms curve

The fitness of experimental data can be observed by using six moisture sorption isotherm (MSI) models [7,14], such as Guggenheim-Anderson-de Boer/GAB (eq. 1), Oswin (eq. 2), Smith (eq. 3), Henderson (eq. 4), Halsey (eq. 5), and Brunauer-Emmett-Teller/BET(eq. 6). The fit goodness to the sorption isotherm models was evaluated using Mean Relative Deviation (MRD) (eq. 7).

$$M_e = \frac{M_0 C K a_w}{(1 - K a_w)(1 - K a_w + C K a_w)} \quad (1)$$

$$M_e = C\left(\frac{a_w}{1 - a_w}\right)^n \quad (2)$$

$$M_e = C_1 + C_2 ln(1 - a_w) \quad (3)$$

$$M_e = \left(-\frac{ln(1 - a_w)}{C}\right)^{1/n} \quad (4)$$

$$M_e = \left(-\frac{C}{ln a_w}\right)^{1/n} \quad (5)$$

$$M_e = \frac{M_0 C a_w}{(1 - a_w) + (C - 1)(1 - a_w) a_w)} \quad (6)$$

$$MRD = \frac{100}{n}\sum_{i=1}^{n}\left|\frac{M_i - M_{pi}}{M_i}\right| \quad (7)$$

Where a represents the equilibrium relative humidity in decimal; M_e is the equilibrium moisture content (g H_2O/g dry solid); M_o, C, C_1, C_2, n are sorption isotherm constants specific to each equation. M_i is experiment moisture content, and M_{pi} is equation moisture content.

2.6. Determination of shelf life

The shelf life of mocatilla chip based on critical moisture content approach was determined using the Labuza equation (eq. 8). The equilibrium moisture content (M_e) was calculated at ambient temperature and RH using the MSI model that gives the lowest MRD.

$$t = \frac{ln\left[\frac{M_e - M_i}{M_e - M_c}\right]}{\frac{k}{x}\frac{A}{W_s}\frac{P_o}{b}} \quad (8)$$

Where,

t = shelf life (days)
M_e = equilibrium moisture content (g H_2O/g dry solid)
M_i = initial moisture content of product (g H_2O/g dry solid)
M_c = critical moisture content (g H_2O/g dry solid)
k/x = permeability of package (g moisture/m^2.day mmHg)
A = surface area of package (m^2)
W_s = product weight in package (g)
Po = pure water vapor pressure at experimental temperature (mmHg)
b = slope of MSI curve in storage operational area.

3. Result and Discussion

3.1. Sorption isotherms model

The moisture content of mocatilla chips changed during storage in a variety of saturated salts (RH between 7% and 90%). Chips stored in the dessicators increased in weight due to water sorption until they reached equilibrium. This indicated an interaction between food material and water molecules in the surrounding air, driven by the difference in RH between the sample and the surrounding air. Equilibrium moisture content is defined as the water content when the water vapor pressure of the material is balanced with the partial water vapor pressure of the environment [14]. The sample equilibrium results achieved within a period of three weeks in saturated saline solutions with various RH are shown in Table 1.

Table 1. Equilibrium moisture content (M_e) of mocatilla chips

Saline Solution	RH (%)	a_w	Me (gH_2O/g solid)
NaOH	7.58	0.0758	0.0165 ± 0.0003
$MgCl_2$	32.44	0.3244	0.0484 ± 0.0006
KI	67.89	0.6789	0.1052 ± 0.0004
NaCl	75.09	0.7509	0.1262 ± 0.0016
KCl	83.6	0.8360	0.1630 ± 0.0034
$BaCl_2$	90.3	0.9030	0.2167 ± 0.0055

Table 1 exhibits that the higher RH value of the solution, the greater value of the equilibrium moisture content. This occurs because high environmental RH causes the partial pressure of the environment water vapor to increase so that water vapor from the environment moves to the material, or there is water migration from the high-pressure steam to low-pressure areas.

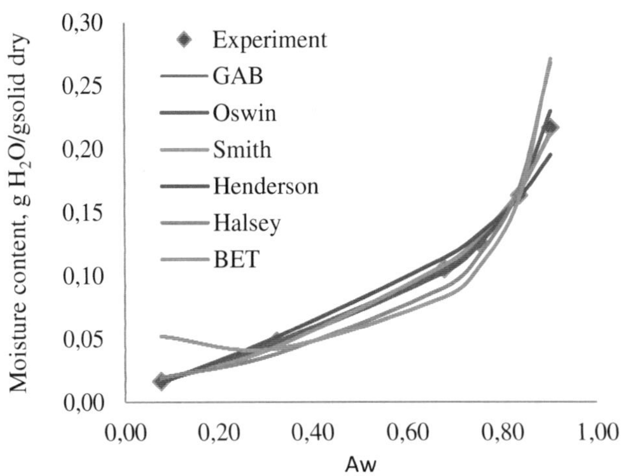

Figure 1. The moisture sorption isotherm curves of mocatilla chips.

The value of equilibrium moisture content (M_e) prediction based on each model (equation 1-6) is obtained by entering a_w values in linear equations and presented in Figure 1. The mocatilla chips showed sigmoid characteristic curves of the isotherms. The type sigmoidal sorption isotherms is the most frequently found type in food products for they contain some ingredients [14,15]. The curves showed a slow raising of moisture content in the first segment (low relative humidity). At low RH, the chips absorbed relatively lower amounts of moisture due to the water in this region represents a strongly bound water. However, at a higher relative humidity (above 67%), a large amount of moisture was absorbed and the curve started sharp raising. It suggested that the chips deteriorate faster above 67% RH. The water properties in this region are close to those of the free water held in voids, capillaries, crevices, and loosely binds to the food materials.

The results of the model accuracy test showed that the GAB model, Oswin model, and Smith model are appropriate to describe the MSI phenomenon of mocatilla chips, Henderson model is rather appropriate, while Halsey model and BET model are not quite right in predicting the isothermic state of water absorption of mocatilla chips.

Table 2. Some models of moisture sorption isotherm equation

Model	Equation	MRD
GAB	$M_e = \dfrac{0.2833a_w}{(1-0.872a_w)(1-0.872a_w+5.997a_w)}$	0.73
Oswin	$M_e = 0.0681\left(\dfrac{a_w}{1-a_w}\right)^{0.545}$	3.82
Smith	$M_e = 0.0099 + 0.0865ln(1-a_w)$	3.48
Henderson	$M_e = \left(-\dfrac{ln(1-a_w)}{20.7014}\right)^{1/1.336}$	6.13
Halsey	$M_e = \left(-\dfrac{0.0201}{lna_w}\right)^{1/1.231}$	14.77
BET	$M_e = \dfrac{-0.7002a_w}{(1-a_w)-27.753(1-a_w)a_w}$	48.54

The GAB model was the most suitable model to describe the sorption isotherm behavior of mocatilla chips. The GAB model was found to be suitable for analyzing more than 50% of meat, fruits and vegetables [7]. Furthermore, this model results a good description of the sorption behavior because it is generated from a further refinement of Langmuir and BET theories of physical adsorption that can describe almost every food product in a wide range of a_w (0 - 0.9). The GAB model is also reported to be appropriate to describe the phenomenon of water sorption isotherm in soybean crackers with an MRD value of 2.51 [17] and potato chips with MRD 5.55 [11]. The different result was reported by Budijanto *et al.* [8] that the Oswin model precisely predicts the actual state of corn tortilla products with an MRD value of 2.33.

3.2. Intial, critical, and equilibrium moisture content

The initial moisture content of mocatilla chips is 2.79% (wet base). This water content is in accordance with SNI No. 2886: 2015 regarding Extrudate Snacks, which limits that the maximum water content is 4 [18]. The condition of the water content shows the characteristics of dry food products. The low moisture content of mocatilla chips is due to the product undergoing a drying process in its production stage. The drying process reduced the water content in the product. Thus, it is expected that the resultant product has a longer shelf life.

The critical moisture content method emphasizes water content and water activity as an expiry criterion. Determination of critical moisture content is done by placing the product openly in a room with high humidity so that moisture absorption occurs until it reaches critical moisture content. Critical moisture content conditions are characterized by products that have experienced physical / chemical / microbiological changes that are no longer acceptable to consumers and are not in accordance with product quality standards. Physical changes that occur after reaching critical water levels are a decrease in the level of product crispness. Chemical changes occur in the form of increased product water content. Microbiological changes occur in the form of growth of molds and fungi [5]. Based on the test results, it was obtained that the critical water content of mocatilla chips was 0.0532 (gH_2O/g solid). In these conditions mocatilla chips physically have a texture that is sluggish and chemically has high water content.

The model with the smallest MRD value is stated as the best model and is used in calculating the estimated shelf life of mocatilla chips. The results of the calculation of MRD values indicate that the GAB equation model has the smallest MRD value that is equal to 0.73, so this model was chosen to be used in calculating the shelf life of mocatilla chips. The equilibrium moisture content of mocatilla chips based on GAB equation (Table 2) at ambient relative humidity in Indonesia about 80.58% ($a_w = 0.8058$) was 0.1471 gH_2O/g solid (www.weather-atlas.com).

3.3. Shelf life

Shelf life is the time interval since a product is made until the quality is not accepted or has lost its special properties. Calculation of shelf life of mocatilla chips is determined by substituting data into the Labuza equation (equation 8). The initial, critical, and equilibrium moisture content, product weight, packaging area and permelability, and slope are collected to calculate shelf life.

Packaging is one of the important factors in influencing the shelf life of food products. The parameters taken into equation are packaging area and permeability. Mocatilla chips was packed in 25g weight using packaging with size of 17 cm x 10 cm resulted packaging area of $0.034m^2$. Product weight per package (Ws) is calculated by weighing the initial weight of the mass of mocatilla chips in one package (W) and then corrected with the initial moisture content to get dry weight. The calculation results that the solid weight on the 25g package was 24.65g, suggested that the product has a solid weight above 98%. Food products with crunchy texture characteristics have a total solidity of more than 90% [19].

Permeability of the package (Water Vapor Transmission Rate/WVTR) is defined as the amount of water vapor that can pass through a package per day at certain atmospheric conditions. Based on the results of the analysis (Table 3), it is known that PET packaging has the highest packaging permeability of 0.5237 $g/m^2/days$, followed by PP packaging of 0.1914 $g/m^2/days$ and aluminum foil packaging of 0.0242 $g/m^2/days$. The higher the packaging permeability value, the higher the transfer of moisture entering the packaging, so that the shelf life of the product becomes shorter. The solution to reduce the entry of water vapor into the product is by using a packaging that has low vapor permeability.

Table 3. Packaging permeability and shelf life mocatilla chips

Parameter	Packaging type		
	PET	PP	Alufo
Packaging thickness, μm	87.8	130.8	108.3
Packaging permeability, $g/m^2/days$	0.5237	0.1914	0.0242
Shelf life time, months	2.86	7.82	61.82

Table 3 shows that the shelf life of the product in the packaging also showed the same pattern with the packaging permeability. The shortest shelf life was found in products that packaged with PET packaging, PP packaging, and aluminum foil packaging, namely 3, 8, and 62 months, respectively. Previous studies of tortillas packed using OPP20 / VMPET / LLDPE25 showed a shelf life of 4.6 months [8]. Corn biscuits that are packaged using aluminum foil packaging have a shelf life of 20.7 months

[20]. The difference in shelf life values is partly due to the natural characteristics of the foodstuff, the model used temperature, RH and permeability of product packaging during storage [21].

4. Conclusion

The critical quality factor of mocatilla chips based on the study results was the reduction in snack crunchiness. The accuracy test of the moisture sorption isotherm model was obtained by MRD of 0.73 for the GAB model. Thus, the GAB equation was used to explain the moisture sorption isotherm phenomenon of mocatilla chips. Combining all variables in the shelf life calculation resulted the shelf life of mocatilla chipswas stored in the PET, PP, and aluminum foil packaging was 3, 8, and 62 months, respectively.

References

[1] Gwirtz J A and Garcia-Casal M N 2014 Processing maize flour and corn meal food products *Ann. N. Y. Acad. Sci.* **1312** 66–75

[2] Sulistyo J and Nakahara K 2014 Cassava Flour Modification by Microorganism *International Symposium on Microbial Technology for Food and Energy Security*

[3] Afifah N, Sholichah E and Yulianti L E 2020 Pengaruh Fortifikasi Olahan Kedelai dan Proses Penggorengan Terhadap Sifat Fisikokimia dan Sensoris Keripik Tortila dari Jagung dan Mocaf (The Effect Of Soybean-Based Products Fortification and Frying Process on Physicochemical and Sensory Properties of T *J. Ris. Teknol. Ind.* **14** 79–87

[4] Phimolsiripol Y and Suppakul P 2016 Techniques in Shelf Life Evaluation of Food Products *Ref. Modul. Food Sci.*

[5] Asiah N, Cempaka L and David W 2018 *Panduan Praktis Pendugaan Umur Simpan Produk Pangan [Practical Guidences of Food Shelf-Life Estimation]* (Jakarta: UB Press)

[6] Nurhayati R, Pratiwi R, Anandito B K, Herawati E R N and Angwar M 2018 Accelerated shelf life testing of chocomix using critical moisture content approach *Reaktor* **18** 63–70

[7] Andrade R D P, Roberto L M and Pérez C E C 2011 Models of sorption isotherms for food: Uses and limitations *Vitae* **18** 325–34

[8] Budijanto S, Sitanggang A B and Kartika Y D 2010 Penentuan Umur Simpan Tortilla dengan Metode Akselerasi berdasarkan Kadar Air Kritis serta Pemodelan Ketetapan Sorpsi Isothermisnya [Shelf Life Study of Tortilla Using Accelerated Shelf Life Testing (ASLT) Method and Its Mathematical Modeling of Moisture] *J. Teknol. dan Ind. Pangan* **21** 165–70

[9] Kurniadi M, Khasanah Y, Kusumaningrum A, Angwar M, Rachmawanti D, Parnanto N H R and Pratiwi L D 2019 Formulation and Shelf Life Prediction of Cookies From Modified Cassava Flour (Mocaf) In Flexible Packaging Formulation and Shelf Life Prediction of Cookies From Modified Cassava Flour (Mocaf) In Flexible Packaging *IOP Conf. Series: Earth and Environmental Science 251* (Tangerang) pp 1–9

[10] Tristantini D, Ramadhan M R and Hanifah A 2019 Shelf life estimation of anti-atherosclerosis herbs using ASLT: Critical water content and sorption isotherms model *AIP Conf. Proc.* **2193**

[11] Alfiyani N, Wulandari N and Adawiyah D R 2019 Validasi Metode Pendugaan Umur Simpan Produk Pangan Renyah dengan Metode Kadar Air Kritis [Validation Method of Shelf Life Estimation of Dry Crisp Food Products by Using Critical Moisture Content Method] *J. Mutu Pangan* **6** 1–8

[12] Greenspan L 1977 Humidity Fixed Points of Binary Saturated Aqueous Solutions *J. Res. Natl. Bur. Stand. (1934).* **81A** 89–96

[13] Lievonen S M and Roos Y H 2002 Water sorption of food models for studies of glass transition and reaction kinetics *J. Food Sci.* **67** 1758–66

[14] Basu S, Shivhare U S and Mujumdar A S 2006 Models for sorption isotherms for foods: A review *Dry. Technol.* **24** 917–30

[15] Blahovec J and Yanniotis S 2009 Modified classification of sorption isotherms *J. Food Eng.* **91** 72–7

[16] Kusnandar F, Adawiyah D R and Fitria M 2010 Pendugaan Umur Simpan Biskuit dengan Metode Akselerasi berdasarkan Pendekatan Kadar Air Kritis [Accelerated Shelf-life Testing of Biscuits Using A Critical Moisture Content Approach] *J. Teknol. dan Ind. Pangan* **XXI** 117–22

[17] Azizah M P N, Hartini S and Cahyanti M N 2016 Isoterm Sorpsi Air dari Kerupuk Kedelai [Moisture Sorption Isotherm of Soy Chips] *Seminar Nasional Kimia dan Pembelajarannya* (Malang) pp 571–81

[18] BSN 2015 *SNI 2886:2015 Makanan Ringan Ekstrudat [SNI 2886:2015: Extruded Snack]* (Jakarta: Badan Standarisasi Nasional Indonesia)

[19] Basuny A M M, Mostafa D M M and Shaker A M 2009 Relationship between chemical composition and sensory evaluation of potato chips made from six potato varieties with emphasis on the quality of fried sunflower oil *World J. Diary Food Sci.* **4** 193–200

[20] Lopulalan C C G 2008 *Kajian Formulasi dan Isothermis Sorpsi Air Biskuit Jagung [Study of Formulation and Moisture Sorption Isotherm of Corn Biscuits)* (Institut Pertanian Bogor)

[21] Steele R 2004 *Understanding and Measuring the Shelf-Life of Food* (Boca Raton Boston New York Washington, DC)

Antioxidant activity of butanolic extract from Madang Gatal (*Schima wallichii* Choisy) Leaves

Galuh Widiyarti[1], Ginanjar Widodo[2], Yulianti Sampora[1], Puspa Dewi Lotulung[1] and Muhammad Hanafi[1]

[1]Research Center For Chemistry, Indonesian Institute of Sciences, Kawasan Puspiptek, Serpong, South Tangerang City, Banten 15314, Indonesia.
[2]Department of Chemistry, Mathematics and Natural Science Faculty, University of Jendral Soedirman, Purwokerto, Indonesia

Corresponding author: galu001@lipi.go.id, galuh.laksmono@gmail.com

Abstract. Antioxidant activity of the butanolic extract from Madang Gatal (*Schima wallichii* Choisy) leaves was studied in the research. Antioxidant activity was analyzed using 2,2-diphenyl-1-picrylhydrazyl (DPPH) radical oxidation method. Furthermore, the extract was purified using column chromatography to yield 4 partially purified fractions (B1-B4). The extract and B2 fraction possess strong antioxidant action with IC_{50} value 12.55 and 24.50 μg/ml, respectively. The B1 and B3 fractions have moderate antioxidant activity with IC_{50} value 70.55 and 90.26 μg/ml, while B4 was not active as antioxidant with IC_{50} value of 228.28. The target compound of the fraction most active as an antioxidant (B2) was separated by preparative Thin Layer Chromatography (TLC). The isolate was then identified using Fourier Transform Infra-Red (FTIR) and Liquid Chromatography Mass Spectroscopy (LCMS). The IR spectra displayed that the bioactive compound of B2 was predicted as flavonoid with MS 290.25 g/mol. The major compound contained in a putative compound known as catechin with an IC_{50} value was 17.97 μg/ml.

1. Introduction

The environment is highly polluted by free radicals which may contribute to the aging process of tissue and cause various chronic and degenerative diseases in humans such as cancer and cardiovascular problems. Excessive amount of free radicals are harmful to human because they can interact with DNA and cell membrane which are important components of cell leading to mitochondrial dysfunction or cell death. The risk can be reduced by utilizing of free radical scavenger compounds known as antioxidants [1-3]. Antioxidants act as antiradical by donating their hydrogen atoms to radical compounds. The function of antioxidants are as free radical neutralizer, thus the aging process can be inhibited [4,5].

In addition, antioxidant ability in preventing cellular damage and scavenging free radicals could be related to its potential to prevent malignancies. Thus, antioxidant activity analysis is considered as one of screening processes for investigating potential candidates for natural antioxidant and anticancer agent [2]. A various number of Indonesian plants have been used traditional medicine to prevent chronic and degenerative diseases due to their remarkable potential as antioxidant. Madang gatal (*Schima wallichii* Choisy), also locally known as puspa or cheloni, belongs to the genus of *Schima* and tea family (Theaceae) [6]. Madang gatal is widely found in Indonesia and several countries in Asia, however, its scientific information and utilization for folk medication is still limited. The stem of Madang gatal is commonly used to produce high-quality carpentry wood, while the tree's leaves are employed as a natural diet of primates [6,7]. In Indonesia, the plant has been traditionally used as a treatment for insomnia and hypertension.

Previous studies reported that Madang gatal extract has anti-inflammatory, antipyretic, analgesic [8], antimicrobial [9] and anti-mutagenic [10] activities. Kaemferol-3-O-ramnosida isolated from this plant known to exert anti-malarial [11], anticancer [12,13], and anti-plasmodial properties [14], while rotenone was found to possess antimicrobial and antifungal. In addition, another reported compound, theanine, potentially shows psychoactive properties for overcoming anxiety, depression, stress, insomnia, sleep disorders, hypertension and some symptoms of schizophrenia [15].

In this current research, the antioxidant activity of the butanolic extract from Madang gatal leaves will be examined. The bioactive compounds contained in the extract were purified and isolated. Further, the most active fraction as an antioxidant was isolated using column chromatography and continued by preparative TLC. The isolate was then identified using TLC, spectroscopy FTIR, and LCMS. Antioxidant activity was performed on the basis of the radical scavenging effect of the stable DPPH radical formula [17].

2. Materials and Methods

2.1. Materials

Madang gatal leaves were collected from PUSPIPTEK provincial garden. This plant was identified by a botanist of the Biology Laboratory, Research Center for Biology LIPI, Cibinong West Java, Indonesia and voucher specimens have been stored at the Biology Laboratory. All technical solvents such as methanol, n-hexane, ethyl acetate and butanol were re-distilled according to standard procedure. The butanolic extract obtained by partitioned of the metanolic extract using n-hexane, ethyl acetate, and butanol, successively.

Some analytical grades obtained from *E. Merck* were used without further purification. Silica gel plates (Kiesel gel 60_{F254} 0.25 mm) and H_2SO_4 for TLC analysis, methanol and acetate acid for mass spectrometry (MS) analysis, DPPH for antioxidant activity analysis, and silica gel 60 (230-400 mesh) for column chromatography to purify bioactive compound of the extract. The equipment used in this study were FTIR Shimadzu prestige 21 using KBr pellets to determine the functional groups and LCMS Mariner Biospectrometry Electrospray Ionization (ESI) System to determine the MS of the isolate.

2.2. Methods

The bioactive compounds from the Madang gatal extract was isolated by using chromatography column techniques with silica gel 60 as the stationary phase. N-hexane and a gradient of ethyl acetate to 100 % were selected as liquid-liquid extraction solvents to obtain a number of fractions. Purifications of the bioactive compounds were conducted on fraction which had the highest antioxidant activity by preparative TLC. The functional group of the isolate was then identified on FTIR Shimadzu prestige 21 using KBr pellets, while MS was analyzed by LCMS spectroscopy.

Antioxidant activity was evaluated using the DPPH method. All samples are prepared at the concentration of 1000 μg/ml. After that, the series solutions of 10-200 μg/ml were prepared to squeeze successively from the stock solution. Quercetin (10-25 μg/ml) was performed as a standard antioxidant. Subsequently, the series solution was diluted with methanol so the volume of the solution became 2000 μL at each concentration. Five hundred μL of DPPH solution was added to sample test and incubated for 30 min in the dark. Further, the absorbance was measured by using UV-Vis Cary 60 spectrophotometer at a wavelength of 517 nm. The percentage of DPPH inhibition and IC_{50} value were calculated [16,17].

3. Results and Discussion

In our previous study, extraction of Madang gatal leaves produced % yield of methanol (22.18%), butanol (20%), aqueous (19.67%), ethyl acetate (2.92%) and n-hexane (0, 06%) extract, sequentially [19]. This current study examined the antioxidant activity of Madang gatal butanolic extract by scavenging the free radicals 1,1-diphenyl-2-picryl-hydrazyl (DPPH). DPPH is converted to 1,1-diphenyl-2-picryl-hydrazine in the present of antioxidant compound, indicated by the reduction of the DPPH solution color from purple into a yellowed-colored DPPH-H molecule product. The color change was proportional to the number of radicals captured and can be measured by visible light spectrophotometer at wavelength 515-517 nm [16-18]. The antioxidant activity test was compared to quercetin as a standard antioxidant which is known as a strong antioxidant with IC_{50} less than 10 μg/ml. Hence, the tested sample solutions will be known to have the same, less, or higher activity than standard materials [17].

The result showed that Madang gatal butanol extract was active as an antioxidant indicated by changing of the DPPH solution color, from purple to yellow. The IC_{50} value of butanol extract in scavenging DPPH was 12.55 μg/ml. The IC_{50} value lower than 50 μg/ml indicates that the butanol extract possess a robust antioxidant activity [16-18], but markedly lower than standard quercetin with IC_{50} value of 9.35 μg/ml. Accordingly, Madang gatal butanol extract was found to be an effective antioxidant agent.

The extract was purified next by silica gel column chromatography eluting with n-hexane, a gradient of ethyl acetate to 100%. The separation by column chromatography yielded 35 fractions (F1– F35), and then combined into 4 purified fractions, which were B1(F1-F13), B2 (F14-F15), B3 (F16-F22), and B4 (F23-F35). The B1-B3 fractions have antioxidant activity, while B4 was not active as antioxidant indicated its IC_{50} value of 228.28 μg/ml. B1 and B3 fractions with IC_{50} value 90.26 and 70.55 μg/mL were categorized as moderate antioxidants, while B2 was categorized as a strong antioxidant with IC_{50} value of 24.50 μg/ml. The antioxidant activity of the fractions is shown in Figure 1.

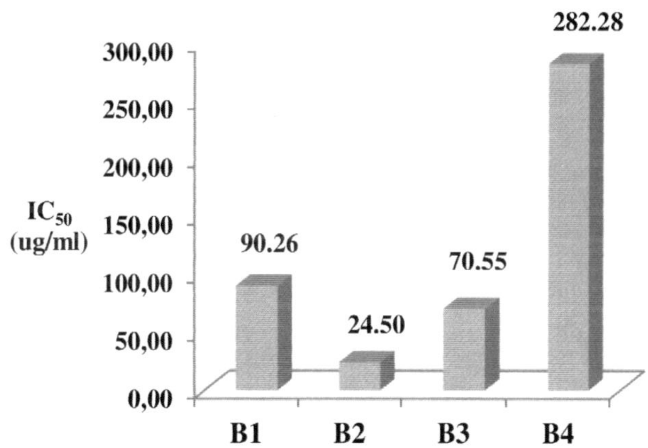

Figure 1. The IC_{50} value of antioxidant activity from butanolic extract fractions of Madang gatal leaves.

The column chromatography has not yet produced pure compounds, therefore the next purification needs to be done. B2 which most active as an antioxidant was subjected to purified further by preparative TLC [20]. Preparative TLC produced 15.5 mg of isolate and then analyzed by using FTIR and LCMC spectrophotometers. The isolate was thought to have functional groups that are similar to flavonoid compounds. The previous research conducted by Barliana *et al.* showed that the main compound contained in *S. wallichii* was a flavonoid compound [14]. The phytochemical analysis of the Madang gatal leaves extracts from the previous study indicated the presence of flavonoids [19]. One of the main flavonoid compounds in theaceae plants are theanin and catechins [21].

The IR spectrum of B2 has similarities to the catechin IR spectrum. The compound exhibited absorption bands at 3246.20-3506.59 cm^{-1} indicating the presence of hydroxyl (OH) groups. The vibration of the carbonyl (C=O) group of the compound was indicated at 1072.42 cm^{-1}. A wave number of 2922.16 cm^{-1}, 678.94-837.11 cm^{-1}, and 1608.63-1699.29 cm^{-1} were indicating the presence of C-H sp2 stretched, C-H aromatic, C=C stretched groups, respectively. Whilst catechins exhibited absorption bands at 3353.17 cm^{-1} and 1611.10 cm^{-1}indicating the presence of hydroxyl (OH) and C=C stretched groups. The IR spectrum similarities of B2 with catechins were presented in Figure 2. The molecular structure of catechin that one of the main flavonoids in teacheae was showed in Figure 3.

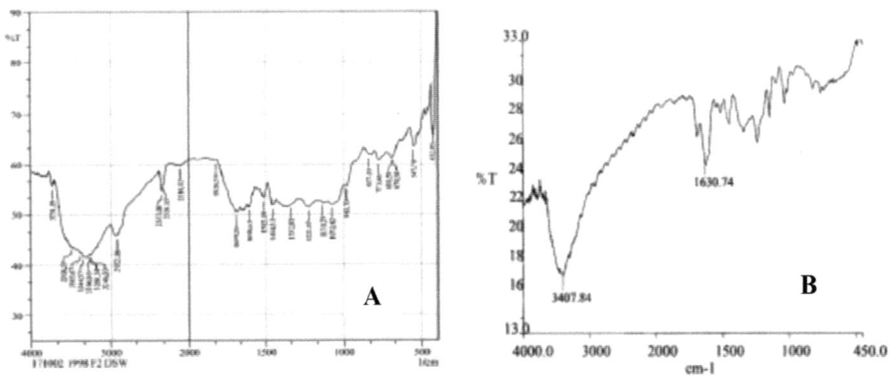

Figure 2. The similarities of IR spectra of the putative compound contained in B2 fraction (A) and catechin compound as reference material (B).

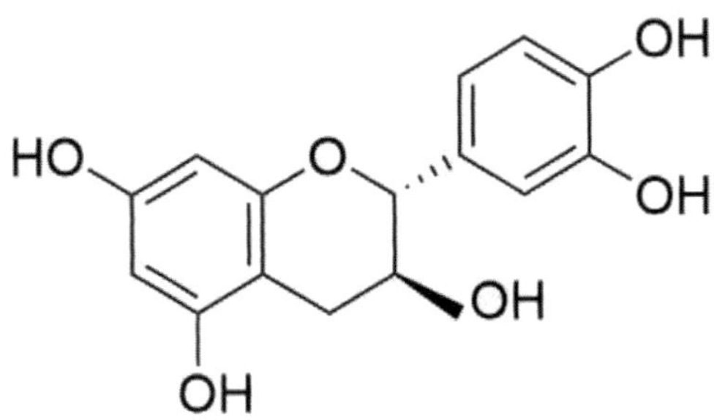

Figure 3. Molecular struture of catechin.

MS of the isolate from B2 identified using LCMS was presented in Figure 4. The chromatogram illustrated 1 main peak at the retention time of 1.68 minutes and MS of 290.25 g/mol was recognized as catechin, according to MS of catechin based on literature is 290.30 g/mol. Catechin and theanine were major compounds commonly found in theaceae plants [21, 22]. Catechine has been known as a powerful antioxidant for chronic and degerenetive disease prevention and health promotion due to its antioxidant, antidiabetic and anti-carcinogenic effects [23].

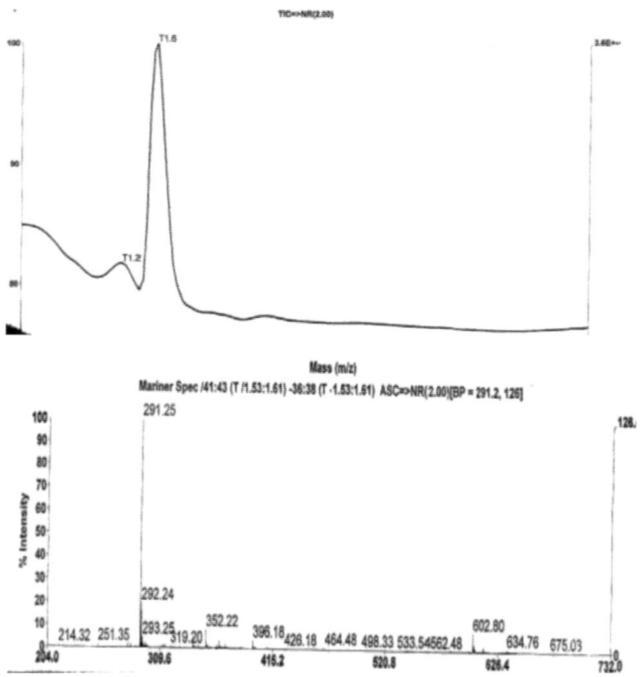

Figure 4. MS of isolate from B2 fraction

The isolate has powerful antioxidant with IC_{50} value 17.97 μg/ml.

4. Conclusion

The current report demonstrated that butanol extract of Madang gatal leaves exhibited a strong antioxidant property with the IC_{50} value of 12.55 μg/ml. Under the optimum purification and isolation condition, the extract produced 1 isolate (15.5 mg), which was predicted as a flavonoid. IR and mass spectrometry (MS) analysis displayed that the major compound of the isolate from B2 fraction might be a catechin compound which has some functional group similarities with MS 290.25 g/mol. The isolate performed a strong antioxidant activity with IC_{50} value of 17.97 μg/ml. Taken together, Madang gatal leaves potentially be developed as a natural antioxidant for preventing human diseases. Further well-design study using ^{1}H and ^{13}C-NMR analysis should be carried out to identify the isolate based on observed spectral.

Acknowledgement

The authors expressed their acknowledgement to Indonesian Ministry for Research and Technology, Higher Education (Kemenristek-DIKTI) for its financial support in this research and to Research Center for Chemistry, Indonesian Institute of Sciences for support this research. We also thank to Mr. Suratman for his help collecting Madang gatal leaves samples, Dr. Hartiwi Diastuti and Purwati, MSi for their suggestions for this study.

References

[1] Halliwell B 2015 *International life science institute (ILSI) Europe Concise Monograph Series* 1-36
[2] Borek C 2017 *Journal of Restorative Medicine* 53-8
[3] Newman D J and Cragg G M 2016 *Journal of Natural Products* 629-61
[4] Sharma N, Samarakoon K W and Gyawali R 2014 *Molecules* **19**(9) 14567–81.
[5] Sarbadhikary S B, Bhowmik S, Datta B D and Mandal N C *2015 International Journal of Current Microbiology and Applied Science* **4**(8) 643 – 55

[6] Orwa C, Mutua A, Kindt R, Jamnadass R, and Simons A 2009 Agroforestree Database: A Tree Reference and Selection Guide Version 4.0, [online], Diunduh dari http://www.worldagroforestry.org/af/treedb/, Diakses pada 25 Oktober 2016.
[7] Martawijaya A, Kartasujana I, Mandang Y I, Prawira S A and Kadir K 1989 *Atlas Kayu Indonesia 2* Bogor: Balitbang Kehutanan Dephut.
[8] Das S and Ghosh L K 2013 *Pharmacologia* 400-3
[9] Dewanjee S, Maiti A, Kundu M, and Mandal S C 2008 *Journal of Pharmaceutical Sciences* **6**(2) 121-3
[10] Subarnas A, Hadiansyah C, Gurmaya K J and Muhtadi A 2003 *Biotika* **2** 7-13
[11] Barliana M I, Diantini A, and Abdullah R 2012 Aktivitas Prosiding of InSINas: Bandung, 28-29 November 2012.
[12] Diantini A A, Subarnas K, Lestari E, Halimah Y, Susilawati, Supriyatna, Julaeha E, Achmad T H, Suradji E W, Yamazaki C, Kobayashi K, Koyama H, and Abdullah R 2012 *Oncology Letters* **3** 1069-72
[13] Halimah E, Diantini A, Destiani D P, Pradipta I S, Sastramihardja H S, Lestari K, Subarnas A, Abdullah R, and Koyama H 2014 *Biomedical Reports* **3** 115-7
[14] Barliana M I, Suradji E W, Abdulah R, Diantini A, Hatabu T, Shimada J N, Subarnas A, and Koyama H 2014 *Biomedical Reports* **2** 579-583.
[15] Subba B and Ram R 2014 *Journal of Engineering Science and Research Technology* **3**(8) 175-80
[16] Molyneux P 2004 *Technology* **26**(2) 211-219.
[17] Yen G C, Chen H Y 1995 Journal of Agricultural and Food Chemistry **43**(1) 27–32
[18] Baravalia Y, Vaghasiya Y, Chanda S 2012 *Iranian Journal of Phar-maceutical Research* **11**(3) 851–61
[19] Widiyarti G, Supiani S, Tiara Y 2018 *The Journal of Tropical Life Science* **8**(2) 151–7
[20] Paudel R R, Subba B 2014 *Pharmacy* **11**(1) 7–13.
[21] Mazza G and Oomar B D 1998 *Herbs, botanicals, and teas* (Boca Ranton: CRC Press)
[22] Deng WW, Ogita S, Ashihara H 2010 *Plant Physiology and Biochemistry* **48**(1) 70–72
[23] Williams J, Kellett J, Roach PD 2016 *Beverages* **2**(13) 1–16

IOP Conf. Series: Materials Science and Engineering 1011 (2021) 012025 doi:10.1088/1757-899X/1011/1/012025

Cold water swelling starch: methods to prepare and recent applications

Achmat Sarifudin[1*], Enny Sholichah[1], Woro Setiaboma[1] and Sunanta Tongta[2]

[1]Research Center for Appropriate Technology Development, Indonesian Institute of Sciences (Pusbang TTG-LIPI), Jl. K.S. Tubun No. 5 Subang 41213, West Java, Indonesia.

[2]School of Food Technology, Institute of Agricultural Technology, Suranaree University of Technology, Muang, Nakhon Ratchasima 30000, Thailand.

*Corresponding author and main contributor: achmatsarifudin@gmail.com

Abstract. Cold water swelling starch (CWSS) offers advantage of unique properties over its origin. Among methods available to prepare CWSS, heating starch in ethanol can be considered to be the simplest one. The unique features of CWSS are depending on its starch origin and degree of conversion; therefore, its application should be specific depending on the feature of CWSS. This review summarizes methods to produce CWSS, advantage of producing CWSS by heating starch in ethanol method, the most important features of CWSS that influence its applicability, and exploration on CWSS applications. We expect that this review will be beneficial for food industries that utilize modified starch as ingredient in their products as well as researchers who seek potential applications of CWSS.

1. Introduction

Starch is one of the most abundant biopolymers built from two major components namely amylose and amylopectin. Amylose is a macromolecule primarily linear glucose chains linked each other by α-(1-4) glycosidic bond while amylopectin is a highly branched molecule with α-(1-4) linked D-glucose as backbones and branches of glucose chains interlinked with backbones by α-(1-6) glycosidic bond [1]. The proportion as well packing arrangement of amylose and amylopectin within the starch structure influence the properties of the native starch.

Native starches are inherently unsuitable for most applications due to limited characteristics such as insoluble in cold water, loss of viscosity at high cooking temperature, and low thickening power post cooking [2]. Therefore, most starch need to be modified either by chemical or physical modification prior to application in food industry. Modified starch can be used in many food products as thickeners, gelling agents and encapsulating agents. It also can be used for other non-food applications such as in papermaking industry. Some modified starch can also be used as a wet-end additive in coating binders. In textile industry, modified starch is an important ingredient for warp sizing of textiles also glass fibre sizing [3].

Due to certain safety considerations such as the requirement to produce no chemical residue after starch modification process, some food industry prefer to use physically modified starch over chemical modified ones. The most common argument is that physical modified starch is considered safer for human consumption [4]. Instaneous gelatinized starch is an example of physically modified starch which can be produced either by pre-gelatinized starch (PGS) and cold water swelling starch (CWSS) methods.

PGS can be prepared by heating starch in excess of water at above the gelatinization temperature of starch while CWSS can be produced by certain methods to swell starch granules at some extent [5]. The CWSS gives more superior properties compared to PGS. The CWSS offers greater solubility, viscosity, smoother texture and more tolerance for further processing [6].

2. Methods to produce cold water swelling starch

Effort to develop starch products that is soluble in cold water has been started since the mid of 1950. Knowledge on the starch granule morphology and pasting behavior of native starch brought out the ideas to produce instantaneous soluble starch. Powell E L [7] reported that pre-gelatinized starch could be prepared by drum drying, extrusion, and conventional cooking followed by spray drying processes. The pre-gelatinized starch product exhibits inferior properties such as graininess, less sheen and less flexible for further processing; thus the product has lower quality than the freshly cooked-up product. However, method of pre-gelatinized starch cooking is still implemented to prepare certain food products such as instant porridges.

Lindqvist I [8] was the first who revealed the ability of starch to gelatinize in cold water after it was treated with potassium iodide (KI) and potassium thiocyanate (KSCN). Pitchon E, et al. [9] introduced a process to prepare CWSS using a special spray drying technique. The starch slurry was atomized in an enclosed chamber from one nozzle, and then at the same time, the other nozzle injected a high-pressure steam to cook up the starch. The product showed indented spheres and swelled well upon rehydration. The major disadvantages of this technique were the high capital cost to construct the spray dryer and the high operational cost of spray dryer.

The first patent of CWSS preparation process was filed by Eastman J E and Moore C O [10]. In their method, normal starch was slurried in an aqueous alcohol and then it is subjected to high temperature in a pressurized autoclave. The CWSS obtained by this method exhibited higher solubility than that of normal starch. However, this method could not be applied for waxy starch because the starch completely gelatinized during the conversion process. Eastman J E [11] successfully polished up his previous method to convert waxy starch to become CWSS. The CWSS product obtained by this method was claimed to be chemically unmodified. The CWSS product can be set to become a sliceable gel without cooking or chilling when it was blended with aqueous sugar syrup. The CWSS product was also claimed to be useful for ingredient of many instant food products such as pie fillings, jellies, demouldable desserts, and puddings.

Rajagopalan S and Seib P A [12] successfully patented another method to prepare CWSS. In their method, starch slurry of a mixture of starch, water and polyhydric alcohol was heated in an atmospheric pressure. The successful keys of the method depended on the type of starch, type of polyhydric alcohol, temperature and the proportions of starch, water, and polyhydric alcohol. Aqueous propan-1,2-diol was the most preferable medium, although aqueous ethylene glycol, glycerol and any other four positional isomers of butandiol were also useful. Aqueous butandiol was the most effective reagent to convert hydroxypropylated starch to CWSS, while aqueous glycerol was the most effective reagent when the starting material was potato starch. This finding was significant in terms of energy conservation for CWSS processing. They claimed that the method can be applied to produce CWSS from many starch sources such as cereal, tuber, root, and legume starches. They also reported that the CWSS products had cold water solubility up to 95%.

Another patent related to CWSS preparation method was released by Jane J and Seib P A [13]. They claimed that the new method was capable to prepare CWSS from waxy rice or high amylose starches which could not be achieved by another previous method. In their method, they used a mixture of ethanol and alkali. The alkali (primarily NaOH) was functioned to swell the starch granules; meanwhile, the ethanol was used to prevent the starch granule from disintegration. The treated starch was then neutralized with HCl, washed and dried at 80°C. They found that higher percentage of used NaOH will produce CWSS with higher solubility.

A new approach in CWSS preparation method was introduced by Wang J, et al. [14] in which they used polyethylene glycol to gelatinize the starch. In this method, polyethylene glycol acts as a grafting agent to create micro pits on the starch granule surface. Therefore, allowing water molecules to penetrate inward and swell the starch granule.

3. Advantage of producing CWSS by heating starch in ethanol method

Simple CWSS preparation method was demonstrated by Zhang B, et al. [15]. A mix of starch and aqueous alcohol was heated in a rotary evaporator in order to facilitate uniform mixing and at the same time it is functioned to evaporate the reactant. The development was mainly intended to avoid the use of harsh chemical such as NaOH. Unfortunately, this method was not applicable to prepare CWSS from waxy starch because the starch granules underwent rapid swelling and gelling upon conversion. Dries D M, et al. [16] prepared CWSS by heating a mixture of water-ethanol in a leak-proof Schott bottle. The method is relatively simple to obtain CWSS in a lab scale. Also heating starch in ethanol can be considered to be the simplest and green process for preparing CWSS. After the conversion process the ethanol can be completely removed from the CWSS product by dehydration [15, 16].

4. The most important features of CWSS that influence its applications

4.1. Solubility and water absorption in cold water

Solubility in cold water is one of the most essential properties of CWSS. Pre-gelatinized starch and CWSS are both known as instantaneous soluble starch product; however, CWSS has lower solubility than pre-gelatinized starch. Majzoobi M, et al. [17] reported that the solubility of CWSS from maize starch was 6.24% while that of pre-gelatinized starch from maize starch was 8.45%. Decreasing or increasing the pH of solution tends to increase the solubility of CWSS [5]. Moreover, increasing the solubilization temperature also increases the solubility of CWSS [18].

The water absorption capacity of CWSS reflects the ability of CWSS to absorb and hold water within its matrix. Pre-gelatinized starch shows higher cold water solubility than CWSS, but CWSS exhibits higher cold water absorption capacity than pre-gelatinized starch. Jane J, et al. [19] reported that CWSS from maize starch had water absorption capacity of 12.33 g/g while pre-gelatinized starch was 9.85 g/g. Hedayati S, et al. [5] also reported similar result in that the water absorption capacity of CWSS from maize starch reached 13.35 g/g which was higher than that of pre-gelatinized starch (10.04 g/g). The water adsorption capacity of CWSS and pre-gelatinized starch increased at high pH and decreased at low pH [5]. CWSS that still preserve its granular form and CWSS that losses its granular form have almost similar water absorption capacity values, however, both CWSS have different mechanism to absorb water [20]. Upon water absorption, CWSS with granular form absorb water via fissures on the granule surface meanwhile, water is entrapped within the three-dimensional structure of starch components of non-granular CWSS.

Packing arrangement of amylopectin which constructs the native crystallinity of starch is also expressed in the physical properties of the starch. The degree of conversion of CWSS is affected by the native crystallinity of the raw starch. Chen J and Jane J [21] investigated the effect of native starch crystallinity during CWSS conversion by using the cold water solubility as indicator. The cold water solubility of normal maize which is a typical of A-type starch exhibited lower than cold solubility of Hylon 5 which has B-type crystalline, respectively, 22% and 90%. Imberty A and Perez S [22] recorded that B-type starches has space in the unit cell whereas A-type starches is more densely packed. Therefore the structure of B-type is more susceptible to any treatments [23-25]. From their observation Chen J and Jane J [21] supposed that this was because of the effect of crystalline structures of the raw starch which brings consequence the CWSS prepared from B-type starches has higher cold water solubility than those prepared from A-type crystalline starches.

Table 1. Cold water solubility of CWSS prepared from different starch sources which have different native crystalline type [21].

Sample	Native crystalline type	% CWSS solubility
Normal maize	A	22.5±3.5*
Hylon 5	B	90.3±2.1

*Data shown as mean ± standard deviation of duplicate samples

4.2. V-type crystalline structure of amylose-ethanol complex

The V-type crystalline structure is formed when guest molecules resides in the cavity of amylose chain as observed by X-ray diffraction technique. This structure may present as an artifact of a complex between amylose and ethanol during CWSS conversion process. This is because ethanol can be removed from the cavity of amylose by dehydration means [16].

Table 2. Amylose contents of native starches was determined using the Megazyme assay kit and the peak degrees of polymerization (DPpeak) of amylose chain was determined using high performance size exclusion chromatography (HPSEC) [26].

Starch sources	Amylose content (% dry matter)	DP peak of amylose chain
Maize*	23.7	1710
Rice	18.1	1520
Cassava	17.3	4240
Potato	14.8	3520
Pea	26.8	2150

*This value was previously published in [16]

The role of amylose during CWSS conversion has been investigated since early finding of this modified starch [15, 16, 27-29]. Most of them agreed that linear AM chains are the main component responsible for V-type crystal formation in CWSS structure. In the recent paper published by Dries D M, et al. [26], they reported the role of amylose in term of amount and chain length (in term of degree of polymerization, DP) of the amylose during CWSS conversion. They found that V-type crystalline never exceeded the AM contents of the starches. However, they did not find any quantitative relation between the levels of V_H-type crystalline with the amylose contents. In contrast, they found a qualitative relation between the chain length of amylose and V-type crystalline in which the V_H-type crystalline in the CWSS decreased with increasing amylose DP. The higher the amylose DP, the more strongly V-type crystal formation is hindered. The molecular weight-dependency is well-known in crystallization of synthetic peak polymers [30] and it is explained by the increased occurrence of entanglements for polymers of higher molecular weight. In addition, Dries D M, et al. [16] emphasized that these entanglements are part of the amorphous regions, which explains why starches with a higher amylose DP have lower V-type crystallinity.

Different with amylose that involved in the V type crystal formation, the amylopectin structure which constructs the native crystalline was loss, in some part, during CWSS conversion. In 1986, Jane J-L, et al. [31] proposed a mechanism to explain the structural transition during formation of CWSS. They hypothesized after native crystal melting; the alcohol rapidly induces single helical formation and V-type crystalline with alcohol being located within the single helix cavities and possibly interstices. Their hypothesis is based on their observation that the endothermic heat of gelatinized starch in aqueous alcohol was lower than in pure water. Jane et al. (1986) also reported that formation of V_H-type

IOP Conf. Series: Materials Science and Engineering 1011 (2021) 012025 doi:10.1088/1757-899X/1011/1/012025

crystalline is mainly assigned equally by the presence of amylose and amylopectin based on the comparison of the X-ray diffraction pattern of their sample with a reference sample. In fact this hypothesis cannot be applied for all starch types. Chen J and Jane J [28] found that no V-type crystalline was exist in CWSS from waxy maize starch. Another report by Zhang B, et al. [15] stated that all CWSS prepared from A-, and B-type sources displayed V-type crystalline except for partially converted waxy maize starches. This fact lead to another hypothesis proposed by Dries D M, et al. [16] that crystallinity transition during CWSS formation requires amylose during nucleation of V-type crystalline and the amylopectin contribution is triggered by amylose.

Table 3. Degree of A-, B, and VH-type crystallinity (%) for native and CWSS determined by wide angle X-ray diffraction (WAXD) [26].

Starch source	Crystalline type	% crystallinity in native starch	% crystallinity in CWSS
Maize*	A	27	0
	V_H	2	18
Rice	A	33	0
	V_H	2	16
Cassava	A	40	0
	V_H	0	10
Potato	B	38	0
	V_H	0	10
Pea	A	13	0
	B	12	0
	V_H	0	13

*data for maize starch were previously published in [16].

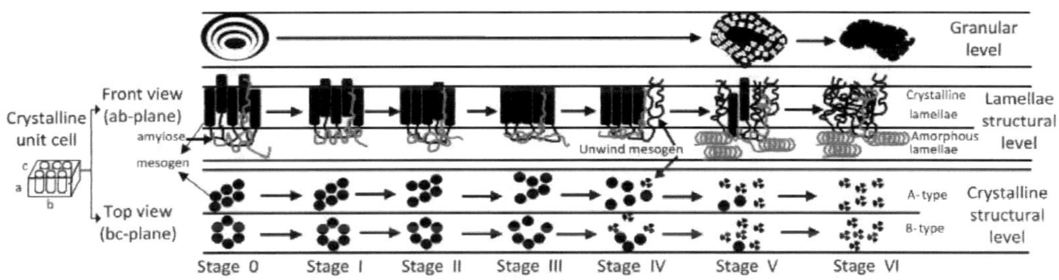

Figure 1. The proposed model of gradual structural transformations of cold water swelling starch prepared by heating in ethanol method [32] .

A model developed by Jane J, et al. [27] described the transformations of CWSS as a single step process at the crystalline structural level, while related phenomena at higher structural levels were not considered. Sarifudin A, et al. [32] developed a model that includes the transformations of starch at different structural levels during CWSS conversion. The model of was developed based on the real time observation of structural transformation at crystalline and lamellae levels. They found that when the temperature increased the thermal energy of the system started to rise. It was sufficient to mobilize the mesogen toward a more stable conformation, as indicated by the increase in the crystalline lamellae thickness. The ability of crystalline lamellae to realign into the most perfect register is illustrated by stage III. Above the maximum stress resilience temperature, the crystalline lamella starts to be disrupted which is illustrated by stage IV. Stage VI illustrates the loss of birefringence of the CWSS granules heated at high temperature (100 °C), but they were still in granular form. This might be attributed to the role of amylose chains and amylose-ethanol complexes to preserve the integrity of the CWSS granules [16, 26].

4.3. Morphology of dried and hydrated CWSS

Conversion to CWSS indeed changes the morphology of native starch. Singh N, et al. [33] reported native corn starch granules appeared less smooth, rounded and angular shaped, meanwhile potato starch granules showed smooth, some were irregular or cuboidal shapes and others were larger oval, and few small granules. The size of corn starch granules were range between 5-7 μm for small granules and 15-18 μm for large granules and for potato starch granules have size between 15-20 μm for small granules and 20-45 μm for large granules. When native starches was subjected to CWSS treatment they underwent morphological changes [34]. They observed that the CWSS of potato starch granules swelled to higher extent more than corn starch granules indicating the effect CWSS conversion on the starch morphology was visually more pronounce for potato starch than that for corn starch. However, the morphological properties of CWSS prepared from different preparation methods do not show much difference between them. Chen J and Jane J [21] that used alcoholic-alkaline treatment for preparing CWSS also observed that there were indentation on the CWSS potato starch granules but the shapes were not rupture. Zhang B, et al. [15] that used rotary evaporation at high temperature for CWSS preparation also noticed that CWSS from potato starch showed bigger size than that of maize starch when observed under light microscope. All reports mentioned that observation of CWSS by using SEM technique indicated that shrinkage in the centre region of the granules can be due to internal fragmentation.

Sample	Corn starch granules	Potato starch granules
Native	A1	B1
CWSS	A2	B2

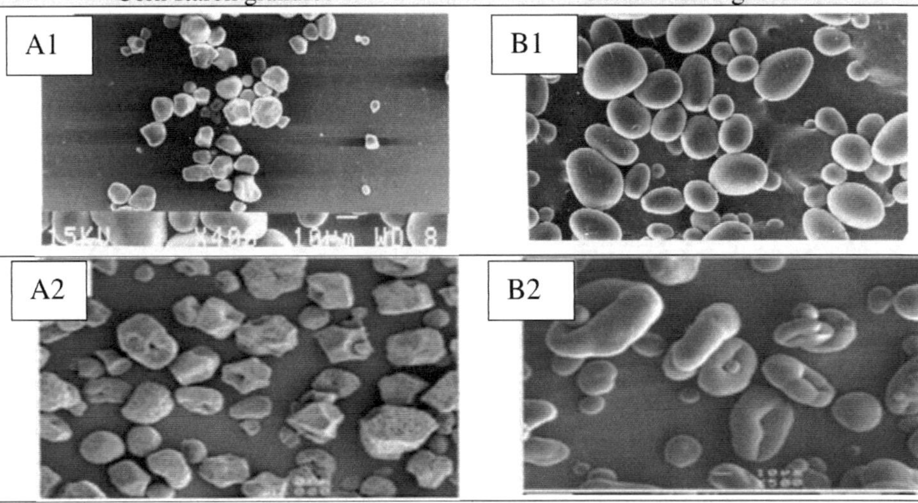

Figure 2. Native starch and CWSS under scanning electron microscope [15, 33].

Zhang B, et al. [15] reported that native corn starch appeared to have shape polygonal with a distinct Maltese cross in each granule under polarized light microscope. Native potato starch showed smooth, round to oval shaped and it also showed distinct birefringence. As the native starch converted to CWSS, the granules swollen thus increasing the granule size which was correlated with the amount of water adsorbed within the CWSS granules [15]. The CWSS granules started to burst from the hilum region (dark area in the center of most CWSS granules) with wrinkling pattern on the surface. Under polarized light microscope showed majority of granules loss their Maltese cross. However, a diffuse birefringence pattern can still be observed in the outer part of granules. This might be an indication that their native crystalline at the outer layer can still be maintained after CWSS conversion [15, 16].

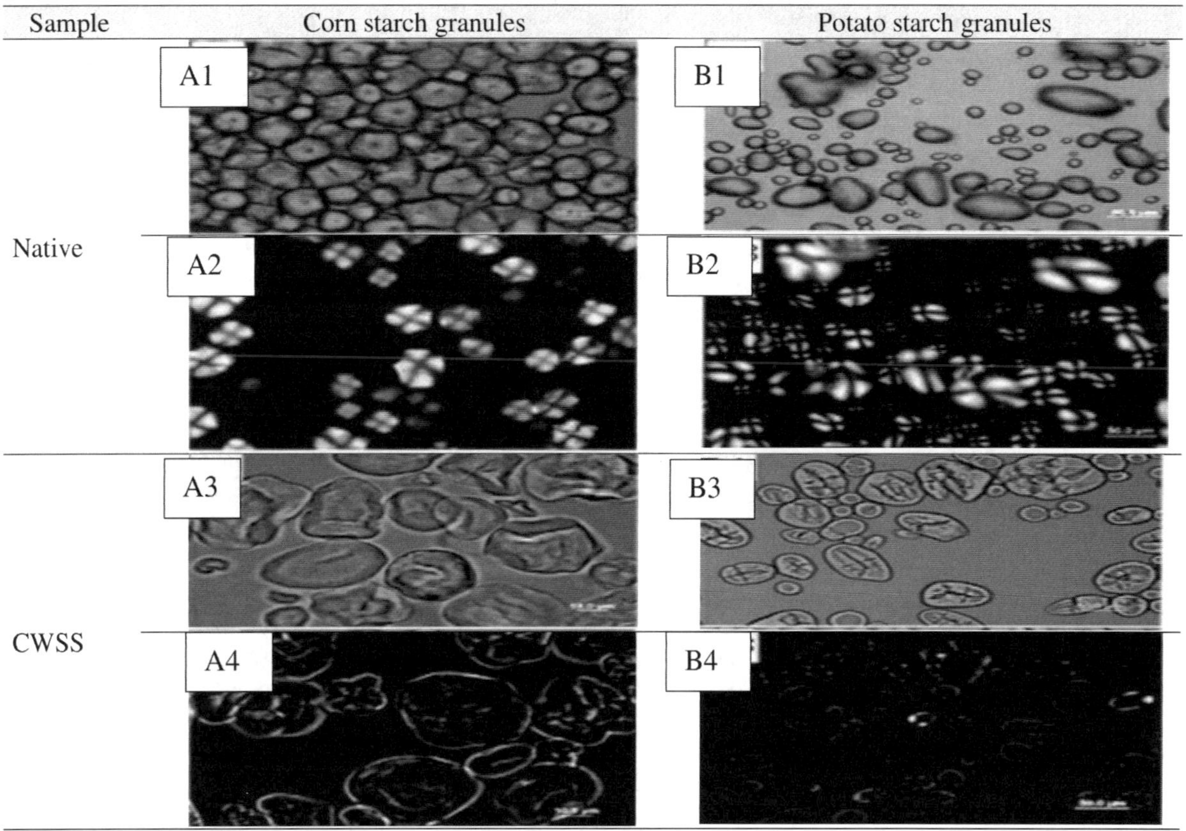

Figure 3. Native starch and CWSS under normal light microscope (A1, B1, A3, B3) and polarized light microscope (A2, B2, A4, B4) [15].

4.4. Pore characteristics of CWSS

Porous starch is a modified starch product which is prepared to enhance the porosity of native starch, thus increasing its water absorption capacity [35, 36]. Another type of starch that exhibits high water absorption capacity is cold water swelling starch (CWSS). Many studies have reported the capability of CWSS to absorb water at low temperatures [16, 29]. CWSS can absorb water up to 12 times more than its weight [17]. Meanwhile, porous starch can only maximally absorb water up to 3 times more than its weight [35]. Therefore, it is advantageous to use CWSS over porous starch in terms of their water absorption capacity.

The capability of CWSS to absorb high amount of water can be an indication that it has high porosity. Model of Jane J, et al. [37] explained that the structure of native starch drastically changes after it is converted to CWSS. The crystalline structure turned to amorphous structure. The amorphous structure provides more free space to reside guess molecules within it [38].

Sarifudin A, et al. [20] reported that CWSS that still has granular form such as CWSS from maize and potato starches contains non-rigid and slit-shaped pores. Meanwhile, CWSS that doesn't maintain their granular form lost their functional pore. For granular CWSS, water penetrated the granules through the fissures, hydrated the amorphous regions, melted the V-type crystalline structure, and was held within the CWSS granules upon water absorption. For non-granular CWSS, water hydrated the amorphous and V-type crystalline structures, and was entrapped within a three-dimensional network of starch components upon contact with water.

5. Recent applications of CWSS

In the first time of CWSS finding, it was developed in tandem with pre-gelatinized starch to overcome the low solubility of native starch [10]. Due to their capability to adsorb cold water and swell promptly, giving appearance and texture of a cooked-like starch. Therefore, they are commonly used for ready-to-be-consumed product [10]. These modified starches are used as ingredients of instant foods such as pie fillings, jellies, demouldable desserts, and puddings Eastman J E [11].

CWSS has been used to encapsulate active ingredients [39, 40]. They reported that the encapsulated product properties by CWSS were influenced by the encapsulation method. Chen J and Jane J [40] used water as a medium during encapsulating atrazine by CWSS. They reported that chemical interactions between atrazine and starch components did not occur; instead, B-type diffraction appeared as an indication of starch gel retrogradation. Dries D M, et al. [39] used multiple encapsulation stages involving water-ethanol mixture to encapsulate ascorbyl palmitate using CWSS. The method was applied to retain the V-type crystalline structure in the final product.

Boonwatcharapan Y, et al. [41] has investigated alcohol-alkaline-treated rice starch as a tablet disintegrant. They reported that the modified rice starch exhibited higher water solubility and swelling capacity compared to that of native rice starch. They also suggested that the modified rice starch was a good disintegrant for direct-compressed tablet formulations, especially in the presence of water insoluble fillers. The most recent investigation suggested that CWSS is potential for excipients of medicinal tablet [42]. They reported that tablet prepared from CWSS are potential to be used in controlled release system. tablet prepared from granular CWSS showed high friability and low crushing strength indexes, and dispersed and released active ingredients rapidly upon contact with water. Therefore, it is suitable for fast release system. Moreover, tablet from non-granular CWSS exhibited low friability and high crushing strength indexes. Upon hydration, the tablet of non-granular CWSS containing lauric acid eroded gradually and released the active ingredient during tablet erosion. Meanwhile, the tablet of non-granular CWSS containing ascorbic acid swelled slowly during hydration, and the active ingredient diffused out gradually from the swelled tablet. Therefore, the non-granular CWSS is suitable for ingredient of tablet for sustained release system [42].

6. Conclusions

CWSS can be prepared by mixing between gelatinization inducer and anti-swelling agent in certain proportion. However, heating starch in ethanol method is the most simplest method to prepare CWSS since it doesn't require to do neutralization step also the reactant (ethanol) can be removed completely from the product during dehydration process. The most important features of CWSS are correlated to the capability of CWSS to absorb water and swell instaneously. CWSS which is commonly used for ingredient of instant product, recently, it is used for encapsulating agent of active ingredients and also binder of medicinal tablet for sustained release system.

References

[1] Buléon, A., Colonna, P., Planchot, V., and Ball, S. 1998 Starch granules: structure and biosynthesis *International Journal of Biological Macromolecules* **23** 85-112

[2] Chiu, C.-w. and Solarek, D. 2009 *Chapter 17 - Modification of Starches* ed(s) San Diego Academic Press 629-655

[3] Xie, S. X., Liu, Q., and Cui, S. W. 2005 *Starch Modification and Applications* S. W. Cui ed(s) Boca Raton, Florida CRC Press, Taylor & Francis Group

[4] Kaur, B., Ariffin, F., Bhat, R., and Karim, A. A. 2012 Progress in starch modification in the last decade *Food Hydrocolloids* **26** 398-404

[5] Hedayati, S., Shahidi, F., Koocheki, A., Farahnaky, A., and Majzoobi, M. 2016 Physical properties of pregelatinized and granular cold water swelling maize starches at different pH values *International Journal of Biological Macromolecules* **91** 730-735

[6] Light, J. M. 1990 Modified food starches: Why, where, and how *Cereal Foods World* **35** 1081-1084

[7] Powell, E. L. 1967 *Production and use of pregelatinized starch* L. Whistler and E. F. Paschall ed(s) New York Academic Press 2 523
[8] Lindqvist, I. 1979 Cold gelatinization of starch *Starch - Stärke* **31** 195-200
[9] Pitchon, E., O'Rourke, J. D., and Joseph, T. H. 1981 Process for cooking or gelatinizing materials U.S.P.T.O 4.280.851
[10] Eastman, J. E. and Moore, C. O. 1984 Cold soluble water granular starch for gelled food compositions USPTO 4.465.702
[11] Eastman, J. E. 1987 Cold water swelling starch composition USPTO 4.634.596
[12] Rajagopalan, S. and Seib, P. A. 1991 Process for the preparation of granular cold water-soluble starch. USPTO
[13] Jane, J. and Seib, P. A. 1991 Preparation of granular cold water swelling/soluble starches by alcoholic-alkali treatments USPTO 5.057.157
[14] Wang, J., Zhai, W., and Zheng, W. 2011 Preparation of granular cold-water-soluble corn starch by surface modification with poly(ethylene glycol) *Starch - Stärke* **63** 625-631
[15] Zhang, B., Dhital, S., Haque, E., and Gidley, M. J. 2012 Preparation and characterization of gelatinized granular starches from aqueous ethanol treatments *Carbohydrate Polymers* **90** 1587-1594
[16] Dries, D. M., Gomand, S. V., Goderis, B., and Delcour, J. A. 2014 Structural and thermal transitions during the conversion from native to granular cold-water swelling maize starch *Carbohydrate Polymers* **114** 196-205
[17] Majzoobi, M., Kaveh, Z., Blanchard, C. L., and Farahnaky, A. 2015 Physical properties of pregelatinized and granular cold water swelling maize starches in presence of acetic acid *Food Hydrocolloids* **51** 375-382
[18] Li, W. *et al.* 2014 Physically modified common buckwheat starch and their physicochemical and structural properties *Food Hydrocolloids* **40** 237-244
[19] Jane, J., Craig, S. A. S., Seib, P. A., and Hoseney, R. C. 1986 Characterization of granular cold water-soluble starch *Starch - Stärke* **38** 258-263
[20] Sarifudin, A., Keeratiburana, T., Soontaranon, S., Tangsathitkulchai, C., and Tongta, S. 2020 Pore characteristics and structural properties of ethanol-treated starch in relation to water absorption capacity *LWT* **129** 109555
[21] Chen, J. and Jane, J. 1994 Preparation of granular cold-water-soluble starches by alcoholic-alkaline treatment *Cereal Chemistry* **7** 618-622
[22] Imberty, A. and Perez, S. 1988 A revisit to the three dimensional structure of B-tpe starch. *Biopolymers* **27** 1205
[23] Gidley, M. J. 1987 Factors affecting the crystalline type (A-C) of native starches and model compounds: A rationalisation of observed effects in term of polymorphic structures. *Carbohydrate Research* **161** 301
[24] Lii, C. Y. and Lee, B. L. 1993 Heating A-, B-, C-type starches in aqueous sodium chloride: Effects of sodium chloride concentration and moisture content on differential scanning calorimetry thermograms. *Cereal Chem.* **70** 188
[25] Zobel, H. F. 1992 *Starch granule structure.* R. J. Alexander and H. F. Zobel ed(s) St. Paul, Minnesota American Association of Cereal Chemist 1
[26] Dries, D. M., Gomand, S. V., Delcour, J. A., and Goderis, B. 2016 V-type crystal formation in starch by aqueous ethanol treatment: The effect of amylose degree of polymerization *Food Hydrocolloids* **61** 649-661
[27] Jane, J., Craig, S. A. S., Seib, P. A., and Hoseney, R. C. 1986 Characterization of granular cold water-soluble starch *Starch/Staerke* **38** 258-263
[28] Chen, J. and Jane, J. 1994 Properties of granular cold-water-soluble starches prepared by alcoholic-alkaline treatments *Cereal Chemistry* **71** 623-626
[29] Rajagopalan, S. and Seib, P. A. 1992 Properties of granular cold-water-soluble starches prepared at atmospheric pressure *Journal of Cereal Science* **16** 29-40

[30] Mathot, V. B. F. and Pijpers, M. F. J. 1984 Crystallization and melting behavior of polyethylene fractions obtained by various fractionation methods. *Polymer Bulletin* **11** 297-304

[31] Jane, J.-L., Craig, S. A. S., Seib, P. A., and Carl Hoseney, R. 1986 A granular cold water-soluble starch gives a V-type X-ray diffraction pattern *Carbohydrate Research* **150** c5-c6

[32] Sarifudin, A., Soontaranon, S., Rugmai, S., and Tongta, S. 2019 Structural transformations at different organizational levels of ethanol-treated starch during heating *International Journal of Biological Macromolecules* **132** 1131-1139

[33] Singh, N., Singh, J., Kaur, L., Singh Sodhi, N., and Singh Gill, B. 2003 Morphological, thermal and rheological properties of starches from different botanical sources *Food Chemistry* **81** 219-231

[34] Singh, J. and Singh, N. 2003 Studies on the morphological and rheological properties of granular cold water soluble corn and potato starches *Food Hydrocolloids* **17** 63-72

[35] Wang, H., Lv, J., Jiang, S., Niu, B., Pang, M., and Jiang, S. 2016 Preparation and characterization of porous corn starch and its adsorption toward grape seed proanthocyanidins *Starch/Stärke* **68** 1254-1263

[36] Weirong, Y. and Huiyuan, Y. 2002 Adsorbent characteristics of porous starch *Starch/Stärke* **54** 260-263

[37] Jane, J., Craig, S. A. S., Seib, P. A., and Hoseney, R. C. 1986 Characterization of granular cold water-soluble starch *Starch/Staerke* **38** 258-263

[38] Turnbull, D. and Cohen, M. H. 1961 Free-Volume Model of the Amorphous Phase: Glass Transition *The Journal of Chemical Physics* **34** 120-125

[39] Dries, D. M., Gomand, S. V., Pycarelle, S. C., Smet, M., Goderis, B., and Delcour, J. A. 2017 Development of an infusion method for encapsulating ascorbyl palmitate in V-type granular cold-water swelling starch *Carbohydrate Polymers* **165** 229-237

[40] Chen, J. and Jane, J. 1995 Effectiveness of granular cold-water-soluble starch as a controlled release matrix *Cereal Chemistry* **72** 265-268

[41] Boonwatcharapan, Y., Srisuk, P., Palladino, P., Sutthiparinyanont, S., and Chitropas, P. 2016 Preparation and evaluation of alcohol-alkaline-treated rice starch as a tablet disintegrant *Tropical Journal of Pharmaceutical Research* **15** 221-229

[42] Sarifudin, A., Soontaranon, S., Peerapattana, J., and Tongta, S. 2020 Mechanical strength, structural and hydration properties of ethanol-treated starch tablets and their impact on the release of active ingredients *International Journal of Biological Macromolecules* **149** 541-551

IOP Conf. Series: Materials Science and Engineering 1011 (2021) 012026 doi:10.1088/1757-899X/1011/1/012026

Encapsulation of proteins with high/low pI in PLGA-particles for oral delivery

Riyona Desvy Pratiwi*, Dian Fitria Agustiyanti, Fitri Agustiani, Mega Ferdina Warsito and Apon Zaenal Mustopa

Research Center for Biotechnology, Indonesian Institute of Sciences, Cibinong Science Center, Jalan Raya Bogor Km. 46 16911, Indonesia

*Corresponding author email: riyona.desvy.pratiwi@lipi.go.id

Abstract. Polylactic co-glycolic acid (PLGA) is a biocompatible and biodegradable polymer used for controlled drug delivery of small molecules or macromolecules including protein. Proteins highly requires an advanced drug delivery system because of its sensitivity towards many factors such as temperature, hygroscopicity, and biological system. Particularly for oral dosage form, protein must be encapsulated in extremely low pH resistant particle. Protein's intrinsic characterization determines effectivity of the encapsulation. The recent method of PLGA particle preparation was more suitable for proteins with low pI. PLGA showed higher effectivity to encapsulate protein with low pI such as BSA (pI = ~ 4.8) than that of high pI for instance lysozyme (pI = ~ 11). In the study, protein release and physico-chemical characterizations of BSA loaded PLGA-particle were performed.

1. Introduction

Polylactic co-glycolic acid (PLGA) is synthetic thermoplastic aliphatic polyester that has biodegradable and biocompatible characteristic. PLGA has been approved for medical applications by the US Food and Drug Administration (FDA) [1]. It has been reported as potential drug delivery agent for various therapeutic agents such as, chemotherapy, antibiotics, antiseptic, anti-inflammatory and antioxidant drugs, and also proteins [2,3].

The US FDA has been approved about 239 therapeutic proteins since 1982 [4]. The majority of commercially available therapeutic proteins are for parenteral administration [5, 6]. This type of administration had several downfall including not acceptable for self administration, painful, the need for sterile conditions, potential risks to health care practitioners from blood-borne pathogens, require specialized storage and cold chain transport for distribution. To date, oral administration is the most convenient and preferable administration route for most patient and it is usually the safest and least expensive option [7]. In addition, though, there are challenges to administer therapeutic protein orally because of multiple physicochemical and biological barriers in the digestive tract [8]. PLGA helps protein to pass through gastrointestinal tract and successfully deliver the active compound to be taken by M Cell of peyer patches [9].

This study used two proteins that have different isoelectric points, i.e. bovine serum albumin (BSA) and lysozyme, as the model proteins. BSA is reported to has isoelectric point (pI) of 4.8, while lysozymes 10.5-11 [10,11]. Physicochemical properties of the protein loaded particles were characterized using particle size analyzer, FTIR, and SEM, while its stability and release in the gastrointestinal tract were conducted in vitro and analyzed using SDS PAGE – semidensitometry.

2. Materials and Methods

2.1. Preparation of PLGA-particle

PLGA (lactide : gycolide 50 : 50; MW 30,000 – 60,000) was dissolved in chloroform and was dissolved in PBS pH 7.4. BSA solution was wisely dripped into PLGA solution, and then immediately sonicated using 40% A for 2 min in ice water bath. Subsequently, BSA – PLGA emulsion (emulsion 1) was dripped into 1% sodium cholate solution, then sonicated at 40% amplitude for 2 min. BSA-PLGA-sodium cholate emulsion (emulsion 2) was also wisely dripped into 0.3% sodium cholate solution and similarly sonicated with the previous condition. The last emulsion was added into 300 ml of 0,3% sodium cholate using 3 ml syringe and then stirrered for overnight at 600 rpm, room temperature to evaporate organic solvent. On the following day, the PLGA-particles were separated by centrifugation at 12,000 rpm, 4°C. The particles were rinsed with dH_2O three times. All materials were purchased from Sigma and Thermo.

2.2. Analysis of particles stability in acidic and neutral condition representing digestive system

Particles pellet was diluted in 0.1 N HCl pH 1.2 and 1× PBS pH 7.4 then incubated at 37°C for 4 h, respectively. Afterwards, the pellet was separated with the released protein by ultracentrifugation at 30,000 rpm, 10 min, 4°C. The supernate containing released protein and the pellet with the remained protein were analyzed using SDS PAGE with Coomasie blue staining. The concentration of protein was analyzed using semi-densitometer with a serial concentration of BSA standard and then calcultaed by using ImageJ software.

2.3. Loading capacity (LC) and entrapment efficiency (EE) of PLGA Particles

Total loaded protein was calculated from released and remained protein in the pellet. Loading capacity (LC) is calculated based on total of loaded protein to total particles produced. Entrapment efficiency (EE) is calculated based on quantification of succesfully loaded protein to total protein used. Calculation following equation by Fernandez-urrusuno et.al. 1999 [12].

$$\text{Entrapment Effeiciency} = \frac{(Total\ protein - unbound\ Protein)}{Total\ protein}\ x\ 100 \quad (1)$$

$$\text{Loading capacity} = \frac{(Total\ protein - unbound\ protein)}{Total\ nanoparticles}\ x\ 100 \quad (2)$$

2.4. Analysis of particles size, PDI and zeta potential

Measurement of particles size, PDI (polydispersity index), and zeta potential were determined using Particle Size Analyzer (Zetasizer Nano Range, Malvern Panalytical). Z-average size obtained by DLS (Dynamic Light Scattering), a mean value for size and width parameter known as PDI (zetasizer nano manual, Malvern 2004)

2.5. Analysis of Fourier Transform Infra-Red (FTIR)

FTIR was analysis of interfaces to investigate the surface adsorption of functional groups on particle. Samples were freeze-dried, then mixed with potassium bromide (KBr). IR spectra in ranged 4000-400 cm^{-1}. IR spectra result by Spectrum Two FT-IR spectometer (Perkin Elmer, USA).

2.6. Analysis of Scanning Electron Microscopy (SEM)

SEM analysis was used to determine dispersion, size, shape and texture of particle surface morphology. A freeze-dried sampel was coated with gold, then observed under scanning electron microscope (JSM IT200, JEOL, Japan).

3. Result and Discussion

3.1. Amount of PLGA and solvent determined hydrodynamic size

Concentration of polymer (PLGA) and organic solvent significantly affect particle size. Table 1 shows large particles were obtained from high concentration or high amount of PLGA. Formula 1 and formula 2 resulted higher particles than formula 3 and 4. Huang and Zhang reported that particles prepared from PLGA >40 mg/ml were found in large size (>500 nm) [13]. Polydispersion (PDI >0.7) was found in the particles with high concentration of PLGA (Table 1) [14]. High concentration of PLGA contributed to increase amount of polymers chains per unit volume of organic solvent leading to more polymer – polymer interaction and forming larger particles and aggregates [15].

PLGA particles are formed via precipitation polymer droplet (organic phase) in an aquoeus phase [13]. Therefore, diffusion rate of the organic solvent into the aqueous solvent predominantly determine particle size [13]. The organic solvent was removed from the aqueous phase by evaporation using stirring method at room temperature for overnight[16]. This method is known simple and mild, howbeit, long stirring time is potential to cause particles coalescence and protein diffussion into the second water layer generating particle aggregation and low EE [17]. A faster organic solvent removal using rotary evaporator is then considered to improve the PLGA-particles, in terms of particles size and EE percentage [17].

Comparing formula 3 and formula 4 containing similar amount of PLGA, yet less organic solvent in formula 4 resulted smaller particles in size (Table 1). In this study, chloroform was used. Chloroform is a water immiscible solvent which is the best solvent for PLGA, however it requires longer time to diffuse to the aquaeous phase causing formation particles in large size. Similar with water immiscible solvent, fully water soluble solvent such as acetone also yield large particle. Song and colleagues reported that partially water soluble solvent, for instance ethyl acetate and propylene carbonate exhibited smaller particles [18].

Moreover, the formula 4 contained higher amount of surfactant, i.e sodium cholate. The surfactant facilitates organic droplets of polymer to diffuse into aquaeous phase and stabilizes the system [15]. Surfactant is beneficial to avoid particles collide and coalesce among themselves. Therefore, the formula 4 also gave monodispersed particles indicated by PDI < 0.7 [18].

Size range in particles are varied depending on the characteristics of the particles themselves. The Nanomedicine European Technology Platform (ETPN) determines that particles (nanomedicine) is not more than 100 nm, but European Commission mentioned that particles, specifically nanomedicine has size up to 1000 nm because it is commonly provided with additional properties for better biological interaction [19,20,21]. Turning to particles obtained in this study, formula 4 resulted BSA – loaded particles in nanosize.

Table 1. Formula and physical characterizations of PLGA-particles.

Component	Formula 1	Formula 2	Formula 3	Formula 4	Formula 4 (lysozyme)
PLGA	200 mg (in 5 ml chloroform)	400 mg (in 5 ml chloroform)	50 mg (in 5 ml chloroform)	50 mg (in 2 ml chloroform)	
Protein (BSA)	400 µl (stock 10 mg/ml)			1 ml (stock 1.8 mg/ml)	
Trehalose 10%	150 µl			100 µl	
Sodium cholate 1%		2 ml			8 ml
Sodium cholate 0.3%			100 ml		
Result	**Formula 1**	**Formula 2**	**Formula 3**	**Formula 4**	**Formula 4**
Yield (mg)	134.5 mg	184.5 mg	70 mg	100 mg	NA
Hydrodynamic size	2666.5 nm ± 849.26	2208.3 nm ± 117.78	NA (> 1000 nm)	512.6 nm ± 130.35	NA
Polydispersity index (PDI)	1	0.85	NA	0.659	NA
Zeta potential	NA	NA	NA	-26.63 mV	NA

3.2. Protein loading and release study

Proteins are loaded into particles through several loading mechanisms depending on pH, ionic strength, denaturant, temperature or 'breathing mechanism' [22]. In this study, pre-load technique in which protein loading and particles preparation are performed at the same time was applied. The protein, BSA dissolved in PBS pH 7.4, was encapsulated into particles by mixing it with PLGA in organic solvent.

Surprisingly, with a few modification in amount of organic solvent and surfactant which yielded significantly smaller size PLGA particles, the amount of encapsulated BSA was remarkably increased reaching loading efficiency up to 88.89 % (Figure 1c lane 9). In addition, encapsulation efficiency (EE) depends on type of PLGA polymer and ionic strength [23,24]. PLGA with low molecular weight (MW) is potential to give high EE. Increase in PLGA MW reduces percentage of EE. Among PLGA 15,000; 25,000; and 30,000 MW, the PLGA 15,000 MW resulted the highest EE, whereas PLGA 30,000 MW gave the lowest EE [23]. In this study, PLGA with 30,000 – 60,000 MW. The EE was quite high because the initial amount of protein was minimized to wasted un-encapsulated protein. However, in line with Fu and colleagues, percentage of LC was slightly low, about 3.2% which was caused by PLGA with high MW. Even so, particles made from PLGA with high MW is stable towards degradation, particularly hydrolysis [23].

The obtained PLGA particles showed capability to protect the encapsulated protein in neutral condition, and slightly released the protein in acidic solution. Figure 1 shows remained and released BSA/lysozyme in HCl pH 1.2 37°C and PBS pH 7.4 37°C representing acidic gastric and neutral esophagus condition, respectively [25]. Figure lane 1 12; 17; and 21 indicate slight release of BSA from PLGA-particles after 2 h incubation in acidic environtment. Figure 1 lane 13; 16; and 20 no release of BSA after 2 h incubation in neutral environtment. The results explain that the PLGA-particles protected BSA in neutral condition, but was slightly degraded in extremely acidic pH. Figure 1 lane 6 also shows slight release of BSA in the acidic and normal condition, yet not valuable compared with amount of loaded BSA.

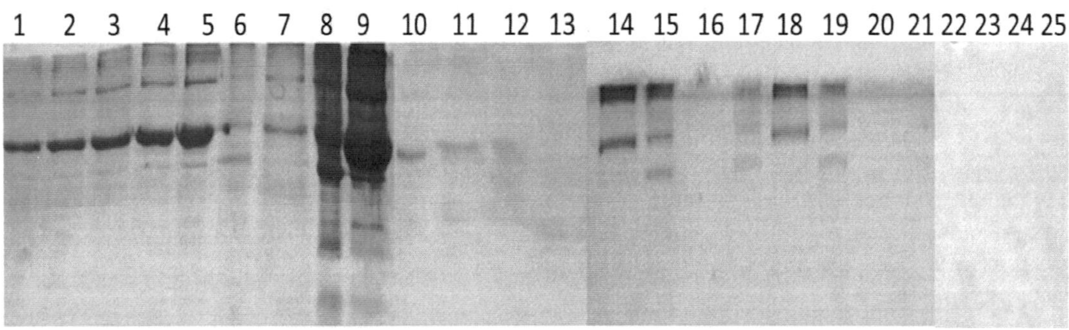

Figure 1. Electroforegram of released and remained proteins in *in vitro* conditions representing gastric and esophagus environment. **Lane 1 – 5:** BSA standard (0.0625; 0.125; 0.25; 0.5; 1 mg/ml); $y = 26520x + 6990.6$, $r^2 = 0.9966$. **Lane 6;12; 17; 21**: BSA released in HCl 0.1N pH 1.2 37°C. **Lane 7; 13; 16; 20**: BSA released in PBS pH 7.4 37°C. **Lane 8;11; 15; 19:** BSA remained after incubation in HCl 0.1N pH 1.2 37°C. **Lane 9; 10;14;18:** BSA remained after incubation in PBS pH 7.4 37°C. **Lane 22**: Lysozyme released in HCl 0.1N pH 1.2 37°C. **Lane 23**: Lysozyme released in PBS pH 7,4 37°C. **Lane 24**: Lysozyme remained after incubation in HCl 0.1N pH 1.2 37°C. **Lane 25**: Lysozyme remained in HCl 0.1N pH 1.2 37°C. **Lane 6 – 9**: Formula 1. **Lane 10 – 13**: Formula 2. **Lane 14 – 17**: Formula 3. **Lane 22 – 25**: Formula 4.

Figure 1 lane 22 - 25 are remained and released profile of lysozyme from the PLGA-particles. No band were found in the lysozyme electroforegram meaning that the protein was not succesfully encapsulated with PLGA using formula 4, the best formula for BSA. Lysozyme is commonly used as model protein for high pI [24]. In the recent study, proteins were dissolved in PBS pH 7.4 and the PLGA-particles were prepared also in neutral pH aqueous solution. pH of the aquaeous phase is the most important factor influencing EE. Proteins are known less ionized at pH near its isoelectric point, so the proteins are less soluble in the aqueous phase compared than that of in the organic phase. Therefore, the protein leakage into the aqueous phase is minimized leading to more effective encapsulation [26]. The applied pH of aqueous phase in this study was much lower towards lysozyme pI than BSA pI and it clearly contributed to no entrapment of lysozyme into the PLGA-particles. Besides, encapsulation of lysozyme in PLGA-particles was strongly dependant on the PLGA MW, in which lysozyme was only entraped into the PLGA-particles made from low MW polymer. Otherwise, BSA entrapment into the PLGA-particles was not affected by the PLGA MW [24].

3.3. FTIR spectroscopy and SEM

BSA - loaded PLGA-particles formula 4 was characterized with FTIR spectroscopy and SEM. Interaction of each component after particles formation and washing was studied using FTIR spectroscopy. Spectra of BSA – loaded PLGA-particles was compared with all components thereof, which were PLGA polymer, sodium cholate, trehalose, and BSA. Major spectrum of sodium cholate and trehalose were not found in BSA – loaded PLGA-particles spectra, as expected. Those components were not incorporated with the particles and removed during the washing steps. The FTIR spectrum show that both components were succesfully cleared.

On the other hand, spectrum of PLGA polymer and BSA – loaded particles were almost similar, except there are two vibration peaks appear at wavenumber 1657 cm^{-1} presenting asymetric stretching vibration of arginin for $CN_3H_5^+$ and 1546 cm^{-1} for asymetric stretching vibration of glutamin, specifically for COO^-.Those are specific vibration peak for amino acids [27]. It decribes that BSA has been bound with PLGA polymer forming BSA – loaded particles (Figure 2).

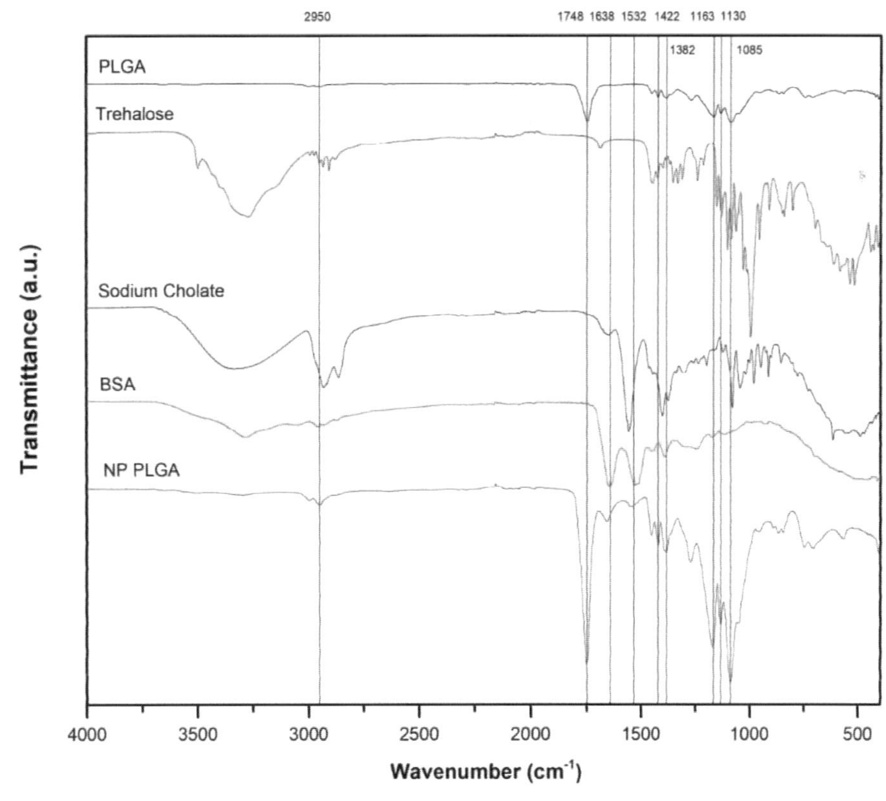

Figure 2. FITR spectrum of BSA – loaded PLGA particles and components thereof.

The other vibration peaks depict functional groups of PLGA. A strong and narrow vibration peak at 1748 cm^{-1} points C=O of PLGA [28]. Fingerprint peaks of PLGA are observed at 1163, 1130, and 1085 cm^{-1} indicating C-O stretching vibration, whereas C-H stretching vibration peaks are shown at 1422 and 1382 cm^{-1} [28-30]. Two small peaks around 2950 cm^{-1} are stretching vibration of C-H from aliphatic CH_2 and -C-H- of PLGA [28-30].

Figure 3 depicts BSA loaded particles under SEM with 13,000 × magnification. As detected with particle size analyzer, the BSA loaded PLGA particles were found approximately 200 – 600 nm, yet some bigger particles which were around 1000 nm in size were also captured.

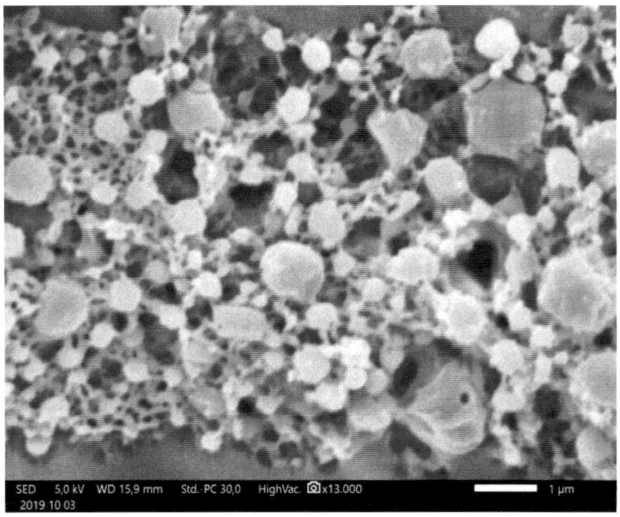

Figure 3. Visualization of BSA – loaded PLGA particles with SEM

4. Conclusion

Preparation of protein loaded PLGA particles is complicated and depends on a number of factors. The particle size was influenced by types of PLGA polymer, PLGA MW, organic solvent, and surfactant. The protein encapsulation into PLGA particles was determined by the pH system or aqueous phase and PLGA MW. The pH was required as near as the protein pI expecting high EE. Moreover, process of organic solvent removal also played significant role controlling particle size and EE percentage. The reported procedure was more suitable for BSA as protein with quite low pI than that of for lysozyme which has high pI. The recent protein delivery system was highly potential and suitable for oral delivery in which it showed capability to protect the protein in extremely acidic solution corresponding gastric environment.

Acknowledgment

The authors would like to thank to Program of Prioritas Nasional (PN) Obat 2019, Research Center for Biotechnology, Indonesian Institute of Science, Research Center for Biology and Research Center for Biomaterial Indonesian Institute of Sciences for facilitating us in analysis with SEM and FTIR, and Ms. Nur Raisah Maddepungeng for her assistance on redrawing the FTIR spectrum using Origin® software.

References

[1] Pandey A, Jain D S and Chakraborty S 2015 *Handb. Polym. Pharm. Technol. 2*, ed V K Thakur and M K Thakur (Beverly: Scrivener Publishing LLC/Wiley) p 151–72

[2] Danhier F, Ansorena E, Silva J M, Coco R, Le Breton A and Préat V 2012 *J. Control. Release* **161** 505–22

[3] Berthet M, Gauthier Y, Lacroix C, Verrier B and Monge C 2017 *Trends Biotechnol.* **35** 770–84

[4] Usmani S S, Bedi G, Samuel J S, Singh S, Kalra S, Kuma P, Ahuja A A, Sharma M, Gautam A and Raghava G P S 2017. *PLoS ONE* **12** 1–12
[5] Ramirez J E V, Sharpe L A and Peppas N A 2017 *Adv Drug Deliv Rev.* **114** 116–31
[6] Ye C and Venkatraman S 2019 *Ther. Deliv.* **10** 269—72
[7] Dinda A K, Bhat M, Srivastava S, Kottarath S K and Prashant C K 2016 *Vaccine* **34** 3076–81
[8] Mamo B 2015 *Asian J. Biomed. Pharm. Sci.* **5** 1–12
[9] Malathi S, Nandhakumar P, Pandiyan V, Webster T J and Balasubramanian S 2015 *Int. J. Nanomed.* **10** 2207
[10] Zeng X and Ruckenstein E 1998 *J. Membrane Sci.* **148** 195–205
[11] Deng Q Y, Zhou C R and Luo B H 2006 *Pharm. Biol.* **44** 336–42
[12] Fernandez-Urrusuno R, Calvo P, Remunan-Lopez C, Vila-Jato J L and Alonso M J 1999 *Pharm. Res.* **16**(10) 1575 – 81
[13] Huang W and Zhang C 2018 *Biotech. J.* **13**(1) 1 – 25
[14] Rahmawanty D, Anwar E and Bahtiar A 2015 *Jurnal Ilmu Kefarmasian Indonesia* **13(**1) 1693 – 831
[15] Reddy P R V, Acharya S R and Acharya N S 2015 *Asian J. Pharm.* **9** 152 – 61
[16] Kwon B S, Lee H E, Kim D H, Kang H K, Kang J S, Lee S and Choi S J 2007 *I. Ind. Eng. Chem.* **13**(6) 1043 - 6
[17] Mishra N, Tiwati S, Vaidya B, Agrawal G P and Vyas S P 2011 *J. Drug Target* 19(1) : 67 - 8
[18] Song C, Lee H S, Choung Y, Cho K I, Ahn Y and Choi E J 2006 *Physicochem. Eng. Aspects* **276** 162 - 7
[19] Tomellini R, Faure D and Panzer O 2005 *European Technology Platform, Strategic Research Agenda for Nanomedicine, Nanomedicine Nanotechnology for Health*
[20] Wagner V, Husing B and Bock A K 2008 *JRC Scientific and Technical Reports – European Commission* 1-113
[21] Lohcharoenkal W, Wang L, Chen Y C and Rojanasakul Y 2014 *Biomed. Res. Int.* 1-12
[22] Diaz A C and Sunna A 2018 *Genes* **9**(7) 1 – 30
[23] Fu X, Ping Q and Gao Y 2005 *J. Microencapsul.* **22**(7) 705 - 14
[24] Blanco D and Alonso M J 1998 *European J. Pharm. Biopharm.* **45**(3) 285 - 94
[25] D'Souza S 2014 *Adv. Pharm.* **304757** 1 - 12
[26] Swed A, Cordonnier T, Fleury F and Boury F *J. Nanomed. Nanotechnol.* **5**(241) 1- 8
[27] Barth A 2007 *BBA – Bioenergetics* **1767**(9) 1073 – 101
[28] Arasoglu T, Derman S and Mansuroglu B 2016 *Nanotechnology* **27**(025103) 1 – 12
[29] Dompeipen E J 2017 *Majalah BIAM* **13**(01) 31-41
[30] Nandiyanto A B D, Oktiani R and Ragadhita R 2019 *IJoST* **4**(1) 97 - 118

IOP Conf. Series: Materials Science and Engineering 1011 (2021) 012027 doi:10.1088/1757-899X/1011/1/012027

A review on factors affecting chitosan nanoparticles formation

Mega Ferdina Warsito[1*] and Fitri Agustiani[2]

[1]Research Center for Biotechnology, Indonesian Institute of Sciences (LIPI), Komplek CSC-LIPI Jl. Raya Bogor Km 46, Cibinong-Bogor 16911, West Java, Indonesia.
[2]School of Biotechnology, IPB University, Bogor, West Java, Indonesia

*Corresponding author email: mferdina@gmail.com

Abstract. Chitosan has been widely used as an excipient in the pharmaceutical industries, due to its biodegradable, biocompatible, low toxicity, and mucoadhesive properties. Chitosan nanoparticles have been extensively investigated for delivery of drugs, herbal products, proteins or peptides and genes. The particle size of chitosan nanoparticles has an important effect on their properties in its pharmaceutical application. Smaller particle size can entrap higher concentration of therapeutic agents, improve drug stability and its bioavailability, and provide sustained delivery. Although diverse efforts have been made to obtain the chitosan nanoparticles and its potential pharmaceutical applications, optimization of the fabricating conditions and the comprehensive properties of the resultant chitosan nanoparticles is still an ongoing important study subject. In this review, we will describe several factors that affect chitosan nanoparticle formation specifically in ionic gelation method, such as chitosan characteristic, i.e. degree of deacetylation, molecular weight, and the ratio of chitosan-crosslinker, type and concentration of crosslinker, mixing procedure, and condition. We will also give an overview of the characterization process of the chitosan nanoparticles.

1. Introduction

Chitosan is an amino polysaccharide biopolymer derive from N-deacetylation processes of chitin [1] and consist of β-1,4-linked glucosamine and N-acetylglucosamine residues [2]. Chitosan has diverse biological activities, such as antimicrobial [3-5], antioxidant [6,7], anticancer [8,9], anti-inflammation [10,11], immune stimulant [12] and wound healing properties [13]. These activities are assume due to the free amino groups in its molecule which had positive charge at acidic pH, hence it can interact easily with polyanionic molecules, such as proteins, DNA, or phospholipids [2]. Chitosan had renewable, sustainable, nontoxic [14], biodegradable and biocompatible characteristic [15-17]. It can encapsulate various compounds, either hydrophobic or hydrophilic [18]. Therefore, chitosan is widely developed as potential drugs carrier.

The development of chitosan nanoparticles (CNPs) for active pharmaceutical ingredients (APIs) encapsulation has been developed rapidly. CNPs is a potential drug carrier, as it can controls drug release, improves the solubility and stability of the drugs, enhances efficacy, and reduces toxicity. The small particle size makes it easier to pass through biological barriers and improve drug targeting to enhance the efficacy of the drug. CNPs had positive charge in its surface and mucoadhesive properties that allow the particles to adhere on the mucus membranes and release the drug molecules [19]. CNPs also enhance the penetration of the APIs through the epithelium by opening the tight junctions, which will facilitates paracellular and transcellular transport [20].

This review will give an overview regarding parameters that affect the production of CNPs and its characterization, since the fabrication of nanosized particles is pivotal for the successful administration of the APIs. It will focus especially on the production of CNPs through ionic gelation process.

2. Factors affecting chitosan nanoparticle formation

CNPs can be prepared by several methods, i.e., ionic gelation, emulsification cross-linking, reverse micellization, spray-drying, and nanoprecipitation [21-23], modified ionic gelation with radical polymerization, emulsion droplet coalescence, emulsion solvent diffusion and desolvation [15]. Among them, ionic gelation is the most common method used in CNPs production, because of its ease of formation, fast, low cost [21], and absence of organic solvent [24].

In ionic gelation method, the CNPs are form (Figure 1) through complex formation of the positively charged amino groups in the chitosan molecule and the negatively charged groups of polyanion or crosslinker [25]. The amino group of chitosan will be protonated to form $-NH_3^+$ when dissolve in aqueous acidic solution. When the cationic charges chitosan added with polyanionic solution dropwise under constant stirring, it will form spherical hydrogel particles known as CNPs [21]. Parameters that reported affect the formation of CNPs in the ionic gelation method are molecular weight (MW), degree of deacetylation (DDA) and concentration of the chitosan, the type and amount of crosslinking agent, mixing procedure, and the mixing condition, such as pH and temperature.

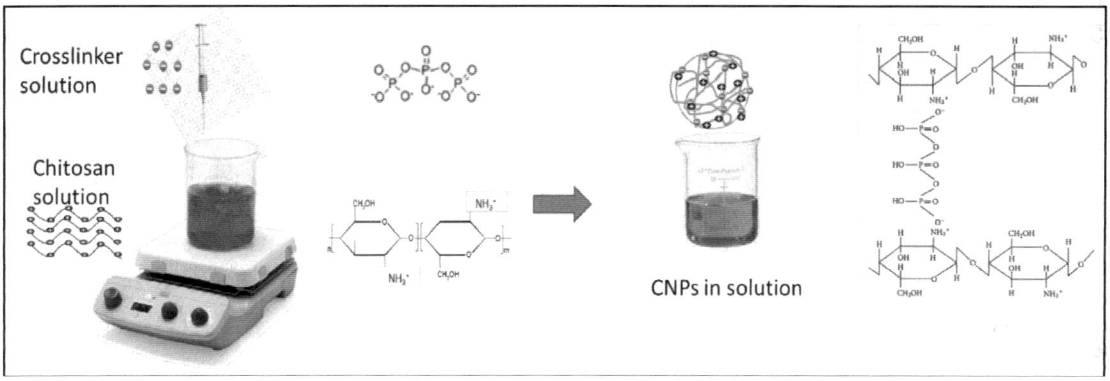

Figure 1. The illustration of CNPs Synthesis through ionic gelation using TPP as crosslinker.

2.1. Chitosan

DDA and MW of chitosan affect the properties and functionality of CNPs [26]. DDA value correlate with the amount of free amino group present in the chitosan [27]. The free amino group will be interacting with the anion groups in the crosslinker to form CNPs. Chitosan with high MW and DDA reported to form CNPs with lower drug release rate [28], while chitosan with a low DDA tend to degrade faster, hence they had faster drug release profile. One of factor that affecting chitosan degradation is variation of the acetamide distribution in the molecule [29].

Chitosan has linear nonbranched structure, with MW varied from 300 to over 1000 kDa. In acid solution, chitosan with higher MW had higher viscosity and behaves as a pseudoplastic material [30]. Xu and Du [31] reported that chitosan with higher MW had greater encapsulation capacity, because the molecule had longer chains which can entrap more drug molecule. CNPs produce using high MW chitosan also had lower drug release rate.

The DDA value of chitosan affecting the particle size [30, 31], zeta potential [30], encapsulation efficiency and release rate of CNPs [31]. The DDA value indicate the percentage of the deacetylated amino groups in the chitosan molecules, and it is increased as the function of the percentage of amino groups to acetylated groups. The density of the positive charge in the CNPs surface is correlate with DDA value and it is affecting the zeta potential. The potency of the chitosan molecules to cross-link with TPP was higher when the positive charge increase, which eventually led to smaller CNPs [25, 30,

IOP Conf. Series: Materials Science and Engineering 1011 (2021) 012027 doi:10.1088/1757-899X/1011/1/012027

31]. Chitosan which had higher DDA value with same MW provides more compact CNPs due to the greater number of ammonium groups interact with polyanion, hence it had lower CNPs permeability which results in slow release rate. Higher DDA value also reported to increased encapsulation efficiency [31].

Sreekumar et al. [2] reported that chitosan concentration also had impact on the average of hydrodynamic diameter of the CNPs at a molar ratio of 1.5 and this effect is stronger when the chitosan had high DDA value. But when a lower molar ratio was used, i.e. 1, the influence of chitosan concentration on the average of the CNPs diameter is stronger when a low DDA chitosan is used. It also showed that chitosan with low (10%) or high (60%) DDA had less control over the CNPs production process. Hence, the CNPs production with average diameters can be controlled by simply changing the chitosan and TPP concentration to get constant chitosan-TPP ratio, if the DDA value of the chitosan was in the range of 20–50%.

2.2. Crosslinker

Chitosan is a linear amino-polysaccharide consisting of 2-deoxy-d-glucosamine and 2-deoxy-N-acetyl-d-glucosamine units linked by glycosidic β-(1→4) bonds [32]. It has amine and hydroxyl groups as the active sites to form linkages comprise of Schiff base formation, amide, or ester bonding, which lead to the formation of chitosan hydrogels [33]. Chitosan has pH sensitive behavior due to the amine groups in its molecule. Hence, chitosan dissolves and depolymerizes in acidic solution. The depolymerization is a major downside for controlled oral delivery, but it can be circumvent using crosslinking agents.

Cross-linking affects polymer properties such as dimensional stability, mechanical properties, chemical stability, swelling, aqueous permeability, and solubility [34, 35]. Crosslinkers that are used in the formation of CNPs via ionic gelation method called polyanions. The anion groups of crosslinker have electrostatic interaction with the amine group of the chitosan, which will weaken the chitosan surface charge and reduce its solubility leading to the spontaneous CNPs formation. These polyanions can be classified into three groups (a) low MW ions, i.e., sodium pyrophosphate [SPP], tripolyphosphate, tetrapolyphosphate, octapolyphosphate, and hexametaphosphate; (b) hydrophobic, i.e., sodium alginate and k-carrageenan; and (c) high MW ions, i.e., octyl, dodecyl, hexadecyl, and cetyl stearyl sulfate [30]. The polyanions type reported to affect the production time of the CNPs, particle size, polydispersity index (PDI), drug loading capacity, stability, mucoadhesiveness, and drug release profile.

It was reported that the polyanions variation, i.e. MW and binding site, affects the size, PDI, and loading capacities of the CNPs. Thandapani et al. [23] compared two type of crosslinkers, i.e., sodium tripolyphophate (STPP) and sodium hexametaphosphate (SHMP), in the preparation of CNPs. It showed that CNPs-STPP had smaller particle size compare to CNPs-SHMP[1,23]. It was because SHMP had six negative charges in neutral and slightly basic solution which present readily available interaction with amino groups of chitosan compare to STTP. SHMP also had stronger ionic complexation and higher interaction with chitosan because it had more binding sites [23]. It was also assumed that HMP is core-ionotropic crosslinker, while TPP acts in the surface. Hence, TPP wields electrostatic interaction to the loose cationic chitosan layers directly underneath CNPs surfaces leading to the subsidence of the particle size, while the core layers of the chitosan will be denser and harder to compress by HMP electrostatic effect thus it forms bigger CNPs. In comparation, polyphosphoric acid (PPA) that work as both core/shell crosslinker had smaller CNPs compare to CNPs-TTP and CNPs-HMP. Triwulandari et al. [36] reported that CNPs-STTP had the smallest particle size compared to CNPs-alginate, CNPs-carrageenan, and CNPs-sodium dodecyl sulfate (SDS). CNPs-SDS had the largest particle size because of the influence of the MW of the polyanion, while STTP has the smallest MW. It was also reported that the PDI value of the CNPs formed using different polyanions show similar result, except for particles formed using SDS, which had higher tendency to aggregate. Saeed et al. [1] and Sang et al. [37] reported that CNPs prepared using polyphosphoric acid (PPA), hexametaphosphate (HMP), and TPP had different encapsulation efficiency. CNPs-PPA and CNPs-HMP have higher encapsulation efficiency compared to CNPs-TPP when used to encapsulate model drugs, i.e. methylene blue and doxorubicin. It was also reported that CNPs-PPA and CNPs-HMP had slower drug release rate compare to CNPs-TPP.

Crosslinker MW also affect the time needed in cross-linking process. CNPs-sulfate and CNPs-citrate had faster reaction compare to CNPs-TPP, it was assumed due to its smaller molecular size. But CNPs-TPP had stronger mechanical strength than that of CNPs-sulfate or CNPs-citrate. CNPs-sulfate and CNPs-citrate swelled and segregated in the simulated gastric fluid (SGF), it led to the release of drug molecules which is completed in 5 hours. While in simulated intestinal fluid (SIF), they kept in the shrink stage, therefore the drug released profile was slower. In comparation, CNPs-TPP swelling and drug release was reported to be insensitive toward media pH [38].

2.3. Mixing procedure

Colloidal systems have tendency to aggregate because of the attraction force between colloidal particles may surpass the electrostatic repulsion energy [39]. The formation of aggregates can greatly obstruct the redispersion process of the lyophilized samples and affecting the particle size [40, 41]. The homogeneity of the mixture and turbulent force in the suspension is directly correlate with stirring process [42]. But high stirring rate could cause a less uniform energy distribution that lead to a bigger particle size [43].

Several articles reported that the increasing stirring speed resulted in to the formation of smaller particle size [23,25, 44]. It was assumed that the increasing shearing force accelerates the dispersion of TPP in chitosan solution and resulted in monodisperse particles. Thandapani et al. [23] reported that increasing stirring speed from 200 rpm to 600 rpm drastically reduced the particle size of the CNPs from 210 to 50.75 nm for CNPs-TPP and 333 to 122.4 nm for CNPs-SHMP. Stirring speed above 800 rpm showed no significant change in the particle size, while agitation speed below 400 rpm resulted in bigger CNPs size. But higher stirring speed, around 1000 rpm, caused particle aggregation because intense stirring could destroy the repulsive force between particles [44].

There are several alternatives of chitosan-TPP mixing method, i.e., dropwise addition, single-shot mixing and dilution. Among them, dilution method showed the most noticeable particle size gradual increase, whereas dropwise addition and single shot mixing induced particle size sharp increase. Dropwise addition showed the biggest abrupt increase. The probable reasons for these results was still unclear. The dilution method assumed to offers the most precise control over the formation of particle size with the average size. The PDI-values of CNPs sample from three mixing methods were not different significantly, i.e., 0.19 for the dilution method, 0.17 for the single shot and 0.13 for the dropwise addition methods [45].

2.4. Mixing condition

Shu and Zhu [46] reported that the electrostatic interaction between the polyanions and chitosan during the formation of CNPs was depend on the pH of the mixing solution. The electrostatic interactions only occur in certain pH region which correspond with the anion's natural characteristics, for examples 1.0–7.5 for CNPs-sulphate, 4.5–7.5 for CNPs-citrate and 1.9–7.5 for CNPs-TPP.

Liu and Gao [47] reported that pH also affecting particle size and zeta potential of CNPs. The particle size growing rapidly from pH 1 to 3.5 and then decreased slowly from pH 3.5 to 5.5. The effect of pH to zeta potential was similar, with a slightly higher transition at pH 4, the zeta potential value increased from pH 1 to 4, then reduced slowly from pH 4 to 5.5. It was because in acidic condition the amine groups of chitosan were protonated that lead to strong charge repulsion and molecule extension [48,49]. In low pH condition, TPP molecule also got protonated and had low charging density. Hence, the chitosan molecules cannot crosslink sufficiently with TPP to form stable CNPs in acidic condition. It was concluded that the stable CNPs with a suitable size was form at pH 3.5. At higher pH value, the charging degree of TPP molecule enhanced and neutralize the chitosan charge to a greater extent, which lead to the shrinkage of CNPs size. At pH value of 4.5 to 6.0, the CNPs had tendency to decrease its particle size and zeta potential, thus form relatively unstable particles. It suggested that the chitosan positive surface charge is shielding because of reorganization of the molecule structure or/and adsorption of other negative charged ions at low pH.

Torres et al. [49] reported that the CNPs protein adsorption capacity was depends on the solution pH. The BSA adsorption capacity of CNPs-glutaraldehyde decreased with the increase of pH, from 6 to 7.5, because of the electrostatic interaction. The optimum adsorption capacity was reported at its isoelectric point, pH about 4.8, because at this pH value, the substrate has a high charge density which will increase the adsorption capacity. The lysozyme adsorption capacity also reaching maximum at around the isoelectric point of the lysozyme. This result was different with Katas et al. [50], which reported that it showed optimum binding capacity at above its the isoelectric point, although the crosslinker used also different. Katas et al. [50] reported that the entrapment efficiency (EE) of BSA in the CNPs-dextran was affected by solution pH and the optimum pH was at 5.5. When protein dissolved at pH above its isoelectric point, it tends to negatively charge and had ionic interaction with positively charged. amino groups which favors the entrapment of BSA into the CNPs. Thus, further research is needed to confirm these findings.

Several studies reported that the CNPs formation also being affected by temperature [23, 51, 44, 52] Gomathi reported that when CNPs-SHMP was produce using stirring at 600 rpm at the concentration of 5mg/mL it obtained the smallest CNPs when produced at ambient temperature (25°C) compare to higher temperature range (40, 50, 60 °C). CNPs obtained in higher temperature were easier to agglomerate as the result of the increased size and yield. Kamat et al, 2016 also reported that lower temperature produced smaller particle size. CNPs was smaller when produce at 4 °C with chitosan concentration of 0.2 and 0.4 mg/mL, while when the concentration 0.8 mg/mL the particle size was not different significantly when it was compare to CNPs produced at 27 ◦C and 35 ◦C. CNPs produced at 4 °C were smaller (60–80nm) and with narrower size distribution, compare to the one produced at 35 ◦C which had larger size (130–190nm) and broader size distribution, it was also easy to agglomerated.

Tsai reported that mixing procedure also had effect to the optimum temperature to obtain smallest CNPs. In the CNPs production by mechanical shearing at 1000 rpm, the best temperature to produce the smallest CNPs were at 45 °C (145 nm), followed by 4 °C (150 nm), and 25 °C (163 nm). But it may be because the larger CNPs was removed during centrifugation thus the smaller CNPs remained than the other CNPs form in the temperature of 4 °C or from 25 °C ones. Meanwhile in ultrasonic radiation method, the CNPs average diameter decreased as the function of the increasing temperature. The viscosity of the solution decreasing as the temperature increase thus aid the sporadic cavitation effect wield on the chitosan molecules producing smaller MW fragments.

Report from Gomathi and Tsai were contradicted with the finding by Fan. Fan reported that CNPs produced from 0.5 mg/ml low MW chitosan solution with agitation at 700 rpm, the increasing temperature from 10 ◦C to 60 ◦C produced lower CNPs size, but when the temperature was higher than 60 ◦C the decrease of particle size was negligible. The homogeneity also getting better with the increasing temperature. It showed that when the temperature was 10 ◦C the PDI value was 0.112 and it get lower than 0.05 at 25 ◦C and 70 ◦C. Thus, further study is needed to confirm the effect of several factor at once on the formation of CNPs.

3. Characterization of chitosan nanoparticles

The characterization of CNPs is an important step to curb the evaluation of the particle's property. Several parameters that commonly used in CNPs characterization are particle size and surface charge, morphological and surface characteristics, encapsulation efficiency, loading capacity, and in vitro release profile (Table 1). These parameter could give insight to the stability, physicochemical performance, and also cell uptake of the CNPs [53, 54]

Particle size can be evaluated using non-imaging and/or imaging-based technique, the option depends on the predicted size and population of the CNPs. Non imaging technique is based on light scattering, either Static Light Scattering (SLS) and/or Dynamic Light Scattering (DLS). These methods are fast, precise, sensitive, and utilizable for a wide range particle size measurement [55], [55], but the downside of these technique is they cannot provide information about particle morphology and the sample need to be meticulously prepared. It is advised to be complement non-imaging techniques with imaging technique and vice versa [56]. Imaging technique based on electron microscopy can be used to evaluate

CNPs shape and morphology, beside its particle size [57, 58]. There are two commonly used electron microscope used for this analysis, i.e., scanning electron microscopy (SEM) and transmission electron microscopy (TEM). SEM gives image of the sample surface, while TEM gives better resolution and image of the inner structure of the CNPs. The CNPs surface and morphology obtained from imaging techniques can be used as one of parameter to deduce the dissolution behavior or in vivo responses [59]. But the electron microscopy technique come with several disadvantages such as longer preparation time, expensive, cannot provide information about the particle size distribution and destructive sample preparation.

Table 1. Assessment techniques for physicochemical characterization of CNPs.

Characterization technique	Parameters
Imaging based technique	
SEM	CNPs size, size distribution, surface structure, stability
TEM	CNPs size and size monodispersity, shape and inner structure, stability
Non-imaging technique	
DLS	CNPs size and size distribution
SLS	CNPs size, mass-weighted particle size, particle shape and structure
Spectroscopy based technique	
UV-Vis spectroscopy	Optical properties, size, concentration, agglomeration state, hints on CNPs shape
Raman spectroscopy	Determine the interactions between the API and its excipients
FTIR	Determine surface composition, ligand binding, the interactions between the API and its excipients
Zeta potential	Surface charge
Thermal analysis	
DSC	Identify modification and physical qualities of the materials
TGA	Determine the precise nature of the thermal transitions of the samples
X-Ray technique	
XRD	Determine the crystallinity of the sample, Size (structural properties); the atomic and molecular level structure of the material

The CNPs surface charge gives an overview about the physical stability and redispersibility of the formulation as well as their in vivo performance. It can predict the formulation long-term storage stability profile. Zeta potentials is the net charge of the particles, which reflects the electrostatic repulsion or attraction degree between charged particles. High zeta potential value, above ± 30 mV, indicates a stable dispersion system because of the resistance of the particles to aggregate or high repulsion forces [60]. Zeta potential also can be used to study the interaction of CNPs with the biological system.

The efficiency of the CNPs preparation and dose calculation can be done through the determination of several parameter such as CNPs yield, encapsulation efficiency (EE), and loading capacity (LC). The drug content of the CNPs should be determined using a suitable analytical method, either based on spectrophotometry or chromatography method. Then the drug EE, LC, and nanoparticles yield are determined by the following equations:

$$EE\ (\%) = \frac{Amouunt\ of\ drug\ in\ definite\ mass\ of\ the\ prepared\ CNP}{Theoritical\ amount\ of\ drug\ in\ the\ same\ mass} \times 100 \qquad (1)$$

$$LC\ (\%) = \frac{Amount\ of\ drug\ in\ definite\ mass\ of\ the\ prepared\ CNP}{Total\ mass\ of\ the\ CNP} \times 100 \qquad (2)$$

$$Yield\ (\%) = \frac{Total\ weight\ of\ the\ CNP}{Total\ weight\ of\ drug\ and\ polymers\ used} \times 100 \quad (3)$$

Differential scanning calorimetry (DSC) evaluate the modification of the material structure in pharmaceutical formulation based on thermal analysis as a function of temperature. It measures the heat flow, endothermic and exothermic events, i.e., heat uptake or heat emission. It also can be used to determine the physical qualities of the materials. While thermogravimetric analysis (TGA) measures sample mass during the heating process and used to determine the precise nature of the thermal transitions. It is a very convenient method for discrimination of hydrates or substance decomposition from its melting point, polymorphic transitions, and loss of water. It can also be used to determine temperature-dependent weight measurement [56].

Spectroscopic methods, such as X-ray diffraction (XRD), infrared and Raman also often used in CNPs characterization. XRD give information about the atomic and molecular level structure of the material. It also can characterize crystalline size, shape, and lattice distortion of the material. But for liquid sample, it needs to be prepared in the dry state to eliminate the moisture. Both IR and Raman are vibrational spectroscopic techniques. IR spectroscopies operate in IR wavelength region, while Raman is in the visible, near IR or near-ultraviolet (UV) wavelengths. IR can be used to determine the interactions between the API and its excipients, based on the ability of the molecules to absorb certain frequencies, either the transition energy of the vibrating bond or functional group in the molecule. Raman technique detect the vibration, rotation and other low-frequency transitions of the materials [56].

In vitro drug release test determines the profile of the drug released from CNPs in the simulation media. The release mechanism generally depends on the location of the drug molecules in the CNPs, whether in the encapsulated in the particle core or adsorbed in the CNPs surface. The encapsulated drug needs to diffuse from chitosan matrix or released after the degradation and erosion of the chitosan to determine its release profile [55]. Other parameter that also affecting drug release profile are chitosan MW and DDA, CNPs particle size, cross-linking degree of chitosan-polyanion, other excipients in the formulation, polarity, pH, and enzymes in the dissolution media [21, 61].

4. Conclusion

The most common method for CNPs production is ionotropic gelation, because of its simplicity, rapid, and lower cost. The CNPs hydrogel production through ionic gelation method is based on ionic interaction between the cationic groups of chitosan and the anionic groups of the crosslinker. Parameters that affecting the CNPs formation are chitosan MW, DDA and concentration, the type and amount of crosslinking agent, mixing procedure, and production condition, such as pH and temperature. The quality of the CNPs is evaluated based on several parameters, such as particle size, zeta potential, morphological and surface characteristics, encapsulation and loading efficiency, and drug release profile.

References

[1] Saeed RM, Dmour I and Taha MO 2020 *Front Bioeng Biotechnol.* **8** pp. 1–21
[2] Sreekumar S, Goycoolea FM, Moerschbacher BM and Rivera-rodriguez GR 2018 *Sci Rep.* 1–11
[3] Malinowska-Pańczyk E, Staroszczyk H, Gottfried K, Kołodziejska I and Wojtasz-Pająk A 2015 *Polimery/Polymers.* **60** pp. 735–741
[4] Goy RC, Morais STB and Assis OBG 2016 *Brazilian J Pharmacogn.* **26** pp. 122–127
[5] Takahashi T, Imai M, Suzuki I and Sawai J 2008 *Biochem Eng J.* **40** pp. 485–491
[6] Xie W, Xu P and Liu Q 2001 *Bioorganic Med Chem Lett.* **11** pp. 1699–1701
[7] Li H, Xu Q, Chen Y and Wan A 2014 *J Biomed Mater Res - Part A.* **102** pp. 911–916
[8] Xu Y, Wen Z and Xu Z 2009 *Anticancer Res.* **29** pp. 5103–5109
[9] Qi L and Xu Z 2006 *Bioorg Med Chem Lett.* **16** pp. 4243–4245

[10] Davydova VN, Kalitnik AA, Markov PA, Volod'ko A V., Popov S V. and Ermak IM 2016 *Appl Biochem Microbiol.* **52** pp. 476–482
[11] Oliveira MI, Santos SG, Oliveira MJ, Torres AL and Barbosa MA 2012 *Eur Cell Mater.* **24** pp. 133–136
[12] Peluso G, Petillo O, Ranieri M, Santin M, Ambrosio L, Calabró D, Avallone and Balsamo G 1994 *Biomaterials.* **15** pp. 1215–1220
[13] Azad AK, Sermsintham N, Chandrkrachang S and Stevens WF 2004 *J Biomed Mater Res B Appl Biomater.* **69** pp. 216–222
[14] Kim S 2018 *Int J Polym Sci.* **1708172** pp. 1–16
[15] Naskar S, Sharma S and Kuotsu K 2019 *J Drug Deliv Sci Technol.* **49** pp. 66–81
[16] Riva R, Ragelle H, des Rieux A, Duhem N, Jérôme C and Préat V 2011 . In: Jayakumar R, Prabaharan M and Muzzarelli RAA (eds) Chitosan for Biomaterials II. Springer Berlin Heidelberg, Berlin, Heidelberg, pp 19–44
[17] Silva MM, Calado R, Marto J, Bettencourt A, Almeida AJ and Gonçalves LMD 2017 *Mar Drugs.* **15** pp. 370
[18] Zhao LM, Shi LE, Zhang ZL, Chen JM, Shi DD, Yang J and Tang ZX 2011 *Brazilian J Chem Eng.* **28** pp. 353–362
[19] Shi X-Y and Fan X-G 2002 *J China Pharm Univ.* **33** pp. 169–172
[20] Mohammed MA, Syeda JTM, Wasan KM and Wasan EK 2017 *Pharmaceutics.* **9**
[21] Agnihotri SA, Mallikarjuna NN and Aminabhavi TM 2004 *J Control Release.* **24** pp. 5–28
[22] Rai VR, Bai JA, Rosales-Martínez P, Cornejo-Mazón M, Arroyo-Maya IJ and Hernández-Sánchez H 2019 . In: Nanotechnology Applications in the Food Industry. pp 427–440
[23] Thandapani G, Supriya Prasad P, Sudha PN and Sukumaran A 2017 *Int J Biol Macromol.* **104** pp. 1794–1806
[24] Neves Borgheti-Cardoso L, Vicentini F, Cunha Filho M and Gelfuso G 2017 . In: Handbook of Composites from Renewable Materials. pp 285–308
[25] Al-Nemrawi NK, Alsharif SSM and Dave RH 2018 *Int J Appl Pharm.* **10** pp. 60–65
[26] Elgadir MA, Uddin MS, Ferdosh S, Adam A, Chowdhury AJK and Sarker MZI 2015 *J Food Drug Anal.* **23** pp. 619–629
[27] Rodrigues FC, Devi NG, Koteshwara KB and Thakur G 2020 *IOP Conf Ser Mater Sci Eng.* **872** pp. 1–6
[28] Ko JA, Park HJ, Hwang SJ, Park JB and Lee JS 2002 *Int J Pharm.* **249** pp. 165–174
[29] Elgadir MA, Uddin MS, Ferdosh S, Adam A, Chowdhury AJK and Sarker MZI 2015 *J Food Drug Anal.* **23** pp. 619–629
[30] Islam N, Dmour I and Taha MO 2019 *Heliyon.* **5** pp. 1–9
[31] Xu Y and Du Y 2003 *Int J Pharm.* **250** pp. 215–226
[32] Barbosa HFG alves, Attjioui M, Ferreira APG, Dockal ER, Gueddari NE El, Moerschbacher BM and Cavalheiro ÉTG 2017 *Molecules.* **22** pp. 1–19
[33] Croisier F and Jérôme C 2013 *Eur Polym J.* **49** pp. 780–792
[34] Xu J, Strandman S, Zhu JXX, Barralet J and Cerruti M 2015 *Biomaterials.* **37** pp. 395–404
[35] Mitra RN, Han Z, Merwin M, Al Taai M, Conley SM and Naash MI 2014 *ChemMedChem.* **9** pp. 189–196
[36] Triwulandari E, Fahmiati S, Sampora Y, Meliana Y, Ghozali M and Sondari D 2018 *AIP Conf Proc.* **2024** pp. 1–6
[37] Sang Z, Qian J, Han J, Deng X, Shen J, Li G and Xie Y 2020 *Carbohydr Polym.* **230** pp. 1–26
[38] Shu XZ and Zhu KJ 2002 *Int J Pharm.* **233** pp. 217–225
[39] Vauthier C, Cabane B and Labarre D 2008 *Eur J Pharm Biopharm.* **69** pp. 466–475
[40] Bozdag S, Dillen K, Vandervoort J and Ludwig A 2005 *J Pharm Pharmacol.* **57** pp. 699–707
[41] Abdelwahed W, Degobert G, Stainmesse S and Fessi H 2006 *Adv Drug Deliv Rev.* **58** pp. 1688–1713

[42] Villegas-Peralta Y, López-Cervantes J, Madera Santana TJ, Sánchez-Duarte RG, Sánchez-Machado DI, Martínez-Macías M del R and Correa-Murrieta MA 2020 *Polym Bull.* 1–20
[43] Fashli Razak, Vlasta Chyzna, Noreen Morris, Alan Murphy and James Kennedy 2017 *Int J Adv Sci Res Manag.* **2** pp. 13–24
[44] Fan W, Yan W, Xu Z and Ni H 2012 *Colloids Surfaces B Biointerfaces.* **90** pp. 21–27
[45] Sawtarie N, Cai Y and Lapitsky Y 2017 *Colloids Surfaces B Biointerfaces.* **157** pp. 110–117
[46] Shu XZ and Zhu KJ 2001 *J Microencapsul.* **18** pp. 237–245
[47] Liu H and Gao C 2009 *Polym Adv Technol.* **20** pp. 613–619
[48] George M and Abraham TE 2006 *J Control Release.* **114** pp. 1–14
[49] Torres MA, Beppu MM and Santana CC 2007 *Brazilian J Chem Eng.* **24** pp. 325–336
[50] Katas H, Hussain Z and Awang SA 2013 *J Nanomater.* **536291** pp. 1–10
[51] Tsai ML, Bai SW and Chen RH 2008 *Carbohydr Polym.* **71** pp. 448–457
[52] Kamat V, Bodas D and Paknikar K 2016 *Sci Rep.* **6** pp. 8–11
[53] Truong NP, Whittaker MR, Mak CW and Davis TP 2015 *Expert Opin Drug Deliv.* **12** pp. 129–142
[54] Zhao J, Lu H, Wong S, Lu M, Xiao P and Stenzel MH 2017 *Polym Chem.* **8** pp. 3317–3326
[55] Kaasalainen M, Aseyev V, von Haartman E, Karaman DŞ, Mäkilä E, Tenhu H, Rosenholm J and Salonen J 2017 *Nanoscale Res Lett.* **12** pp. 74
[56] Peltonen L 2018 *Adv Drug Deliv Rev.* **131** pp. 101–115
[57] Sander JRG, Bucar D-K, Henry RF, Zhang GGZ and MacGillivray LR 2010 *Angew Chem Int Ed Engl.* **49** pp. 7284–7288
[58] van Sebille M, van der Maaten LJP, Xie L, Jarolimek K, Santbergen R and van Swaaij RACMM, Leifer K, Zeman M 2015 *Nanoscale.* **7** pp. 20593–20606
[59] Ahmed TA, El-Say KM, Mahmoud MF, Samy AM and Badawi AA 2012 *AAPS PharmSciTech.* **13** pp. 760–771
[60] Ahmed TA and Aljaeid BM 2016 . Drug Des. Devel. Ther. 10:483–507
[61] Felt O, Buri P and Gurny R 1998 *Drug Dev Ind Pharm.* **24** pp. 979–993

Catalytic conversion of ethanol to butanol over magnesium oxide catalysts

Stanley Austin Susanto[1,2], Rika Tri Yunarti[2], Robert Ronal Widjaya[1], Yati Maryati[1], Adid Adep Dwiatmoko[1], Fauzan Aulia[1] and Nino Rinaldi[1]

[1]Research Center for Chemistry, Indonesian Institute of Sciences, Kawasan Puspiptek, Serpong, Tangerang Selatan 15314, Indonesia
[2]Department of Chemistry, Faculty of Mathematics and Natural Sciences, Universitas Indonesia, Kampus UI Depok, Depok 16424, Indonesia

*Corresponding author email: data.chemisorption@gmail.com

Abstract. Several types of magnesium oxide (MgO) has been studied as catalyst for catalytic conversion of ethanol into butanol. The reactions have been conducted at 350 °C in batch reactor. MgO catalysts have been characterized by XRD, CO_2-TPD, and N_2-physisorption. It was found that of the highest yield of butanol was obtained from MgO synthesized from precipitation method. The highest butanol yield was 2.4% at our experimental conditions. This was attributed to its surface area and basicity.

1. Introduction

Ethanol is a chemical compound that can be produced from biomass and is usually produced by fermentation of hexose sugar. Ethanol is a versatile raw material that can be used as a fuel additive or solvent [1]. Ethanol can be used as a fuel like gasoline. However, ethanol is not qualified as an effective transportation fuel since it is hydrophilic, soluble in water, corrosive, and highly volatile [2]. In addition, ethanol also cannot be used in high proportion since it produces low energy per unit volume compared to gasoline [3]. This is related to boiling point of ethanol which is lower than n-butanol thus the combustion of ethanol in the motor engine will be faster [4]. To obtain such a high proportion, ethanol can be catalytically converted to important chemical compound such as n-butanol since this compound has low vapor pressure, low solubility in water, and is less corrosive [2].

Catalytic conversion of ethanol to n-butanol involves Guerbet reaction, where a primary aliphatic alcohol is converted into its β-alkylated dimer alcohol with loss of one equivalent of water. The reaction is useful to increase the value of short chain alcohol since it can provide high product selectivity and a controlled synthesis process [1]. The conversion reaction involves aldol reaction in alkaline condition [5] thus requiring the addition of alkaline/base catalyst.

Metal oxide such as MgO is the catalyst that show intrinsic activity against n-butanol production because of the dominance of alkaline and acid active sites [1]. Other methods such as calcination of magnesium precursor and surfactant mixture can improve the performance of MgO catalyst since surfactants can control the morphology, particle size, crystallinity, and surface area of the catalyst thus increasing the concentration of basic sites in MgO [8]. Synthesis of MgO catalyst can be conducted by precipitation method, direct calcination method with magnesium precursor and calcination of magnesium precursor and surfactant mixture. These methods produce different chemical and physical characters of MgO. In this study we studied the effect of MgO preparation methods and evaluate its performance on catalytic reaction of ethanol conversion to *n*-butanol.

2. Experimental

2.1. Catalyst preparation

Commercial magnesium oxide (Merck) was dried in oven at 110 °C then grounded to produce MgO and named MgO(m). Magnesium oxide was synthesized using two calcination methods, that is direct calcination of magnesium precursor (MgO(k)) and calcination of magnesium precursor and surfactant mixture (MgO(s)). For direct calcination of magnesium precursor, $Mg(NO_3)_2.6H_2O$ was calcinated at 700 °C for 4 hours [8]. Calcination of magnesium precursor and surfactant mixture method was conducted by preparing $MgCl_2$ 1 M and added Hexadecyl Trimethyl Ammonium Bromide (C16TAB) 0.125 M and Sodium dodecyl sulfate (SDS) 0.125 M. After that, the mixture was stirred and dropwise with NaOH 2 M until pH of mixture is 11-12. Next, the mixture was centrifuged at 500 rpm for 5 minutes. The mixture was then vacuum filtered and washed with warm water, dried in an oven at 100 °C and calcinated at 700 °C for four hours. $MgCl_2$ 1 M was prepared as magnesium precursor then stirred and dropwise with NaOH 2 M until pH of mixture is 11-12. After that, the mixture was then vacuum filtered and washed with warm water, dried in an oven at 100 °C, calcinated at 700 °C for four hours, and named MgO(p).

2.2. Catalyst characterizations

Specific surface area analysis was conducted with 0.5 g MgO from various preparation methods with BET and BJH methods. XRD analysis was conducted with 0.5 g MgO from various preparation methods and MOx/MgO(p) (M = Ni, Ru) catalyst were grounded first then tested in sample holder with Cu-Kα (1,542 Å) radiation and angle range 2θ from 10° to 80°. CO_2-TPD analysis for basicity was conducted with 0.05 g MgO from various preparation methods and MOx/MgO(p). The sample was flowed with helium gas while heated to 400 °C and dwelled for one hour, then conducted cooling down to room temperature and CO2 gas adsorption for 30 minutes, after that gas flow was changed with helium gas for 15 minutes at room temperature, then temperature was raised to 600 °C and dwelled for 10 minutes, and the sample was conducted cooling to room temperature.

2.3. Catalytic reaction

The catalytic reaction was carried out by adding 20 mL ethanol with no catalyst, MgO catalysts, and MOx/MgO(p) catalysts into a 100 mL autoclave which was placed in a batch reactor at 350 °C for four hours at atmospheric pressure. The reaction results are filtered to separate the catalyst and the product obtained. GC-FID analysis was conducted for samples with manual gas chromatography specification of Agilent 7890A with a Carbowax/20M column (30 m × 320 μm × 25 μm) and 25 μL of toluene was needed as an internal standard to samples.

3. Results and Discussion

In Table 1, of the four MgOs, MgO(m) and MgO(k) can be classified as mesoporous (2 – 50 nm) and MgO(p) and MgO(s) as microporous (< 2 nm) [10]. In the table, MgO(k) has larger surface area, pore size, and pore volume than MgO(m), while MgO(p) is the catalyst having largest surface area and pore volume of the four MgOs having been analyzed although having small pore size. MgO(s) also has large surface area with smallest pore size and large pore volume. The results of surface area of four catalysts are proportional to pore volume of catalysts but are inversely proportional to pore size of catalysts. The trend of surface area of MgO is very likely in accordance with [8].

Table 1. N_2-physisorption of MgO catalyst.

Catalyst	BET Surface Area (m2/g)	Pore Size (nm)	Pore Volume (cm3/g)
MgO(m)	0.57	19.5	0.00280
MgO(k)	5.82	25.0	0.036
MgO(p)	107.6	1.75	0.288
MgO(s)	60.61	1.58	0.101

Figure 1 shows the dominant peak pattern of MgO appears at an angle of 2θ 42.9° and 62.3° with Miller indices respectively of (200) and (220) while the other peaks appear at an angle of 2θ 36.9°; 74.7°; and 78.6° with Miller indices respectively of (111), (311), and (222) [11]. This diffraction has the value of *hkl* being always even or odd. According to the selection rule for the Miller indices, this *hkl* value will have a Bravais lattice of face-centered cubic thus MgO has a face-centered cubic structure. From Figure 1, the diffraction pattern of MgO(k) has a high peak intensity and is characteristic to MgO(m). When MgO is synthesized from precipitation method, the intensity of peaks decreases drastically because of small size of crystallite from MgO(p) resulting in low-density white powder MgO (p). However, these peaks still have characteristics of MgO(m). The same thing happens when MgO is synthesized from calcination of magnesium precursor and surfactant mixture. The pattern has low-intensity peaks as well as peaks of MgO(p) and even peak at 2θ 36.9° are barely visible. The decrease in the intensity of the peaks is also observed in the report [8] on the precipitation method and calcination of magnesium precursor and surfactant mixtures and is related to the crystallite size of MgO.

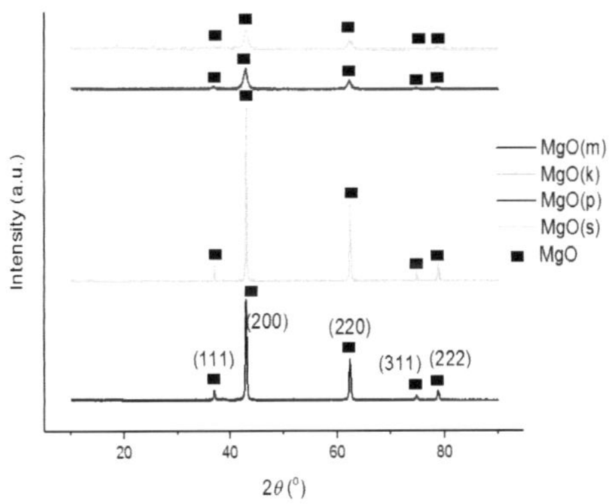

Figure 1. XRD pattern of commercial MgO (MgO(k)) and MgO synthesized.

In this research, crystallite size of MgO from various methods was estimated using Scherrer equation on main peak (200). The trend based on Table 2 is suitable with SAA analysis results where the larger the pore size/crystallite size of MgO molecule in the catalyst, the higher the intensity of XRD. Based on this, MgO(p) has lowest crystallinity and pore size as well as those observed by [8].

Table 2. Crystallite size of MgO catalysts.

Catalyst	Crystallite Size (nm)
MgO(m)	51.9
MgO(k)	71.4
MgO(p)	17.3
MgO(s)	17.6

Characterization of MgO catalysts and MO_x/MgO(p) catalyst with chemisorption was conducted with CO_2-Temperature Programmed Desorption (CO_2-TPD) to measure the amount of adsorbed CO_2 by catalyst chemically and basic property of catalyst [13]. In Figure 2 and Table 3, it is shown that MgO(m) has high basicity at low temperature and decreases when temperature gets higher. Meanwhile, MgO(k) has high basicity at high temperature but the intensity is much lower than other MgO basicities at same temperature. Furthermore, MgO(p) has high basicity both at low, medium, and high temperatures thus the catalyst is suitable for reactions at higher temperature. MgO(s) has high basicity at low and high temperatures but not at medium temperature. Table 3 shows that "weak" base site has CO_2 desorption between 27 to 147 °C, "medium" base site has CO_2 desorption between 147 to 377 °C, and "strong" base site has CO_2 desorption above 377 °C [13]. Therefore, MgO(m) will have the most dominant "weak" base site, then MgO(k) has the most dominant (low intensity) "strong" base site, then MgO(p) has the most dominant "medium" base site, and MgO(s) has the most dominant "weak" base site. Of this classification, the order/trend of basicity strength (from weakest) of synthesized MgOs is: MgO(k) < MgO(s) ≈ MgO(m) < MgO(p). Despite this trend, it turns out that the basicity does not really match to surface area of catalysts. MgO(p) does have the largest surface area thus producing highest basicity whereas for MgO(m) surface area is much smaller than MgO(s) or MgO(k).

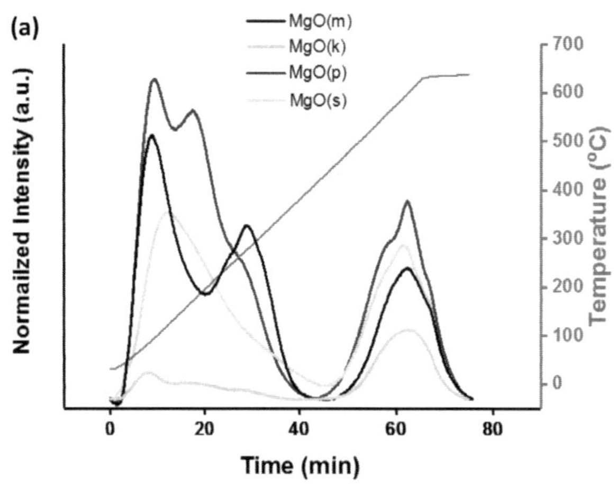

Figure 2. CO_2-TPD for MgO catalysts.

Table 3. Basicity (mmol/g) of MgO catalysts.

Catalyst	"Weak" Base Site (mmol/g)	"Medium" Base Site (mmol/g)	"Strong" Base Site (mmol/g)	Basicity Total (mmol/g)
MgO(m)	0.14	0.11	0.08	0.33
MgO(k)	0.01	0.01	0.04	0.06
MgO(p)	0.13	0.19	0.13	0.45
MgO(s)	0.18	-	0.10	0.28

Table 4 shows that even without catalyst, the percentage of ethanol conversion is 25%. According to [14,15], ethanol undergoes decomposition into gases such as ethylene, water vapor, and acetaldehyde thus generating residual pressure. Percent conversion of ethanol increases when using catalyst MgO(k) rather than MgO(m) which shows the surface area of both catalysts playing important and proportional roles. Interestingly, percent conversion of ethanol with MgO(p) only produces 23% even though MgO(p) has the largest surface area and pore volume. Although all four methods produce MgO, different methods will produce MgO with different characteristics, especially basicity of four MgOs produced.

Table 4. Conversion of ethanol and yield of n-butanol.

Catalyst	Conversion (%)	Yield of Butanol (%)
No catalyst*	25.01	0.067
MgO(m)	25.10	0.249
MgO(k)	25.71	1.352
MgO(p)	23.30	2.362
MgO(s)	25.14	0.352

Table 4 shows that it is shown that the reaction of ethanol with MgO(p) produces highest percent yield of *n*-butanol from the reaction of ethanol with MgO from various methods while lowest percent yield of *n*-butanol is obtained when ethanol is reacted without catalyst. This is in accordance with analysis of CO_2-TPD graph of MgO(p) at "weak" and "medium" base sites which results in MgO(p) having high catalytic activity at 350 °C. This result is also in accordance with lowest crystallite size of MgO(p) where the smaller size of crystallite catalyst, the higher probability of the MgO(p) molecule to react effectively with ethanol to produce high yield [17],[18]. This result is also supported by large surface area and pore volume of catalyst MgO(p). Interestingly, percent yield for MgO(s) is lower than MgO(k). It seems that crystallite size and surface area of catalyst MgO(s) and MgO(k) have less role for percent yield of *n*-butanol. CTAB and SDS surfactant mixture has an important role in the catalytic reaction of the ethanol. The same thing happens in percent yield of *n*-butanol with catalyst MgO(m) and MgO(k) where surface area is more role playing than crystallite size of both catalysts.

4. Conclusions

Catalytic Guerbet reaction involving ethanol occurs with MgO catalyst prepared from various methods and results percent conversion of ethanol for MgO(m), MgO(k), MgO(p), dan MgO(s) respectively, that is: 25.10%; 25.71%; 23.30%; and 25.14% and percent *yield* of *n*-butanol from these catalysts are: 0.25%; 1.35%; 2.36%; and 0.35%. Factors such as surface area, basicity, and MgO molecule crystallite size on catalyst affect the catalytic activity of the reaction. Batch reactor used for the reaction also affects the catalytic activity of the reaction.

Acknowledgments

This work was supported by the Research Collaboration grant funded by Research and Technology Center, PT Pertamina. This work was also supported by the BSLN Program 2019 of the Ministry of Science, Technology and Higher Education.

References

[1] Pang J F *et al.* 2016 *J. Catal.* **344** 184-93
[2] Hanspal, Sabra, dissertation, University of Virginia, Virginia, 2016.
[3] Cimino S, Lisi L and Romanucci S 2017 *Catal. Today* **304** 58-63
[4] Ndaba B, Chiyanzu S and Marx I 2015 *Biotechnol. Rep.* **8** 1-9
[5] Smith J G 2011 *Organic chemistry* 3rd Ed. (New York:McGraw-Hill)
[6] Tsuchida T, Sakuma S, Takeguchi T and Ueda W 2006 *Ind. Eng. Chem. Res.* **45** 8634-42
[7] Apuzzo J, Cimino S and Lisi L 2018 *RSC Adv.* **8** 25846-55
[8] Simanjutak F, Hubertson S, Lim S R, Ahn B S, Kim H S and Lee H 2014 *Appl. Catal. A Gen.* **484** 33-8
[9] Fereshteh M and Rezaei M 2009 *Powder Technol.* **196** 85-8
[10] Rouquerol J, *et al.* 1994 *IUPAC* **66** 1739-58
[11] Dercz, G., K. Prusik, L. Pająk, R. Pielaszek, J. J. Malinowski, W. Pudło. Materials Science-Poland, 27, 201-207 (2009).
[12] Li B, Zhang Y, Zhao Y, Wu Z and Zhang Z 2007 *Mater. Sci. Eng. A* **452-453** 302-5

[13] Peng S Y *et al.* 2014 *Catal. Sci. Technol.* **4** 1925-30
[14] Park J, Zhu R S and Lin M C 2002 *J. Chem. Phys.* **117** 3224-31
[15] Li J, Kazakov A and Frederick L 2004 *J. Phys. Chem. A* **108** 7671-80
[16] Quesada J, Faba L, Díaz E and Ordóñez S 2018 *Catal. Today* **563** 64-72
[17] Hermann C, Haas J and Fetting F 1987 *Appl. Catal.* **35** 299-310
[18] Mguni L L, Meijboom R and Jalama K 2012 *Int. J. Mater. Metallur. Eng.* **6** 375-9

IOP Conf. Series: Materials Science and Engineering 1011 (2021) 012029 doi:10.1088/1757-899X/1011/1/012029

Proficiency testing for determination of pesticides residues in black tea: Comparison of three robust statistical approaches based on ISO 13528 to estimate the consensus values for small number of participants

Nurhani Aryana

Research Center for Chemistry, Indonesian Institute of Sciences, Building 452 Kawasan Puspiptek Serpong, Tangerang Selatan, Indonesia

Corresponding author email: nurh020@lipi.go.id

Abstract. In Indonesia, proficiency testing (PT) provider for determination of pesticides residues in tea is still limited. In addition, many proficiency testing schemes still use consensus value from participants' results as the assigned value rather than reference assigned value due to some technical and economical issues. However, this mode of assigned value may lack of reliability when the number of participants are small and the test results are highly varied. In these conditions, many PT providers were not able to perform the statistical data evaluation causing the needed PT became unavailable. This study is aimed to investigate the suitable statistical approach in estimating the consensus value for small number of participants in PT for determination of pesticides residues in black tea. Three robust statistical approaches recommended by ISO 13528: variants of Algorithm A, Hampel method, and M estimator were compared to evaluate the participants' data from the PT schemes organized by Research Center of Chemistry-LIPI in 2014 and 2015. All of the robust statistical approaches showed to be able to assess the performance of participants. The Hampel method and M estimator were the most adequate robust statistical approaches in obtaining the closest consensus values compared to the homogeneity test results, stability monitoring results, and the assigned valued in the real PT scheme, for endosulfan sulphate and bifenthrin, respectively.

1. Introduction

The amount of pesticides residues in tea commodities is routinely checked during trading to ensure that the quality of tea meets the requirements set by the food safety standards or the respective national regulations. However, an accurate measurement occasionally cannot be easily performed since some of the maximum residue limit (MRL) of the residual pesticides are at trace level. Laboratories for testing pesticides should incorporate all the management and technical requirements as specified in ISO/IEC 17025 in order to ascertain reliable analytical data in this challenging work [1]. One of the mandatory requirements for a laboratory in monitoring the validity of its results according to ISO/IEC 17025 is by participation in a proficiency testing (PT) program. PT is one of the key elements in the implementation of an appropriate quality assurance program and performance monitoring procedure for chemical analysis laboratories [2].

Metrology in chemistry laboratory-Indonesian Institute of Sciences (MiC-LIPI) has organized two PT schemes for determination of pesticides residues in black tea. Both of the PT schemes were using candidate reference materials pesticides residue in black tea, PTH14, as the test samples. The test parameters were endosulfan sulphate and bifenthrin. The first PT scheme was conducted in 2014 and was participated by three laboratories from Indonesia. The second PT scheme was conducted in 2015 and was participated by six laboratories from Myanmar and two laboratories from India. According to ISO 13528 – Statistical methods for use in proficiency testing, where the number of participants is small, the assigned value should ideally be determined using a metrologically valid procedure and independent of the participants [3]. In these ideal conditions, the PT scheme can be rolled even with just one participant. The metrologically valid procedure can be achieved by using a certified reference material or a primary method. The assigned values in the PT schemes were derived from analytical measurements by using high sensitivity methods: gas chromatography-μ-electron capture detector (GC-μECD) method, gas chromatography-mass spectrometry (GCMS) method, and gas chromatography-isotope dilution mass spectrometry (GC-IDMS) method.

In Indonesia, many PT schemes were still operated by the consensus values derived from the participants' results. Apart from technical issues, one of the reasons often mentioned is that PTs operated with consensus value would be cheaper than those using reference value obtained from a *priori* characterization measurements [4]. It has been reported that over 90 % of the proficiency testing programmes rely on consensus values [5]. However, when the PT scheme was only participated by small number of participants and the test results were highly varied, the performance evaluation by using consensus value from the participants' results may become unreliable. In these conditions, many PT providers were not able to perform the statistical data evaluation causing the needed PT became unavailable. To overcome these issues, in this study, some data treatment processes for PT with small number of participants were investigated including outlier identification methods and three robust statistical approaches recommended by ISO 13528 for calculating consensus value from participants' results: variants of Algorithm A, Hampel method, and M estimator. The participants' data combined from the first and second PT schemes for determination of pesticides residue in black tea were used in this study. The effectiveness of the robust statistical approach was expressed as the percentage of participants with acceptable performance score. The results in this study were further verified by using the homogeneity test results performed by gas chromatography-μ-electron capture detector (GC-μECD) method, stability monitoring results performed by gas chromatography-isotope dilution mass spectrometry (GC-IDMS) method, and the assigned values from the real PT scheme.

2. Experimental section

2.1. Preparation of the test samples
The test samples were prepared as candidate certified reference materials in accordance with ISO/IEC 17034 [6] and ISO Guide 35 [7]. The procedure for the preparation of the test samples were described elsewhere [8]. The homogeneity and stability of the test samples were checked in accordance with ISO 13528 [3].

2.2. Analysis by the participants
The PT schemes were organized in accordance with ISO/IEC 17043 [9] and were subscribed by total of 11 participants from Indonesia, Myanmar, and India. Each participant received one bottle of the test sample together with the technical instruction and the result report form. The participants were asked to determine preferable the mass fraction of the test parameters endosulfan sulphate and bifenthrin (in three significant figures, in μg/kg) by using the participants' routine methods.

2.3. Statistical approaches
The following procedures describe the data treatment processes for PT with small number of participants, including outlier identification methods, calculation of consensus value from participants' results, and performance evaluation of participants' results. All of the statistical data evaluation were done in accordance with ISO/IEC 17043 [9] and ISO 13528 [3] by using the Microsoft Excel.

2.3.1. Outlier identification. Three identification methods were used in this study to detect any outlier in the data set: Grubb's test, Hampel's test, and trimmed average. The most commonly used and recommended test by ISO for outliers is the Grubb's test since it is easy to apply and operated using the mean and standard deviation of the data [10]. The Grubb's test was performed to the highest and lowest values in the set of test results. Grubb's statistics for the highest value and the lowest value are given as equation (1) and equation (2), respectively.

$$G_p = \frac{x_p - \bar{X}}{s} \tag{1}$$

$$G_p = \frac{\bar{X} - x_l}{s} \tag{2}$$

where G_p indicates the Grubb's statistics, x_p indicates the highest value, x_l indicates the lowest value, $\bar{X}$ indicates the mean value, and s indicates the standard deviation. G_p value was compared with the critical value for the p number of test results.

The Hampel's test [3] was performed by equation (3).

$$d_i = |x_i - med(x_i)| \tag{3}$$

where d_i indicates the absolute differences, x_i is the ith data of the test results (for $i = 1$ to p number of test results), and $med(x_i)$ indicates the median of the test results. The outlier value from the data set can be obtained by comparing the d_i value from the condition given in equation (4).

$$|d_i| \leq 4.5\, med(d_i) \tag{4}$$

where $med(d_i)$ is the median of the absolute differences.

The trimmed average method [11] was performed by calculating the mean $\bar{X}$ and standard deviation s from all data. The values outside of the likelihood interval ($\bar{X} - s$ or $\bar{X} + s$) then were discarded.

2.3.2. Estimation of the consensus values from participants' results. The first robust statistical approach used to estimate the consensus values from the participants' results was variants of Algorithm A [3]. The algorithm A was performed with no iteration of location, using the median (equation 5) as a location estimate.

$$x^* = med(x_i) \tag{5}$$

The robust standard deviation of participants' results (s^*) was calculated with Q estimator [3] and the s^* was not updated during iteration. The Q_n for a data set was calculated with p reported results. Firstly, the absolute differences ($p(p$-$1)/2$) were calculated as equation (6),

$$d_{ij} = |x_i - x_j| \text{ for } i = 1, 2, ..p - 1 \text{ and } j = i + 1, i + 2, ..p \tag{6}$$

The ordered differences (d_{ij}) were denoted by $d_{\{1\}}, d_{\{2\}}, ...d_{\{p(p-1)/2\}}$. Then, the number distinct pairs (k) chosen from h objects was calculated by using equation (7),

$$k = \frac{h(h-1)}{2} \tag{7}$$

where

$$h = \begin{cases} p/2 & p \text{ even} \\ (p-1)/2 & p \text{ odd} \end{cases} \tag{8}$$

The Q_n was calculated as equation (9),

$$Q_n = 2.2219 d_{\{k\}} b_p \tag{9}$$

where b_p was selected from Table C.2 in ISO 13528 for a particular number p of data points.

The second robust statistical approach used to estimate the consensus values from participants' results was the Hampel estimator. The Hampel estimate of location was provided by an iterative reweighting scheme [3], started with calculating the initial robust mean (x^*) as calculated in equation (5). The robust standard deviation of participants' results (s^*) was calculated as the scaled median absolute deviation (*MADe*) by using equation (10),

$$MADe = 1.483\, med|x_i - x^*| \tag{10}$$

where $i = 1, 2, \ldots p$, with p being the number of test results.

For each data point x_i, q_i was calculated as equation (11). Then, for each q_i, the weight w_i was calculated as equation (12).

$$q_i = \left|\frac{x_i - x^*}{s^*}\right| \tag{11}$$

$$w_i = \begin{cases} 0 & |q| > 4.5 \\ (4.5 - q/q) & 3 < |q| \leq 4.5 \\ 1.5/q & 1.5 < |q| \leq 3.0 \\ q & |q| \leq 1.5 \end{cases} \tag{12}$$

The x^* was then recalculated as equation (13),

$$x^* = \frac{\sum_{i=1}^{p} w_i x_i}{\sum_{i=1}^{p} w_i} \tag{13}$$

The steps in equation (11) to equation (13) were repeated until x^* converges. Convergence can be assumed when the change in x^* from one iteration to the next was less than $0.01s^*/p^{1/2}$, corresponding to approximately 1 % of the standard error in x^*.

The third robust statistical approach used to estimate the consensus values from participants' results was the M estimator. According to Rousseeuw et al. [12], the robust mean (x^*) and robust standard deviation (s^*) are estimated by the M estimator as equation (14) and equation (15),

$$\frac{1}{n}\sum_{i=1}^{p} \psi\left(\frac{x_i - T_n}{S_n}\right) = 0 \tag{14}$$

$$\psi(x) = \frac{e^x - 1}{e^x + 1} \tag{15}$$

where T_n is the location x^* estimator and S_n is the scale s^* estimator.

The initial robust mean (x^*) was calculated as equation (5). Next, T_n was iteratively calculated using equation (14) computed using a Newton-Raphson algorithm. The robust standard deviation (s^*) was calculated as Q_n, following equation (6)-(9).

2.3.3. Performance evaluation. Before selecting the criterion for performance evaluation, the standard uncertainty of the consensus value ($u(x_{pt})$) was compared to the standard deviation of proficiency assessment (σ_{pt}). The standard uncertainty of the consensus value may be considered to be negligible and need not be included in the interpretation of the results of the round of proficiency testing when the criterion in equation (16) was met [3].

$$u(x_{pt}) < 0.3\sigma_{pt} \tag{16}$$

When this criterion was met, laboratory performance was assessed by using z-score as shown in equation (17). When the criterion in equation (16) was not met, the standard uncertainty of the consensus value need to be included in the interpretation of the results of proficiency testing and the laboratory performances was assessed by using z'-score as shown in equation (18),

$$z_i = \frac{(x_i - x_{pt})}{\sigma_{pt}} \tag{17}$$

$$z_i' = \frac{x_i - x_{pt}}{\left(\sigma_{pt}^2 + u^2(x_{pt})\right)^{1/2}} \tag{18}$$

where x_i is the submitted participants result.

The performance scores calculated by the standard deviation for proficiency assessment (σ_{pt}) derived from the robust standard deviation of participants' results (s^*) were compared with the standard deviation from Horwitz Equation as calculated by equation (19),

$$\sigma_{pt} = 0.22c \quad (19)$$

3. Results and discussion

From the total of 11 registered participants, there were 8 and 10 laboratories submitted the test results for endosulfan sulphate and bifenthrin, respectively. In order to see the scattering of the data, the participants' test results are presented in figure 1 and 2, sorted by increasing value with the red line represents the median value. Figure 1 and 2 show wide range of test results for both of endosulfan sulphate and bifenthrin. In the PT with small number of participants, it is important to treat extreme results correctly if they are not caused by a known gross error or miscalculation [13]. Information about the distribution patterns of the data (such as normality) is needed before applying any statistical analysis tools since typical statistical tests incorporate assumptions about the underlying distribution of data [9]. The histograms were drawn to assess the distribution of the data set and detect visually any outliers. The histograms of the test results of endosulfan sulphate and bifenthrin show tailed distributions and the presence of some test results which can be considered as outliers.

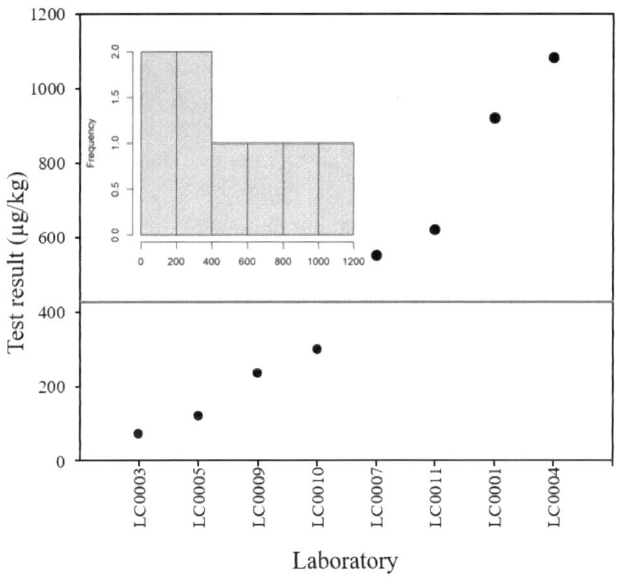

Figure 1. The participants' test results and the histograms of the data in the PT for determination of endosulfan sulphate in black tea.

When the number of participants is small, extreme results cannot usually be identified as outliers by known statistical tests because of low power of these tests [13]. Outlier tests are seldom completely satisfactory, especially when many outliers may be present [14]. In this study, three outlier identification methods were employed to identify the outliers: Grubb's test, Hampel's test, and trimmed average. Outlier identification of the test results are summarized in table 1 and 2 for endosulfan sulphate and bifenthrin, respectively. The Grubb's and Hampel's tests compare the difference between an extreme value and the centre of the data with the variability of the data and identify the extreme value as an outlier if the ratio is too large [15]. For endosulfan sulphate, the G_p value for the lowest test result was 1.12 and the G_p value for the highest test result was 1.60. These G_p values were smaller than the critical value of Grubb's statistic ($G_{p\,crit}$ = 2.03, n=8, at confidence level 95 %), therefore, there was no outlier can be detected by the Grubb's test. For bifenthrin, the G_p value for the lowest test result was 1.26 and the G_p value for the highest test result was 1.70. These G_p values were also smaller than the critical value of Grubb's statistic ($G_{p\ crit}$ = 2.18, n=10, at confidence level 95 %), therefore, there was no outlier can be detected by the Grubb's test. According to ISO 13528, the Grubb's test is most useful for p>10 and can fail when there are multiple outliers [3]. There was also no outlier detected by the Hampel's test since the test results for endosulfan sulphate and bifenthrin have not met the criterion in equation (4).

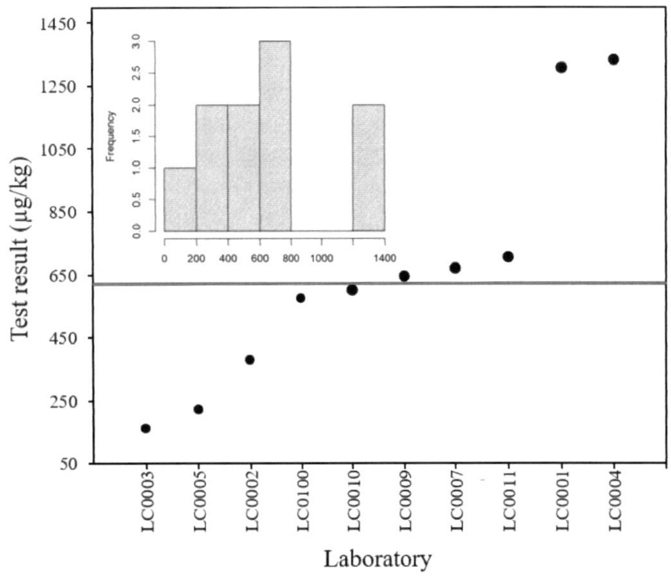

Figure 2. The participants' test results and the histograms of the data in the PT for determination of bifenthrin in black tea.

Table 1. Identification of outliers using different methods in the PT for determination of endosulfan sulphate in black tea.

Lab Code	Test result (μg/kg)	Grubb's test	Hampel's test		Trimmed average
			$\lvert d_i \rvert$	Outlier test	
LC0001	919	OK	493	OK	$$
LC0003	71	OK	355	OK	$$
LC0004	1082	OK	656	OK	$$
LC0005	121	OK	305	OK	OK
LC0007	552	OK	126	OK	OK
LC0009	235	OK	191	OK	OK
LC0010	300	OK	126	OK	OK
LC0011	620	OK	194	OK	OK

OK indicates inlier test result
$$ indicates outlier test result

The values of *4.5med(di)* were 1123 μg/kg and 740 μg/kg, for endosulfan sulphate and bifenthrin, respectively. Outliers finally can be detected by the trimmed average method as shown in table 1 and 2. For endosulfan sulphate, the outlier-corrected average value was change from 488 μg/kg to 366 μg/kg when the outliers detected by trimmed average method were excluded. For bifenthrin, the outlier-corrected average value was change from 660 μg/kg to 596 μg/kg when the outliers detected by trimmed average method were excluded. The trimmed average method was the most suitable method to detect the outliers in this study.

Table 2. Identification of outliers using different methods in the PT for determination of bifenthrin in black tea.

Lab Code	Test result (µg/kg)	Grubb's test	Hampel's test		Trimmed average
			$\|d_i\|$	Outlier test	
LC0001	1307	OK	684	OK	$$
LC0002	377	OK	246	OK	OK
LC0003	162	OK	461	OK	$$
LC0004	1333	OK	710	OK	$$
LC0005	221	OK	402	OK	$$
LC0007	671	OK	48	OK	OK
LC0009	646	OK	23	OK	OK
LC0010	600	OK	23	OK	OK
LC0011	706	OK	83	OK	OK
LC0100	573	OK	50	OK	OK

OK indicates inlier test result
$$ indicates outlier test result

According to IUPAC/ CITAC Guide – Selection and use of proficiency testing schemes for a limited number of participants [13], the assigned value of the PT items cannot be calculated safely from the measurement results obtained by the participants as a consensus value. The consensus value and the robust standard deviation from participants' results, in principle, cannot be used for a reliable assessment of an individual laboratory performance when the number of participants is small. Methods of obtaining traceable assigned values are to be used wherever possible to provide comparable PT results. In this study, the consensus values are used only for comparison of PT results as a statistical sample.

In general, the consensus value from participants' results can be obtained by central tendency estimators such as the arithmetic mean or robust mean [5]. The arithmetic mean is the minimum variance estimator, but is sensitive to the influence of outliers. Meanwhile, robust statistics is a modern approach to the outlier problem in which the influence of outliers and heavy tails is downweighted. Robust statistics have been shown to be applicable to analytical data in number of situations [16]. However, when the number of participants is small (p<20), the choice of statistical approaches should be taken carefully since most of them become increasingly unreliable and deviations from a normal distribution are harder to identify. Among many types of robust statistics, the median, *MADe*, and Q/Hampel methods were selected in this study since these methods can tolerate a very large proportion of outliers.

When the data were calculated by classical mean and standard deviation without exclusion of outliers, the obtained consensus values and standard deviation of participants' results were 488 µg/kg and 370 µg/kg, for endosulfan sulphate, and 660 µg/kg and 244 µg/kg, for bifenthrin. These standard deviation were very large to be used in the calculation of performance scores. In this condition, limit on the largest standard deviation that will be used should be placed so that results that are not fit for purpose will receive an action signal. According to ISO 13528, there are two spesific recommendations for robust estimates of dispersion in very small data where $p \geq 4$: M-estimate of standard deviation based on a logarithmic weighting function as recommended by Rousseeuw [12], and algorithm A with no iteration of location, using the median as a location estimate. Based on the outlier identification results, the proportion of outliers were expected to be more than 20 %, therefore, s^* was replaced with the Q estimator and was not updated during iteration. In this study the Hampel method was also employed in estimating the consensus values since this method has high breakdown point and high efficiency.

Table 3. The obtained consensus value of participants' results by using three robust statistical approaches for endosulfan sulphate.

Statistical approach	x^* (µg/kg)	u (µg/kg)	U (µg/kg, k=2)	U_{rel} (%)	s^* (µg/kg)	σ_{pt} (µg/kg)	0.3 σ_{pt} (µg/kg)
Variants of Algorithm A	426	107	214	50	243	94	28
Hampel method	482	164	327	68	370	106	32
M estimator	459	107	214	47	243	101	30

Table 4. The obtained consensus value of participamts' results by using three robust statistical approaches for bifenthrin.

Statistical approach	x^* (µg/kg)	u (µg/kg)	U (µg/kg, k=2)	U_{rel} (%)	s^* (µg/kg)	σ_{pt} (µg/kg)	0.3 σ_{pt} (µg/kg)
Variants of Algorithm A	623	124	248	40	314	137	41
Hampel method	596	96	193	32	244	131	39
M estimator	690	124	248	36	314	152	46

Table 3 and 4 present the obtained consensus values from participants' results, for endosulfan sulphate and bifenthrin, respectively. All of the test results of endosulfan sulphate and bifenthrin were included in the calculation of the consensus values by variants of algorithm A and Hampel method. Meanwhile, the outliers detected by the trimmed average method were excluded in the calculation of the consensus values by M estimator.

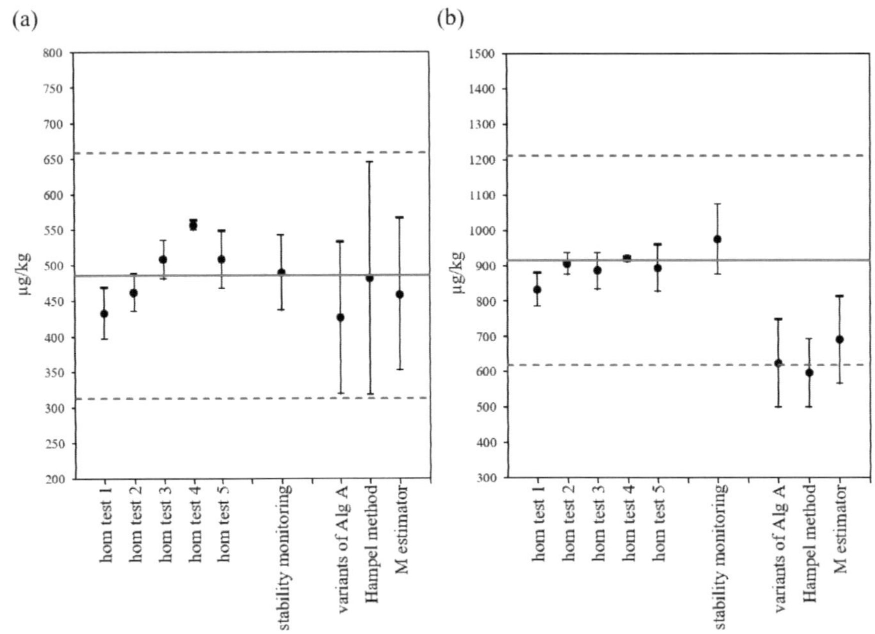

Figure 3. Comparison between the obtained consensus values of participants' results with the homogeneity test results derived from GC-μECD measurements, stability monitoring results derived from GC-IDMS measurements, and the assigned values in the real PT scheme of (a) endosulfan sulphate and (b) bifenthrin. The solid red line indicates the assigned value in the real PT scheme. The dotted blue lines indicate range of tolerance (|z'-score|≤2.00).

Figure 3 shows the comparison between the obtained consensus values of participants' results with the homogeneity test results derived from GC-μECD measurements, stability monitoring results derived from GC-IDMS measurements, and the assigned values in the real PT scheme. For endosulfan sulphate, the consensus values of participants' results obtained from three robust statistical approaches were inside the range of tolerances and close to the homogeneity and stability test results and the assigned value in the real PT scheme. Meanwhile, for bifenthrin, the obtained consensus values of participants' results were close to the low limit of the range of tolerance. This may be caused by some technical issues in the sample preparation and analysis of the target analyte.

As shown in table 3 and 4, the standard uncertainties of the consensus values obtained from the calculation of three robust statistical approaches have not met the criterion set in equation (16), for both of test parameters. Therefore, the performance scores of participants' results were evaluated by z'-score. The standard deviation for proficiency assessment (σ_{pt}) was taken from the Horwitz equation since performance score evaluation by using σ_{pt} from s^* cannot detect the outliers as action signal. Table 5 and 6 presents the performance scores of participants in terms of z'-score, for endosulfan sulphate and bifenthrin, respectively. Participants' performance was ranked into three categories based on ISO/IEC 17043 [9], where $|z'| \leq 2.00$ is considered to be satisfactory (OK), $2.00 < |z'| < 3.00$ is considered to be questionable ($), and $|z'| \geq 3.00$ is considered to be unsatisfactory ($$).

Table 5. The performance scores of participants results in terms of z'-score for endosulfan sulphate calculated by variants of Algorithm A, Hampel method, and M estimator.

Lab Code	Test result (µg/kg)	Variants of Alg A		Hampel method		M estimator	
LC0001	919	3.46	$$	2.24	$$	3.13	$$
LC0003	71	-2.49	$	-2.11	$	-2.63	$
LC0004	1082	4.61	$$	3.08	$$	4.23	$$
LC0005	121	-2.14	$	-1.85	OK	-2.29	$
LC0007	552	0.88	OK	0.36	OK	0.63	OK
LC0009	235	-1.34	OK	-1.27	OK	-1.52	OK
LC0010	300	-0.88	OK	-0.93	OK	-1.08	OK
LC0011	620	1.36	OK	0.71	OK	1.09	OK

Table 6. The performance scores of participants results in terms of z'-score for bifenthrin calculated by variants of Algorithm A, Hampel method, and M estimator.

Lab Code	Test result (µg/kg)	Variants of Alg A		Hampel method		M estimator	
LC0001	1307	3.70	$$	4.37	$$	3.15	$$
LC0002	377	-1.33	OK	-1.34	OK	-1.60	OK
LC0003	162	-2.49	$	-2.67	$	-2.69	$
LC0004	1333	3.84	$$	4.53	$$	3.28	$$
LC0005	221	-2.18	$	-2.30	$	-2.39	$
LC0007	671	0.26	OK	0.46	OK	-0.09	OK
LC0009	646	0.12	OK	0.31	OK	-0.22	OK
LC0010	600	-0.12	OK	0.03	OK	-0.46	OK
LC0011	706	0.45	OK	0.68	OK	0.08	OK
LC0100	573	-0.27	OK	-0.14	OK	-0.59	OK

Table 7. Percentage of acceptable z'score of endosulfan sulphate and bifenthrin calculated by the consensus values from participants' results derived by three robust statistical approaches.

Test parameter	Variants of Algorithm A	Hampel method	M estimator
Endosulfan sulphate	50 %	62.5 %	50 %
Bifenthrin	60 %	60 %	60 %

All of the three robust statistical approaches can be used to evaluate the PT results and were able to detect the acceptable, warning, and action signals. For endosulfan sulphate, the highest percentage of acceptable z'-score was achieved by the Hampel method (table 7). The consensus value of endosulfan sulphate obtained by the Hampel method was also the closest to the homogeneity test results, stability monitoring result, and the assigned value in the real PT scheme. For bifenthrin, three robust statistical approaches give same percentage of satisfactory results (table 7). The consensus value of bifenthrin obtained by the M estimator was the closest to the homogeneity test results, stability monitoring result, and the assigned value in the real PT scheme.

4. Conclusion

The effectiveness of three robust statistical approaches recommended by ISO 13528 for small number of participants: variants of algorithm A, Hampel method, and M estimator were identified. The calculation of consensus value by variants of algorithm A and Hampel method were done without outlier exclusion. Meanwhile, for M estimator, the outlier identified by trimmed average were excluded before calculation of the consensus value. The robust standard deviation of participants' results (s^*) were calculated by Q estimator and the scaled median absolute deviation (*MADe*). However, the obtained s^* cannot detect the outliers as warning or action signal since the variation was too high. The standard deviation for proficiency assessment (σ_{pt}) for performance score evaluation was finally taken from the Horwitz equation. All of the obtained consensus values and σ_{pt} showed to be able to detect the acceptable, action, and warning signals. The highest effectiveness in achieving the acceptable z'-score for endosulfan sulphate was achieved by the consensus value from the Hampel method. Meanwhile, for bifenthrin, all of the robust statistical approaches gave same effectiveness in assessing the performance of participants.

References

[1] International Organization for Standardization 2017 *ISO/IEC 17025 - General requirements for the competence of testing and calibration laboratories* (Switzerland: ISO)
[2] Yarita T, Otake T, Aoyagi Y, Kuroiwa T, Numata M and Takatsu A 2015 *Talanta* **132** 269–77
[3] International Organization for Standardization 2015 *ISO 13528 - Statistical methods for use in proficiency testing by interlaboratory comparison* (Switzerland: ISO)
[4] Baldan A, Van Der Veen A M H, Prauß D, Recknagel A, Boley N, Evans S and Woods D 2001 *Accredit. Qual. Assur.* **6** 164–7
[5] Wong S K 2005 *Accredit. Qual. Assur.* **10** 409–14
[6] International Organization for Standardization 2016 *ISO 17034 - General requirements for the competence of reference material producers* (Switzerland: ISO)
[7] International Organization for Standardization 2017 *ISO Guide 35 - Reference materials — Guidance for characterization and assessment of homogeneity and stability* (Switzerland: ISO)
[8] Aryana N, Aristiawan Y, Putri D and Styarini D 2016 *Indones. J. Chem.* **16** 72–9
[9] International Organization for Standardization 2010 *ISO/IEC 17043 - Conformity assessment — General requirements for proficiency testing* (Switzerland: ISO)
[10] Aslam M 2020 *J. King Saud. Univ. Sci.* 1 5
[11] Einax J 1995 *Chemometrics in environmental chemistry: applications* (New York: Springer)
[12] Rousseeuw P J and Verboven S 2002 *Computational Statistics & Data Analysis* **40** 741-58

[13] Kuselman I and Fajgelj A 2010 *Pure Appl. Chem.* **82** 1099–135
[14] Chakravarty S, Mohanty A, Ghosh B, Tarafdar M, Aggarwal S G and Gupta P K 2014 *Mapan - J. Metrol. Soc. India* **29** 87–95
[15] Coucke W and Soumali M R 2017 *Biochem. Medica* **27** 37–48
[16] Lawn R E, Thompson M, Thompson M and Walker R F 1997 *Proficiency Testing in Analytical Chemistry* (Cambridge: Royal Society of Chemistry Publishing)

Enzymatic biodelignification of Corncob by Laccase (Lac) from *Cerrena* Sp.B.Md.T.A.1.

Muryanto Muryanto[1*], Rifqa Muharramah[2], Syamsul Falah[2] and Asep Hidayat[3]

[1]Research Center for Chemistry, Indonesian Institute of Sciences, Kawasan PUSPIPTEK Serpong, Tangerang Selatan, Indonesia, 15314
[2]Department of Biochemistry, Faculty of Mathematics and Natural Sciences, Bogor Agricultural University. Gd. D Lt. 5 Kampus IPB Dramaga, Bogor, 16680
[3]Forest Microbiology Research Group, Forest Research and Development Centre, the Ministry of Environment and Forestry, Jalan Gunung Batu No.5 Bogor, 16610, Indonesia

*corresponding author's : moeryanto.mrt@gmail.com

Abstract. Corn *(Zea mays)* is a commodity of food crops that are important after rice in Indonesia. Corn plantation produces waste in the form of corncob. Corncob is lignocellulosic biomass that can converted to cellulose, even sugar or another chemical like ethanol. The first step for utilization of corncob is delignification. The purpose of this study was to isolate and purify the enzyme laccase from white rot fungus *Cerrena* sp.B.Md.T.A.1 that inoculation in GDP media used for biodelignification of corncobs. Enzymatic biodelignification conduct with the variation of substrate was 5%, 10%, and 15% (w/v), while the time variation were 6 hours and 24 hours. The percentage of lignin concentration before pretreatment reached is 37.25%. The highest reduction of lignin content was achieved in 15% substrate for 24 hours incubation time around 42.71%.

1. Introduction

Ethanol is one of a group of chemical compounds from alcohols, that contain a hydroxyl group bonded to a carbon atom. The use of ethanol is very wide such as for solvents, disinfectants, antiseptics, and also to be developed as an alternative fuel. Ethanol can be produced from fermentation of sugar. The substrate of bioethanol can used foodstuffs such as sweet potato [1], cassava [2], and corn[3], that known as first generation bioethanol. But this raw material interferes with food security, another raw material for bioethanol production was lignocellulosic biomass. Lignocellulosic bioethanol or second-generation bioethanol used lignocellulosic biomass that contain lignin, cellulose and hemicellulose. The cellulose in lignocellulosic can converted to sugar and then fermented into ethanol. Some resource of lignocellulosic biomass are usually derived from agricultural waste such as corncob [4–6], Rice Husk [7] and oil palm empty fruit bunch [8,9].

Corn *(Zea mays)* is a commodity of food crops after rice. Corn plantation/industries produce waste in the form of corncob. Corncob is composed of lignin, cellulose, and hemicellulose. Corncob contain lignin 26.47%, hemicellulose 24.39%, and cellulose 31.69% [6]. Lignin is a part of plant cell walls and natural polymer after cellulose. Lignin serves as a water, nutrient, and metabolic carrier in plant cells. Lignin has a complex and heterogeneous structure that binds cellulose and hemicellulose in plant tissues so it is difficult to be degraded, providing a sturdy structure for plants, and providing protection against insects and pathogens [10].

High lignin content causes a lack of cellulose utilization. Lignin inhibit enzyme for degradation cellulose. One way to overcome the levels of lignin in lignocellulose waste is biodelignification. Biodelignification can conduct by extracellular lignoclytic enzyme from white rot fungi of the Basidiomycetes class [11,12]. The extracellular lignolytic enzyme is laccase (LAC), lignin peroxidase (LiP), and manganese peroxidase (MnP) [13]. The use of fungi in delignification has some disadvantages that require a longer time (at least 3 to 7 weeks) than the enzymatic pretreatment that makes it less profitable to use. The advantages of enzyme delignification are shorter processing time, did not require nutrient supplementation and no sugar consumption by fungi [14].

Cerrena sp. is a white rot fungus from a class of Basidiomycetes. These fungi have the ability to degrades lignin. *Cerrena* sp. B. Md. T.A.1. isolated from Mount Rinjani National Park has the potential to produce lignolytic enzymes such as laccase (Lac) and manganese peroxidase (MnP) [15]. The laccase (Lac) has a higher enzyme activity compared to manganese peroxidase (MnP). This paper describes the isolation and purify the enzyme laccase from white rot fungus *Cerrena sp*. B.Md.T.A.1. to perform biodelignification of corncob.

2. Material and Methods

2.1. Materials

The strain used in this research are *Cerrena* sp. B.Md.T.A.1. Corncob was obtained from Indonesian Center for Agricultural Post Harvest R&D, Ministry of Agriculture. The corncob was reduced to the size below 1 cm granule then stored in a plastic bag. Other ingredients used include potato dextrose agar, ABTS (2.2'-Azino-bis(3-ethbenthiazolin-6-sulfonic acid), ammonium sulfate $(NH_4)_2SO_4$, acetate buffer 0.5 M, H_2SO_4 and $CaCO_3$ are for synthesis and analytical grade was purchased from Merck.

2.2. Methods

2.2.1. Cultivation of fungi. Medium for inoculation was made by Potato Dextrose Agar (PDA) containing glucose (20 g L^{-1}), potato (200 g L^{-1}), and agar (15 g L^{-1}). *Cerrena* sp. B.Md.T.A.1. was taken from culture and transferred to a petri dish containing a PDA medium, then incubated at room temperature for 7 days.

2.2.2. Enzyme production and extraction. Cerrena sp. B.Md.T.A.1. from PDA was transferred to 20 ml sterile growth medium containing Potato Dextrose Broth (PDB). The cultivation was incubated for 6 days at 30 ° then filtered and to take filtrate for purification.

2.2.3. Purification of laccase. The purification of laccase was done in the time of maximum laccase activity. The extraction was conducted by grinding and followed by centrifugation at 10000 rpm for 20 min. Crude laccase extract are precipitated by ammonium sulfate salts. Salt is added gradually to crude laccase extract to gain saturation of 0-60%. The solution precipitate overnight at a temperature of 4°c, then centrifuged at 10000 rpm for 30 minutes. The sludge then dissolved into the 0.5 M acetate buffer pH 5.0 (10% w/v) for measure enzyme activity. The supernatants were decanted and the precipitant was dissolved in 0.5 M sodium acetic buffer (pH 5). This solution was stored at 4°C until further use.

2.2.4. Enzyme assay. Laccase activity was determined spectrophotometrically using 1 mM ABTS (2.2 '-Azino-bis (3-ethbenzthiazolin-6-sulfonic acid) in 0.5 M sodium acetate buffer, pH 5 and measuring at 420 nm ($\varepsilon420 = 36{,}000\ mol^{-1}\ cm^{-1}$). One unit of enzyme activity was defined as the amount of enzyme oxidizing 1 nmol ABTS min^{-1}. Enzyme activity was measured based on the following equation:

$$\text{Enzyme activity (U/mL)} = \frac{(A_t - A_0) x V_{tot}(mL) x 10^6}{\varepsilon_{maks}\ x\ d\ x\ V_{enz}(mL) x\ t} \qquad \text{(Eq. 1)}$$

At = Absorbance at T-minute
Ao = absorbance in the 0 minute (value 0 (autozero measurement))
Ɛmax = Absorption of molar tetrametoxy Azobismetilen Quinon (36000 $M^{-1}cm^{-1}$)
d = Cuvette thickness (cm)
Vtot = Total test Volume (mL)
Venz = total added enzyme volume (mL)
t = Incubation time (minutes)

2.2.5. Analysis of corncob component. The lignin content in corncob was determined by NREL methods [16]. Corncob (300 mg, dry weight) was subjected to acid hydrolysis for lignin, cellulose, and hemicellulose content analysis. After hydrolysis, acid-insoluble lignin (AIL) was weighed using Sartorius BS224S, and acid-soluble lignin (ASL) was measured using UV/Vis Spectrophotometer (Optizen 2120 UV) at 205 nm. Total lignin was obtained from the sum of AIL and ASL. Ash content was measure by gravimetric after hydrolysis process. Hydrolizate was filtered use filter paper, residue and filter paper inserted to crucible cup and furnace at 575 ° C for ± 3 hours. After finished the crucible is cooled in the desiccator ± 15 minutes then weighed for measure ash content. Lignin content was measured according to the following equation:

$$\text{AIL } [\%] = \frac{W_{ks} - W_k - A}{W_s} \text{ X } 100\ \% \qquad \text{(Eq. 2)}$$

$$\text{ASL } [\%] = \frac{\left(Abs\ x \frac{df}{110}\right) x \left(\frac{87}{1000}\right)}{W_s\ [g]} \text{ x } 100\% \qquad \text{(Eq. 3)}$$

AIL : Acid Insoluble Lignin [%]
ASL : Acid Soluble Lignin [%]
W_s : Dry sample weight (without moisture) [g]
W_k : Filter paper weight [g]
W_{ks} : Filter paper weight and residue dried [g]
A : Ash weight [g]
Abs : Absorbance
df : Dilution Factor

2.2.6. Biodelignification of corncob. Corncob material was treated by extracted laccase from *Cerrena* sp. B.Md.T.A.1 in solid substrate culture. The substrate in variation 5%, 10% and 15% (w/v) was added by extracted laccase in the total volume 40 ml. Biodelignification was conducted in shaker incubator at 40°C, 100 rpm for 6 and 24 hours. Controls are done under the same conditions without using laccase. All experiments are made 2 times repeated. After treatment, the corncob carried out for measure lignin content after biodelignification.

3. Result and Discussion

3.1. Activity of enzyme laccase

Enzyme purification is an important step prior characterization. Purification was carried out by the salting out method using ammonium sulfate salt against the extract crude enzyme. The principle of ammonium sulfate precipitation is based on water-binding competition between salt and protein. Ammonium sulfate salt can separate enzymes from other proteins that did not precipitated during centrifugation. The advantage of using ammonium sulfate is that it has a high solubility of around 840 g/liter and exothermic process. Dissolved ammonium sulfate will ionize into an $NH_4{}^{+}$and $SO_4{}^{2-}$ ions. This salt ion will attract water that binds to protein, so that protein solubility decreases then forms an aggregate and settles [17].

Ammonium sulfate concentrations are added to get the most diverse enzyme specific activities for each microorganism [18]. According to research conducted by Hidayat [15] the ammonium sulphate deposition fraction 40-60% produces the highest laccase activity. Therefore, the ammonium sulfate used in this study was 0-60%.

The measurement of laccase activity from fractionation was determined using ABTS. The change reaction rate in ABTS substrate into ABTS radical cation becomes a determinant of laccase activity in the optimum point. The measurement time interval on the laccase enzyme occurring at 0-15 minutes to reach the optimum point, exceeding that time will decrease the acceleration of the laccase reaction [15,19]. The results of laccase activity in this research was obtained at 0-15-minute intervals reached 242.76 U/mL. This result was slightly lower than Hidayat's result in previous research that obtained 280 U/mL [15].

3.2. Lignin content of corncob
Corncob is one of the lignocellulosic biomass, that contain high lignin content. Lignin content in this study was 37.25%, while ash content was 1.54% (table 1). The lignin content in corncob was lower than rice straw and wheat straw, but higher than empty fruit bunch and sugarcane bagasse [20]. Due to the height lignin content, the delignification must be carried out prior utilize corncobs. The process of biodelignification (enzymatic pretreatment) is by mixing corn cobs with the enzyme laccase to degrade the lignin content in corncobs [21]. The ability of an enzyme in hydrolyzing lignocellulosic material is influenced by several factors, such as lignin content, hemicellulose, and the level of crystallization of cellulose [22].

Table 1. Content of lignin and ash corncob.

Biodelignification	Variations		Lignin Content	Ash Content
	t (hour)	% substrate [w/v]	[%]	[%]
untreated	-	-	37.25	1.54
pretreated	6 hours	5%	24.53	1.39
		10%	23.40	0.80
		15%	22.47	0.97
		5%	23.08	0.92
	24 hours	10%	21.89	0.66
		15%	21.34	0.94

3.3. Biodelignification of corncob
Biodelignification is one of the environmentally friendly lignin degradation techniques by microorganism such as fungi. The use of the enzyme laccase from the fungus *Cerrena* sp. B.Md.TA.1 aims to reduce the incubation time, thereby speeding up the biodelignification process. Laccase enzymes play an important role in lignin degradation. Laccase enzymes can induce oxidation reactions, producing unstable reactive phenoxy radicals that can lead to polymerization and depolymerization reactions [23].

Figure 1 showed the biodelignification result in corncobs using laccase from *Cerrena* sp. B.Md.T.A.1. The result showed that Laccase could reduce lignin content in corncob. The longer the processing time the higher the delignification. This is caused by the optimal work of the laccase enzyme in degrade lignin. The highest delignification was obtained on a substrate of 15% (w/v) for 24 hours incubation time reached 42.71%. Similarly, the increasing percentage of substrate content the higher the biodelignification that occurs.

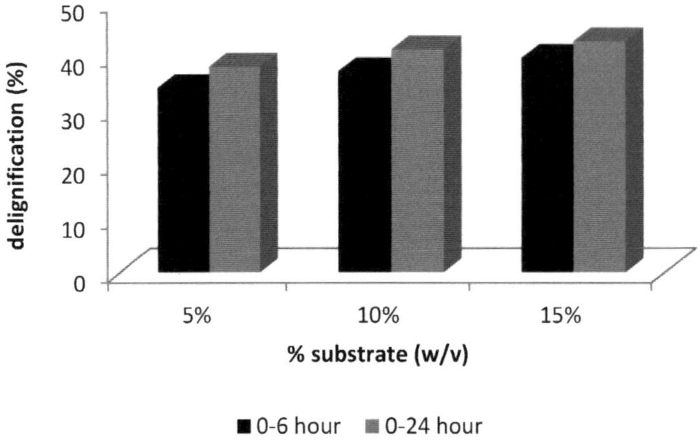

Figure 1. Biodelignification of corncob.

Delignification produced by the laccase is still smaller than delignification using chemicals. In our previous studies result, corn cobs after the delignification process using NaOH achieved a lignin reduction of 60.22%, but the biodelignification result closely to pretreatment process using spent black liquor [6]. Although enzymatic biodelignification results is lower than chemical delignification, this process didn't produce another waste after process.

4. Conclusion

The activity of laccase obtained from isolation and purification using ammonium sulfate with 0-60% saturation reached 242.76 U/mL. The laccase from *Cerrena* sp. B.Md.T.A.1 can be used for enzymatic biodelignification of corncob. The laccase *Cerrena* sp. B.Md.T.A.1 used in delignification of 15% corn cobs for 24 hours can reduce the lignin content up to 42.71%. This process is promising for the process of removing lignin in lignocellulosic biomass without chemical and environmentally friendly.

Acknowledgments

The authors gratefully acknowledge to RC Chemistry Indonesian Institute of Sciences, and Forest Research and Development Centre, the Ministry of Environment and Forestry for support this research.

References

[1] Wang X, Tian S, Lou H and Zhao R 2020 A reliable method for predicting bioethanol yield of different varieties of sweet potato by dry matter content *Grain Oil Sci. Technol.* **3** 110-116
[2] Hanif M, Mahlia T M I, Aditiya H B, Chong W T and Nasruddin 2016 Techno-economic and environmental assessment of bioethanol production from high starch and root yield Sri Kanji 1 cassava in Malaysia *Energy Reports* **2** 246–53
[3] Mojović L, Nikolić S, Rakin M and Vukasinović M 2006 Production of bioethanol from corn meal hydrolyzates *Fuel* **85** 1750–5
[4] Zheng T, Yu H, Liu S, Jiang J and Wang K 2020 Achieving high ethanol yield by co-feeding corncob residues and tea-seed cake at high-solids simultaneous saccharification and fermentation *Renew. Energy* **145** 858–66
[5] Apiraksakorn J, Sukklang S and Boonmee M 2017 The evaluation of bio-energy produced from ethanol fermentation using corncob dust hydrolysate *Energy Procedia* **138** 145–50

[6] Muryanto M, Handayani E M and Irawan Y 2020 Black Liquor Pretreatment of Concorb and Enzymatic Hydrolysis for Lignocellulosic Bioethanol *IOP Conf. Ser. Mater. Sci. Eng.* **854** 12006
[7] Ebrahimi M, Caparanga A R, Ordono E E, Villaflores O B and Pouriman M 2017 Effect of ammonium carbonate pretreatment on the enzymatic digestibility, structural characteristics of rice husk and bioethanol production via simultaneous saccharification and fermentation process with Saccharomyces cerevisiae Hansen 2055 *Ind. Crops Prod.* **101** 84–91
[8] Muryanto, Triwahyuni E, Abimayu H, Cahyono A, Cahyono E T, Sudiyani Y 2015 Alkaline delignification of oil palm empty fruit bunch using black liquor from pretreatment *Procedia Chemistry* **16** 99–105
[9] Triwahyuni E, Muryanto, Sudiyani Y and Abimanyu H 2015 The effect of substrate loading on simultaneous saccharification and fermentation process for bioethanol production from oil palm empty fruit bunches *Energy Procedia* **68** 138–146
[10] Ahmed Z, Banu H, Rahman M M, Akhter F and M. Shamsul Haque 2001 Microbial Activity on the Degradation of Lignocellulosic Polysaccharides *J. Biol. Sci.* **1** 993–7
[11] Valadares F, Gonçalves T A, Damasio A, Milagres A M F, Squina F M, Segato F and Ferraz A 2019 The secretome of two representative lignocellulose-decay basidiomycetes growing on sugarcane bagasse solid-state cultures *Enzyme Microb. Technol.* **130** 109370
[12] Machado A da S and Ferraz A 2017 Biological pretreatment of sugarcane bagasse with basidiomycetes producing varied patterns of biodegradation *Bioresour. Technol.* **225** 17–22
[13] Shin S K, Ko Y J, Hyeon J E and Han S O 2019 Studies of advanced lignin valorization based on various types of lignolytic enzymes and microbes *Bioresour. Technol.* **289** 121728
[14] Liguori R and Faraco V 2016 Biological processes for advancing lignocellulosic waste biorefinery by advocating circular economy *Bioresour. Technol.* **215** 13–20
[15] Hidayat A, Ningrum A C and Falah S 2019 Decolorization of anthraquinone, azo and triphenylmethane dyes by laccase from newly isolated fungus, Cerrena sp. BMD.TA.1 *IOP Conf. Ser. Earth Environ. Sci.* **308** 12019
[16] Raymond R and Tina E 1996 *Determination of Carbohydrates in Biomass by High Performance Liquid Chromatography LAP-002* vol LAP-002
[17] Sinatari H M The use of thermophilic compost cellulase in Bayat Klaten Village uses ammonium sulfate fractionation *Chem Info Journal; Vol 1, No 1*
[18] Zimmerman J 2002 Protein purification techniques (2nd ed.), Roe, S. (ed.); and protein purification applications (2nd ed.), Roe, S. (ed.) *Biochem. Mol. Biol. Educ.* **30** 72
[19] Falah S, Sari N M and Hidayat A 2018 Decolorization of remazol Brilliant blue R by laccase of newly isolated Leiotrametes flavida strain ZUL62 from Bangka heath forest, Indonesia *Biodiversitas* **19** 588–94
[20] Baruah J, Nath B K, Sharma R, Kumar S, Deka R C, Baruah D C and Kalita E 2018 Recent Trends in the Pretreatment of Lignocellulosic Biomass for Value-Added Products *Front. Energy Res.* **6** 141
[21] Yesiladalı S K, Pekin G, Bermek H, Arslan-Alaton İ, Orhon D and Tamerler C 2006 Bioremediation of Textile Azo Dyes by Trichophyton rubrum LSK-27 *World J. Microbiol. Biotechnol.* **22** 1027–31
[22] Robak K and Balcerek M 2018 Review of Second Generation Bioethanol Production from Residual Biomass *Food Technol. Biotechnol.* **56** 174–87
[23] Ahmad M, Taylor C R, Pink D, Burton K, Eastwood D, Bending G D and Bugg T D H 2010 Development of novel assays for lignin degradation: comparative analysis of bacterial and fungal lignin degraders *Mol. Biosyst.* **6** 815–21

IOP Conf. Series: Materials Science and Engineering 1011 (2021) 012031 doi:10.1088/1757-899X/1011/1/012031

Alternative Use of Modified Cassava Flour (Mocaf) as Carrier of Natural Folic Acid in Infant Biscuits for Complementary Feeding

Agustine Susilowati*, Aspiyanto, Hakiki Melanie, Yati Maryati, Puspa Dewi Lotulung, Anastasia Fitria Devi, and Setyani Budiari

Research Centre for Chemistry, Indonesian Institute of Scieces, Building 452, Kawasan PUSPIPTEK, Serpong, Tangerang Selatan, Banten 15314, Indonesia

*E-mail : agustine_1408@yahoo.co.id

Abstract. Modified cassava flour (mocaf) is an alternative food ingredient used in preparation of biscuits as a binder and a carrier of folic acid fortificant. This work aimed to evaluate the mocaf addition with natural folic acid fortificants in two infant biscuit formulations. Mocaf was added at concentration of 0, 0.3, 0.6, 1.2, and 2.4% (w/w, control formula of infant biscuit) with folic acid fortificant A (mixture of nixtamalized corn, soy tempeh and fermented broccoli) and B (mixture of nixtamalized corn, mung beans tempeh and fermented broccoli). The optimum condition was achieved at mocaf concentration of 0.3% and resulted in infant biscuits A and B with folic acid content of 212.23 μg/mL and 438.74 μg/mL, dissolved protein 27.41 mg/mL and 28.88 mg/mL, total sugars of 358.20 mg/mL and 322.72 mg/mL, reducing sugars of 35.10 mg/mL and 39.78 mg/mL, total solids of 94.82% and 92.43%, respectively. The formulations of biscuits A and B increased folic acid content by 80.05% and 199.55% compared to the counterpart without mocaf addition. The predominant folic acid monomers had molecular weight of 442 m/z. Volatile compounds were identified as acetic acid, furan, and fatty acids. Biscuits A and B displayed average particle size of 69.07 μm and 69.78 μm with hardness of 8.44 N and 16.94 N, and lightness of 35.68 and 36.86, respectively.

1. Introduction

Infant biscuits with folic acid fortificant have potential for complementary feeding (CF). Infant biscuits for CF are made to adapt the digestive system of infants. These biscuits are meant to complement breast milk or infant formula, so thus infants can learn to hold and bite their foods. Infant biscuits are designed for easy dilution into porridge. These biscuits are usually weighed 20 g per piece [1].

Folic acid is necessary for proper growth, particularly during pregnancy, so thus the baby can produce adequate red blood cells and prevent Neural Tube Defects (NTDs) [2], whilst the expecting mother can avoid anemia [3]. Folic acid (L-glutamic acid, pteroyl-L-glutamic acid), also known as Vitamin B9, consists of three parts: pteridine heterocyclic, para-aminobenzoate (p-ABA) and L-glutamic acid [4]. Folic acid is sensitive to oxidative degradation, which is triggered by light, high temperature, and mechanical treatment [5].

Mocaf, added in the formulation of infant biscuit to increase crispiness, can also act as a carrier or a binder of folic acid. Mocaf is produced through fermentation of thin cassava slices by lactic acid bacteria (LAB), followed by drying and grinding. Mocaf addition is a potential gelling agent since it increases viscosity and imparts gelation, with good hydration power [6]. The potential role of mocaf as a binder and a carrier of bioactives in biscuits has not been explored yet. Mocaf is expected to be a good binder for folic acid due to smaller size of starch granule.

The objective of this experiment study was to determine the optimum mocaf concentration as a binder and a carrier of folic acid in infant biscuits, as observed through analysis of folic acid monomer. Physicochemical properties of the resultant biscuits were also analysed.

2. Materials and Methods

2.1. Materials and equipments

Materials used in this experimental work were mocaf (a gift from Research Center for Biotechnology, LIPI), dried yellow corn (*Zea mays* var. indentata) from a local corn plantation (Tangerang Selatan), soy beans, mung beans, broccoli, sucrose salt, whole milk powder, vanilla extract, unsalted butter, baking powder, wheat flour, and corn starch were purchased from local market. *Rhizopus oligosporus* strain C_1 and Kombucha culture were the collection of Research Center for Chemistry, LIPI.

Main equipments implemented in this experimental activity were analytical balance (Fujitsu, Japan), autoclave (CHENG YI, LS – 50 L, China), homogenizer (Ultra-Turrax, Ika Labortechnik, T50, Jane & Kunkel, Germany), blender (local, National), container, incubator (local), electrical oven, and cabinet dryer (local). Instruments for analysis were UV-Vis Spectrophotometer (Model RF-550, Shimadzu, Japan), Liquid Chromatography-tandem Mass Spectrometry (LC-MS) (Mariner Biospectrometry) with LC (Hitachi L 6200), Particle Size Analyzer (Horiba Laser Scattering Particle Size Distribution Analyzer LA-960), GC-MS (Shimadzu, Japan) and Texture Analysis (TMS-Pro, Stable Micro System, USA).

2.2. Experimental Design

Mocaf was added to the control formula of infant bicuits at concentration of 0, 0.3, 0.6, 1.2, and 2.4% (w/w). The biscuits contained either fortificant A or B which was equivalent to 1,000 μg of folic. Fortificant A was a mixture of nixtamalized yellow corn, soy tempeh, and fermented broccoli. Fortificant B consisted of the same ingredients except the use of mung beans instead of soy tempeh.

The infant biscuits were prepared according to Indonesian National Standard (SNI) [20] for infant foods. The requirement of folic acid is 20% from the whole vitamin requirement according to recommended daily intake (RDA) or approximately 400 μg. The weight of one piece of biscuit ranged from 20 to 25 gr, with equivalent energy of 20 kcal. The biscuits were predicted to contain folic acid of 100– 50 μg per piece of biscuit.

Analyses were performed on both fortificants and resultant infant biscuits included total solid (gravimetric method), dissolved protein (spectrophotometry) [7], total sugar (phenol sulphate method), reducing sugars (Somogyi Nelson method) [8], and folic acid (spectrophotometry) contents [9]. Identification on folic acid monomer was conducted using LCMS (Mariner Biospectrometry) tandem with LC (Hitachi L 6200) [10]. Other analyses included particle size distribution (Horiba LS-PSA LA-960) [11], volatile compounds (Shimadzu, Japan) [12], hardness (TMS Pro Texture Analyser), and colour measurement (Konika Minolta CR-410) [13].

2.3. Procedure

Folic acid fortificants were prepared by mixing either soy bean (fortificant A) or mung bean (fortificant B), fermented broccoli and nixtamalized yellow corn at mass ratio of 1:1:1, followed by homogenation, drying, size reduction, and sieving through 60-80 mesh until fine powder of fortificant A and B were obtained [14].

The biscuits were prepared from wheat flour (33.62%), corn starch (11.2%), sucrose (15.17%), vanilla (0.79%), salt (0.79%), whole milk powder (9.86%), unsalted butter (1.86%), baking powder (0.86%), and water (11.73%) (w/w, control formula). Fortificant was added at concentration equivalent to 1,000 μg of folic acid per 100 g of control formula. Mocaf was added at concentrations of 0, 0.3, 0.6, 1.2, and 2.4 % (w/w, control formula).

Butter, sucrose, and milk powder was beaten using an electrical mixer until fluffy. Fortificants A or B, salt, vanilla, wheat flour, corn starch, baking soda, and water were added and gently mixed with spatula. The dough was then moulded and baked in oven at 150 °C for 15–20 minutes. The biscuits were subsequently cooled and packed.

3. Results and Discussions

3.1. Characteristics of materials

Fortificant B (130.54 μg/mL) had higher content of folic acid than fortificant A (101.89 μg/mL) (Table 1). Fortificant B also showed higher total sugars, dissolved protein, and total solids. It, however, contained lower reducing sugar than fortificant A. Higher folic acid content in fortificant A was possibly due to the use of soy tempeh which contained 503.60 μg/mL of folic acid. While fortificant B used mung beans tempeh that contained lower folic acid (381,5 μg/mL) [15].

During the fermentation of either soy or mung beans by *Rhizophus oligosporus* strain C1, the protein degraded to amino acids, particularly glutamic acid, which was a component for folic acid formation. The results of protease activity on the soy or mung bean substrates was expressed as dissolved protein [14]. Meanwhile, broccoli was fermented by Kombucha culture using sucrose as source of carbon in order to produce folic acid through synthesis of de novo [16, 17]. The remaining reducing sugars achieved from this synthese was soluble in biomass as a source of energy and contributed to the taste of the product.

Nixtamalised yellow corn not only gave a better binding of folic acid, but also acted as a source of carbohydrate and imparted texture [18]. Folic acid is sensitive on mechanical treatment, light and temperature [5], and processed foods might experience reduction of folic acid. Therefore, folic acid fortification will help to supply adequate folic acid in the final products. The control infant biscuit already showed folic acid content of 109.09 μg/mL, which obtained from the baking ingredients i.e., wheat flour, corn starch, sugars, powder milk, and unsalted butter

Table 1. The properties of fortificants A and B, control biscuits, and mocaf

Materials	Components				
	Folic acid (μg/mL)	Total solids (%)	Dissolved protein (mg/mL)	Reducing sugars (mg/mL)	Total sugars (mg/mL)
Fortificant A*	101.89	92.91	5.80	28.00	78.28
Fortificant B**	130.54	93.07	6.00	19.14	89.65
Control biscuit	109.09	69.17	12.30	30.48	143.96
Mocaf	22.26	96.13	1.17	5.50	96.13

* A, mixture of nixtamalized yellow corn, soy tempeh and fermented broccoli
** B, mixture of nixtamalized yellow corn and mung beans tempeh and fermented broccoli

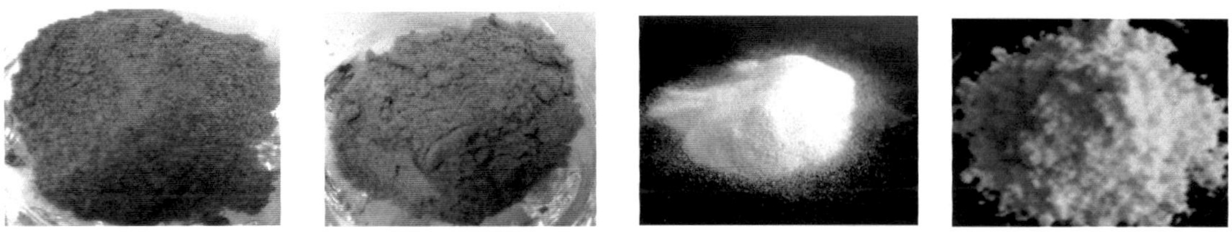

(a) (b) (c) (d)

Figure 1. (a) Fortificant A, (b) fortificant B, (c) control formula of infant biscuits, (d) mocaf.

3.2. Effect of formulation on infant biscuits

3.2.1. Folic acid (µg/mL) and dissolved protein (mg/mL) contents. The addition of mocaf to the formulation of either biscuit A or B, in general, increased the folic acid content. The increase of folic acid content, however, was not linearly correlated with the increase of mocaf addition. Biscuit B displayed higher content of folic acid compared to biscuit A across percentages of mocaf additions (Figure 1a). This was due to higher concentration of folic acid in fortificant B (130.54 µg/mL) than that of fortificant A (101.89 µg/mL). Besides, the interaction of mocaf and other ingredients in each biscuit might be different. Mocaf in biscuit B was possibly better in binding folic acid and thus the folic acid could handle the baking process better. Therefore, more folic acid remained in biscuit B after the exposure to baking temperature (150 °C for 20 minutes).

Mocaf addition led to increase of available carbohydrate for Maillard reaction. At higher mocaf concentration, the Maillard reaction [19] between sugars (carbohydrate) and protein was accelerated, and thus more melanoidin was produced. This might also contribute to a drop of final folic acid content in the biscuit at higher mocaf concentrations. The highest folic acid content was achieved at mocaf addition of 0.3% for both biscuits, which resulted in final folic acid content of 212.23 µg/g and 438.74 µg/g for biscuit A and B, respectively. Compared to the control biscuits (biscuits with folic acid fortification at 1,000 µg/g and without mocaf addition), the mocaf addition was able to increase folic acid content in biscuit by 80.05% and 199.55%.

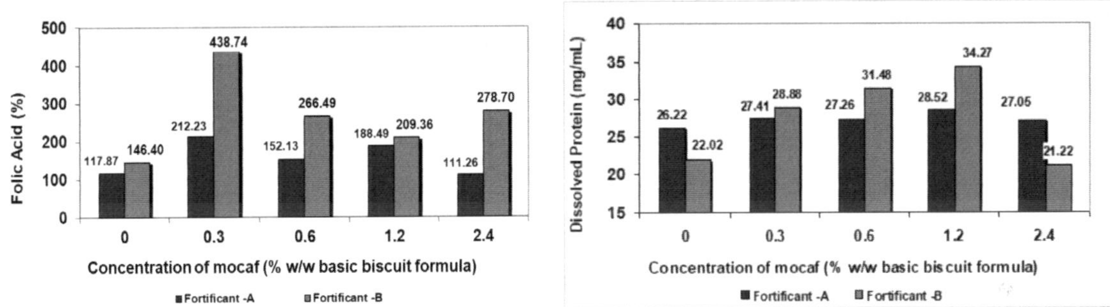

(a) (b)

Figure 2. Relationship between type of fortificant and mocaf concentration on recovery (a) folic acid and (b) dissolved protein from infant biscuit as source of folic acid.

Increasing mocaf concentration in biscuits also resulted in fluctuative values of dissolved protein. Dissolved protein is the whole proteins and its derivate, particularly amino acids and folic acid structured by glutamic acid, pteridine heterocyclic and para-aminobenzoate (p-ABA) [4]. The highest values of dissolved protein in biscuits A and B were 28.52 mg/mL and 34.27 mg/mL respectively, which were both reached at mocaf concentration of 1.2%. Compared to the control positive biscuits, the mocaf addition at 1.2% increased dissolved protein by 8.77% and 55.63%. The initial dissolved protein concentration in fortificants A and B (5.8 and 6.0 mg/mL, respectively) might contributed to the different dissolve protein contents of the final biscuits. In addition, the rate of Maillard reaction and protein denaturation in each biscuit was possibly different due to different fortificant ingredients although the mocaf addition was the same. However, both biscuits experienced a decrease of dissolved protein content at mocaf concentration of 2.4 mg/mL. The Maillard reaction at this mocaf concentration was possibly too fast and resulted in more melanoidin compounds that were undetected by the Lowry method [7].

3.2.2. Total sugars (mg/mL), reducing sugars (mg/mL) and total solids (%). Mocaf addition, in general, increased the total sugars of biscuit A. The increase of total sugars in biscuit B, however, did not follow a smooth pattern (Figure 2a). The highest total sugars of biscuits A (408.99 mg/mL) and B (429.97 mg/mL) were achieved at mocaf concentration of 2.4 and 0.6 % (w/w), respectively. Total sugars of biscuit A was most of the time higher than that of biscuit B across percentage of mocaf addition, except at 0.6% mocaf addition. The initial fortificant analysis showed that fortificant A had a lower (78.28 mg/mL) total sugars compared to fortificant B (89.65 mg/mL). This might explain the reason beyond a lower total sugar of biscuit A at 0.6% mocaf addition. More available sugars in the formulation of biscuit B implied more intensive Maillard reaction took place in biscuit B compared to that of biscuit A. Therefore, less sugars remained in biscuit B for most of mocaf concentrations as revealed by the phenol sulphate method [8].

The reducing sugars of both biscuits, in general, increased with increased mocaf concentrations. The highest amounts of reducing sugars for biscuits A and B was 42.67 mg/mL and 49.94 mg/mL, respectively, at mocaf concentration of 2.4% (Figure 2b). Compared to the control biscuits, the mocaf addition increased the reducing sugars in biscuit A and B up to 111.87% and 66.46%, respectively.

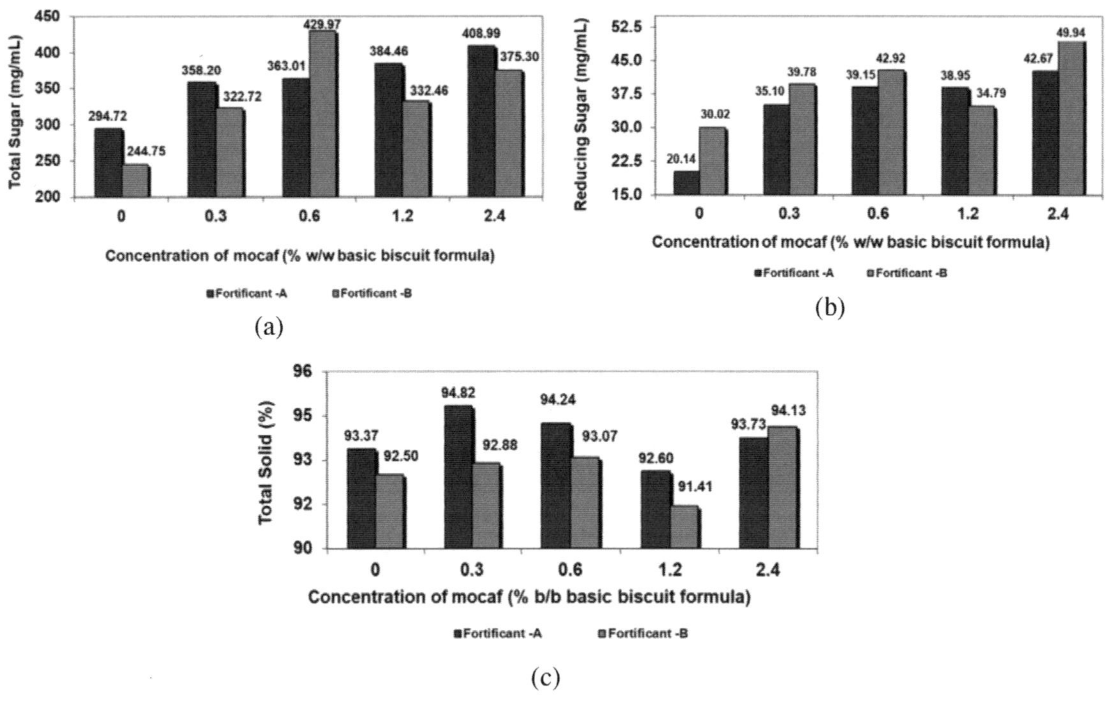

Figure 3. Relationship between type of fortificants and mocaf concentrations on (a) total sugars, (b) reducing sugars, and (c) total solids of infant biscuit as source of folic acid.

The increasing percentages of mocaf addition in both biscuits did not give a clear pattern. Particularly at mocaf concentration of 1.2% which gave a lower total solid content for both biscuits compared to the control biscuits. Interaction between ingredients during baking process might lead to reduction of total solids due to water evaporation and shrinkage of materials. The highest total solids of biscuits A and B were 94.82% and 94.13%, respectively. These values were achieved at different mocaf concentrations. At 0.3%, mocaf displayed its best interaction with fortificant A and other ingredients. This left unbound water which later evaporated during baking. For biscuit B, the highest total solid content was achieved at 2.45. This indicated that the interaction between mocaf, fortificant, and other ingredient in each biscuit was unique.

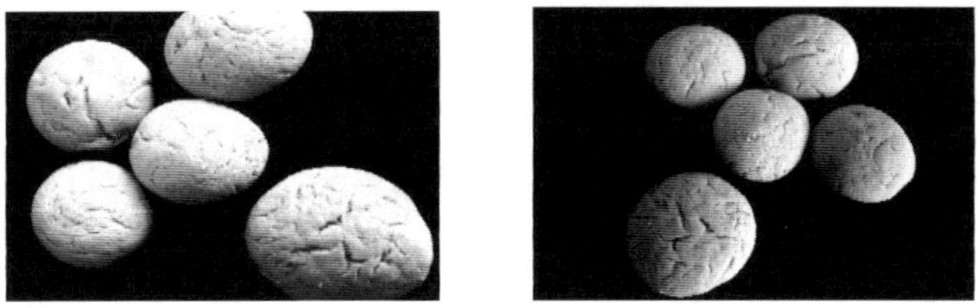

(a) (b)

Figure 4. (a) Infant biscuit A and (b) infant biscuit B by adding mocaf 0.3% (w/w, control formula of biscuit).

3.2.3. Characteristic of Volatile Compounds for Infant Biscuit. Volatile analysis was performed on biscuits A and B which were added with 0.3% (w/w) since they displayed the best folic acid recovery. Chromatogram of biscuit A (Figure 3a) shows 25 peaks which were dominated (46.1%) by acetic acid and ethyl/methyl ester acetic acid. Other peaks represented furan and furan derivatives (18.64%) e.g., furfural, ethylhydroxylamine, furanmethanol, furanone, hydroxymethylfurfural and rhamnose. Fatty acids (11.93%) consisted of undecanoic acid, tridecanoic acid, pentadecanoic acid, hexadecanoic acid, 9-octadecenoic acid, methyl stearate, and butanoic acid. Other methyl ester (9.68%) were mercaptoacetone, cyclopentanol, 2-propanamine, benzaldehyde, n-hexane and oxirane.

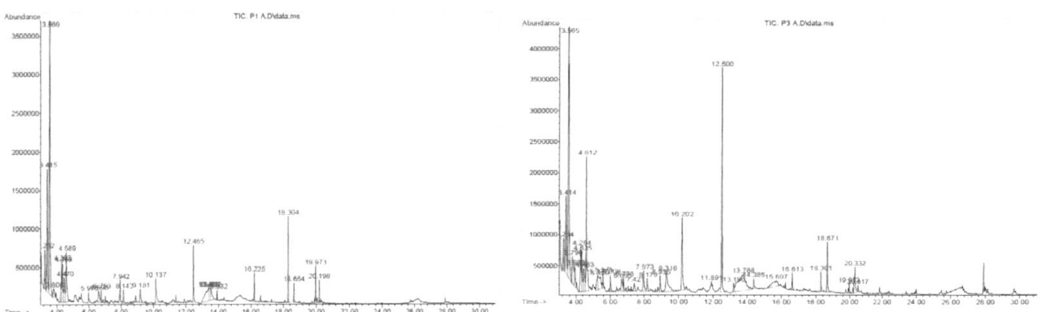

(a) (b)

Figure 5. Volatile compounds in (a) infant biscuit of A and (b) infant biscuit of B at mocaf concentration 0.3% (w/w, control formula of infant biscuit).

Chromatogram of biscuit B (Figure 3b) displays 33 peaks that were dominated (37.12%) by acetic acid and ethyl/methyl ester acetic acid. Furan and its derivatives (29.77%) consisted of furfural, 2-furan methanol, 4H-pyran-4-one, (3H)-furanone, and 5-hydroxymethylfurfural. Fatty acids (7.8%) comprised of methyl ester, dodecanoic acid, tetradecanoic acid, hexadecanoic acid, and 10-octadecenoic acid. Alcohol and its derivates (15.35%) were dihydroxyacetone, 2-Cyclopenten-1-one, mercaptoacetone, 2-propen-1-ol, 2-butanone, etraethoxyethylene, benzaldehyde, cyclopropyl carbinol, and trihydroxy-2-butoxy.

3.2.4. Identification of folic acid monomer. Folic acid has MW of 441 Da. The LCMS analysis of folic acid standard showed the retention time (RT) was 3.27 min (Figure 4a) and MW of 441.37 (Figure 4b). The chromatogram of biscuit A had RT of 1.44 min (Figure 4c). The spectra (Figure 4d) demonstrated six folic acid monomers with MW of 441.12–441.83 Da, dominated by folic acid monomer with MW of 441.83 Da. The LCMS analysis of biscuit B showed a RT of 2.12 min (Figure

IOP Conf. Series: Materials Science and Engineering 1011 (2021) 012031 doi:10.1088/1757-899X/1011/1/012031

4e) with eight folic acid monomers with MW 440.90–441.98 Da, dominated by three folic acid monomers with MW of 441.09, 441.44, and 441.59 Da (Figure 4f).

Qualitative comparison between chromatograms revealed that biscuit B had more folic acid monomers than biscuit A. This was most likely due to differences of fortificants and the further interactions between fortificants and other biscuit ingredients during baking.

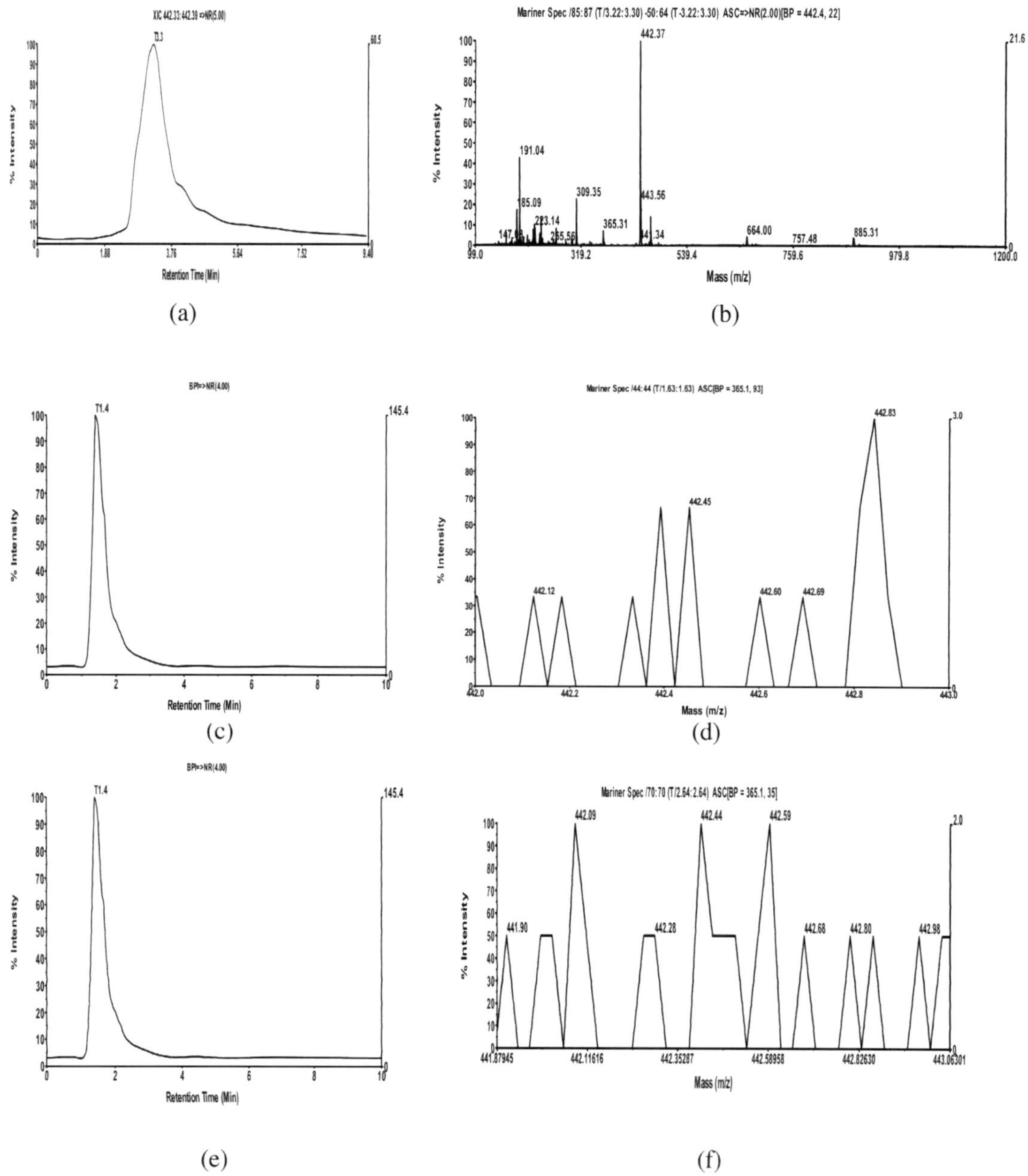

Figure 6. Chromatograms and spectra of standard folic acid (a and b), biscuit A (c and d), and biscuit B (e and f). The selected biscuits A and B were prepared with 0.3% (w/w) mocaf addition.

3.2.5. Particle size and particle size distribution for infant biscuits. The particles analysis of biscuit A showed the D10, D50, and D90 was 8.32, 25.81, and 195.34 μm, respectively (Figure 5a) with mean diameter size of 69.07 μm. Meanwhile the D10, D50, and D90 of biscuit B was 8.33, 26.37, and 196.38 μm, with mean diameter size of 69.78 μm. In general, biscuit B demonstrated larger particle size than biscuit A. This matter is possibility caused by difference in fortificant and interaction between formula and baking process, which resulted in different particle agglomeration.

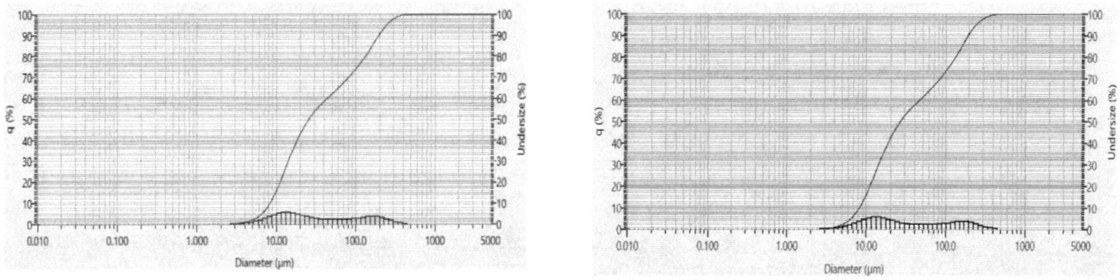

(a) (b)

Figure 7. Particle size distribution of biscuit A (a) and biscuit B (b), both were prepared with 0.3% (w/w) mocaf addition.

3.2.6. Identification of crispiness and sharpeness of biscuit. Both hardness and lightness analysis were performed on biscuits with 0.3% (w/w) mocaf addition. Biscuit B showed higher hardness (16.94 N) compared to biscuit A (8.44 N) eventhough total solids of biscuit A (94.82%) was higher compared with total solids of biscuit B (92.88%). The hardness of biscuit B which was two fold that of biscuit B was possibly due to the nature of fortificant B. Fortificant B was more sticky than fortificant A. Thus, the dough of biscuit B was harder.

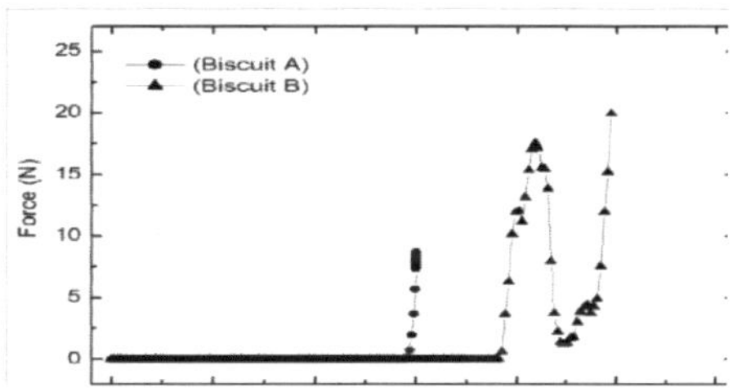

Figure 8. Level of hardness from infant biscuit of A and infant biscuit of B at mocaf concentration 0.3% (w/w, control formula of biscuit).

The colour analysis of both biscuits are presented as Table 2. Biscuit B showed lower ΔE (14.95±0.32) than that of biscuit A (25.31±0.61). Value of ΔE demonstrated how different the colour of material from standard white tile. Both biscuits were very much different from standard white tile, as diplayed by their ΔE that exceeded 5.

Table 2. The three stimulus of colour parameters of biscuits A and B and their corresponding ΔE against standard white tile.

Type of biscuit	*L**	*a**	*b**	*ΔE*
Biscuit A	65.45 ± 1.09	11.51 ± 0.26	35.68 ± 0.68	25.31 ± 0.61
Biscuit B	81.18 ± 2.52	5.88 ± 0.59	36.86 ± 0.55	14.95 ± 0.32

The bigger *ΔE* value of biscuit A indicated a faster Maillard reaction that resulted in melanoidin and other advanced glycation end products [19]. This was also reflected by green-red coordinate (*a*) in Table 2; the *a* value of biscuit A was higher than that of biscuit B. Although biscuit B had higher dissolved protein and reducing sugars than the counterpart biscuit A, the total sugars of biscuit B (322.72 mg/mL) was lower compared to biscuit A (358.20 mg/mL). This possibly lead to a lower rate of Maillard reaction in biscuit B, and thus generated a higher lightness.

4. Conclusions

Mocaf has potential use as folic acid binder which also impart texture in infant biscuits. The optimum mocaf addition that resulted in the best folic acid recovery was achieved at 0.3% (w/w) for both biscuits A and B. These biscuits were prepared to result in 1,000 μg/g folic acid with different fortificants, namely fortificant A and B. The formulations produced infant biscuits A and B with folic acid of 212.23 and 438.74 μg/mL, dissolved proteins of 27.41 and 28.88 mg/mL, total sugars of 358.20 and 322.72 mg/mL, reducing sugars of 35.10 and 39.78 mg/mL, and total solids of 94.82 and 92.43%, respectively. Compared to control biscuits (biscuits with 1,000 μg/g folic acid without mocaf addition), the 0.3% mocaf addition was able to recover folic acid in biscuits A and B as much as 80.05% and 199.55%, respectively.

References

[1] Aini QN and Wirawani Y 2013 Kontribusi MP-ASI Biskuit Substitusi Tepung Garut, Kedelai dan Ubi Jalar Kuning Terhadap Kecukupan Protein, Vitamin A, Kalsium dan Zink Pada Bayi. *Journal of Nutrition College*. 2(4) : 458-66.

[2] Berry J R, Li Z, Erickson J D, Li S, Moore CA, Wang H 1999 Prevention of neural tube defects with folic acid in China. *N. Engl. J Med*. 341(20) :1485-90

[3] McKenzie S B 1999 *Megaloblastic and Nonmegaloblastic Macrocytic Anemias Hematology* (Baltimore: Williams adn Wilkins) pp 179-99

[4] Quilivan E P, Roje S, Basset G, Shahar-Hill Y, Gregory J F, and Hanson A D 2003 The Folate precursor ϙ-aminobenzoate is reversible converted to is glucose ester in the plant cytosol. *J. Bio. Chem*. 278 : 20731-7.

[5] Delchier N, Ringling C, Maingonnat J F, Rychlik M, Renard C M G C. 2014. Mechanisms of folate losses during processing : diffusion vs. heat degradation. *Food Chem*. 157 *:* 439 – 47. Accessed at 25 April 2017.

[6] Rahayu E S 2010 *Lactic Acid Bacteria and Their Role in Food and Health: Current Research In Indonesia*. Skripsi Sarjana (Yogyakarta : UGM)

[7] Lowry O H, Rosebrough N J, Farr A L and Randall R J 1951 Protein measurement with Folin phenol reagent. *J Biol. Chem*. 193(1) : 265-276

[8] AOAC 1995 *Official Methods of Analysis of Association of Official Analytical Chemists* (Washington DC : AOAC)

[9] Ruengsitagoon W and Hattanat N 2012 *Simple Spectrophotometric Method for Determination of Folic Acid*. Conf. The 4th Annual Northeast Pharmacy Research (Thailand : Khon Kaen University)

[10] Eichhorn P and Knepper T P 2001 Electrospray ionization mass spectrometric studies on the amphoteric surfactant cocamidopropylbetaine. *J Mass Spectrom*. 36(6) : 677-84

[11] Backman 2018 *Laser diffraction measurement for non sperical particles Backman Coulter Life Sciences* (USA: Backman Coulter Inc)

[12] Hussain S Z and Maqbol K 2014 GC-MS: Principle, technique and its application in food science *Int J Curr Sci* 13 : 116-26.

[13] Pathare PB, Opara UL, and Al-Said FA 2013 Colour measurement and analysis in fresh and processed foods: a review. *Food and Bioprocress Techno*. 6(1) : 36-60.

[14] Susilowati A, Iskandar Y M, Melanie H, Maryati Y, Lotulung P D, and Aryani D G 2015 Pengembangan Konsentrat Sayuran Hijau dan Kacang-Kacangan Terfermentasi pada Jagung

IOP Conf. Series: Materials Science and Engineering 1011 (2021) 012031 doi:10.1088/1757-899X/1011/1/012031

(Zea Mays L.) Pramasak sebagai Sumber Asam Folat untuk Formula Pangan Pintar Laporan Hasil Penelitian Program Tematik Kedeputian IPT Tahun Anggaran 2015 (Tangerang Selatan : Pusat Penelitian Kimia LIPI)

[15] Susilowati A, Maryati Y, Lotulung P D. 2018. Proses emulsifikasi dan pengeringan pasta campuran sumber asam folat alami dan idensifikasinya dalam perolehan serbuk suplemen untuk wanita hamil. *Jurnal Sains dan Kesehatan*, 1 (9) p- ISSN: 2303-0267, e-ISSN 2407-6082

[16] De Crécy-Lagard V, El Yacoubi B, De la Garza R D, Noiriel A, and Hanson A D 2007 Comparative genomics of bacterial and plant folate synthesis and salvage : predictions and validations. *BMC Genomics*. 8 : 245.

[17] Rossi M, Amaretti A and Raimondi S 2011 Folate production by probiotic bacteria Folate Production by Probiotic Bacteria. *Nutrients*. 3(1) : 118 – 134

[18] Agustine S, Puspa D L, Yati M and Aspiyanto 2018 Modification procces in nixtamalization of folic acid–rich dent corn (*Zea mays identata*) and its identification as smart food fortificant. *Matecconf*. 154, 01017 (2018), p 1 – 9.

[19] Manley C H 1993 Process Flavors and Precursor Systems. *ACS Symp Series*, 16–25. Bourne M C 2002 *Concept and Measurement Food Texture and Viscosity* (New York : Elsevier)

[20] SNI 2007 Makanan Pendamping Air Susu Ibu (MP-ASI) Surat Keputusan Menteri Kesehatan Republik Indonesia No. 224/Menkes/SK II/2007 (Jakarta : Ministry of Health).

IOP Conf. Series: Materials Science and Engineering 1011 (2021) 012032 doi:10.1088/1757-899X/1011/1/012032

The effect of particle size and brewing time of Ginger *(Zingiber officinale)* powder to the characteristic and acceptance of the herbal product

L Ratnawati*, D Desnilasari, L E Yulianti, D Kristanti, D P Putri, E Sholichah, C E Andriansyah and A Herminiati

Research Center for Appropriate Technology, Indonesian Institute of Sciences, JL KS Tubun No.5, Subang, West Java, Indonesia

*Corresponding author and main contributor: lia.romeo@gmail.com

Abstract. Ginger is one of the famous herbals that traditionally medicinal used in Indonesia. In this study, we investigated the effect of particle size and brewing time of ginger powder to the characteristic (moisture content, extractable matter, color and antioxidant activity) and acceptance of the herbal product. This study conducted using a completely randomized design with 2 factors namely particle size (40, 20 and 10 mesh) and brewing time (5, 10, 15, and 20 minutes). The result showed that particle size significantly ($p<0.05$) affected the moisture content. It is also significantly affected the extractable matter, color, and antioxidant activity. The brewing time was significantly ($p<0.05$) affected the extractable matter, color, antioxidant activity, and acceptance of the panelist to the ginger extract as herbal product. The largest particle size with the longest brewing tends to produce the highest extractable matter *i.e* 15.7%. The larger particle size causes a decrease in the L-values, an increase in the a-values, and a decrease in the b-values. The shortest brewing time with 20 mesh tends to produce the highest antioxidant activity *i.e* 77.97%. In concluding the ginger powder with a 20-mesh particle size and a 5 minutes brewing time was chosen to be the best treatment based on its antioxidant activity and sensory evaluation. It has the characteristics of extractable matter, L, a, b, antioxidant activity and acceptance including color, aroma, taste, after taste and overall namely 10.04%, 30.40, 1.15, 4.07, 77.97%, 3.2, 3.27, 3.1, 3.3 and 3.33, respectively.

1. Introduction

Ginger (*Zingiber officinale*), a member of *Zingiberaceae* is widely used as a spice in foods and beverages. Furthermore, it has long been used in traditional medicine as a cure for some disease. Ginger contains active phenolic compounds such as gingerol, shogoal and paradol that have antioxidant [1], anti-cancer [2], anti-inflammatory [3], anti-angiogenesis [4] and anti-artherosclerotic properties [5,6].

Harvest time of ginger is continuously throughout the year, this causes an abundance of fresh ginger on the market. Fresh ginger is perishable and has a short shelf life, so the utilization of it was not optimum. Because has many health benefits and public interest on herbal product was increased, nowadays ginger has been processed in various types of food products such as syrup, instant powder, candy and snacks. Ginger powder is one type of processed ginger. Powdered product is a processed food product that has good solubility in water, easy to serve and has a long shelf life due to its low moisture content. The ginger powder can be utilized for development of different commercial products like cookies, candy, tea, tinctures, sodas, jam, beer and syrup.

Despite of their potential applications, available studies on the effect of particle size and brewing time of ginger powder is limited. Makanjuola [7] studied about influence of particle size and extraction solvent on antioxidant properties of extracts of tea, ginger and tea-ginger blend. The study suggests that particle size influences the extraction of antioxidants. The powder with the lowest particle size of 0.425 mm tends to produce aqueous extracts of tea, ginger and tea-ginger with highest antioxidant content. Zhao *et al.* [8] reported that superfine grinding of ginger had many significance characteristics: surface area of powder was increased; had the good fluidity, water holding capacity, water solubility index and protein solubility; and was well suited to manufacture instant and convenient foods.

The aim in this study was to investigate the effect of particle size and brewing time of ginger powder on characteristics and acceptance of the herbal product. The characteristics were analyzed include moisture content, extractable matter, color, antioxidant activity and sensory evaluation.

2. Materials and Methods

2.1. Materials

Ginger *(Zingiber officinale* var. Amarum*)* simplicia was obtained from Herbal Anugerah Alam (Yogyakarta). Tea bag was purchased from sunshine55.id (Jakarta). 2,2-diphenyl-2-picrylhydrazyl (DPPH) and ascorbic acid for standard were obtained from Sigma-Aldrich Corp. (St. Louis, Mo., USA). Analytical grade ethanol was purchased from Merck Ltd. (Merck Pte Ltd. Singapore). Distilled water was used throughout and obtained from water distiller (Nuve ND12, Turkey). Filter paper was used Whatman No. 41 (GE Healthcare, UK).

2.2. Ginger powder preparation and extraction

Ginger simplicia were crushed using dry mill (Philips HR 2115). The ginger powder were passed through sieves of the following sizes : 40, 20 and 10 mesh. Then, the sieved ginger powder was analyzed for moisture content. The sieved ginger powder was put in a tea bag (3 g) and sealed. The ginger powder in a tea bag were stored in PP plastic for further analysis.

The brewing was prepared by placing 3 g of ginger powder (in a tea bag) in 150 ml of distilled water at boiling temperature (98±2°C) and brewing for 5, 10, 15 and 20 minutes. The samples with different particle size and brewing time were then analysis for extractable matter, color and antioxidant activity. The selected samples based on the antioxidant activity were sensory evaluated.

2.3. Ginger powder and extract characterization

2.3.1. Moisture content. Moisture content of ginger powder was measured using gravimetric method according to procedure described by Indonesian National Standard (SNI 01-2891-1992) [9]. Briefly, 2 g of samples were dried in a air oven (Memmert, Germany) at 105°C until constant weight. The result was expressed in percentage of evaporated water.

2.3.2. Extractable Matter. Extractable matter was analyzed according to procedure described by World Health Organization (WHO) [10] with modifications. Briefly, 3 g of samples in a tea bag brewed using 150 ml of distilled water with the brewing time according to the treatment. Brewing samples were then filtered using whatman paper No.41. Transfer the filtrate to a tared beaker glass and evaporate to dryness on an oven (Memmert, Germany). Dry at 105 °C for 12 hours, cool in a desiccator for 30 minutes, then weigh without delay. Extractable matter was calculated using Equation 1 as follows:

$$Extractable\ matter\ (\%) = \frac{dry\ weight\ extract}{dry\ weight\ sample}\ x\ 100\% \qquad (1)$$

2.3.3. Color Evaluation. The color of samples was measured according to procedure described by Raja et al. [11]. The ginger extract was measured using a chromameter (NH310, China). The lightness (L=0-100 (black-white)), greenness to redness (-a to +a) and blueness to yellowness (–b to +b) were evaluated.

IOP Conf. Series: Materials Science and Engineering 1011 (2021) 012032 doi:10.1088/1757-899X/1011/1/012032

2.3.4. Antioxidant activity. The antioxidant activity of samples was evaluated by using the 2,2-diphenyl-2-picrylhydrazyl (DPPH) according to the method described by Shen et al. [12] with some modifications. Briefly, 2 ml of DPPH (0.1 mM) already dissolved in ethanol (absolute, 100%) were mixed with 2 ml ginger extract and 1 ml ethanol (absolute, 100%). After incubation at room temperature for 30 minutes at 37°C, the absorbance was measured at 517 nm using an UV-Vis spectrophotometer (UV 1900, Shimadzu, Japan). Ascorbic acid was used as standard. The results were expressed as the percentage of inhibition of the DPPH radical and calculated using Equation 2 as follows:

$$Antioxidant\ activity\ (\%) = \frac{Absorbance\ of\ blanko - Absorbance\ of\ sample}{Absorbance\ of\ blanko}\ x\ 100\% \qquad (2)$$

2.3.5. Sensory evaluation. Sensory evaluation was conducted according to the method described by Hariharan and Mahendran [13]. Sensory evaluation was carried out using 20-30 ml of samples from brewing results which have placed in a glass then they served to 30 untrained panelists. The panelists filled on the organoleptic test sheets to give score preference for each sample based on a numerical scale (1-5), starting from very dislike (1) to very like (5). The scoring includes preferences for color, taste, aroma, aftertaste and overall.

2.4. Statistical analysis

An Analysis of Variance (ANOVA) and Duncan test (confidence level, α:0.05) were performed on the obtained results in order to establish significant differences. SPSS version 13 was used for data treatment and statistcal analysis. All measurements were carried out in triplicate and the results were expressed as the mean value ± standard deviation (SD).

3. Result and Discussion

3.1. Moisture content

The moisture content of ginger powder was ranged between 10.11-11.40 % (figure 1). The moisture content of ginger powder in the present study had fulfilled Indonesian National Standard (SNI 3933-1994) : Ginger which required the maximum moisture content in dried ginger is 12 % [14].

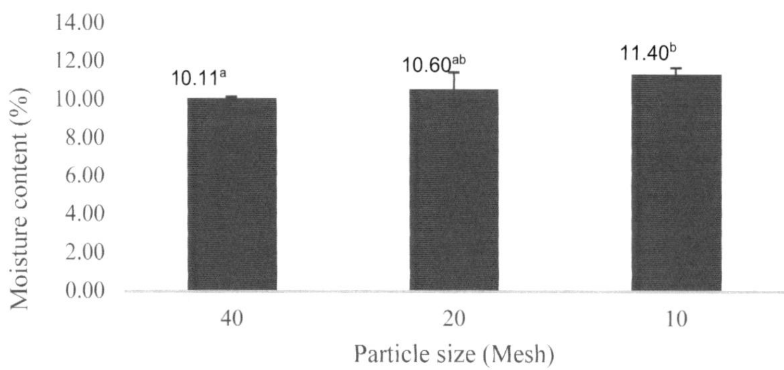

Figure 1. Moisture content of ginger powder with different particle size

The moisture content of ginger powder with 40 mesh particle size was significantly different ($p<0.05$) from ginger powder with 10 mesh particle size, but not significantly different ($p>0.05$) from ginger powder with 20 mesh particle size. As could be seen in Figure 1, the moisture content of ginger powder with larger particle size (10 mesh) had higher moisture content than smaller particles size (40 mesh). This result in line with the previous studies conducted by Raja *et al.* [11], stated that *Carica* leaf powder with larger particle size also had higher moisture content. The larger particle size of ginger powder may still have water trapped in the plant tissue. In addition, ginger powder with smaller particle size, the evaporation of water vapor will be more optimal when drying due to the greater surface area. In the drying process, the particle size is closely related to surface area. The larger particle size would reduce the surface area so that the water evaporation not optimal.

3.2. Extractable matter and color evaluation

Extractable matter is the number of active constituents extracted with solvents from a given amount of herbal material [10]. The larger particle size and the longer the brewing time increases the extractable matter. The difference in particle size and brewing time were significantly affected extractable matter ($p<0.05$) (table 1). The highest extractable matter (15.70%) was obtained from treatment with 10 mesh particle size and 20 minutes brewing time. Meanwhile, the lowest extractable matter (5.65%) was found in samples with treatment 40 mesh particle size and 5 minutes brewing time. Previous study by Sembiring *et al.* [15] suggested that the longer the extraction time the higher extractable matter. The extractable matter of red ginger (*Zingiber officinale* var. Rubrum) extract increased with increasing temperature and extraction time, it was optimal at 95°C for 15 minutes [16].

Based on table 1, particle size was proportional to the extractable matter value. The smaller particle size, the lower extractable matter value. It is due to the possibility that during extraction of dried ginger with finer particle sizes, the material extracted was carbohydrate or starch. Ajayi *et al.* [17] reported that carbohydrate content dried ginger was higher than raw ginger, it was 16.7% and 68.7%, respectively. In the process of analyzing extractable matter, there are several stages of filtering using Whatman filter paper No.41. It caused the extracted carbohydrate/starch was restrained at filter paper. Therefore, the component that really extracted was the active component and other substances that are soluble in water.

Table 1. Extractable matter, L, a and b-values of ginger extracts with different particle size and brewing time

Particle Size (mesh)	Brewing Time (minutes)	Extractable Matter (%)	L	a	b
40	5	5.65 ± 1.03^{Aa}	30.07 ± 0.01^{Bc}	1.19 ± 0.01^{Aa}	4.19 ± 0.03^{Cc}
	10	5.91 ± 0.95^{Aab}	30.36 ± 0.03^{Bc}	1.15 ± 0.03^{Ab}	4.21 ± 0.01^{Cb}
	15	7.15 ± 1.13^{Abc}	29.70 ± 0.02^{Ba}	1.36 ± 0.03^{Ad}	4.16 ± 0.01^{Cb}
	20	8.28 ± 0.44^{Ac}	29.97 ± 0.02^{Bb}	1.29 ± 0.01^{Ac}	4.03 ± 0.00^{Ca}
20	5	10.04 ± 1.77^{Ba}	30.40 ± 0.02^{Bc}	1.15 ± 0.01^{Aa}	4.07 ± 0.03^{Ac}
	10	10.36 ± 0.86^{Bab}	29.87 ± 0.04^{Bc}	1.33 ± 0.02^{Ab}	4.02 ± 0.01^{Ab}
	15	10.82 ± 0.92^{Bbc}	29.92 ± 0.00^{Ba}	1.29 ± 0.03^{Ad}	4.02 ± 0.02^{Ab}
	20	11.55 ± 0.82^{Bc}	30.00 ± 0.01^{Bb}	1.25 ± 0.03^{Ac}	4.03 ± 0.03^{Aa}
10	5	12.57 ± 0.95^{Ca}	29.70 ± 0.23^{Ac}	1.39 ± 0.02^{Ba}	4.11 ± 0.02^{Bc}
	10	14.27 ± 0.67^{Cab}	29.86 ± 0.01^{Ac}	1.37 ± 0.02^{Bb}	4.06 ± 0.03^{Bb}
	15	15.23 ± 0.51^{Cbc}	29.42 ± 0.07^{Aa}	1.51 ± 0.05^{Bd}	4.12 ± 0.01^{Bb}
	20	15.70 ± 1.01^{Cc}	29.37 ± 0.03^{Ab}	1.44 ± 0.02^{Bc}	4.00 ± 0.02^{Ba}

The capital letters denote significant difference ($p<0.05$) between the different particle size. The small letters denote significant difference ($p<0.05$) between different brewing time. Data are means of triplicates (n=3) = SD

The L-values indicates the sample lightness, the greater value showed that lighter the color of sample. The a-values indicates the greenness to redness color. The positive value indicates the redder color. The b-values shows a blueness to yellowness color. The positive value indicates yellower color. The L-values of ginger extracts with various particle size and brewing times were significantly different ($p<0.05$). Table 1 shows that the treatment of 20 mesh particle size and 5 minutes brewing time had the highest L-values (30.40), while the treatment of 10 mesh particle size and 20 minutes brewing time had the lowest L-values (29.30). It is shows that the ginger extracts from smaller particle and faster brewing time were much lighter than those from larger particle and longer brewing time. This results in accordance to the previous studies conducted by Ibrahim *et al.* [16], stated that the lightness value of ginger juice tends to decrease (dark) along with the increase in extraction temperature and extraction time.

Based on the instrumental measurements, ginger extracts from larger particle and longer brewing time were much redder than finer particle and faster brewing time as shown by their high and positive a-values (table 1). The a-values of ginger extracts with various particle size and brewing times were significantly different ($p<0.05$). The a-values tends to increase with increasing temperature and extraction time [16]. The red color in ginger extracts comes from Edulan II compounds which are classified as carotenoid groups [18]. In contrary, the b-values of ginger extract from larger particle sizes and longer brewing time were showed decreased. The particle size and brewing time were significantly affected on b-values of ginger extracts ($p<0.05$). It is shows that smaller particle and faster brewing time produce a yellower ginger extract. The yellow color comes from oleoresin contained in ginger. Ginger oleoresin is bright yellow, yellow to dark brown. Ginger oleoresin components consist of gingerol, zingiberen, shogaol, essential oils and resins [16]

3.3. Antioxidant Activity

The antioxidant activity of ginger extracts was evaluated using DPPH assay and the results are shown in table 2.

Table 2. Antioxidant activity of ginger extracts.

Particle Size (mesh)	Brewing Time (minutes)	Antioxidant activity (%)
40	5	74.91 ± 0.16^{Ba}
	10	71.60 ± 1.58^{Bb}
	15	69.08 ± 0.90^{Bc}
	20	64.29 ± 0.40^{Bd}
20	5	77.97 ± 0.92^{Ba}
	10	75.37 ± 1.26^{Bb}
	15	63.79 ± 2.39^{Bc}
	20	60.79 ± 2.49^{Bd}
10	5	66.44 ± 1.01^{Aa}
	10	65.40 ± 1.56^{Ab}
	15	64.38 ± 0.43^{Ac}
	20	65.18 ± 0.26^{Ad}

The capital letters denote significant difference ($p<0.05$) between the different particle size. The small letters denote significant difference ($p<0.05$) between different brewing time. Data are means of triplicates (n=3) = SD

The antioxidant activity of ginger extracts was showed in ranged 60.79-77.97%. It is shows that the antioxidant activity of ginger extracts in this study had high activity. This refers to the statement of Wulansari and Chairul [19], if the antioxidant activity using the DPPH method has results above 50%, it is indicated as high antioxidant activity; 20–50% shows moderate antioxidant activity; while the result below 20% indicates low antioxidant activity. Other research by Eleazu *et al.* [20] reported that 75% inhibition of DPPH radical for their best ginger variety, but Shukla *et al.* [2] reported that higher value 74.33 to 91.25% for ginger sungro-sung variety.

The data showed that the particle size of ginger in this study significantly affected the antioxidant activity of the ginger extracts (table 2). The biggest the particle size has the lowest antioxidant activity of the ginger powder. This was indicated that the smaller particle would increase the contact surface area. The contact surface area ginger powder with solvent would be maximize the extraction of active compound [21]. Makanjuola [7] reported that size reduction of plant substrates before extraction maximizes the surface area, which in turn enhances the mass transfer of active compound from plant material to the solvent. The antioxidant activity of 40 mesh ginger powder was not significantly different with 20 mesh ginger powder. This also has been reported by Makanjuola [7] that although it is a known fact that a reduction in particle size could always lead to increased extraction efficiency, however, a critical particle size is reached such that any further reduction in the particle size-could lead to no further increase-or a reduction-in extraction efficiency. The brewing time in this research showed siginificantly different to the the antioxidant activity. The result showed that the longer brewing time could decreasing gradually the antioxidant activity. The antioxidant activity, total phenol content, and gingerol content of ginger (*Zingiber officinale* Rosc.) decreased with increasing the time of thermal processing [18]. The longer the contact time of material with heat is possibility cause the bioactive compound damage. Other research with *Centella asiatica* by Ariffin *et al.* [22] with treatment were brewing time in 1 to 20 minutes (at 100°C) showed that the antioxidant activity increased gradually until 10 minutes later become stable. The increasing temperature and extraction time of red ginger (*Zingiber officinale* var. Rubrum) extract affect the increasing of antioxidant activity and total phenol, it was optimal at 95°C for 15 minutes [16].

3.4. Sensory Evaluation

The sensory evaluation of ginger extracts with different brewing time are shown in Table 3.

Table 3. Sensory evaluation of ginger extracts with the different brewing time

Particle Size (mesh)	Breewing Time (minutes)	Color	Aroma	Taste	Aftertaste	Overall Acceptance
20	5	3.20^{a}	3.27^{ab}	3.10^{a}	3.30^{a}	3.33^{ab}
	10	3.43^{a}	3.33^{ab}	3.30^{a}	3.30^{a}	3.43^{ab}
	15	3.10^{a}	3.03^{a}	2.93^{a}	3.07^{a}	3.07^{a}
	20	3.87^{b}	3.63^{b}	3.33^{a}	3.53^{a}	3.63^{b}

Notes: Different uppercase letters in the same column indicate significant differences ($p<0.05$).

Based on the extractable matter value and the antioxidant activity of ginger extract, the best particular size for extraction was 20-mesh. This data showed that color parameter of ginger extracts with different brewing time was not significantly different, except the ginger extracts with 20 minutes brewing time was significantly different from other treatments. The panelist preferable for color parameters were ranged from 3.10-3.87 (rather like). The treatment with 20 minutes brewing time had highest score on color parameter. From table 3 could be seen that aroma of ginger extracts was not significantly different, except the samples with 15 minutes brewing time was significantly different from the samples with 20 minutes brewing time. The panelists rated the aroma in the range of 3.03-3.63 (rather like). The treatment with 20 minutes brewing time had highest score on aroma. The major of hydrocarbon groups that influenced the aroma of ginger were α-pinene, camphene, β-phellandrene and ar-curcumene. In addition, the presence of geranyl acetate from volatile oil that have a ginger-like odor contributing to the most ginger aroma [18].

The taste score for samples were ranged between 2.93-3.33 (dislike to rather like) (table 3). From the statistical analysis, the brewing time was not significantly affected ($p>0.05$) on the taste. The highest taste score was 20 minutes brewing time treatments. As could be seen from Table 3, that aftertaste of ginger extracts was not significantly different ($p>0.05$). The aftertaste score for samples were ranged 3.07-3.53 (rather like). The highest aftertaste score was 20 minutes brewing time treatments. The taste and aftertaste in this sensory evaluation was to describe the pungency of ginger extracts. The compounds responsible for the pungency taste of ginger are zingerone, 6-gingerol, 8-gingerol, 10-gingerol, shogaol and 1-dehydro-10-gingerdione [23; 24].

Table 3 showed that overall acceptance of ginger extracts with different brewing time was not significantly different, except the ginger extracts with 15 minutes brewing time was significantly different from the ginger extracts with 20 minutes brewing time. The overall acceptance score for samples were ranged 3.07-3.63 (rather like) with the highest score for overall acceptance was 20 minutes brewing time. This result supported by data of the extractable matter and a-values (table 1). The ginger extract which brewed for 20 minutes had the highest of extractable matter and a-values. This result indicated that the brewing of ginger extract for 20 minutes produced a brownish yellow extract with the stronger aroma and taste of ginger.

Although the panelist acceptance of ginger extract with a 20-minute brewing time had the highest score, the selection of the best treatment was based on the highest antioxidant activity, namely the 20-mesh particle size and 5 minutes brewing time. It is also based on the results of sensory evaluation between ginger extract of 5 and 20 minutes of brewing time were not significantly different for the taste, aroma, aftertaste and overall acceptance.

4. Conclusion

The particle size significantly affected the moisture content, extractable matter, color, and antioxidant activity. The brewing time was significantly affected the extractable matter, color, antioxidant activity, and acceptance of the panelist to the ginger extract as herbal product. The largest particle size with the longest brewing tends to produce the highest extractable matter. The larger particle size causes a decrease in the L-values, an increase in the a-values, and a decrease in the b-values. The shortest brewing time with 20 mesh tends to produce the highest antioxidant activity. In concluding the ginger powder with a 20-mesh particle size and a 5 minutes brewing time was chosen to be the best treatment based on its antioxidant activity and sensory evaluation.

Acknowledgments

The authors are grateful to Research Center for Appropriate Technology for providing the experimental facilities and to the Indonesian Institute of Sciences for financial support.

References

[1] Stoilova I, Krastanov A, Stoyanova A, Denev P and Gargova S 2007 Antioxidant activity of a ginger extract *Food Chem* **102** 764-770

[2] Shukla Y and Singh M 2007 Cancer preventive properties of ginger: a brief review *Food Chem Toxicol* **45** 683-690

[3] Mashhadi N S, Ghiasvand R, Askari G, Hariri M, Darvishi L and Mofid M R 2013 Anti-oxidative and anti-inflammatory effects of ginger in health and physical activity: review of current evidence *Int J Prev Med* **4** 36-42

[4] Kim E C, Min J K, Kim T Y, Lee S J, Yang H O, Han S, Kim Y M and Kwon Y G 2005 [6]-Gingerol, a pungent ingredient of ginger, inhibits angiogenesis in vitro and in vivo. *Biochem Bioph Res Co* **335** 300-308

[5] Boroujeni H R, Gharipour M, Samani M A and Boroujeni H M 2016 The protective effects of ginger on the development of coronary atherosclerosis: an experimental animal study *Der Pharmacia Lettre* **8** 105-109

[6] Gunathilake K D P P and Rupasinghe H P V 2015 Recent perspective on the medicinal potential of ginger *Botanics: Targets and Therapy* **5** 55-63

[7] Makanjuola S A 2017 Influence of particle size and extraction solvent on antioxidant properties of extracts of tea, ginger and tea-ginger blend *Food Sci Nutr* 1-7

[8] Zhao X, Yang Z, Gai G and Yang Y 2009 Effect of superfine grinding on properties of ginger powder *J Food Eng* **91** 217-222

[9] Badan Standarisasi Nasional (BSN) SNI 01-2891-1992 Cara uji makanan dan minuman (How to test food and drinks) Jakarta

[10] World Health Organization (WHO) 1998 Quality control methods for herbals materials Switzerland ISBN 978 92 4 150073 9

[11] Raja K S, Taip F S, Azmi M M Z, Shisir M R I 2019 Effect of pre-treatment and different drying methods on the physicochemical properties of *Carica papaya* L. leaf powder *J Saudi Agr Sci* **18** 150-156

[12] Shen Q, Zhang B, Xu R, Wang Y, Ding X, Li P 2010 Antioxidant activity in vitro of the selenium-contained protein from the Se-enriched Bifidobacterium animalis 01 *Anaerobe* **16** 380-386

[13] Hariharan G and Mahendran T 2016 Physicho-chemical, sensory and microbial evaluation of ginger-lime ready-to-serve (RTS) functional beverage, sweetened by palmyra sugar candy *Imperial J Interdisciplinary Res* **2** 1545-1552

[14] Badan Standarisasi Nasional (BSN) SNI 3933-1994: Ginger Jakarta

[15] Sembiring B B, Ma'mun, Ginting E I 2006 Pengaruh kehalusan bahan dan lama ekstraksi terhadap mutu ekstrak temulawak (*Curcuma xanthorriza* Roxb) (Effect of fineness of the ingredients and extraction time on the quality of temulawak extract (*Curcuma xanthorriza* Roxb)) *Bul Littro* **17** 53-58

[16] Ibrahim A M, Yunianta, Sriherfyna F H 2015 Effect of Temperature and Extraction Time on Physicochemical Properties of Red Ginger (*Zingiber officinale* var. Rubrum) Extract with The Additional of Honey Combination as Sweetener for Functional Drink J Pangan dan Agroindustri **3** 530-541

[17] Ajayi O A, Ola O O, Akinwunmi O O 2017 Effect of drying method on nutritional composition, sensory and antimicrobial properties of Ginger (*Zingiber officinale*) *Int. Food Res J.* **24** 614-620

[18] Purnomo H, Jays F and Widjanarko B 2010 The effects of type and time of thermal processing on ginger (*Zingiber officinale* Roscoe) rhizome antioxidant compounds and its quality) *Int. Food Res J* **17** 335-347

[19] Wulansari D, Chairul 2011 Penapisan aktivitas antioksidan dan beberapa tumbuhan obat Indonesia menggunakan radikal 2,2-diphenyl-1 picrylhydrazyl (DPPH) (Antioxidant screening activity of several Indonesian medicinal plants using 2,2-diphenyl 1-1 picrylhidrazyl (DPPH)). *Majalah Obat Tradisional* **16** 22 – 25

[20] Eleazu C, Amadi C, Iwo G, Nwoso P, Ironua C 2013 Chemical Composition and Free Radical Scavenging Activities of 10 Elite Accessions of Ginger (*Zingiber officinale* Roscoe) *J. Clin. Toxicol.* **3** 155–159

[21] Maulida R, Guntarti A 2015 The influence of particle size of black rice (*Oryza sativa* L.) on extract yield and total anthocyanin content *Pharmaciana* **5** 9–16

[22] Ariffin F, Chew S H, Bhupinder K, Karim A A, Huda N 2011 Antioxidant capacity and phenolic composition of fermented *Centella asiatica* herbal teas *J Sci Food Agric* **91** 2731–2739

[23] Jayashree E, Visvanathan R, John Z T 2014 Quality of dry ginger (*Zingiber officinale*) by different drying methods *J Food Sci Tecnol* **51** 3190-3198

[24] Tiwari S, Saha S N, Tiwari A, Patel Y K 2020 Ginger Physico Chemical Properties and It's Consumer Products processing: A Review *Int Arch App Sci Technol* **11** 159-169

Synthesis layered material of MnO_2 and TiO_2 as the support of molybdenum catalyst for bio-crude oil upgrading process

N Rinaldi*, A A Dwiatmoko and A Kristiani

Research Center for Chemistry, Indonesian Institute of Sciences. Kawasan PUSPIPTEK, Serpong, Tangerang Selatan, Banten 15314, Indonesia

*Corresponding author: nino.rinaldi@lipi.go.id

Abstract. Layered materials are attractive for many applications because of their unique structure consisting of stacked sheets that can be controlled and modified. In this work, various synthetic layered materials (SLMs) *i.e.,* Na-MnO_2, H-MnO_2, and Cs-TiO_2 have been successfully prepared as the support for molybdenum (Mo)-based catalyst and applied for bio-crude oil hydrodeoxygenation (HDO) process. A commercial Mo/Al_2O_3 catalysts will also be utilized as a benchmark. The physical and chemical properties of the catalysts were characterized by using an X-ray diffractometer, X-ray fluorescence, surface area analyzer, NH_3-temperature-programmed desorption, and scanning electron microscopy. The characterization results showed that the interlayer ion exchange of Na^+ ions with H^+ ions was successful for MnO_2 support. The impregnation of the MoO_3 into SLMs increased the surface area, pore-volume, average pore size, and total acidity. The HDO of bio-crude oil using MoO_3/Na-MnO_2 catalyst produced a higher HHV value of 38.75 MJ/Kg and a DOD value of 82.75% compared to other prepared catalysts or even the MoO_3 commercial catalyst.

1. Introduction

The global energy demand has been increased due to the increase in population growth and economic developments. On the other hand, the availability and sustainability of fossil fuels have led to major global concerns. It encourages the need for sustainable energy sources and liquid fuels or bio-crude oil produced from biomass become promising fuels in the future [1-3]. However, direct bio-crude oil cannot be directly used because of its high corrosiveness, high viscosity, high oxygen and water contents, and low high heating value (HHV). Thus, bio-crude oil needs upgrading to remove the high oxygen content (10-40 wt%) which affects the homogeneity, polarity, high heating value (HHV), viscosity, and acidity of the oil [4-13].

Catalytic hydrodeoxygenation (HDO) is an important process to remove oxygen from the bio-crude oil and produce a high-grade oil product equivalent to fuels [13-17]. Supported MoO_3 catalyst is already known its effectiveness in the hydrodeoxygenation (HDO) reaction that can selectively cut C-O bonds in various derivatives of organic compounds to produce olefin and aromatic hydrocarbons with high activity and selectivity [18]. The reactivity and stability of different supported MoO_3 catalysts (SiO_2, γ-Al_2O_3, TiO_2, ZrO_2, and CeO_2) were also investigated for HDO of m-cresol. ZrO_2 and TiO_2 support promoted the catalytic activity and improved stability of MoO_3 catalyst compared with unsupported MoO_3 catalyst [19]. Therefore, the support role of the catalyst is also important that different support could significantly affect the catalyst activity.

Layered materials still developed for many applications due to their unique structure consisting of stacked sheets [20]. These materials can be modified their structural, textural, and compositional depend on their applications. Layered materials allow a fine-tuning and control of the catalytic reactivity [21,22], therefore very interesting for many catalytic applications. Nakamura *et al.* [23] reported monolayered titanium, tungsten, and manganese oxides synthesized by exfoliation of the nanocrystals through aqueous solution processes at room temperature. The results showed that the exfoliation of the nanocrystals generated the monolayered structures without changes in the lateral size. This monolayered nanodots with morphological control and tuning properties can be applied as a building block of composites at a molecular level and comprehensive applications. Honda *et al.* [24] have discovered the synthesis path of MnO_2 synthetic layered material (SLM) and TiO_2 synthetic layered material to enhance their ability as the catalyst.

Due to synthetically layered material (SLM) has significant potential to develop on the catalyst system with high porosity and activity [21], therefore, in this research, various SLMs, such as Na-MnO_2, H-MnO_2, and Cs-TiO_2 were prepared as the support of MoO_3 catalyst. The physicochemical properties of SLM before and after supported by MoO_3 supported were characterized. The catalytic performance of prepared catalysts was investigated on the catalytic HDO reaction of bio-crude oil. The supported MoO_3 commercial catalyst was also utilized as a comparison.

2. Experimental

2.1. Materials

The chemicals employed in this research were manganese chloride tetrahydrate ($MnCl_2.4H_2O$), Aldrich), disodium dihydrogen ethylene-diamine tetraacetate (EDTA, Merck), sodium hydroxide (NaOH, Merck), rutile titanium oxide (TiO_2, Aldrich), cesium carbonate (Cs_2CO_3, Aldrich), hydrochloric acid (HCl, Merck), ammonium heptamolybdate heptahydrate ($(NH_4)_6Mo_7O_{24}.4H_2O$, Merck), H_2 gas, carbon disulfide (CS_2, Merck), ethyl acetate ($C_4H_8O_2$, Merck), and MoO_3/Al_2O_3 commercial catalyst (15% wt MoO_3). The bio-crude oil used in this work was produced from pyrolysis of palm kernel shell, a naturally abundant lignocellulosic biomass waste obtained from a local palm oil plant in Sumatera.

2.2. Synthesis of Na-MnO_2 SLM

Na-MnO_2 SLM was synthesized by the chelation-mediated approach reported before Honda *et al.* [24]. An amount of typically 5 mmol/dm^3 (0.58 g) EDTA and 20 mmol/dm^3 (1.58 g) $MnCl_2.4H_2O$ into a polypropylene container which contains 400 mL of distilled water, then stir until all powder dissolved. After dissolved, 200 mmol/dm^3 (3.2 g) NaOH in 400 mL distilled water was prepared and added into the solution. The solution was maintained under stirring for five days at room temperature. The black precipitate obtained was filtered and dried at room temperature.

2.3. Synthesis of H-MnO_2 SLM

The H-MnO_2 SLM synthesized by the ion-exchange method. 1 g of the synthesized Na-MnO_2 SLM was added into 100 mL of 1M HCl solution. The mixture was stirred for one day at room temperature. The precipitates were collected and dried at room temperature [24].

2.4. Synthesis of Cs-TiO2 SLM

Synthesis of Cs-TiO_2 SLM was carried out in solid-state conditions reported by Honda *et al.* [24]. 1.5 g of TiO_2 rutile powder and 1 g of Cs_2CO_3 are mixed and calcined at 800 °C for 20 hours in argon gas flow. Then, the powder was cooled until room temperature.

2.5. Impregnation of MoO_3

The SLMs were impregnated with MoO_3 through the wet impregnation method from the ammonium heptamolybdate (AHM) precursor [25,26]. A total of 13% wt MoO_3 (0.275 g) AHM was dissolved in 5 mL distilled water then added to the solution contained 1.5 g of the synthesized layer material in 10 mL distilled water. The mixture was stirred and heated around 80 °C until dry. The powder results were dried overnight in oven 100 °C and calcined at 550 °C for 3 hours.

2.6. Catalyst characterization

X-Ray Diffractometer (XRD) of the catalysts was analyzed using MAX Science MXP3V, with Cu Kα radiation at 40 kV and 25 mA, and a secondary graphite monochromator.

Brunauer-Emmett-Teller (BET) adsorption method of the catalysts was conducted using Tristar II 3020 Micromeritics Instrument to obtain the specific surface area of the samples through nitrogen adsorption-desorption isotherms, performing at -196 °C on liquid nitrogen after degassing the sample at 200 °C for 2 h.

NH_3-Temperature Programmed-Desorption (NH_3-TPD) of the catalysts was conducted using Micromeritics 2900. The analysis was carried out from 100 to 450 °C at a heating rate of 10 °C/min with a flow rate of Helium (He) gas of 15 mL/min. After each TPD measurements, the amount of ammonia adsorbed was determined from the calibration curve obtained from varying volumes of ammonia in He. The total acidities of the catalysts were measured by comparing the NH_3-TPD spectra to that of the reference material.

Scanning electron microscopy (SEM) images were obtained on a JEOL JSM-IT300 electron microscope. The samples are fixed on the holder with conductive adhesive and coating with gold for 90 s to improve the conductivity of the samples.

X-Ray Fluorescence (XRF) of the catalysts was performed using Thermo Niton XL3t. About 1 gr sample is placed on the sample holder and was measured on the inert condition using He gas.

2.7. Catalyst sulfidation

All prepared catalysts were sulfided before the HDO reaction by the same procedure as reported by Jiang *et al.* [26,27]. 0.3 g of catalyst was placed into the U tube reactor and flowed by N_2 gas at a pressure of 1 bar. The temperature was maintained at 150 °C for 1.5 h. Then, the catalyst was cooled to room temperature. The sulfidation was continued with 1 bar of H_2 gas that flows over CS_2 at 400°C for 2 h. The catalyst was cooled to room temperature and applied for hydrodeoxygenation (HDO) of bio-crude oil.

2.8. Catalytic performance test

0 g of bio-crude oil and 3% w/w of catalyst were placed into an autoclave reactor, HDO reaction was carried out at 300°C for 4 h under 60 bar of H_2 [26]. The MoO_3/Al_2O_3 commercial catalyst was used as a benchmark. The liquid product was collected and analyzed using a CHON analyzer to measure the high heating value (HHV). The HHV results were compared with the HHV of commercial diesel fuel.

3. Result and discussion

3.1. Characterization of catalysts

The crystal structure of supports and supported Mo catalysts was observed using XRD to ensure the diffraction pattern of the synthesized catalyst were appropriate to the diffraction pattern of the target material. The diffraction pattern of prepared catalysts is shown in figure 1.

For Na-MnO_2 SLM (figure 1.1-a), the dominant peaks at 2θ of 12.40°, 24.90°, and 35.60° (ascribed to the diffraction by the (001), (002), (200) planes) might be described to typical of birnessite-type of sodium manganite which is compatibility with JCPDS 43-1456. H-MnO_2 SLM (figure 1.1-b) shows some dominant peaks which are similar to Na-MnO_2 with a lower intensity. Some peaks diminished and a new peak observed at 2θ of 23.1°. It indicates that Na^+ ions in the interlayer of Na-MnO_2 SLM have been exchanged by H^+ ions from HCl and produced H-MnO_2 SLM. Moreover, the XRD pattern of Cs-TiO_2 SLM shows the 2θ of 10.2°, 20.5°, 27.9°, and 30.5° (ascribed to the diffraction by the (020), (040), (130), (011) planes) which revealed to $Cs_2Ti_6O_{13}$ according to JCPDS 38-0170. The diffraction pattern also observed in the region of 2θ = 27.4°; 36.0°; 54.3° with miller indexes (110), (101) and (211) under JCPDS 21-1276 which is a diffraction pattern of rutile TiO_2 that from material precursors. Although Cs-TiO_2 SLM was obtained with TiO_2 as the main layer and Cs ions found in between layers [28], it is suggested that only a small formation of Cs-TiO_2 SLM was formed, due to the diffraction pattern of rutile TiO_2 still be observed (figure 1.1-c). All XRD pattern results were obtained in figure 1 is the same observed on Honda *et al.* [24], it is described that preparation of synthesis layer materials of Na-MnO_2, H-MnO_2, and Cs-TiO_2 are obtained successfully. For impregnated MoO_3 in SLMs support (figure 1.2), the peaks observed at 2θ = 12.8°, 23.3°, 25.3°, and 27.3° (ascribed to the diffraction by the (020), (110), (040), (021) planes) indicates the existence of orthorhombic phase of MoO_3 (JCPDS 35-0609). The presence of MoO_3 peaks on the SLMs shows that the impregnation of the MoO_3 into SLMs was successful. However, this result is also described that Mo structure on SLM supports is mostly in MoO_3 crystal and indicated poor dispersion of Mo on the SLM support [29].

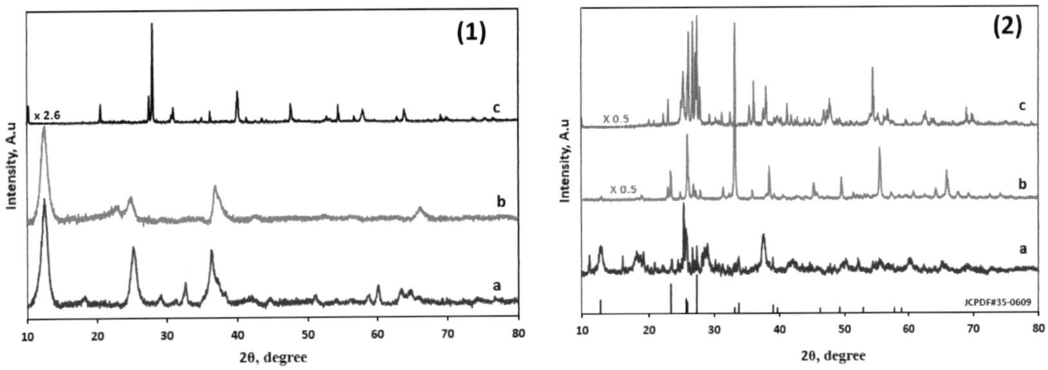

Figure. 1. X-ray diffraction results of (1) Synthetic layer materials (a) Na-MnO_2 (b) H-MnO_2 and (c) Cs-TiO_2; (2) MoO_3 prepared catalysts (a) MoO_3/Na-MnO_2 (b) MoO_3/H-MnO_2 and (c) MoO_3/Cs-TiO_2

The surface area, pore-volume, and average pore diameter of the supports and supported Mo catalysts are summarized in table 1. The impregnation of MoO_3 increased the surface area, pore-volume, and average pore size of SLMs. It is suggested that some of MoO_3 is not dispersed on the pore of SLM support after impregnation but also modified the stacked sheets of SLM acting as the pillar [30-32], thus increasing surface area and pore diameter. Moreover, the average pore diameter of Na-MnO_2 (17.384 nm) is lower than H-MnO_2 (12.803 nm). It indicates the success of ion exchange of Na^+ ions with H^+ ions in the interlayer of MnO_2 which decreased the distance between layers.

Table 1. Surface properties and total acidity of prepared catalysts.

Sample	Surface area (m^2/g)	Pore volume (cm^3/g)	Average Pore diameter (nm)	Total Acidity (mmol/g)
Na-MnO_2	6.273	0.027	17.384	$1.03x10^{-4}$
MoO_3/Na-MnO_2	11.980	0.060	20.098	$3.03x10^{-4}$
H-MnO_2	10.039	0.032	12.803	$2.64x10^{-4}$
MoO_3/H-MnO_2	14.353	0.069	19.174	$6.12x10^{-5}$
Cs-TiO_2	1.736	0.012	26.869	$1.56x10^{-4}$
MoO_3/Cs-TiO_2	5.425	0.030	22.044	$7.30x10^{-5}$

The acidity properties of catalysts were evaluated through NH_3-TPD analysis as also shown in table 1. The total acidity of Na-MnO_2 ($1.025x10^{-4}$ mmol/g) is higher than H-MnO_2 ($2.64x10^{-4}$ mmol/g). It means the interlayer ion exchange of Na^+ ions with H^+ ions that contribute to the increase of the acidity sites was successful [30]. The impregnation of MoO_3 into MnO_2 SLM support also increased the total acidity significantly from 1.025x10^{-4} to 3.03x10^{-4} mmol/g. On the other hand, the impregnation of MoO_3 into H-MnO_2 and Cs-TiO_2 SLMs support decreased the total acidity from 2.64x10^{-4} to 6.12x10^{-5} mmol/g and from 1.56x10^{-4} to 7.30x10^{-5} mmol/g, respectively. This decrease in total acidity after impregnation is suggested that the acidity of H-MnO_2 and Cs-TiO_2 SLMs is higher compare to MoO_3 metal. Therefore, during the impregnation proses, some of the MoO_3 metal was exchanged ion H^+ and Cs^{2+} and acting as the pillar between the stacked sheets, thus decrease the acidity. This result is similar to the surface area and average pore diameter in table 1 that also showed a decrease after the impregnation.

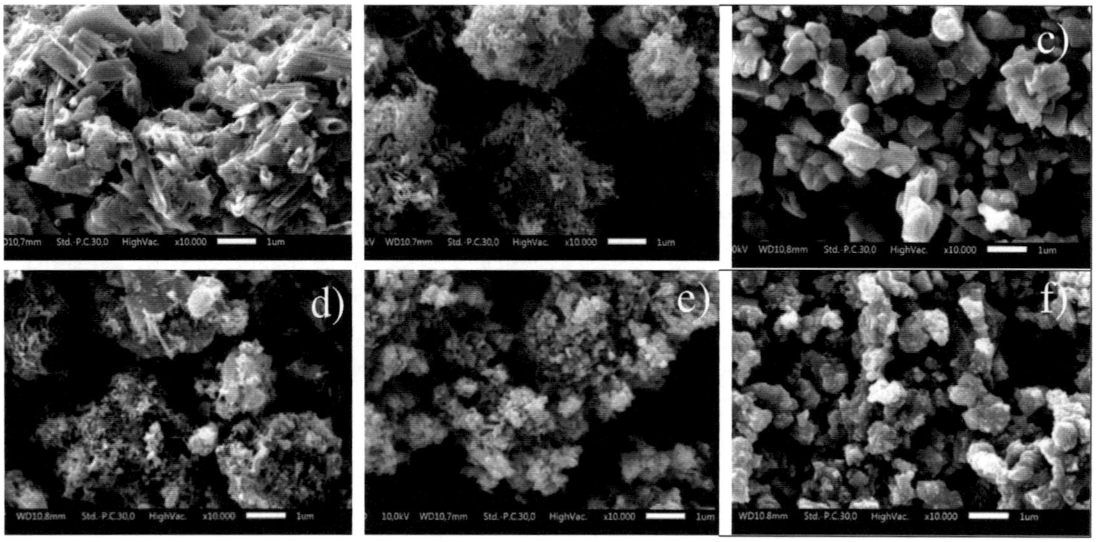

Figure 2. Surface morphology of (a) Na-MnO_2, (b) H-MnO_2, (c) Cs-TiO_2, (d) MoO_3/Na-MnO_2, (e) MoO_3/H-MnO_2, and (f) MoO_3/Cs-TiO_2 (Bar scale: 1μm)

SEM analysis was carried out to identify surface morphology and the distribution of components on the synthesized catalysts as shown in figure 2. It can be seen that the surface morphology of Na-MnO_2 SLM (figure 2a) is mostly hollow tubes structure and stacked. Moreover, after ion-exchanged with H^+, the surface morphology of H-MnO_2 (figure 2b) is different from Na-MnO_2, the structure of H-MnO_2 SLM becomes small agglomerate particles composed of the thin layers stack. This morphological change confirms that the interlayer exchange Na^+ ions with H^+ ions were successful. For Cs-TiO_2 SLM, the surface structure is mostly small distributed particles. The addition of MoO_3 into the Na-MnO_2 and H-MnO_2 supports (figure 2d and 2e) also changed the surface morphology. All surface structures of MoO_3/Na-MnO_2 and MoO_3/H-MnO_2 become similar to the H-MnO_2 structure and even smaller agglomerate particles. It indicates that MoO_3 particles filled the pores and decreased the particle size. On the other hand, Cs-TiO_2 does not show significant differences in the surface morphology after impregnated by MoO_3, but it seems to become smaller particles after the impregnation.

XRF analysis is conducted to determine the composition of metal elements in prepared catalysts both qualitatively and quantitatively. The results of the qualitative measurement of chemical composition are shown in table 2. It can be observed that the composition of Na in the H-MnO_2 sample disappeared. It confirms that Na^+ ions in the Na-MnO_2 sample are exchanged successfully by H^+ ions. The impregnation of MoO_3 shows a decrease in the Mn composition and the appearance of Mo metal content in the material. The presence of Mo shows that the MoO_3 metal was successfully impregnated into the SLM supports. However, it is different for the Cs-TiO_2 sample. The amount of Mo is slightly small observed than Na-MnO_2 and H-MnO_2 after the impregnation. It is indicated that only a minor amount of MoO_3

could disperse and exchange with Cs ion as the pillar on Cs-TiO_2 SLM support. This phenomenon is also supported by the SEM analysis result (figure 2), which showed that the surface structure of Cs-TiO_2 SLM is not significantly altered after the impregnation.

Table 2. Element compositions by XRF analysis.

Sample	% Atomic weight				
	Na	Mn	Cs	Ti	Mo
Na-MnO_2	1.5	98.2	-	-	-
MoO_3/Na-MnO_2	0.5	69.2	-	-	29.6
H-MnO_2	-	98.2	-	-	-
MoO_3/H-MnO_2	-	78	-	-	21.7
Cs-TiO_2	-	-	36.3	43.8	-
MoO_3/Cs-TiO_2	-	-	31.8	40.9	9.6

3.2. Catalytic performance of MoO_3 supported catalysts

HDO reaction of bio-crude oil was conducted in a batch reactor and products were analyzed using the CHON analyzer to determine a high heating value (HHV) and degree of deoxygenation (DOD) of the product. The results were summarized in Table 3.

Table 3. CHON analysis results.

Samples	%. wt				HHV
	N	C	H	O	(MJ/Kg)
Bio-crude oil	0.45	58.18	7.79	33.58	24.8
MoO_3/Al_2O_3	0.59	72.84	7.95	18.62	32.7
MoO_3/Na-MnO_2	0.75	82.55	8.30	8.40	38.3
MoO_3/H-MnO_2	0.56	70.58	7.80	21.06	31.3
$MoO_3/CsTiO_2$	2.31	71.86	7.41	18.42	31.6
Diesel oil	-	87.00	13.00	-	44.91

All of the catalyst samples show the decrease in oxygen content of bio-crude oil from 33.58% to below 22% after HDO. On the other hand, the carbon content of bio-crude oil increased from 58.18% to more than 70% after HDO. Consequently, HHV of bio-crude oil increased after HDO using MoO_3/Na-MnO2, MoO_3/H-MnO_2, and MoO_3/Cs-TiO_2, which is from 24.8 MJ/Kg to 38.3 MJ/Kg. 31.3 MJ/Kg and 31.6 MJ/Kg, respectively. It indicates that the HDO process was successfully diminished the oxygen content and improve the quality of bio-crude oil even though the quality obtained was still below diesel fuel. The commercial MoO_3/Al_2O_3 catalyst was used as a benchmark in this study. It is observed that our prepared catalysts were closed to the activity of MoO_3/Al_2O_3 commercial catalyst or even higher, especially on the MoO_3/Na-MnO_2 catalyst. The HHV value of the MoO_3/Na-MnO_2 catalyst is superior compared with the commercial catalyst.

The DOD value indicates the percentage of oxygen lost during the HDO reaction takes place. The higher DOD value means a higher catalytic activity on breaking the C-O bond in the deoxygenation reaction [33]. Table 4 shows the DOD value of bio-crude oil products produced by HDO. It is observed that the MoO_3/Na-MnO_2 catalyst produced the highest decrease in the O/C ratio with a DOD value of 82.75%. Meanwhile, MoO_3/Cs-TiO_2 catalyst and commercial catalyst MoO_3/Al_2O_3 showed the same value of 55.17%. The MoO_3/H-MnO_2 catalyst showed the lowest DOD value of 48.28%. This decrease in the O/C ratio confirms that the deoxygenation reaction was successful. It is shown that the MoO_3/Na-MnO_2 SLM catalyst was superior on the catalytic performance compared to the other prepared catalysts instead of the commercial MoO_3 catalyst.

Table 4. DOD value of HDO products.

Sample	O/C	H/C	DOD (%)
Bio-crude oil	0.58	1.32	-
MoO_3/Al_2O_3	0.26	0.86	55.17
MoO_3/Na-MnO_2	0.10	0.64	82.75
MoO_3/H-MnO_2	0.30	0.92	48.28
MoO_3/Cs-TiO_2	0.26	0.86	55.17

Topsøe *et al.* [34] describe there are several properties of the MoO_3 catalyst are affected on the hydrotreating reaction of crude oil, such as surface area, pore diameter, acidity, porous catalyst, dispersion of active metal on the surface support, the interaction between active metal catalysts and the surface support, and additive component [34-40]. Based on the characterization data above, it is suggested that the total acidity of catalyst is one of the crucial points to obtain a high catalyst activity next to the surface area, pore-volume, and pore diameter. Surface area, pore-volume, and pore diameter of MoO_3/Na-MnO_2 is similar to MoO_3/H-MnO_2 (table 1), and the surface morphology of both catalysts also showed as the porous material with small agglomerate particles composed of the thin layers stack (figure 2). However, the MoO_3/Na-MnO_2 catalyst is more acid compare to other catalysts (table 1). Table 4 showed that O/C and H/C values of MoO_3/Na-MnO_2 are very narrow compared to the other catalysts, especially the MoO_3 commercial catalyst, which is indicated that the ability of deoxygenation and hydrogenation reactions is more active of MoO_3/Na-MnO_2 catalyst. The activity catalyst on the hydrotreating process could be increased by using the support with a high acidity [34-37]. Thus, the acidity factor is concluded as significant properties to obtain a high HHV and DOD values after the HDO reaction of bio-crude oil in this study.

4. Conclusion

Synthetic layered materials of Na-MnO_2, H-MnO_2, and Cs-TiO_2 have been successfully prepared and characterized as support for MoO_3 catalyst. MoO_3/Na-MnO_2 catalyst showed superior in the HDO activity compared with other catalysts, especially the MoO_3 commercial catalyst. The highest HHV and DOD values were obtained of 38.3 MJ/Kg and 82.75%, respectively, although this value was still lower than diesel fuel properties. It is suggested that the acidity factor is the primary characteristic next to the surface area, pore-volume, and average pore diameter to obtain a high HHV and DOD values after the HDO reaction of bio-crude oil.

Acknowledgment

The authors would like to express their great gratitude to Yuya Oaki, Department of Applied Chemistry, Keio University, Japan during fruitful research discussions about the prospecting of synthesis layer material.

References

[1] Lian X, Xue Y, Zhao Z, Xu G, Han S and Yu H 2017 *Int. J. Energy Res.* **41** 1798.
[2] Milina M, Mitchell S and Pérez-Ramírez J 2014 *Catalysis Today* **235** 176.
[3] Xiu S and Shahbazi A 2012 *Renewable and Sustainable Energy Reviews* **16(7)** 4406.
[4] Luque R, Herrero-Davila L, Campelo JM, Clark JH, Hidalgo JM, Luna D, Marinas JM and Romero AA 2008 *Energy & Environmental Science* **1(5)** 542.
[5] Zhang Q, Chang J, Wang T J and Xu Y 2007 *Energy Conversion and Management* **48(1)** 87.
[6] Zhang L, Liu R H, Yin R Z and Mei Y F 2013 *Renewable and Sustainable Energy Reviews* **24** 66.
[7] Sipilä K, Kuoppala E, Fagernäs L and Oasmaa A 1998 *Biomass and Bioenergy* **14(2)** 103.
[8] Zacher A H, Olarte M V, Santosa D M, Elliott D C and Jones S B 2014 *Green Chemistry* **16(2)**

491.
[9] Black S and Ferrell J R 2016 *Energy & Fuels* **30(2)** 1071.
[10] Lu Q, Li W Z and Zhu X F 2009 *Energy Conversion and Management* **50(5)** 1376.
[11] Murugappan K, Mukarakate C, Budhi S, Shetty M, Nimlos M R and Román-Leshkov Y 2016 *Green Chem* **18** 5548.
[12] Wang H, Male J and Wang Y 2013 *ACS Catal.* **3** 1047.
[13] Tavizón-Pozos J A, Santolalla-Vargas C E, Valdés-Martínez O U and Heredia J A R 2019 *Catalysts* **9** 550.
[14] Mortensen M, Grunwaldt J D , Jensen P A, Knudsen K G and Jensen A D 2011 *Applied Catalysis A: General* **407** 1.
[15] Linck M, Felix L, Marker T and Roberts M 2014 *Integrated Biomass Hydropyrolysis and Hydrotreating: A Brief Review* **vol 3** issue 6 Wiley Interdiscip. Rev. Energy Environ. pp 575.
[16] Kim S, Kwon E E, Kim Y T, Jung S, Kim H J, Huber G W and Lee J 2019 *Green Chem.* **21** 3715.
[17] Jin W, Pastor-Perez L, Shen D, Sepulveda-Escribano A, Gu S, and Reina T R 2019 *ChemCatChem.* **11(3)** 924.
[18] Prasomsri T, Nimmanwudipong T and Román-Leshkov Y 2013 *Energy Environ. Sci.* **6** 1732.
[19] Shetty M, Murugappan K, Prasomsri T, Green W H and Román-Leshkov Y 2015 *J. Catal.* **331** 86.
[20] Centi G and Perathoner S 2008 *Microporous and Mesoporous Materials* **107** 3.
[21] Auerbach S M, Carrado K A and Dutta P K 2004 *Handbook of Layered Materials,* New York Marcel Dekker. Inc.
[22] Fleischmann S, Spencer M A and Augustyn V 2020 *Chem. Mater.* **32(8)** 3325.
[23] Nakamura K, Oaki Y, and Imai H 2013 *J. Am. Chem. Soc.* **135** 4501.
[24] Honda M, Oaki Y and Imai H 2014 *Chem. Mater.* **26** 3579.
[25] Rinaldi N, Kubota T, Okamoto Y 2010 *Appl. Catal. A : Gen.* **374** 228.
[26] Rinaldi N, Simanungkalit S P and Kristiani A 2017 *AIP Conference Proceedings* **1904** (020078) 1.
[27] Jiang M, Wang B, Yao Y, Li Z, Ma X, Qin S and Sun Q 2013 *Catal. Sci. Technol.* **3** 2793.
[28] Saito M, Akeboshi Y, Ohno N, Kuwano J, Shiroishi H and Uchimoto Y. 2009 *ECS Transactions* **16(25)** 97.
[29] Okamoto Y, Arima Y, Nakai K, Umeno S, Katada N, Yoshida H, Tanaka T, Yamada M, Akai Y, Segawa K, Nishijima A, Matsumoto H, Niwa M and Uchijima T 1998 *Appl. Catal. A: Gen.* **170(2)** 315.
[30] Occelli M L and Rennard R J *1988 Catal. Today* **2** 309.
[31] Khan K, Tareen A K, Aslam M, Wang R, Zhang Y, Mahmood A, Ouyang Z, Zhang H and Guo Z 2020 *J. Mater. Chem. C* **8** 387.
[32] Ötvös S B, Pálinkó I and Fülöp F 2019 *Catal. Sci. Technol.* **9** 47.
[33] Basu P 2010 *Biomass Gasification and Pyrolysis Practical Design* **1st edition** Academic Press pp ISBN 978-0-12-374988-8.
[34] Topsøe H, Clausen B S, and Massoth F E 1996 *Hydrotreating Catalysis*, Springer, Berlin.
[35] Song C 2003 *Catal. Today* **86** 211.
[36] Whitehurst D D, Isoda T, Mochida I 1998 *Adv. Catal.* **42** 345.
[37] Kabe T, Ishihara A, Qian W 1999 *Hydrodesulfurization and Hydrodenitrogenation*, Kodansha, Tokyo.
[38] Eijsbouts S, 1997 *Appl. Catal. A* **158** 53.
[39] Sun M, Nicosia D, and Prins R 2003 *Catal. Today* **86** 173.
[40] van Veen J A R, Gerkema E, van der Kraan A M, Knoester A 1987 *J. Chem. Soc. Chem. Commun.* 1684.

IOP Conf. Series: Materials Science and Engineering 1011 (2021) 012034 doi:10.1088/1757-899X/1011/1/012034

Physicochemical and microbiological properties of Pudding Powder as a complementary food during storage

D Kristanti* and A Herminiati

Research Center for Appropriate Technology, Indonesian Institute of Sciences Jl. KS. Tubun No. 5 Subang 41213

*Corresponding author and main contributor: dita.kristanti@gmail.com

Abstract. The product evaluation during storage is important because it is related to food safety. The pudding powder as a complementary food was made from skim milk powder, sugar, mocaf, and carrageenan which enriched with inulin and fortified with micronutrient premix. The pudding powder were packed in polyethylene plastic (PE) and aluminium foil (AF) pouches to evaluate the physicochemical and microbiology properties at 15°C, 30°C, and 45°C for 4 weeks. The data were analyzed using IBM SPSS Statistics 20 and analysis of variance (ANOVA) with Duncan's means test (significance level of $p<0.05$). The moisture content, protein, and lightness of all pudding powder were significantly decreased, while a value and Total Plate Count (TPC) were significantly increased after storage. The storage with AF pouches were not significantly affect the weight of pudding powder after first week storage. The water activity (a_w) of AF45 was the highest among others. It affects the decreasing of protein, increasing of TPC, browning and caking on this pudding powder. The best storage condition of pudding powder for both pouches was at temperature 15°C.

1. Introduction

The inulin-enriched pudding powder for complementary food has been developed to support the growth and development of children over one year old [1], [2]. The sensory evaluation assessment of pudding as complementary food showed a good response from both mother and children's under 1-5 years old [1]. However, the iron and zinc contents did not meet the SNI requirement for instant complementary food powder [1]. Thus, the product development has been carried out to increase mineral content, especially iron. The iron and zinc content of pudding powder has been increased by tripled with mineral fortification which contain 50 mg/kg of iron, 30 mg/kg of zinc, 4 mg/kg of vitamin B2, 2.5 mg/kg of vitamin B1, and 2 mg/kg of folic acid [2]. The physicochemical properties of pudding powder with mineral fortified as a complementary food were 4.40% of moisture content, 1.00% of ash, 15.74% of protein, 0.12% of fat, 78.74% of carbohydrate, 419.13 mg/100mg of calcium. 2.04 mg/100g of iron, 7.18 mg/100g of zinc, 71.65 of lightness, -0.35 of a value, 11.05 of b value, and 105 cP of viscosity [2].

The complementary food must meet the nutritional needs and it is safe for infants. The safety of complementary food is related to contamination of chemical, metals, and microbial during preparation process and storage. Evaluation of the physicochemical and microbiological properties of the product during storage is essential as it is related to food safety and consumer acceptance. Moisture content and total microbes are the important characteristics of product during storage that related to food safety.

Furthermore, the color properties are essential characteristic of product during storage period that related to consumer acceptance.

The research on evaluation of physicochemical and microbiological properties in a complementary food during storage has been carried out. The effect of instant baby food storage has been done on the moisture content, peroxide value, and free fatty acids value, which kept in double polyethylene bag at 20-25°C for 3 months [3]. Danarsi et al. [4] had conducted on the effect of storage of instant porridge baby with the snakehead fish and pumpkin flour substitution on microbiological quality that stored in airtight plastic bag at 15°C for 4 weeks. The evaluation on the physicochemical characteristic of liquid and powder infant formula that stored at 10-40°C for 6 months had also been carried out [5], [6]. The storage affects the moisture content, color properties, surface protein, and total microbial of the complementary food [3]–[6].

The factors affect the physicochemical and microbiological properties changes during storage are type of packaging, storage temperature, relative humidity, and storage time. In this study, the pudding powder were packed in polyethylene plastic (PE) and aluminium foil (AF) pouches then stored at 15°C, 30°C, and 45°C for 4 weeks. The physicochemical and microbiological characteristics of powder pudding during storage were evaluated in this study. The evaluation was including moisture content, weight, protein, water activity, color properties, and total plate count.

2. Material and Method

The materials used in this study were skim milk powder from local store in Bandung, mocaf from Indomocaf in Bandung, sugar from local market in Subang, carrageenan, Orafti inulin (Synergy1-Oligofructose), and micronutrient premix SNI Enrichment Mix from Muhlenchemie.

2.1. Preparations and Storage Conditions of CF Pudding Powder

The pudding powder was prepared according to Kristanti et al. [2]. The pudding powder were packed in polyethylene plastic (PE) and aluminium foil (AF) pouches at 15°C, 30°C, and 45°C for 4 weeks. The physicochemical and microbiology properties evaluated were moisture content, water activity (a_w), weight, protein, color properties, and total plate count (TPC).

2.2. Physicochemical Properties Analysis

The moisture content of pudding powder was analysed by the gravimetric method [7]. The water activity (a_w) of pudding powder was determined by Smart Water Activity Meter HD-3A with WSC-4 water activity sensor. The weight of pudding powder was determined using analytical balance (FV-220C). The protein content was analysed by the Dumas combustion method using DuMAster Buchi D-480, Switzerland. The color of CF pudding powder was determined by Chromameter Konica Minolta CR-400 using CIE method. The color scales of CIE were estimated as: L = lightness 0 - 100 is white color; a = red color 0 – 60 and green color 0 – (-60); b = yellow color 0 – 60 and blue color 0 – (-60). The total plate count (TPC) method was evaluated by AOAC [7]. The sampling and analysis were done every week during storage time.

2.3. Statistical Analysis.

Three replications of analysis from pudding powders were used in all of the quality measurements. The data were analyzed using IBM SPSS Statistics 20 and analysis of variance (ANOVA) with Duncan's means test (significance level of $p<0.05$). All data were presented as the mean with standard deviation.

3. Result and Discussion

The physicochemical properties of pudding powder as complementary food in the initial observation (week 0) were 3.67% of moisture content, 15.50% of protein, 0.42 of water activity (a_w), 32.16 g of weight powder, 73.62 of lightness, 0.18 of a value, 11.08 of b value, and 0.92×10^4 CFU of total plate count. These results are slightly different from our previous studies. The physicochemical properties of pudding powder as a complementary food were 4.40% moisture content, 15.74% of protein, 71.65 of lightness, -0.35 of a value, and 11.05 of b value [2]. The differences in physicochemical properties in pudding flour occur due to the different source of mocaf used in the formulation.

The moisture content of pudding powders was significantly decreased during storage in 4 weeks, except for PE30 which was increased. The PE45 sample was the pudding powder that had a highest decline in moisture content, it formed an agglomerate. The moisture content of pudding powder during storage was presented in Figure 1. The moisture content of spray dried yogurt powder and mango soy fortified yogurt powder stored in aluminium limited polyethylene and high density polypropylene pouches were increase during storage time [8], [9]. The moisture content of instant baby food was increase during storage time at 20-25°C in double polyethylene bag [3]. Other research showed that moisture content spray dried camel milk powder was slight decline during storage time in plastic petri dish polystyrene at relative humidity (RH) levels of 11%, 22%, and 32% [10]. The research of Masum et al. [6] showed that the storage of instant milk formula powder at 40°C for 15 day caused the very low of moisture content and the powder was caked. The factors may contribute to moisture content decrease during storage are permeability of packaging, storage temperature, and storage time.

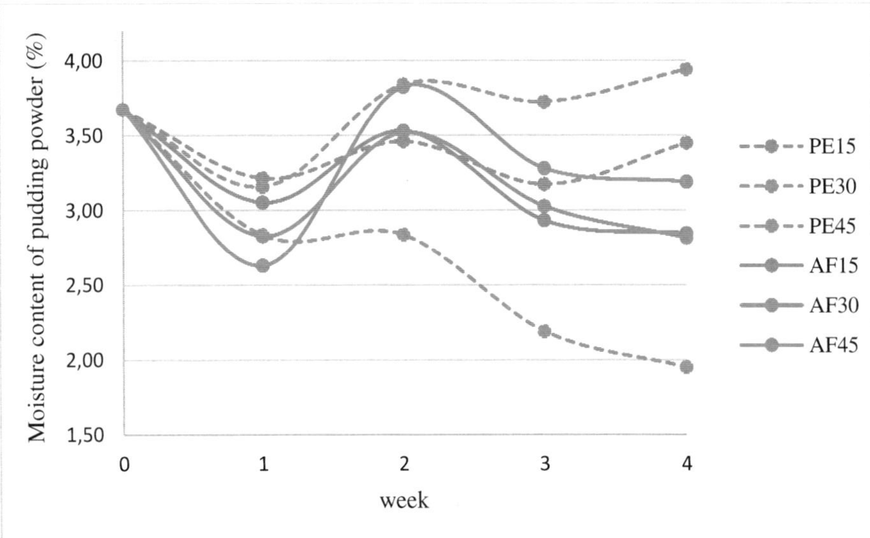

Figure 1. The moisture content of pudding powder as complementary food during storage

The weight of pudding powder stored in aluminium foils were significantly increase in first week then it was stable until final week 4, except for AF45 sample that stable during storage period. Furthermore, the weight of pudding stored in polyethylene plastics were significantly increase during storage, except for PE45 sample which was drastically declined (Figure 2). These results study was not in accordance with previous study. The storage of dried persimmons at temperature of -20°C, 5°C, 12°C, and 25°C for 70 days was not affected by their weights and moisture contents [11]. The weight rising of PE15 and PE30 may cause by the increasing of moisture content during storage. The drastically decline of PE45 sample were due to drastically decrease of moisture content during storage (Figure 1).

The storage condition affects the water activity of pudding powder during storage (Table 1). The water activity of PE15, PE30, and AF45 significantly increased after storage. However, the a_w of PE45, AF15, and AF30 were significantly decreased after storage. The increasing of the a_w in AF45 was the highest among others. The changes in water activity of the pudding powder are caused by changes in moisture content during the storage period. The water activity of spray dried bovine colostrum powder stored in aluminium laminated polyethylene and polyethylene pouches increased during storage at temperature 4°C, 25°C, and 50°C for 90 days [12]. The increasing of water activity in Yu et al. [12] study was accompanied by an increasing of moisture content. Other research showed that the storage of commercial goat milk powder and spray dried yogurt powder did not affect the water activity [8], [13].

The permeability of packaging affects the characteristics of food products during storage, especially moisture content and water activity. Plastics was relatively permeable to small molecules such as gases, water vapour, organic vapours, and liquids [14], but aluminium paper foil was not permeable for the oxygen [15]. The permeability of polyethylene was higher than the permeability of aluminium foil, there were 0.05 and 0.004 gH_2O/day.m^2.mmHg [16]. Generally, the high packaging permeability makes the characteristic of food products more easily changed so that the shelf life is short.

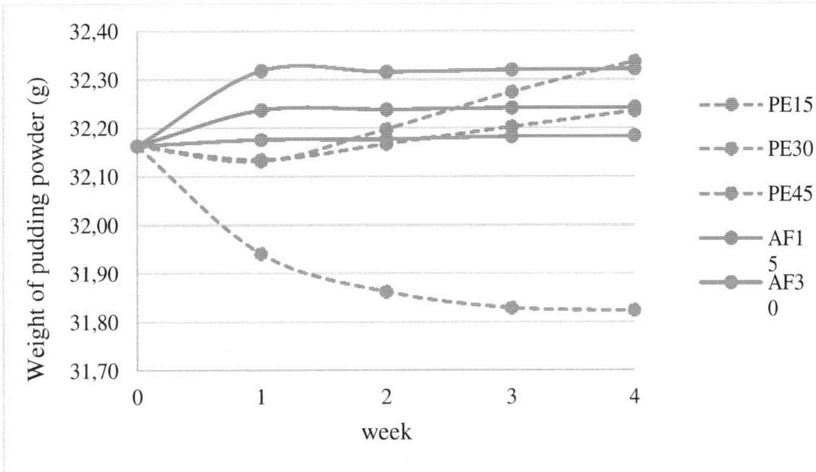

Figure 2. The weight of pudding powder as complementary food during storage.

Table 1. The water activity (a_w) of pudding powder as complementary food during storage/

Sample	Week				
	0	1	2	3	4
PE15	0.42 ± 0.00^{Ba}	0.42 ± 0.00^{Ba}	0.42 ± 0.00^{Bb}	0.44 ± 0.01^{Ab}	0.44 ± 0.01^{Ab}
PE30	0.42 ± 0.00^{Ba}	0.40 ± 0.00^{Db}	0.41 ± 0.00^{Cab}	0.43 ± 0.00^{Ab}	0.43 ± 0.00^{Ac}
PE45	0.42 ± 0.00^{Ba}	0.34 ± 0.00^{CDd}	0.34 ± 0.01^{Ce}	0.33 ± 0.01^{De}	0.35 ± 0.01^{Ee}
AF15	0.42 ± 0.00^{ABa}	0.40 ± 0.00^{Cb}	0.40 ± 0.00^{Cc}	0.41 ± 0.00^{BCc}	0.40 ± 0.00^{Cd}
AF30	0.42 ± 0.00^{Ba}	0.39 ± 0.00^{Cc}	0.39 ± 0.00^{Cd}	0.39 ± 0.01^{Cd}	0.39 ± 0.00^{Cd}
AF45	0.42 ± 0.00^{Ca}	0.40 ± 0.03^{Eb}	0.48 ± 0.00^{Ba}	0.48 ± 0.00^{Ba}	0.49 ± 0.00^{Aa}

Values are expressed as mean ± standard deviation. Means in the same column (lowercase letters) and row (uppercase letters) with different letters were significantly different at $p < 0.05$

The protein is one of the parameters that can be affected by temperature changes during storage. The protein of pudding powder during storage showed in Figure 3. The protein content of PE15 sample was relative stabile until the second week of storage. It was decreased after the third week storage. Other protein samples were significantly declined after four-week storage. Furthermore, the highest decline of protein after four-week storage occurred in AF45. The decrease in protein content in the pudding powder is probably caused by Maillard reaction or protein denaturation during storage. The surface protein content of infant milk formula powder was decline after storage at 22 °C and 40 °C for 180 days [6]. The denaturation protein enthalpy (ΔH) of dried egg white was decrease after storage at 25 and 40 for four months that indicated a partially unfolded protein conformation [17].

The lowest protein occurred in AF45 sample. The water activity of the pudding powder may plays a role in protein damage. The AF45 sample had a highest water activity, which is 0.49 (Table 1). The high water activity affects the loss rate of amino acids such as lysine when the storage temperature above the glass transition temperature [18]. The storage of infant milk formula powder decreased the glass transition temperature that affect the increasing of caking [6]. The research of Masum et al. [6] showed that the glass transition temperature of infant milk formula powder in the initial observation (week 0) was 70-80°C then decline to 10-30°C and <0°C in storage conditions at 20°C and 40°C, 54% RH for 90 days, respectively.

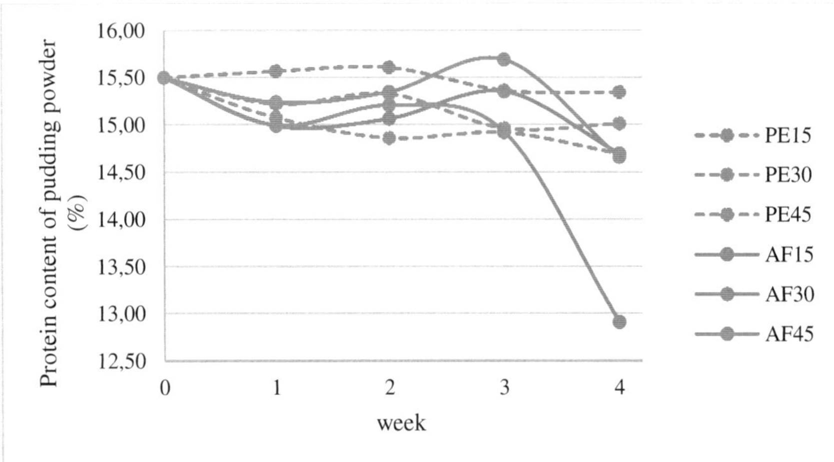

Figure 3. The protein of pudding powder as complementary food during storage

The color properties of pudding powder as complementary food present in Table 2. The lightness of pudding powder was significantly decreased in all sample. The lightness in final observation were not significantly different. The a value of the PE pouches samples increased, while the AF pouches decreased, but the value were not significantly different. The b value of PE15, PE30, and AF15 significantly decrease, but it increased for AF30 even though it was significantly different, and the AF45 was significantly increased. The results of previous studies stated that the lightness of mango soy fortified yogurt powder and liquid infant formula were decreased, while the a and b value were increased after storage [5], [9]. The research study of Ho et al. [10] showed that the storage affect the increasing of b value and the decreasing of a value, but not affect the lightness of spray dried camel milk powder. The storage caused the increasing of color different, b value, and Maillard Browning Index (MBI) value in the high protein powder products [8], [12], [17], [19]. Other research showed that the lightness and a value of mango milk powder were increased, while the b value was decreased [20].

The factors affect the lightness of food product during storage are type of packaging, storage temperature, sugar and protein content of product, water activity, and storage time. The lowest of lightness and highest of b value were occur in the AF45 sample. This result indicated that the AF45 sample was darker and yellowish colored. Moreover, the AF45 was caked. The pudding powder had a browning reaction during storage period. The discoloration of pudding powder during storage lead by reaction between sugar and amino acids, its browning reaction of the Maillard reaction effect. Furthermore, water activity of the pudding powder also plays a role in Maillard reaction. The water activity contributes for the development of browning via non-enzymatic reaction, its Maillard reaction [21]. The high-water activity of product leads browning reaction. The AF45 sample was 0.49 (Table 1). The decreasing of the glass transition temperature during storage period of infant milk formula powder caused caking [6].

IOP Conf. Series: Materials Science and Engineering 1011 (2021) 012034 doi:10.1088/1757-899X/1011/1/012034

Table 2. The color properties of pudding powder as complementary food during storage

Sample		Week				
		0	1	2	3	4
L	PE15	73.77 ± 0.14^{Bb}	81.54 ± 0.40^{Aa}	71.27 ± 0.02^{Eb}	71.57 ± 0.24^{Da}	72.40 ± 0.34^{Cb}
	PE30	73.30 ± 0.00^{Bc}	80.69 ± 0.41^{Ac}	71.17 ± 0.67^{Db}	70.19 ± 0.25^{Ec}	72.31 ± 0.17^{Cb}
	PE45	72.69 ± 0.10^{Bd}	80.95 ± 0.25^{Abc}	71.74 ± 0.10^{Da}	70.60 ± 0.02^{Eb}	72.33 ± 0.17^{Cb}
	AF15	74.04 ± 0.34^{Ba}	81.55 ± 0.01^{Aa}	71.30 ± 0.03^{Db}	70.79 ± 0.06^{Eb}	72.65 ± 0.10^{Ca}
	AF30	74.18 ± 0.17^{Ba}	81.19 ± 0.02^{Aab}	70.32 ± 0.11^{Ec}	70.67 ± 0.08^{Db}	72.33 ± 0.04^{Cb}
	AF45	73.74 ± 0.03^{Bb}	81.19 ± 0.51^{Aab}	70.17 ± 0.09^{Cc}	67.91 ± 0.38^{Ed}	68.69 ± 0.90^{Db}
a	PE15	0.13 ± 0.01^{Ccd}	0.29 ± 0.06^{Ac}	0.14 ± 0.04^{Cb}	0.11 ± 0.01^{Cb}	0.22 ± 0.03^{Bc}
	PE30	0.12 ± 0.07^{Bd}	0.37 ± 0.02^{Ab}	0.11 ± 0.07^{Bb}	0.15 ± 0.04^{Bb}	0.42 ± 0.02^{Ab}
	PE45	0.20 ± 0.02^{Bab}	0.42 ± 0.04^{Aab}	0.13 ± 0.02^{Cb}	0.23 ± 0.05^{Bb}	0.46 ± 0.03^{Ab}
	AF15	0.21 ± 0.05^{Aab}	0.21 ± 0.04^{Ad}	0.02 ± 0.05^{Bc}	-0.05 ± 0.04^{Cc}	0.15 ± 0.02^{Ac}
	AF30	0.25 ± 0.02^{ABa}	0.29 ± 0.05^{Ac}	0.04 ± 0.03^{Dc}	0.10 ± 0.01^{Cb}	0.24 ± 0.04^{Bc}
	AF45	0.18 ± 0.06^{Ebc}	0.45 ± 0.02^{Da}	0.83 ± 0.01^{Ca}	1.71 ± 0.17^{Ba}	2.70 ± 0.25^{Aa}
b	PE15	10.99 ± 0.16^{Cb}	11.78 ± 0.11^{Aab}	11.19 ± 0.10^{Bbc}	10.90 ± 0.03^{CDb}	10.77 ± 0.03^{Dc}
	PE30	11.09 ± 0.07^{ABab}	11.29 ± 0.02^{Ab}	10.95 ± 0.14^{BCc}	10.53 ± 0.41^{Db}	10.71 ± 0.11^{CDc}
	PE45	10.98 ± 0.03^{Bb}	12.09 ± 0.46^{Aab}	11.32 ± 0.44^{Bb}	10.83 ± 0.51^{Bb}	11.40 ± 0.32^{Bb}
	AF15	11.18 ± 0.09^{Ba}	11.71 ± 0.03^{Aab}	11.24 ± 0.10^{Bbc}	11.25 ± 0.03^{Bb}	11.06 ± 0.03^{Cbc}
	AF30	11.20 ± 0.18^{Ba}	11.76 ± 0.10^{Aab}	11.55 ± 0.22^{ABb}	10.85 ± 0.28^{Db}	11.32 ± 0.08^{BCb}
	AF45	11.04 ± 0.04^{Dab}	11.66 ± 0.81^{Da}	13.25 ± 0.17^{Ca}	14.96 ± 0.98^{Ba}	16.42 ± 0.49^{Aa}

Values are expressed as mean ± standard deviation. Means in the same column (lowercase letters) and row (uppercase letters) with different letters were significantly different at $p < 0.05$

The total plate count (TPC) in all pudding powder samples significantly increased after four-week storage (Figure 4). The TPC of pudding powder stored at temperature 15°C and 30°C in both polyethylene and aluminium pouch were relatively stable until third week storage. Furthermore, the TPC of these pudding powder in the fourth week storage were significantly increased. However, the TPC of AF45 sample particularly increased in the third week storage. In addition, the TPC of PE45 sample since the first week storage and continued to drastically increase until the fourth week storage. The TPC at the first week storage meet the SNI requirement for instant complementary food powder, but at maximum limits.

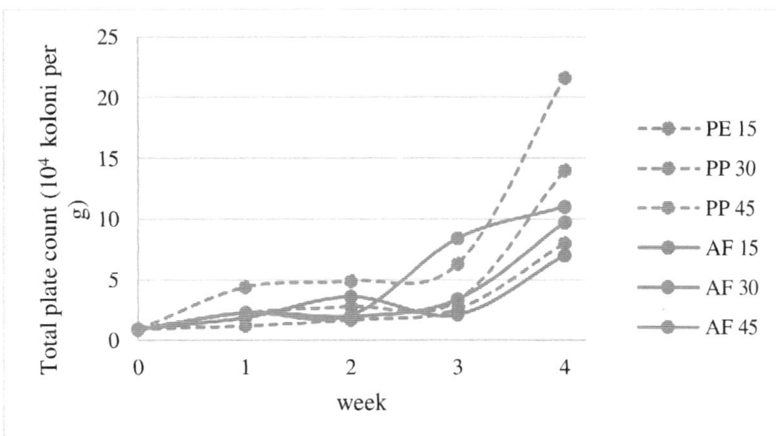

Figure 4. The total plate count of pudding powder as complementary food during storage

These results were consistent with the previous study. The TPC of the snakehead fish and pumpkin flour in instant porridge as complementary food increased during storage in airtight plastic bag at 15°C for four weeks [4]. Another research showed that TPC of the mango milk powder decreased during storage in tin can, metalized polyester, 4-ply lamites, and polystryrene bag at 5°C and 15°C for 8 months [20]. The increasing of TPC during storage is probably caused by the high of water activity in pudding powder. The water activity of pudding powder in this study were in the range of 0.34 -0.49. The high value of a_w affects the eminent probability of microbial growth, resulted in the increasing of total bacteria in pudding powder.

4. Conclusion

The storage of pudding powder in polyethylene plastic and aluminium foil pouch at 15°C, 30°C, and 45°**C** for 4 weeks affect the moisture content, weight, protein, water activity, color properties, and TPC. The moisture content, protein, and lightness of pudding powder in both polyethylene plastic and aluminium foil pouches were decreased, while the a value and TPC were increased after storage. The storage with aluminium foil pouches were not significantly affect the weight of pudding powder after first week storage. The storage of pudding powder in aluminium foil pouch at 45°C had a highest of water activity than others in which lead to a browning reaction during storage. The best storage condition of pudding powder for both polyethylene plastic and aluminium foil pouches were at temperature 15°C. Furthermore, it is necessary to conduct sensory evaluation during storage in order to determine the consumer acceptance level.

Acknowledgments

The authors are thankful to Insentif Riset Sistem Inovasi Nasional (INSINAS) 2019 Program from Ministry of Research, Technology and Higher Education of the Republic of Indonesia with contract number 065/P/RPL-LIPI/INSINAS/II/2019 for funding this research. The authors also are thankful to Research Center for Appropriate Technology, Indonesian Institute of Sciences for the provision of research facilities.

References

[1] D. Kristanti dan A. Herminiati, "Characteristics of physical, chemical, and organoleptic properties of inulin-enriched pudding as a complementary food," in *IOP Conference Series: Earth and Environmental Science*, 2019.
[2] D. Kristanti dan A. Herminiati, "Physicochemical properties of pudding powder as a complementary food fortified with the essential mineral," *AIP Conf. Proc.*, vol. 2175, no. November, hal. 020053-6-020053-6, 2019.
[3] S. Raza, M. N. Safdar, A. Mumtaz, N. Siddiqui, K. Nasim, dan M. Amjad, "Preparation and quality evaluation of nutritious instant baby food from indigenous sources.," *Pakistan J. Agric. Res.*, vol. 22, no. 1/2, hal. 50–55, 2009.
[4] C. S. Danarsi dan E. R. Noer, "Pengaruh Lama Penyimpanan terhadap Mutu Mikrobiologi Makanan Pendamping Air Susu Ibu (MP-ASI) Bubur Instan dengan Substitusi Tepung Ikan Gabus dan Tepung Labu Kuning," *J. Nutr. Coll.*, vol. 5, no. 2, hal. 58–63, 2016.
[5] C. W. Seo, S. Hong, Y. K. Shin, dan S. H. Kang, "Physicochemical properties of liquid infant formula stored at different temperatures," *Korean J. Food Sci. Anim. Resour.*, vol. 38, no. 5, hal. 995–1007, 2018.
[6] A. K. M. Masum, J. Chandrapala, T. Huppertz, B. Adhikari, dan B. Zisu, "Effect of storage conditions on the physicochemical properties of infant milk formula powders containing different lactose-to-maltodextrin ratios," *Food Chem.*, vol. 319, no. October 2019, hal. 126591, 2020.
[7] AOAC, *Official Methods of Analysis of The Association of Official Analytical Chemist, 14 ed.* Airilington: AOAC Inc, 1995.
[8] B. Koç, M. S. Yilmazer, P. Balkir, dan F. K. Ertekin, "Moisture sorption isotherms and storage

stability of spray-dried yogurt powder," *Dry. Technol.*, vol. 28, no. 6, hal. 816–822, 2010.

[9] P. Kumar dan H. N. Mishra, "Storage stability of mango soy fortified yoghurt powder in two different packaging materials: HDPP and ALP," *J. Food Eng.*, vol. 65, no. 4, hal. 569–576, 2004.

[10] T. M. Ho, S. Chan, A. J. E. Yago, R. Shravya, B. R. Bhandari, dan N. Bansal, "Changes in physicochemical properties of spray-dried camel milk powder over accelerated storage," *Food Chem.*, vol. 295, no. September 2018, hal. 224–233, 2019.

[11] J. E. Hyun, J. Y. Kim, E. M. Kim, J. C. Kim, dan S. Y. Lee, "Changes in Microbiological and Physicochemical Quality of Dried Persimmons (Diospyros kaki Thunb.) Stored at Various Temperatures," *J. Food Qual.*, vol. 2019, 2019.

[12] H. Yu, Y. Zheng, dan Y. Li, "Shelf life and storage stability of spray-dried bovine colostrum powders under different storage conditions," *J. Food Sci. Technol.*, vol. 52, no. 2, hal. 944–951, 2013.

[13] B. I. Davis, A. Siddique, dan Y. W. Park, "Effects of Different Storage Time and Temperature on Physicochemical Properties and Fatty Acid Profiles of Commercial Powder Goat Milk Products," *Adv. Dairy Res.*, vol. 5, no. 4, 2017.

[14] V. Siracusa, "Food packaging permeability behaviour: A report," *Int. J. Polym. Sci.*, vol. 2012, 2012.

[15] L. Kubík dan S. Zeman, "Determination of oxygen permeability and mechanical properties of the aluminium paper foils," *Res. Agric. Eng.*, vol. 60, no. 4, hal. 172–179, 2014.

[16] L. D. Pratiwi, "Determination of Cookies,s Shelf Life Sources of Protein and Energy from Mixed Flour Based Mocaf by Using Packaging Variations," Universitas Negeri Sebelas Maret Surakarta, 2017.

[17] W. Katekhong dan S. Charoenrein, "Color and gelling properties of dried egg white: Effect of drying methods and storage conditions," *Int. J. Food Prop.*, vol. 20, no. 9, hal. 2157–2168, 2017.

[18] A. S. Pereyra Gonzales, G. B. Naranjo, G. E. Leiva, dan L. S. Malec, "Maillard reaction kinetics in milk powder: Effect of water activity at mild temperatures," *Int. Dairy J.*, vol. 20, no. 1, hal. 40–45, 2010.

[19] S. Chudy, J. Pikul, M. Rudzinka, dan A. Makowska, "The Effect of Storage on Physicochemical Properties of Spray-Dried Milk, Egg and Milk-Egg Mixture," vol. 22, no. 1, hal. 17–26, 2015.

[20] A. K. Chauhan dan V. Patil, "Effect of packaging material on storage ability of mango milk powder and the quality of reconstituted mango milk drink," *Powder Technol.*, vol. 239, hal. 86–93, 2013.

[21] K. Rossini, C. P. Z. Noreña, dan A. Brandelli, "Changes in the color of white chocolate during storage: Potential roles of lipid oxidation and non-enzymatic browning reactions," *J. Food Sci. Technol.*, vol. 48, no. 3, hal. 305–311, 2011.

IOP Conf. Series: Materials Science and Engineering 1011 (2021) 012035 doi:10.1088/1757-899X/1011/1/012035

Antioxidant compound from *Myristica fatua* Houtt Leaves

S Fajriah[1]*, M Megawati[1], A Darmawan[1], S Hudiyono[2], S Kosela[2] and M Hanafi[1]

[1]Research Center for Chemistry, Indonesian Institute of Sciences
Kawasan PUSPIPTEK, Serpong, Tangerang Selatan, Banten 15314, Indonesia
[2]Department of Chemistry, Faculty of Mathematics and Natural Sciences, Universitas Indonesia, Kampus UI Depok, Jawa Barat 16424, Indonesia

*Corresponding author: sofafajriah@gmail.com

Abstract. Antioxidant plays an important role in reducing the risk of chronic diseases. By using DPPH radical scavenging assay-guided fractionation, a bioactive compound has been isolated from the methanol extract of the leaves of *Myristica fatua* Houtt and subsequently identified as methyl 3,4-dihydroxybenzoic (MDB). The structure of MDB was confirmed by UV, FT-IR, MS, 1- and 2-D NMR spectroscopy. This compound exhibit antioxidant activity together with previous compounds *i.e* (7S,8R,8′S,7′S)-7,7′-bis(3-hydroxy-5-methoxyphenyl)-8,8′-dimethylbuthane-7,7′-diol and 3”-hydroxy demethyldactyloidin with IC_{50} values 7.96, 101.1 and 15.4 μg/mL, respectively. These findings demonstrate that MDB can be considered as a natural antioxidant.

1. Introduction

Antioxidant have been widely reported to plays an important role in the prevention of aging and degenerative diseases such as cancer, atherosclerosis, and Alzheimer’s, which are closely related with the production of reactive oxygen and nitrogen species [1]. Free radicals and other reactive oxygen species (ROS), such as superoxide anion, hydroxyl radical and hydrogen peroxide are an entire class of highly reactive molecules derived from the normal metabolism of oxygen or from exogenous factors and agents [2]. Plants are believed to play a vital role in the prevention or progression of the degenerative diseases, which rich sources of natural antioxidants [3]. Therefore, there is a great deal of interest in natural antioxidants in view of their health implications such as *Myristica fatua* Houtt leaves from Indonesian medicinal plants.

Myristica fatua Houtt is a plant species originating from India, Borneo, Sulawesi, Maluku, Philippines, New Guinea, Solomon Islands, New Hybrid, Fiji, Samoa, Vanuatu and New Caledonia [4]. *M. fatua* extracts possessed several pharmacological activities, such as antioxidant, antibacterial, anti-parasitic and anti-proliferative against A549, DLD-1, DU145 and MCF-7 cancer cells [5-8]. The secondary metabolites of *M. fatua* mainly are lignan and polyketide such as malabaricon A, malabaricon C, licarin A, (7S,8R,8′S,7′S)-7,7′-bis(3-hydroxy-5-methoxyphenyl)-8,8′-dimethylbuthane-7,7′-diol and 3”-hydroxy demethyldactyloidin [8-9]. Essential oils (α-chopaene and caryophillene) are also found in *M. fatua* [10]. Our preliminary study from the methanol extract of *M. fatua* leaves showed a potency as an antioxidant (DPPH) with IC_{50} 26.5 μg/mL. Thus, the present study was undertaken to isolate and identify methyl 3,4-dihydroxybenzoic (MDB) from *M. fatua* leaves, their antioxidant activity was also examined using DPPH method including (7S,8R,8′S,7′S)-7,7′-bis(3-hydroxy-5-methoxyphenyl)-8,8′-dimethylbuthane-7,7′-diol and 3”-hydroxy demethyldactyloidin from previous research [9].

2. Research methods

2.1. Plant material

M. fatua leaves was collected from Mekongga forest, Southeast Sulawesi, Indonesia. The plant was identified at Research Center for Biology, Indonesian Institute of Sciences (LIPI) with voucher specimen UHA-55.

2.2. General experiments ormatting author names

The ^{1}H, ^{13}C, HMQC, and HMBC NMR spectra were measured JEOL JNM-ECA 500 MHz spectrometer using chloroform-d ($CDCl_3$). Tetramethylsilane (TMS) was used as an internal standard. The molecular weight was obtained, using LC-MS Mariner Biospectrometry with ESI (Electrospray Ionisation) system. FTIR and UV/Vis spectra were recorded using Shimadzu Type IR Prestige-21 and Agilent Technologies (Cary 60 UV-Vis), respectively.

2.3. Extraction and Isolation

The dried powder of the leaves (1.37 kg) was extracted three times with methanol by maceration at room temperature for 3 X 10 L. The mixture was filtered and the filtrate was concentrated on a rotary evaporator under reduced pressure at 45 °C to afford 99.6 g of extract. A successive partition of methanol extract (45.6 g) using Vacuum Liquid Chromatography yielding n-hexane (66.1 mg), ethyl acetate (EtOAc) (35.7 g) and residue (2 g) fractions. Subsequently, these fractions were subjected to an assay of antioxidant activity.

The bioassay-guided fractionation was only carried out with ethyl acetate fraction because it was the most active antioxidant. The EtOAc fraction was subjected to silica gel chromatography (200-300 mesh, Merck). Elution was performed with a gradient elution of n-hexane, ethyl acetate and methanol to obtain 12 primary fractions (Fr. 1- Fr. 12). All the fractions were assayed for antioxidant activity by DPPH method (Table. 1). The most active fraction Fr. 5 (3.0 g) eluted with gradient elution (n-hexane-ethyl acetate-methanol) to obtain 10 sub-fractions (Fr. 5-1 until Fr. 5-10). Sub-fractions Fr. 5-3 and Fr. 5-4 further purified using centrifugal chromatography and Sephadex LH-20 to obtained (7S,8R,8′S,7′S)-7,7′-bis(3-hydroxy-5-methoxyphenyl)-8,8′-dimethylbuthane-7,7′-diol and 3”-hydroxy demethyldactyloidin [9], but their antioxidant activity have not been done. Then, Fr. 5-5 (16.53 mg) was further purified by Sephadex LH-20 (Sigma) eluted with dichloromethane-methanol (1:1) to obtain compound 3 (4 mg). Compound **3** was further identified using spectroscopy methods and examined the antioxidant activity of compound **1**-**3**.

2.4. DPPH free radical scavenger activity assay [11]

The antioxidant assay was performed on the scavenging effect of DPPH. Briefly, 4 mg of sample was dissolved in 4 mL DMSO to obtain 1000 µg/mL as sample mother solution. The sample solution was diluted with ethanol to obtain various sample concentrations (10, 20, 50, 100, and 200 µg/mL). The test sample was mixed with the ethanol solution of 300 µM DPPH (Sigma) in 90-well microtiter plate and incubated at 37°C for 30 minutes. The absorption was measured with a spectrophotometer at λ 515 nm. Inhibition percentage of the sample was compared to control (DMSO). Gallic acid used as positive controls. Percentage of inhibition was calculated by the equation: [1-(B/A)] x 100%; whereas A is absorbance in the absence of sample (blank) and B is absorbance in the presence of sample. IC_{50} value denotes the concentration of sample required to scavenge 50% of the DPPH free radical.

3. Results and Discussion

A rapid, simple and inexpensive method to evaluate antioxidant activity involves the use of the free radical, 2,2-diphenyl-1-picrylhydrazyl (DPPH). DPPH radical scavenging activity assay is commonly used to test the ability of compounds to act as free radical scavengers or hydrogen donors and to evaluate antioxidant activity [2]. It has also been used to quantify antioxidants in complex biological systems in recent years. The methanol extract of *M. fatua* leaves was screened for antioxidant activity with IC_{50}

26.5 μg/mL. This extract was subjected to activity-guided repeated fractionation on Kieselgel 60G with n-hexane, ethyl acetate, methanol as mobile phase using Vacuum Liquid Chromatography technique. The ethyl acetate fraction exhibited the highest radical scavenging activity with IC_{50} 58.73 μg/mL (Table 1). A further bio-guided fractionation of the ethyl acetate fraction on silica gel led to twelve fractions with fraction 5 (F-5) has the highest antioxidant activity (IC_{50} 16.5 μg/mL, Table 1). Further purification using column chromatography technique afforded ten sub-fractions. Sub-fractions Fr. 5-3 and Fr.5-4 had been purified on the previous report and obtained (7S, 8R, 8′S, 7′S) 7,7′-bis(3-hydroxy-5-methoxyphenyl)-8,8′-dimethylbuthane-7,7′-diol (**1**) and 3”-hydroxy demethyldactyloidin (**2**) [9], but their antioxidant activity have not been reported. Sub-fraction Fr. 5-3 further purified on Sephadex LH-20 and obtained compound **3**. This compound identified using spectroscopy methods and evaluated its antioxidant activity.

Table 1. Antioxidant activity of fractions of *M. fatua* leaves

No	Sample	IC_{50} (μg/mL)	No	Sample	IC_{50} (μg/mL)
1	n-hexane fraction	80.88	9	F-6	20.43
2	Ethyl acetate fraction	58.73	10	F-7	19.24
3	Residue	>100	11	F-8	35.87
4	F-1	>100	12	F-9	77.79
5	F-2	>100	13	F-10	>100
6	F-3	67.62	14	F-11	>100
7	F-4	48.46	15	F-12	
8	F-5	16.05			

The compound **3** (4 mg) isolated from the subfraction Fr. 5-5 from ethyl acetate fraction of *M. fatua* leaves with a yellow light oil. Based on UV-Vis spectrophotometer measurement, compound 3 showed maximum wavelength (λ) at 205 and 224 nm similar to aromatic/benzene ring peaks [12]. In FTIR spectrum, compound 3 showed the absorbance peak at wave number 3352 cm^{-1} as vibration of the -OH group supported by bending vibration at wave number 1300 cm^{-1}. Vibration of aliphatic group showed at 2939 and 2860 cm^{-1}. The wave number at ν 1444 cm^{-1} is the bending vibration of the aromatic C-H, while the C-O-C bond absorbs at 1228 and 1039 cm^{-1}. In addition, the sharp peak at ν 1685 cm^{-1} is the vibration of the carbonyl ester group. From LC-MS chromatogram, compound **3** showed peak (100%) $[M + H]^+$ = 169.33 with m / z 168.33 at retention time 5.3 min.

The ^{1}H-NMR spectrum (500 MHz in CD_3OD) of compound 3 showed the presence of an ABX system aromatic ring at chemical shift δ_H 7.40 (1H, d, J = 7.79 and 1.53 Hz, H-6), 7.42 (1H, d, J = 1.95 Hz, H-2) and 6.80 ppm (1H, d, J = 7.79 Hz, H-5). The value of the *J* coupling constant of 1.95 and 7.79 Hz at δH 7.40 ppm (H-6) showed the *meta* position against the proton at a chemical shift of 7.42 ppm (1H, H-2, d, J = 1.53 Hz) and the *ortho* position against the proton at a chemical shift of 6.80 (1H, H-5, d, J = 7.79 Hz). Furthermore, a methoxy group also appeared at δ_H3.3 ppm (3H, s, H-8).

^{13}C-NMR spectra of compound **3** (125 MHz in CD_3OD) showed eight carbons consist of a methoxy group at chemical shift δ_C 52.3 (C-8); three methine aromatics at δ_C 115.9 (C-2); 117.5 (C-5); 123.7 (C-6); and four quaternary carbons at δ_C 168.9 (C-7); 122.7 (C-1); 146.3 (C-3); 151.8 (C-4). A quaternary carbon at δ_C 168.9 ppm indicated the presence of a carbonyl ester [13].

Based on ^{1}H-NMR, ^{13}C-NMR and LCMS data, it could be assumed that compound **3** has a C6-C1 aromatic skeleton. This was supported by HMQC and HMBC correlation data. In the HMQC spectrum, compound **3** indicates the presence of a methoxy in a chemical shift of δ_H3.3 ppm bound to carbon in C-8; and 3 aromatic protons (δ_H 7.42, 6.8 and 7.40 ppm) bounded to C-2, C-5 and C-6, respectively. From the HMBC spectrum, we can see the correlation of H-2 with C-1, C-3, and C-4; H-5 correlated with C-3 and C-4; and H-6 correlated with C-4, and C-5. A methoxy at δ_H 3.3 ppm correlated with a carbonyl at C-7 indicated a methyl ester group (Fig. 1).

Based on spectroscopic data of UV/Vis, FTIR, LCMS, FT-NMR, it can be concluded that compound **3** is a benzoate derivative, methyl 3,4-dihydroxybenzoate. This compound is first report from *M. fatua* leaves.

Table 2. NMR data of compound 3 and methyl 3,4-dihydroxybenzoic.

No	^{1}H-NMR (CD_3OD) (δ_H (ppm), multiplicity, *J* (Hz))	^{13}C-NMR* (δ_C, ppm)	Methyl 3,4-dihydroxybenzoic (Miyazawa *et al.*, 2003) (aseton d_6) (δ_H (ppm), multiplicity, *J* (Hz))	(δ_C, ppm)
1	-	122.7	-	122
2	7.42 (*d*, 2.60)	115.9	7.39 (*d*, 2)	117
3	-	146.3	-	145
4	-	151.8	-	150
5	6.80 (*d*, 7.79)	117.5	6.80 (*d*, 8.3)	115
6	7.40 (*dd*, 2.60 & 7.79)	123.7	7.34 (*dd*, 2.0; 8.3 Hz)	123
7	-	168.9	-	166
8	3.83	52.3	3.80 (*s*)	51

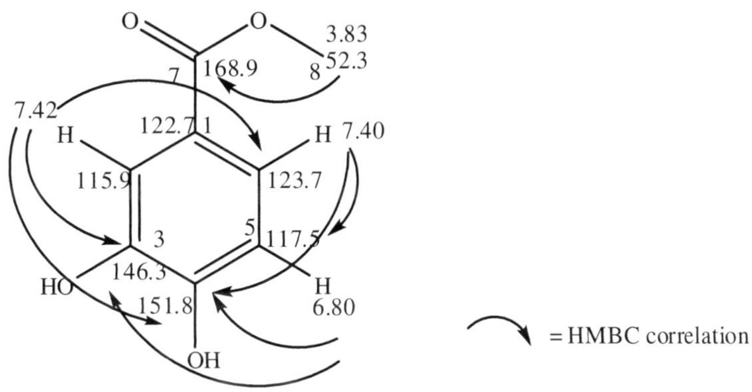

Figure 1. HMBC correlation of compound **3.**

The antioxidant activities of compounds **1**-**3** showed in Table 3. Compound **3** has highest antioxidant activity with IC_{50} 7.94 µg/mL, but lower than gallic acid as positive control.

Table 3. Antioxidant activity of compounds 1-3.

No	Sample	IC_{50} (µg/mL)
1	methyl 3,4-dihydroxybenzoate (**1**)	7.96
2	(7S, 8R, 8′S, 7′S) 7,7′-bis(3-hydroxy-5-methoxyphenyl)-8,8′-dimethylbuthane-7,7′-diol (**2**)	101.07
3	3”-hydroxy demethyldactyloidin (**3**)	15.42
4	Gallic acid (positive control)	2.30

Gallic acid was known as a strong radical scavenger compared to a range of phenolic acids. The number of electron donor hydroxy and methoxy substitutions will increase the stability of the phenoxy radicals so that it affected the radical-scavenging activity of phenolic acids. Therefore, gallic acid with three hydroxyl groups and a carboxyl group was observed to be more active than methyl 3,4-dihydroxybenzoate with two hydroxyl groups and a carboxylic methyl ester group. The methoxy group of methyl 3,4-dihydroxybenzoate may hinder the scavenging effect of the hydroxyl groups by intra- or intermolecular hydrogen bonding [14].

Acknowledgements
The authors are thankful to RISTEKDIKTI for financial support and to Puspa Dewi N. Lotulung for her encouragement in the MS measurement. The first author is main contributor of this research.

References
[1] Parejo I, Viladomat F, Bastida J, Schmeda-Hirschmann G, Burillo J and Codina C 2004 *J. Agric. Food Chem.* **52** 1890-7
[2] Loizzo M R, Said A, Tundis R, Hawas UW, Rashed K, Menichini F, Frega N G and Menichini F 2009 *Plant Foods Hu Nutr.* **64** 264-70
[3] Harrish R, Divakar S, Srivastava and Shivanandappa T 2005 *J. Agric. Food Chem.* **53** 7709-14
[4] Lim T K 2012 *Edible Medicinal and Non-Medicinal Plants. In: Fruit* (New York: Springer).
[5] Desrivot J, Waikedre J, Cabalion P, Herrenknecht C, Bories C, Hocquemiller R and Fournet A. 2007 *J. Ethnopharmacol.* **112** 7-12
[6] Billo M, Fournet A, Cabalion P, Waikedre J, Bories C, Loiseau P, Prina E, de Arias A R, Yaluff G, Fourneau C and Hocquemiller R 2005 *J. Ethnopharmacol.* **96** 195-200
[7] Viveka M R and Chandrashekar K R 2016 *Asian J Pharm. Clin. Res.* **9**(4) p235-239.
[8] Pandey R, Mahar R, Hasanain M, Shukla SK, Sarkar J, Rameshkumar KB and Kumar B 2016 *Food Chem.* **211** 483-93
[9] Fajriah S, Darmawan A, Megawati, Hudiyono S, Kosela S and Hanafi M. 2017 *Phytochem. Lett.* **20** 36-9
[10] Jamal Y and Agusta A. 2004 *Berita Biologi* **7**(3) 155-8
[11] Yen G C and Chen H Y 1995 *J. Agric. Food Chem.* **43** 27-32
[12] Pretsch E, Bühlmann P and Affolteer C 2000 *Structure determination of organic compounds—Tables of spectra data.* (Berlin: Springer)
[13] Miyazawa M, Oshima T, Koshio K, Itsuzaki Y and Anzai J 2003 *J. Agric. Food Chem.* **51** 6953-6
[14] Asnaashari M, Farhoosh R and Sharif A 2014 *Food Chem.* **159** 439-44

IOP Conf. Series: Materials Science and Engineering 1011 (2021) 012036 doi:10.1088/1757-899X/1011/1/012036

Bitterness reduction of Green Chiretta (*Andrographis paniculata*) leaves and its functionality

Kam Natania* and Kezia Haniel

Food Processing Laboratory, Food Technology Department, Faculty of Science and Technology, Universitas Pelita Harapan, Lippo Karawaci, Indonesia

*Email: Natania.fti@uph.edu

Abstract. Green chiretta leaves, known for its bitterness, is an Indonesian native traditional herb, used for its functional and medicinal properties. This research aimed to reduce the bitterness of green chiretta leaves using adsorbent technique and to observe how it might affect the functional properties of the chiretta leaves. The debittering process was done by utilizing diatomite, bentonite, and attapulgite as the adsorbent, together with 5, 10, and 20 minutes of boiling time. The leaves were then subjected to bitterness and phytochemical analysis. The bitterness was analyzed using scalar methods, the antioxidant properties were analyzed by measuring the IC50 value against DPPH solution, and the phytochemical compounds (total phenolic, flavonoid, alkaloid, tannin, and terpenoid) was analyzed quantitatively. Different adsorbent will adsorb different phytochemical compound. As the boiling time increases, the number of phytochemical compounds also reduced, hence resulting in a decrease of bitterness and IC50 value. This study found that the use of attapulgite with 20 minutes of boiling reduce the bitterness up to (90%) of the leaves due to its ability to adsorb terpenoid content while still retained their antioxidant properties better than bentonite or diatomite.

1. Introduction

Bitter taste though undesirable in the plant, is often associated with the functional and medicinal properties of the plant. Green chiretta or sambiloto, known as the "king of bitters", is an herbal plant that has been used as traditional medicine due to its many functional properties. However due to its extreme bitterness, the consumption of the green chiretta is not favorable. Green chiretta is known to contain many Andrographolide compound, a diterpenoid lactone, which also known to possess medicinal properties [1,2]. Therefore, the removal of bitterness might potentially reduce or remove the functional properties of green chiretta leaves. There are some traditional methods used to reduce the bitter taste of vegetables, which includes boiling with the addition of salt, sugar, vinegar, and many more. Other than that, various method has been attempted to reduce bitter taste in foods.

The adsorbent is a solid material that is able to trap molecules and form a thin layer [3]. Some of the adsorbents that are commonly used are activated attapulgite, bentonite clay, and diatomaceous earth. All of these adsorbents are silicon based adsorbent but are differentiated by the principal minerals that are attached to the base. These minerals give different adsorbing characteristics and capacity [3,4].

A research conducted by Amran *et al.* in 2018, shown that the use of adsorbent is able to reduce the bitterness of *Carica papaya* leaves. The adsorbent resins used are Amberlite IRA-67, Diaion WA30, and Diaion WA21J. All of the resins are able to adsorb the saponin component that causes the bitter taste significantly [5]. Another research also removed the bitter taste of orange juices by using amberlite

IOP Conf. Series: Materials Science and Engineering 1011 (2021) 012036 doi:10.1088/1757-899X/1011/1/012036

as an adsorbent. The result proves that the limonin (bitter component) can be reduced up to 90% to 96% at temperature 30°C and 40°C, respectively [6].

In this research, activated attapulgite, bentonite clay and diatomaceous earth were used as adsorbents to remove or reduce the bitterness of green chiretta leaves. Heat treatment was applied to increase the efficiency of bitterness removal. The efficiency of bitterness removal will be correlated with the change in phytochemical compounds and functional properties (radical scavenging activities) before and after adsorbent treatment.

2. Research and Methodology

This research aimed to observe and determine the best type of adsorbent and contact time in terms of bitterness removal and the change in phytochemical properties of green chiretta leaf.

2.1. Material and methods

The materials used in this research were green chiretta leaves obtain from local market (uncurled with approximate length 3-5 cm), activated attapulgite, bentonite clay and diatomite. For phytochemical analysis, the materials used were ethanol, 10% Folin-Ciocalteu reagent, 7.5% Na_2CO_3, 2% $AlCl_3$, DPPH reagent, methanol, 10% acetic acid, 1% ammonium hydroxide solution, petroleum ether, gallic acid, tannic acid, and quercetin. The equipment used in this research were freeze dryer, rotary evaporator and UV-Vis spectrophotometer.

The research aimed to determine the best combination of adsorbent type and treatment in terms of bitterness and phytochemical properties of papaya leaves. The types of adsorbent used were activated attapulgite, bentonite clay and diatomite with various boiling times (5, 10 and 20 minutes). The extraction process and the physicochemical analysis included a Radical Scavenging activities, Total Tannins, Flavonoid, Total Phenolic Content and Alkaloid Content [6-9].

2.2. Panelist training

Panelist training was conducted to improve the ability of the panelist and their consistency to evaluate the sensory characteristics of the products. Panelists were introduced to the terminology, attributes, and intensity of the bitter taste from caffeinated solution as the reference taste. There were 6 training sessions conducted twice a week in the form of Focus Group Discussion (FGD). The duration of each session was 30 - 45 minutes.

The bitterness test was conducted using scalar test [10]. The intensity for bitterness ranges from 0-15 score which is 0 score means the determined taste was not detected, while 15 score means that the determined taste was highly detected. The scalar test was conducted in the standard sensory booth. The panelists were given samples, a questionnaire, and water for rinsing. Each sample was evaluated for two replications. The collected data were processed using IBM SPSS 22.0 using Analysis of Variance (ANOVA).

2.3. Adsorption treatments

Fresh green chiretta leaves were added with different adsorbent and boiled using traditional stove with 3 different boiling time i.e. 5, 10, and 20 minutes. There were 3 different adsorbents used, which were bentonite clay, activated attapulgite, and diatomite with ratios of absorbents and green chiretta leaves used is 3:1. Green chiretta leaves were then analyzed for its bitterness and phytochemical compounds.

3. Results and Discussion

3.1. Effect of adsorbent and boiling time towards bitterness intensity

Based on a statistical analysis of sensory evaluation, the adsorbent, boiling time, and the interaction between adsorbent and boiling time give significant effect ($p<0.05$) on the bitterness of the green chiretta leaves. Table 1 shows the sensory result of the samples while Figure 1. shows the percentage reduction of bitterness intensity in the treated samples compared to the control. According to literature, papaya leaf has a bitterness intensity of 11.4±0.42, while control green chiretta leaves in this research has bitterness intensity of 13.95±0.40 when evaluated by the scalar method [3]. This result indicates that green chiretta leaves are bitter than papaya leaves. Other than that, green chiretta leaves have a bitterness value of 2.86±1.74 x 103 units/g or equivalent to 1.43±0.87% w/w of quinine hydrochloride. The bitterness value is expressed in units comparable to the bitterness of a solution that contains 1 g of quinine hydrochloride in 2000 ml [11].

Table 1. Bitterness intensity.

Sample	Bitterness Intensity
Control	13.95±0.40[g]
Diatomite 5 minutes	13.18±0.57[g]
Diatomite 10 minutes	11.98±0.60[f]
Diatomite 20 minutes	9.86±0.42[e]
Bentonite 5 minutes	12.85±0.58[g]
Bentonite 10 minutes	10.17±0.77[e]
Bentonite 20 minutes	6.49±0.46[d]
Attapulgite 5 minutes	4.47±0.30[c]
Attapulgite 10 minutes	3.57±0.30[b]
Attapulgite 20 minutes	1.37±0.28[a]

Note: control = 13.95±0.40
Different notations (letters) indicates a significant difference ($p<0.05$)

Figure 1 displays that all adsorbents are able to significantly reduce the bitter taste of the chiretta leaves. The adsorbents are able to reduce the bitter taste the most were activated attapulgite, while the least was diatomite. It can also be seen that as the boiling time increases, the bitterness intensity decreases. The most effective boiling time that can reduce the bitter taste is 20 minutes, while the least is effective is 5 minutes. Figure 1 demonstrates that the application of attapulgite for 20 minutes was able to reduce the bitterness up to 90.18%, while diatomite with boiling for 20 minutes was only able to reduce 29.32% of bitterness, and bentonite with boiling for 20 minutes reduced 53.48% of bitterness in the leaves.

Figure 1. Bitterness reduction of green chiretta leaves.

It is known that the compound responsible for the bitter taste in green chiretta leaves is Andrographolide, which is classified as a diterpenoid lactone [1]. Terpenoid compounds are naturally hydrophobic; thus, the adsorbents capability of adsorbing terpenoid compounds depend on their hydrophobicity properties [3]. Activated attapulgite has a hydrophobic surface, while bentonite clay and diatomaceous earth have a highly hydrophilic surface. Since activated attapulgite has the highest ability to adsorb terpenoid due to its hydrophobicity properties, the samples treated by activated attapulgite were the least bitter compared to others [12]. Diatomaceous earth (diatomite) adsorbs the least terpenoid compounds followed by bentonite clay, therefore the bitterness of the samples follows the adsorbent capability trend.

3.2. Effect of adsorbent and boiling time towards total phytochemical compounds

The use of adsorbents and methods of application also affect the amounts of phytochemical compounds present in leaves. Phytochemical compounds can be defined as the compounds naturally found in plants that have biological activities. Dietary phytochemicals can be classified into 6 different groups, *i.e.* carotenoids, phenolics, alkaloids, nitrogen-containing compounds and organosulfur compounds [8,13]. The bioactive constituents found in green chiretta are terpenoids, alkaloids, saponins, tannins, flavonoids, phenols, and cardiac glycosides [2,14]. In this current research, the phytochemical compounds (phenolics, flavonoids, tannins and alkaloids) were analyzed before and after treatment with the adsorbent.

Table 2. Total phytochemical compounds of Green Chiretta leaves after adsorption treatment.

	Terpenoid (%)	Phenolic (mg GAE/g extract	Flavonoid (mg QE/g extract)	Alkaloid (%)	Tannin (mg TAE/g extract)
Control	17.92±0.28	174.58 ±4.13	61.59±0.44	46.46±1.26	36.71±0.57
Diatomite 5 minutes	16.74±0.50^{g}	159.40±6.10^{e}	55.08±0.30^{e}	27.83±0.79^{g}	34.02±0.51^{e}
Diatomite 10 minutes	16.85±0.56^{g}	147.28±9.43^{d}	49.70±0.80^{d}	22.11±0.70^{e}	32.41±0.68^{e}
Diatomite 20 minutes	12.65±0.03^{e}	119.62±6.52^{a}	47.88±0.21^{c}	16.16±0.15^{c}	23.91±1.33^{bc}
Bentonite 5 minutes	15.49±0.47^{f}	136.95±2.90^{c}	50.60±0.91^{d}	20.11±0.92^{d}	34.13±0.68^{e}
Bentonite 10 minutes	12.78±0.12^{e}	127.42±1.66^{b}	42.16±1.13^{b}	13.49±0.76^{b}	25.88±1.15^{cd}
Bentonite 20 minutes	9.68±0.13^{d}	121.94±2.13^{ab}	38.71±0.49^{a}	10.03±0.20^{a}	22.88±1.33^{ab}
Attapulgite 5 minutes	7.61±0.12^{c}	155.79±1.66^{e}	54.20±0.31^{e}	23.96±0.02^{f}	33.85±0.56^{e}
Attapulgite 10 minutes	6.71±0.03^{b}	147.78±8.85^{d}	50.75±0.21^{d}	20.38±0.93^{d}	26.33±0.62^{d}
Attapulgite 20 minutes	3.81±0.02^{a}	118.73±10.25^{a}	47.49±0.07^{c}	15.45±0.68^{c}	21.88±0.51^{a}

*Different notations (letters) indicates a significant difference ($p<0.05$)

Table 2 shows that all adsorbents significantly reduced the phytochemical content of the chiretta leaves after treatment. As the boiling time increases, most of the phytochemical content of the leaves was also significantly reduced. However different adsorbents have different susceptibility toward the phytochemical compound. Bentonite has the highest CEC (Cation Exchange Capacity) and specific surface area among the adsorbents used, followed by attapulgite and diatomite respectively. Montmorillonite, which is the main component of bentonite, has a specific surface area range from 760-810 m2/g and CEC ranging from 70-120 meq/100 g [15, 3]. In contrast, the acid and heat activated attapulgite have a surface area of 399 m2/g and 210 m2/g respectively, with CEC ranging from 20-30 meq/100 g [3, 4]. As for diatomaceous earth, the specific surface area ranges from 0.7-180 m2/g with CEC of 22.21 meq/100 g [3, 16].

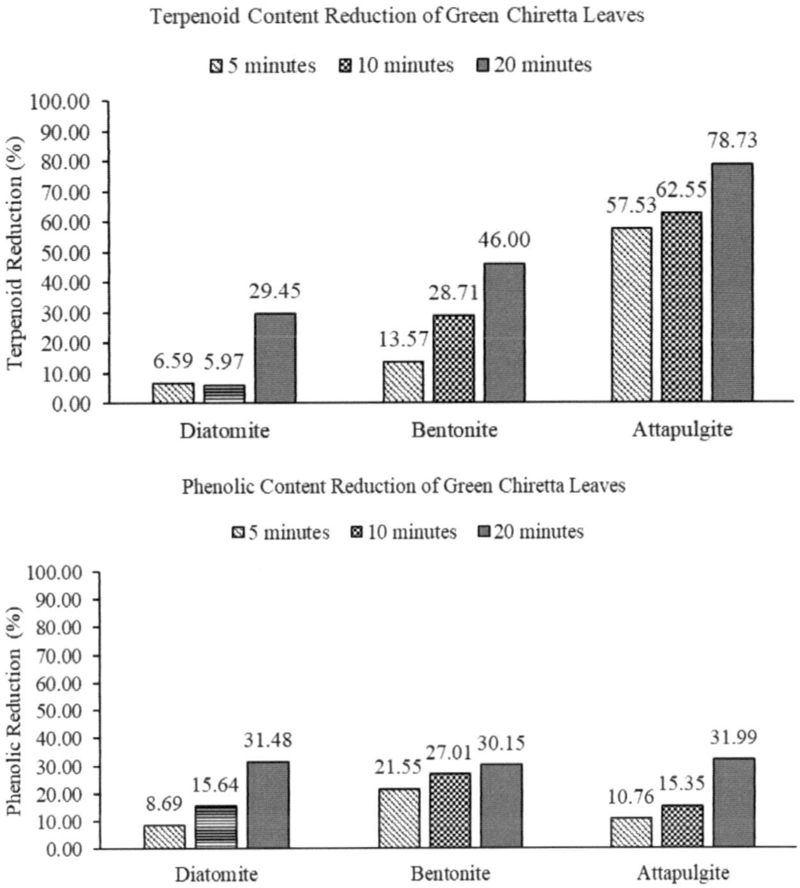

Figure 2. Percentage reduction of terpenoid and phenolic content of Green Chiretta leaves after adsorbent treatment.

Terpenoid is a chemical compound that has carbon backbones made up of isoprene. Terpenoid can be further classified into several groups based on their number of isoprene, such as hemiterpene, monoterpene, sesquiterpene, diterpene, sesterpene, triterpene, and tetraterpene [6, 2]. Terpenoid in green chiretta leaves, plays a major role in providing functional properties. The main component in green chiretta leaves is andrographolide, which is a labdane diterpenes. Activated attapulgite which has the highest surface hydrophobicity, is able to adsorb more terpenoid compound, than bentonite and diatomite. Attapulgite itself is a natural clay that contains hydrous magnesium aluminum silicate. The activation of the clay enhances the adsorption ability and surface hydrophobicity thus making it efficient in removing terpenoid compounds [17, 18].

The reduction of the terpenoid content in attapulgite might affected the functionality of the green chiretta leaves. Andrographolide provides many of functional properties, such as anticancer, anti-inflammatory, antihyperglycemic, antimicrobial, and many more [14, 21], and this bitter compound has correlations with the functional properties of the green chiretta leaves. Zhang, *et.al.*, 2018, reported several bitter andrographolide families that responsible for the bitter taste of the green chiretta *i.e.* neoandrographolide, 14-deoxyandrographolide, and dehydroandrographolide. Therefore, the removal of terpenoid might affected some of the functional properties of the green chiretta, however the research conducted by Zhang, also showed possibilities of other unknown bitter-related substances that also possess some sort of functional properties [21].

Phenolics are products of secondary metabolism in plant that contain one or more aromatic rings with one or more hydroxyl groups in its structure. In plants, phenolic contributes to the reproduction and growth, defense mechanism, and color of the plant. Phenolic compounds can be divided into several groups such as phenolic acids, flavonoids, tannins, stilbenes, and coumarins [8]. Bentonite and diatomite are more efficient in removing phenolic and flavonoid compounds better than attapulgite. The result of the alkaloid content is in accordance with the specific surface area of the adsorbents used. Higher specific surface area enables the adsorbent to adsorb more compounds. Many studies have been conducted to evaluate plant alkaloid modified bentonite towards pesticide [19]. These researches imply that bentonite is able to react with alkaloid and thus may reduce the alkaloid content in the leaves. Modified bentonite can adsorb 0.14 mmol/g of alkaloid at the equilibrium of 40 hours of contact time [20].

3.3. Effect of adsorbent and boiling time towards radical scavenging activity

Radical Scavenging Activity (IC_{50}) is defined as the concentration needed to scavenge 50% of free radicals. Higher IC_{50} value indicates lower radical scavenging activity and antioxidant properties. All treated samples showed that the application of adsorbent reduces the antioxidant activities of green chiretta to a certain level.

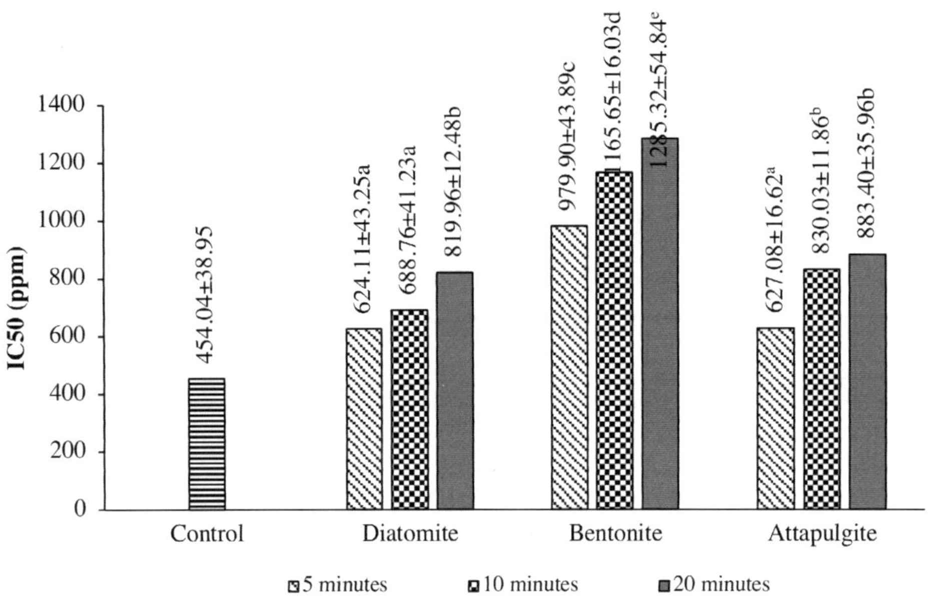

Figure 3. Radical scavenging activities (IC_{50}) values of Green Chiretta after adsorbent treatment.

All of the phytochemical compounds contribute to the antioxidant activity of the green chiretta leaves; thus, the removal of the phytochemical compounds might decrease the antioxidant activity. Phenolic compounds can act as antioxidants due to their ability to donate hydrogens and chelate metal ions similarly both flavonoids and tannins can act as antioxidants by scavenging free radicals [8,13]. Several studies had also shown the contribution of alkaloids towards antioxidant activity [13,22]. From this research, it was observed that the leaves treated with adsorbent that removed the most phenolic content showed the least antioxidant activity. Leaves that were treated using diatomite had the least phenolic content and lowest antioxidant activity. Figure 3 shows that green chiretta leaves that were boiled with active attapulgite were able to retain the highest antioxidant properties. While leaves that were boiled with bentonite had the lowest antioxidant activity.

3.4. Correlation between radical scavenging activities, bitterness, and phytochemical compounds

The mean value of IC_{50}, percentage changes in bitterness, and phytochemical compounds were analyzed statistically with Pearson's correlation to know the correlation between the parameter analysis. Table 3 shows the correlation between IC_{50} value, bitterness, phytochemical compounds

Table 3. Correlation between IC_{50}, bitterness, and phytochemical compounds.

	IC_{50}	Bitterness	Phenolic	Flavonoid	Tannin	Terpenoid	Alkaloid
Bitterness	-0.076	1	0.239	0.138	0.557	0.989**	0.310
IC_{50}	1	-0.076	-0.707*	-0.919**	-0.587	-0.140	-0.888**

Note: * correlation is significant at the 0.05 level
** correlation is significant at the 0.01 level

Table 3 shows that only terpenoid were found to have a significant positive correlation with bitterness, compared to other phytochemical compounds. This result further supports the literature stating that the terpenoid compound (Andrographolide) contributes to the bitterness in green chiretta leaves [1]. Hence, the adsorbent that result in the least bitter leaves, was the one that can adsorb the most terpenoid compound. Among the adsorbents, attapulgite was able to adsorb the most terpenoid compound due to its hydrophobic surface. Thus, the adsorbent was able to reduce the highest percentage of bitterness. On the other hand, diatomite and bentonite adsorb less terpenoid compound due to its high hydrophilic surface.

The loss of bitterness did not have a significant correlation with the antioxidant activity of the green chiretta leaves. The antioxidant activity was significantly correlated with the amount of phenolic, flavonoid, and alkaloid compounds. The negative correlation indicates that the reduction of those phytochemical compounds, results in the increase of IC_{50} value or lower antioxidant properties. The result is coherent with the theory that state phenolic contents are proven to have strong positive relation with antioxidant properties of fruits and vegetables [13]. Flavonoid is also part of phenolic compounds and thus give antioxidant potential. Other than that, alkaloids also have been proven to give an antioxidant effect higher than phenol [22,8]. Attapulgite was the adsorbent that can reduce the highest amount of bitterness but has less adsorption rate toward phenolic, flavonoid, and alkaloid compound. Taken together, the green ciretta leaves treated with attapulgite still regain their antioxidant compound better than other adsorbents while having the highest reduction in the bitterness intensity.

4. Conclusion

Taken together, all of the adsorbents (diatomite, bentonite, and attapulgite), boiling time (5, 10 and 20 minutes), and interaction between adsorbents and boiling time are affecting the bitterness and functional properties of green chiretta leaves significantly ($p<0.05$). There is no significant correlation between bitterness and IC_{50}, however bitterness was found to be significantly correlated with terpenoid content. While IC_{50} value has significant correlation with phenolic, flavonoid, and alkaloid content. Bentonite clay was able to adsorb the highest phenolic, flavonoid, and alkaloid content of the leaves due to its high specific surface area. While activated attapulgite was able to adsorb the highest terpenoid content due to its hydrophobic surface. Activated attapulgite with 20 minutes of boiling was able to reduce the highest bitterness of the leaves due to its ability to adsorb the most terpenoid content (90%). Despite all that, the green chiretta leaves treated with activated attapulgite for 20 minutes also retained their antioxidant properties, with IC_{50} 883.40, better than bentonite (IC_{50} 1285.32).

Further studies on optimization of the adsorption method and their effects toward other functional properties of green chiretta such as antimicrobial, anti-inflammatory, and antidiabetic can also be analyzed.

References

[1] Gosh B K, Datta A K, Mandal A and Halder S 2012 An overview on *Andrographis paniculata* (Burm. F.) Nees *Int. J. Res. Ayurveda Pharm.* **3**(6) 752-60

[2] Nagajothi S, Mekala P, Raja A and Senthilkumar A 2018 *Andrographis paniculata*: qualitative and quantitative phytochemical analysis *J. Pharmacogn. Phytochem.* **7**(4) 1251-3

[3] Schoonheydt R and Johnston C T 2006 Chapter 3 surface and interface chemistry of clay minerals *Dev. Clay Sci.* **1** 87-113

[4] Boudriche L, Calvet R, Hamdi B and Balard H 2011 Effect of acid treatment on surface properties evolution of attapulgite clay: An application of inverse gas chromatography *Colloids Surf. A Physicochem. and Eng. Asp.* **39**(1) 45-9

[5] Amra S N S, Abidin N Z, Hashim H and Zubair S I 2018 Saponin bitterness reduction of *Carica papaya* leaf extracts through adsorption of weakly basic ion exchange resins *J Food Qual.* **2018** 5602729

[6] Malik S K, Ahmad M and Khan F 2017 Qualitative and quantitative estimation of terpenoid contents in some important plants of Punjab, Pakistan *Pakistan J. Sci.* **69**(2) 150-5

[7] Mohan A C 2017 Phytochemical analysis and screening of total flavonoid, tannin and phenolic contents in *Croton bonplandianum* leaf and stem *World J. Phar. Res.* **6**(4) 1066-76

[8] Agati G, Azzarelo E, Pollastri S and Tattini M 2012 Flavonoids as antioxidants in plants: Location and functional significance *Plant Sci.* **196** 67-76

[9] Azwanida NN 2015 A Review on the extraction methods use in medicinal plants, principle, streght, and limitation *Med. Aromat. Plants* **4** 196

[10] Meilgaard M C, Carr B T and Van G 2007 *Sensory Evaluation Techniques, Fourth Edition* (Boca Raton: CRC Press)

[11] Nas G and Karatas S 2017 Removal of bitterness by using of amberlite in orange juices *INTERNATIONAL JOURNAL OF ELECTRONICS, MECHANICAL AND MECHATRONICS ENGINEERING* **7**(2) 1419-32

[12] Sedai B, Khatiwada B K, Mortazavian H and Blum F D 2016 Development of superhydrophobicity in fluorosilane-treated diatomaceous earth polymer coatings *Appl. Surf. Sci.* **386** 178-86

[13] Minatel I O, Vanz Borges C and Ferreira M I 2017 *Phenolic Compounds: Functional Properties, Impact of Processing and Bioavailability Phenolic Compounds - Biological Activity* Ed. 1 (London: Intech Open Science) chapter 1 pp 1-2.

[14] Nyeem M A B, Mannan MA, Nurruzaman M, Kamrujjaman KM and Das SK 2017 Indigenous king of bitter *(Andrographis paniculata)*: a review *J. Med. Plants Stud.* **5**(2) 318-24

[15] Kobayashi I, Owada H, Ishii T and Lizuka A Evaluation of specific surface area of bentonite-engineered barriers for Kozeny-Carman law," *Soils Found.* **57**(5) 683-97

[16] Gu Q Y, Wu G and Lu X N 2012 Adsorption of volatile phenol in coking wastewater by diatomite," *Adv. Mater. Res.* **573-574** 648-53

[17] Haden W L and Schwint A 1967 Attapulgite: its properties and applications *Ind. Eng. Chem.* **59**(9) 58-69

[18] Deng Y, Wang L, Hu X, Liu B, Wei Z, Yang S and Sun C 2012 Highly efficient removal of tannic acid from aqueous solution by chitosan-coated attapulgite *Chem. Eng. J.* **181-182** 300-6

[19] Tha-in S, Anh D H and Dumri K The enhanced carbamate adsorption of modified bentonite with *Coscinium fenestratum Int. J. Environ. Sci. Dev.* **4**(4) 415-8

[20] Oliveira M F, Da Silva M and Vieira M G 2019 Equilibrium and kinetic studies of caffeine adsorption from aqueous solutions on thermally modified Verde-lodo bentonite *App. Clay Sci.* **168** 366-73

[21] Zhang X, Wu H, Yu X, Luo H, Lu Y, Yang H, Li X, Li Z, Tang L and Wang Z 2018 Determination of bitterness of Andrographis herba based on electronic tongue technology and discovery of the key compounds of bitter substances *Molecules* **23**(12) 3362

[22] Gan J, Feng Y, He Z, Li X and Zhang H 2017 Correlations between antioxidant activity and alkaloids and phenols of Maca *(Lepidium meyenii) J. Food Qual.* **2017** 3185945

[23] Sunday AJ, Obodezie O, Uford I, Abubaka M and Garba M 2010 Quality control tests on *Andrographis paniculata* Nees (Family: Acanthaceae)–an Indian 'Wonder'plant grown in Nigeria *Trop. J. Pharm. Res.* **9**(4) 387-94

IOP Conf. Series: Materials Science and Engineering 1011 (2021) 012037 doi:10.1088/1757-899X/1011/1/012037

Characterization of solid perfume based on Cocoa Butter with Jasmine Oil as fragrance

Melati Septiyanti[1,a)], Yenny Meliana[1,b)], Novia Suryani[2,c)] and Hendrawati[2,d)]

[1]Indonesian Institute of Sciences, Research Center for Chemistry, Serpong –Tangerang Selatan 15314, Indonesia
[2]Chemistry Department, Faculty of Science and Technology, Universitas Islam Negeri Syarif Hidayatullah, Jakarta, Indonesia

Corresponding author: [a)]mela003@lipi.go.id
[b)]yenn001@lipi.go.id
[c)]noviaaass@gmail.com
[d)]hendrawati.wibowo@uinjkt.ac.id

Abstract. Cocoa butter can be obtained from by-product of chocolate powder processing from the result of mechanical pressing using hydraulic press. As a source of vitamin E, cocoa butter in cosmetic has some advantages which are to soften and moisturize the skin. Meanwhile, jasmine can be used as a fragrance in perfume and cosmetics. In this study, cocoa butter was used as raw material for solid perfume. Its optimal concentration in the product was determined. Solid perfume was made by melting cocoa butter and beeswax at 90°C followed by adding jasmine oil as fragrance. The formulation variation of solid perfume was done with cocoa butter concentration 10%, 20%, 30%, 40% and 50% (w/w) and addition of patchouli oil as fragrance fixative agent. The following tests were carried out on the products: functional group analysis, hardness test, homogeneity test, physical stability test, melting point test, antioxidant test, and organoleptic test. The result showed that the variation of cocoa butter concentration affected the physical characteristics of the solid perfume. The optimum product of solid perfume based on panelists' organoleptic test was achieved with cocoa butter concentration of 30%. It had an organoleptic score of 3.63, antioxidant activity with IC_{50} value of 201.98μg/mL, hardness penetration 14.16 mm/second, melting point 90°C and pH value 4.

1. INTRODUCTION

Cocoa butter, known as theobroma oil, is a natural fat obtained from cocoa beans processing by mechanical pressing using a hydraulic press or by chemical processing such as extraction. Cocoa butter is widely used as basic ingredient in food, pharmaceutical and cosmetics industries. It contains compounds that are beneficial for skin, namely stearic acid, palmitic acid, oleic acid also vitamins A, C and E. The addition of cocoa butter in cosmetics can moisturize, smoothen, and soften the skin. Cocoa butter has melting point around 30-35°C which is lower than the temperature of the human body thus it seeps into the skin quickly [1]. Cacao butter consists of fatty acid mostly oleic acid, triacylglycerols and γ-tocopherol [2]. Oleic acid can disrupt skin barriers thus it can enhance permeability on stratum corneum layer [3]. The fat content prevents evaporation on the skin and provides lubrication for skin surface which maintains the lipid content of the skin thus it has a good texture. Cocoa butter also has similar lipid compositions with skin lipid composition [4].

Many cosmetic products use cocoa butter as raw material such as cream, sunscreen, lipstick, soap and shampoo [5]. In this study, cocoa butter was used as raw material of solid perfume. Solid perfume was made by mixing the wax base, carrier oil, and essential oil as fragrances followed heating process. The liquid product was poured into a container and allowed to harden at room temperature. Hardiyati *et al.*, has made solid perfume using the essential oils aroma of vanilla, jasmine and sweet orange. The ingredients of solid perfume consisted of carrageenan, stearic acid and cetyl alcohol as wax base, triethanolamine, acrylonitrile butadiene styrene (ABS), preservatives and patchouli essential oils as fixatives [6].

Essential oils are widely used for aromatherapy through respiration which allows to penetrate into the human body. Cosmetics with aromatherapy purpose used essential oils in skin, body, face and hair products. It is also safe to use as raw material for perfume [7], [8]. One of the essential oils that often used as fragrances or perfume is jasmine oil. Besides the nice aroma, jasmine oil also has a pharmacological function, namely increase breathing speed, increase oxygen absorption in the blood, reduce depression and improve human mood [9].

Studies related to solid perfume made from cocoa butter are still rarely found. Therefore, in this study, optimization of solid perfume formulation was carried out using cocoa butter as raw material and jasmine oil as fragrance. The solid form aims for practical use and storage as well as increase its stability. In this study, characterization of cacao butter as raw material was conducted by antioxidant test, saponification numbers and acid numbers test and analysis of the chemical components of jasmine oil essential oils. Characterization carried out on solid perfume products included functional group analysis, pH test, hardness level test, homogeneity test, stability test and organoleptic test or hedonic test.

2. MATERIALS AND METHODS

2.1. Materials

The materials used in this study namely, beeswax technical grade (brataco, Indonesia), jasmine oil (brataco, indonesia), avocado oil (brataco, Indonesia), patchouli oil from Chemical Research Center LIPI, Indonesia and cocoa butter from Biotechnology Research Center LIPI, Indonesia.

2.2. Methods

2.2.1. The Process of Making Solid Perfume. Cocoa butter and beeswax were melted at 90oC. Jasmine oil as fragrance was mixed with avocado oil and then added to wax compound. After that, patchouli oil was added gradually. The mixture was stirred until homogeneous for 30 minutes then cooled and allowed to harden at room temperature. The composition of solid perfume is shown in Table 1.

Table 1. Variations in Solid Perfume Composition (% weight).

Materials	F1	F2	F3	F4	F5	F6	F7	F8	F9	F10
Cocoa butter	10	20	30	40	50	10	20	30	40	50
Beeswax	50	40	30	20	10	50	40	30	20	10
Avocado Oil	10	10	10	10	10	9.9	9.9	9.9	9.9	9.9
Jasmine Oil	30	30	30	30	30	30	30	30	30	30
Patchouli Oil	-	-	-	-	-	0.1	0.1	0.1	0.1	0.1

2.2.2. Characterization. Cocoa butter as raw material was tested for acid number based on SNI 3748:2009, saponification number based on SNI 3748:2009 and antioxidant test using 2,2-diphenyl-1-picrylhydrazyl (DPPH) method. Jasmine essential oil and patchouli oil were analyzed for the chemical compounds by Gas Chromatography-Mass Spectrometry (GC-MS) Agilent 7890B with MSD 5977A and refractive index by refractometer Atago RX-5000α. Solid perfume products were tested for their physical properties that were pH test using universal pH indicator, hardness level using penetrometer, melting point using the capillary pipe method and analysis of chemical functional groups using Fourier Transform Infra-Red (FTIR) Shimadzu IR Prestige-21.

3. RESULTS AND DISCUSSION

3.1. Acid Number and Saponification Number of Cocoa butter

Characterization of acid number and saponification number of cocoa butter raw material was carried out to determine the quality of the material compared with the SNI 3748:2009 quality standard. Characterization of acid number and saponification number of cocoa butter are shown in Table 2.

Table 2. Acid Number and Saponification Number of Cocoa Butter.

Parameter	Result	SNI Standard
Acid Number (%)	1.19	Max 1.75
Saponification Number (mg/KOH/gr fat)	192.28	181-198

Based on the results, cocoa butter used in this study had good quality with acid and saponification number below the SNI standard' threshold. The presence of excess free fatty acids in cocoa products were undesirable because it indicated low quality product due to moist and unclean storage [10]. The saponification number indicates the amount of fatty acids. Cocoa butter with saponification fat ranging from 190-200 mgKOH/gr contained fatty acids with moderate molecular weight [11].

3.2. Analysis of Essential Oils' Chemical Compounds

Essential oils used in this study were jasmine oil as fragrance and patchouli oil as fixative or scent binder. Samples were analyzed using GCMS to determine the content of chemical compounds in each essential oil. Chromatogram images along with details of jasmine oil compounds are shown in Figure 1 and Table 3.

Based on Edris et al., study, there were about 47 compound components from the extraction of Jasmine Jasminum Sambac flowers from Egypt. The dominant components included benzyl acetate 14.2%; indole 13.4%; E-E-a-farnesene 13.1%; Z-3-hexenyl benzoate 9.4%; benzyl alcohol 8.4%; and linalool 6.3%. Hidayat et al., found 38 components from Jasminum Sambac extraction and the dominant components were benzyl acetate 15.78%; linalil acetate 10.32%; cis jasmone 10.04%; Z-jasmone 8.32%; and linalool 6.10%. The composition of chemical compounds in the same plant can be differ depends on the level of impurity, plant characteristic and plant origin [12], [13].

In this study, commercial jasmine oil was used. There were several similar compounds with the previous study such as linalool, benzyl acetate and benzyl alcohol. The dominant compound was amylcinnamaldehyde, which was a derivative of cinnamic acid. Various derivatives of cinnamic acid were used as treatment material for skin, hair, antimicrobial ingredients and widely used as raw materials for perfumes. Amylcinnamaldehyde compound characterized as a pale yellow liquid with a floral aroma that gave the impression of jasmine aroma in cosmetic products [14].

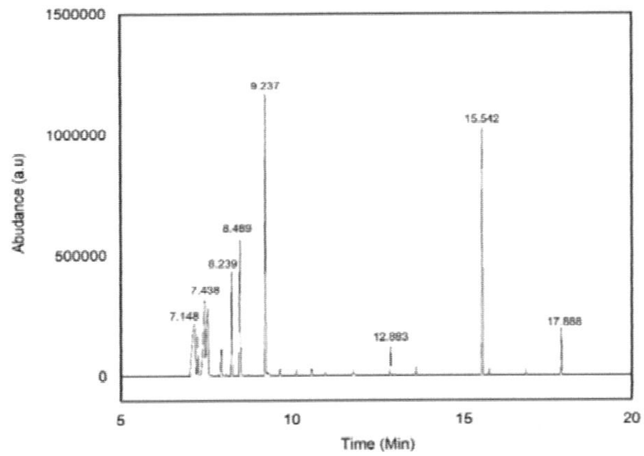

Figure 1. GC-MS Chromatogram of Jasmine Oil

Table 3. Chemical Compounds of Jasmine Oil

Retention Time	% Area	Compound Name
7.148	14.43	2-propanol, 1.1'-oxybis-
7.229	2.72	Benzyl alcohol
7.438	11.92	1-propanol,2-(2-hydroxypropoxy)-
8.239	6.15	Linalool
8.489	9.99	Phenylethyl alcohol
9.237	18.88	Benzyl acetate
12.883	1.84	Benzemethanol, alpha, 4-dimethyl
15.542	18.97	Cinnamaldehyde, alpha-pentyl
17.888	3.13	Benzene, 1-(1,1-dimethylethyl)-3,5-dimethyl-2,4,6-trinitro-

The results of the patchouli oil chemical compounds characterization are shown in Figure 2 and Table 4. The dominant compositions based on characterization were alpha.-Guaiene, delta.-Guaiene and Patchouli alcohol. The resulting patchouli oil composition characterization was similar to Ermaya et al (2019) where the patchouli oil content consisted of Patchouli alcohol 42.75%, Delta-Guaiene 28.3%, Azulene 20.48%, Trans Caryophellene 11.84%, Seychellence 10.77%, Nephtalene 8.02%, Cycloheptane 6.03 % and Caryophyllene 5.73% [15].

Table 4. Chemical Compounds in Patchouli Oil

Retention Time	% Area	Compound Name
12.359	3.16	Beta-Patchoulene
12.862	3.05	Caryophyllene
13.090	14.31	alpha.-Guaiene
13.191	5.69	Seychellene
13.354	7.24	Alpha-Patchoulene
13.947	21.78	delta.-Guaiene
15.849	32.57	Patchouli alcohol
16.344	3.32	Pogostone

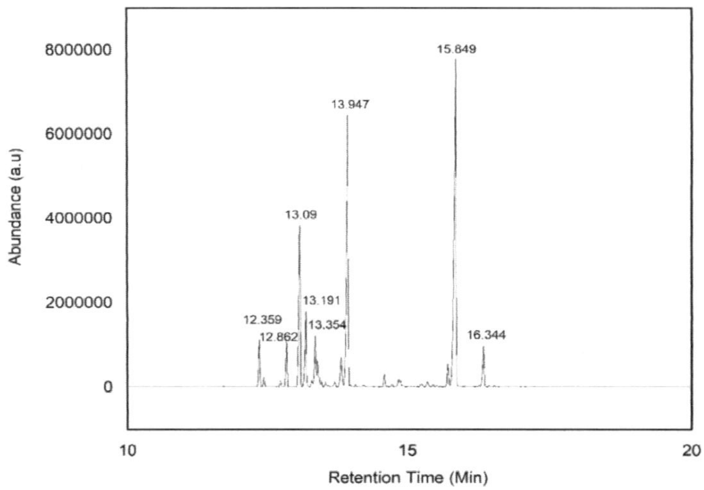

Figure 2. GC-MS Chromatogram of Patchouli Oil

3.3. Essential Oil Refractive Index

Refractive index is one of the parameters of quality analysis of essential oil based on SNI 06-2385-2006. Refractive index of Jasmine oil was about 1.47028-1.47038. This value was similar to Ginting 2019 with refractive index of Jasmine oil 1,478 whereas in general the value of Jasmine oil refractive index was about 1470-1492 [16].

3.4. Chemical Functional Group of Solid Perfume

The functional group of solid perfume and the raw materials were analyzed using FTIR and the results of the analysis are shown in Figure 3. Based on the analysis results, there were absorbance at 3377 cm^{-1} on solid perfumes that corresponded to the -OH functional groups. The cocoa butter showed the wavenumber at 3610 cm^{-1}, jasmine oil at 3369 cm^{-1} and patchouli oil at 3606.89 cm^{-1} where these absorptions indicated the -OH group. Other peaks occurred that showed $-CH_3$ and $-CH_2$ groups in solid perfume at wavenumbers 2862 cm^{-1} and 2927 cm^{-1}, cocoa butter at 2860.43 cm^{-1} and 2927.94 cm^{-1}, jasmine oil at 2881.65 cm^{-1} and 2953 cm^{-1}, and patchouli oil at 2941 cm^{-1}. The presence of wavenumber in solid perfume at 1739 cm^{-1}, cocoa butter at 1735.93 cm^{-1}, jasmine oil at 1680 cm^{-1}, and oil at 1732 cm^{-1} corresponded to the content of -CO aldehyde group in each sample.

Based on the results of FTIR analysis, solid perfume was produced through the physical mixing process of various raw materials, of which there were no new functional groups that indicated the formation of new compound. The functional groups found in raw materials that were cocoa butter, jasmine oil and patchouli oil were also found in solid perfume product.

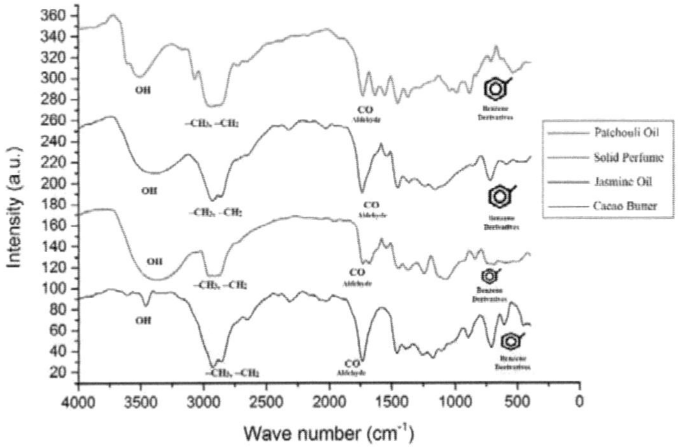

Figure 3. FTIR Spectra Functional Groups of Solid Perfume and Raw Materials

3.5. pH Test of Solid Perfume Formulation
pH standard for topical preparation in contact with skin was about 4-8 [17]. The pH value was expected not to be too acidic because it can cause irritation and not too alkaline because it can cause scaly skin. The resulting solid perfume was tested using universal pH paper and had pH value of 4. This pH value was considered safe for topical preparation for human skin application. Moreover, the content of this solid perfume was cocoa butter which contained fatty acids that similar to fatty acids composition on the skin [1].

3.6. Homogeneity of Solid Perfume Formulations
Homogeneity test of solid perfume was conducted by applying the samples to a flat glass surface as shown in Figure 4. The results of this test showed there were no coarse grains on the surface of the glass. According to Nurany et al (2018) homogeneous product had no coarse grain on the glass surface that had been smeared by the product [18].

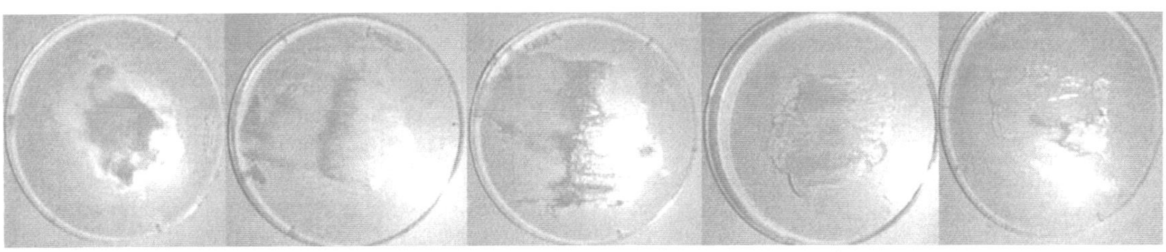

(a) (b) (c) (d) **(e)**

Figure 4. Solid Perfume Homogeneity Test with Cocoa butter Concentration (a) 10% (b) 20% (c) 30%, (d) 40% and (e) 50%

3.7. Hardness Level of Solid Perfume Formulation
The effect of concentration variations of cocoa butter on solid perfume products hardness level can be seen in Figure 5.

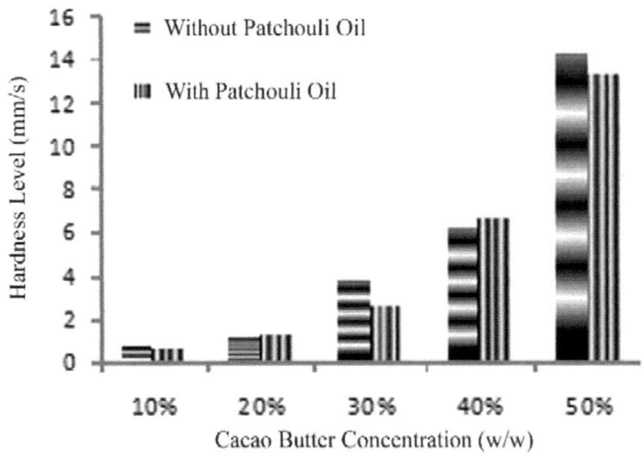

Figure 5. Level Hardness of Solid Perfume.

As shown in Figure 5, the penetration value would be higher when more cocoa butter was added. The highest penetration rate was at 50% cocoa butter concentration by value 13.23-14.16 mm/sec and the lowest was at 10% cocoa butter concentration by value 0.46-0.66 mm/sec. The addition of cocoa butter made the solid perfume product soften. The presence of cocoa butter in formulation was an important factor to adjust the texture and melting point. The cocoa butter affected the composition of solid perfume crystals into a more stable polymorphic arrangement. More amount of cocoa butter increased the consistency and density properties [19]. Cocoa butter can filled the space between the mixed ingredients which caused lubrication process that lead to decrease of rheological properties, texture and thermal properties [20].

3.8. Melting Point of Solid Perfume Formulation

The melting point is one of quality testing for solid perfume products. This test represents the stability of product quality during manufacture, storage, and usage processes [21]. Melting point results for all solid perfume variations are shown in Table 5.

Table 5. Melting Point for Solid Perfume Products.

Sample	Melting Point (°C)
F1	100
F2	100
F3	90
F4	80
F5	74
F6	115
F7	100
F8	100
F9	80
F10	85

Melting point from solid perfume products generally ranged above 70°C. Solid perfume with the highest melting point was the formulation with the lowest cocoa butter concentration by value 10%. The increased of melting point proportional with the increased of product hardness [22]. This corresponded to the decreased of rheological properties along with increased cacao butter concentration. It caused a decrease in the solid perfume products' melting point at high cocoa butter concentration.

3.9. Antioxidant Activity Test of Solid Perfume
Cocoa butter and solid perfume product were tested for antioxidants by DPPH method using UV-Vis Spectrophotometer with a maximum wavelength of 515 nm and used vitamin C as positive control. The results of antioxidant activity tests are shown in Table 6. IC_{50} value in cocoa butter was 168 µg/mL and in solid perfume was 201.98 µg/mL, this value was categorized as weak antioxidant. Based on Molyneux (2004), very strong antioxidants were those that have IC_{50} values less than 50 µg/mL, strong antioxidants with IC_{50} 50-100 µg/ml, medium antioxidants with IC_{50} 101-150 µg/ml and low antioxidants with IC_{50} 151-200 µg/mL [23]. Pure cocoa products had high antioxidant value due to the presence of polyphenol content, while cocoa butter was a by-product of the cocoa bean process, thus the antioxidant activity of cocoa butter was lower than pure cocoa [24].

Teble 6. Antioxidant activity of Solid Perfume and Cocoa butter

Sample	IC_{50} (µg/ml)
Cocoa butter	168
Solid Perfume	201.98
Vitamin C (positive control)	44.6

3.10. Stability Test of Solid Perfume Product
The stability test aimed to evaluate the product characteristics by observing the physical endurance of the product under certain condition. The test on solid perfume was carried out at room temperature by observing changes in texture, color and aroma for 4 weeks. The results of solid perfume products stability are shown in Table 7.

Table 7. Physical Stability of Solid Perfume for 4 Weeks

Solid Perfume	Weeks	Observation		
		Texture	Color	Aroma
F1	1-4	Solid, hard, and difficult to apply	Yellowish white, stable until the fourth week	Typical jasmine, stable until the fourth week
F2	1-4	Solid, hard, and difficult to apply	Yellowish white, stable until the fourth week	Typical jasmine, stable until the fourth week
F3	1-4	Solid, a little soft and easy to apply	Yellowish white, stable until the fourth week	Typical jasmine, stable until the fourth week
F4	1-4	Creamy, a little soft and uncomfortable to apply	Yellowish white, stable until the fourth week	Typical jasmine, stable until the fourth week
F5	1-4	Creamy, soft and uncomfortable to apply	Yellowish white, stable until the fourth week	Typical jasmine, stable until the fourth week
F6	1-4	Solid, hard, and difficult to apply	Yellowish white, stable until the fourth week	Typical jasmine and a little patchouli aroma, stable until the fourth week
F7	1-4	Solid, hard, and difficult to apply	Yellowish white, stable until the fourth week	Typical jasmine and a little patchouli aroma, stable until the fourth week

Solid Perfume	Weeks	Observation		
		Texture	Color	Aroma
F8	1-4	Solid, a little soft and easy to apply	Yellowish white, stable until the fourth week	Typical jasmine and a little patchouli aroma, stable until the fourth week
F9	1-4	cream, a little soft and uncomfortable to apply	Yellowish white, stable until the fourth week	Typical jasmine and a little patchouli aroma, stable until the fourth week
F10	1-4	cream, soft and uncomfortable to apply	Yellowish white, stable until the fourth week	Typical jasmine and a little patchouli aroma, stable until the fourth week

Overall, the observation for 4 weeks on all formulations had good stability for color and aroma parameters, but had a different texture consistency. The sensory of preparations with high cocoa butter concentration had soft texture which was uncomfortable when it applied to the skin. Product with high beeswax concentration had hard texture which was difficult to apply. The addition of waxes such as beeswax and cocoa butter greatly affected the consistency of the final product [22]. The addition of patchouli oil gave its distinctive aroma and intensify the scent of jasmine oil. It because the function of patchouli oil was to hold other scents thus the fragrance didn't disappear quickly and long last [23].

3.11. Organoleptic Test of Solid Perfume Products

Organoleptic test was carried out on 10 formulations. Organoleptic test refers to SNI 01-2346-2006 about organoleptic and/or sensory testing. Organoleptic test was a method of testing using the five human senses. This test involved 30 untrained panelists and they were given an assessment with the parameters written on the questionnaire. The test was done by selecting the assessment score with range 1-5 where each number indicated the level of preference for the tested solid perfume product. Organoleptic test results are shown in Table 8.

Table 8. Organoleptic Test of Solid Perfume.

Sample	Texture	Color	Aroma	Appearance	Moisture	Comfort	Homogeneity	General
F1	2.80	3.37	3.23	2.87	2.40	2.63	3.40	2.87
F2	3.07	3.50	3.50	3.40	2.77	3.17	3.97	3.03
F3	3.60	3.73	3.33	3.67	3.47	3.67	4.03	3.63
F4	3.23	3.13	2.93	3.03	3.17	3,00	3.40	2.90
F5	1.80	2.03	1.87	1.57	2.00	1,83	2.17	1.50
F6	2.70	3.37	3.67	2.87	2.23	2,67	3.83	2.87
F7	3.00	3.63	3.47	3.27	2.93	3,07	3.73	3.20
F8	3.60	3.80	3.70	3.70	3.70	3,67	4.10	3.37
F9	3.47	3.20	3.03	3.37	3.47	3,30	3.80	3.30
F10	1.50	1.60	2.07	1.67	1.73	1,83	2.43	1.67

The most preferable solid perfume formula based on organoleptic test was formula F3, with cocoa butter concentration 30% without the addition of patchouli oil by preference value 3.63. Panelists did not like solid perfume with cocoa butter concentration 50% which had the lowest preference value of 1.50.

Based on statistical test using one-way ANOVA, there was a significant difference with probability value of 0,000 ($P < 0.005$) which showed that the addition of cocoa butter concentration in solid perfume affected the level of panelist preference. This caused by consistency that was very dependent on the ratio of beeswax and cocoa butter. Organoleptic test process involved a sensory process and also a cognitive process. This sensory analysis can identify the level of acceptance of a product which was the basis for product development and a reference to the concentration limits of the used ingredients [25].

4. CONCLUSION

The concentration of cocoa butter significantly affected the characteristics of the final product. The process only involved the physical process because there were no new chemical groups formed in the final product. The most preferred formulation based on organoleptic test was product with 30% cocoa butter concentration with a preference level score of 3.63. Solid perfume with 30% cocoa butter concentration had characteristics of penetration level of 14.16 mm/sec, melting point 90°C, pH value of 4 and IC_{50} 201.98 μg/mL. Solid perfume products showed homogeneity and physical stability during 4 weeks observation at room temperature. The optimum formulation of this solid perfume was expected to provide new form aside of liquid conventional perfumes.

ACKNOWLEDGMENTS

Authors are gratefully acknowledging Program Riset Inovatif-Produktif (RISPRO) Komersial, Indonesia Endowment Fund for Education (LPDP), Ministry of Finance Republic of Indonesia for funding this research. Universitas Islam Negeri Syarif Hidayatullah, Faculty of Science and Technology, Chemistry Department and Research Center for Chemistry, Indonesian Institute of Sciences (LIPI) for laboratory and analysis facilities.

References

[1] S. Ramlah, "Characteristics of Quality and Effect of Polyphenol Addition to Cocoa Butter Based Hand Body Lotion to Skin," *J. Ind. Has. Perkeb.*, vol. 12, no. 2, pp. 29–39, 2017.

[2] M. Lipp *et al.*, "Composition of genuine cocoa butter and cocoa butter equivalents," *J. Food Compos. Anal.*, vol. 14, no. 4, pp. 399–408, 2001.

[3] T. K. Lin, L. Zhong, and J. L. Santiago, "Anti-inflammatory and skin barrier repair effects of topical application of some plant oils," *Int. J. Mol. Sci.*, vol. 19, no. 1, 2018.

[4] E. Gyedu-akoto, F. Amon-armah, and E. O. K. Oddoye, "Production and Marketing of Cocoa Butter and Shea Butter Based Body Pomadesas A Small Scale Business in Ghana," *Asian J. Bus. Manag.*, vol. 03, no. 05, pp. 385–391, 2015.

[5] E. Y. Ristanti, S. Ramlah, and D. Indriana, "Anti-Aging Properties Of Cream Made With Cocoa Polyphenol , Aloe Vera (Aloe Barbadensis) And Seaweed (Euchema Cottoni) As Active Agents Sifat Anti-Aging Dari Krim Berbahan Aktif Polifenol Kakao , Aloe Vera (Aloe Barbadensis) Dan Rumput Laut (Euchema C," *J. Ind. Has. Perkeb.*, vol. 13, no. 1, pp. 43–51, 2018.

[6] I. Hardiyati *et al.*, "Pembuatan dan Evaluasi Parfum Padat dari Minyak atsiri Vanilla (Vanilla planifola), Melati (Jasminum sambac (L.) Ait, Jeruk Manis (Citrus sinesis (L.) Osbeck) Dalam Kemasan Bros," *J. Ilm. Kedokt.*, vol. 6, no. 3, pp. 101–106, 2019.

[7] B. Ali, N. A. Al-wabel, S. Shams, A. Ahamad, S. A. Khan, and F. Anwar, "Essential Oils Used in Aromatheraphy: A Systemic Review," *Asian Pac. J. Trop. Biomed.*, vol. 5, no. 8, pp. 601–611, 2015.

[8] P. N. Waithaka, E. M. Gathuru, B. M. Githaiga, and J. Mbithe, "Making of Perfumes from Essential Oils Extracted from Lavender Plant Collected Making of Perfumes from Essential Oils Extracted from Lavender Plant Collected from Egerton University , Main Campus Njoro , Kenya," *Pyrex J. Biomed. Res.*, vol. 2, no. 6, pp. 35–40, 2016.

[9] T. Hongratanaworakit, "Stimulating Effect of Aromatherapy Massage with Jasmine Oil," *Nat. Prod. Commun.*, vol. 5, no. 1, pp. 157–162, 2010.

[10] T. Aziz, V. F. Sitorus, and B. A. Rumapea, "Pengaruh Pelarut Heksana dan Etanol, waktu Ekstraksi Terhadap Hasil Ekstraksi Minyak Coklat," *J. Tek. Kim.*, vol. 16, no. 2, pp. 48–54, 2009.
[11] L. Junaidi, A. Sidibyo, T. F. Hutajulu, and D. Abdurakhman, "Pengaruh Perlakuan Suhu Ekstraksi Terhadap Karakteristik Mutu Lemak Kakao," *J. Agro-Based Ind.*, vol. 25, no. 2, pp. 24–34, 2008.
[12] A. E. Edris, R. Chizzola, and C. Franz, "Isolation and characterization of the volatile aroma compounds from the concrete headspace and the absolute of Jasminum sambac (L .) Ait . (Oleaceae) flowers grown in Egypt," *Eur. Food Res. Technol.*, vol. 226, pp. 621–626, 2008.
[13] N. Hidayat, I. Atsari, and D. Alfiana, "Extraction of Jasmine (Jasminum sambac) Oils (Study of Solvent Type and Extraction Time)," *J. Ind.*, vol. 4, no. 2, pp. 82–88, 2015.
[14] A. G. Krzyzak, K. Sloczynska, J. Popiol, P. Koczurkiewicz, H. Marona, and E. Pekala, "Cinnamic acid derivatives in cosmetics : current use and future prospects," *Internastional J. Cosmet. Sci.*, vol. 40, pp. 356–366, 2018.
[15] D. Ermaya, S. Sari, A. Patria, F. Hidayat, and F. Razi, "Identification of patchouli oil chemical components as the results on distillation using GC-MS," in *IOP Conference Series: Earth and Environmental Science*, 2019, pp. 1–6.
[16] M. D. Ginting, F. Iskandar, I. Iriany, and O. Bani, "Extraction of Jasmine Essential Oil: The Effect of Ration Between Jasmine Flower to N-Hexane as Solvent (m/v), Time and Temperature," *J. Tek. Kim. USU*, vol. 8, no. 1, pp. 42–47, 2019.
[17] Y. Padmadisastra, A. Syaugi, and S. Anggia, "Formulasi Sediaan Salep Antikeloidal yang Mengandung Ekstrak Terfasilitasi Panas Microwave dari Herba Pegagan (Centella asiatica (L.) Urban)," in *Seminar Kebudayaan Indonesia Malaysia*, 2007, pp. 1–11.
[18] A. Nurany, A. Sri, S. Amal, and S. A. Estikomah, "Formulasi Sediaan Lipstik Ekstrak Bunga Rosella (Hibiscus sabdariffa) Sebagai Pewarna dan Minyak Zaitun (Olive oil) Sebagai Emolien," *Pharmasipha*, vol. 2, no. 1, pp. 1–9, 2018.
[19] N. Baldino and D. Gabriele, "The influence of formulation and cooling rate on the rheological properties of chocolate," *Eur. Food Res. Technol.*, vol. 231, pp. 821–828, 2010.
[20] R. V. Rios, M. Pessanha, P. F. Almeida, and S. C. S. Lannes, "Application of fats in some food products," *Food Sci. Technol.*, vol. 34, no. 1, pp. 3–15, 2014.
[21] K. S. Vishwakarma, S. S. Ram, S. A. Ahmed, S. Sadiq, S. D. Mallick, and T. R. Sree, "Formulation and Evaluation of Natural Lisptick from Coloured Pigments of Beta Vulgaris Taproot," *Res. Rev. J. Pharm. Pharm. Sci.*, vol. 3, no. 3, pp. 65–71, 2014.
[22] O. Perdanakusuma and Z. Wulandari, "Optimasi Proses Pembuatan Lipstik dengan Penambahan Berbagai Konsentrasi Malam Lebah," *J. Teknol. Ind. Pertan.*, vol. 14, no. 3, pp. 95–100, 2003.
[23] P. Molyneux, "The use of the stable free radical diphenylpicryl- hydrazyl (DPPH) for estimating antioxidant activity," *Songklanakarin J. Sci. Technol.*, vol. 26, no. 2, pp. 211–219, 2004.
[24] M. Natsume *et al.*, "Analyses of Polyphenols in Cacao Liquor, Cocoa, and Chocolate by Normal-Phase and Reversed-Phase HPLC," *Biosci. Biotechnol. Biochem.*, vol. 64, no. 12, pp. 2581–2587, 2000.
[25] T. D. L. Ángeles, M. Tayupanta, M. Espadero, M. Mancheno, and S. Peña, "Sensory analysis of cosmetic formulations made with essential oils of Aristeguietia glutinosa (matico) and Ocotea quixos (ishpingo)," *Int. J. Phytocosmetics Nat. Ingredients*, vol. 5, no. 5, pp. 1–5, 2018.

IOP Conf. Series: Materials Science and Engineering 1011 (2021) 012038 doi:10.1088/1757-899X/1011/1/012038

Effect of solvent concentration in sago starch fractionation

Dewi Sondari[1], Witta Kartika Restu[2]*, Yulianti Sampora[2], Yenni Apriliany Devy[2], Teo Duri Yosta[2] and Anna Muawanah[3]

[1]Technology Process of Biomass and Bioremediation Group, Research Center for Biomaterial, Indonesian Institutes of Sciences (LIPI), Kawasan Cibinong Science Center, Jl. Raya Bogor Km.46, Cibinong, Bogor 16911, West Java, Indonesia
[2]Polymer Chemistry Group, Research Center for Chemistry, Indonesian Institutes of Sciences (LIPI), Kawasan PUSPIPTEK Serpong, South Tangerang 15314, Banten, Indonesia
[3]Department of Chemistry, Syarif Hidayatullah State Islamic University, J Jl. Ir H. Juanda No.95, Cempaka Putih, Ciputat, South Tangerang 15412, Banten, Indonesia

*E-mail: witt001@lipi.go.id

Abstract. The research studied the effect of solvent concentration in the fractionation process from sago starch in order to achieve the amylose fraction. The amylose is aimed to be applied as a functional material such as for edible film. The fractionation process is carried out by utilizing N-butanol as the solvent with a variety of concentration (10, 12.5 and 15%). The highest result is achieved from the concentration of butanol of 15%. It resulted in a yield of 31.19%, a 30.78% solubility, and swelling ability of 2.37%. The spectra from Fourier transform infrared showed the same specific peaks of the obtained amylose. However, each of them shows different intensity means the fractionation process affect the resulted amylose.

1. Introduction

Sago has high carbohydrate content that can be used as a substitute for wheat flour and it is also well used as a thickener that is better than tapioca [1,2]. Starch is a large polysaccharide and is mainly composed of linear amylose and highly branched amylopectin. Amylose is a long, straight, unbranched, and hydrophobic chain polymer with α- (1,4) -D-glucose bonds [3,4]. Amylopectin has the same backbone as amylose, but it also branched by α-(1,6) linkages. Sago starch contains 70-80% more amylopectin, while the amylose content is 15-30%. These components determine the gelatinization, gel stability, solubility, and swelling power. High percentages of amylose give starch less stickiness and low gelatinization temperature. The molecular weights of amylose and amylopectin have been estimated to be about 105 and 108 Da, respectively [5].

Amylose plays an important role in starch gelation, but insight in its main role has been vague by the presence of amylopectin [6,7]. Therefore, this makes amylose content an important quality parameter for starch-based products [8]. Separation or fractionation of amylose and amylopectin has been a field of interest for some decades, but only several successes have been achieved. Schoch described starch separation for the first time [9]. Some authors provided alternative paths and optimized this method, in which the purity of the obtained amylose and amylopectin fractions becomes the main consideration [10–13]. High purity starch fractions are obtained by precipitating the amylose fraction of amylopectin with guest molecules.

Since the previous studies showed that the amylose is able to make inclusion complexes, so that its solubility is reduced in an aqueous dispersion. Then, the insoluble amylose-guest molecule is recovered as a precipitate. Some guest molecules, such as methanol, butanol, and pentasol have been employed to precipitate free amyloses from aqueous dispersions. However, butanol has been extensively used due to its ability in selecting the amyloses compared to other guest molecules. To the best of our knowledge, there is limited study regarding the amylose from fractionation process of sago starch. Therefore, in this study, butanol with varying concentration was used as a guest molecule to create amylose-guest molecule complexes in the fractionation process of sago starch. The obtained amylose fraction from the sago starch were characterized and evaluated their behaviors.

2. Materials and Method

2.1. Materials

Sago (Metroxylon sp.) starch with approximately 34% amylose and 66% amylopectin was used as the main material. Glycerol (analytical grade) was purchased from Sigma–Aldrich Co., Ltd. (St. Louis, MO, USA). Ethanol and N-butanol technical grade were used. All other reagents used were analytical grade.

2.2. Preparation of sago starch composition

The amylose content was analysed by the spectroscopic method. The solution was measured its absorbance by a UV-Vis spectrophotometer at a wavelength of 625 nm and the levels was determined using the standard curve equation of amylose solution standard. The soxhletation method was used to determine the fat content according to SNI 01-2891-1992.

2.3. Preparation of sago starch fractionation

Sago starch fractionation was carried out by modifying the method of Mizukami *et al.* [14]. 30 gram of sago starch was mixed with 2 litre of distilled water and then heated to 85°C for 60 min while stirring by a stirrer under nitrogen atmosphere. The water-soluble fraction was separated from the suspension by centrifugation at 8000 rpm for 10 min. The separated water-soluble fraction was added with butanol with variations of 10, 12.5, and 15 % of the water-soluble fraction volume, then incubated for 24 hr. The resulting precipitate is an amylose fraction and then centrifuged at 8000 rpm for 10 min. The precipitate was washed with ethanol 3 times, then washed also with petroleum ether 3 times, followed by drying with $CaCl_2$ then put in a desiccator. The yield of obtained amylose was calculated as follow:

$$\text{Yield of amylose (\%)} = \frac{\text{weight of obatained amylose}}{\text{initial weight of sago starch}} \times 100\% \quad (1)$$

2.4. Swelling power (SP) and water solubility index (WSI)

Swelling power (SP) of the obtained amylose was determined according to the method of Senanayake *et al.* [15]. 100 mg amylose was weighed directly into a screw-cap test tube and 10 mL distilled water was added. The capped tubes were placed on a vortex mixer for 10 sec, then incubated at 85 °C water bath with frequent mixing about 10 sec after 5, 15 and 25 min. The tubes were cooled at room temperature and centrifuged at 2000 rpm for 30 min. The supernatant was removed and the remaining sediment in the tube was weighed (Ws). The supernatant was dried to constant weight (W_1) in a drying oven at 105 °C. The water swelling power was calculated as follows:

$$\text{SP (\%)} = \frac{Ws}{[0.1 \text{ x } (100\% - \text{WSI}]} \text{ x } 100\% \quad (2)$$

where

$$\text{WSI} = W1/0.1 \text{ x } 100\% \quad (3)$$

2.5. Fourier transform infrared spectroscopy analysis (FTIR)

FTIR spectra were recorded using FTIR ATR (Tensor II, Bruker, USA). The resolution was 4 cm-1 and there was a total of 40 scans at wave number of 4000-400 cm^{-1}.

3. Results and Discussion

3.1. Sago starch composition

The sago starch composition is given in Table 1. The moisture content of the sago starch was 11.46 % which is typical of commercial starches. The moisture content of most unmodified starches is around 12 %. All commercial starches contain small quantities of inorganic materials. The inorganic material generally derived from the way the starch was isolated and also from the used water in the process.

Table 1. Component of sago starch.

Component	Composition (%)
Amylose content	33.73
Amylopectin content	66.27
Moisture	11.46
Crude fat	1.8
Crude protein	0.0017

3.2. Sago starch fractionation

Sago starch with the content of amylose of about 33% and water content of 11% was prepared in fractionation process. The fractionation process was carried out by utilizing butanol as the solvent to dissolve the amylopectin fraction. This study varied the concentration of butanol namely 10, 12.5, and 15 % in order to achieve the highest yield of amylose. Yield is an important parameter in order to know the level of efficiency of a processing process. The greater the yield indicates the more effective and efficient the process carried out. The yield of each variation is shown in Table 2.

Table 2. Yield of amylose fraction.

Butanol concentration (%)	Yield of amylose fraction (%)
10	25.29
12.5	28.24
15	31.19

Amylose fractionation in this study was carried out by the modification method. The fractionation process was initially carried out using hot water, because hot water is able to dissolve the starch completely. The starch solubility increases with the increasing temperature of the hot water that used as the solvent. Then the separation was done using butanol as a solvent to dissolve amylopectin. Table 1 shows that the greater the concentration of butanol used, the higher the yield of the obtained amylose fraction from sago starch. The concentration of 15% butanol resulted in the yield of amylose fraction of 31.19%. This result was higher compared to other study which gained the yield of 16.36 and 19.36 % for the butanol concentration of 10 and 20% [16]. Other study also showed that the increasing in yield was in line with the increasing of butanol concentration. This is due to the butanol as solvent to dissolve amylopectin fraction so that resulted in high content of amylose. The use of hot water as the initial part in the fractionation proses of starch plays an important role to change the structure and function of starch granules.

Amylose is a starch component that has a straight chain and soluble in water, while amylopectin has a branched chain, not soluble in water but soluble in N-butanol. This is because amylose consist of a straight D-glucose chain that binds to α-1,4. It is also influenced by hydrogen bonds that occur between hydroxyl groups in amylose and water. When starch is heated in hot water at gelatinization temperature, heat energy causes hydrogen bonds in starch to weaken and it provide the water to get into the granules and allows a slight dissolution and exchange of amylose molecules into the water. This emerges due to the presence of excess butanol can dissolve amylopectin, so that amylose can be separated from amylopectin in the form of sediment [10].

3.3. Swelling power and water solubility

Swelling power is the power of amylose to expand, which occurs due to non-covalent bonds between amylose molecules. Weak hydrogen bonds in the amorphous region will be broken during heating, resulting in water hydration. Amylose will continue to expand so that the viscosity increases to the maximum hydration volume that can be achieved [17]. From data in Table 3, it was stated that the concentration of butanol significantly affected the swelling power and solubility of the amylose fraction of sago starch with a range of 40.18% (butanol 10%) to 30.78% (butanol 15%). These results were lower compared to other study, which the swelling ability reached more than 40% depending on the temperature as its parameter.

Table 3. Swelling power and solubility of amylose fraction.

Butanol concentration (%)	Swelling power (%)	Solubility (%)
10	3.89	40.18
12.5	3.39	37.33
15	2.37	30.78

Table 3 shows that the increasing concentration of butanol up to 5% significantly decreases the swelling power and solubility of the amylose fraction of sago starch. This might be due to the utilization of butanol as solvent decreasing the mobility of the water fraction, thereby inhibiting the amylose restructure in the amorphous phase. The highest value of swelling power and solubility is produced at 10% butanol concentration and the smallest is at 15% butanol concentration.

The amylose restructuring process which is inhibited in the amorphous region during heating and caused the inter- and intra-molecules of hydrogen bonds to become denser, making it difficult for water molecules to enter, since solubility is related to the ease of water molecules to interact with molecules in amylose. The increase in swelling power values is also suggested due to the hydrophilic nature of the amylose so as to bind to hydrogen in water molecules [18]. While the occurring partial hydrolysis during heating in the fractionation process resulted in low molecular weight so that the ability to expand is limited.

3.4. Fourier transform infrared spectroscopy analysis (FTIR)

FTIR analyses were applied in order to study the changes in chemical structure among the obtained amyloses. The changes of functional group in chemical structure can be known from the characteristic of absorption peak. Figure 1 shows the spectra of sago starch, similar as in previous study [19,20]. In the spectra for sago starch the presence of -C-O-C was confirmed by the absorption peaks at 1628-1676 cm-1, the absorption peak at 1436-1479 cm-1 was attributed to -OH bending.

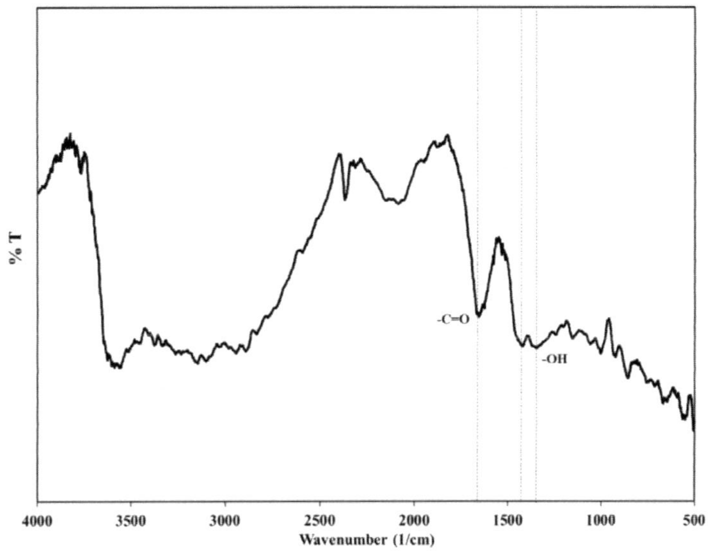

Figure 1. FTIR spectra of sago starch.

Figure 2 shows the FTIR spectrums of biocomposite films. The comparison of each absorption peak from the designated samples is summarized in Table 4.

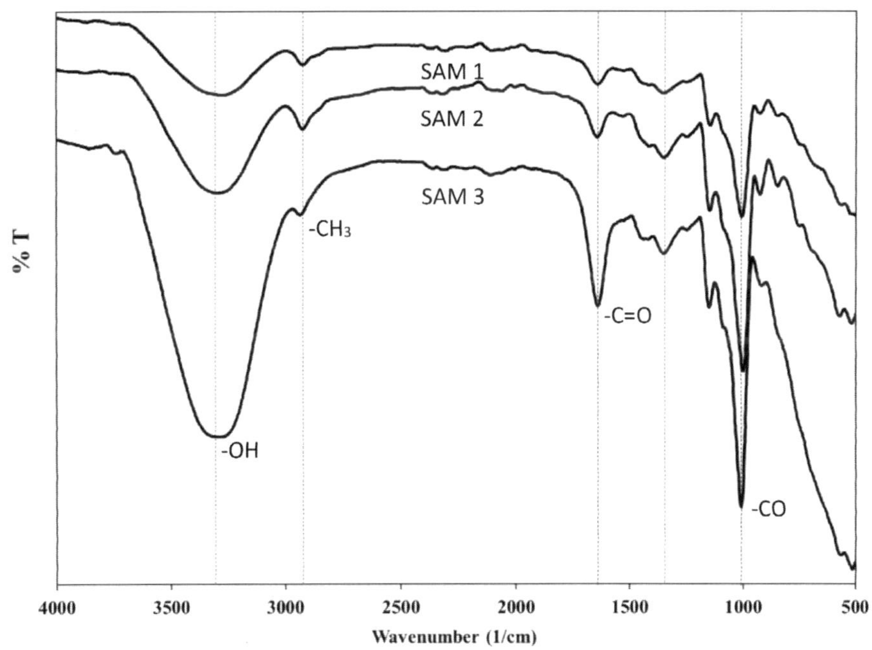

Figure 2. FTIR spectra of obtained amylose.

Table 4. FTIR characteristic of starch films.

Samples (cm^{-1})			Characteristics
SAM 1	SAM 2	SAM 3	
3253-3304	3248-3344	3242-3354	OH stretching
2908-2937	2912-2937	2933-2941	CH stretching
1625-1656	1627-1660	1627-1652	C=O group
1311-1382	1348-1354	1346-1355	OH bending
1002-1010	999-1012	1001-1020	C-O stretching

FTIR spectroscopy was used to examine the influence of butanol concentration in the fractionation process toward the functional group of the obtained amylose from sago starch. This is because if two or more chemicals are mixed, the physical blends towards chemical interactions are reflected by the changes in the characteristic of spectra peaks. The infrared spectra of each amylose are shown in Figure 2. All the results show certain peaks similar to the amylose studied by other groups [21]. It can be seen from all obtained infrared spectra that there was no significant peak change, but difference in intensity of the peak. This might be due to the effect of different concentration of butanol as a solvent for the precipitation of amylose-butanol complexes.

4. Conclusion

The starch separation procedure was successfully synthesized and amylose fraction was obtained from sago starch. The variation of butanol as the solvent resulted in different yield of the obtained amylose. This also implied in the swelling ability and water solubility of each obtained amylose. All amyloses separated from the sago starch had the same functional group with different intensity.

Acknowledgments

Authors wish to acknowledge the Indonesian Institute of Sciences (LIPI).

Funding

This work was supported by the Ministry of Research Technology and Higher Education: [Grant Number 09/P/RPLLIPI/INSINAS-2/VII/2019].

References

[1] Sumardiono S, Rakhmawati RB and Pudjihastuti I 2016 Physicochemical and rheological properties of sago (Metroxylon Sagu) starch modified with lactic acid hydrolysis and UV rotary drying. *ASEAN J. Chem. Eng.* **18** 41–53

[2] Syahariza ZA and Yong HY 2017 Evaluation of rheological and textural properties of texture-modified rice porridge using tapioca and sago starch as thickener. *J. Food Meas. Charact.* **11** 1586–91

[3] Garcia MC, Franco CML, Soares Júnior MS and Caliari M 201 Structural characteristics and gelatinization properties of sour cassava starch. *J. Therm. Anal. Calorim.* **123** 919–26.

[4] Stawski D 2008 New determination method of amylose content in potato starch. *Food Chem.* **110** 777–81

[5] Roger P, Bello-Perez LA and Colonna P 1999 Contribution of amylose and amylopectin to the light scattering behaviour of starches in aqueous solution. *Polymer (Guildf)* **40** 6897–909

[6] Lemos PVF, Barbosa LS, Ramos IG, et al 2019 Characterization of amylose and amylopectin

fractions separated from potato, banana, corn, and cassava starches. *Int. J. Biol. Macromol.* **132** 32–42
[7] Li H, Prakash S, Nicholson TM, Fitzgerald MA and Gilbert RG 2016 The importance of amylose and amylopectin fine structure for textural properties of cooked rice grains. *Food Chem.* **196** 702–11
[8] Copeland L, Blazek J, Salman H and Tang MC 2019 Form and functionality of starch. *Food Hydrocoll.* **23** 1527–34
[9] Schoch TJ 1942 Fractionation of starch by selective precipitation with Butanol. *J. Am. Chem. Soc.* **64** 2957–61
[10] Montgomery EM and Senti FR 1955 Separation of amylose from amylopectin of starch by an extraction-sedimentation procedure. *J. Polym. Sci.* **XXVIII** 1–9
[11] Mua JP and Jackson DS 1998 Retrogradation and gel textural attributes of corn starch amylose and amylopectin fractions. *J. Cereal Sci.* **27** 157–66
[12] Klucinec JD and Thompson DB 1998 Fractionation of high-Amylose maize starches by differential alcohol precipitation and chromatography of the fractions. *Cereal Chem.* **75** 887-96
[13] Majzoobi M, Rowe AJ, Connock M, Hill SE and Harding SE 2003 Partial fractionation of wheat starch amylose and amylopectin using zonal ultracentrifugation. *Carbohydr. Polym.* **52** 269–74
[14] Mizukami H, Takeda Y and Hizukuri S 2009 The structure of the hot-water soluble components in the starch granules of new Japanese rice cultivars. *Carbohydr. Polym.* **38** 329–35
[15] Senanayake S, Gunaratne A, Ranaweera K and Bamunuarachchi A 2013 Effect of heat moisture treatment conditions on swelling power and water soluble index of different cultivars of sweet potato (Ipomea batatas (L). Lam) starch. *ISRN Agron.* **2013** 502457
[16] Yuliasih I, Irawadi TT, Sailah I, Pranamuda H, Setyowati K and Sunarti TC 2017 Pengaruh Proses Fraksinasi Pati Sagu terhadap Karakteristik Fraksi Amilosanya. *J. Tek. Ind. Pert.* **17** 29–36
[17] Daramola B and Osanyinlusi SA 2006 Investigation on modification of cassava starch using active components of ginger roots (Zingiber officinale Roscoe). *African J. Biotechnol.* **5** 917–20
[18] Zhu F 2014 Structure, physicochemical properties, modifications, and uses of sorghum starch. *Compr. Rev. Food Sci. Food Saf.* **13** 597–610
[19] Sondari D, Falah F, Suryaningrum R, Sari FP, Septefani AA, Restu WK, *et al* 2019 Biofilm Based on Modified Sago Starch : Preparation and Characterization. *Reaktor* **19** 125–30
[20] Sondari D 2019 Modification of sago starch for edible coating. *IOP Conf Ser Mater Sci Eng* **543** 012013
[21] Zhang H, Hortal M, Jordá-Beneyto M, Rosa E, Lara-Lledo M and Lorente I 2017 ZnO-PLA nanocomposite coated paper for antimicrobial packaging application. *LWT - Food. Sci. Technol.* **78** 250–7

IOP Conf. Series: Materials Science and Engineering 1011 (2021) 012039 doi:10.1088/1757-899X/1011/1/012039

Antifungal of Cashew (*Anacardium occidentale* Linn) leaves, nutshells, and peduncle bagasse ashes extracts against sooty mould fungi (*Capnodium* sp)

W R Andayanie[1]*, M Lukito[1] and F Chasanatun[2]

[1]Department of Agrotechnology, Agriculture Faculty, Merdeka Madiun University, Jl. Serayu 79, Madiun, East Java,63133, Indonesia
[2]Department of Primary Teaching, Education Faculty, Madiun PGRI University, Indonesia

*E-mail: wuye.andayanie@gmail.com

Abstract. Different concentrations of cashew (*Anacardium occidentale* Linn) leaves, nutshells, and peduncle bagasse ashes extract evaluated for growth inhibition of *Capnodium sp* and the antifungal activity. Results revealed that different concentration of cashew leaves, nutshells, and peduncle bagasse ashes extracts showed values difference of absorbance (DA) ranging from 0.00 to 0.42. All the concentrations of extracts showed significant inhibition in the spore germination of sooty mould. The highest levels caused maximum inhibition in the spore germination followed by lower concentrations of cashew nut shells, peduncle bagasse and leaf extracts respectively. Cashew nut shells extract had a high level of total polyphenolic, flavonoid content, pH and low values of titratable acidity, which could be attributed to its potent antifungal activity.

1. Introduction

Bemisia tabaci (Genn.) (Hemiptera: Aleyrodidae) is a type of the cosmopolitan insect pest that it can develop in tropical and subtropical climate zones. This insect produces black sooty moulds fungus that grows on honeydew excretion after sucking the sap of soybean leaf. *B tabaci* causes the reduction of soybean production by producing honeydew. The main components of honeydew are sugars, free amino acids, proteins, minerals and other organic compounds [1].

Honeydew triggered the growth of sooty mould (*Capnodium sp*) fungi on the leaf surface and caused chlorosis, which disrupts the process of leaf photosynthesis and soybean value. Several studies reported that the sooty mould could spread from one plant to another by airborne and water splash spores [2-4]. The incidence of sooty mould infection has increased tremendously; as a result, the use of insecticide synthetic against *B. tabaci*. The typical symptom of sooty mould showed the slight black spot to the dark blackspot covers the entire surfaces of the leaves with black mycelia. Therefore, sooty mould has recognized as the main plant pathogenic group.

The use of fungicides in soybean farming increases grain yields and economic profit. However, these often cause negative effects, viz. phytotoxicity and residues in food. More recently, the botanical insecticide used for managing *B. tabaci*, which recommended for *B. tabaci* as the low risk of the toxic residual effect [5]. Cashew nutshell extract was also reported as antiviral against *Cowpea mild mottle virus* on soybean plants and botanical herbicide on the cultivation of black grass jelly (*Mesona palustris* BL) [6-8]. Various plant extracts as inhibitory effect on the mycelial growth and conidial germination of the phytopathogenic fungus *Fusarium*. *Anacardium occidentale* L (Anacardiaceae) peduncle bagasse ash and leaf extract were effective against plant pathogenic Fusarium strain [9]. The ethanolic extracts of cashew nutshell (CNS) and peduncle bagasse ash (CPBA) extract also reported as antifungal activity against *Aspergillus flavus*, *A. fumigatus*, *A. niger, Curvularia* sp, *Colletotrichum gloesporioides* and *Lasiodiplodia theobromae* [10]. Compound derived from CNS and CPBA sources have shown non-phytotoxic on the germination [11,12]. The majority of studies show that CNS extract may have a toxic effect on the mycelial growth of phytopathogenic fungus. Still of these, very few have the ability of leaves, nutshells and peduncle bagasse ashes extract to suppress sooty moulds fungus on soybean. These extracts may also be potential alternatives of fungicidal for managing the sooty moulds fungus, which make them suitable fungicides for organic agriculture. The present study aimed to determine the percentage growth inhibition of *Capnodium sp* with leaves, nutshells and peduncles extracts of cashew (*Anacardium occidentale* Linn) and evaluate the antifungal activity of extracts using phytochemical screening. This research is important to recognize of the cashew tree as a good fungicidal source against *Capnodium* sp on soybeans. In addition, the result of this research can be potentially used as reference to best concentration of parts of the cashew tree (leaves, nut shells and apples) in the control of sooty moulds on soybean.

2. Materials and Methods

2.1. Preparation of plant extract and stock culture

Fresh leaves, nutshells, peduncle bagasse ashes samples of *Anacardium occidentale* L. were washed thoroughly with sterile distilled water and cut into small pieces. Each sample was dried in the oven at 36 °C for 48 h to the ground using a blender to a fine powder. Approximately, 500 g of each powder sample was macerated with n-hexane (0.5 g/ml) and stirred for one h. After that, the extracts were incubated and filtered through double-layered cheesecloth. The filtrate was evaporated using rotary vacuum evaporator at 45 °C±5 with pressure (550 mm Hg). It was made different concentration at levels of 500 ppm, 1000 ppm, 1500 ppm, 2000 ppm. The extract sterilized for 24 h using UV rays.

The *Capnodium* sp isolates were maintained and subcultured via spore suspension on Potato Dextrose Agar. The plates were incubated at 28°C for seven days. Spores were counted and used at a concentration of 3×10^6 spores/ml.

2.2 Spore germination assay

Leaves, nutshells and peduncle bagasse ashes extracts of cashew (*Anacardium occidentale* Linn) were dipped about 2 cm apart with different concentrations on each microscopic slide (30 µl per slide) in triplicate, respectively. Then 20 µl of homogenous spore suspension from the 7 days old cultures (3×10^6 spores/ml) was added to each 30 µl the extract.

Antifungal activity was detected using a microtiter plate. Performed using 200 µl a reaction mixture containing: 80 µl of homogemous spore suspensions from the seven days old cultures (10^5spore suspensions /ml), 100 µl Czapek Dox Broth and 20 µL of plant extract were pipetted with different extract and concentrations into each well of the microtiter plate. One plate row was filled with untreated spore suspension in Czapek Dox Broth as a negative control. The optical density (OD) in each well was calculated after 48 h incubation by a Spectrostar Nano microplate reader at a wavelength of 405 nm.

The percentage of spore germination was assessed after 24 h at 15 °C by mounting 10 µl of spore suspension on a slide glass and counting the number of germinated spores using a haemocytometer grid at 4 × 10-2 mm^2. The percentage of spore germination was calculated for the tenth wells and averaged. Each treatment was repeated four times. The percentage of spore germination is calculated using formula:

$$\text{\% of spore germination} = \frac{\text{No. of spore germination}}{\text{Total no. of spore examinated}} \text{ x } 100\% \qquad (1)$$

2.3 Determination of total phenolic and flavonoid content

The content of total phenolics (TPC) of the different concentration of leaves, nutshells, and peduncle bagasse ashes extracts of cashew was determined by the method of Emelike *et al.* (2017) with slight modification [13]. Exactly 1 mg of each extract dissolved in methanol solution (1 ml). Each sample (250 µl) mixed with 2.5 ml of 10 % Folin-Ciocalteu's reagent by manual shaking for 30 s. 2.0 ml of 7 % Na_2CO_3 added. After that, the reaction mixture was shaken for 10 min on a vortex and incubated in the dark for 60 min at room temperature. The solution was mixed, and absorbance was measured against ultra-pure water as blank at λ 750 nm with a UV visible spectrophotometer (Varian Cary 100 Conc UV-Vis) Blank consists of all reagents except the extract.

The content of total phenolics was expressed as gallic acid (mg GAE.g^{-1}dry weight) of lyophilized plant extract. The standard phenolic compound was prepared 0.00625–0.1 µg ml^{-1}, curve equation: r_2 = 0.9911. Total flavonoids content (TFC) was expressed as quercetin (mg QE. Mg^{-1}dry weight) of the extract. The standard flavonoid compound curve equation: r_2 = 0.9951.

2.4. Determination of Titratable Acidity and pH analysis

The acidity of the sample was determined by titration of a diluted sample with a 0.02 N alkaline sodium hydroxide solution to a phenolphthalein. Phenolphthalein (1%) was used as the indicator (4 drops in 20 mL of each sample before starting the analysis) to determine the endpoint in acid-base titrations. Sodium hydroxide was added dropwise with constant swirling until the solution turned pink throughout. The volume of base required to reach the equivalence point was used to calculate the acidity of the extracts expressed in meq of sodium hydroxide per unit sample weight (meq NaOH/g). The pH value of each extract was determined with a Hamilton pH electrode sensor. All measurements were repeated twice within a period of 7 days.

2.5. Experimental design

A randomized was used for experimental design. Analysis of variance (ANOVA) formulas by Fisher's protected LSD test at $P \leq 0.05$ were used for means values. The Probability values were determined by 0.05 and 0.01 significance level, at 95 % and 99 % confidence limits respectively. The experimental data of total phenolic content, total flavonoid content, pH of leaves, nutshells, and peduncle bagasse ashes extracts obtained were expressed as an average.

3. Result and Discussions

3.1. Antifungal screening

3.1.1. Different absorbance of extracts Spore germination assay. After 48 h incubation, at the different concentration of leaves, nutshells, and peduncle bagasse ashes extracts showed values difference of absorbance of (DA) ranging from 0.00 to 0.42. These values were significantly different from the control value of DA (Table 1).

Table 1. The values of difference absorbance of extracts with different concentration.

Concentration / Extracts	Difference Absorbance after 48 h incubation**					
	Control[a]	Fungicide[b]	500 ppm	1000 ppm	1500 ppm	2000 ppm
Cashew nutshells	0.41 [a]	0.00 [a]	0.13 [b]	0.09 [b]	0.05 [c]	0.00 [c]
Cashew leaves	0.40 [a]	0.01 [a]	0.25 [a]	0.19 [a]	0.17 [a]	0.11 [a]
Cashew peduncle bagasse ashes	0.42 [a]	0.00 [a]	0.21 [ab]	0.15 [ab]	0.07 [b]	0.05 [b]

Mean of four replicates; **; [a] = no treated (positive control);[b] = negative control; Means in the same column followed by same letter are not statistically different at P =0.05 according to the Fisher LSD Method.

In the present assay, out of the different concentration of leaves, nutshell, and peduncle bagasse ashes extracts of cashew, the CNS extract at concentration 2000 ppm showed the same trend of difference absorbance (DA) as the fungicide value. It has been either shown that plant extracts with a DA < 0.04 were not or very low conidia germination of the phytopathogenic fungus *Fusarium oxysporum* [9].

3.1.2. Spore germination

From the analysis of data in Table 2, we infer that the different concentration of leaves, nut shell, and peduncle bagasse ash extracts of cashew caused significant inhibition of spore germination. The highest inhibition on spore germination was obtained at 2000 ppm in leaves, nutshell, and peduncle bagasse ash extracts of cashew against sooty mould fungi. It was followed by 1500 ppm, 1000 ppm and 500 ppm of extracts. In addition, there was significant difference among treatments (P = 0.01). The antifungal activity of extracts was better than conventional fungicide. The extract of the cashew nutshell (CNS) at 2000 ppm proved the most effective in reducing the spore germination. The fungus did not have sufficient nutrients to spore germination. Analysis of spore germination showed a high correlation between the difference of absorbance (DA) at 48 h and the percentage of spore germination observed after 24 h at concentration of 1500 ppm and 2000 ppm, respectively. The effect of CNS extract was observed on spore germination at various concentrations ranging from 3.80 to 43.94 %. The compound of CNS tended to inactivate enzymes related to the synthesis of basic metabolites of fungus and very effective in inhibiting spore germination at a concentration of 200 ppm. The control treatment was found the lowest inhibition on spore germination of sooty mould.

Table 2. The percentage of spore germination of cashew leaves, nut shells, and peduncle bagasse ashes extracts

Concentration / Extracts	Spore germination (%) after 24 h incubation**				
	Control	500 ppm	1000 ppm	1500 ppm	2000 ppm
Cashew nutshells	96.61 (78.15)	43.94 (46.32)	28.20 (30.05)	12.01 (22.19)	3.80 (7.26)
Cashew leaves	90.57 (70.32)	81.17 (69.44)	66.28 (49.26)	48.26 (44.35)	20.58 (25.19)
Cashew peduncle bagasse ashes	94.74 (75.11)**	60.05 (55.72)	50.73 (42.80)	35.22 (38.76)	16.70 (19.54)
	SE.diff		C.D (P=0.05)	C.D (P=0.01)	
Fungicides	2.12		1.75	2.18	2.12
Concentration	0.87		1.59	2.35	0.87
Fungicide x conc.	1.55		3.06	3.94	1.55

Mean of four replicates; ** Figures in parentheses are arc Sin√% age transformed value and are statistically identical.

The potential effect of cashew nut shell liquid (CNSL) at a concentration of 320 µg mL-1, resulted in significant inhibition of the mycelial growth of *C. gloeosporioides* and *L. theobromae* on papaya fruit [14]. The fungus inactivated enzyme-producing energy by the toxic substance of CNSL. Therefore, after contact with conidia or mycelia did not have sufficient nutrients to mycelial growth. Additionally, the ash of cashew peduncle bagasse has also great potential to be a powerful non-toxic fungicidal agent against particular *Fusarium* species [11].

3.2. The content of total phenolics (TPC) and flavonoid content (TFC), acidity and PH analysis

The content of total phenolic and flavonoid content was expressed as mg GAE/g DW and mg QE/g DW, respectively. There were varied narrowly among concentration 1500 ppm and 2000 ppm) of cashew nutshell extract. Results presented in Table 3 that the average of total phenolics, flavonoids contents were 67.28 mg GAE. g^{-1}, 68.03 mg GAE. g^{-1} and 15.79 mg QE.g^{-1}, 16.06 mg QE.g^{-1} of CNS extract at a concentration of 1500 ppm and 2000 ppm, respectively. While cashew nutshells extract at a concentration of 2000 ppm showed the lowest spore germination activity after 24 h incubation. The extract has a low value of acidity and high value of pH. Cashew nutshells extract with high of total phenolic, and flavonoid content generally also indicated an excellent antifungal activity against sooty mould.

Table 3. Total phenolic content (TPC), Total flavonoid content (TFC), and pH of leaves, nut shells, and peduncle bagasse ashes extracts at the different concentrations.

Treatments	**Conc (ppm)**	**TPC** *mg GAE.g^{-1} DW*	**TFC** *mg QE. mg^{-1}DW*	**Acidity** *meq NaOH/g*	**pH**
No treated (positive control)		-	-	-	-
Fungicide (negative control)		-	-	-	-
Cashew nutshells	500	11.24	5.43	0.687	4.17
	1000	35.62	9.80	0.507	4.85
	1500	67.28	15.79	0.480	5.12
	2000	68.03	16.06	0.412	7.56
Cashew leaves	500	3.05	1.95	0.860	4.15
	1000	6.99	3.58	0.895	3.87
	1500	10.40	5.32	0.541	5.95
	2000	15.11	6.56	0.598	5.60
Cashew peduncle bagasse ashes	500	18.78	7.12	0.575	6.03
	1000	22.40	7.25	0.586	5.91
	1500	30.89	9.08	1.015	5.32
	2000	52.24	10.15	8.172	6.11

These results are in agreement with previous studies [5,9,12] which found that bioactive polyphenol compounds, singly or in combination, could not stimulate fungal and viral growth. The compounds may have destroyed of the cell membrane fungi.

4. Conclusion

All the concentrations of extracts showed significant inhibition in the spore germination of sooty mould. Cashew nut extract at a concentration of 2000 ppm showed the lowest spore germination activity. The extract had a high level of total polyphenolic, flavonoid content, pH and a low value of titratable acidity which could be attributed to its potent antifungal activity.

Acknowledgments

The authors would like to thank the DRPM Ministry of Research, Technology and Higher Education of the Republic of Indonesia who has funded this research and publication through the Leading Higher

References

[1] El-Gali Z I and Mohammed M S 2017 Sooty mold on lemon trees-first recording from El-Beida city *Rev. of Plant Stud.* **4** 1-7
[2] Hilje L and Morales F J 2008 Whitefly bioecology and management in Latin America. In: Capinera, J (ed) Encyclopedia of entomology (Dordrecht: Springer) 4250-60.
[3] Vieira S S, Bueno R C O F, Bueno A F, Boff M I C, Gobbi A L 2013. Different timing of whitefly control and soybean yield *Cienc. Rural* **43** 247-53
[4] Andayanie W R and Ermawati N 2019 Developmental Effect of Cashew Nut Shell Extract

Against Nymphal Instar of Silver Leaf Whitefly (*Bemisia tabaci)* Genn *IOP Conf. Series: Earth and Enviromental Science* **293** 012039

[5] Andayanie W R, Nuriana W and Ermawati N 2019 Bioactive compounds and their their antifeedant activity of cashew nut (*Anacardium occidentale* L.) shell extract against *Bemisia tabaci*, (Gennadius, 1889) (Hemiptera: Aleyrodidae) *Acta Agric. Slov.* **113** 281-8

[6] Andayanie W R, Adinurani P G, Nuriana W and Ermawaty N 2019 The Plant Defence Inducer Activity of *Anacardium occidentale* Linn., *Azadiracta indica* A. Juss. and *Zingiber officinale* Rosc. Extracts against *Cowpea Mild Mottle Virus* infecting Soybean *AIP Conf. Proc.* **2175** 020033

[7] Andayanie W R, Nuriana W and Ermawati N 2019 Antiviral activity of cashew nutshell extract against Cowpea mild mottle virus on soybean *Jurnal Hama dan Penyakit Tumbuhan Tropika* **19** 170-178

[8] Andayanie W R, Nuriana W and Iswati R 2019 Use tillage system and botanical herbicide of cashew nut shell extract on losses nutrient and organic matter in the sloping land *AIP Conf. Proc.* **2024** 020018

[9] Rongai D, Milano F and Sciò E 2012. Inhibitory Effect of Plant Extracts on Conidial Germination of the Phytopathogenic Fungus *Fusarium oxysporum Am. J. Plant Sci.* **3** 1693-8

[10] Rajesh Kannan V, Sumathi C S, Balasubramanian V and Ramesh N 2009 Elementary chemical profiling and antifungal properties of cashew (*Anacardium occidentale* L.) Nuts *Botany Research International* **2** 253-7

[11] Pires dos Santos R, Marinho R M, Araújo Sá R, Martins J L, Teixeira E H, Chagas de Sousa Alves F, Ramos V S C, Soares de Sousa G and Cavada B S 2011 Compositional analysis of cashew (*Anacardium occidentale* L.) peduncle bagasse ash and its in vitro antifungal activity against Fusarium species. *Rev. Bras. Biocienc.* **9** 200-5

[12] Udoh, Philip I, Aladenika, Tunrayo S, Eleazar, Idara C; Onyebueke, Adaze E, Azubuike, Chinonyelum N, Okwuosa, Nwachukwu C, Ibezim, Ekpereka N, Esimone and Charles O 2019 Antifungal properties of methanolic leaf extract of Anacardium of occidentale L. against Fusarial isolates from human and plant origin. *Archives* **1** 117-35

[13] Emelike N J T, Akusu M O, Ujong A E 2017 Antioxidant and physicochemical properties of oil extracted from cashew (*Anacardium occidentale* L) kernels *Int. J. Food Sci. Nutr.* **2** 122-8

[14] Garcia N Z T, Barbosa G F, Matias R, Pedrinho D R, Bono J A M and Martini D 2018. Antifungal potential of cashew nut shell liquid in the control of plant pathogens. *Biosci. J., Uberlândia* **34** 95-103

Anticancer activity of Lampung Robusta Coffee extract fractions

Husniati Husniati[1a], Devi Oktiani[1], Tati Suhartati[2], Valentino Budi Pratama[2], Anastasia Fitria Devi[3] and Nina Artanti[3]

[1]Center for Research and Standardization of Industry Bandar Lampung, Indonesia.
[2]Department of Chemistry, University of Lampung, Lampung, Indonesia.
[3]Research Centre for Chemistry, Indonesian Institute of Sciences (LIPI), Banten, Indonesia

[a]E-mail: husniati@kemenperin.go.id

Abstract. Coffee is a popular beverage that is consumed worldwide. Habitual coffee drinking is associated with cancer risk reduction including pre- and post-menopausal breast cancer. The invert correlation between coffee intake and breast cancer risk is weak, therefore coffee is potential for functional beverage. Coffee should be taken for several times per day to get its benefit. Compounds which give the cancer reduction effect are still debatable, they can be compounds with antioxidant or non-antioxidant properties. Robusta coffee from Lampung has a special meaning for Indonesian coffee lovers due to its strong aroma and flavour with affordable price. In order to observe the anticancer activity of Lampung Robusta coffee, its aqueous methanol extract was passed through a chromatography column containing polyamide with aqueous methanol as the eluent with a gradient of 10%. A total of 54 fractions were obtained and labelled as A to K fractions. Chlorogenic acid was detected in fractions A and B. Meanwhile, it was not detected in other fractions including G fraction. This G fraction, however, was the only one to show cytotoxicity against MCF-7 breast cancer cell line. This result suggests that this fraction contains potential compounds with anticancer activity.

1. Introduction

One of the main plantation commodities in Indonesian is coffee. Lampung is the second biggest coffee producer province in Indonesia after South Sumatera. Lampung has a very potential role in national coffee production, in 2017 it produced 107,219 ton of total 716,089 ton Indonesia coffee production for the year [1].

Coffee contains various compounds such as caffeine, chlorogenic acid, protein, and aromatic volatile compounds [2]. Most of the research and scientific publications only focus on caffeine in coffee, while characteristics and benefit of other compounds such as the chlorogenic acid (CGA) was not widely known and studied. CGA is an ester formed from caffeic acid and quinic acid, it is one of the most abundant polyphenol compounds, consists of a group of phenolic secondary metabolites, it isolated from the leaves and fruits of dicotyledonous plants [3]. CGA soluble phenolic compounds are conjugated forms of hydroxycinnamic acids with quinic acid, which accumulate in coffee plants [4].

There are several isomers of CGA for each subgroup of coffee plant varieties and each coffee extract contains different CGA [5]. Qualitative and quantitative differences of CGA are not only among species but also within species, in the coffee berries [4]. The CGA composition of the green coffee bean is also specifically affected by the way of processing [6].

CGA is an important and biologically active dietary polyphenol, playing several important and therapeutic roles such as antioxidant activity, antibacterial, hepatoprotective, cardioprotective, anti-inflammatory, antipyretic, neuroprotective, antiobesity, antiviral, antimicrobial, antihypertension, free radicals scavenger and a central nervous system (CNS) stimulator [4], The CGA extract from coffee bean has a potential activity as an anticancer and antioxidant [7-11] .

Previous research on human liver hepatocellular cancer cell, HepG2 and HepG3 cell concluded that CGA is activating and inhibiting some important pathways in cancer metabolism [12,13]. CGA as cancer treatment on Hep G2 cell indicated that it is less toxic then doxorubicin, the commercial chemotherapy medication for cancer treatment [13]. The IC_{50} value of CGA on Hep-G2 cell is 727μM (μmol/L) [14].

Previous studies on the effect of CGA on colon cancer concluded that CGA reduced colon cancer-cell proliferation [15-16]. Previous research was using CGA complex named CGA7, a decaffeinated water-soluble green coffee bean extract for human and mouse cancer line, colon cancer HCT-116 cells treatment, the overall findings indicated that CGA7 complex, a potential anticancer molecule found in green coffee beans, is most likely safe [17].

This research focuses on the variation of mobile phase solvent composition in product purification of CGA and coffee bean extract fractions and anticancer activity of extracts. This research applied fractionation based on mobile phase polarity. The variation of mobile phase composition is composed from polar to nonpolar organic solvent with gradual addition of nonpolar solvent. The objective of this variation is to select and separate compounds based on compound polarity, from high polar compound to less polar compound. This research studies correlation of the solvent to the extraction products characteristics. The different solvent composition used in chromatographic separation process. Each extract is tested for CGA contents and anticancer activity on MCF-7 cell. MCF-7 breast cancer cell line was chosen for anticancer activity assay. This cell line was available at Research Center for Chemistry (LIPI) and displayed good response toward various botanical extracts. MCF-7 cell is a human breast cancer cell line with estrogenic, progesterone and glucocorticoid receptors [18].

2. Material and Method

2.1. Materials

The material in this research is organic green Robusta coffee beans from Padang Cermin, Pesawaran, a city in Lampung province Indonesia. The coffee beans were determined by Herbarium Bogoriense, Research center for Biology-LIPI Bogor, as *Coffea canephora Pierre ex A. Frochne*r. The solvent for extraction and chromatographic separation process was methanol distillated technical/industrial grade solvents, while for spectrophotometric analysis used pro-analysis grade solvents. Silica Thin Layer Chromatography (TLC) for separation process.

MCF-7 breast cancer cell from Research Center for Chemistry (LIPI) for testing anticancer activity.

2.2. Methods

2.2.1. Extractions. Ground coffee was extracted with methanol [18]. The ratio of ground coffee and solvent was 1:4. The extract was concentrated with rotary vacuum evaporator to obtain the methanol extract. The methanol extract was later purified by nonpolar solvent n-hexane using separation device. During this process nonpolar compounds were removed from the extract. The process continued with purification using semi polar solvent ethyl acetate, which removed the semi polar compounds from extract. The remaining compounds in extract were polar compounds. CGA was possibly one of them.

The separation process continued with the polar mobile phase in film chromatographic column. The different compositions of polar mobile phase were applied in this experiment in order to obtain the different active compounds. The mobile phase were water methanol solutions with different volume ratio (Table 1) start from 100% water without methanol. The methanol addition reduced water polarity. This experiment started from 100% water because the main objective was to get the CGA extract in early the beginning or early fraction. The chromatographic method was performed in a column with polyamide as the stationary phase while demineralized water and methanol mixture was the mobile phase with 10% gradient (Table 1).

Table 1. Mobile phase composition.

Sample Fraction (Code)	Water (%)	Methanol (%)
A	100	0
B	90	10
C	80	20
D	70	30
E	60	40
F	50	50
G	40	60
H	30	70
I	20	80
J	10	90
K	0	100

2.2.2. Detection of CGA. Each fraction from chromatographic column was coded according to Table 1 and analysed using thin layer chromatography (TLC) for CGA qualitative testing. The later test was performed using LAAN-A-LC-E008 manual Shimadzu Application NN. L.3016 and Agilent Application Note 2016.

2.2.3. Test for anticancer activity. Anticancer activity was performed using MCF-7 breast cancer cells in Roswell Park Memorial Institute (RPMI) solution ($1x10^4$ cells/ml) [19]. The cells were transferred into a 96-well plate (100 μL per well) and incubated for 4 h at 37 °C to let the cells grew. Samples (10 μL each) with above mentioned codes (Table 1) were added. The cells were incubated for 24 h and then analyzed under microscope. Solvent control (positive cell growth) was prepared by transferring and incubating 100 μL cancer cells in RPMI, in solvent control 10 μL of solvent only was added into the cells without sample addition.

3. Result

The presence of CGA and anticancer activity in each sample fraction is presented in Table 2. This table also mentioned about the colour of the fractions as determined visually. For better visualisation, the compiled pictures of all fractions are displayed as Figure 1.

Figure 1. Different fraction coffee extracts, coded according to Table 1.

IOP Conf. Series: Materials Science and Engineering 1011 (2021) 012040 doi:10.1088/1757-899X/1011/1/012040

Table 2. The presence of CGA and anticancer activity of coffee sample fractions, along with their color as perceived by naked eyes.

Sample Fraction (Code)	CGA (+/–)[a]	Anticancer Activity (+/–)[b]	Color
A	+	–	Dark Brown
B	+	–	Brown
C	–	–	Light Brown
D	–	–	Brown Yellow
E	–	–	Green
F	–	–	Brown
G	–	+	Brown Yellow
H	–	–	Light Yellow
I	–	–	Clear (colorless)
J	–	–	Clear (colorless)
K	–	–	Clear (colorless)

Note: [a] (+) =detected, (-) = not detected based on TLC (see example in Figure2)
[b] (+) =detected, (-) = not detected based on anticancer analysis (see example in Figure 3)

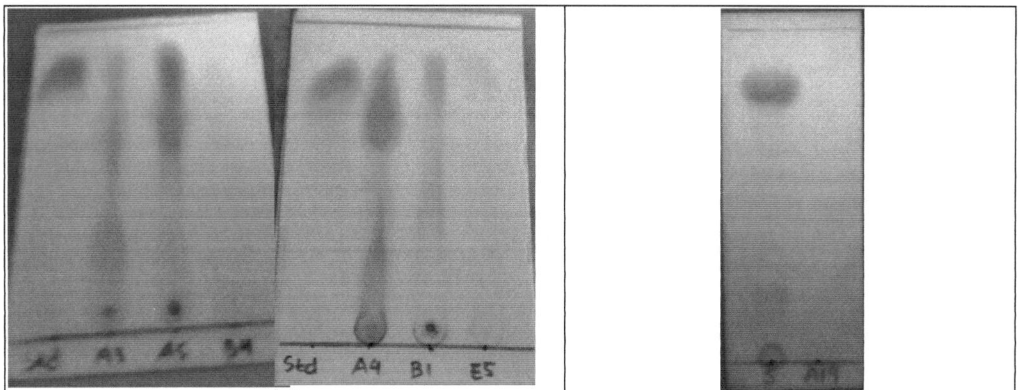

(a) (b)

Figure 2. Compounds identification by TLC compare to CGA standard solution. (a) Fraction A-E, (b) fraction G.

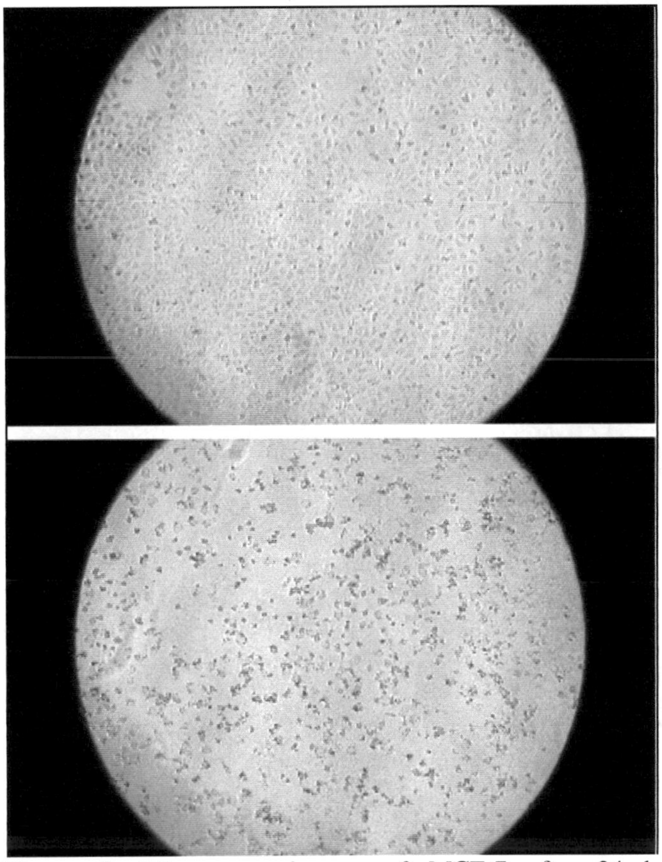

Figure 3. Microscopy images of MCF-7 after 24 h incubation. Without sample (above) and with sample G (below). The below image displays death of most MCF-7 cells.

3. Discussion

This research applied separation based on polarity using chromatographic column. In chromatographic methods, separation process distributes components and separates between two phases, stationary phase and mobile phase which move in a definite direction. The mobile phase leaves the column and this is called effluent. Based on molecule structure theory CGA is a polar compound, therefore the separation process was started polar solvent, followed by removing semi polar and non polar impurities. This was meant to purify the extract from nonpolar and semi polar compounds. The purified extract was fractionated with chromatographic column of water-methanol mixture to obtain 10 different fractions, coded according to Table 1.

The results obtained in accordance with the theory, the CGA was detected in the early fractions, which were A and B fractions. As predicted, CGA was not detected in the later fractions, which referred to C to K fractions (Table 2 and Figure 2).

Qualitative separation based on visual technique is a preliminary process for collecting fraction which dissolved in solvent from polyamide column. The next step is verification compounds by identify chemical compounds and determine compound purity using Silica thin layer chromatography (TLC), visual analysis on TLC spot, it is a quick and simple chemical analysis method [1].

The determination of CGA absence was conducted by visual determination on the spot by TLC analysis compare to CGA standard solution. Picture above (Figure 2) mentioned TLC analysis on CGA standard solution, the TLC analysis of sample from fraction A-E indicates the positively availability of CGA substance in the sample A3, A4, A5, and B1, as a conclusion all fraction A and B is positively contain CGA, while the test on sample E5 indicates there was no CGA substance in fraction E5. The test on sample also G indicates that there was no CGA substance in sample G.

The separation in this fractionation process is based on polarity produced different samples, in each fraction indicated that there were different chemical compounds in every fraction. These differences were possibly due to different polarity of water-methanol mixtures, which were used as the mobile phase. The most polar compounds were in the beginning fractions, hereinafter were the less polar compounds.

Each fraction was tested against MCF-7 breast cancer cell line in order to study its anticancer activity. The G fraction was the only sample showed cytotoxicity against MCF-7 cell line although it did not show any presence of CGA based on the thin layer chromatography analysis. Previous studies on anticancer activity of CGA are scarce. Lee and Zu (2006) found that human breast cancer cell cultures MCF-7 which were treated with CGA 20 μM displayed inhibited DNA methylation approximately by 80% [22]. Meanwhile Yamagata et al. (2018) observed that human lung cancer cells which were cultured with CGA 50 μM displayed reduced cell proliferation by 40% [23]. Our samples, particularly fractions A and B, which contained CGA, did not indicate any anticancer characteristics. There are some possibilities beyond these negative results. First, anticancer activity of CGA is dose and time dependent [24]. The concentration of CGA in both fractions A and B were too low to show any anticancer effect. Second, there are compounds other than CGA in both fractions A and B with anticancer effects. In addition to CGA, it is suggested that several compounds from coffee such as caffeine, ferulic acid, and kahweol, which have anticancer characteristic as displayed in either in vitro or in vivo studies [25]. The G fraction in this study needs to be investigated in the future research to identify the potential anticancer compound.

4. Conclusion

Separation of coffee extract using a polyamide column chromatography with gradient water methanol solvent produced fractions containing different composition of compounds. TLC analysis showed that fractions A and B contain CGA substances. Fraction A-F and H-K did not show anticancer activity. Fraction G showed no CGA present was the only sample that showed anticancer activity against MCF-7 breast cancer cell line.

Acknowledgments

This research was supported by Center of Research and Standardization of Industry (Baristand) in Bandar Lampung and Industrial Research and Development Body (BPPI) of Ministry of Industry. We also would like to show our gratitude to Department of Chemistry (University of Lampung), Integrated Laboratory and Technological Innovation Center (LTIST, University of Lampung), and Research Center for Chemistry (Indonesian Institute of Sciences, LIPI).

References

[1] Sub Directorate of Estate Crops Statistics, Statistics Indonesia (BPS) 2019 *Indonesian coffee statistics Indonesia 2018*, Jakarta: Badan Pusat Statistik (BPS) - Statistics Indonesia; 2018. https://www.bps.go.id/publication, Accessed : 25 August 2020, 08:00 a.m.

[2] Mulato S, Widyotomo S and Lestari H 2001 Water extraction of caffeine in Robusta coffee using a fixed bed column *Coffee and Cocoa Research Journal* **17**(2) 97–109

[3] Kim JK and Park SU 2019 Chlorogenic acid and its role in biological functions: an up to date *EXCLI J* **18** 310–6

[4] Campa C, Kochko A D and Jacques R J 2008 Chlorogenic acids: diversity in green beans of wild coffee species *Adv. Plant Physiol.* **10** 421–37

[5] Naveed M, Hejazi V, Abbas M, Kamboh A A, Khan G J, Shumazid M, Ahmad F, Babazadeh D, Fangfang X, Modarresi-Ghazani F, WenHua L and XiaoHui Z 2018 Chlorogenic acid (CGA): a pharmacological review and call for further research. *Biomed Pharmacother* **97** 67-74

[6] Selmer D, Kleinwachter M and Bytof G 2015 Metabolic responses of coffee beans during processing and their impact on coffee flavor. In: Schwan RF, Fleet GH, and Afoakwa EO *Cocoa and Coffee Fermentation* (Florida: CRC Press) pp 432–76

[7] Miller E G, Gonzales A P, Orr A M, Binnie W H and Sunahara G I 2000 The anticancer activity of coffee beans. In: Parliament TH Ho C, and Schieberle P *Caffeinated Beverages*, chapter 7, ACS Symposium Series vol 754. (USA: American Chemical Society) pp 56-63

[8] Yashin A, Yashin Y, Wang J Y and Nemzer B 2013 Antioxidant and antiradical activity of coffee (Review) *Antioxidants* **2** 230–45

[9] Farah A and Lima J 2018 Consumption of chlorogenic acids through coffee and health implications (Review) *Beverages* **5**(1) 11

[11] Lukitasari M, Nugroho D A and Widodo N 2018 Chlorogenic acid: the conceivable chemosensitizer leading to cancer growth suppression. *JEBIM* **23** 1–6

[12] Tajik N, Tajik M, Mack I and Enck P 2017 The potential effect of chlorogenic acid, the main phenolic components in coffee, on health: a comprehensive review of the literature *Eur. J. Nutr.* **56**(7) 2215–44

[13] Moreno-Ceballos M, Arroyave J C, Cortes-Mancera F M and Röthlisberger S 2019 Chemopreventive effect of coffee against colorectal cancer and hepatocellular carcinoma *Int. J. Food Prop.* **22**(1) 536–55

[14] Sukohar A and Muhartono M 2015 Comparative effects of chlorogenic acid and doxorubicin in against expression of caspase3 in cell lines Hep-G2 *J. Chem. Pharm. Res.* **7**(1) 187–92

[15] Sukohar A, Herawati H, Witarto A B, Setiawan S, Wirakusumah F F and Sastramihardja H S 2013 Role of chlorogenic acid from Lampung Robusta coffee against gene expression of Mirna (Micro RNA) 146 A on hepatocellular carcinoma cells *Int. J. Res. Phar. Nano Sci.* **2**(6) 776 – 84

[16] Ekbatan S S, Li X, Ghorbani M, Azadi B and Kubow S 2018 Chlorogenic acid and its microbial metabolites exert anti-proliferative effects, S-phase cell-cycle arrest and apoptosis in human colon cancer caco-2 cells. *Int. J. Mol. Sci.* **19**(723) 1–13

[17] Buldak R J, Hejmo T, Osowski M, Buldak L, Kukla M, Polaniak R and Birkner E 2018 The impact of coffee and its selected bioactive compounds on the development and progression of colorectal cancer in vivo and in vitro *Molecules* **23**(3309) 11–25

[18] Goutomchandra K, Sudeep H V, Venkatesh B J, and Prasad K S 2017 Chlorogenic acid complex (CGA 7), standardized extract from green coffee beans exerts anticancer effects against cultured human colon cancer HCT-116 cells. *Food Science and Human Wellness* **6**(3) 147–53

[19] Sukohar A, Wirakusumah F F and Sastramihardja H S 2011 Isolation and characterization cytotoxic compounds caffeine and chlorogenic acid seeds of Lampung Coffee Robusta. *Jurnal Medika Planta* **1**(4) 11–26

[20] Forster G M, Raina K, Kumar A, Kumar S, Agarwal R, Chen M H, Bauer J E, McClung A M and Ryan E P 2013 Rice varietal differences in bioactive bran components for inhibition of colorectal cancer cell growth *Food Chem.* **141** 1545–52

[21] Khopkar SM 2003 *Konsep Dasar Kimia Analitik* translated by A. Saptorahardjo. Jakarta: Universitas Indonesia pp 84-311

[22] Lee WJ and Zhu BT 2006 Inhibition of DNA methylation by caffeic acid and chlorogenic acid, two common catechol-containing coffee polyphenols *Carcinogenesis* **27**(2) 269-77

[23] Yamagata K, Izawa Y, Onodera D and Tagami M 2018 Chlorogenic acid regulates apoptosis and stem cell marker-related gene expression in A549 human lung cancer cells *Mol. Cell. Biochem.* **441** 9-19

[24] Goutomchandra K, Sudeep H V, Venkatesh B J and Prasad K S 2017 Chlorogenic acid complex (CGA 7), standardized extract from green coffee beans exerts anticancer effects against cultured human colon cancer HCT-116 cells *Food Science and Human Wellness* **6**(3) 147–53

[25] Jabir N R, Islam M T, Tabrez S, Shakil S, Zaidi S K, Khan F R, da Silva Araújo L, de Meneses A P M, de Oliveira Santos J V and de Carvalho Melo-Cavalcante A A 2018 An insight towards anticancer potential of major coffee constituents *BioFactors* **44**(4) 315-26

Metabolite Detection and Antibacterial Activity of Fungal Endophytic Extracts Isolated from Brotowali (*Tinospora crispa*) Plants using TLC-Bioautography Assay

A Fathoni[1,2b], S. Hudiyono[1a], E Budianto[1], A H Cahyana[1] and A Agusta[2]

[1] Department of Chemistry, Faculty of Mathematics and Natural Sciences (FMIPA), Universitas Indonesia, Depok 16424, Indonesia
[2] Research Center for Chemistry, Indonesian Institute of Sciences (LIPI), Cibinong Science Center, Cibinong, Bogor 16911, Indonesia

Corresponding author: [a]sumi.hudiyono@sci.ui.ac.id; [b]ahmad.fathoni81@ui.ac.id

Abstract. Endophytic fungi may be a potent source of bioactive compounds and have a vast repertoire of diverse metabolites. One source of endophytic fungi host plants is a medicinal plant such as Brotowali or *Tinospora crispa*. Research corresponding to the antibacterial activity of endophytic fungi extracts isolated from *T.crispa* taken from several regions in West Java has never been reported yet. While antibacterial methods to evaluate their activity as well as to identify the chemical compounds is a TLC-bioautography assay. This research is aimed to investigate the components of endophytic fungal extracts and their antibacterial activity using the TLC-bioautography assay method. In this study, eighty isolates of endophytic fungi have been successfully isolated from plant tissues of *T.crispa* from several regions in West Java. Antibacterial activity by using TLC (dot-blot) plates revealed that the fungal extracts could inhibit *S. aureus* (77 extracts) and *E. coli* (35 extracts) with inhibition zones ranging from 8-30 mm. The endophytic fungi extracts that showed potent antibacterial activity against Gram-positive bacteria (*S.aureus*) were sixteen extracts, while nine extracts have good inhibition activity against Gram-negative bacteria (*E.coli*). Among the fungal extracts that have excellent ability against both Gram-positive and negative was namely TcBt1Bd-10 extract. Based on metabolites analysis using the TLC method, the possibility of the chemical compounds that played a role in antibacterial activity of the extracts, including phenolics, flavonoids, alkaloids, and terpenoids.

1. Introduction

Endophytic fungi may be an excellent source of bioactive compounds and have a vast repertoire of diverse metabolites [1]. Endophytic fungi are known as symbiotically associated with the plant tissues, and they induce without symptoms to their hosts [2]. Due to they are capable of synthesizing bioactive metabolites that can be used for their host plant defence against fungal and bacterial pathogenic, so endophytic fungi also have known as "the chemical synthesizers inside plants" [3]. The fungal bioactive compounds produced have important pharmacological activities such as antioxidant, anticancer, immunomodulatory, antivirus, antituberculosis, antiparasite, and insecticides [4]. One source of endophytic fungi is a medicinal plant such as Brotowali or *Tinospora crispa*. The extracts of *T.crispa* plants have known as the potential multiple pharmacological and therapeutic activities due to the presence of various chemical compounds in the herb [5]. Chromatography methods can separate various bioactive compounds. Thin-layer chromatography analysis performed with biological activity detection is termed bioautography. TLC-bioautography assay is a fast and inexpensive method [6]. Bioautography assay belongs to screening methods commonly used for antimicrobial activity detection [7] and enables the search for biologically active substances in complex mixtures [8] such as endophytic fungi, plant, and actinomycetes extracts [9, 10, 11]. The principle of this method is the separation and detection of components of extracts which are carried out directly on a TLC plate. The extracts were transferred on a TLC plate and developed with a suitable solvent, then observed using both viewed under UV-light and sprayed staining-reagents to identify the secondary metabolites. The determination of antibacterial activity was also carried out using TLC dot-blot and TLC-bioautography assay, then dipped in bacterial culture, and bacteria grew directly on the layers during the incubation time. Inhibition zones are formed in the places where the antibacterial component is located. This study aims to investigate the components of endophytic fungal extracts and their antibacterial activity using the TLC-bioautography assay method.

2. Materials and Methods

2.1. Material
Samples of fresh and healthy parts of *T.crispa* plants (stem, leaves, and petiole) were collected from Bogor, Bekasi, and Bandung, West Java, Indonesia. Identification of the plant specimen was carried out at Herbarium Bogoriense, Research Center for Biology, Indonesian Institute of Sciences (LIPI).

2.2. Staining-reagents
Dragendorff (Merck) reagent, 0.25% vanillin (Sigma-Aldrich) solution in 10% sulphuric acid-ethanol, 1% $Ce(SO_4)_2$ (Merck) solution in 10% sulphuric acid-ethanol, folin-ciocalteu (Merck) reagent. After sprayed vanillin-sulfuric acid or cerium (IV) sulfate, the TLC plates (Silica gel 60 F_{254}, Merck) were heated at 110 °C for optimal color visualization. All reagents and solvents used in this research were analytical grades.

2.3. Bacterial Isolates for Antibacterial Testing
Bacterial isolates, namely *S. aureus* InaCC-B4 and *E. coli* InaCC-B5, were used for the antibacterial test. They are microbial collections of Indonesian Culture Collection (InaCC), Microbiology Division, Research Center for Biology, Indonesian Institute of Sciences (LIPI).

2.4. Isolation of Endophytic Fungi
Fungal endophytic isolation was carried out by surface sterilization, according to Praptiwi et al. [9] with minor modification. Healthy and fresh parts of plants (stems, leaves, and petiole) collected from the field were stored at a temperature of 4°C. These samples were cleaned under tap water and dipped in 70% ethanol for 1 minute, then dipped in 5.3% sodium hypochlorite for 2.5-3.5 minutes and finally dipped in 70% ethanol for 30 seconds. Samples were dried under aseptic conditions. The sterilized samples were cut aseptically into around size of 1×1 cm^2, and then, placed on top of the Corn Meal Malt Agar (CMMA) growth medium containing 17 g/L corn meal agar (Fluka, Sigma-Aldrich), 20 g/L malt extract (Bacto™), and 2 g/L yeast extract (Bacto™) were supplemented with 0.05 g/L chloramphenicol (Sigma), and then incubated at room temperature for 3-5 days. The emerging colonies were subcultured several times on Potato dextrose agar (PDA, Difco™) until a single isolate was obtained. Pure isolates were preserved in 10% glycerol and stored at -80°C at the Indonesian Culture Collection (InaCC), Indonesian Institute of Sciences (LIPI).

2.5. Cultivation and Extraction of Endophytic Fungi
Each fungal isolate was cultured on potato dextrose broth (PDB, Difco™) (@200 ml in culture flask of 500 ml) and incubated in a dark room at a temperature of 25-26°C for 21 days. After the incubation period is completed, both growth media and fungal biomass were extracted with ethyl acetate (EtOAc)

thrice. The extract solvent was removed under reduced pressure, and the dry extracts were stored in the glass vials at -30°C until used.

2.6. Chemical Compounds Analysis: Thin Layer Chromatography (TLC) Detection
Chemical compounds analysis of fungal extracts were performed on silica gel Thin-layer chromatography (TLC) plates (silica gel 60 F_{254}, Merck). The dried extract was prepared in 10 mg/ml, and ten microliters of extract were transferred on the TLC plate and developed in CH_2Cl_2: MeOH (10:1). Separated chemical compounds were visualized under UV light of 254 and 366 nm (Camag) and followed by spraying with staining-reagent dragendorff, or vanillin-sulfuric acid, or cerium (IV) sulfate, or folin-ciocalteu.

2.7. Screening of Antibacterial Activity: TLC Dot-Blot Assay
Antibacterial Activity screening by using the TLC Dot-Blot assay was carried out to assess of antibacterial potency of fungal extracts. Ten microliters of fungal extracts were loaded on the TLC plates (Silica gel 60 F_{254}, Merck) and then dried at room temperature. The TLC Plate then immersed into the suspension of bacterial culture, then incubated the plate in a humid chamber for 18-24 hours at a temperature of 37 °C. After the incubation period was finished, the plates were sprayed with iodonitrotetrazolium p-violet (INT, Sigma). The inhibition of bacteria growth was viewed by the formation of a clear zone around the active extract. Then the inhibition zone of the active extract was measured as inhibition zone diameter. Then, the active extracts were further evaluated the antibacterial activity by using a TLC-direct bioautography assay.

2.8. TLC-Direct Bioautography Assay
Ten microliters of the selected extracts were loaded on the TLC plates and then dried at room temperature. Then the active extract on the TLC plate developed with mobile phase CH_2Cl_2: MeOH (10: 1) and dried at room temperature. The TLC Plate then immersed into the suspension of bacterial culture, then incubated the plate in a humid chamber for 18-24 hours at a temperature of 37 °C. After the incubation period was finished, the plates were sprayed with iodonitrotetrazolium p-violet (INT, Sigma). The inhibition of bacteria growth was viewed by the formation of a clear zone around the active band of the chemical compounds contained in the extracts. Then inhibition band of the active compounds was measured as Retention factor (Rf).

3. Results and Disscussion
Eighty isolates of endophytic fungi were successfully obtained from healthy and fresh tissues of *T.crispa* plants. Fungal endophytic isolates were associated with *T.crispa* plants, i.e., 30, 26, and 24 isolates from stem leaves and petioles, respectively (Table 1). Another study, Sun et al. [12] reported that host plant species and tissue types significantly affected the endophytic fungi community, where the overall colonization rate of endophytic fungi was significantly higher in branches than in leaves. This study showed that the host plant, *T.crispa*, was taken from Bogor and obtained the most endophytic fungal isolates, probably due to differences in geographic location and abiotic factors. Several endophytic fungi isolated from *T.crispa* after cultivated in PDB media for 21 days (Figure 1) with distinct exudate colors in broth media showed the different and diverse fungal species. According to a previous study, the rate of endophytic fungal colonization, diversity, and community composition was influenced by host plant species, genotypic, tissue types, geographic location, and abiotic factors [13, 12, 14, 15, 16]. The composition of endophytic fungal communities influenced by their host plant species and tissue types, and some endophytic fungi showed significant host and tissue preferences [17].

Table 1. Endophytic Fungi associated with *T.crispa*

No. Sample	Isolate Name	Origin plant from	Plant part	Weight of EtOAc extracts* (mg)	No. Sample	Isolate Name	Origin plant from	Plant part	Weight of EtOAc extracts* (mg)
1	TcBt1Bd-02	Bandung	Stem	69.5	41	TcTd2Be-04	Bekasi	Petiole	27.2
2	TcBt1Bd-05	Bandung	Stem	99.9	42	TcBt1Bo-01	Bogor	Stem	66.5
3	TcBt1Bd-07	Bandung	Stem	131.0	43	TcBt1Bo-02	Bogor	Stem	154.3
4	TcBt1Bd-08	Bandung	Stem	182.1	44	TcBt1Bo-03	Bogor	Stem	112.4
5	TcBt1Bd-09	Bandung	Stem	108.4	45	TcBt1Bo-04	Bogor	Stem	82.2
6	TcBt1Bd-10	Bandung	Stem	62.2	46	TcBt1Bo-05	Bogor	Stem	122.7
7	TcBt2Bd-01	Bandung	Stem	226.2	47	TcBt1Bo-06	Bogor	Stem	233.1
8	TcBt2Bd-03	Bandung	Stem	72.0	48	TcBt1Bo-07	Bogor	Stem	241.0
9	TcBt2Bd-04	Bandung	Stem	85.9	49	TcBt1Bo-08	Bogor	Stem	73.9
10	TcBt2Bd-06	Bandung	Stem	33.0	50	TcBt1Bo-09	Bogor	Stem	40.0
11	TcDn1Bd-01	Bandung	Leaf	102.5	51	TcBt1Bo-10	Bogor	Stem	134.0
12	TcDn1Bd-02	Bandung	Leaf	52.2	52	TcBt2Bo-01	Bogor	Stem	91.8
13	TcDn1Bd-03	Bandung	Leaf	37.9	53	TcBt2Bo-02	Bogor	Stem	23.7
14	TcDn2Bd-01	Bandung	Leaf	18.8	54	TcBt2Bo-03	Bogor	Stem	134.8
15	TcDn2Bd-02	Bandung	Leaf	38.0	55	TcBt2Bo-04	Bogor	Stem	60.9
16	TcDn2Bd-03	Bandung	Leaf	70.8	56	TcDn1Bo-01	Bogor	Leaf	31.0
17	TcDn2Bd-04	Bandung	Leaf	69.6	57	TcDn1Bo-02	Bogor	Leaf	32.3
18	TcTd1Bd-01	Bandung	Petiole	90.5	58	TcDn1Bo-03	Bogor	Leaf	27.1
19	TcTd1Bd-02	Bandung	Petiole	32.8	59	TcDn1Bo-04	Bogor	Leaf	89.5
20	TcTd1Bd-03	Bandung	Petiole	104.1	60	TcDn1Bo-05	Bogor	Leaf	14.7
21	TcTd2Bd-01	Bandung	Petiole	92.0	61	TCDn2Bo-01	Bogor	Leaf	31.0
22	TcTd2Bd-02	Bandung	Petiole	178.3	62	TcDn2Bo-02A	Bogor	Leaf	61.0
23	TcTd2Bd-03	Bandung	Petiole	118.0	63	TCDn2Bo-03	Bogor	Leaf	19.2
24	TcTd2Bd-04	Bandung	Petiole	98.0	64	TCDn2Bo-04	Bogor	Leaf	100.4
25	TcTd2Bd-05	Bandung	Petiole	13.5	65	TCDn2Bo-05	Bogor	Leaf	130.7
26	TcTd2Bd-06	Bandung	Petiole	139.2	66	TcDn2Bo-06	Bogor	Leaf	31.1
27	TcBt1Be-01	Bekasi	Stem	126.8	67	TcDn2Bo-07	Bogor	Leaf	50.0
28	TcBt1Be-03	Bekasi	Stem	79.0	68	TcDn2Bo-08A	Bogor	Leaf	151.9
29	TcBt2Be-01	Bekasi	Stem	17.1	69	TcDn2Bo-09	Bogor	Leaf	76.2
30	TcBt2Be-05	Bekasi	Stem	29.3	70	TcTd1Bo-01	Bogor	Petiole	38.2
31	TcBt2Be-06	Bekasi	Stem	73.1	71	TcTd1Bo-02	Bogor	Petiole	24.4
32	TcBt2Be-07	Bekasi	Stem	79.2	72	TcTd1Bo-03	Bogor	Petiole	18.2
33	TcDn1Be-01	Bekasi	Leaf	33.9	73	TcTd1Bo-04	Bogor	Petiole	35.8
34	TcDn2Be-01	Bekasi	Leaf	57.1	74	TcTd2Bo-01A	Bogor	Petiole	24.8
35	TcDn2Be-02	Bekasi	Leaf	34.1	75	TcTd2Bo-02	Bogor	Petiole	28.8
36	TcDn2Be-03	Bekasi	Leaf	52.1	76	TcTd2Bo-03	Bogor	Petiole	58.5
37	TcDn2Be-04	Bekasi	Leaf	15.1	77	TcTd2Bo-04	Bogor	Petiole	67.0
38	TcTd1Be-02	Bekasi	Petiole	383.8	78	TcTd2Bo-05	Bogor	Petiole	132.3
39	TcTd2Be-02	Bekasi	Petiole	33.0	79	TcTd2Bo-06	Bogor	Petiole	104.9
40	TcTd2Be-03	Bekasi	Petiole	224.0	80	TcTd2Bo-07	Bogor	Petiole	68.9

Remark : (*) Extracts produced by 200 ml endophytic fungi cultured in PDB media, static condition, for 21 days.

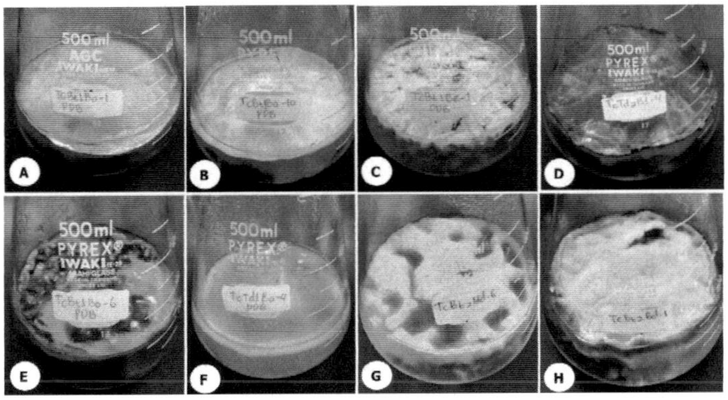

Figure 1. Some Endophytic Fungi isolated from *T.crispa* plant (cultured in PDB media, 21 days). A: TCBt1Bo-1, B: TCBt1Bo-10, C: TCBt1Be-1, D: TCTd2Bd-4, E: TCBt1Bo-6, F: TCTd1Bo-4, G: TCBt2Bd-6, H: TCBt2Bd-1.

In this study, several flavonoids without chemical treatment appeared red color under visible light, and some terpenoids and coumarins showed blue fluorescent and dark zones in the chromatogram that viewed under UV of 254 nm (Figure 2A and 2B). While, some flavonoids viewed under 366 nm shown yellow or dark yellow, or brown fluorescence (Figure 2C). The spray-reagents used to allow the composition of the fungal extracts. General staining-reagent and very well for the detection of alkaloid compounds is cerium (IV) sulfate that shown brown color (Figure 2D). Another general spray-reagent and suitable for the detection of hydroxyl or carbonyl compounds is vanillin-sulfuric acid that appeared diverse colors, i.e., red (phenolics), and green (terpenoids) (Figure 2E). Staining-reagent by using dragendorff, alkaloids appear as a stable orange color (visible) (Figure 2F). Folin-ciocalteu reagent used for phenolic compound detection that shown dark-blue color (Figure 2H). The standard of metabolites used berberine (alkaloid), phloroglucinol (phenolic), and quercetin (flavonoid) appeared brown (without treatment, and viewed under UV 254 nm), red (after sprayed vanillin), and dark zone (after sprayed folin-ciocalteu) (unpublished data) (Table 2). Phenolics compound (gallic acid) in other studies showed a blue or dark zone [24], but the phenolic standard (phloroglucinol) did not appear dark or blue (unpublished data). While dark or blue color appeared for the detection of flavonoid (quercetin standard) after sprayed folin-ciocalteu. From this study, folin-ciocalteu possible allow as staining-reagent for flavonoid compound detection by the TLC method.

In the previous study reported that endophytic fungi extract isolated from *T.crispa* could have produced diverse metabolites such as flavonoids, phenolics, terpenoids, and coumarins. While an alkaloid (cytochalasin D) and phenolic (phloroglucinol) produced by an endophytic fungus that isolated from *Albertisia papuana* leaves [24, 25]. Praptiwi et al. [26] and Agusta et al. [27] reported that fungal endophytic associated with *Uncaria gambir* stems could produce anthraquinone compounds ((+)-1,1'-bislunatin and (+)-2,2'-epicytoskyrin A). Two alkaloids are produced by endophytic fungus isolated from *Annona squamosa* plant [28]. Fungal metabolites such as steroids were produced by endophytic fungus isolated from *Cynodon dactylon* leaves [29], and some sesquiterpene quinones were isolated from a fungal strain endophytic fungus associated with *Platycladus orientalis* plant [30]. Bioactive compounds such as some coumarins were isolated from an endophytic fungus that associated with the leaves of the Chinese mangrove *Rhizophora mucronate* [31].

Table 2. Metabolite detection by using UV light and spray-reagent.

UV light /Staining Reagent	Detection for	Colors (metabolites)	Literatures
UV light (254 nm)	Compounds with extended conjugation (aromatic compounds)	Blue-white (coumarins) Dark zones (terpenoids with conjugated double bonds)	[18, 19]
		Dark zones (flavonoid: quercetin standard, Sigma)	Unpublished data
UV light (365 or 366 nm)	Compounds with extended conjugation (aromatic compounds)	Yellow, or brown, or blue, or blue-green fluorescence (coumarins) Yellow, or orange, or red fluorescence (all anthracene derivatives) Blue, or blue-green, or violet fluorescence (alkaloids) Yellow or yellowish-blue or dark yellow, or orange, or brown, or green, or blue fluorescence (flavonoids)	[18, 19, 20]
		Yellow (alkaloid : berberine standard, Merck)	Unpublished data
Cerium (IV) sulfate	General staining-reagent, very well for alkaloids	Brown (alkaloid: cytochalasin D)	Unpublished data
		Yellowish-brown (alkaloid: berberine standard, Sigma)	Unpublished data
		Orange (flavonoid: quercetin standard)	
Vanillin-H_2SO_4	General staining-reagent, very well for hydroxyl or carbonyl compounds	Blue-violet, or red (saponins) Red or yellow or brown or blue-green (flavonoid glycoside) Blue, or dark blue, or dark green, or brownish-green, bluish-green, light orange, purple, maroon, lavender, brown (terpenoids)	[19, 20, 21]
		Red (phenolic: phloroglucinol standard)	Unpublished data
Dragendroff	Alkaloid staining-reagent	Orange or reddish-brown (alkaloids)	[22, 23]
Folin-ciocalteu	Phenolic staining-reagent	Blue or dark zones (phenolics)	[24]
		Dark zones (flavonoid: quercetin standard)	Unpublished data

Antibacterial activity by using TLC Dot-Blot assay revealed that six endophytic fungi extracts associated with *T. crispa* showed potent antibacterial activity (the inhibition zone diameter range of 17-30 mm or > 16 mm) against both *S. aureus* (Gram-positive) and *E. coli* (Gram-negative), while some extracts had strong activity only against Gram-positive or Gram-negative bacteria (Table 3). Then all of the active extracts were further evaluated for antibacterial activity test by using the TLC-direct bioautography assay. This study revealed that the alkaloids (Rf: 0.800), phenolics (Rf: 0.125-0.500), and flavonoids (0.500-0.625) produced by endophytic fungi No. 27 and 28 showed antibacterial against only Gram-positive bacteria (Figure 2 and 3). A previous study reported that bioactive metabolites such as alkaloid, pseurotin A, produced by fungal endophytic isolated from *Bauhinia guianensis* plant had potent activity against *B.subtilis*, *E.coli*, *P.aeruginosa*, and *S.aureus* [33]. While, phenolic compound, 4-(2,4,7-trioxa-bicyclo [4.1.0] heptane-3-yl) phenol, was produced by endophytic fungus isolated from *Mangifera indica* plant has strong not only as an antibacterial against Gram-positive and negative bacteria but also antifungal activity [34].

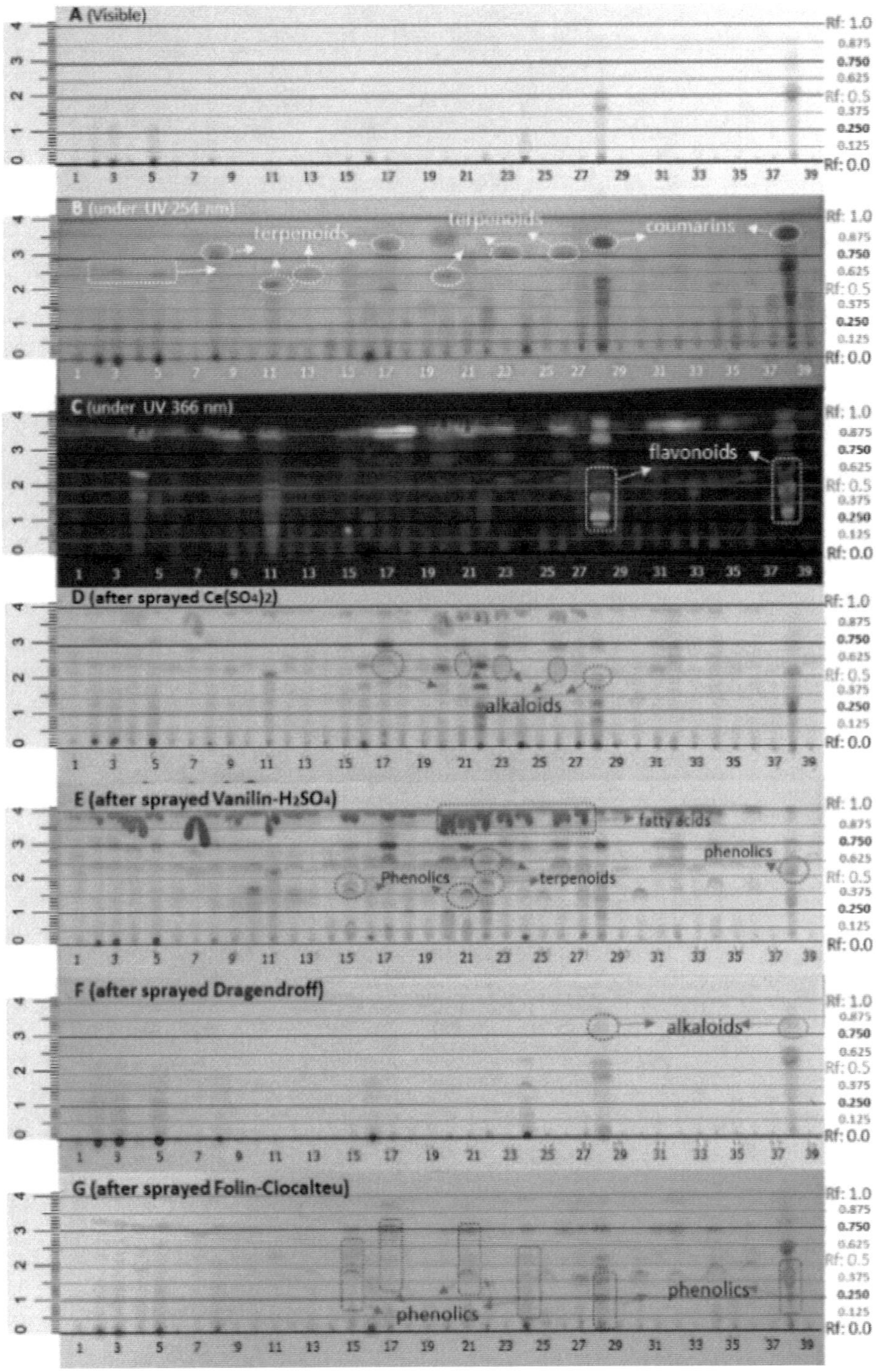

Figure 2. Some chromatograms of fungal endophytic extracts (no.1-39). A: under visible light, B: under UV light (254 nm), C: under UV light (366 nm), D: after sprayed staining-reagent (cerium sulfate), E: after sprayed staining-reagent (vanillin-sulfuric acid), F: after sprayed staining-reagent (dragendroff), G: after sprayed staining-reagent (folin-ciocalteu). Mobile phase: CH_2Cl_2-MeOH (10 : 1). Stationary phase: Silica gel 60 GF_{254} (Merck). Metabolite analysis based on band color according to the previous studies [18, 19, 20, 22, 23, 24].

Table 3. Screening Antibacterial Activity of Fungal Endophytic Extracts.

No Sample	Isolate Name	Region origin	Plant Parts	IZD against *S.aureus* (mm)	Category of Antibacterial against *S.aureus*	IZD against *E.coli* (mm)	Category of Antibacterial against *E.coli*
6	TcBt1Bd-10	Bandung	Stem	30	Strong	30	Strong
1	TcBt1Bd-02	Bandung	Stem	22	Strong	23	Strong
31	TcBt2Be-06	Bekasi	Stem	22	Strong	20	Strong
37	TcDn2Be-04	Bekasi	Leaf	21	Strong	18	Strong
8	TcBt2Bd-03	Bandung	Stem	19	Strong	17	Strong
68	TcDn2Bo-08A	Bogor	Leaf	19	Strong	17	Strong
67	TcDn2Bo-07	Bogor	Leaf	22	Strong	10	Weak
80	TcTd2Bo-07	Bogor	Petiole	22	Strong	NA	Not active
32	TcBt2Be-07	Bekasi	Stem	20	Strong	14	Moderate
47	TcBt1Bo-06	Bogor	Stem	20	Strong	15	Moderate
28	TcBt1Be-03	Bekasi	Stem	19	Strong	NA	Not active
51	TcBt1Bo-10	Bogor	Stem	19	Strong	15	Moderate
53	TcBt2Bo-02	Bogor	Stem	19	Strong	NA	Not active
57	TcDn1Bo-02	Bogor	Leaf	19	Strong	NA	Not active
10	TcBt2Bd-06	Bandung	Stem	17	Strong	NA	Not active
38	TcTd1Be-02	Bekasi	Petiole	17	Strong	NA	Not active
16	TcDn2Bd-03	Bandung	Leaf	15	Moderate	20	Strong
22	TcTd2Bd-02	Bandung	petiole	14	Moderate	19	Strong
45	TcBt1Bo-04	Bogor	Stem	14	Moderate	20	Strong

Remark: IZD: Inhibition Zone Diameter. Selected data (strong activity against *S. aureus* and/or *E.coli*). Antibacterial activity category based on the diameter of inhibition zone: 7-10 mm (weak); 11-15 mm (moderate); >16 mm (strong) [32].

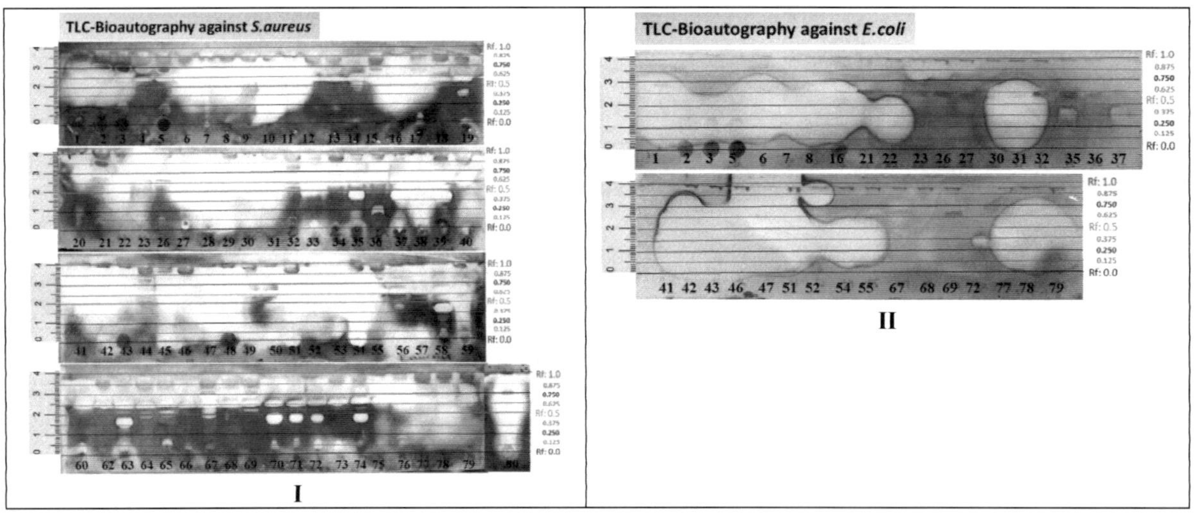

Figure 3. Bioautogram of selected extracts (active extracts against *S.aureus* and *E.coli*). Mobile phase: CH_2Cl_2-MeOH (10 : 1). Stasionary phase: Silica gel 60 GF_{254} (Merck). I: active extracts against *S.aureus*, II: active extracts against *E.coli.*

In addition, some flavonoids such as 3-methoxy flavone, nobiletin, formononetin, scopoletin, and daidzein contained in endophytic fungi that have antioxidant and antibacterial activities [35]. One of the endophytic extracts that good antibacterial activity against both Gram-positive and negative bacteria is isolate extract No. 22 showed the terpenoid compounds with Rf of 0.400 and 0.625 as bioactive compounds. In another study, some terpenoid compounds such as conidiogenon B and conidiogenol had excellent antibacterial Gram-positive and negative bacteria [36]. Fungal extract No. 6 was the best extract as antibacterial with Rf around of 0.125-0.750. Identification of bioactive metabolites produced by fungal isolate No. 6 is not very clear, because the band color of compounds is not obviously visible (Figure 2). The limitation of component analysis by using TLC methods for the compound with unclear color bands, but this method provides a fundamental separation of compounds as well as bioassay-guided isolation for compounds. Future research will involve the determination of further investigation of antibacterial activity by using microdilution method, and the chemical characterization and structural elucidation of bioactive compounds.

4. Conclusions

The study of metabolite detection and the antibacterial activity of fungal endophytic extracts that living associated with *T.crispa* using the TLC-dot blot assay can be summarized that six endophytic fungi extracts showed strong antibacterial activity against both Gram-positive and negative bacteria. One extract had an excellent antibacterial activity (isolate No.6: TcBt1Bd-10). According to metabolites analysis using the TLC method along with TLC-Bioautography assay, the possibility of the chemical compounds that played a role in antibacterial activity of the extracts, including phenolics, flavonoids, alkaloids, and terpenoids. This study showed that fungal endophytic extracts associated with *T.crispa* might be used as promising sources of antibacterial. Further study needs to be done to isolate and purify the antibacterial compounds in the active extract and further investigation of antibacterial evaluation by using the microdilution method.

Acknowledgments

This study was financially supported by PUTI-Doktor 2020, Universitas Indonesia. The authors are very grateful to Beasiswa Saintek Pusat Pendidikan dan Pelatihan Kementerian Riset dan Teknologi/Badan Riset dan Inovasi Nasional which supported the first author's graduate scholarship. The authors would like to thank Ersaliany Nurul P. Qodrie and Andi Saptaji Kamal for their helping and supporting assistance on the laboratory works.

Author Contributions

All authors wrote the manuscript and were equally working as main contributors.

References

[1] Deshmukh SK, Verekar SA and Bhave SV 2014 *Front. Microbiol.* **5** 1-44
[2] Petrini O 1991 *Endophytes of Tree Leaves*, in *Microbial Ecology of Leaves* ed J Andrew and S Hirano (New York: Springer US) pp 179-197
[3] Owen NL and Hundley N 2004 *Sci. Prog.* **87**(2) 79-99
[4] Hussain H, Kliche-Spory C, Al-Harrasi A, Al-Rawahi A, Abbas G, Green IR, Schulz B, Krohn K and Shah A 2014 *Asian Pac. J. Trop. Med.* **7**(S1) S224-S227
[5] Ahmad W, Jantan I and Bukhari SNA 2016 *Front. Pharmacol.* **7** 1-19
[6] Botz L 2019 *Bioassays - Bioautography* (Amasterdam: Elsevier Inc) pp 253-260
[7] Choma IM and Grzelak EM 2011 *J. Chromatogr. A* **1218**(19) 2684-2691
[8] Owen L, White AW and Laird K 2019 *Phytochem. Anal.* **30**(2) 121-131
[9] Praptiwi, Raunsai M, Wulansari D, Fathoni A and Agusta A 2018 *J. Appl. Pharm. Sci.* **8**(8) 69-74
[10] Choma I and Jesionek W 2015 *Chromatography.* **2**(2) 225-238
[11] Praptiwi, Fathoni A, Putri AL, Wulansari D and Agusta A 2019 *AIP Conf. Proc.* **2120** 080011

[12] Sun X, Ding Q, Hyde K and Guo L 2020 *Fungal Ecol.* **5** 624–632
[13] Arfi Y, Buée M, Marchand C, Levasseur A and Record E 2012 *FEMS Microbiol. Ecol.* **79** 433-444
[14] Lau M, Arnold A and Johnson N 2013 *Fungal Ecol.* **6** 365–378
[15] Silva FdA, Liotti RG, Boleti APdA, Reis EdM, Passos MBS, dos Santos EL, Sampaio OM, Januário AH, Branco CLB, da Silva GF, de Mendonça EAF and Soares MA 2018 *PLoS One* **13**(4) e0195874
[16] Li J, Sun X, Chen L and Guo L 2016 *Mycology* **7**180–190
[17] Li JL, Sun X, Zheng Y, Lü PP, Wang YL and Guo LD 2020 *MycoKeys* **62** 75-95
[18] Duric K, Kovac-Besovic EE, Niksic H, Muratovic S and Sofic E 2015 *Bosn. J. Basic Med. Sci.* **15**(2) 9-14
[19] Wagner H and Bladt S 1996 *Plant Drug Analysis: A Thin Layer Chromatography Atlas* (Berlin: Springer-Verlag)
[20] Karthika K, Paulsamy S 2015 *Indian J. Pharm. Sci.* **77**(1) 111–116
[21] Gocan S 2005 *Terpenoids: Analysis by TLC,* in *Encyclopedia of Chromatograpy* ed J Cazes (United States: Taylor & Francis) pp 1676-1680
[22] Pascual ME, Carretero ME, Slowing KV and Villar A 2002 *Pharm. Biol.* **40**(2) 139-143
[23] Brechú-Franco AE, Laguna-Hernández G, De-La-Cruz-Chacón I and González-Esquinca AR 2016 *Eur. J. Histochem.* **60**(1) 55-58
[24] Murningsih T, Praptiwi, Liana and Fathoni A 2019 *Nusant. Biosci.* **11**(1) 44-48
[25] Fathoni A, Ilyas M, Cahyana AH and Agusta A 2013 *Ber. Biol.* **12**(3) 307-314
[26] Praptiwi P, Jamal Y, Fathoni A, Nurkanto A and Agusta A 2013 *Microbiol. Indones.* **7**(4):159-166
[27] Agusta A, Wulansari D, Jamal Y, Nurkanto A, Praptiwi and Fathoni A 2015 *Microbiol. Indones.* **9**(1) 35-43
[28] Yunianto P, Rusman Y, Saepudin E, Suwarso W and Sumaryono W 2014 *Pak. J. Biol. Sci.* **17**(5) 667-674
[29] Ma YM, Li Y, Liu JY, Song YC and Tan RX 2004 *Fitoterapia* **75**(5) 451-456
[30] De Souza JJ, Vieira IJC, Rodrigues-Filho E and Braz-Filho R 2011 *Molecules* **16**(12) 10604-10618
[31] Xu J, Kjer J, Sendker J, Wray V, Guan H, Edrada RU, Müller WEG, Bayer M, Lin W, Wud J and Proksch P 2009 *Bioorganic Med. Chem.* **17**(20) 7362-7367
[32] Harun A, Rahim NA, Jalil MA, Rosdi NAM, Daud S, Harith SS, So'ad SM and Hassan NM 2016 *Malaysian J. Sci.* **35**(2) 259-274
[33] Pinheiro E, Carvalho J, Santos D, Diellem C, Feitosa A and Marinho P 2013 *Nat. Prod. Res.* **27** 1633–1638
[34] Subban K, Subramani R and Johnpaul M 2013 *Nat. Prod. Res.* **27**(16) 1445-1449
[35] Tang Z, Wang Y, Yang J, Xiao Y, Cai Y,Wan Y, Chen H, Yao H, Shan Z, Li C and Wang G 2020 *PeerJ* **8** e8978
[36] Gao S, Li X, Zhang Y, Li C and Wang B 2011 *Chem. Biodivers* **8** 1748-1753

IOP Conf. Series: Materials Science and Engineering 1011 (2021) 012042 doi:10.1088/1757-899X/1011/1/012042

Antihyperlipidemia activity of ethanol extract the stem bark of *Flacourtia rukam* on propylthiouracil-induced albino rats, *Rattus noverticus* (Wistar strain)

Muharni Muharni[1]*, Julinar Julinar[1], Heni Yohandini[1], Fitrya Fitrya[2] and Rima Melati[2]

[1]Department of Chemistry, Faculty of Mathematics and Natural Sciences, University of Sriwijaya. Jl. Raya Palembang Prabumulih KM 32 Indralaya 30662 South Sumatera Indonesia

[2]Department of Pharmacy, Faculty of Mathematics and Natural Sciences, University of Sriwijaya. Jl. Raya Palembang Prabumulih KM 32 Indralaya 30662 South Sumatera Indonesia

*E-mail: muharnikimia@unsri.ac.id

Abstract. *Flacourtia rukam* belongs to one of the traditional medicinal plants used by the people, especially in South Sumatra for the treatment of hypertension. One of the triggers for hypertension is high levels of cholesterol in the blood. The *in vivo* research of the anti-hyperlipidemia activity of ethanol extract the stem bark of *F. rukam* on Propylthiouracil - induced albino rats *Rattus noverticus* (Wistar Strain) has been done. The animal test was separated into 6 groups i.e. normal group, negative, positive (simvastatin), and 3 treatment groups with a variation extract dose of 100, 200, and 400 mg/kg BW. The measurement of total cholesterol, triglyceride, LDL and HDL levels using the photometric enzymatic method. The result shows that ethanol extract of *F. rukam* gives the highest anti hyperlipidemia activity at 400 mg/kg BW with decreased total cholesterol percentage of 56.28%, triglycerides 57.34 %, LDL 54.29%, and increased HDL 58.77%. Statistical analysis shows that the treatment groups I, II, and III can reduce the level of total cholesterol, triglycerides, and higher-level LDL. The effective dose (ED_{50}) is 327.83 mg/kg BW. Based on this result, the traditional plant can be used in hyperlipidemia treatment.

1. Introduction

Flacourtia rukam is one of the traditional medicines has been used for the treatment of various diseases. The fruit can be used to diarrhea and dysentery treatment, the juice of the leaves was used to treat swollen eyelids, the roots were used by a postpartum woman as an antiseptic [1] while the stem bark using as an antihypertensive treatment [2]. One of the triggers of hypertension is high cholesterol in the blood where causes various diseases such as coronary heart and stroke diseases [3]. Phytochemical tests the ethanol extract stem bark of *F. rukam* shown contain triterpenoid, steroid, flavonoid, and phenolic compounds [4,5]. Steroids and phenolics compounds have the effect of reducing cholesterol levels [6]. Flavonoids also can reduce hyperlipidemia levels in the blood [7]. The increased cholesterol levels in the blood are mostly occurred in people with hypertension, and increased lipid level in blood might be an important risk factor for the development of Cardiovascular disease.

Hyperlipidemia treatment can be using synthetic drugs, such as simvastatin, gemfibrozil, cholestyramine, and niacin, but the use of synthetic drugs usually caused undesirable side effects, while natural compounds usually have lower side effects than synthetic compounds. Besides that, natural compounds also can treat other diseases [8]. In our previous research, the stem bark of *F. rukam* also can reduce cholesterol levels in the blood while *in vitro* anti-cholesterol activity the ethanol extracts the stem bark of *F. rukam* obtained IC_{50}values 209.092 µg/mL [9]. This research will report the potential of ethanol extract the stem bark of *F. rukam* as anti-hyperlipidemia using male albino rats Wistar strain using the enzymatic colorimetric method.

2. Methodology

2.1. Material

Sample of the stem bark of *F. rukam* used in this study was the same as we used in previous studies [4]. Thirty albino rats employed as animal model with body weight (BW) of 150-250 g were obtained from animal holdings of Bio Science Research Laboratory, Palembang city. The experiments were performed after accepting an approval letter of the protocol ethical clearance from ethics the health research review committee of Sriwijaya University (No. 074/kepkrsmhfkunsri/2020) and were carried out in accordance with current guidelines for the care of laboratory animals. The animal models were acclimatized for seven days and fed with feed and water standard [10].

2.2. Methods

2.2.1. Animal treatment. One kg of *F. rukam* was distillated in 96% ethanol using distillation apparatus distillated ethanol. The maceration for 3 days and then concentrated using rotary evaporator. Before the treatment, the body weight, total cholesterol, triglyceride, LDL, and HDL levels of each rats were measured. Animals test were divided into 6 groups, which are normal, negative control, positive control, and 3 treatment groups (Dose I, Dose II, Dose III). The normal control group were given standard foods, each given 2 mL every day with oral administration. Meanwhile, other groups were induced with solution of propylthiouracil (PTU) and high-fat foods (contained of 40% duck egg yolk, 50% cooking oil, and 10% butter). This induction was done for 14 days. On the day 15, the blood was taken to determine the total cholesterol, triglyceride, HDL, and LDL levels of the rat after induction.

The rats under hyperlipidemia were given treatment: the normal and negative control groups were given Na CMC solution 1% and positive control were given simvastatin solution at a dose of 0.193 mg/ 200 g BW. For the treatment group was given a solution of ethanol extract the stembark of *F. rukam* at Dose of 100; 200; and 400 mg/kg BW. The dosage was given orally with a volume of 2 mL/200 g BW of rats for 14 days. On the day 15, blood was taken to determine the total cholesterol, triglyceride, HDL, and LDL levels of rats by using a UV-Vis spectrophotometer at the maximum wavelength respectively using a biosystem analyzer [11]. Blood of tested rats was taken from plexus retro orbitalis of their eyes by microhematocrit. During the process, the bodyweight of rats was weighed before induction, after induction and after treatment.

Table 1. The treatment of group animal test.

Groups	Treatment
Normal	Na CMC 1%
Control (-)	Propylthiouracil 0.1% + food high-fat + 1% Na CMC
Control (+)	Propylthiouracil 0.1% + food high-fat + Simvastatin 0.193 mg/200 g BW
Dose I	Propylthiouracil 0.1% + food high-fat + ethanol extract 100 mg/kg BW
Dose II	Propylthiouracil 0.1% + food high-fat + ethanol extract 200 mg/kg BW
Dose III	Propylthiouracil 0.1% + food high-fat + ethanol extract 400 mg/kg BW

2.2.2. Measurement of plasma lipid level. Two ml of blood was collected in EDTA tubes and centrifuged for 20 minutes at 3500 rpm for 15 min at 4°C to obtain blood serum. The serum was used to measured total cholesterol, triglyceride, LDL, and HDL concentrations by using commercial kits [12].

2.2.3. Determination of serum cholesterol. The blood serum was analyzed for total cholesterol levels using the CHOD-PAP (Cholesterol Oxidase peroxidase aminoantipyrine) method). Hundred μL of cholesterol reagent solution was added to ten μL of blood serum and mixed well. The mixture was then incubated for 20 minutes at room temperature. The absorbance was measured at λ_{max} 500 nm. Distilled water was used as a blank, while cholesterol standard was served as a standard.

2.2.4. Determination of serum triglyceride. Serum triglyceride was measured using the Glyserol Peroxidase Phosphat Acid (GPO-PAP) method. Blood serum (10 μL) was taken in a test tube, the 1000 μL solution of triglyceride reagent kit QCA (Quimica Clinica Aplicada) was added. The solution was mixed well and incubated for 20 minutes at room temperature. The absorbance was read at λ_{max} 500 nm. Distilled water was used as a blank, while cholesterol standard was served as a standard.

2.2.5 Determination of LDL and HDL. Serum (10 μL) was added to 1000 μL of precipitation reagent (polyvinyl sulphate), then the solution was mixed and incubated for 15 minutes at 37 °C. Centrifugation was done for 20 min at 3500 rpm. The clear supernatant was used for the determination of LDL and measured the absorbance of the sample against reagent blank at λ_{max} 546 nm.

Determination of serum HDL level was carried out by CHOD-PAP method. Sample serum (10 μL) was placed in a test tube and 1000 μL of cholesterol reagents solution was added. The mixture was shaken thoroughly and left to stand for 10 min at 20-25 °C. The absorbance was measured at 500 nm. HDL standard was used as a standard.

2.3. Effective Dose Determination (ED_{50})
The effective dose (ED_{50}) was calculated using linear regression between decrease of total cholesterol, triglycerides, LDL levels and concentration of extract used.

2.4. Analysis Data
Analysis Statistical of data was calculated with Shapiro-Wilk normality test. If the data is normally distributed then paired simple t-test and One-Way ANOVA were tested using the SPSS 21.0 for Windows program where that $p < 0.05$ means there is a significant difference. If the data is not normally distributed then proceed with the Kruskal-Wallis non parametric test

3. Results and discussion
In vivo anti-hyperlipidemia activity ethanol extract of *F. rukam* using albino rats Wistar strain. The measurement of total cholesterol, triglycerides, LDL, and HDL levels were done after acclimatization process. The normal condition of total cholesterol in rats was 10 - 54 mg/dL [13]. Animal tests were fed with high-fat diets and 0.01% PTU solution to increase the total cholesterol levels. Induction of propylthiouracil will inhibit thyroid hormone production lead to hyperthyroidism where will affect lipoprotein metabolism increasing cholesterol levels, especially LDL [14]. The total level of cholesterol, triglyceride, LDL, and HDL were measured on day 15 (Table 1). The data in Table 2 shows the total cholesterol after acclimatization is in the normal range that is 35.49 ± 1.36 mg / dL – 38.93 ± 1.22 mg/dL.

3.1. Total cholesterol
The results (Figure 1) show increased total cholesterol levels for the treatment group with average cholesterol values, 80.75 ± 5.37 - 83.60 ± 0.56 mg / dL, while the normal group is 40.55 ± 1.48. This data shows that the induction process has been succeeded in increasing cholesterol levels above normal. The rats in hyperlipidemia condition were given suspension treatment of ethanol extract of *F. rukam* at

doses of 100; 200; and 400 mg/Kg BW. On the day 15, the total level of cholesterol, triglycerides, LDL, and HDL were measured.

Figure 1 shows the decrease in total cholesterol levels for the treatment groups of 100; 200; and 400 mg/dL, and the highest decrease was given by the treatment of dose III with value of 33.74 ± 3.86. Meanwhile, simvastatin as positive control showed a decrease in total cholesterol level of 34.04 ± 4.07 mg/dL which is almost the same as the treatment dose of 400 mg/dL. The statistical analysis of the paired t-test on the total cholesterol level data, between groups before and after treatment showed a positive control, groups I, II, and III showed significant difference ($p < 0.05$), whereas in the negative control group showed there was no significant difference ($p > 0.05$). Statistical analysis also shows. there was no significant difference between the treatment dose of 400 mg/kg BW with positive control.

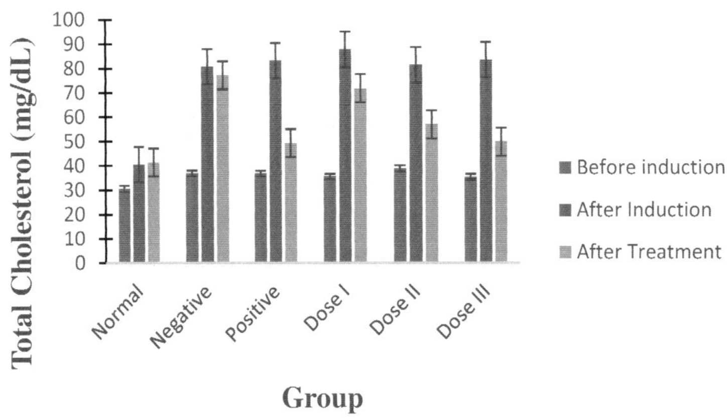

Figure 1. Total cholesterol of animal test before and after induction and after treatment.

Table 2. Percent decrease in total cholesterol.

Dose	Decrease in total cholesterol (%)	ED_{50} (mg/Kg BW)
I	17.38	
II	34.38	336.43
III	56.28	

Dose 1: 100 mg/kg BW ; Dose II : 200 mg/ kg BW ; Dose III : 400 mg /kg BW

Furthermore, the determination of the effective dose (ED_{50}) value based on the linear regression the percent of decreasing the total cholesterol to the extract dose [15]. Based on the regression equation obtained ED_{50} of 336.43 mg/kg BW (Table 2).

3.2. Triglyceride

The triglycerides value of normal albino rats is 25-145 mg/dL. The rats induced by high fat diet and PTU showed significant ($p<0.05$) increment in triglyceride levels, but in the normal group show not significant ($p>0.05$ increases in triglyceride levels. The given ethanol extract at doses 100; 200; and 400 mg/kg BW and positive control (simvastatin 1 mg/kg BW) exhibited a significant decrease in triglyceride levels ($p < 0.05$) compared to before treatment. Furthermore, it was calculated the value of the percent decrease in levels triglycerides (Figure 2).

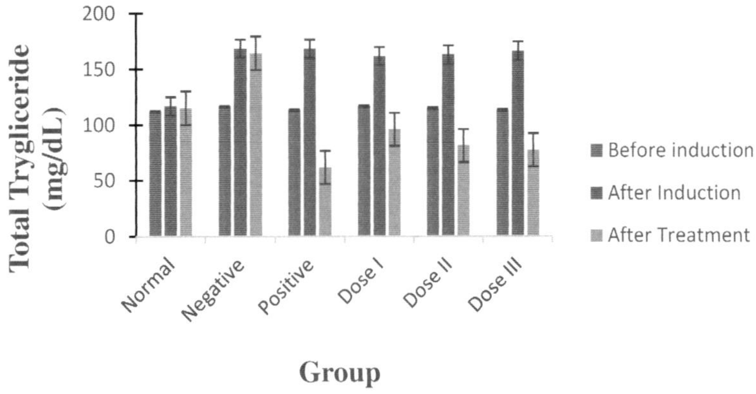

Figure 2. Triglyceride of animal test before and after induction and after treatment.

Table 2. Percent decrease in triglyceride.

Dose	%PKTG	ED_{50} (mg/KgBW)
I	15.34 %	
II	39.58 %	328.09
III	57.34 %	

Dose I: 100 mg/kg BW ; Dose II: 200 mg/ kg BW ; Dose III : 400 mg /kg BW

The statistical analysis of the paired t-test on the triglyceride level, between groups before and after treatment showed a positive control, groups I, II, and III showed significant difference ($p<0.05$), whereas in the negative control group possessed no significant difference ($p> 0.05$). The statistical analysis showed significant decrease of triglyceride between groups, and treatment groups I, II, and III could reduce triglyceride levels significantly ($P<0.05$). The ED_{50} value (Table 2) of triglyceride was determined to base on the linear regression between the dose of ethanol extract and the percent decrease in triglyceride levels. Based on linear regression ED_{50} the decreased of triglycerides is 328.09 mg/kg BW.

3.3. LDL Measurement

Normal LDL levels in rats is 7 - 27.2 mg/dL. high-fat diet and PTU treatment were significantly increased LDL levels compared to before induction ($p<0.05$). The average increase in LDL levels was (32.72 ± 2.23 - 38.82 ± 0.61 mg / dL). The decreased of LDL levels after being treated showed significant difference ($p<0.05$) compared before treatment. In contrast, normal and negative control groups displayed no significant difference ($p>0.05$) on LDL level (Figure 3). The average decrease in LDL levels in the normal group only 0.353 ± 0.249 mg / dL, whereas in the negative control group only 0.94 $\pm$ 0.66 mg /dL. The decrease in LDL levels for positive control also showed a significant decrease, where simvastatin can inhibit the biosynthesis of cholesterol in the liver, causing an increase in LDL receptor work in binding LDL particles in the liver.

The administration ethanol extract at a dose of 400 mg/kg BW also possessed an extreme difference ($p<0.05$) in reducing LDL levels with a decreased percentage is 54.29% (Table 3). The statistical analysis of the paired t-test on the LDL level, between groups before and after treatment showed a positive control, groups I, II, and III showed significant difference ($p <0.05$), whereas in the negative control group showed there was no significant difference ($p> 0.05$). The statistical analysis showed that were significant differences ($p <0.05$) in the decrease of LDL between groups, and treatment groups I,

II, and III could markedly reduce LDL levels (P <0.05). The calculation of ED_{50} LDL (Table 3) based on linear regression between concentrations ethanol extract and percent decrease in LDL levels and obtained ED_{50} LDL value 360.69 mg/kg BW.

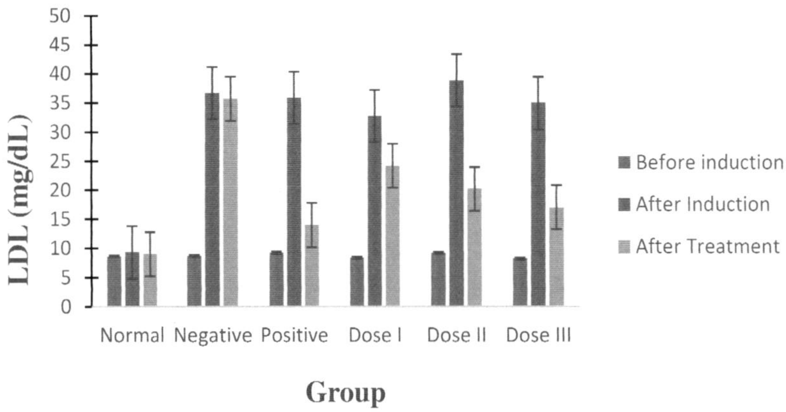

Figure 3. LDL of animal test before and after induction and after treatment.

Table 3. Percent decrease of LDL.

Dose	Decrease in LDL (%)	ED_{50} (mg/KgBW)
I	13.36 %	
II	31.49 %	360.69
III	54.29 %	

Dose 1: 100 mg/kg BW Dose II : 200 mg/ kg BW Dose III : 400 mg /kg BW

3.4. Determination of HDL

HDL (High Density Lipoprotein) is a protective lipoprotein that reduces the risk of coronary heart disease. Normal levels of HDL in rats is ≥35 mg/dL [16]. The measurement of HDL levels after acclimation showed that HDL levels of rats were in the normal range (32.41 – 35.15 mg/dL). PTU and high-fat diet treatment markedly down-regulated HDL levels (10.12 - 12.78 mg/dL). Low levels of HDL can cause a risk of coronary heart disease.

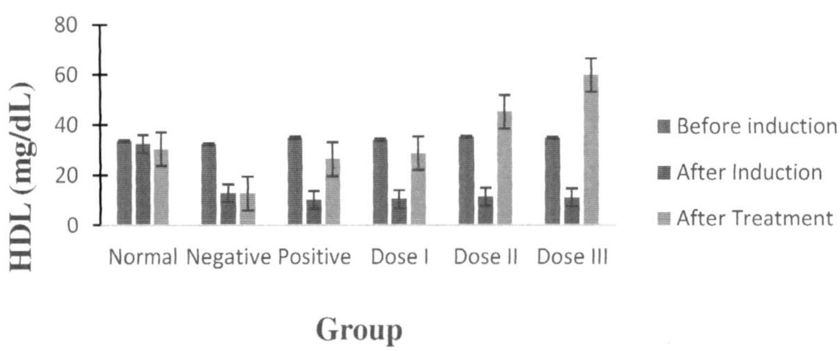

Figure 4. HDL of animal test before and after induction and after treatment

The treatment with *F. rukam* ethanolic extract showed a significant increase in HDL levels (p <0.05) between before and after the treatment dose and the highest increase was indicated by the dose of 400 mg/KgBW. The positive control where the treatment with simvastatin solution did not show a significant increase in HDL levels (p> 0.05) because simvastatin does not affect HDL metabolism in the blood

(Figure 4). The statistical analysis of the paired t-test on the HDL level, between groups before and after treatment showed a positive control, groups positive control, I, II, and III showed significant difference ($p < 0.05$), and more height compared positive control, whereas in the negative control group showed there was no significant difference ($p > 0.05$). The statistical analysis showed that were significant differences ($p < 0.05$) in the decrease of triglyceride between groups, and treatment groups I, II, and III could reduce total cholesterol levels significantly ($P < 0.05$). Determination of ED_{50} HDL values (Table 4) from linear regression between doses of ethanol extract the *F. rukam* and percent increase in HDL levels based on linear regression equations between extract doses with percent increase HDL obtained ED_{50} HDL 327.83 mg/kg BW.

Table 4. Percent increase in HDL.

Dose	Increase in HDL (%)	ED_{50} (mg/KgBW)
I	12.55 %	
II	36.76 %	327.83
III	58.77 %	

Dose 1: 100 mg/kg BW ; Dose II : 200 mg/ kg BW ; Dose III : 400 mg /kg BW

The given extract can reduce total cholesterol, triglyceride, LDL levels, and significantly increased HDL levels, where the effect of decreasing the level of total cholesterol, triglycerides, and LDL is greatest in the treatment group dose of 400 mg/kg. The greater the extract dose given, the greater the effect in reducing total cholesterol, triglyceride, LDL levels, and in increasing HDL levels. Base on this data ethanol extracts of *F. rukam* activity as anti-hyperlipidemia.

The anti-hyperlipidemia properties of the ethanol extract were related to the content of flavonoids, steroids, triterpenoids, and phenolics [4]. Flavonoids can reduce total cholesterol and LDL levels in the blood. flavonoids are also able to reduce triglyceride levels and increase HDL and another report states β-sitosterol glycoside compound work by inhibiting the absorption of cholesterol in the digestive tract [17,18].

3.5. The body weights

The weighing body weight of rats was carried out before induction, after induction, and after treatment. Statistical analysis using paired t-test showed a significant difference ($p < 0.05$) between the bodyweight of rats before induce with after induction. This increase in body weight is due to the increase in the amount of fat and attached to adipose tissue, especially in the abdominal cavity. The Fat Deposited will cause triglyceride levels to increase in adipose tissue and the blood and cell size will increase so that there is an increase in body weight. The results measurements of body weight shown in Table 5.

Tabel 5. Body weight animal test.

Group	Body weight (g)		
	Before induction	After induction*	After treatment
Normal	155.72 ± 9.65	163.71 ± 7,26	78.25 ± 9.56
Negative	144.73 ± 0.08	174.85 ± 18.35	191.36 ± 31.06
Positive	166.94 ± 9.07	194.61 ± 5.02	183.21 ± 5.55
Dose I	181.90 ± 4.46	253.79 ± 6.98	235.27 ± 10.92
Dose II	147.63 ± 8.67	173.81 ± 6,18	156.56 ± 5.38
Dose III	171.08 ± 9.67	197.83 ± 27.09	186.97 ± 13.41

* = p<0,05 a significant difference to before induction

After the treatment, the positive control group, and the treatment group doses I, II, and III demonstrated significant difference weight loss ($p<0.05$) compared with before treatment, whereas for the normal group and negative control showed increase body weight significant difference ($p<0.05$) (Table 5). This is presumably because the content of flavonoids, steroids, and phenolic in ethanol extracts stem bark of *F. rukam* causes a decrease in body fat levels so that the weight of the animal test will also go down.

4. Conclusion

Treatment groups are showing significant reduction of total cholesterol, triglyceride and LDL ($P<0.05$) and significant increment of average HDL ($P<0.05$) compared with the negative control. The ethanol extracts from the stem bark of *F. rukam* have an anti-hyperlipidemia activity that not only can reduce total cholesterol, triglyceride, and LDL levels, but also can increase HDL level in male rats with Effective dose (ED_{50}) 328.09 mg/kg BW.

Acknowledgments

The authors are thankful to the Ministry pf Research and Technology of the Republic of Indonesia Basic Research Funding 2020. We also thank Department of Chemistry, University of Sriwijaya on providing facilities to carry out the research work.

References

[1] Heyne K 1987 *Tumbuhan Berguna Indonesia.* Jilid III. Jakarta: Yayasan Sarana Wana Jaya :390
[2] Yustian I, Muharni, Sukarmi S, Arbi and Zulaika 2012 Special research on the exploration of ethnomedicine and local community medicinal plants in Indonesia (ethnic Musi II). Palembang Ministry of Health of the Republic of Indonesia
[3] Adisakwattana S, Moonrat S, Srichairat S, Chanasit C, Tirapongporn H, Chanathong B, Ngamukote S,l Makynen K and Sapwarobol S 2010 Lipid-lowering mechanisms of grape seed extract (*Vitis vinifera* L) and its antihyperlipidemic activity *J. Med. Plant Res.* **4** 2113-20
[4] Muharni, Elfita, Yohandini H, Julinar, Yasrina and Miranti 2019 Chemical constituents from the stem bark of *F. rukam* Zool & Mor. And their antioxidant activities. *Sains Malays.* **48**(**9**) 1899–906
[5] Muharni, Fitrya, Nurmaliana and Ristojo. 2016 *Skrining fitokimia aktifitas antioksidan dan antibakteri dari tumbuhan obat tradisional etnis Musi*, Balai Besar Obat Dan Jamu Kementrian Kesehatan Republik Indonesia, Palembang, Indonesia
[6] Choi JM, Lee EO, Lee HY, Kim KH, Ahn KS, Shim BS, Kim N, Song MC, Baek NI, Kim AH 2007 Identification of campesterol from *Chrysanthenum coronarium* L. and its antiangiogenic activities *Phytother. Res.* **21** 954-9
[7] Zeka K, Ruparelia, Arroo RRJ, BB, Budriesi R, Micucci M 2017 Flavonoids and their metabolites: prevention in cardiovascular diseases and diabetes *Diseases* **5**(3): 19
[8] Karimi A, Majlesi M, Rafieian-Kopaei M 2015 Herbal versus synthetic drugs; beliefs and facts *J. Nephropharmacol.* **4**(1): 27–30
[9] Muharni, Heni Y, Julinar 2019 *Analisa zat aktif dari tumbuhan rukam (Flacourtia rukam) serta prospeknya sebagai kandidat obat herbal alami baru untuk penyakit hipertensi*, Laporan Penelitian Dasar Ristek Dikti 2019, Universitas Sriwijaya
[10] Kshirsagar A, Purnima, A 2008 Evaluation of Calotropis gigantea flower extract of alchohol induced hepatotoxicity *J. Cell Tissue Res.* **4**(19) 1551–56
[11] Aberare OL, Okuonghae P, Mukoro N, Dirisu JO, Osazuwa F, Odigie E and Omoregie R 2011 Triglycerides, total cholesterol, high density lipoprotein cholesterol and low density lipoprotein cholesterol in rats exposed to premium motor spirit fumes *N. Am. J. Med. Sci.* **3**(6) 277–280
[12] Sheu MJ, Hsieh YY, Lai CH, Chang CC and Wu CH 2013 Antihyperlipidemic and Antioxidant Effects of C-phycocyanin in Golden Syrian Hamsters Fed with a Hypercholesterolemic Diet *J.*

Tradit. Complement. Med. **3**(1) 41–7

[13] Harini MD and Astirin OP 2009 Blood cholesterol level of hypercholesterolemia tat (Rattus norvegicus) after VCO treatment *Nusantara Bioscience* **1**(2) 53-8

[14] Perla FM, Prelati M, Lavorato M, Visicchio D and Anania C 2017 The role of lipid and lipoprotein metabolism in non-alcoholic fatty liver disease *Children (Basel)* **4**(6) 46

[15] Adams SP, Sekhon SS and Wright JM 2014 Rosuvastatin for lowering lipids *Cochrane Database Syst. Rev.* **11**: CD010254

[16] Nurhidajah, Astuti R and Nurrahman 2019 Black rice potential in HDL and LDL profile in sprague dawley rat with high cholesterol diet *IOP Conf. Series: Earth and Environmental Science* **292** 012019

[17] Bao L, Hu L, Zhang Y and Wang Y 2016 Hypolipidemic effects of flavonoids extracted from *Lomatogonium rotatum* Exp *Ther. Med.* **11**(4) 1417-24

[18] Lin X, Ma L, Racette SB, Spearie CLA and Ostlund Jr RE 2009, Phytosterol glycosides reduce cholesterol absorption in humans *Am. J. Physiol. Gastrointest. Liver Physiol.* **296**(4) 931–35

IOP Conf. Series: Materials Science and Engineering 1011 (2021) 012043 doi:10.1088/1757-899X/1011/1/012043

Effect of Ag and Mn doping for methylene blue photodegradation performance

Jessica, M Ibadurrohman and Slamet*

Department of Chemical Engineering, Faculty of Engineering, Universitas Indonesia, Kampus UI Depok, Depok 15424, Indonesia

*Corresponding author: slamet@che.ui.ac.id

Abstract. Currently, wastewater management is ineffective, operationally expensive, and difficult to be implemented in Indonesia. The photocatalyst is a technology that may answer this problem. Titania (TiO2) is a high-activity, inexpensive, stable, and non-toxic photocatalyst material. However, TiO2 has large band-gap energy and rapid recombination. In this research, TiO2 was doped by Ag (a noble metal) and Mn (a transition metal) to compare the performance of photocatalysts in organic pollutant degradation, using methylene blue. Ag/TiO2 catalyst was synthesized by Photo Assisted Deposition (PAD) method, while Mn/TiO2 was synthesized by impregnation method. The physicochemical properties of both catalysts were assessed by SEM-EDX and UV-Vis DRS. SEM-EDX indicated that both methods successfully doped Ag and Mn into TiO2. Besides, UV-Vis DRS confirmed the presence of metal dopants that caused the narrowing of the bandgap. Ag/TiO2 catalyst increased the degradation of methylene blue by up to 97% in 30 minutes. Meanwhile, the Mn/TiO2 catalyst inhibited the degradation of methylene blue. The increased activity by Ag dopants is due to the Localized Electromagnetic Field (LEMF) and Schottky barrier phenomena which increase charge separation. Meanwhile, decreased activity by Mn dopants is due to the shading effect and Mn electronic structure that can facilitate recombination at high loading. Ag/TiO2 catalyst also improved photodegradation performance of TiO2 when irradiated by visible light.

1. Introduction

Clean water is a human's basic need. However, clean water provision is still problematic in Indonesia. It is estimated that 25.1% of Indonesia's rivers are heavily polluted [1]. The main sources of pollution come from industrial waste and household waste, which dispose of dangerous pollutants like synthetic dyes. Methylene blue is a commonly used synthetic dye which is difficult to degrade because it has a stable heterocyclic aromatic structure. Photocatalyst, one of the Advanced Oxidation Process (AOP) technologies, is a possible solution to degrade stable organic compounds like methylene blue. Titania (TiO_2) is a high-activity, inexpensive, stable, and non-toxic photocatalyst material. However, compared to other AOP technologies such as Fenton and FeGAC/ H_2O_2, the operational cost of photocatalyst is more expensive because TiO_2 has large bandgap and slow kinetics [2]. Moreover, WWTPs generally store large amounts of waste in outdoor ponds, so it is preferable if the catalyst can be activated by solar light. Solar radiation contains only 5% of UV, while the rest is an electromagnetic wave with lower frequency. Therefore, current efforts are directed to increase catalyst ability to capture photons with lower energy and reduce the rate of recombination.

Based on previous studies, it was found that the performance of TiO_2 under solar radiation can be increased by doping metal. Different metal dopants serve this purpose in different ways. TiO_2 that is doped by noble metals such as Ag (silver) improve TiO_2 activity through plasmonic resonance that allows metal dopant to be activated by absorbing visible light [3]. While transition metal dopant allows catalyst activation under lower frequency radiation by introducing new energy levels below the conduction band and narrowing the bandgap. Previous studies that synthesized Ag/TiO_2 to degrade organic pollutants under visible light were prepared using the sol-gel and impregnation method [4-5]. Meanwhile, in current research, Ag/TiO_2 was prepared using Photo Assisted Deposition (PAD). This method was selected due to its better dispersion and performance compared to Ag/TiO_2 that was prepared with other methods [6]. As the price of noble metals is quite expensive, a comparative study with other cheaper metal dopants, such as Mn (manganese) is needed. Several studies found that the addition of Mn metal can increase photocatalyst activity under visible light [7-8]. When compared to transition metals with similar electronic configurations, such as Fe^{2+}, Ru^{3+}, and Os^{3+}, the Mn^{2+} exhibited the best activity as well [9]. However, previous studies have not done comparative studies between the transition metal and noble metal directly.

This study aims to compare the performance of Mn metal dopants with Ag noble metal dopants to obtain a cheaper yet effective catalyst. The photocatalyst performance of various metal dopants on TiO_2 was tested by comparing the degradation of methylene blue. Ag/TiO_2 catalyst was synthesized by Photo Assisted Deposition (PAD) method, while Mn/TiO_2 was synthesized by impregnation method. The purpose of using different methods of synthesis was to obtain metal-doped-TiO_2 by using the most suitable method that yields optimal results.

2. Material and Method

All materials were of analytical grade and used without any further preparation. TiO_2 Evonik P25, anhydrous methanol (CH_3OH) 96%, nitric acid (HNO_3) 65%, silver nitrate solution ($AgNO_3$), methylene blue 1000 ppm, and Mangan acetate tetrahydrate ($Mn(CH_3COOH).4H_2O$) from Sigma Aldrich were purchased from local supplier. Distilled water was used in all experiments.

2.1. Synthesis of Ag/TiO_2

Ag/TiO_2 was synthesized with Photo Assisted Deposition (PAD) method inside a photoreactor. The first step was to make TiO_2 slurry by mixing 1 grams TiO_2 P25 with 100 mL distilled water and add nitric acid until the pH reached 3. The sol was stirred for 15 minutes and sonicated for 30 minutes. Then, $AgNO_3$ solution was made by mixing 0.016 gram (1% Ag) of $AgNO_3$ crystal with 100 mL distilled water until it dissolved. The solution was then stirred with TiO_2 slurry for 30 minutes. After that, 45 mL anhydrous methanol was added to the slurry while being stirred. Then, the mixture was put inside a photoreactor for 6 hours of continuous irradiation and stirring. The slurry was then separated using a centrifuge with rotation speed of 3000 rpm for 30 minutes. Then, the solid obtained was dried on hot plate with temperature of 150^{0}C and calcinated inside a furnace with temperature of 300^{0}C for 1 hour.

2.2. Synthesis of Mn/TiO_2

Mn/TiO_2 was synthesized with impregnation method. The first step was to make Mangan acetate solution by dissolving 0.0446 gram (1% Mn) of Mangan acetate tetrahydrate crystal in 50 mL distilled water. Then, 1 gram of TiO_2 P25 was added to the solution. It was sonicated for 1 hour at 60^{0}C. The catalyst was then dried on hot plate with temperature of 110^{0}C. Lastly, it was calcinated inside a furnace at 350^{0}C for 4 hours.

2.3. Structural and Optical Characterization

Structural and optical characterization was carried out to the catalyst. UV-Vis DRS characterization was done to obtain the band gap value of the catalyst. While, the SEM-EDX (Scanning Electron Microscopy with Energy Dispersive X-ray Spectroscopy) characterization was to analyze the final composition, distribution of constituent material, and surface morphology of the photocatalyst.

2.4. Methylene Blue Degradation Test

Photocatalytic degradation of MB ($C_{16}H_{18}ClN_3S$) in aqueous solution (10 ppm) was carried out inside a reactor with 20-watt UV lamps and visible-light (Gold star, 20 watt) irradiation. The set-up of the reactor can be seen at Fig.1 that depicts the lamps arrangement with solution and reflector. About 50 mg of photocatalyst was added into a 100 ml of aqueous MB (10 ppm) solution. Prior to irradiation, the suspensions were stirred in the dark for 30 min to ensure establishment of an adsorption–desorption equilibrium among the photocatalyst and MB. Every 30 minutes, 5 mL of sample was taken. The test was carried out for 2 hours. After recovering the catalyst by centrifuging, subsequent variations in the MB concentration were analyzed by an ultraviolet–visible spectrophotometer and the absorbance at 664 nm (λmax for MB) was recorded.

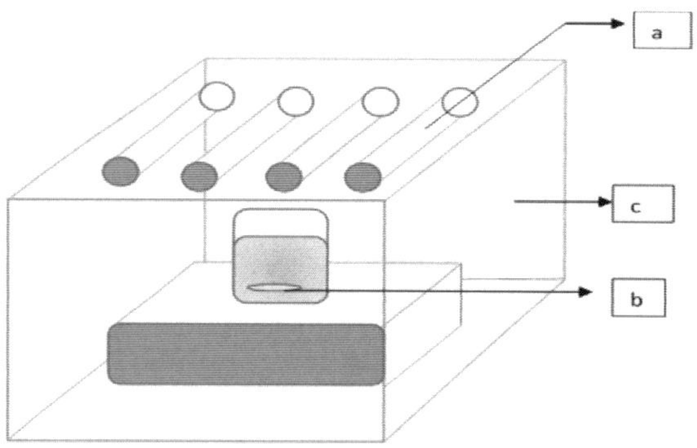

Figure 1. Photoreactor set up with (a) lamp (b) magnetic stirrer and (c) reflector

3. Result and Discussion

3.1. Catalyst Characterization

The SEM results can be observed at Figure 2, that show undoped and doped TiO_2. It can be seen that doped TiO_2 did not change the morphology significantly. All of them looked like small aggregates resembling cotton.

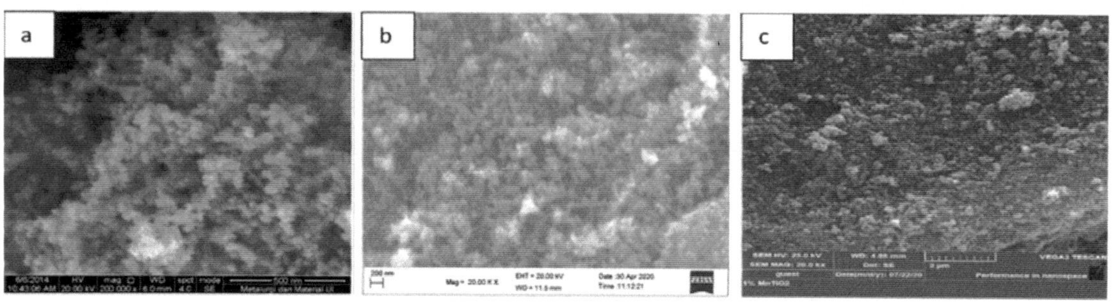

Figure 2. Morphology of catalyst (a)TiO_2 P25 (b)1 % Ag/TiO_2(c)1% Mn/TiO_2

Figure 3 demonstrated that the physical appearance of Mn/TiO_2 was considerably different from Ag/TiO_2 and TiO_2. The color changed and became darker. Even though Ag/TiO_2 and Mn/TiO_2 were synthesized with different methods, the EDX result indicated that both have successfully doped metal with the desired content, which is 1% wt. Besides, UV-Vis DRS confirmed the presence of metal dopants that caused the narrowing of the bandgap. Even though the metal was loaded to TiO_2 with the same amount, the absorbance shift observed is different. The TiO_2 catalyst that was loaded with Mn caused a greater shift compared to Ag.

IOP Conf. Series: Materials Science and Engineering 1011 (2021) 012043 doi:10.1088/1757-899X/1011/1/012043

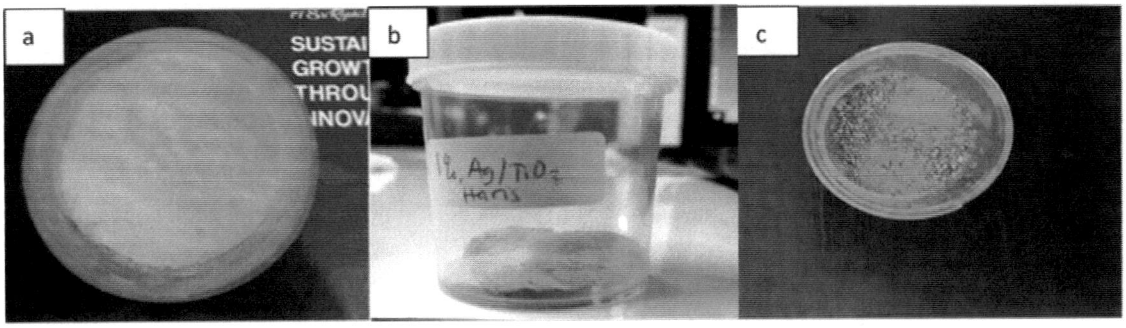

Figure 3. Physical appearance of catalyst (a)TiO_2 P25 (b)1 % Ag/TiO_2(c)1% Mn/TiO_2

Table 1. EDX Result of Ag/TiO_2 and Mn/TiO_2

Elements	Ag/TiO_2 content (%)	Elements	Mn/TiO_2 content (%)
Ti	60.81	Ti	57.05
O	38.08	O	41.99
Ag	1.11	Mn	0.96

Bandgap was calculated using the Kubelka-Munk equation. The result is presented in fig.4. The bandgap of TiO_2 was shifted from 3.05 eV to 3.01 eV for Ag/TiO_2 and 2.87 eV for Mn/TiO_2. This shift was also observed in previous study when doping TiO_2 with 3% mol Mn and with 5% mol Ag [10-11]. It is consistently observed that at the same loading, Mn will give greater shift than Ag.

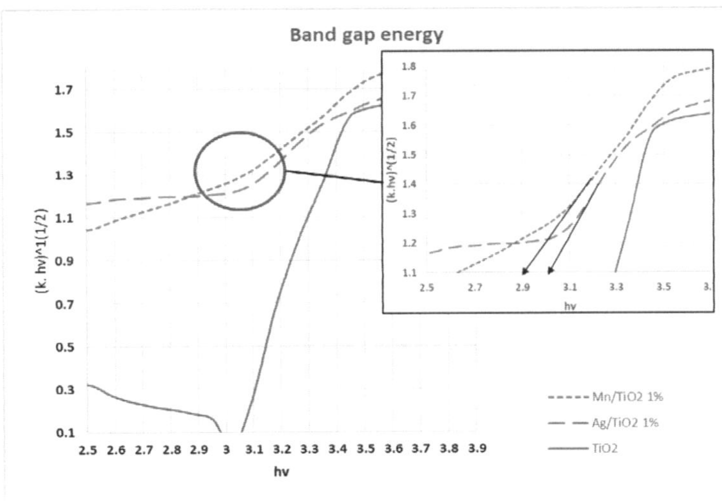

Figure 4. Band gap energy.

3.2. Methylene Blue Degradation

Metal doping caused alteration in photocatalyst activity. In this research, Mn and Ag were used to dope TiO_2. Their photocatalytic activity was measured by observing methylene blue degradation under UV light for 2 hours. The result is presented in figure 4. Doping of Ag (1% wt) increased photocatalytic activity, while the doping of Mn (1% wt) caused an adverse effect. In 30 minutes, Ag/TiO_2 had managed to degrade 97% of methylene blue. This increase indicated the increase in hydroxyl radical production.

In methylene blue degradation, hydroxyl radicals are the most predominant active species [11]. Ag metal doping helped this process in several ways, which all of them are related to its characteristic as a noble metal. Ag metal doping can effectively create the Schottky barrier that prolongs separation between electron and holes [3]. It can also increase the amounts of excited electrons through the electromagnetic field. This process is called localized electromagnetic field (LEMF) which was already observed in Ag/TiO_2 [12].

While in this study Mn/TiO_2 produced inhibition effect to photocatalyst, this is not always the case. It has been observed that, in a small amount of doping, Mn/TiO_2 increased photocatalytic activity under UV and visible light [13-14]. It is suggested that the result might come from the fact that the loading amount exceeded its optimum result. It is also possible than Mn metal caused significant shading effect that prevented photon reaching semiconductor effectively, as can be seen from the color of the catalyst. Mn metal has a unique orbital, which can trap both holes and electrons [15]. Therefore, it is hypothesized that: when Mn doping is small, the orbitals tend to act as electron trappers. While the amount increase, Mn orbital can trap both hole and electron simultaneously that increase the chance of recombination.

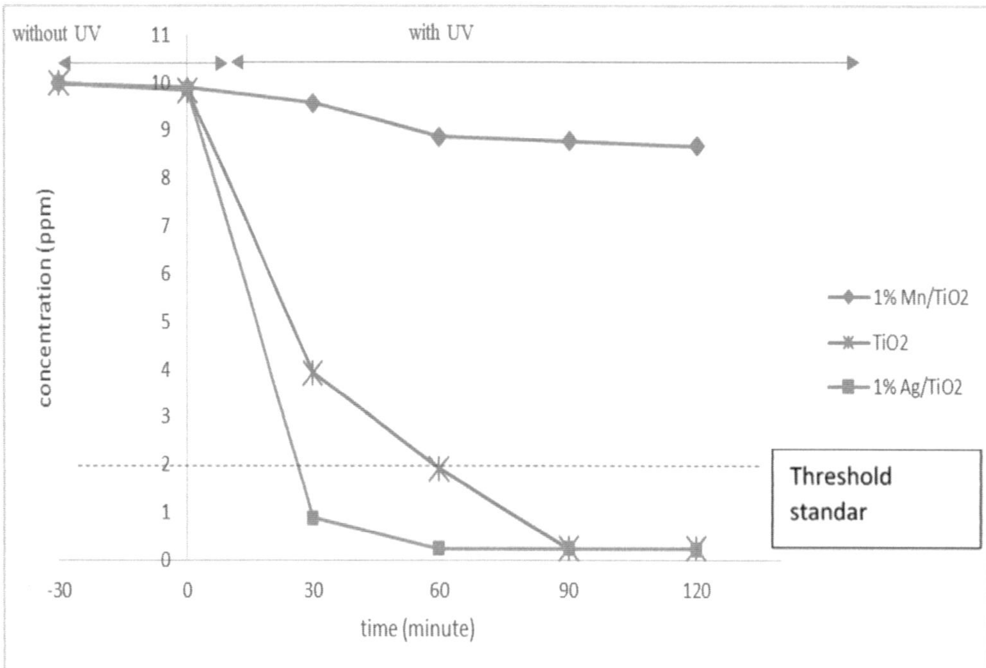

Figure 5. Effect of Dopant to MB degradation.

Proceeding further, Ag/TiO_2 which exhibited increased performance was tested to degrade MB under visible light irradiation. Table 2 presents the result of the degradation of 2 hours. In comparison to performance under UV light, the MB degradation under visible light was much slower. In 2 hours, both doped and Ag-doped TiO2 has degraded>99% MB under UV light, while the result under visible light was less than 70%. However, Ag/TiO2 has significantly higher activity under visible light, which might come from its plasmonic activity.

Table 2. Percent of MB degradation under visible light after 2 hours.

Catalyst	% MB Degradation after 2 hours
TiO_2	48.65
Ag/TiO_2	69.89

4. Conclusion

The purpose of this study is to compare the effect of metal doping for methylene blue photodegradation. The direction of this research is to obtain cheaper metal doping that can be activated under solar light. However, the present work is not successful to prove that a cheaper transition metal like Mn can substitute noble metal doping like Ag. From this study, Ag/TiO_2 catalyst increased activity for the degradation of methylene blue by up to 97% in 30 minutes. Meanwhile, Mn/TiO_2 catalyst inhibited activity for the degradation of methylene blue. The increased activity by Ag doping is due to the Localized Electromagnetic Field (LEMF) and Schottky barrier phenomena which increase the amount and time of charge separation. Meanwhile, decreased activity by Mn doping is due to the shading effect and Mn electronic structure that can facilitate recombination at high loading. Ag/TiO_2 catalyst also improved photodegradation performance of TiO_2 when irradiated by visible light.

Acknowledgement

Authors would like to express gratitude to Universitas Indonesia's PUTI Prosiding Universitas Indonesia 2020, No. Contract NKB-1107/UN2.RST/HKP.05.00/2020, that has funded this study.

References

[1] Kementerian Lingkungan Hidup dan Kehutanan 2019 *Statistik Indonesia 2019* (Jakarta: Badan Pusat Statistik)
[2] Affam A C, Chaudhuri M and Kutty S R 2018 *Bull. Chem. React. Eng. Catal.* **13 (1)** 179-186
[3] Zaleska-Medynska A 2018 *Metal Oxide-Based Photocatalysis* (Amsterdam: Elsevier)
[4] Ali T, Ahmed A, Alam U, Uddin I, Tripathi P and Muneer M 2018 *Mater. Chem. Phys.* **212** 325-335
[5] Duan Y, Zhang M, Wang L, Wang F, Yang L, Li X and Wang C 2017 *Appl. Catal., B* **204** 67-77
[6] Mohamed R M, Shironita S, Mori K and Yamashita H 2009 *Res. Chem. Intermed.* **35(3)** 305-312
[7] Chauhan R, Kumar A and Chaudhary R P 2012 *Spectrochimica Acta A* **98** 256-264
[8] Deng Q, Xia X, Guo M, Gao Y and Shao G 2011 *Mater. Lett.* **65** 2051–54.
[9] Kim D, Im Y, Jeong K M, Um M H, & Kang M 2014 *Bull. Korea Chem. Soc.* **35(8)** 2295–98
[10] Devi L G and Kumar S G 2010 *Appl. Surf. Sci.* **257(7)** 2779–90
[11] Harikishore M, Sandhyarania M, Venkateswarlua K, Nellaippan T and Rameshbabu N 2014 *Procedia Mater.Sci.* **6** 557-566.
[12] Ali T, Ahmed A, Alam U, Uddin I, Tripathi P and Muneer M 2018 *Mater. Chem. Phys.* **212** 325-335
[13] Awazu K, Fujimaki M, Rockstuhl C, Tominaga J, Murakami H, Ohki Y, Yoshida N and Watanabe T. 2008 *J. Am.Chem. Soc.* **130 (5)** 1676–80
[14] Papadimitriou V C, Stefanopoulos V G, Romanias M N, Papagiannakopoulos P, Sambani K, Tudose V and Kiriakidis G 2011 *Thin solid films* **520(4)** 1195-1201
[15] Umar K, Aris A, Ahmad H, Parveen T, Jaafar J, Majid Z A, Reddy A V B and Talib J 2016 *J. Anal. Sci. Technol* **7 (1)**
[16] Pereira A J, Gracia L, Beltran A, Lisboa-Filho P N, Silva J H and Andres J 2011 *J. Phys. Chem. C* **116(15)** 8753–62

IOP Conf. Series: Materials Science and Engineering 1011 (2021) 012044 doi:10.1088/1757-899X/1011/1/012044

Saturation time test of membrane to supply clean water using reverse osmosis with pressure stage in rejection

Muhammad Zakwan, Anton Prayoga and Setijo Bismo*

Department of Chemical Engineering, Faculty of Engineering, Universitas Indonesia, Depok, West Java, 16424, Indonesia.

*Corresponding author: setijo.bismo@ui.ac.id

Abstract. The tsunami in Pandeglang had resulted in high salinity of water and caused water scarcity. Furthermore, there were difficulties accessing water due to dry season in 4 sub-districts in Pandeglang in 2019. Reverse osmosis (RO) is a suitable technology to provide clean water, however, the advanced research is required to undrestand the method and determine the optimum pressure of multi-pass RO to produce clean water with a long saturation time of the membrane. The experimental research used RO multi-pass was operated in Tanjung Lame by various pressures 6-3, 6-4, and 7-4 on the first and second stage to observe the effect of flux decrease and TDS increase throughout 72 hours. Used P1 = 6 bar/ P2 = 3 bar showed a better result compared to P1 = 6/ P2 = 4 and P1 = 7/ P2 = 4 in producing clean water. Provided clean water at 20 ppm has been done by RO multi-pass, which permeates the first stage used as a feed in the second stage. P1 = 6 bar and P2 = 3 bar give the saturation time of membrane 72-80 hour on the first stage and 72-131 hour on the second stage.

1. Introduction

The high salinity of the water is one of the causes of the scarcity of clean water. Water scarcity continues to challenge the world's population, especially those living areas [2], such as Pandeglang-Indonesia, where it experienced difficulty to access clean water.

Technology Reverse Osmosis (RO) is a promising technology to produce clean water in areas where water scarcity is required [4]. RO is gaining worldwide acceptance in both water treatment and desalination applications. The use of RO in supplying clean water and purifying water from bacteria is effective without adding the ozone process [7]. One of the challenges in using membrane is how to prolong membrane lifespan. Fouling is the main factor caused by pore-clogging or adsorption of solutes on the layer [5, 6]. There are several factors that affect the fouling of membranes, including pressure, recovery, and salt passage. Scaling of the RO membrane is the result of precipitation of saturated salts onto the layer's surface. Cherryan in Ninik (2009) reported that molecules in solutes rejected and accumulated with the result of the clogging of pore and triggering polarization of concentration phenomena. The phenomena of polarization-concentration increases with increased recovery. The phenomena caused TDS (Total Dissolve Salute) to permeate increased significantly [9]. Pressure also affects fouling. The pressure is needed to push water through the membrane is called transmembrane pressure. Transmembrane pressure is the differential pressure between the sides of the membrane [9]. Mulder in Ninik (2009) explained that the flux increase due to the rise of transmembrane pressure. However, it occurs on feed with low salinity or low TDS. Below is the equation of transmembrane pressure in RO.

$$pressure\ drop\ (dp) = P_f - P_c \quad (1)$$

$$transmembrane\ pressure\ (TMP) = \frac{(P_f + P_c)}{2} - P_p \quad (2)$$

The fouling process are also affected by salt passage and recovery. As mentioned before, recovery accelerates the fouling of the membrane if recovery increased. The salt-passage also acts similarly, which means if much salt goes through the membrane, fouling increases. Thus, the membrane should be cleaned. Based on the technical note, clean the membrane when flux reduces until 10-15% from the initial while salt passage about 5-10% [5].

There are many strategies to extend the lifespan of the membrane. The cross-flow system's operating membrane has better performance on extending the membrane's lifetime than the dead-end system because the flux in the cross-flow system can wipe off all molecules or ions, which clog the pore of the membrane to the drain side [8]. The multi-pass configuration of the membrane RO can extend the lifetime of the membrane. The method is the product produced by the first membrane, then feed to the second membrane. Instead of using a multi-pass configuration, appropriate pretreatment can alleviate the fouling, which affects the lifespan membrane [1]. People have widely used conventional pretreatment to pre-treat seawater and brackish water RO, such as acid addition, coagulation addition, and media filtration until cartridge filtration. Nevertheless, using a new trend in pretreatments, such as MF, UF, and NF, has been preferred [3]. Based on Jia Xu (2006), using UF as pretreatment in the UF-RO system showed the system performed well as pretreatment [10].

This current study aims to test how long the household membrane's saturation time in supplying clean water from brackish water on low pressure of multi-pass configuration. The correlation between pressure, flux, and TDS from the graphic shows the optimum pressure with a long membrane's saturation time. Then, the lifespan of the membrane is calculated by the obtained graphic.

2. Materials and method

The experiments used brackish water, which available in a well of Tanjung Lame, Pandelang Banten. The experimental setting used was a multi-pass configuration of RO, which uses UF as pretreatment.

There are six types of pump 36V diaphragm, which were employed by this research. The tank, which has dimension 2500 L of feed, was placed above the machine at 3-meter height. It means the pressure used to pump feed to UF comes from the height of the tank. While in the first stage, RO is ULP-3013-400, and the second stage of the RO model is WW-2012-100. Both membranes usually are for residential water. Before the valve in rejection, there are two pressure-indicator to know the value of pressure. RO system is operated by various pressures in the first stage 6 and 7 bar and the second stage using 3 and 4 bar. The data of decrease flux and increase of TDS carried out every 2 hours. Then compare the data to get optimum pressure. Figure 1 shows a schematic diagram of the experimental settings for this study.

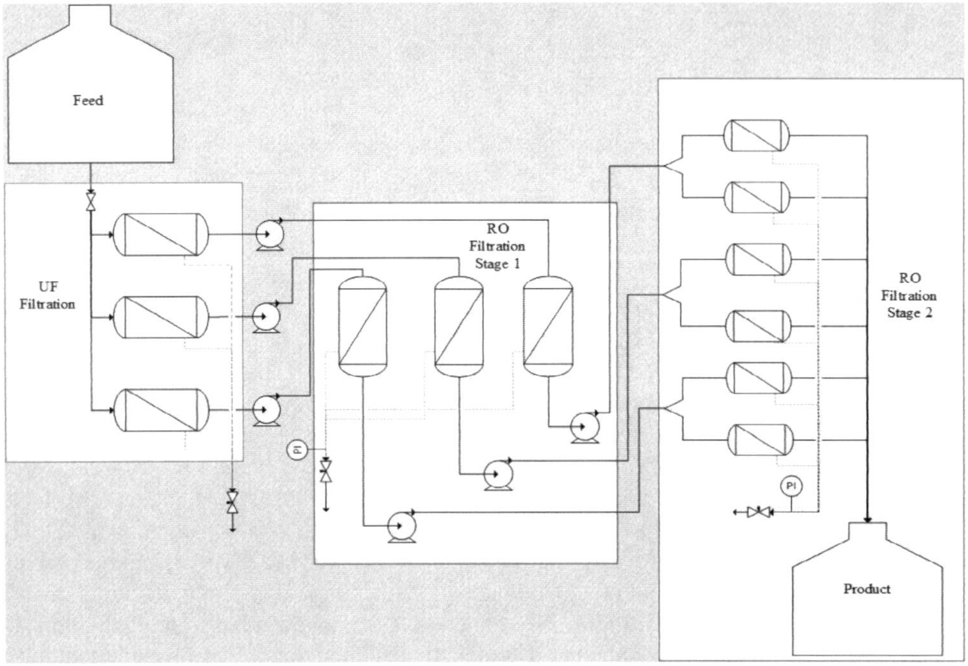

Figure 1. The schematic diagram of the experimental for producing clean water.

3. Result and discussion

3.1. Compared the use of pressure in decreased flux and increased TDS

Table 1 shows the data that had been obtained by this research. The effect of decreasing flux and increasing TDS are evaluated at the variation of pressure P1 = 6 bar and P2 = 3 bar for first experiment, P1 = 6 bar and P2 = 4 bar for second experiment and P1 = 7 bar and P2 = 4 bar for third experiment. Comparing varying pressures were carried out for 4 hours with a constant flux of feed 4.90 Litres per minute. It can see from the result of the operation condition of pressure having the best result. Below is table 1 shows a comparison of varying pressure.

Table 1. The effect of varying pressure at the flux decreases and TDS increases

Time (Hour)	$Pressure_1$	$Pressure_2$	Q-Feed	$Q\text{-}P_1$	$Q\text{-}P_2$	TDS P_1	TDS P_2	Salt $Rejection_1$	Salt $Rejection_2$
2	6	3	4.90	1.50	1.15	116	16	96.11	99.46
4	6	3	4.90	1.50	1.15	115	15	96.14	99.50
2	6	4	4.90	1.35	1.30	118	59	96.04	98.02
4	6	4	4.90	1.35	1.30	119	60	96.01	97.99
2	7	4	4.90	1.20	1.05	142	74	95.23	97.52

Q-feed Flux-in (Litre per minute (LPM))
Q-P1 Flux permeate the first stage (LPM)
Q-P2 Flux permeate the second stage (LPM)
P1 Pressure in the first stage (bar)
P2 Pressure in the second stage (bar)
TDS Total Dissolved Solid (ppm)
Salt Rejection (%)

Table 2. The flux decrease and TDS increase at P_1 = 6 bar and P_2 = 3 bar

Time (Hour)	$Pressure_1$	$Pressure_2$	Q-Feed	Q-P_1	Q-P_2	TDS P_1	TDS P_2	Salt $Rejection_1$	Salt $Rejection_2$
2	6	3	4.90	1.50	1.15	116	16	96.11	99.46
4	6	3	4.90	1.50	1.15	115	15	96.14	99.50
6	6	3	4.90	1.45	1.15	115	16	96.14	99.46
8	6	3	4.90	1.45	1.10	116	15	96.11	99.50
10	6	3	4.90	1.45	1.10	116	16	96.11	99.46
12	6	3	4.90	1.45	1.10	116	16	96.11	99.46
14	6	3	4.90	1.50	1.10	116	16	96.11	99.46
16	6	3	4.90	1.45	1.05	116	17	96.11	99.43
18	6	3	4.90	1.45	1.05	116	16	96.11	99.46
20	6	3	4.90	1.45	1.05	116	16	96.11	99.46
22	6	3	4.90	1.45	1.05	116	17	96.11	99.43
24	6	3	4.90	1.45	1.05	116	17	96.11	99.43
26	6	3	4.90	1.45	1.05	116	16	96.11	99.46
28	6	3	4.90	1.40	1.05	116	17	96.11	99.43
30	6	3	4.90	1.45	1.05	116	17	96.11	99.43
32	6	3	4.90	1.45	1.05	117	17	96.07	99.43
34	6	3	4.90	1.40	1.05	117	17	96.07	99.43
36	6	3	4.90	1.40	1.05	117	17	96.07	99.43
38	6	3	4.90	1.40	1.05	116	18	96.11	99.40
40	6	3	4.90	1.35	1.05	117	18	96.07	99.40
42	6	3	4.90	1.35	1.05	117	18	96.07	99.40
44	6	3	4.90	1.35	1.05	117	18	96.07	99.40
46	6	3	4.90	1.35	1.05	117	19	96.07	99.36
48	6	3	4.90	1.35	1.05	117	19	96.07	99.36
50	6	3	4.90	1.35	1.05	116	19	96.11	99.36
52	6	3	4.90	1.35	1.05	117	19	96.07	99.36
54	6	3	4.90	1.35	1.05	117	18	96.07	99.40
56	6	3	4.90	1.35	1.05	117	19	96.07	99.36
58	6	3	4.90	1.35	1.05	119	19	96.01	99.36
60	6	3	4.90	1.35	1.05	118	19	96.04	99.36
62	6	3	4.90	1.35	1.05	119	20	96.01	99.33
64	6	3	4.90	1.35	1.00	118	19	96.04	99.36
66	6	3	4.90	1.35	1.00	119	19	96.01	99.36
68	6	3	4.90	1.35	1.00	119	19	96.01	99.36
70	6	3	4.90	1.30	1.00	119	19	96.01	99.36
72	6	3	4.90	1.30	1.00	119	20	96.01	99.33

Based on Table 1, can be seen that the use of variation pressure P1 = 6 bar and P2 = 3 bar, also P1 = 6 bar and P2 = 4 bar has low decreasing of flux compared to P1 = 7 bar and P2 = 4 bar. In the first experiment, P1 = 6 bar and P2 = 3 bar gave the constant value of decreasing flux (Q-P1 and Q-P2). These values indicate for 4 hours, the membrane has good performance, especially in rejecting salt. For the second experiment, the use of P_1 = 6 bar has a lower flux product compared to the first. Based on the theory, this is not possible because applying the same pressure provides the same propulsion and flux. This phenomenon is possible for two reasons. The first used the same membrane, which may clog the membrane's pores and result in less water diffusion. However, the possibility is slighter because the membrane is only running for 4 hours. Besides, it is proved by a constant salt rejection approximately 96% equivalent to the first. The second reason is an error measurement.

Furthermore, P1 = 6 bar in the second experiment gives constant salt rejection of about 96%, and the first experiment's value. The pressure 4 bar in the second stage reduces salt-rejection from 98.02% to 97.99% and increases TDS about 16 to 59 ppm. The phenomena indicate additional salt that passes through the membrane and occurring phenomena of polarization concentration in the membrane [6].

At pressure P1 = 7 bar and P2 = 4 bar, flux has a lower value than the first and second experiments. Phenomena polarization concentration occurring in the third experiment is greater than other experiments, instead of using the same membrane. Based on the value of TDS, the use of P1 = 7 bar and P2 = 4 bar give total dissolve solid higher than other experiments. High-pressure forces salt going through the membrane in consequence to clog the pore of the membrane. These salts are monovalent and bivalent ionic such as Na^+, Cl^-, Mg^{2+}, and Ca^{2+} [4]. More pores are clogged by salt, resulting in a lower diffusion of water through the membrane. Used P1 = 7 bar and P2 = 4 bar also gives a high decrease in flux and salt rejection. Based on the effect of decreasing of flux and increasing of TDS that were evaluated by varying pressure showed the operating pressure which has long saturation time of membrane is P1 = 6 bar and P2 = 3 bar. The saturation test of the membrane carried out using these operating pressure where is shown in Table 2.

3.2. Estimated the saturation time of membrane using method decreased of flux and increased of TDS
After comparing those pressure to get optimum pressure, evaluate and estimate the membrane's saturation time. Figure 2 below shows the effect of pressure on decreasing flux (Q) and increasing TDS at 72 hours. The data of the graphic below is shown in Table 2.

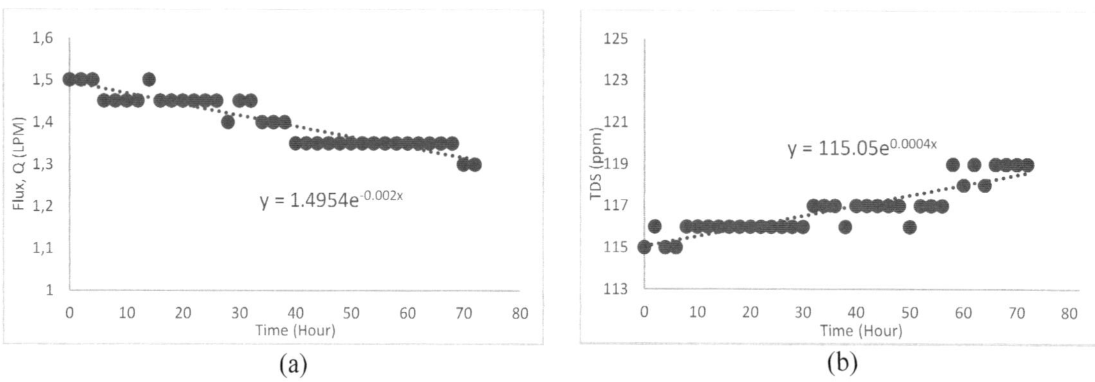

Figure 2. The effect of pressure P_1 = 6 bar on (a) decreased flux (Q-P1) and (b) increased TDS.

Based on Figure 2 (a), flux decreases from 1.5 liters per minute to 1.3 liters per minute. Total flux decreases are 0.2 liters per minute for 72 hours or in percent about 13% because dissolve solid started to accumulate in the membrane. Based on the graphic shown in Figure 2 (a) has the equation $y = 1.495\ x\ e^{-0,002x}$, where e is the Euler number. Assumed the membrane should be changed or cleaned when the decreasing of flux is 15%, then the saturation time of the membrane was calculated and obtained 80 hours. Therefore, based on the decrease flux method, the membrane should be changed with the range of time about 72-80 hours. While in Figure 2 (b), total dissolved solids (TDS) increases from the range of 115-116 to 119 ppm. The increase of TDS because of the accumulation of salt by pressure. The total increase of TDS for 72 hours, about 3% with the equation $y = 115.05\ x\ e^{0.0004x}$. Based on a technical note, clean the membrane if salt-rejection decreases 10%, the increased TDS about 413 ppm. Therefore, using that equation, the membrane should be cleaned after 3,188 hours.

Figure 3 shows the effect of pressure with P_2 = 3 bar on (a) decreasing of flux (Q-P2) and (b) increasing of TDS.

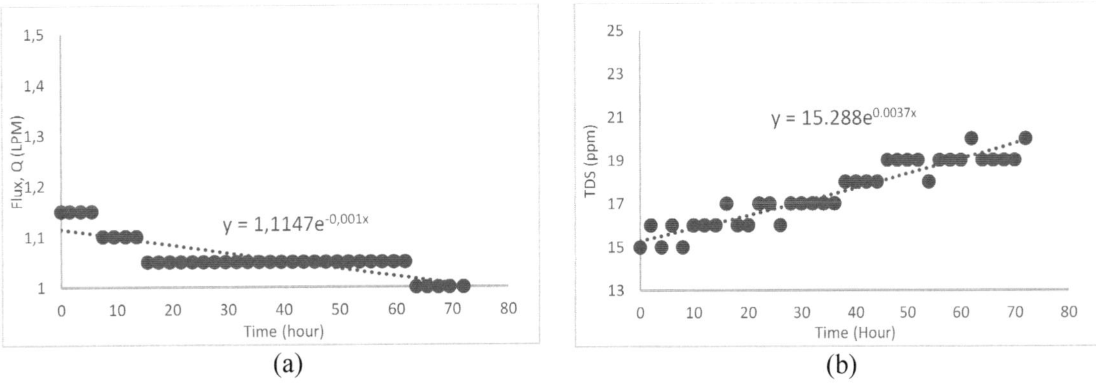

Figure 3. The effect of Pressure P_2 = 3 bar on (a) the decreased flux (Q-P1) and (b) increased TDS.

Figure 3 shows that the flux produced is lower than the first stage and TDS in the second stage lower than the first stage. The product produced in the first stage had used for the second stage lead product purer from contaminants. Figure 3 (a) at a range of 15-60 hours, shows constant flux. It indicates that the layer had not experienced saturation significantly. The total decrease in flux for 72 hours is about 13%. The graph in Fig 3.a has the equation$y = 1.115\, x\, e^{-0,001x}$. Assumed cleaning the membrane after decreasing flux by 15%. Use the equation to calculate the membrane's saturation time and obtain a range of about 72-131 hours. While in Fig. 3.b has shown that the graphic has an equation$y = 15.29\, x\, e^{0.0037x}$. Assumed cleaning the membrane when salt-rejection decreases 10%, it increases TDS at about 313 ppm. Therefore, the saturation time of the membrane for 816 hours. Compared to the calculation at the first stage and the second stage, calculating the membrane's saturation time using the decreased flux method gives a better result than TDS increased. Based on calculating the membrane's saturation time, which was conducted by two methods above, the decreasing flux more representative of the existing theory in which the saturation time of the membrane in the first stage should be faster than the second membrane.

4. Conclusion

This cuurent study revealed that the membrane's saturation time calculation used the decreased flux method more representative than the TDS increase. The use of pressure that has long saturation time of membrane might be ranked as P_1 = 6 bar and P_2 = 3 bar > P_1 = 6 bar and P_2 = 4 bar > P_1 = 7 bar and P_2 = 4 bar. Using P_1 = 6 bar /and P_2 = 3 bar, the saturation time of membrane at the first stage (ULP-3013-400) has range 72-80 hours, and the second stage (WW-2012-100) has 72-131 hours.

Acknowledgments

The researcher thank KEGIATAN UI PEDULI and the PUTI Proceeding 2020 (The Grants of International Indexed Publication for the lecturers of University Indonesia) No. PENG-1/UN2.RST/PPM.00.00/2020, organized by the Directorate of Research and Community Service (DRPM) – University Indonesia. Furthermore, there is no competing interest or conflict of financial interests.

References

[1] Choi J S and Kim J T 2015 Modeling of full-scale reverse osmosis desalination system: influence of operational parameters *J. Ind. Eng. Chem.* **21** 261-8.

[2] ElAzhar F, ElAzhar M, Hafsi M, Taky M and Elmidaoui A 2018 Desalination of brackish water using low-pressure nano filtration membranes comparison and simulation *J. Chem. Pharm.* **10(1)** 119-25.
[3] Greenlee L F, Lawler D.F, Freeman B.D, Marrot B and Moulin P 2009 Reverse osmosis desalination: water sources, technology, and today's challenges. *Water Res* ***43*****(9)** 2317-48.
[4] Oh H J, Hwang T M and Lee S 2009 A simplified simulation model of RO systems for seawater desalination *Desalination* **238** 128-39.
[5] Kucera J 2015 *Reverse Osmosis: Design, Process, and Application for Engineers 2nd edition* Beverly: Scrivener Publishing LLC.
[6] Malaeb L and Ayoub G M 2011 Reverse osmosis technology for water treatment: state of the art review *Desalination* **267(1)** 1-8.
[7] Bodzek M, Konieczny K and Rajca M 2019 Membranes in water and wastewater disinfection – review *Arch. Environ. Prot.* **45(1)** 3-18.
[8] Ninik P 2009 *Thesis: Optimization Reverse Osmosis Process in Recovery and Concentrated Flavour Component from Liquid Waste of Crab Meat Pasteurization* (Bogor: IPB).
[9] Rajindar S 2015 *Membrane Technology and Engineering for Water Purification: Application, System Design and Operation 2nd edition* Oxford: Butterworth-Heinemann Elsevier.
[10] Xu J, Ruan G, Chu X, Yao Y, Su B and Gao C 2007 A pilot study of UF pretreatment without any chemicals for SWRO desalination in China *Desalination* **207(1-3)** 216-26.

IOP Conf. Series: Materials Science and Engineering 1011 (2021) 012045 doi:10.1088/1757-899X/1011/1/012045

Extraction and identification of Mitragynine from the Kratom Leaf (*Mitragyna speciosa*) using HFC-134a subcritical system

Arthur A Lelono[1]*, Ita L Latifah [1], Herlan Herdiawan [1] and Rina Wahyu Cahyani [2]

[1]Research Center for Chemistry, Indonesian Institute of Sciences, Building 452, Puspiptek Serpong Area, South Tangerang, Banten, Indonesia
[2]Dipterocarps Forest Ecosystem Research and Development Center, Ministry of Environment and Forestry, Samarinda East Kalimantan

*Corresponding author: aalelono@gmail.com

Abstract. Kratom (*Mitragyna speciosa*) is tree that useful as herbal medicine commonly found in Indonesia and other Southeast Asia Countries. Kratom has been consumed by chewing it or brewed like tea, it is useful to treated fever, muscle ache and diarrhea pain. Sedatives and stimulant effect on the central nervous system are side effect of Kratom leaves infusion. It contains mitragynine that have strong affinity on opioid receptor, therefor classified as new psychoactive substances (NPS). Others compounds in Kratom is 7-hydroxymitragynine, paynantheine, speciogy nine and speciociliatine. The aim of this study was to study the mitragynine and other secondary metabolites on the Kratom using subcritical HFC R134a extraction method and identified with TLC, LCMS/MS and GCMS. The HFC-134a subcritical system method is based on the advantage strength and physical properties of refrigerant R134a as solvent. Subcritical fluid extraction has potentially alternative method to isolate bioactive compound of the herbal plants. The result was 0.7878 g extract Kratom was produced, TLC spot at Rf 0.49, GCMS peak at RT at 18.327 and LCMS/MS peak of RT 7.12 with mass spectrum at 399.2m/z, 174.1 m/z and 159.1 m/z belong to mitragynine.

1. Introduction

Kratom (*Mitragyna speciosa*) is tree indigenous Southeast Asia, one of them is Indonesia [1] especially grow in Putusibau, West Borneo [2]. The Kratom commonly used by way of chewing it, brewed like tea even made as cigarette to local society [3]. The Kratom leaf used as remedies such as, fever, relieve muscle pain, reduce appetite and diarrhea [1,4]. In Malaysia antiseptic substance from Kratom was used to treat postpartum recovery. Herbal medicine considered more safer for consumption than synthetic chemical-based medicine. Some terms usually used to mention Kratom as herbal medicine composition were Krypton, K2 or herbs and spices [5].

The studies on pharmacological effects of Kratom have been conducted such as analgesic activity, stimulant, anti-inflammatory, antioxidant, antinociceptive, antibacterial, and antidepressants [6-10]. The Kratom was known having ability as stimulant in low dose [11] and sedative effect in high dose.

IOP Conf. Series: Materials Science and Engineering 1011 (2021) 012045 doi:10.1088/1757-899X/1011/1/012045

Potential of sedative effect from ethanolic extraction reported bigger than those of diazepam, i.e. 27.9 mg/20 g BW [12], these effect mimic the act of opium which works suppress pain and used to reduce opioid with drawl syndrome [13], while depressant effect and stimulant in the center nervous system caused from side effect of Kratom [12]. This effect was mediated through monoaminergic receptors and opioid [14].

The Kratom leaf or the Puric leaf has been reported containing some alkaloid compounds i.e mitragynine 66.2%, paynantheine 8.6%, speciogynine 6.6%, 7-hydroxymitragynine 2%, and speciociliatine 0.8% [11].

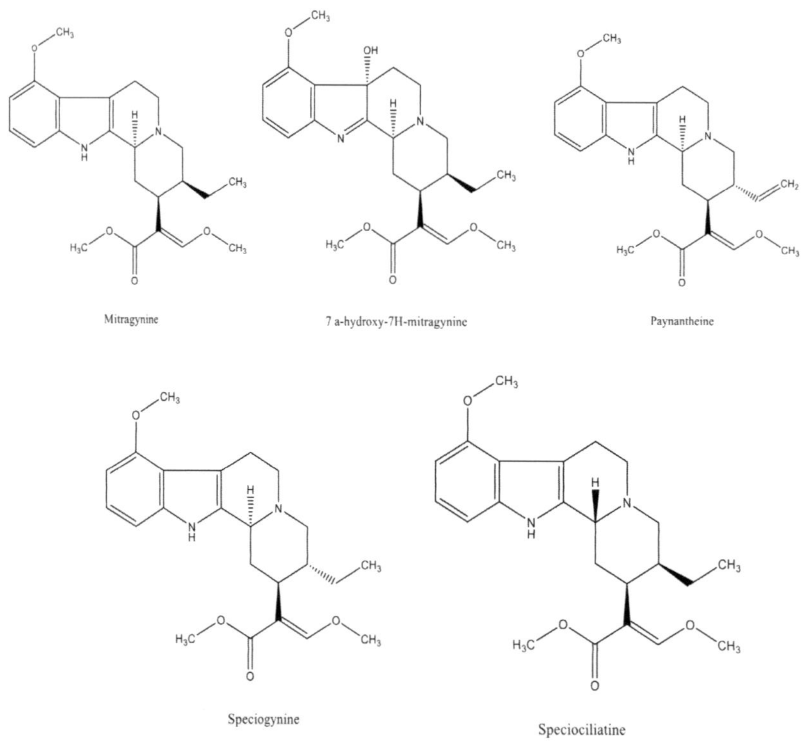

Figure 1. Chemical structure of alkaloid compound in the Kratom leaf

Further studies are required to optimize secondary metabolite contents from The Kratom leaf. Therefore, in this study was conducted the extraction Kratom using subcritical HFC R134a extraction method. The method has been reported by Lelono *et al* [15] on the isolation of artemisinin from *Artemisia annua,* which is proven more effective than convention extraction method. This method has advantage of unique strength and physical properties from pure substance or mixed at critical pressure and temperature in equilibrium phase. Nowadays R134a refrigerant would be used as solvent in natural product extraction process. Because R134a has good strength, non-toxic, non-flammable, relatively stable, environmentally friendly, high purities and solubility and recovered solvent [16].

Tetrafluoroethane (HFC 134a) is safe to use in extraction technology. It works using low pressure and temperature. Extraction is possible using room temperature, lower temperatures but also higher temperatures. R134a is a non-polar fluorocarbon-based solvent that reaches a supercritical state at

101.2° C and 40.6 bar (compared to 31.0° C and 83.3 bar for carbon dioxide, CO_2) [17]. In the application, the use of HFC134a-based extractors has many advantages compared to other extraction methods, including: fast process; high final yield of the extraction (> 85%), low level of solvent wasted; ease of capacity expansion, and lower extraction costs compared to other extraction methods [15].

2. Materials and Methods

2.1. Materials

2.1.1. Plant material. The shredded and powdered dried leaf of Kratom (Mitragyna speciosa) were provided by UPT. KPHL GERBANG BARITO UNIT IX, The Forestry Service, Central Kalimantan, Indonesia.

2.1.2. Reagents and chemicals. N-hexane, ethyl acetate, methanol from Merck (p.a), LCMS/MS reagents grade from Merck, GCMS reagents grade from Merck, Sulfuric acid 5%, Refrigerant R134a Dupont (USA) as refrigerant (solvent), silica gel.

2.1.3. Apparatus. HFC R134a extractor unit, evaporator, column chromatography, silica gel 60 F_{254} TLC plates from Merck, GC-MS Agilent 7890B from Agilent, LC-MS/MS Waters Xevo G2-XS QTof from Agilent, UV-Vis lamp, Cimarex hotplate and Camag TLC chamber.

2.2. Method

2.2.1. Sample preparation. Dried leaves of Kratom were shredded and powdered to increase surface area to optimize result of extract product. Powdered sample was weighed (500 g). Sample was placed into both vessel of extractor, 250 g sample per vessel.

2.2.2. Extraction. Sample was extracted using HFC R134a extractor. Capacity was 5 liters, consist of two reactor extract chambers, collector extract chamber, pressure vacuum pump and chiller, both of reactor extract chamber must be cooled to 20 ºC about 10 minutes prior to extraction. Releasing (HFC 134a) tube and inlet reactor extract chamber were turned (anticlockwise) to open it, pressure vacuum pump was pushed to flow about 2.5 kg HFC-134a into one of reactor chamber. Temperature, pressure and time were recorded as initial time of extraction process. Extraction was conducted for 3 cycles with 1 hour of extraction per cycle. To obtain the product, releasing the solvent out extract into collector extract chamber, which was heated up to 30ºC. Extract was dried using air blower in room temperature. Extract was identified using Thin layer Chromatography (TLC), Gas Chromatography-Mass Spectrometry (GCMS) and Liquid Chromatography-Mass Spectrometry/Mass Spectrometry (LC-MS/MS).

2.2.3. Thin Layer Chromatography (TLC). TLC silica gel 60 F_{254} were used for separation. Mobile phase was n-hexane : ethyl acetate solution (9:1). Approximately 1 mg sample was diluted in 1 mL methanol was applied as bands. Chromatograms were developed to 4/5 of the plate height, then dried, examined under UV lamp (254 nm).

2.2.4. Gas Chromatography-Mass Spectrometry (GCMS). Preparation for GCMS analysis was required about 10 mg extract in 1 mL chloroform and diluted. Sample was filtered prior for injection. Sample was injected for 1 µL using autosampler at 270ºC with increased level temperature 10ºC/minutes to 300ºC. Column's flow rate was 1 mL/minutes.

2.2.5. Liquid Chromatography-Mass Spectrometry/Mass Spectrometry (LC-MS/MS).
Approximately 10 mg sample was diluted in 1 mL methanol. Total of 5 μL was injected to Waters Xevo G2-XS QTof, equipped with ESI Source, pump and Masslynx PC Workstation. Qtop with positive ionization mode referred as Mass Spectrometry (MS). Parameter ESI that used involved capillary temperature 120°C and atomizer gas 5L/hours, voltage 2 kV. Full scan module from m/z 50-1200. Eluent was used H_2O (A) and acetonitrile (B) grade LC-MS. Eluent's flow rate was 0,3 mL/minutes.

3. Results and Discussions
The extraction was conducted three times at one hours of extraction time to maximize the yield content as compared to the studies reported on extraction of artemisinin using HFC-134a subcritical system in Lelono *et al* [15]. HFC-134a Subcritical extraction of Kratom has brown paste extract yielding 0.7878g from 500 gram of Kratom leaves, shown in figure 2. It was analyze using TLC, LCMS/MS and GCMS analysis method.

Figure 2. Kratom HFC-134a extract

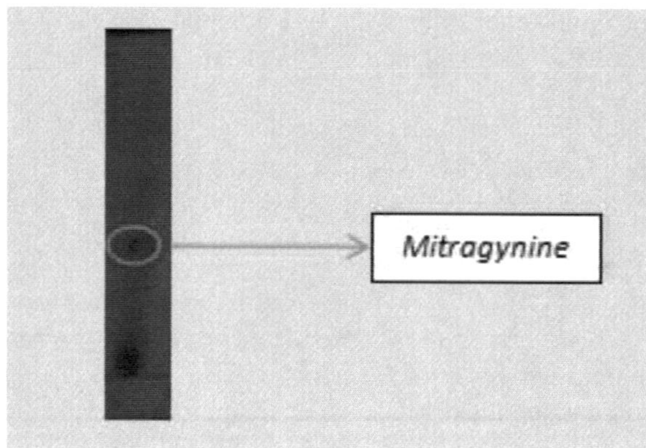

Figure 3. TLC Chromatogram of Kratom extract

Thin Layer Chromatography was conducted as preliminary assay since it is easy, quick and cheap method to identified samples. Chromatogram of Rf sample compared with Rf standard of sample [18]. This method was prepared with gradient eluent n-hexane : ethyl acetate (9:1). Chromatogram was examined under UV lamp (254 nm). Mitragynine was spotted in Rf 0.48 shown in figure 3. According to study that conducted by Kowalczuk AP *et al* [19] Rf 0.49 was mitragynine.

Qualitative test of Kratom HFC-134a extract was conducted using GCMS and LCMS/MS. This was preliminaries step to provide an overview about compounds in Kratom HFC-134a extract. Elsa *et al* [20] have identified active compounds in Kratom with GCMS that mitragynine was separated at retention time (RT) 18.327 that shown in figure 4. Using the database, it was confirmed that peak at RT 18.327 was mitragynine.

Table 1. List of secondary metabolites compounds in Kratom HFC-134a extract from LCMS/MS data.

No	Name Compounds	Mass Fragment (m/z)	Retention Time (min)
1	Mitragynine	399.220	7.12
2	7-hidroxymitragynine	415.215	7.53
3	Isorhynchophylline	385.204	6.12
4	3-Tert-butyl-4-methoxyphenol	181.115	8.94

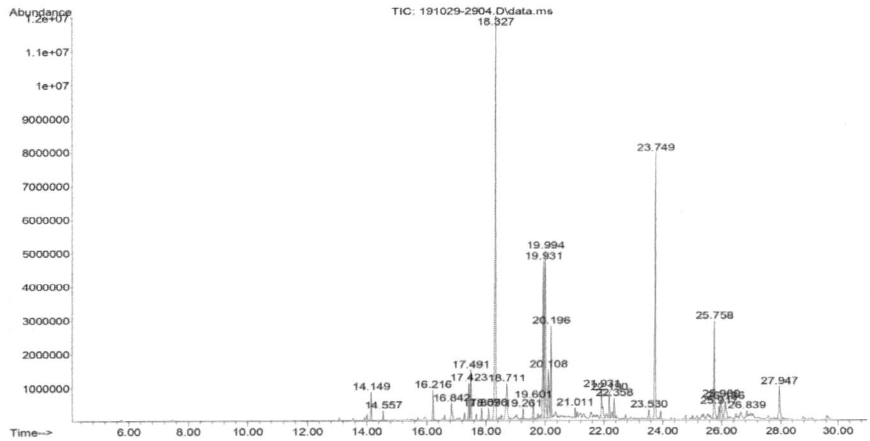

Figure 4. GCMS chromatogram of extract Kratom, mitragynine peak at RT 18.327.

Kratom HFC-134a extract was identified used LCMS/MS. Mitragynine showed retention peak at RT 7.12 minutes with mass fragment m/z 399.220 show in Figure 5 and others secondary metabolites compounds was also identified, that shown in Table 1. This is the first report of the extraction of myrtagynine using HFC-134a subcritical system. Kratom contains a number of indole alkaloids that are believed to be the primary contributors to its psychoactive effects. Chief among these is mitragynine, which typically constitutes 1–2% of the dry leaf mass and up to approximately two-thirds of the total alkaloid content [21]. Mitragynine was appeared at RT 7.12 with fragment m/z 399.2, 174.1 and 159.1 [22]. Mitragynine is alkaloid group, naturally synthesized from some organism including animal, plant, bacteria and fungi. Mitragynine ($C_{23}H_{30}N_2O_4$) have molecular weigh 389.5 g/mol, the most interested to study from The Kratom leaf is sedative effect and stimulant causing who consumed it. Mitragynine is reported as alkaloid compound of Kratom be expected have opioid effect 13 times higher than morphine [14]. Other metabolites which also interesting in Kratom is 7-hydroxymytraginine, which was reported presence in the Kratom HFC-134a extract.

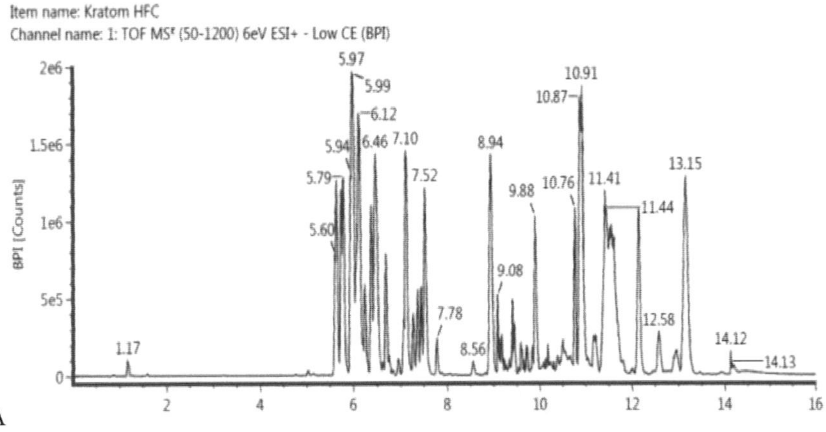

A

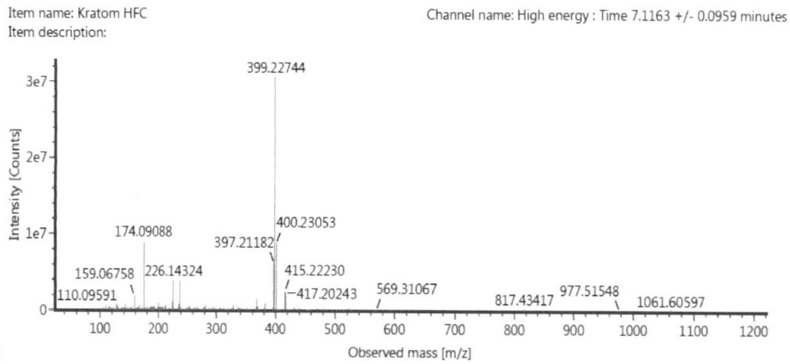

B

Figure 5. LCMS/MS chromatogram of extract Kratom, mitragynine peak at RT 7.12 (A) and LCMS/MS mass chromatogram of extract Kratom, mass fragment of mitragynine at 399.220 m/z (B).

4. Conclusion

Extraction of Kratom leaves using HFC-134a give 0.15% of extract from dry leaves. The chromatography analysis of TLC Rf 0.48, GC-MS at RT 18.327 and LCMS/MS at RT 7.12 and mass fragment 399.220 m/z confirmed that was mitragynine is presence in the Kratom HFC-134a extract. Subcritical HFC R134a extraction method have good ability and efficient for extract natural plants. This is the first report of extraction of mytragynine from Kratom using HFC-134a subcritical system.

References

[1] Hassan Z, Muzaimi M, Navaratnam V, Yusoff NHM, Suhaimi FW, Vadivelu R, Vickknasingam BK, Amato D, von Hörsten S, Ismail NIW, Jayabalan N, Hazim AI, Mansor SM, Müller CP 2013 *Neurosci. Biobehav. Rev.* **37** 138-51
[2] Novindiriana D, Wijianto B and Andrie M 2013 *Jurnal Mahasiswa Farmasi Fakultas Kedokteran UNTAN* **3** 1-13
[3] Sanagia MM, Ayoba MFM, O'Mara N, Ibrahima WAW and Hussaina I 2013 *Pharma Chem.* **5** 131-8
[4] Jansen KL and Prast CJ 1988 *J. Ethnopharmacol.* **2** 115-9
[5] Singh D, Müller CP, Vicknasingam BK 2014 *Drug Alcohol Depend.* **139** 132-7
[6] Luliana S, Robiyanto R, Islamy MR 2018 *Pharmaceutical Sciences and Research* **5** 58-64
[7] Idayu N F, Hidayat MT, Moklas MAM, Sharida F, Nurul Raudzhan AR, Shamma AR and Apryani E 2011 *Phytomedicine* **18** 402-7
[8] Parthasarathy S, Azizi JB, Ramanathan S, Ismail S, Sasidharan S, Said MIM, and Mansor SM 2009 *Molecules* **14** 3964-74
[9] Mossadeq WMS, Sulaiman MR, Tengku Mohamad TA, Chiong HS, Zakaria ZA, Jabit ML, Baharuldin MTH and Israf DA 2009 *Med. Princ. Pract.* **18** 378-84
[10] Reanmongkol W, Keawpradub N and Sawangjaroen K 2007 *Soklanakarin J. Sci. Tech.* **29** 39-48
[11] Kikura-Hanajiri R, Kawamura M, Maruyama T, Kitajima M, Takayama H and Goda Y 2009 *Forensic Toxicol.* **27** 67-74
[12] Ridayani Y 2013 *Jurnal Mahasiswa Farmasi Fakultas Kedokteran UNTAN* **3** 1-9
[13] Carpenter JM, Criddle CA, Craig HK, Ali Z, Zhang, Khan IA, Sufka KJ 2016 *Fitoterapia* **109** 87-90
[14] Warner ML, Kaufman NC and Grundmann O 2015 *Int. J. Legal Med.* **130** 127-38
[15] Lelono A, Simanungkalit S, Umarudin and Herdiawan H 2018 *J. Trop. Pharm. Chem.* **4** 114-21

IOP Conf. Series: Materials Science and Engineering 1011 (2021) 012045 doi:10.1088/1757-899X/1011/1/012045

[16] Lagalante AF, Clarke AM and Bruno TJ 1998 *J. Phys. Chem. B* **102** 8889-92
[17] Setapar SHM, Khatoon A, Ahmad A, Yunus MC, Zaini MAA 2014 *Asian J. Chem.* **26** 5911–16
[18] Sasidharan S, Chen Y, Saravanan D, Sundram KM and Yoga Latha L 2011 *Afr. J. Tradit. Complement. Altern. Med.* **8** 1-10
[19] Kowalczuk AP, Lozak A and Zjawiony JK 2013 *J. Forensic Sci.* **65** 1-20
[20] Elsa L, Yuwono M and Prawita A 2016 *Jurnal Biosains Pascasarjana.* **18** 3
[21] Kruegel AC, Uprety R, Grinnell SG, Langreck C, Pekarskaya EA, Le Rouzic V, Ansonoff M, Gassaway MM, Pintar JE, Pasternak GW, Javitch JA, Majumdar S and Sames D. 2019. *ACS Cent. Sci.* **5** 992–1001
[22] Casey CR, Conley T, Heise A, Thomas T and Ayres PR 2015 *J. Regulatory Science* **2** 1-14

IOP Conf. Series: Materials Science and Engineering 1011 (2021) 012046 doi:10.1088/1757-899X/1011/1/012046

Candlenut oil encapsulation with Hidroxypropyl Methylcellulose (HPMC) for body lotion application

Silvya Yusri, Tiffany and Hery Sutanto*

Department of Chemical Engineering, Faculty of Life Sciences and Technology, Swiss German University, Tangerang, Banten, Indonesia.

*Corresponding author : hery.sutanto@sgu.ac.id

Abstract. Candlenut oil obtained by cold press extraction method was encapsulated in order to preserve the unsaturated fatty acids content from thermal oxidation. The encapsulation of candlenut oil was performed using hydroxypropil mrthylcellulose (HPMC) and maltodextrin as the encapsulating agent due to their filming ability. The solution of wall materials was added with the candlenut oil and emulsified, followed by spray drying process. Three different ratios of oil to wall materials (2:3, 1:1, and 3:2) were evaluated. Based on the peroxide values, encapsulation efficiency, wettability, and moisture content of the microcapsules, the sample with a ratio of 1:1 had better preservation ability towards candlenut oil and therefore used as the main ingredient for body lotion development. Three different concentrations of encapsulated candlenut oil (10%, 6%, and 2%) within the lotion base were observed and analyzed in terms of antioxidant activity by DPPH assay, and overall physical likeness. The result suggested that the wall materials might hinder the reaction between candlenut oil and DPPH radical compounds. The body lotion sample with the lowest content of encapsulated candlenut oil showed the highest antioxidant activity (IC_{50} = 52.31 ppm). Furthermore, according to the ANOVA test result, there was no significant difference in overall physical likeness between the three samples (p-value $\geq$ 0.05).

1. Introduction

Vegetable oils are liquid vegetable fats that are frequently extracted from several diverse parts of vegetation, such as plant seedlings, fruits, or seeds, which contain high concentration of saturated and unsaturated fatty acids, especially triglycerides, that have beneficial influence especially on the skin. Due to this beneficial influence, unsaturated fatty acids are being one of the most important components in dermatology and cosmetology, and becoming frequently used in formulations of dermatological and cosmetic products [1].

As unsaturated fatty acids have numerous benefits, the demand for unsaturated fatty acids have been increasing. However, unsaturated fatty acids can only be found in certain fish and fish oils, and vegetable oils such as olive oil, which are relatively expensive in Indonesia as they are usually imported from another countries. Therefore, other sources of unsaturated fatty acids are needed in order to offer other lower cost alternatives and sustainable resources of natural unsaturated fatty acids, such as candlenut oil.

In Indonesia, candlenut is mainly used as spice to elevate the taste of food. Therefore, the economic value of candlenut is low due to its usage mainly as a cooking ingredient. However, based on the previous study regarding the extraction of candlenut oil done by Cendikiawan (2016), it has been proven that candlenut oil contains high concentration of unsaturated fatty acids which consist of 20.39%

Omega-3; 38.10% Omega-6 and 26.67% Omega-9 [2]. With this high concentration of unsaturated fatty acids, candlenut oil can be applied as a natural source of antioxidant and a main ingredient in various food and beverages, cosmetics, and health care products, thus increasing the economic value of candlenut in Indonesia.

Although candlenut oil is a good source of fatty acids, the thermal sensitivity of candlenut oil degrades the quality of fatty acids that it contains and therefore reduces its concentration [3]. According to Tjhin (2014), the concentration of Omega-3 significantly decreases at the temperature of 220°C with a mixing time of more than 3 hours while it is given various concentration of oxygen sparged. Due to this characteristic, it is going to be difficult to apply candlenut oil as one of the ingredients during production process as most of the processes might involve heat. Based on the study of formulating Omega-3-rich baby rusk from candlenut oil done by Apelia (2013), the concentration of Omega-3 after the baking process reduces significantly compared to the concentration before the baking process (from 2119 mg/100g to 18.15 mg/100g Omega-3) [4]. Another study done by Lukmanto (2013), shows that the concentration of unsaturated fatty acids decreases up to more than 90% during the production of ice cream inserted with candlenut oil due to the heat introduction during the process [5].

Thus, in order to overcome this problem, encapsulation could be the alternative solution for the thermal instability of candlenut oil. Hydroxypropyl methylcellulose (HPMC) is chosen as the encapsulation material for candlenut oil, in regard to its nature as essential oil and moisturizer, as they have the ability to encapsulate fatty acids. HPMC has good water-solubility, emulsifying-ability, and filming-ability, which is already used as microencapsulating agent [6]. Thus, HPMC might be compatible as coating material of candlenut oil.

Therefore, candlenut oil which contains a high concentration of unsaturated fatty acids, are going to be studied as the main ingredient for body lotion as one of the topical cosmetics in order to fulfil the increasing demand of natural fatty acids for cosmetics, to increase the economic value of candlenut in Indonesia, and possibly offer lower costs and provide a sustainable source of natural unsaturated fatty acids. This research aimed to obtain the best ratio between HPMC and candlenut oil in application to topical lotion. Several parameters, such as peroxide values, encapsulation efficiency, wettability, and moisture content of the microcapsules were evaluated to obtain the best ratio of encapsulation agents to candlenut oil. The best ratio of encapsulation agents to candlenut oil was used in the formulation of candlenut oil in the body lotion.The antioxidant activity and overall physical likeness were measured to obtain the best formulation of candlenut oil as a natural fatty acids ingredient in body lotion.

2. Materials and methods

2.1. Candlenut Oil Extraction & Unsaturated Fatty Acids Analysis

Candlenut oil was extracted from candlenut kernels using a cold press expeller under low temperature (≤ 15°C) to preserve the unsaturated fatty acids content. Then, the extracted candlenut oil was filtered using a filter cloth to remove any solid residual remain after the extraction. In order to obtain high purity of candlenut oil, the remaining solid residual that could not be filtered with filter cloth was settled by centrifugation and separated from the candlenut oil. The unsaturated fatty acids (UFAs) content was analyzed using gas chromatography (GC).

2.2. Encapsulation of Candlenut Oil

The candlenut oil was encapsulated with HPMC in which the ratio between HPMC and candlenut oil will be varied into three different concentrations as shown in Table 1.

The sample was pulverized using spray dryer with inlet temperature 200°C and outlet temperature 80°C. All of the samples were then analyzed by measuring the percentage of encapsulation efficiency (%EE), thermal oxidative stability (PV), moisture content, and wettability of the samples.

Table 1. Concentration of HPMC, maltodextrin, and candlenut oil for encapsulation process.

Sample	Candlenut Oil (mL)	HPMC (g)	Maltodextrin (g)	Soy Lecithin (g)	Distilled Water (mL)	Temperature (°C)
A	30	30	15	0.75	1000	5
B	30	15	15	0.75	1000	5
C	45	15	15	1.12	1000	5

2.3. Body Lotion Application

The best candlenut oil sample was used in the formulation of the body lotion. The formulation of the product was varied into three different concentrations as shown in Table 2.

Table 2. Concentration of encapsulated candlenut oil in body lotion.

Sample	Encapsulated Candlenut Oil (%)	Lotion Base (g)
A	10	80
B	6	80
C	2	80

The antioxidant activity of each sample was measured in order to determine which sample has higher antioxidant activity. Sensory analysis was also conducted in order to ascertain the overall physical likeness of the products.

3. Result & Discussion

3.1. Candlenut Oil Extraction

The candlenut oil was extracted using cold press expeller machine to prevent any reduction of unsaturated fatty acids content within the candlenut kernels. The yields of extracted candlenut oil are shown in Table 3.

Table 3. Yield of extracted candlenut oil.

	This research	Isnaeni (2017)	Cendikiawan (2016)
Weight of Candlenut Kernels (g)	20,000	10,000	10,000
Crude Candlenut Oil (mL)	6,500	2,250	3,928
Refined Candlenut Oil (mL)	2,980	1,230	1,467
Yield (mL oil / g kernels)	0.149	0.123	0.150

The yield of extracted candlenut oil obtained in this research was comparably similar with the yield obtained by Cendikiawan (2016) [2], nevertheless the present yield was slightly higher compared with the yield obtained by Isnaeni (2017) [7]. The differences of yield obtained from the present and previous researches might be due to the freshness of candlenut kernels, leakage during the extraction process, and/or mishandling during the process that might decrease the amount of oil obtained.

3.2. Unsaturated Fatty Acids Content of Candlenut Oil

The unsaturated fatty acids content of candlenut oil was analyzed by using gas chromatography (GC). The result of the analysis is shown in Table 4. The total unsaturated fatty acids content of candlenut oil was 90.80% in which 25.58% Omega-3, 40.86% Omega-6, and 23.91% Omega-9 were contained. In comparison with the unsaturated fatty acids content from the previous researches, the Omega-3 content obtained from this research was higher. Additionally, the Omega-6 content was significantly higher compared with Isnaeni (2017) [7], though it was comparably similar to Cendikiawan (2016) [2].

Nevertheless, the content of Omega-9 content obtained in this research was the lowest compared to the previous researches. The differences in the omega content might be due to the difference in origin where the candlenuts were produced and the freshness of the candlenuts.

Table 4. Omega content of candlenut oil.

Parameter	Unit	This research	Isnaeni (2017)	Cendikiawan (2016)
Omega-3	mg/100g	25,583.5	21,742.1	20,390.00
Omega-6	mg/100g	40,864.6	26,009.5	38,100.00
Omega-9	mg/100g	23,906.8	42,767.5	26,670.00

3.3. Encapsulation of Candlenut Oil
The candlenut oil was encapsulated using HPMC and maltodextrin as the wall materials. HPMC is modified cellulose that has the ability to emulsify and formed a layer which has already been utilized as a coating material for drugs [6,8]. Three different formulations of candlenut oil encapsulation were observed in this research to ascertain which formulation had better encapsulation ability towards candlenut oil. The formulation differed in the ratio of candlenut oil to wall materials in which 2:3 for sample A, 1:1 for sample B, and 3:2 for sample C.

3.4. Encapsulated Candlenut Oil Samples Analysis
Peroxide Value. Lipid oxidation occurs when unsaturated fatty acids react with oxygen which resulting in fatty acid hydroperoxides as the primary product [9]. The concentration of hydroperoxides from the primary oxidation of oil can be measured by Peroxide Value (PV) calculation. Therefore, PV was calculated in order to ascertain the encapsulation ability of HPMC to prevent the candlenut oil from rapid oxidation when expose to heat.

As shown in Figure 1 below, the peroxide values of all of the samples were fluctuated except for the non-encapsulated candlenut oil. At day zero, all of the samples have zero peroxide values as they have not been exposed to heat. When the oxidation process begin, the peroxide values of all of the samples were slowly increasing over time which occurs due to the formation of hydroperoxides from the oxidation of unsaturated fatty acids [10]. At day four, the peroxide values of all of the encapsulated candlenut oil samples decreased unexpectedly due to the formation of the secondary products of the lipid oxidation. The hydroperoxides formed during lipid oxidation are broken down into a variety of volatile and non-volatile compounds, such as aldehydes and ketones [10,11].

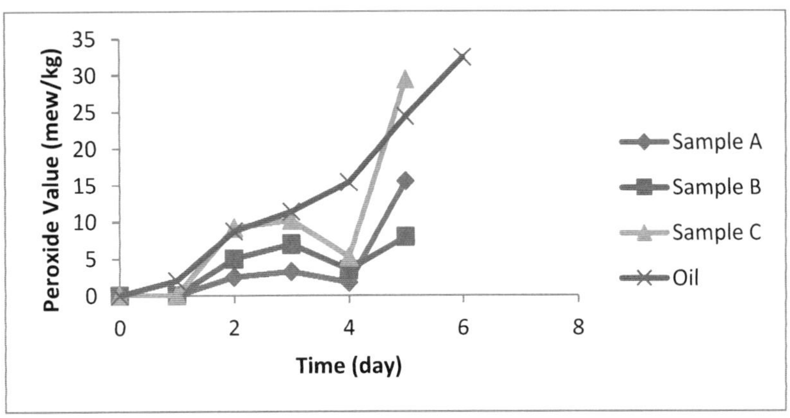

Figure 1. Peroxide value of samples over time.

From the graph, it can be seen that the slope of non-encapsulated candlenut oil sample was higher compared with all of the encapsulated candlenut oil samples, which means that lipid oxidation occurs more rapidly in non-encapsulated candlenut oil sample compared with the encapsulated candlenut oil samples. Thus, it can be concluded that the encapsulation process was able to prevent the oxidation of candlenut oil. Among the three samples, sample B proves to be more proficient in preventing the oxidation of candlenut oil because the hydroperoxides formed during the oxidation process was the lowest

3.5. Encapsulation Efficiency (%EE)
In this research, the encapsulation efficiency indicates the capability of HPMC to encapsulate candlenut oil. The measurement of encapsulation efficiency was done by extracting the free oil on the surface of the encapsulated samples with organic solvent (hexane). The result is shown in Table 5 below.

Table 5. Encapsulation efficiency (%) of samples

Sample	Encapsulation Efficiency (%)
A	40.05 ± 0.0040
B	54.23 ± 0.0057
C	34.92 ± 0.0105

Comparing the three samples, sample B has the highest encapsulation efficiency of 54.23% compared to sample A and sample C with encapsulation efficiency of 40.05% and 34.92% respectively. The increase of encapsulation efficiency is affected by the size of droplets emulsion, which means the higher the stability of the emulsion will result in higher retention time of the oil [12-14]. This theory can be applied for sample C that contained higher oil concentration than wall materials causing the decrease of emulsion stability and thinning out the wall materials leading to a poor retention of oil and rapid lipid oxidation [6,15].

3.6. Wettability
Wettability is a measure of the ability of microcapsules to rehydrate [16]. The wettability of each sample was displayed in Table 6.

Table 6. Wettability of samples.

Sample	Time (minutes)
A	18.19 ± 2.9769
B	4.32 ± 1.2940
C	2.02 ± 0.7212

Comparing the result of each sample, sample A took the longest time to become entirely wet, which was 18.19 minutes while sample B and sample C took 4.32 minutes and 2.02 minutes respectively. This result is similar with Karim et al. (2016) [17] which indicates that lesser polymer content (HPMC) will result in shorter wetting time of the sample as it will facilitate the accessibility and penetration of water into the powder samples. Furthermore, the short amount of time that sample B and sample C took might also be due to the agglomeration of the powders which could increase the penetration of liquid through the pores of the powder particles [17].

3.7. Moisture Content
Moisture content of the samples was measured in order to determine the physical properties of the powders. Moisture content of the samples may ascertain the quality of the samples.

Table 7. Moisture content of samples

Sample	Moisture Content (%)
A	8.32 ± 0.2656
B	6.61 ± 0.1028
C	5.67 ± 0.3846

It can be seen from Table 7 that sample A has the highest value of moisture content compared to sample B and C. Moisture content of the powders is affected by the viscosity of the emulsion which is highly depended on total solids within the emulsion [18]. Higher total solids within the emulsion will produce higher viscosity emulsion. Higher moisture content of powders will also affect the quality of the sample as moisture might enhance lipid oxidation of the oil resulting in the formation of off-flavors molecules and degradation of samples [17].

3.8. Development of Body Lotion with Encapsulated candlenut Oil

Based on the results of encapsulated candlenut oil analyses, sample B was chosen as the main ingredient for the development of body lotion because sample B had better encapsulation efficiency and low peroxide value.

In this research, three different formulations were applied for the development of body lotion in order to select which sample had better antioxidant activity and overall physical acceptance. The formulation differed in the concentration of encapsulated candlenut oil within the lotion base in which 10% of encapsulated candlenut oil for sample A, 6% for sample B, and 2% for sample C. By considering the final physical texture and fragrance of the products, the respective concentration of encapsulated candlenut oil within the lotion base was chosen.

3.9. Sensory Analysis

The sensory analysis of the products was done in order to ascertain the overall physical likeness of the consumers towards the products. Hedonic test of certain attributes of the products was performed by 30 untrained panelists. The assessed attributes were: texture, spreadability, oiliness, and fragrance. The result of sensory analysis of the sample is shown in Figure 2.

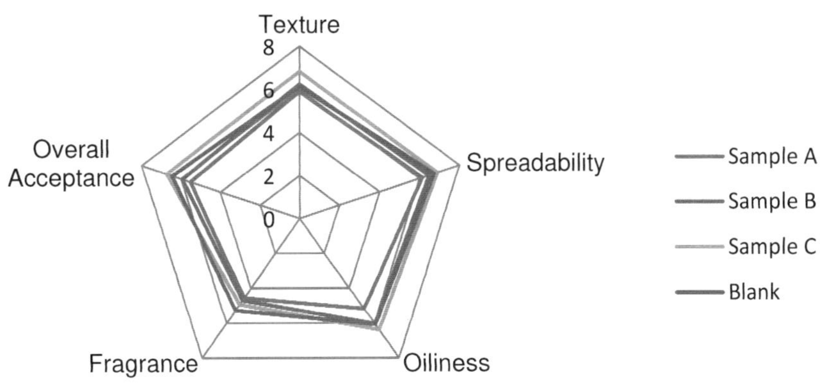

Figure 2. Quantitative result of sensory analysis.

The result of the sensory test was then analyzed using ANOVA in order to determine whether there was any significant difference between the samples. According to the ANOVA test result, there was no significant difference between each sample ($p \geq 0.05$) meaning that despite the difference in encapsulated candlenut oil concentration, the physical attributes of the samples produced similar results.

This result showed similar scores for each sample in physical attributes and overall acceptance. The similar results and insignificant difference between each sample might be due to the use of commercial lotion base in this formulation, as the concentrations of the encapsulated candlenut oil were probably not high enough to alter any physical attributes of the products. Furthermore, as HPMC was used as the wall material for the encapsulation of candlenut oil, the products produced an unfavorable smell in which, despite the difference in concentration of encapsulated candlenut oil, the smell still can be strongly detected from each sample.

3.10. Antioxidant Activity

Antioxidant activity of the body lotion sample was measured using DPPH assay method. Based on the Table 9 below, the IC_{50} of the blank sample (lotion base only) was the highest among the other samples. The high value of IC_{50} of the blank sample might be due to the lack of antioxidant compounds within the lotion base. However, among the samples that contained encapsulated candlenut oil, sample C with the lowest concentration of encapsulated candlenut oil had the lowest IC_{50} value compared to the other samples. According to Rakmai et al. (2017), wall materials might prevent the antioxidant compounds within the core material reacting with the DPPH radical compounds [19]. Therefore, the high IC_{50} value of sample A and sample B might be due to the wall materials blocking the reaction between the antioxidant compounds and DPPH radical compounds. Nevertheless, these results indicated that the addition of encapsulated candlenut oil into the lotion base presumably improved the antioxidant activity of the lotion as the IC_{50} values of each samples were lower than the blank sample.

Table 9. IC_{50} values of samples.

Sample	IC_{50} (ppm)
A	53.70
B	64.27
C	52.31
Blank	91.47

4. Conclusion

The encapsulation process of candlenut oil using HPMC and maltodextrin was successful. From the three different concentrations, sample B with 1:1 ratio of wall materials to core material, was proven to be the most effective to preserve the quality of candlenut oil in terms of peroxide value and encapsulation efficiency. The development of the body lotion using encapsulated candlenut oil was varied in three different concentrations which were 10%; 6%; and 2%. The antioxidant activity of the samples was higher than the commercial lotion base. Based on the sensory analysis, none of the samples had better overall physical likeness towards the untrained panelists.

References

[1] Zielińska A and Nowak I. 2014 Chemik **68**(2), pp 103–110
[2] Cendikiawan, L. H. 2016 *Effect of Candlenut Oil Addition to Physicochemical Properties of Gummy Candy* (Tangerang: Swiss German University)
[3] Tjhin, N. 2014 *Effect of Oxygen Concentration and Heat Treatment to The Stability of Omega-3 Content in Candlenut Oil* (Tangerang: Swiss German University)
[4] Apelia, I. 2013 *Formulation of Baby Rusks Enriched with Omega-3 From Candlenut Oil Extract.* (Tangerang: Swiss German University)
[5] Lukmanto, M. O. 2013 *Effect of Candlenut Oil Addition to Physicochemical Properties of Ice Cream* (Tangerang: Swiss German University)
[6] Wu, K. G., Chai, X. H. and Chen, Y. 2005 Chinese Journal of Chemistry **23**(11), pp 1569–1572
[7] Isnaeni, R. 2017 *Effect of Candlenut Oil Extract to Physicochemical Properties of Tofu*

IOP Conf. Series: Materials Science and Engineering 1011 (2021) 012046 doi:10.1088/1757-899X/1011/1/012046

(Tangerang: Swiss German University)
[8] Kolanowski, W., Laufenberg, G. and Kunz, B. 2004 International Journal of Food Sciences and Nutrition **55**(4), pp. 333–343
[9] Verardo, V. et al. 2009 Food Chemistry **114**(2), pp. 472–477
[10] Kaleem, A. et al. 2015 Fuuast J. Biol, **5**(2), pp. 191–196
[11] Decker, E. A., Elias, R. J. and Mcclements, D. J. 2010 Oxidation in foods and beverages and antioxidant applications Volume 2: Management in dofferent industry sectors
[12] Liu, X. D. et al. 2001 Drying Technology **19**(7), pp. 1361–1374
[13] Soottitantawat, A. et al. 2006 Proceedings of the 10th Intl Drying Symposium **68**(7), pp. 2256–2262.
[14] Jafari, S. M. et al. 2008 Drying Technology **26**(7) pp. 816–835
[15] Carneiro, H. C. F. et al. 2013 Journal of Food Engineering **115**(4), pp. 443–451
[16] Fernandes R V, de B Borges S V and Botrel, D. A. 2013 Ciência e Tecnologia de Alimentos **33**, pp. 171–178
[17] Karim F T, *et al.* 2016 *J. Food Process. Preser.* **40**(2) 140–53
[18] Bhandari B, *et al.* 1992 *Food Sci.* **57** 217–21
[19] Rakmai J, *et al.* 2017 *Ind. Crops Prod.* **111** 219–25

Antioxidant activity of endophytic fungi isolated from the stem bark of *Swietenia mahagoni* (L.) Jacq

Elfita Elfita[1]*. Muharni Muharni[1], Heni Yohandini[1] and Fadhillah Fadhillah[2]

[1]Department of Chemistry, Faculty of Mathematics and Natural Sciences, University of Sriwijaya, Inderalaya, Ogan Ilir, South Sumatra, 30662 Indonesia
[2]Magister Program of Chemistry, Department of Chemistry, Faculty of Mathematics and Natural Sciences, University of Sriwijaya, Palembang, South Sumatra, 30662 Indonesia

*Corresponding author: elfita69@gmail.com

Abstract. Endophytic fungi are microbes that live to form colonies in plant tissue without harming their host plants. Each higher plant has the potential to contain several endophytic fungi that produce secondary metabolites as a result of coevolution or genetic recombination of the host plant. Endophytic fungi derived from medicinal plants are candidates for producing bioactive compounds. *Swietenia mahagoni* known as mahoni, is a medicinal plant that has been used to treat various diseases such as to prevent degenerative diseases. Almost all degenerative diseases are caused by free radicals. The endophytic fungus of *S. mahagoni* is a potential source for the discovery of antioxidant compounds. In the present study, we report an antioxidant compound from endophytic fungal isolated from the stem bark of *S. mahagoni*. Antioxidant activity was tested by 2,2-diphenyl-1-picrylhydrazyl (DPPH) method. The compound was identified by the spectroscopic method as phenolic compound. The endophytic fungi were identified by molecularly based on genetic analysis as *Cladosporium tenuissimum*

1. Introduction

Endophytic fungi are microscopic organisms that live in plant tissues such as leaves, roots, fruit, and stems. The fungus lives in mutual symbiosis with its host [1]. Some endophytic fungi produce the same phytochemical compounds as their host plants. Taxol is a well-known anticancer agent found in every species of the yew tree (Taxus spp.). Stierle *et al.* (1993) found the endophytic fungus *Taxomyces andreanae* from the *Taxus brevifolia*, which can produce taxol compounds [2]. Currently, several taxol-producing endophytic fungi have been identified. Podophyllotoxin compound is an anticancer, antiviral, antioxidant, antibacterial, immunostimulant, and antirheumatic. The main supply of podophyllotoxin is from the *Sinopodophyllum*. Yang *et al.* (2003) reported six endophytic fungi obtained from *Sinopodophyllum hexandrum*, *Diphylleia sinensis*, and *Dysosma veitchii*, which can produce podophyllotoxin [2-5].

Sources of natural antioxidants, such as vegetables, fruits, spices, and medicinal plants, are widely known because the presence of phytochemicals, flavonoid and phenolic compounds. Recently, many studies have revealed that endophytic fungi also produced of phenolic compounds which have antioxidant activity. Recently, endophytes have gained as novel sources of medicinally important metabolites including antioxidant, antimalarial, anticancer, antibacterial, and other activities that have attracted much attention from researchers worldwide [6-8].

Swietenia mahagoni has been widely used in various countries including in Asia as a cure for various diseases such as hypertension, diabetes, malaria, HIV, bleeding, nerve disorders, anti-inflammatory, skincare, treating fever, as a tonic, astringent, etc. This plant possesses numerous pharmacological effects including anti-inflammatory, neuropharmacological, anti-HIV, hepatoprotective, antioxidant, immunomodulatory and antimicrobial [9-12]. The ability of endophytic fungi to produce secondary metabolites depends on the host plant and the environment of the host plant to grow. The host plants that are used as drugs with antioxidant effects are candidates for getting endophytic fungi that produce antioxidant compounds [13,15]. In the previous studies, we have reported the isolation of endophytic fungi from the stalks of *S. mahagoni* and their antioxidant activity. The extract of endophytic fungus *Cladosporium tenuissimum* has a good antioxidant activity, but the compound (5-hydroxy-2-oxo-2H-Piran-4-yl) methyl acetate which have isolated from that extract not shows antioxidants activity [12]. In this paper, we reported that *C. tenuissimum* has isolated from *S. mahagoni* stem bark. Based on this, this study aimed to isolate antioxidant compounds from *C. tenuissimum*.

2. Materials and methods

2.1. Sampling and identification
The stem bark of *S. mahagoni* (L.) Jacq were taken from around the Sriwijaya University campus, Indralaya, Ogan Ilir district, South Sumatra. Plant identification was carried out at the Department of Biology, Faculty of Mathematics and Natural Sciences, Sriwijaya University.

2.2. Chemicals
The Chemicals used in this study include Potato Dextrose Agar (PDA), Potato Dextrose Broth (PDB), distilled water, alcohol 70%, KLT kiessel gel 60 F254 20 x 20 cm, silica gel G 60 70-230 mesh, sodium hypochlorite (NaOCl) 2,2-diphenyl-1-picrylhydrazyl (DPPH), dimethyl sulfoxide (DMSO), methanol, ethyl acetate, chloroform, n-hexane (Merck).

2.3. Isolation of endophytic fungal strains
Fresh sample (2 x 4 cm) was rinsed with tap water and sterilized by immersion in alcohol 70% for 3 minutes, followed by immersion in NaOCl 2% and then rinsed three times with sterile water. The sample (0.5 x 1 cm) was placed on three PDA plates containing 150 mg/L of chloramphenicol. The plates were incubated at 27 ± 2 °C for 5 to 14 days. The hyphal tip from the segments were transferred into new PDA plates containing 150 mg/L of chloramphenicol and further incubated for 5 to 14 days to obtain pure culture [12,16]. The culture of endophytic fungal was deposited for culture collection of the Department of Biology, Faculty of Mathematics and Natural Sciences, Sriwijaya University.

2.4. Identification of endophytic fungal C. tenuissimum
Endophytic fungal was identified based on molecular analysis at the Research Center for Biology, Indonesian Institute of Sciences, Cibinong, Indonesia [12].

2.5. Cultivation and isolation of pure compound
Six plugs (5 x 5 mm2) of endophytic fungus culture on PDA were transferred onto culture flasks containing 200 mL PDB and incubated at room temperature under static conditions. After four weeks of incubation, mycelium was separated from the broth culture using a Whatman filter paper. The broth culture was partitioned using ethyl acetate (1:1, triplicate) to obtain ethyl acetate extracts and then evaporated using a rotary evaporator to obtain concentrated ethyl acetate extract [12, 16]. The extract (1.47 g) was column chromatographed on a silica gel column (70-230 mesh) as a stationary phase and eluted with a gradient solvent system. Eluates were placed in vials (10 mL) and analyzed by thin layer chromatographed (TLC). The potential stain was continued to separation and purification by column

chromatographed to obtain pure compound. The identification of the compound was determined by spectroscopic analysis ^{1}H-NMR and ^{13}C-NMR. The stages of isolation of pure compound from endophytic fungal *C. tenuissimum* from the stem bark of the *S. mahagoni* were compared with the stages of previous studies of the same fungus from the *S. mahagoni* leaf stalks [12] shown in Figure 1.

2.6. Antioxidant activity test by DPPH method

Concentrations of DPPH solution 0.05 mM was prepared by dissolving 1.98 mg DPPH in 100 mL methanol. The samples were dissolved to make series concentrations of 1000; 500; 250; 125; 62.5; 31.25; and 15.625 μg/mL. The samples (200 μL) in various concentrations were added with 3.8 mL DPPH 0.5 mM and incubated at room temperature in the dark. After 30 minutes, the absorbance was measured using a UV-Vis spectrophotometer at λ max 517 nm. The antioxidant activity was expressed in IC_{50} values by plotting % inhibition vs concentration series [17,18]. The percentage of the free radical scavenging activity was calculated as:

$$\% \text{ inhibition} = \frac{control\ absorbance - sample\ absorbance}{control\ absorbance} \times 100 \quad (1)$$

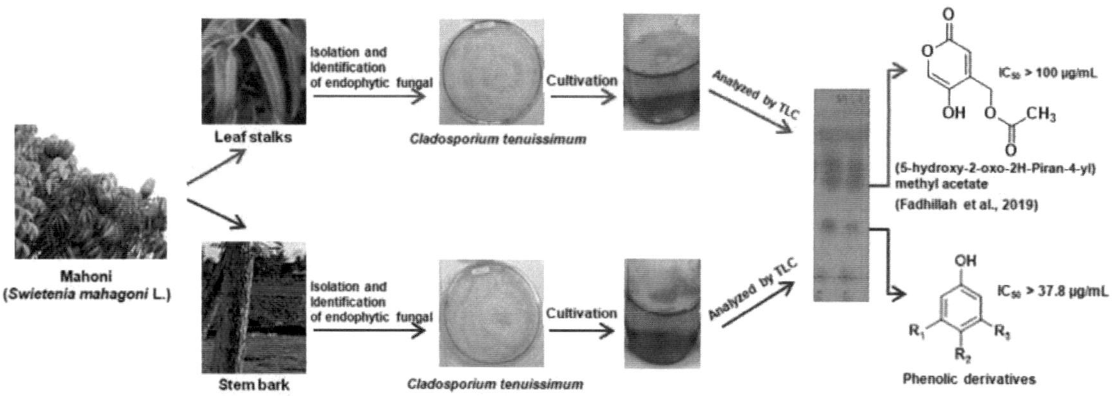

Figure 1. The stages of isolation of pure compound from endophytic fungal *C. tenuissimum* from the stem bark of the *S. mahagoni* (→) were compared with the stages of previous studies of the same fungus from the *S. mahagoni* leaf stalks (→).

3. Results and discussion

Isolation and purification of endophytic fungal from the stem bark of *S. mahagoni* produce endophytic fungi similar to those found in the leaf stalks of *S. mahagoni* [12]. Based on the phylogeny tree's analysis, the endophytic fungus has the highest homology with *C. tenuimum* strain APBSDSF6. The sequence was submitted to GenBank and deposited with the accession number MN379836. Ethyl acetate extract of the fungal *C. tenuissimum* isolated from the stem bark and the leaf stalk of *S. mahagoni* showed the same stain pattern on TLC. Fadhillah *et al.* (2019) have succeeded in isolating one compound, namely (5-hydroxy-2-oxo-2H-Piran-4-yl) methyl acetate. The compound is not an active antioxidant [12]. However, the TLC results showed another purple stain which was possibly demonstrated antioxidant activity.

3.1. Isolation and identification of pure compound

The concentrated of ethyl acetate extract (1.47 g) was separated by column chromatography on silica gel (70-230 mesh, 30 g) and eluted with n-Hexane-ethyl acetate (10:0→0:10) and ethyl acetate-methanol (10:0→0:10), yielded four subfractions (A-D). Fraction C (0.36 g) was re-chromatographed with a solvent system of n-hexane- ethyl acetate= 7:3→0:10 to obtain three fractions (C1–C3). Fraction C3 (71 mg) was further purified as same as previous with n-hexane-ethyl acetate = 2: 8 as mobile phase, to obtain pure compound 1 (38 mg). To determine the structure of compound 1, a 1D-NMR spectrometric analysis was performed.

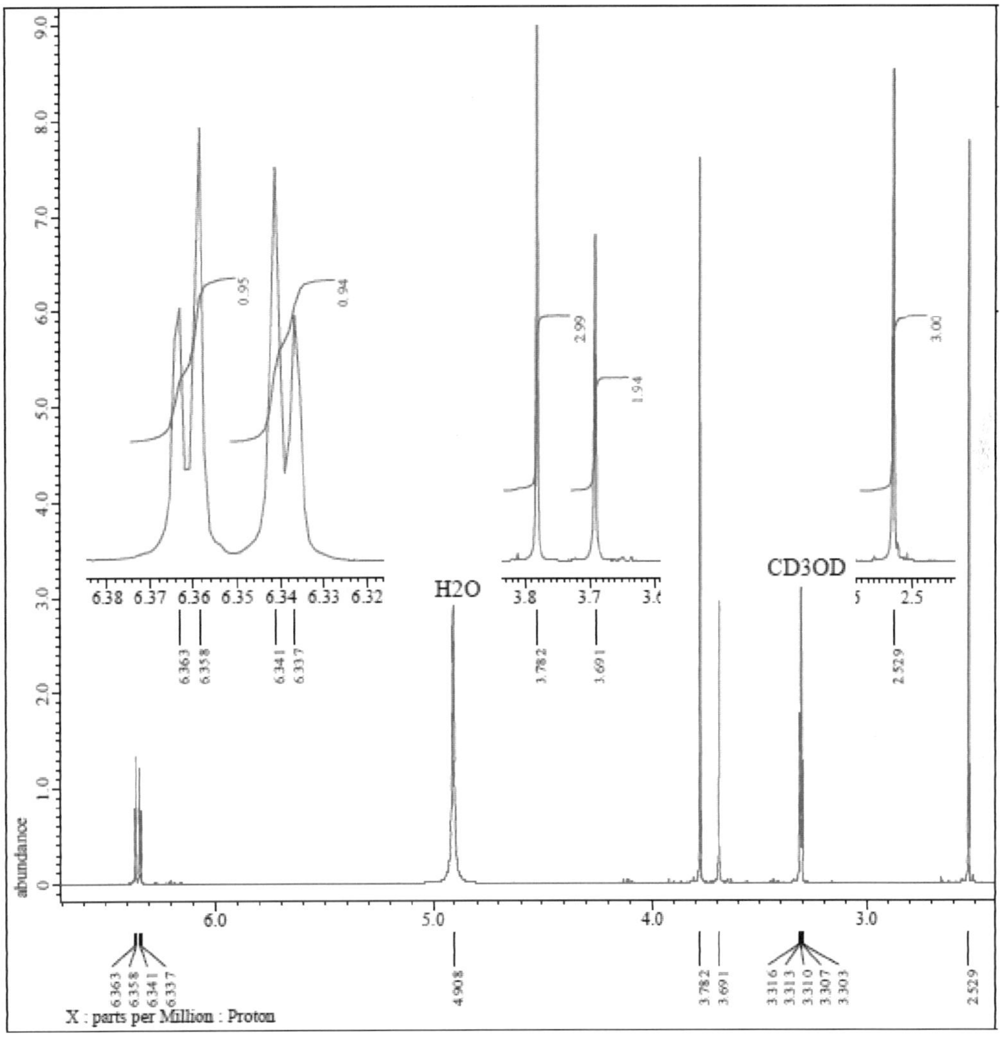

Figure 2. The ^{1}H-NMR spectrum of pure compound from *C. tenuimum* (^{1}H-500 MHz, in CD_3OD).

The spectrum of ^{1}H-NMR (figure 2) showed the presence of five proton signals. The signal indicated two aromatic protons at 6.36 ppm δ_H (1H, d, J = 2.5 Hz) and 6.34 ppm (1H, d, J = 2.5 Hz). The protons by the doublet multiplicity and the coupling constant J = 2.5 Hz indicated the meta position. In the spectrum appeared one signal of methoxy protons at δ_H 3.78 ppm (3H, s), one signal of methylene proton at 3.69 ppm (2H, s), and one signal of methyl proton at δ_H 2.53 ppm (3H, s).

The spectrum of ^{13}C-NMR (Figure 3) showed 11 carbon. The aromatic groups indicated six signals at δ_C 101.17, 110.74, 121.92, 137.85, 160.54, and 163.72 ppm. The Signals at 160.54 and 163.72 ppm indicated the presence of two oxyaryl carbon atoms (= C-O). Based on the ^{1}H-NMR spectrum, there are two protons in the aromatic ring, so the methoxyl and hydroxyl groups were attached to the aromatic ring. The methoxyl carbon atom appears at δ_C 55.85 ppm. The signal in downfield carbons at chemical shift δ_C 206.38 ppm indicated the presence of ketone carbonyl carbon. Thus, the methyl ketone group is attached to the aromatic ring. The presence of this methyl ketone group was supported by the methyl signal's appearance at the spectrum of ^{1}H-NMR δ_H 2.53 ppm with singlet multiplicity and the spectrum of ^{13}C-NMR at δ_C 32.44 ppm. The other signal in downfield carbons at δ_C 175.41 ppm is the area for the carbonyl ester or carbonyl carboxylate groups. The signal is thought to be a carbonyl carboxylate and not an ester, because one methoxyl group was attached to the aromatic ring.

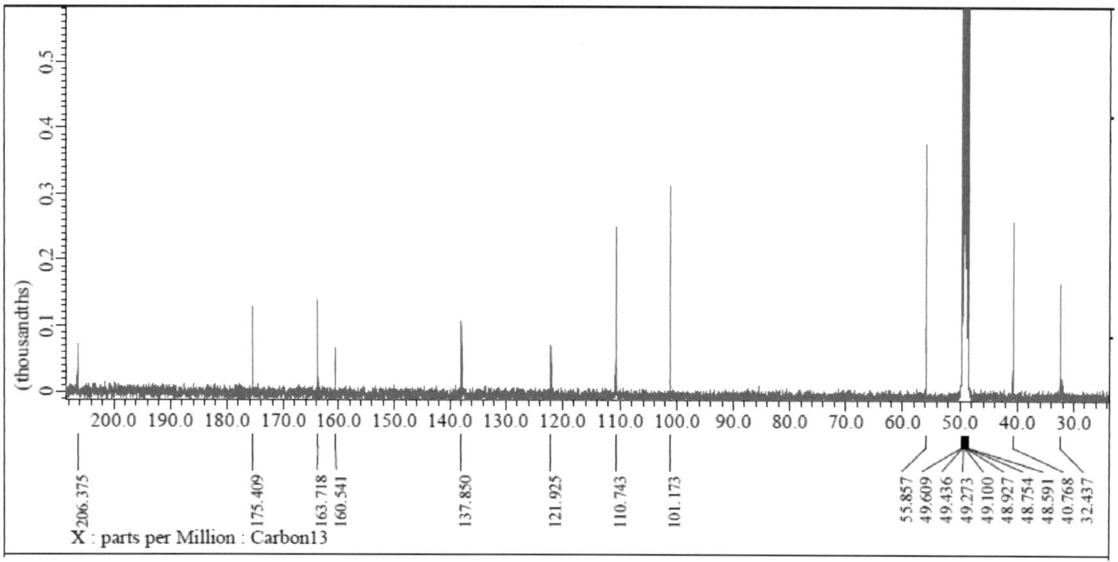

Figure 3. The ^{13}C-NMR spectrum of pure compound from *C. tenuimum* (^{13}C-125 MHz, in CD_3OD).

Based on the analysis of H-NMR and C-NMR spectra, the compound **1** was identified as phenolic derivatives with four substitutes consisting of hydroxyl, methoxyl, methyl ketone, and acetic acid groups. The structure of compound **1** was shown in Figure 4.

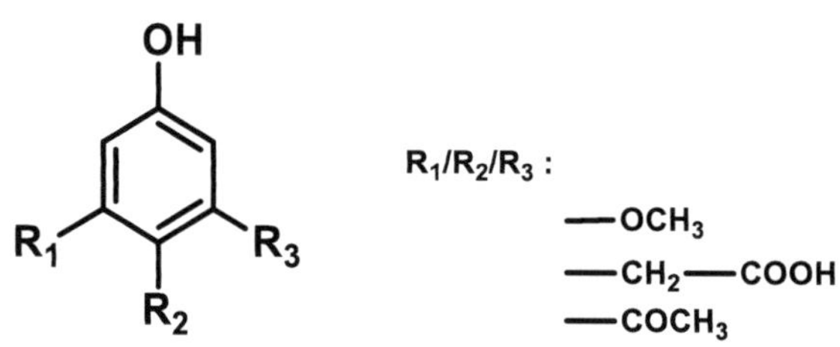

Figure 4. The structure of phenolic derivatives from *C. tenuimum.*

3.2. Antioxidant activity of phenolic compound from C. tenuimum

The antioxidant activity test of ethyl acetate extract and compound **1** were compared with the previous extract and compound (5-hydroxy-2-oxo-2H-Piran-4-yl) methyl acetate) of endophytic fungal *C. tenuissimum* from the *S. mahagoni* leaf stalks [12] as shown in Table 1.

Both of the ethyl acetate extract has good antioxidant activity with the DPPH test method. On the other hand, compound **1** and 5-hydroxy-2-oxo-2H-Piran-4-yl) methyl acetate is not active showed different activity. Based on the structure of the two pure compounds, the abstraction of the proton phenolic compound by DPPH free radicals is much more comfortable and can form new free radicals that can be stabilized by distribution and delocalization electron in the molecule [19, 20]. Khiralla *et al.*

(2015) reported that the phenolic compounds as potential sources of natural antioxidants [21]. In contrast, the (5-hydroxy-2-oxo-2H-Piran-4-yl) methyl acetate compound is inactivated antioxidant because the abstraction of the proton by DPPH free radicals produces new free radicals that are less stabilized. The presence of heteroatoms in the ring seems to inhibit free radical resonance in molecules.

Table 1. The IC_{50} value of ethyl acetate extract and pure compound of endophytic fungal *C. tenuissimum* from the stem bark of *S. mahagoni* and from the leaf stalks of *S. mahagoni* with ascorbic acid as a positive control using DPPH method.

Sample tested	IC50 (µg/mL)	
	Stem bark	Leaf stalk (*)
Ethyl acetate extract	86.2	85.4
Compound **1**	37.8	-
5-hydroxy-2-oxo-2H-Piran-4-yl) methyl acetate	-	> 100
Ascorbic acid	22.2	22.2

(*) [12]

4. Conclusion

The endophytic fungal *C. tenuissimum* has been found in stem bark and the leaf stalk of *S. mahagoni*. However, the chemical contents are the same. This study showed that *C. tenuissimum* could be used as a new source of antioxidants. Further research is needed to isolate other active compounds in *C. tenuissimum*.

Acknowledgments

The authors are thankful to Sriwijaya University, who have funded this research through *Penelitian Unggul Profesi* 2018-2020.

References

[1] Elfita E, Muharni M and Indah T 2011 Secondary metabolite of *Aspergillus fumigatus*, endophytic fungi of the medicinal plant *Garcinia griffithii. Makara J. Sci.* **15** 124-8
[2] Stierle A, Stroble G and Stierle D 1993 Taxol and taxane production by *Taxomyces andreanae*, an endophytic fungus of Pacific yew. *Science* **260** 214–6
[3] Strobel G, Daisy B, Castillo U and Harper J 2004 Natural products from endophytic mikroorganisms. *Journal of Natural Products* **67** 257-68
[4] Zhao J, Zhou L, Wang J, Shan T, Zhong L, Liu X and Gao X 2010 Endophytic fungi for producing bioactive compounds originally from their host plants. *Current Research, Technology and Education Topics in Applied Microbiology and Microbial Biotechnology,* A. Mendez-Vilas (Ed), 567-76
[5] Yang X, Guo S, Zhang L and Shao H 2003 Selection of producing podophyllotoxin endophytic fungi from podophyllin plant. *Nat. Prod. Res. Develop.* **15** 419-22
[6] Elfita E, Munawar M, Muharni M, Pratiwi G and Rahmadania R 2016 A benzoyl compound isolated from endophytic fungi of kandis gajah (*Garcinia griffithii*) and asam kandis (*Garcinia cowa*). *Makara J. Sci.* **20** 167-72
[7] Gunasekaran S, Sathiavelu M and Arunachalam S 2017 *In vitro* antioxidant and antibacterial activity of endophytic fungi isolated from *Mussaenda luteola. J. Appl. Pharm. Sci.* **7** 234-8
[8] Caicedo NH, Davalos AF, Puente PA, Rodriguez AY and Caicedo PA 2019 Antioxidant activity of exo-metabolites produced by *Fusarium oxysporum*: An endophytic fungus isolated from leaves of *Otoba gracilipes. MicrobiologyOpen* **8** 1-7
[9] Metasari S, Muharni M, Elfita E and Yohandini H 2020 Study of antioxidant activities from antihypertension drug plant of the Indralaya area. *Indones. J. Fundamental Appl. Chem.* **5** 22-8

[10] Tan Q and Luo X 2011 Meliaceous limonoids: chemistry and biological activities. *Chem. Rev.* **111** 7437–52

[11] Elfahmi E, Woerdenbag HJ and Kayser O 2014 Jamu: Indonesian traditional herbal medicine toward rational phytopharmakological use. *J. Herb. Med.* **4** 51-73

[12] Fadhillah F, Elfita E, Muharni M, Yohandini H and Widjajanti H 2019 Chemical compound isolated from antioxidant active extract of endophytic fungus *Cladosporium tenuissimumin* in *Swietenia mahagoni* leaf stalks. *Biodiversitas* **20** 2645-50

[13] Jalgaonwala RE, Mohite BV and Mahajan RT 2011 A review: natural products from plant associated endopytic fungi. *Journal of Microbiology and Biotechnology Research* **1** 21-32

[14] Elfita E, Muharni M, Munawar M and Aryani S 2012 Secondary metabolite from endophytic fungi *Aspergillus niger* of the stem bark of kandis gajah (*Garcinia griffithii*). *Indones. J. Chem.* **12** 195 - 200

[15] Cui J, Guo T, Ren Z, Zhang N and Wang M 2015 Diversity and antioxidant activity of culturable endophytic fungi from alpine plants of *Rhodiola crenulata, R. angusta,* and *R. sachalinensis. PLoS ONE* **10** 1-16

[16] Elfita E, Mardiyanto M, Fitrya F, Larasati JE, Julinar J, Widjajanti H and Muharni M 2019 Antibacterial activity of *Cordyline fruticosa* leaf extracts and its endophytic fungi extracts. *Biodiversitas*, 20(12), 3804-3812.

[17] Muharni M, Elfita E, Yohandini H, Julinar J, Yasrina Y and Miranti M 2019 Chemical constituents from stem bark of *Flacourtiarukam* Zoll. & Mor. and their antioxidant activities. *Sains Malays.* **48** 1899-906

[18] Budiono B, Elfita E, Muharni M, Yohandini H and Widjajanti H 2019 Antioxidant activity of *Syzygium samarangense* L. and their endophytic fungi. *Molekul* **14** 48-55

[19] Procházková D, Boušová I and Wilhelmová N 2011 Antioxidant and prooxidant properties of flavonoids. *Fitoterapia* **82** 513-23

[20] Brunetti C, Martina D, Ferdinando MD, Fini A, Pollastri S and Tattini M 2013 Flavonoids as antioxidants and developmental regulators: relative significance in plants and humans. *Int. J. Mol. Sci.* **14** 3540-55

[21] Khiralla A, Mohamed I, Thomas J, Mignard B, Spina R, Yagi S and Laurain-Mattar D 2015 A pilot study of antioxidant potential of endohpytic fungi from some Sundanese medicinal plants. *Asian Pac. J. Trop. Med.* **8** 701–4

IOP Conf. Series: Materials Science and Engineering 1011 (2021) 012048 doi:10.1088/1757-899X/1011/1/012048

Design of hybrid RO/NF permeate staging for the supply of clean water in Tanjung Lame Village

Anton Prayoga, Muhammad Zakwan and Setijo Bismo*

Department of Chemical Engineering, Faculty of Engineering, Universitas Indonesia, Depok, West Java, 16424, Indonesia.

*Corresponding author: setijo.bismo@ui.ac.id

Abstract. The groundwater sources from community wells in Tanjung Lame village, Pandeglang district, are very salty, so they are not suitable for consumption. Reverse osmosis (RO) and nanofiltration (NF) membrane are alternative desalination technologies for treating clean water from saltwater wells. However, the pressure required on a conventional single-stage RO filtration system is very high. The purpose of this study is to design a hybrid RO/NF filtration system that can operate at low pressure. The designed hybrid RO/NF permeate staging has proven to be able to produce clean water from saltwater wells in Tanjung Lame village at low pressure. To obtain optimum results, the designed hybrid RO/NF permeate staging is operated at the first and second stage pressure of 6 and 3 bar, respectively. At that pressure, the salt rejection and TDS of permeate is 97.06% and 85 ppm, respectively. The water recovery rate and permeate flow rate is 31% and 1.55 liters/minute, respectively.

1. Introduction

Tanjung Lame is one of the villages in Ujung Jaya, Sumur District, Pandeglang Regency, Province of Banten. Tanjung Lame village is located on the coast. In the Selat Sunda tsunami of 2019, Tanjung Lame village was one of the affected areas. The location which is very close to the beach and the tsunami has made the ground water from the community wells are very salty. In the rainy season, the TDS of water from community wells can reach 2890 ppm. In the dry season, the TDS of the water can increase due to groundwater evaporation. This makes the source of water from the community wells are not suitable for consumption. One alternative technology for treating clean water from salt water is a reverse osmosis (RO) filtration system.

RO filtration system is one alternative water treatment technology that uses semipermeable membrane to remove impurities such as ions, molecules, and particles that are larger than water. RO filtration systems are easier to operate compared to the evaporation and ion exchange methods, making them easier to apply to the rural areas. However, the pressure required in a conventional single-stage RO system is very high which causes high the electrical energy requirements for pumps [1]. In the case of seawater desalination, the required pump pressure is usually above 40 bar, while water recovery is around 40-50% [2].

Some modifications to RO filtration system have been developed to decrease the pressure, one of which is a two-stage concentrate staging. This configuration succeeded in decreasing operating pressure but the TDS of combined permeate stream was still relatively high. Moreover, the fouling rete of second stage RO higher because the membrane recycles the rejection stream [3]. The purpose of this study is to design the permeate staging configuration. In this RO filtration system, permeate stream on the first RO as a feed on the second RO, so it can reduce operating pressure and the TDS of clean water produced. To increase water recovery rate, NF membrane will be combined with the RO membrane at the first filtration stage. NF membranes have higher flux so that they can increase the permeate flow rate [4]. In some brackish water desalination projects, the NF membrane has replaced the RO membrane due to lower energy consumption and higher flux rates [5]. Thus, the replacement of part of RO membrane with NF membrane is expected to increase the value of water recovery from the designed reverse osmosis filtration system.

2. Materials and method

2.1. Feed water and membrane modules

The experiments used feed water from saltwater well in Tanjung Lame village, Pandeglang district. Water from the well is collected in the feed tank so that the TDS of the feed is constant during the experiment. The TDS of feed is 2890 ppm. The membrane modules used in this experiment are RO TCF membrane ULP-3013-400 type for the first stage RO membrane, RO membrane type WW-2012-100 for the second stage RO membrane, and NF membrane Kusatsu type NF-2812-500 for the first stage NF membrane. There are three and six membrane in the first stage and second stage, respectively.

2.2. Experimental installation and procedure

The experimental setting used in this study was hybrid RO/NF permeate staging filtration system. Figure 1 shows a schematic diagram of the experimental settings for this study.

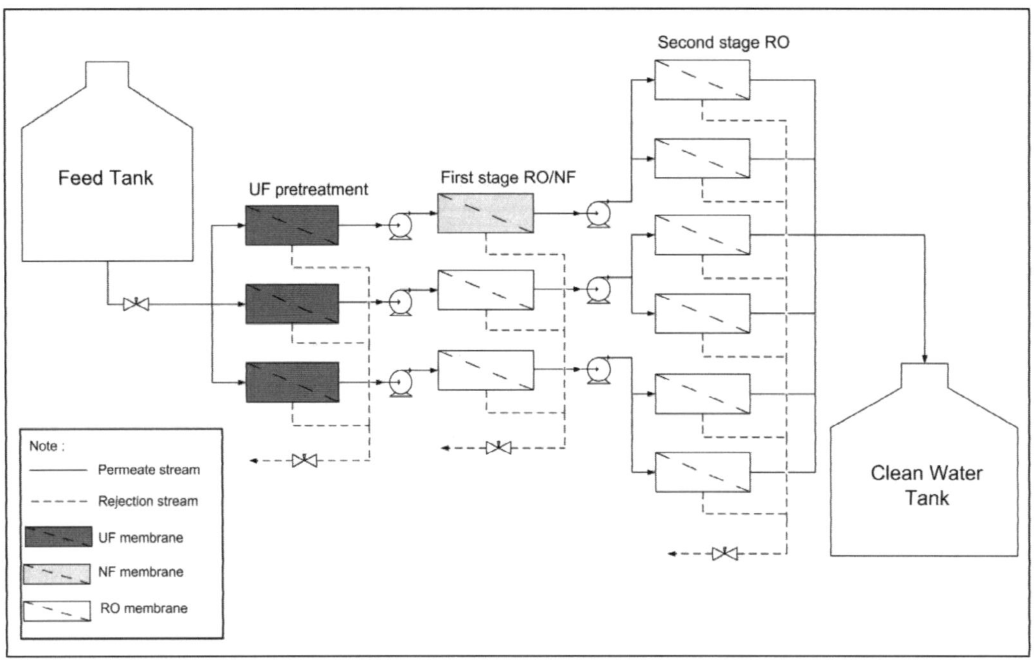

Figure 1. The schematic diagram of experimental setup for hybrid RO/NF filtration system.

In this configuration, three UF membranes are used as pretreatment stage. Permeate stream from the UF membrane is flowed to the first stage pump. The first stage pump, which consists of three pumps, will pump water to the first stage RO/NF membrane. The first stage RO/NF consists of two RO membranes and one NF membrane. Permeate stream from the first RO/NF membrane flow to the second stage pump. The second stage pump, which consists of three pumps, will pump water to the second stage RO membrane. The second stage RO consists of six RO membranes. Rejection stream from each RO

IOP Conf. Series: Materials Science and Engineering 1011 (2021) 012048 doi:10.1088/1757-899X/1011/1/012048

membrane is connected to the valve and pressure gauge to control the pressure on the membrane. The permeate stream from each stage is connected to the rotameter to measure the volumetric flow rate. Feed flow rate is set constant at 5 liters/minute. Permeate stream from the second RO membrane is collected in the clean water tank. Variation of transmembrane pressure used in the first and second stage are 5, 6, 7 bar and 3, 4 bar, respectively.

3. Result and Discussion

Performance of designed hybrid RO/NF permeate staging filtration system is evaluated by the effect of transmembrane pressure on salt rejection and water recovery rate.

3.1. Effect of transmembrane pressure on salt rejection

Salt rejection is one of the parameters to determine the performance of the designed RO/NF filtration system. The salt rejection (in%) is the ratio between the concentration of salt that can be rejected into the rejection stream and the concentration of salt in the feed stream at each stage of filtration. The higher the salt rejection, the better the performance of the RO/NF filtration system. The effect of transmembrane pressure on the salt rejection in this study is shown in Figure 2 and Table 1.

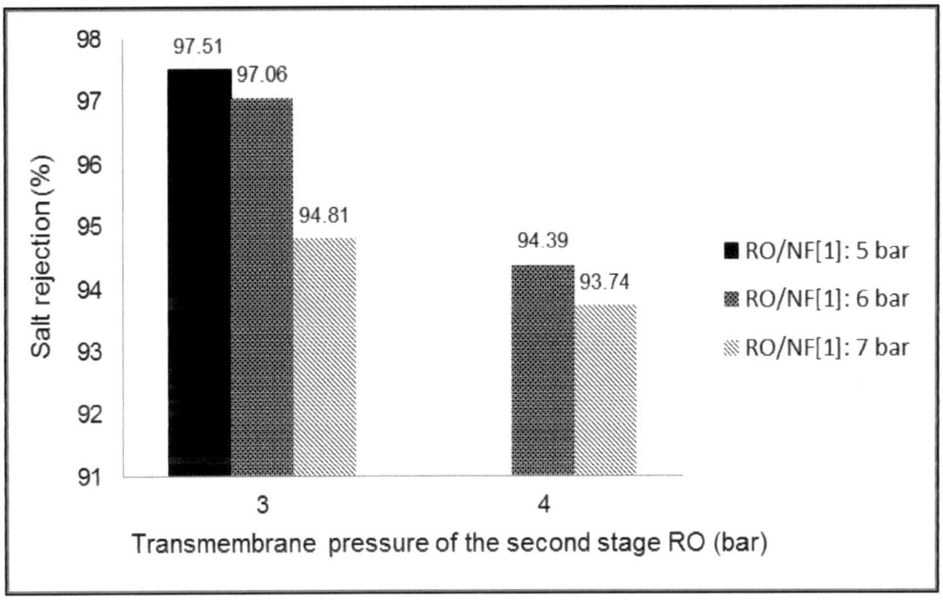

Figure 2. Effect of transmembrane pressure on the salt rejection.

Table 1. Effect of transmembrane pressure on the salt rejection.

Pressure (bar)		Salt rejection (%)			TDS of clean water (ppm)	Volumetric flow rate of clean water (liters/minute)
First stage RO	Second stage RO	First stage RO	Second stage RO	Final value		
5	3	65.50	92.78	97.51	72	1.00
6	3	64.01	91.83	97.06	85	1.55
6	4	63.67	84.57	94.39	162	2.40
7	3	60.90	86.73	94.81	150	1.95
7	4	60.55	84.12	93.74	181	2.60

Figure 2 and Table 1 show that the salt rejection will decrease when the transmembrane pressure increases. This is caused by the variation in pressure used which has passed its optimum pressure to separate the salt. The presence of the NF membrane in the hybrid RO/NF permeate staging makes the required optimum transmembrane pressure lower than the optimum transmembrane pressure in the configuration RO permeate staging. This is because the operating pressure of the NF membrane is lower than the RO membrane [6]. In the case of brackish water desalination using NF membranes, divalent and multivalent ions will be filtered, while monovalent ions will diffuse with the water molecules to the permeate flow [4]. When the transmembrane pressure continues to increase beyond its optimum condition, the salt flow in the feed stream will be forced to diffuse through the semipermeable membrane with a diffusion rate that exceeds the salt flow rate in the rejection steam so that more salt will diffuse towards the permeate stream [7]. When the salt passage increases, the salt rejection will decrease. Transmembrane pressure that exceeds the mechanical strength of the semipermeable membrane on the RO/NF membrane will also cause mechanical damage to the membrane, so that the semipermeable membrane cannot be a good separator between salt and water solution [8]. This will also increase the salt passage value and decrease the salt rejection.

To obtain the maximum salt rejection, the designed hybrid RO/NF filtration system is operated at the first and second stage RO/NF pressure of 5 and 3 bar, respectively. At that pressure, the value of salt rejection and TDS of permeate is 97.51% and 72 ppm, respectively. The salt rejection and TDS of the permeate close to the result at the optimum pressure at the first and second stage RO/NF pressure of 6 and 3 bar, respectively. At that pressure, the value of salt rejection and TDS of permeate is 97.06% and 85 ppm, respectively. The designed RO filtration systems with the configuration of hybrid RO/NF permeate staging has proven to have better performance in salt rejection.

3.2. Effect of transmembrane pressure on water recovery rate

Another parameter that shows the performance of the designed RO/NF filtration system is water recovery rate. The water recovery rate (in%) is the ratio between the permeate flow rate and the feed flow rate at each stage of filtration. The higher water recovery rate, the better the performance of the RO/NF filtration system. The effect of transmembrane pressure on the water recovery rate in this study is shown in Figure 3 and Table 2.

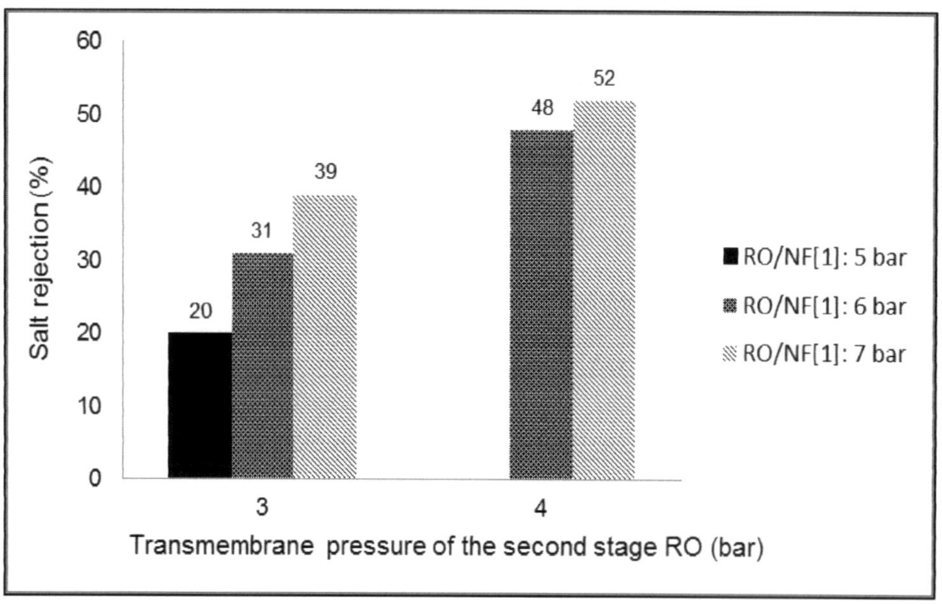

Figure 3. Effect of transmembrane pressure on the water recovery rate.

Table 2. Effect of transmembrane pressure on the water recovery.

Pressure (bar)		Water recovery rate (%)			TDS of clean water (ppm)	Volumetric flow rate of clean water (liters/minute)
First stage RO	Second stage RO	First stage RO	Second stage RO	Final value		
5	3	44.00	45.45	20.00	72	1.00
6	3	66.00	46.97	31.00	85	1.55
6	4	72.00	66.67	48.00	162	2.40
7	3	86.00	45.35	39.00	150	1.95
7	4	90.00	57.78	52.00	181	2.60

The results shown in Figure 3 and Table 2 show that the water recovery rate is directly proportional to the transmembrane pressure. The higher the transmembrane pressure, the more volume of water is forced to diffuse from the feed stream to the permeate stream. As explained in the previous section, this is because the increase in transmembrane pressure will increase the diffusion flux value in the RO/NF membrane so that the water recovery rate in the RO/NF filtration system also increases [7]. When the pressure increases, the permeate flux will also increase until the salt concentration in the permeate flow reaches its equilibrium value in the RO/NF membrane [9].

The presence of the NF membrane in first stage makes the water recovery rate of the first stage filtration higher than the second stage. Because the NF membrane contains a negatively charged hydrophilic group, the NF membrane has a higher water flux than the RO membrane [4]. In this study, the highest water recovery value was obtained at the first and second stage RO/NF transmembrane pressures of 7 and 4 bars respectively, which is 52%. However, in this pressure, the salt rejection is very low. The salt rejection that can be achieved is only 93.74% with the TDS of clean water is 181 ppm. In the pressure of the highest salt rejection, the water recovery rate that can be achieved is only 20%. At the first and second stage RO/NF transmembrane pressures of 6 and 3 bars, respectively, the decrease in the salt rejection produced is not too significant compared to its highest value (less than 0.5%). However, the water recovery rate that can be achieved has increased significantly. The salt rejection under these pressures only decreased by 0.45% from 97.51% to 97.06%. However, the water recovery rate has increased by 11% from 20% to 31%. Thus, the optimum pressure chosen in this study are at the first and second stage RO/NF transmembrane pressures of 6 and 3 bar, respectively. At the optimum operating conditions chosen, the value of water recovery that can be achieved is 31% with a flow rate of clean water products produced at 1.55 liters / minute.

Based on experimental study, the optimum condition for operating the machine is 12 hours per day with a time lag of 30 minutes after the machine is operated for 4 hours. It is aims to avoid pump damage due to overheating pump temperatures. If the machine is operated for 12 hours per day, clean water that can be produced is 1116 liters/day. It is enough to meet the clean water needs of 105 families in Tanjung Lame village. The average clean water requirements for drinking and cooking per family is 10 liters/day, so the average clean water needs for drinking and cooking for Tanjung Lame village is 1050 liters/day [10]. Thus, the volume of clean water produced can meet the needs of the people of Tanjung Lame with water quality that still meets WHO standards.

4. Conclusion

Design of hybrid RO/NF permeate staging configuration consisting of two RO membranes and one NF membrane on the first stage RO/NF and six RO membranes on the second stage RO/NF is proven to be able to produce clean water from saltwater well. The volume of clean water produced is able to meet the needs of the people of Tanjung Lame village, Pandeglang Regency, Banten. To obtain optimum results, the configuration is operated at the first and second stage RO/NF operating pressures of 6 and 3 bar, respectively. At that pressure, the salt rejection and TDS of permeate is 97.06% and 85 ppm, respectively. water recovery rate and permeate flow rate is 31% and 1.55 liters/minute, respectively.

The designed hybrid RO/NF permeate staging has proven to be able to produce clean water from saltwater wells in Tanjung Lame village at low pressure. The volume of clean water produced can meet the needs of the people of Tanjung Lame.

Acknowledgments
These researchers thank the PUTI Prosiding 2020 (The Grants of International Indexed Publication for the lecturers of Universitas Indonesia) No. PENG-1/UN2.RST/PPM.00.00/2020, which was organized by the Directorate of Research and Community Service (DRPM) – Universitas Indonesia. Furthermore, there is no competing interest or conflict of financial interests.

References

[1] Ahunbay M G, Tantekin-Ersolmaz S B and Krantz W B 2018 Energy optimization of a multistage reverse osmosis process for seawater desalination *Desalination* **429** 1–11
[2] Chong T H, Loo S L and Krantz W B 2015 Energy-efficient reverse osmosis desalination process *J. Memb. Sci.* **473** 177–88
[3] Rybar S, Boda R dan Bartels C 2010 Split partial second pass design for SWRO plants *Desalin. Water Treat.* **13** 186–94
[4] Ozaki H, Sharma K, Saktaywin W, Wang D and Yu Y 2000 Application of ultra low pressure reverse osmosis (ULPRO) membrane to water and wastewater *Water Sci. Technol.* **4** 123–135
[5] Sohn J, Valavala R, Han J, Her N and Yoon Y 2011 Pretreatment in reverse osmosis seawater desalination: A short review *Environ. Eng. Res.* **16(**4) 205–12
[6] Lee J J, Woo Y C and Kim H 2014 Effect of driving pressure and recovery rate on the performance of nanofiltration and reverse osmosis membranes for the treatment of the effluent from MBR *Desalin. Water Treat.* 3589-95
[7] FilmTec. Factors Affecting RO Membrane Performance 2007 [Online] Available on: https://www.watertreatmentguide.com/factors_affecting_membrane_performance.htm. [Accessed: 17-Apr-2020]
[8] Abid M F, Zablouk M A and Abid-Alameer A M 2012 Experimental study of dye removal from industrial wastewater by membrane technologies of reverse osmosis and nanofiltration *Iran. J. Environ. Health Sci. Eng.* **9** 17
[9] Amin M, Hafez A, Shaaban A and Hanafy M 2011 Removal of reactive dyes from dye-house effluent using nanofiltration removal of reactive dyes from dye-house effluent using nanofiltration membrane
[10] WHO 2011 How much water is needed in emergencies: Technical Notes on Drinking Water, Sanitation, and Hygiene in Emergencies [online] available on: https://www.who.int/water_sanitation_health/emergencies/WHO_TN_09_How_much_water_is_needed.pdf?ua=1

Degradation of 2-chlorophenol in dielectric barrier discharge reactor

Samantha Suryo[a], Febri Monica Diristiani, Badzlina Khairunizzahrah and Setijo Bismo[b]

Department of Chemical Engineering, Universitas Indonesia, Indonesia

[a)]samantha.suryo@ui.ac.id
[b)]Corresponding author: setijo.bismo@ui.ac.id

Abstract. Liquid waste containing 2-chlorophenol compound has high toxicity, carcinogenic, and poor degradation properties, leading to bioaccumulation in the environment. It also considered as a primary chemical feedstock in the manufacture and widely used in various industry. Conventional methods such as chlorination, adsorption, liquid-liquid extraction, steam distillation, photocatalytic are inefficient and expensive. Plasma technology with a non-thermal dielectric barrier discharge (DBD) plasma reactor can efficiently degrade chlorophenol compounds without producing side effects. Various active species produced in the reactor are electrons and radical compounds •OH, O_3, H_2O_2. This study aims to evaluate the degradation performance of DBD plasma reactors, namely the percentage of degradation, degradation concentrations of 2-chlorophenol, and chemical oxygen demand (COD). Variations were made with pH 4 and 10, while the fixed variables were waste flow rate 50 mL/min, air flow rate 2.5 L/min, and plasmatron voltage 19kV. It was found that degradation process of 2-chlorophenol with DBD reactor in pH 4 and 10 achieve 70.96% and 79.41%, respectively.

1. Introduction

Several industries such as pharmaceutical, coloring, agricultural and domestic release large amounts of toxic and dangerous pollutants into the environment. Most of these pollutants have xenobiotic and recalcitrant properties, which tend to be bioaccumulated and difficult to degrade in the environment. Increasing water pollution from industrial development and diminishing water resources lead to greater demand for clean water and various new regulations emerged [1]. One of the dangerous compounds contained in waste is chlorophenol. Chlorophenol is a serious problem if it is found in soil, sediments, and ground water. These pollutants and their derivatives are very mutagenic, carcinogenic, and immunogenic for humans and other organisms.

Many conventional waste treatment methods such as biological processes do not achieve high purification. The methods of chlorination, adsorption, liquid-liquid extraction, solid phase extraction, steam distillation, photocatalytic also have disadvantages, including slow reaction kinetics so that the degradation process is ineffective. Biological and classical physics-chemical processes are less effective in the processing of chlorophenol. The advanced oxidation process is an effective method for removing chlorophenol. In addition to the large number of AOPs studied, a very promising process is based on the use of electric discharge technology to remove hazardous compounds in wastewater, namely dielectric barrier discharge (DBD) [2].

Non-thermal plasma technology, namely the use of dielectric barrier discharge reactors, is a method of treating waste in environmentally friendly manner. The process is simple, effective, low energy consumption, does not require the use of other chemicals, and does not produce by-product. The DBD reactor has the potential to decompose chlorophenol in liquid waste due to the production of active species such as O_3, •OH, •O, and H_2O_2 in the reactor so that it can decompose organic pollutants. Because it has a high level of oxidation, many complex compounds and toxic organic pollutants can be degraded into non-toxic products and completely degraded to CO_2 and H_2O or to become biodegradable and less toxic intermediates. Compared to other plasma technologies, DBD reactors have several advantages compared to other plasma technologies, including enabling scale-up processes to operate with larger volumes of gas and producing stable micro discharges [3].

This research used DBD plasma reactor with spiral cylinder configuration to degrade chlorophenol compounds contained in liquid waste. Chlorophenol compound has been able to be degraded by as much as 50% at pH 4 with oxidation of hydrogen peroxide [4]. This research was conducted to obtain more effective results in degrading chlorophenol. The effectiveness of the performance of the plasma DBD reactor system can be seen from the degradation of chlorophenol compounds by observing the pH value of the compound, especially in acidic and basic conditions. The influence of pH affects the reactivity of the compound since chlorophenol is dissociating compound. In addition, the effect of the pH value will have an impact on the production of hydroxyl and ozone radicals [5].

2. Materials and Methods

2.1. Materials

The synthetic effluent was prepared with distilled water and 2-chlorophenol. 2-chlorophenol was purchased from Sigma-Aldrich with concentration of 99% suitable for analysis. Solution of 50 mg/L concentration were prepared by adding synthetic effluent to distilled water. Sulfuric acid 2 M was used to make acid solution and natrium hydroxide 2 M was used to make basic solution.

2.2. Experimental design

2.2.1. Reactor. Figure 1 displays the reactor which consists of a high voltage power supply, DBD plasma reactor, pump, compressor, and many other instruments as shown in Table 1.

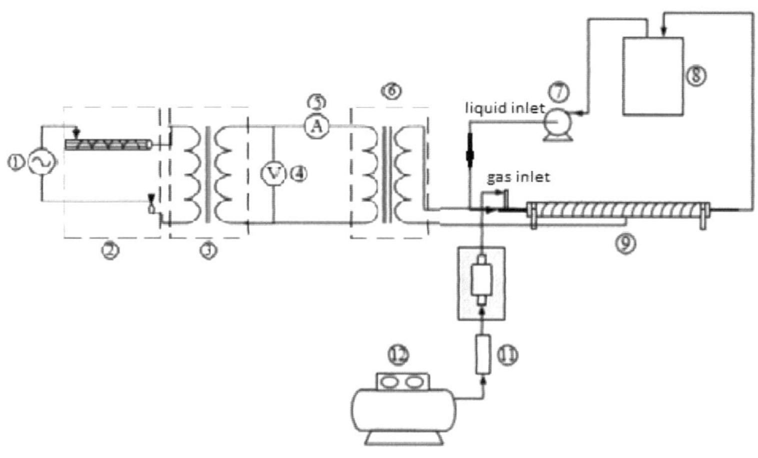

Figure 1. Overall DBD plasma reactor system.

The DBD reactor has a spiral cylindrical coaxial pipe configuration. The reactor has been designed to have several criteria, namely minimal power consumption and high productivity, easy operation and maintenance, and low operating and maintenance costs. The reactor was made of quartz glass as dielectric with a diameter of 22 mm and a length of 40 cm. The electrodes used are spiral-cylinder configuration made of stainless steel (SS304). There are two electrodes *which are* inner and outer electrodes. The inner electrode functions as a high-voltage electrode, while the outer electrode functions as a low-voltage electrode. The inner electrode is a cylinder with a diameter of 9 mm, placed in the centre of the quartz tube and produces a discharge gap of 6.5 mm. The outer electrodes are stainless steel wires wrapped around the outer surface of the quartz tube with the number of turns (N) 30. Information of Figure 1 can be seen in the following Table 1.

Table 1. Information of DBD reactor system.

No.	Instrument	No.	Instrument
1	Power source	**7**	Pump
2	*Slide Regulator*	**8**	Reservoir
3	Transformer	**9**	DBD Plasma Reactor
4	Voltmeter	**10**	Ozonator
5	Amperemeter	**11**	Flowmeter
6	NST (Neon Sign Transformer)	**12**	Compressor

2.2.2. Procedure. The research carried out hydroxyl radical quantification and ozone solubility test before degradation process. Hydroxyl radical quantification used iodometric titration and ozone solubility test used ozone reagent. Chlorophenol solution was filled inside the reservoir. Air was fed to the discharging zone and gas flow meters were used to control the flow rate. Then, the high voltage was applied to produce the plasma between solution and quartz plate. The 500 mL 2-chlorophenol was treated for 120 minutes. The pH value was adjusted to 4 and 10. Degraded samples of 2-chlorophenol were measured with amino-antipyrine reagent. The chemical oxygen demand (COD) was measured with a COD HACH reagent. All experiments were performed at room temperature. All experiments carried out with the fixed variables are air flow rate 2.5 L/min, volume of liquid 500 mL, voltage 19kV, while the independent variables are pH 4 and pH 10. The pH values were chosen at those value because a lot of waste is also present at that pH so there is no need to adjust the pH first before it is degraded.

2.3. Analysis methods

The concentration of 2-chlorophenol was analyzed by spectrophotometer in the wavelength of 513,2 nm dan 510,3 nm respectively. The efficiency of degraded 2-chlorophenol was determined from the following equation:

$$\%_{degradation} = \frac{C_{inlet} - C_{outlet}}{C_{inlet}} x\ 100\% \qquad \text{(Eq. 1)}$$

where C_{inlet} is the initial concentration of chlorophenol and C_{outlet} is the concentration of chlorophenol after 't' minutes.

The degradation reaction of 2-chlorophenol obeyed the apparent first order kinetic law. The equation could be expressed as follow:

$$\ln\left(\frac{C_o}{C_t}\right) = k_{cp} t \qquad \text{(Eq. 2)}$$

where C_o is the initial concentration of chlorophenol, C_t is the final concentration of chlorophenol after degradation, k_{cp} is rate constant, and t is degradation time.

3. Result and Discussion

3.1. Hydroxyl radical quantification

The pH condition is very influential in the formation of hydroxyl radicals because it can affect the kinetics and reaction path. At alkaline pH, the formation of hydroxyl radicals increases significantly,

IOP Conf. Series: Materials Science and Engineering 1011 (2021) 012049 doi:10.1088/1757-899X/1011/1/012049

while the pH of acid, ozone remains in the molecular form so that the formation of hydroxyl radicals at acidic pH is very much different. Figure 2 the amount of H_2O_2 produced at pH 10 is 0.1 mmol while pH 4 is only 0.00125 mmol in 120 minutes. This proves that the more alkaline pH will produce high hydroxyl radicals. This indicate the hydroxyl radical generation is highly related to pH. At high pH, the transfer of one electron directly occurs according to equations (3) and (4) [6].

$$O_3 + OH^- \rightarrow \bullet O_3^- + \bullet OH \quad \text{(Eq. 3)}$$

$$\bullet O_3^- + H_2O \rightarrow \bullet OH + O_2 + OH^- \quad \text{(Eq. 4)}$$

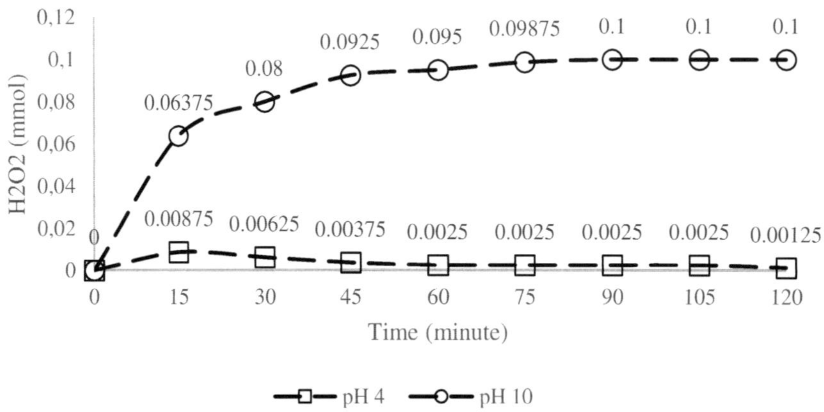

Figure 2. The effect of pH 4 and 10 on the amount of H_2O_2 formed.

3.2. Ozone solubility test

From Figure 3, the concentration of dissolved ozone is strongly influenced by pH. The highest dissolved ozone concentration is found at pH 4. At pH 10, ozone undergoes auto-decomposition to form OH radicals (7). This happens because pH 4 slows the rate of ozone decomposition so that molecular ozone is found. Conversely, the rate of ozone decomposition at pH 10 increases significantly. Alkaline pH There are many hydroxyl ions and hydroxide ions act as initiators in ozone decomposition. There are two possible reaction pathways, equation (5) shows ozone reaction with OH^- and equation (6) shows ozone reaction with HO_2^-. These both anions can lead ozone decomposition to hydroxyl radical production. The reaction can be seen in the equation below [8]:

$$O_3 + OH^- \rightarrow HO_2^- + O_2 \quad \text{(Eq. 5)}$$

$$O_3 + HO_2^- \rightarrow \bullet OH + \bullet O_2^- + O_2 \quad \text{(Eq. 6)}$$

$$O_3 + \bullet O_2^- \rightarrow \bullet O_3^- + O_2 \quad \text{(Eq. 7)}$$

$$\bullet O_3^- \leftrightarrow \bullet O^- + O_2 (pH \geq 8) \quad \text{(Eq. 8)}$$

$$\bullet O^- + H_2O \rightarrow \bullet OH + OH^- \ (pH \geq 8) \quad \text{(Eq. 9)}$$

$$\bullet OH + O_3 \rightarrow HO_4 \bullet \quad \text{(Eq. 10)}$$

$$HO_4 \bullet \rightarrow O_2 + HO_2 \quad \text{(Eq. 11)}$$

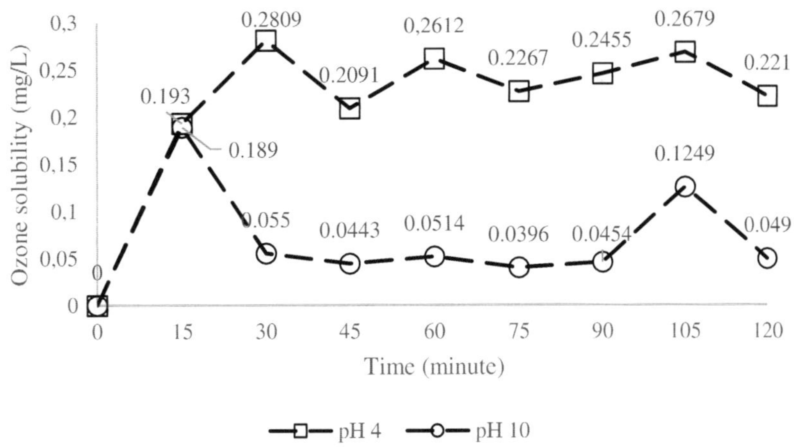

Figure 3. Ozone solubility at pH 4 and 10.

3.3. Degradation of 2-chlorophenol in DBD reactor

The degradation of the compound was carried out using a DBD reactor. This section will provide an analysis of the effect of pH with the flow rate of liquid 50 mL/min, air flow rate 2.5 L/min, and plasmatron voltage 19kV.

3.3.1. Effects of pH. From Figure 4, 2-chlorophenol degradation meets first-order kinetics as shown in equation (2). The highest decrease in concentration occurs in the 5th minute. This is because at the beginning of the reaction, the reactant concentration is still high so that many collisions occur between the active species and organic molecules. After the chlorophenol compound is degraded, direct collisions between active species and chlorophenol are reduced [7].

The degradation mechanism of chlorophenol compound is different under acidic and basic conditions. Ozonation at acidic pH is dominated by reactions with ozone molecules, whereas at alkaline pH it is dominated by OH radicals. Degradation at pH 10 is better than pH 4 as shown in Figure 4 for 2-chlorophenol compounds because of the higher degradation rate at alkaline condition. Degradation at pH 10 occurs faster due to dissociation of chlorophenol to chlorophenolate ions, which then tend to react with ozone and hydroxyl radical. When ozone is transferred from the discharge area to the liquid phase, it will react with 2-chlorophenol. A solution with a pH of 10 makes it easy to dissociate 2-chlorophenol into 2-chlorophenolate ions, which has a higher reaction rate with ozone so that at pH 10 ozone can be consumed at a higher rate. In other words, alkaline pH makes 2- chlorophenol ionized and supports ionic reactions. As a result, at pH 10, 2-chlorophenol experienced a degradation of 79.4% to 10.4 ppm, while pH 4 experienced a degradation of 70.96% to 14.2 ppm.

The difference on the degradation of 2-chlorophenol at pH 10 and pH 4 is possibly due to the reaction between chlorophenol and number of hydroxyl radical and ozone produced under certain pH. The degree of dissociation of chlorophenol to pH that affects the overall reaction. Dissociation increases with increasing pH and deprotonated compounds are easier to oxidize. At pH 10, 2-chlorophenol is in a more deprotonated state than at pH 4. The pKa value of a compound is related to the pH of the solution. When pKa < pH, more compounds are turned into deprotonated states. This is due to the value of pKa 2-chlorophenol which is 8.52. Therefore, 2-chlorophenol at pH 10 is more in a deprotonation state and causes 2-chlorophenol more easily attacked by OH radicals. Likewise, at pH 4, 2-chlorophenol is more abundant in the protonated state or in the molecular form. At pH 4, the reaction is dominated by ozone and there are only a few OH radicals so that the 2-chlorophenol degradation yield at pH 4 is smaller than pH 10.

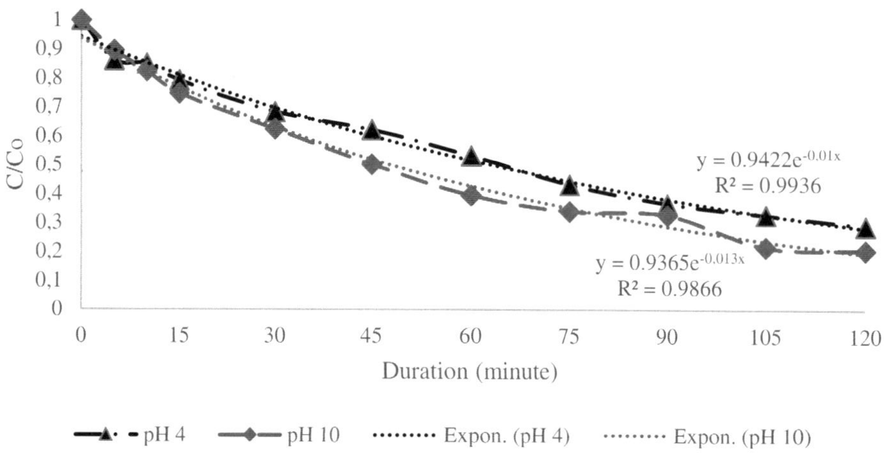

Figure 4. Degradation of 2-chlorophenol.

3.4. COD

After degradation, the next step is the analysis of COD of 2-chlorophenol. The compound was analysed at liquid flow rate of 50 mL/minute and air flow rate of 2.5 L/minute at reaction time of 0, 5, 60, and 120 minutes.

3.4.1. Analysis. COD analysis is carried out with the aim of assessing the total amount of oxygen needed to oxidize organic compounds. The percentage of mineralization of 2-chlorophenol at pH 4 is 38.82% while the pH of 10 removal occurs 44.04%. This is consistent with the results that have been obtained previously that at pH 10 the removal percentage is greater than pH 4. According to the Minister of Environment regulation No. 5 of 2014, waste to be discharged into the environment must meet the COD content standard, which is 100 mg / L (class I) and 300 mg / L (class II) as a standard quality requirement for industrial liquid waste so that it can be declared safe to be disposed. It was found that the processed waste has not met the Ministry of Environment regulation No. 5 of 2014.

4. Summary

Based on the result of the research, the study has shown that the degradation percentage of 2-chlorophenol is greater in pH 10 than pH 4. In alkaline condition, chlorophenols are easily to degrade than in acidic condition. At pH 10 the yield of 2-chlorophenol degradation reached 79.4%. At pH 4 the result of degradation of 2-chlorophenol only reached 70.96%. The difference in degradation yield of 2-chlorophenol is also strongly influenced by reaction mechanism according to the condition of the solution. The slightly degradation differences between pH 4 and 10 indicate that the DBD reactor can successfully degrade 2-chlorophenol compound under acidic and alkaline condition.

Acknowledgements

This work was supported by the PUTI Prosiding (Publikasi Terindeks Internasional Prosiding) No. PENG-1/UN2.RST/PPM.00.00/2020, Universitas Indonesia. The authors stated that there were no competing conflicts of financial interests.

References

[1] Van Aken P, Lambert N, Van den Broeck R, Degrève J and Dewil R 2019 Advances in ozonation and biodegradation processes to enhance chlorophenol abatement in multisubstrate wastewaters: a review. *Environ. Sci. Water Res. Technol.* **5**(3) 444-81

IOP Conf. Series: Materials Science and Engineering 1011 (2021) 012049 doi:10.1088/1757-899X/1011/1/012049

[2] Marković M D, Dojčinović B P, Obradović B M, Nešić J, Natić M M, Tosti T B, *et al.* 2015 Degradation and detoxification of the 4-chlorophenol by non-thermal plasma-influence of homogeneous catalysts *Sep. Purif. Technol.* **154** 246-54
[3] Nawaz M I, Yi C, Asilevi P J, Geng T, Aleem M, Zafar A M, *et al.* 2019 A study of the Performance of dielectric barrier discharge under different conditions for nitrobenzene degradation *Water* **11** 4
[4] De A, Chaudhuri B and Bhattacharjee S 1999 A kinetic study of the oxidation of phenol, o-chlorophenol, and catechol by hydrogen peroxide between 298K and 333K *J. Chem. Technol. Biotechnol.* **74** 162-8
[5] Zhang H, Liu Y, Cheng X, Zhang A, Li X, Liu J, *et al.* 2018 Degradation of phenol in water using a novel gas-liquid two-phase dielectric barrier discharge plasma reactor *Water Air Soil Pollut.* **229** 10
[6] Wang J L and Xu L J 2012 Advanced oxidation processes for wastewater treatment: formation of hydroxyl radical and application *Crit. Rev. Environ. Sci. Technol.* **42**(3) 251-325
[7] Poznyak T, Tapia R, Vivero J and Chairez I 2006 Effect of pH to the decomposition of aqueous phenols mixture by ozone *Sociedad Química de México* 23-35
[8] Von Gunten U 2003 Ozonation of drinking water: Part I. Oxidation kinetics and product formation *Water Res.* **37**(7) 1443-67

IOP Conf. Series: Materials Science and Engineering 1011 (2021) 012050 doi:10.1088/1757-899X/1011/1/012050

The use of ^{1}H-NMR spectroscopy and chemometrics of pattern recognition for authentication of *Curcuma xanthorrhiza* adulterated with *Zingiber montanum*

A Windarsih[1*], K Nisa[1], A W Indrianingsih[1], C Darsih[1], S Handayani[1], M P Wulanjati[1], V T Rosyida[1], T Wijayanti[2] and A Rohman[3]

[1]Research Division for Natural Product Technology (BPTBA), Indonesian Institute of Sciences (LIPI), Yogyakarta, 55861, Indonesia
[2]The National Agency of Drug and Food Control (BPOM) Republic of Indonesia, Jakarta, Indonesia
[3]Department of Pharmaceutical Chemistry, Faculty of Pharmacy, Universitas Gadjah Mada, Yogyakarta, 55281, Indonesia

*Email: anjarwindarsih2@gmail.com

Abstract. *Curcuma xanthorrhiza* is widely used in food and traditional medicine products. Due to its high demand, it is potential to be substituted or mixed with other species having similar appearance, therefore, rapid and reliable analytical method is highly required. The objective of this study was to develop ^{1}H-NMR spectroscopy and chemometrics of pattern recognition as a metabolite fingerprinting technique for authentication of *C. xanthorrhiza* from *Zingiber montanum*. The powdered rhizomes were extracted using combination of methanol-D4 and phosphate buffer pH 6.0 in deuterium oxide (1:1 v/v) containing trimethylsilyl propionic acid (TSP) for chemical shift calibration. The variables extracted from ^{1}H-NMR spectra were used for creating chemometrics models. Chemometrics of partial least square-discriminant analysis (PLS-DA) using 7 principal components (PCs) successfully classified between authentic and adulterated samples of *C. xanthorrhiza* with high value of R2X (0.988), R2Y (0.998), and Q2 (0.993). Moreover, chemometrics of orthogonal projection to latent structures-discriminant analysis (OPLS-DA) using 2 PCs and 4 orthogonal components perfectly discriminated authentic and adulterated samples of *C. xanthorrhiza*. The model showed high R2X (0.965), R2Y (0.976) as well as Q2 (0.946) values. Validation using permutation test confirmed the validity both PLS-DA and OPLS-DA models. It suggested that combination of ^{1}H-NMR and chemometrics method is promising for authentication of medicinal plants.

1. Introduction

C. xanthorrhiza is an herbaceous perennial plant belongs to Zingiberaceae family and it is widely cultivated in several countries in Asia especially in South East Asia. The rhizome is known to have several biological activities such as antioxidant, anti-inflammatory, hepatoprotective, cardioprotective, and antibacterial activity. Some research has also explored regarding its anticancer activity [1-2]. The rhizome has been widely applied in several products including food, cosmetics as well as traditional medicine products [3]. Due to its high demand, the rhizome of *C. xanthorrhiza* especially in powder form is often adulterated with other rhizome which is widely available and having similar physical properties such as *Z. montanum* by unethical player to meet the demand [4]. Adulteration is a serious problem because it is related to the quality, efficacy, and safety of the products. In a powder form, it is obviously difficult to differentiate between authentic and adulterated powder of *C. xanthorrhiza* [5]. The metabolite profile of genus Curcuma (*C. manga, C. zedoaria,* and *C. xanthorrhiza*) and Zingiber (*Z. officinale, Z. montanum,* and *Z. zerumbet*) from three different locations has been determined using ^{1}H-NMR spectroscopy [6]. However, it just measured the pure or individual samples, not in adulterated samples composed of the mixtures between different species. Therefore, the development of analytical method capable of detecting the presence of adulterant in binary mixtures with *C. xanthorrhiza* powdered rhizome is highly required for quality control.

Several analytical methods have been used for authentication of *C. xanthorrhiza* rhizome. Some of the methods focus on measuring the curcuminoid contents such as spectroscopy ultraviolet-visible [7], thin layer chromatography [6-7], ultra-performance liquid chromatography [10], capillary electrophoresis [11], and liquid chromatography tandem mass spectrometry [12]. However, these methods have disadvantages such as require long time, require high amount of solvent, and difficult in samples preparation. ^{1}H-NMR spectroscopy is emerging as an analytical method in metabolite fingerprinting for authentication of medicinal plants [13]. It offers fast analysis, ease in sample preparation, highly robust and highly reproducible. Combined with chemometrics of multivariate analysis which can manage the complex data generated from ^{1}H-NMR spectra, it becomes a powerful analytical technique for authentication [14]. Chemometrics of pattern recognition is aimed for classification and differentiation of samples. It has been widely used to analyze data generated from several instrumental methods including spectroscopy and chromatography for food authentication [14-15]. Therefore, the use of chemometrics pattern recognition is promising and potential to be applied for authentication of medicinal plants.

^{1}H-NMR spectroscopy and pattern recognition chemometrics of principal component analysis (PCA), partial least square-discriminant analysis (PLS-DA), and orthogonal projection to latent structures-discriminant analysis (OPLS-DA) have been successfully used for detection of *C. heyneana* in *C. longa* powdered rhizome [17]. Adulteration of high quality Saffron has also been successfully detected using combination of ^{1}H-NMR spectroscopy and chemometrics of OPLS-DA and bidirectional OPLS-DA [18]. To the best of our knowledge, there is no report regarding the use of ^{1}H-NMR spectroscopy method in combination with chemometrics of pattern recognition for authentication of *C. xanthorrhiza* adulterated with *Z. montanum*. This research was aimed to develop ^{1}H-NMR spectroscopy combined with chemometrics of pattern recognition for authentication of *C. xanthorrhiza*.

2. Material and Methods

2.1. Materials

Rhizomes of *C. xanthorrhiza* and *Z. montanum* were obtained from Yogyakarta, Indonesia. The solvents (methanol-D4, deuterium oxide) and chemicals (KH_2PO_4, Trimethylsilyl propionic acid/TSP) were purchased from Merck.

2.2. Sample preparation

Sample preparation was carried out according to Kim *et al.* [13] with slight modifications. Rhizomes were cleaned and chopped into small sizes approximately of 5 cm in length. Subsequently, rhizomes were dried and ground into powder. The series of adulterated samples of *C. xanthorrhiza* in binary mixtures with *Z. montanum* were prepared using several adulterant concentrations ranged from 0-100 % w/w. Amount of 25.0 mg samples were accurately weighed using analytical balance and placed into 2 mL microtube. Samples were extracted using NMR solvents consist of methanol-D4 and phosphate buffer pH 6.0 in deuterium oxide using ratio of 1:1 v/v. Methanol-D4 was chosen because it has an ability to extract wide range of metabolites. Whereas phosphate buffer was used to maintain the pH of metabolites in order to avoid significant shift in their chemical shift. Trimethylsilyl propionic acid (TSP) was added for chemical shift calibration. The mixtures were vortexed, ultrasonicated for 15 min and centrifuged for 10 min at 13500 rpm. Supernatant was taken and transferred to NMR tube for NMR measurement.

2.3. ^{1}H-NMR spectra measurement

The ^{1}H-NMR spectra acquisition was carried out using JEOL ECZ-R 500 MHz NMR spectrometer. The samples were measured using field strength of 11.74 T with X_offset of 5.0 ppm and 5s of relaxation delay. Spectra were recorded in the range of 0-12 ppm consisting of 128 scans. Each sample was measured in three replicates.

2.4. Data analysis

Data pre-processing was performed prior to chemometrics analysis. ^{1}H-NMR spectra were phase-corrected, then baseline correction was carried out using polynomial fit mode. To obtain data variables for multivariate analysis, binning of the spectra was performed in the chemical shift of 0.50-10.00 ppm for every 0.04 ppm. The binned data were subjected for chemometrics analysis of partial least square-discriminant analysis (PLS-DA) and orthogonal projections to latent structures-discriminant analysis (OPLS-DA).

3. Results and Discussion

The use of NMR spectroscopy for metabolite fingerprinting and metabolite profiling is getting increased in the past few years. ^{1}H-NMR spectroscopy could be used as fingerprint method to differentiate samples because the spectra could be unique for each sample meaning that different sample will have different ^{1}H-NMR spectra. The main factor contributed to the spectra differences is the difference in metabolite contents in every sample. ^{1}H-NMR spectra of *C. xanthorrhiza* differs from the spectra of *Z. montanum* indicating they have different content in metabolite compounds. The secondary metabolites of *C. xanthorrhiza* are xanthorhizol, curcumin, demethoxycurcumin, phelandren, sucrose, borneol, sesquiterpenes, and flavonoids [3] while the secondary metabolites of *Z. montanum* are cassumunarin A, B, C, terpinen-4-ol, alpha and beta pinene, sabinene, myrcene, terpinene, and limonene [19]. The curcumin and demethoxycurcumin are the active compounds in *C. xanthorrhiza* which appeared at the chemical shift of 7.28, 7.17, 6.77, and 3.66 ppm for curcumin and 6.93, 5.88, and 3.66 for demethoxycurcumin [17]. Generally, ^{1}H-NMR spectra of plant metabolites are divided into three categories namely amino acid and organic acid region (0.10-3.00 ppm), carbohydrate region (3.00-5.00 ppm), and aromatic region (6.00-8.00 ppm) [13]. The differences between *C. xanthorrhiza* and *Z. montanum* observed mainly in the region of amino acid and organic acid (0.10-3.00 ppm) which the signals in *C. xanthorrhiza* are more complex with high intensities (Figure 1). It is presumed that the concentration of amino acid and organic acid in *Z. montanum* is low. A sharp signal appeared in the chemical shift of 5.45 ppm which is presumed to be the signal from sucrose [17] could be used as marker signal for *C. xanthorrhiza* because it is specific for *C. xanthorrhiza.* Eventhough the spectra of pure *C. xanthorrhiza* and pure *Z. montanum* could be clearly differentiated, however, in adulterated samples of *C. xanthorrhiza*, the spectra of authentic and adulterated samples were similar each other especially in the adulterant (*Z. montanum*) concentration of 10-40%. As a consequence, it is difficult to state whether the sample is authentic or adulterated if there is no information regarding the samples. Therefore, the use of advanced statistical analysis such as chemometrics of pattern recognition PLS-DA and OPLS-DA is highly required.

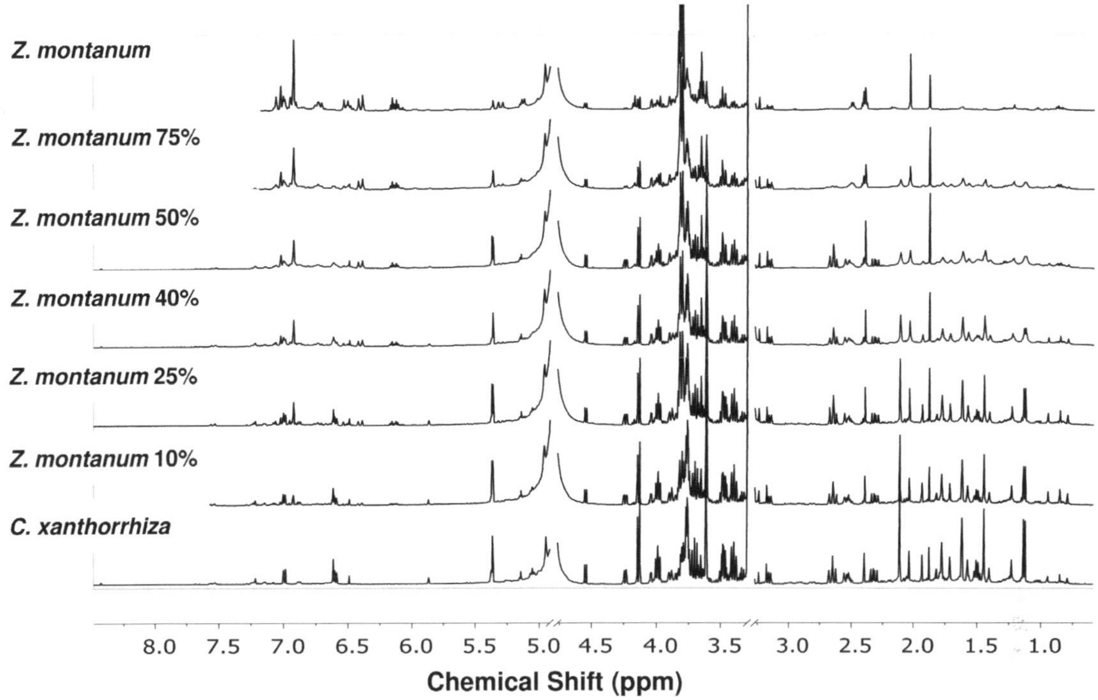

Figure 1. ^{1}H-NMR spectra of pure and adulterated *C. xanthorrhiza* with *Z. montanum* using adulterant concentration from 10-75% displayed in the chemical shift of 0.00-9.00 ppm.

Chemometrics of pattern recognition is divided into unsupervised and supervised pattern recognition. Supervised pattern recognition such as PLS-DA and OPLS-DA has the potential for classification because of their high order in algorithm [20]. PLS-DA using 7 principal components has been successfully used to differentiate between authentic and adulterated powder of *C. xanthorrhiza*. with *Z. montanum*. All samples containing *Z. montanum* were correctly classified as adulterated samples and clearly separated from authentic sample of *C. xanthorrhiza* as can be seen in PLS-DA score plot (Figure 2A). The PLS-DA model showed high value of R2X (0.988), R2Y (0.9980) and Q2 (0.993). The high values of R2X and R2Y (close to 1) indicated good of fitness of the model, whereas the high Q2 value (> 0.50) showed the good predictivity of the model. Another supervised pattern recognition method namely OPLS-DA using 2 principal components and 4 orthogonal component variables were also successfully applied for classification between authentic and adulterated samples of *C. xanthorrhiza*. The model has good of fitness indicated by the obtained value of R2X (0.965) and R2Y (0.976) and also good of predictivity demonstrated by its high Q2 (0.946). OPLS-DA showed better classification compared to PLS-DA model as shown in figure 2B. This might be caused by the different algorithm used in OPLS-DA. The OPLS-DA model used latent variables both in X and Y matrix, therefore resulting better separation than in PLS-DA which used latent variables only in Y matrix.

IOP Conf. Series: Materials Science and Engineering 1011 (2021) 012050 doi:10.1088/1757-899X/1011/1/012050

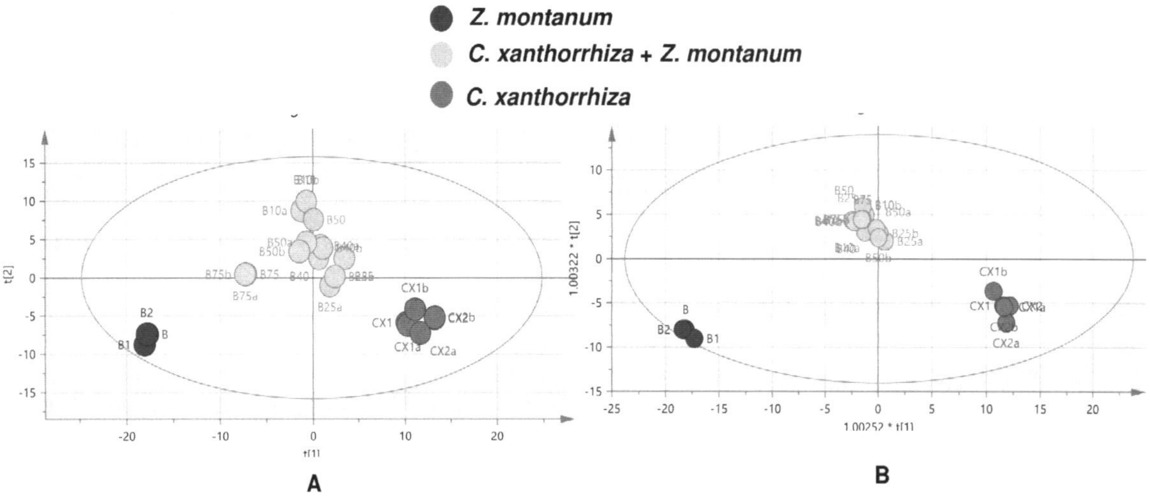

Figure 2. PLS-DA (A) and OPLS-DA (B) model of *C. xanthorrhiza* and adulterated *C. xanthorrhiza* with *Z. montanum.*

The PLS-DA and OPLS-DA models were validated using permutation test to confirm the validity of the developed models. Validation is required to avoid model overpredicting which tend to bias result. Permutation test is the common validation technique used for PLS-DA and OPLS-DA models which compared the value between original and permutated of R2 and Q2. Good permutation test is obtained when the original value of R2 and Q2 (in the right side) have the highest value compare to the all permutated R2 and Q2 (in the left side) as shown in Figure 3. Moreover, the intersection value of R2 and Q2 must between zero or less than zero. Result showed that the original value of R2 and Q2 has the highest value than the permutated ones both in PLS-DA and OPLS-DA models. Moreover, the obtained intersection of R2 and Q2 in PLS-DA model was (0.0, 0.24) and (0.0, -1.17) whereas the intersection of R2 and Q2 in OPLS-DA model was (0.0, 0.196) and (0.0, -0.871), respectively indicating the good validation model.

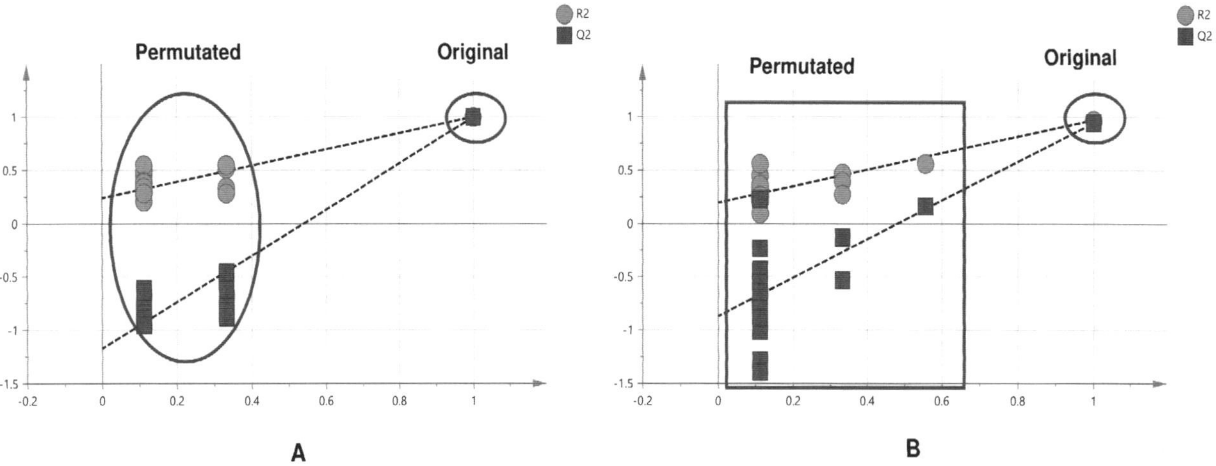

Figure 3. Permutation test of PLS-DA (A) and OPLS-DA (B) models of *C. xanthorrhiza* and adulterated *C. xanthorrhiza* with *Z. montanum.*

4. Conclusion

^{1}H-NMR spectroscopy provided fingerprint spectra which is very useful to differentiate between authentic and adulterated samples of *C. xanthorrhiza* powdered rhizome. Chemometrics of pattern recognition namely PLS-DA and OPLS-DA was successfully classified between authentic and adulterated samples of *C. xanthorrhiza* with *Z. montanum*. It can be concluded that ^{1}H-NMR spectroscopy combined with chemometrics of pattern recognition is potential and promising to be used for authentication of medicinal plants.

Acknowledgements

The authors thank to Lembaga Pengelola Dana Pendidikan (LPDP) for financial support through Prioritas Riset Nasional (PRN) 2020 organized by Ministry of Research and Technology/The National Agency of Research and Innovation.

References

[1] Jantan I, Saputri F C, Naeem Qaisar M and Buang F 2012 Correlation between chemical composition of *Curcuma domestica* and *Curcuma xanthorrhiza* and their antioxidant effect on human low-density lipoprotein oxidation *Evidence-Based Complement. Altern. Med.* **2012** 1–10.

[2] Ramdani E D, Marlupi U D, Sinambela J M and Tjandrawinata R R 2016 A New method of xanthorrhizol isolation from the rhizome extract of *Curcuma xanthorrhiza Sch. Acad. J. Biosci.* **4** 732-737.

[3] Wijayanti T, Riyanto S, Lukitaningsih E and Rohman A 2019 Application of ^{1}H-NMR specta and multivariate analysis for the authentication of *Curcuma xanthorrhiza* from *Zingiber cassumunar Int. J. Appl. Pharm.* **11** 258–263.

[4] Rohman A, Windarsih A, Hossain M A M, Johan M R, Ali M E and Fadzilah N A 2019 Application of near- and mid-infrared spectroscopy combined with chemometrics for discrimination and authentication of herbal products: A review *J. Appl. Pharm. Sci.* **9** 137–147.

[5] Rafi M, Wulansari L, Heryanto R, Darusman L K, Lim L W and Takeuchi T 2015 Curcuminoid's content and fingerprint analysis for authentication and discrimination of *Curcuma xanthorrhiza* from *Curcuma longa* by high-performance liquid chromatography-diode array detector *Food Anal. Methods* **8** 2185–2193.

[6] Indiarto E N 2018 Profil metabolit ^{1}H NMR rimpang genus Curcuma dan Zingiber dari tiga lokasi berbeda *Semin. Nas. Biol. Trop. 2018* Universitas Gadjah Mada 28 Juli 2018.

[7] Kadam P, Yadav K, Kadam P V, Yadav K N, Bhingare C L and Patil M J 2018 Standardization and quantification of curcumin from *Curcuma longa* extract using UV visible spectroscopy and HPLC ~ *1913 ~ J. Pharmacogn. Phytochem.* **7** 1913–1918.

[8] Paramasivam M, Poi R, Banerjee H and Bandyopadhyay A 2009 High-performance thin layer chromatographic method for quantitative determination of curcuminoids in *Curcuma longa* germplasm *Food Chem.* **113** 640–644.

[9] Awin T, Mediani A, Shaari K, Faudzi S M M, Sukari M A H, Lajis N H and Abas F 2016 Phytochemical profiles and biological activities of *Curcuma* species subjected to different drying methods and solvent systems: NMR-based metabolomics approach *Ind. Crops Prod.* **94** 342–352.

[10] Cheng J, Weijun K, Yun L, Jiabo W, Haitao W, Qingmiao L and Xiaohe X 2010 Development and validation of UPLC method for quality control of *Curcuma longa* Linn.: Fast simultaneous quantitation of three curcuminoids *J. Pharm. Biomed. Anal.* **53** 43–49

[11] Lechtenberg M, Quandt B and Nahrstedt A 2004 Quantitative determination of curcuminoids in Curcuma rhizomes and rapid differentiation of *Curcuma domestica* Val. and *Curcuma xanthorrhiza* Roxb. by capillary electrophoresis *Phytochem. Anal.* **15** 152–158

[12] Shen Y, Han C, Chen X, Hou X and Long Z 2013 Simultaneous determination of three Curcuminoids in *Curcuma wenyujin* by liquid chromatography-tandem mass spectrometry combined with pressurized liquid extraction *J. Pharm. Biomed. Anal.* **81**–**82** 146–150.
[13] Kim H K, Choi Y H and Verpoorte R 2010 NMR-based metabolomic analysis of plants *Nat. Protoc.* **5** 536–549.
[14] Worley B and Powers R 2013 Multivariate analysis in metabolomics *Curr. Metabolomics* **1** 92–107.
[15] Erwanto Y, Muttaqien A T, Sugiyono, Sismindari and Rohman A 2016 Use of Fourier transform infrared (FTIR) Spectroscopy and chemometrics for analysis of lard adulteration in "rambak" crackers *Int. J. Food Prop.* **19** 2718–2725.
[16] Fadzillah N A, Rohman A, Salleh R A, Amin I, Shuhaimi M, Farahwahida M Y, Rashidi O, Aizat J M and Khatib A 2017 Authentication of butter from lard adulteration using high-resolution of nuclear magnetic resonance spectroscopy and high-performance liquid chromatography *Int. J. Food Prop.* **20** 2147–2156.
[17] Windarsih A, Rohman A and Swasono R T 2019 Application of 1H-NMR based metabolite fingerprinting and chemometrics for authentication of *Curcuma longa* adulterated with *C. heyneana J. Appl. Res. Med. Aromat. Plants* **13** 1-8.
[18] Petrakis E A, Cagliani L R, Polissiou M G and Consonni R 2015 Evaluation of saffron (*Crocus sativus* L.) adulteration with plant adulterants by ^{1}H NMR metabolite fingerprinting *Food Chem.* **173** 890–896.
[19] Singh C and Chanu S 2015 Ethnobotany, Phytochemistry and pharmacology of *Zingiber cassumunar* Roxb. (Zingiberaceae) ~ *1* ~ *J. Pharmacogn. Phytochem.* **4** 01-06.
[20] Zettl D, Bandoniene D, Meisel T, Wegscheider W and Rantitsch G 2017 Chemometric techniques to protect the traditional Austrian pumpkin seed oil *Eur. J. Lipid Sci. Technol.* **1600468** 1–11.

IOP Conf. Series: Materials Science and Engineering 1011 (2021) 012051 doi:10.1088/1757-899X/1011/1/012051

The antioxidant and antibacterial study of *Canarium indicum* L. latex extract

A W Indrianingsih*, V T Rosyida, W Apriyana, C Darsih, M Angwar, A Windarsih and Batrisya

Research Division for Natural Product Technology, Indonesian Institute of Sciences
Jl. Wonosari Km 31.5, Gading, Playen, Gunungkidul, Yogyakarta 55861, Indonesia

*Corresponding author: anastasia.wheni.i@gmail.com

Abstract. Recently, plant extracts are natural additives that are in great demand. Many biological activities of plant extracts in the fields of food and health, make research on plant extracts quite rapid. *Canarium indicum* L. is one of the most famous fruits in Indonesia. However, usually only the seed of *C. indicum* that was studied such as anti-inflammatory, antirheumatic, and hepatoprotective. In this paper, the antimicrobial and antioxidant activities of *C. indicum* latex were studied. The *C. indicum* extract were prepared by maceration method using ethanol for 48 hours. After the evaporation, the crude extracts were evaluated its biological activity. The antibacterial activity was performed using dilution method and the antioxidant activity was using DPPH assay. The scavenging activity of *C. indicum* latex extracts values in the range of 10.92 – 58.36% with the extract concentration of 100 to 800 ppm, respectively. The antibacterial activity of *C. indicum* latex extract was conducted against Gram-positive bacteria (*S. aureus*) and Gram-negative bacteria (*E. coli*). The inhibition zone *C. indicum* was 6.39±0.06 mm against *S. aureus* and 2.38±0.62 mm against *E. coli* at extract concentration of 10000 ppm. The results obtained indicate that *C. indicum* latex extract is favourable to be applied in the fields of medicine and health.

1. Introduction

Antioxidants are substances or materials that could significantly inhibit oxidative process. Antioxidant can protect cells so that the harmful effect of free radicals does not occur. Several studies reported the bad effect of free radicals to living things include aging, stroke, diabetes, cancer and neurodegeneration [1-3]. Food and pharmaceutical industries usually used synthetic antioxidant, to prolong the product shelf life, especially to inhibit the lipid oxidation. Commonly used synthetic antioxidant is butylated hydroxytoluene, t-butylhydroquinone and butylated hydroxyanisole [4]. Butylated hidroxytoluene was effective in fish products, animal fats, packaging material and low-fat food [5]. However, negative side effects of synthetic antioxidant such as its toxicological effect to various species was become a limitation of the using of synthetic antioxidant [6].

Secondary metabolites produced by plants, herbs, vegetable are usually containing chemical substance such as polyphenols, alkaloids, flavonoids, and terpenoids [7]. Secondary metabolites in plants are protecting their lives from UV radiation, dryness or microbial infection [8]. Recently, interest in secondary metabolites from plants focused on tannin, flavonoids and anthocyanin, that be proven to have an important role in prevention of several diseases [9]. These compounds were having high antioxidant properties [10]. Moreover, polyphenol also have good antibacterial activity, so that could be a candidate for antibiotics [11].

Bacteria as a single cell organism, have an important role in human life. Some of it have been used in various industries such as in pharmaceuticals and food. However, bacteria could be also dangerous to human body, since they can generate many disease and infections [12]. Antibiotics or antimicrobial agents have been seriously needed and saved peoples from bacterial infections [13]. However, several microorganisms have an ability to resist the antibiotic [14], so that new antimicrobial agents with medicinal properties from nature have been studied extensively [15, 16]. Meanwhile, some infection and food spoilage by microbes were becoming problem in public health [17-19]. Several bacteria such as *Bacillus subtilis*, *Escherichia coli* and *Staphylococcus aureus* are identified as the cause of food spoilage [20]. Thus, search on antimicrobial compounds resistant bacteria is an urgent [21].
The genus of Canarium has approximately 100 species worldwide [22]. *Canarium indicum* L. or Kenari in Indonesian language, is one of the species of Canarium. It is an indigenous plant in Indonesia, especially in the eastern region such as Maluku islands [23]. The distribution of *C. indicum* plants spread in various countries, including Indonesia, the Philippines, Papua New Guinea, and Malaysia [24]. The seed of *C. indicum* is usually used as a snack food. The oil from *C. indicum* seed also comparable to other plants such as cashew, hazelnuts, and macadamia [25]. Moreover, the other parts of *C. indicum* such as leaves, stems and exocarp of *C. indicum* seed also had been studies for treating disease and natural food colouring [24]. However, the bioactivity studies about latex of *C. indicum* is still very rare. Because of that, the aim of this study is to determine the antioxidant and antibacterial activity of *C. indicum* latex extract. The results of this study hopefully can be a knowledge for the community, especially for those who have access to *C. indicum* trees.

2. Materials and methods

2.1. Sample collection
The sample of *C. indicum* latex was obtained from Papua Province, Indonesia. The sample was dried with indirect exposure of the sun. For the next extraction process, the dried latex was powdered.

2.2. Preparation of C. indicum latex extract
The powder of *C. indicum* latex was macerated using ethanol 95% for two days at room temperature (1:8 w/v). It was then filtered and evaporated using vacuum rotary evaporator to obtain ethanolic extract of *C. indicum* latex.

2.3. FTIR spectroscopy analysis
The sample was mixing with KBR to make a KBr pellet. The FTIR spectrometer used was Shimadzu 8201 PC (Japan). The spectra were achieved in the range frequency of 4000 to 500 cm^{-1}.

2.4. DPPH radical scavenging activity assay
The radical scavenging activity of *C. indicum* latex extract was performed using DPPH assay according to Indrianingsih *et al* (2020) with slight modification [26]. Several concentration of *C. indicum* latex extract were prepared using methanol as solvent. It was then reacted with DPPH solution (1.01 mM) for 30 minutes at room temperature. The room was in the dark condition to avoid the effect of light on DPPH since it was sensitive to light. The absorbance of the final samples was measured using Elisa Reader at 517 nm. The DPPH radical scavenging activity was calculated by the equation as follows:

$$\text{DPPH radical scavenging activity (\%)} = \left(\frac{A_o - A_1}{A_o}\, x\, 100\right) \quad (1)$$

where A_o is absorbance of the control and A_1 is absorbance of the sample. The assays were carried out in triplicates. Ascorbic acid was used as the standard control for this assay.

2.5. Antibacterial activity assay
The ability of *C. indicum* latex extract to inhibit the bacteria was performed using agar well diffusion method according to literature with some modification [27]. The extract was dissolved in DMSO at

several concentration (500; 1000; 10,000 ppm). One hundred microliter of bacteria suspension was inoculated in nutrient agar plates. The extract solution was then added into the wells in the solid media. After incubation at at 37°C for 24 hours, the diameter of inhibition zone was measured. In this study, ampicillin was used as positive control while DMSO was used as negative control.

3. Results and Discussion

In this study, *C. indicum* latex extract was obtained by maceration process using ethanol. The latex of *C. indicum* was rarely studied, however, the nut of *C. indicum* were extensively studied. Several literatures reported the composition of *C. indicum* nut and kernel. Carbohydrate, vitamin E, and fatty acids (palmitic acid, stearic acid, oleic acid and linoleic acid) were obtained from kernel of *C. indicum* [28]. The yield of *C. indicum* latex extract showed in Table 1 as of 44.01 % (w/w).

Table 1. The yield of *C. indicum* latex extract.

Sample	Dry weight (g)	Extract weight (g)	Yield (%w/w)
C. indicum	61.69	27.16	44.01

The capability of *C. indicum* latex extract to scavenge free radical was evaluated using DPPH radical scavenging assay. This method is the commonly used to perform the antioxidant activity capability of plant extracts. The antioxidant of *C. indicum* latex extract can be seen in Figure 1.

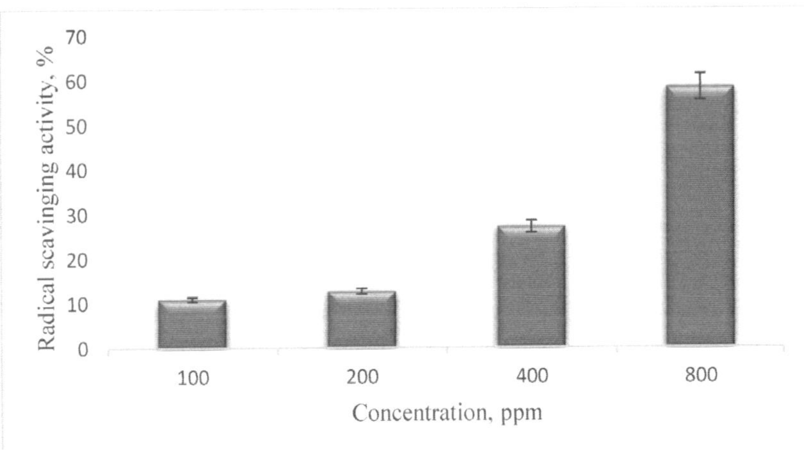

Figure 1. The antioxidant activity of *C. indicum* latex extract using DPPH assay.

Based on the DPPH assay, the higher the concentration of *C. indicum* latex extract, the higher the radical scavenging activity. The *C. indicum* latex extract has high activity as of 10.92% at 100 ppm, 12.74 % at 200 ppm, 27.12% at 400 ppm and 58.36 % at 800 ppm. The IC 50 of *C. indicum* latex extract is 697.45 ppm, meanwhile the antioxidant activity of ascorbic acid as standard as of 76.7 at 25 ppm. A study by Djarkasi *et al.* (2017) reported that *C. indicum* nut contains phenolic and flavonoid compounds so that it possessed a high antioxidant activity. The quantification of phenolic substances in *C. indicum* L. methanolic extract as of 7.4 to 8.8 mg GAE/g. Meanwhile, the vitamin E content of *C. indicum* nut around 439.36 ppm [29]. It had positive correlation between the antioxidant activity with phenolic and vitamin E contents. It also presented that tannins extracted from the leaves of *Canarium album* L showed potent antioxidant activity with IC50 of 56.8 μg/ml using DPPH assay [30]. A research also isolated the pure compounds from *Canarium album* L. that were hyperin, ellagic acid and brevifolin that also had good antioxidant activity [31].

In this research, antibacterial activity of *C. indicum* latex extract was evaluated using agar well diffusion method. The capability of *C. indicum* latex extract to inhibit the bacteria was evaluated by its inhibition zone diameter against Gram-positive bacteria *S. aureus* and Gram-negative bacteria *E. coli*. Ampicillin was used as the positive control. The results of antibacterial activity of *C. indicum* latex extract are presented in Table 2. At the concentration of 500 and 1000 ppm, *C. indicum* latex extract did not show inhibition activity of bacterial growth for both *E. coli* and *S. aureus*. At concentration 10,000 ppm, *C. indicum* latex extract inhibited *E. coli* growth with inhibition zone diameter as 2.38±0.62 mm and inhibited S. aureus growth with inhibition zone as of 6.39±0.06 mm. Meanwhile DMSO that was used as negative control showed no inhibition zone diameter against the bacteria. *C. indicum* latex extract showed moderate antibacterial activity against *E. coli* and good antibacterial activity against *S. aureus*.

Table 2. Antibacterial activity of *C. indicum* latex extract against *E. coli* and *S. aureus*.

Bacterial strain	Sample concentration (ppm)	Well 1	Well 2	Mean
E. coli	500	-	-	-
	1000	-	-	-
	10,000	2.815	1.935	2.38±0.62
S. aureus	500	-	-	-
	1000	-	-	-
	10,000	6.435	6.345	6.39±0.06
Ampicilin	500	3.07	3.11	3.09±0.03
DMSO	-	-	-	-

Literature studies showed that the research on antibacterial or antimicrobial properties from *C indicum* extract are still very rare. However, some literature has studied the bioactivity of the genus of Canarium. It was about only 12% of total Canarium species that had been studied for its biological activities included pharmacological activity [30]. Antibacterial activities against *Vibrio cholera* with MIC as of 0.62 mg/ml was reported from the dichloromethane extract of *Canarium schweinfurthii* while its ethyl acetate extract was active against *P. vulgaris* and *S. aureus* with MIC value as of 5 mg/ml and 10 mg/ml, respectively. The ethanol extract of *C. schweinfurthii* was also active against *V. cholera* and *P. vulgaris* with MIC values of 0.62mg/ml and 10mg/ml respectively [32]. The essential oil of *C. schweinfurthii* also active against *S. aureus*, *S. pyogens* and *S. enterica* with an inhibition zone of 18 mm, 25 mm and 27 mm, respectively [33]. Another species of Canarium, that was *Canarium patentinervium* also inhibited significantly against *S. aureus*, *B. cereus* and *P. aeruginosa* from its ethanol and hexane extract of barks and leaves [34]. Some mechanism of how the antibacterial peptides destroy the bacteria was reported by a literature, that is the antimicrobial agents could bind to bacterial DNA after reach the inner cells through membrane. The membrane itself can disrupt or not was possible.

The functional group of *C. indicum* latex extract were analysed using FTIR spectroscopy. The FTIR spectra between 4000 to 500 cm^{-1} represent the chemical component on its functional groups of *C. indicum* latex ethanolic extract (Figure 2; Table 3.).

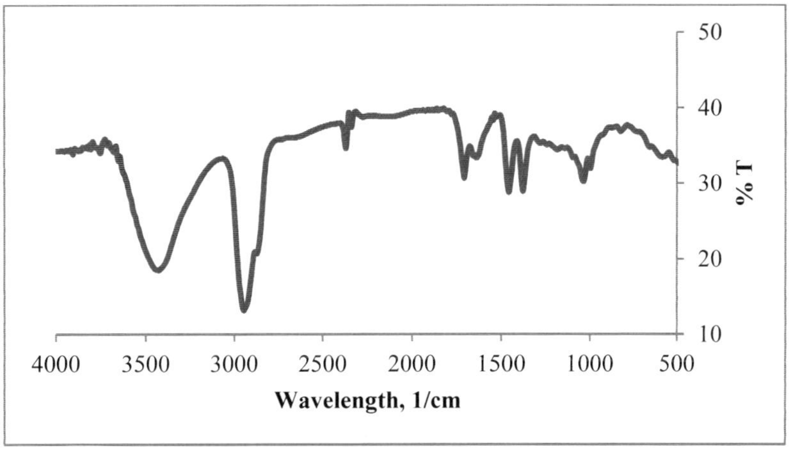

Figure 2. The FTIR spectra of *C. indicum.*

The functional groups detected in C. *indicum* latex extract were consist of OH (could be come from phenol and alcohols), C-H of alkane, C = C of aromatic, and C = O from aldehyde, ketone, ester, or carboxylic acid. The OH (hydroxyl) group was observed by absorption peak at 3425.58 cm^{-1} (broad), CH alkane was observed by absorption peak at around 2862.36-2924.58 cm-1 with high intensity, C = C group was observed by absorption peak at 1635.64 cm^{-1} (sharp) and C = O was indicated by the absorption peak at 1705.1 cm^{-1} (sharp).

Table 3. Interpretation of FTIR spectra of *C. indicum* latex.

Functional group vibration	Wavenumber (1/cm)	Intensity
O-H, phenol, alcohols	3425.6	broad
C-H, alkane	2862.4-2924.6	sharp
C=C, aromatics	1635.6	sharp
C=O, aldehyde, keton, ester, carboxcylic acid	1705.1	sharp

The FTIR spectra of *C. indicum* showed that it possibly contained phenolic compounds. Another study revealed that several secondary bioactive metabolites from the genus Canarium L. contained terpenes include monoterpenes, carotenoids, sterols, triterpenes, coumarins, furans, carboxylic acids and phenol [30]. Single compounds also had been isolated from *Canarium album* L and the study of its biological activity showed that they had good antioxidant activity [31]. Djarkasi *et al.* (2017) also reported that *C. indicum* nut contain several amino acids such as arginine, leucine, aspartate and glutamate [29]. The various amino acids in *C. indicum* nut indicated that it can be had maintain human health since amino acids had the function to regulate human metabolism [35]. Another study showed that besides antioxidant and antibacterial capability, the extract of Canarium species also had an ability as an anti-inflammatory agent [28], an anti-atherosclerosis agent [36] and to maintain blood glucose levels [37].

4. Conclusion

In conclusion, *C. indicum latex* collected from Papua Province, Indonesia exhibit a potent antioxidant and antibacterial activity. *C. indicum* latex extract showed moderate antibacterial activity against *E. coli* and good antibacterial activity against *S. aureus*. The antioxidant activity of *C. indicum* latex extract was moderate compare to ascorbic acid as positive control. It is likely that the antioxidant and antibacterial activity was attributed to the phenolic compounds of *C. indicum* latex extract. More functional use and bioactivity of the *C. indicum* latex extract still need to further be performed.

Acknowledgements
This research was supported by the National Research Priority research grant (PRN) from Indonesia Endowment Fund for Education (LPDP) and National Research and Innovation Agency (Kemenristek/BRIN) of Indonesia.

References
[1] Spence J D 2006 *Stroke* **37**(9) 2430–2435
[2] Calabrese V and Maines M 2006 *Antioxid. Redox Sign.* **8**(3-4) 362–364
[3] Droge W and Schipper H M 2007 *Aging Cell* **6**(3) 361–370
[4] Mishra P K, Shukla R, Singh P, Prakash B and Dubey N K 2012 *Ind. Crops Prod.* **36** 74–80
[5] Gordon M and Kourimska L 1995 *Food Chem.* **52**(2) 175–177
[6] Botterweck A, Verhagen H, Goldbohm R, Kleinjans J and van den Brandt P 2000 *Food Chem. Toxicol* **38**(7) 599–605
[7] Croteau R, Kutchan TM, and Lewis NG 2000 *Biochemistry and molecular biology of plants* **24** 1250–1319
[8] Ramakrishna A and Ravishankar G A 2011 *Plant Signal Behav.* **6** 1720–1731
[9] Khlifi D, Sghaier R M, Amouri S, Laouini D, Hamdi M and Bouajila J 2013 *Food Chem. Toxicol.* **55** 202–208
[10] Manach C, Scalbert A, Morand C, Remesy C and Jimenez L 2004 *Am. J. Clin. Nutr.* **79** 727–747
[11] Martins S, Amorim E L, Sobrinho T J P, Saraiva A M, Pisciottano M N, Aguilar C N and Mussatto SI 2013 *Ind. Crops Prod.* **41** 306–311
[12] Hlila M B, Majouli K, Jannet H B, Aouni M, Mastouri M and Selmi B 2017 *Asian Pac. J. Trop. Biomed.* **7** 629–632
[13] Lee C R, Cho I H, Jeong B C and Lee S H 2013 *Int. J. Environ. Res. Public Health* **10** 4274–4305
[14] Marchese A, Barbieri R, Sanches-Silva A, Daglia M, Nabavi S F, Jafari N J, Izadi M, Ajami M, Nabavi S M 2016 *Trends Food Sci. Tech.* **52** 49–56
[15] Nagaraj R, Prakash M and Karmegam N 2014 *Int. J. Curr. Res. Biosci. Plant Biol.* **1** 53–57
[16] Sofowora, A, Ogunbodede E, Onayade A 2013 *Afr. J. Tradit. Complement. Altern. Med.* **10** 210–229
[17] Mostafa A A, Al-Askar A A, Almaary K S, Dawoud T M, Sholkamy EN, Bakri MM 2018 *Saudi J. Biol. Sci.* **25** 361–362
[18] Rabbani M, Etemadifar Z, Karamifard F and Borhani M S 2016 *Comp. Clin. Pathol.* **25** 59–65
[19] Sokmen A, Gulluce M, Akpulat H A, Daferera D, Tepe B, Polissiou, M, Sokmen M and Sahin F 2004 *Food Control* **15**(8) 627–634
[20] Betts GD, Linton P and Betteridge R J 1999 *Food Control* **10**(1) 27–33
[21] Michael N A and Stuart B L 2007 *Cell* **128**(6) 1037–1050
[22] Chua H P, Nicholas D and Yahya M N A 2015 *J. Trop. Agric. Food Sci.* **43** 1–10
[23] Bai S H, Darby I, Nevenimo T, Hannet G, Hannet D, Poienou M, Grant E, Brooks P, Walton D, Randall B and Wallace H M 2017 *PLoS One* **12**(9): e0184279
[24] Rahman H, Anggadiredja K, Gusdinar T, Sitompul J P and Ryadin A R 2019 *Jurnal Fitofarmaka Indonesia* **6**(1) 325–333
[25] Esteban L, Munio M M, Robles A, Hita E, Jimenez M J, Gonzales P A, Camacho B and Molina E 2009 *Biochem. Eng. J.* **44** 271–279
[26] Indrianingsih A W, Rosyida VT, Apriyana W, Hayati S N, Darsih C, Nisa K and Ratih D 2020 *J. Appl. Pharm.* **10**(07) 037–042
[27] Wulanjati M P, Indrianingsih A W, Darsih C and Apriyana W 2019 *IOP Conf. Ser.: Earth Environ. Sci.* **462** 012028
[28] Leakey R, Fuller S, Treloar T, Stevenson L, Hunter D, Nevenimo T, Binifa J and Moxon J 2008 *Agroforest. Syst.* **73** 77–87
[29] Djarkasi G S S, Lalujan L, Nurali E J N, Tuju T J D, Rawung D and Sumual M F 2017 *Pharm. Chem. J.* **4**(4) 79-84

[30] Mogana R and Wiart C 2011 *J. Pharm. Res.* **4**(8) 2482–2489
[31] Ito M, Shimura H, Watanabe N, Tamai M, Hanada K and Takahashi A 1990 *Chem. Pharm. Bull.* **38** 2201–2203
[32] Moshi M, Innocent E, Masimba P J, Otieno D F, Weisheit A, Mbabazi P, Lynes M, Meachem K, Hamilton A and Urassa I 2009 *Tanzan. J. Health Res.* **11** 23–28
[33] Obame L, Koudou J, Kumulungui B S, Ismael H N, Prosper E, Aboubakar S O and Alfred S T 2007 *Afr. J. Biotechnol.* **6** 2319–2323
[34] Mogana R, Khoo T J and Wiart C 2011 *Biotechnol. Res. Int.* **2011** 768673
[35] Wu G 2013 *Amino Acids* **45**(3) 407–411
[36] Shakirin F H, Azlan A, Ismail A, Amom Z and Yuon L C 2012 *Oxid. Med. Cell Longev.* **2012** 840973
[37] Yeh Y, Chiang A and Hsieh S 2017 *Nutrients* **9** 1123

IOP Conf. Series: Materials Science and Engineering 1011 (2021) 012052 doi:10.1088/1757-899X/1011/1/012052

In vitro antioxidant activity of macroalgae *Sargassum duplicatum* and *Palmaria palmata* extracts collected from Sepanjang Beach, Gunungkidul, Yogyakarta

C Darsih*, A W Indrianingsih, C D Poeloengasih, D J Prasetyo and N Indirayati

Research Unit for Natural Product Technology, Indonesian Institute of Sciences, Gunungkidul, Yogyakarta, Indonesia

*Email: cici.darsih@gmail.com

Abstract. Marine macroalgae are considered as important sources for bioactive phytochemicals. In this research, two potential marine macroalgae, i.e. *Sargassum duplicatum* and *Palmaria palmata* taken from Sepanjang beach, Gunungkidul, Yogyakarta, Indonesia were explored as sources of phenolic compound, and their antioxidant activities were evaluated. Three different solvents, *i.e.* chloroform, ethyl acetate, and methanol were applied as extraction medium. For each species, its total phenolic content (TPC) and antioxidant activity were determined using Folin-Ciocalteu reagent and DPPH scavenging assay, respectively. The result showed that methanolic extracts exhibited the highest yield in both species. The highest TPC (726.54 mg GAE/g) was obtained from *S. duplicatum* extracted using ethyl acetate. The antioxidant activity from *S. duplicatum* and *P. palmata* have similar value with IC_{50} values of 790.34 µg/mL and 789.29 µg/mL. The scavenging activities of all extracts related to the total polyphenol content, whereas the polarity of solvent affected the resulted yield and TPC of the extracts. However, our results demonstrated that all extracts were inactive against DPPH radicals ($IC_{50} > 250$ µg/mL).

1. Introduction

Marine macroalgae are considered as important sources for bioactive phytochemical with promising application in drugs, cosmetics, and food. Macroalgae were classified into four major groups, *i.e.* green algae (Chlorophyceae), red algae (Rhodophyceae), blue-green (Cyanophyceae), and brown algae (Phaeophyceae) [1]. Metabolites on macroalgae such as polyphenols, polysaccharides, carotenoids, flavonoids, pigments, and fatty acids have been proven exhibit biological activities [2,3]. Moreover, macroalgae also contain vitamins, protein, inorganic elements, and carbohydrates [3,4]. These metabolites possess biological activities such as antifungal, antibacterial, anti-inflammatory, antidiabetic, antioxidant, antiviral, and antimalarial [5-9]. The various metabolites of macroalgae and its concentration were affected by environmental, season, age of macroalgae, and geographical location [10].

The metabolites from brown alga *S. duplicatum* have been reported showing various pharmacological activities such as antioxidant, antibacterial, anti-inflammatory, anticancer, and tyrosinase inhibition [11-15]. Meanwhile, some studies reported that metabolites from red alga *P. palmata* or known as dulse exhibit biological activities such as antioxidant, angiotensin-converting enzyme (ACE) inhibition, anti-inflammatory, and inhibit primary human neutrophil activation [16-18]. Both of these macroalgae can be found in Sepanjang Beach Gunungkidul, Yogyakarta. In Gunungkidul, *P. palmata* traditionally consumed as a snack food, whereas *S. duplicatum* has not been widely used yet by the Gunungkidul community. The aim of this study was to determine the total phenolic content (TPC) and the antioxidant activity of *S. duplicatum* and *P. palmata* extracts because study about both of macroalgae and its biological activity were rarely conducted. As a valuable compound for preservation and protection against oxidation, the study about antioxidant is to be concerned.

2. Materials and Method

2.1. Materials
The fresh macroalgae *S. duplicatum* and *P. palmata* (Figure 1) were collected from Sepanjang Beach, Gunungkidul, Yogyakarta, Indonesia on November 2018. The samples were then identified in Plant Systematic Laboratory, Biology Faculty, Gadjah Mada University. The fresh macroalgae were rinsed using seawater followed by distilled water to remove debris and sands. Samples were dried in room temperature for seven days, ground, packed in plastic bags, and stored at room temperature until further analysis.

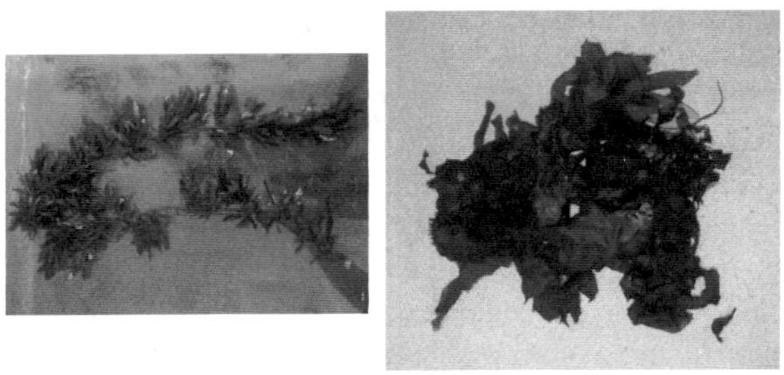

(a) (b)

Figure 1. Macroalgae *S. duplicatum* (a) and *P. palmata* (b).

2.2. Extraction of samples
In this work, three different solvents were used, *i.e.* chloroform, ethyl acetate, and methanol with the ratio of 1:10 (w/v). The powder samples were sonicated at room temperature for 15 minutes, followed by maceration in room temperature for 24 hours. The filtrate and the solid residues were filtered using Whatman 1. The crude extracts of *S. duplicatum* and *P. palmata* were obtained by evaporating the filtrate.

2.3. Total Phenolic Content (TPC)
The TPC of crude extracts from *S. duplicatum* and *P. palmata* were determined using Folin-Ciocalteu reagent. The extract samples (500 µL) were added with 500 µL of the Folin-Ciocalteu reagent and mixed thoroughly. The samples were added with 1.5 mL of 20% sodium carbonate. The final volume was added up to 10 mL with distilled water. The mixture was incubated for two hours in room temperature. The absorbance of the samples was measured using ELISA reader at 765 nm. Gallic acid was used as a standard [19].

2.4. Assay for antioxidant activity
The antioxidant activity of *S. duplicatum* and *P. palmata* extracts was determined using DPPH method. The sample solutions consisted 20µL of crude extracts and 180 µL DPPH solution (40 mM in methanol) were

incubated in the dark at room temperature for 30 minutes. A microplate reader was used to measure the absorbance of the solutions at 517 nm. Ascorbic acid was used as a positive control and absolute methanol as a blank solution. The percentage activity of crude extracts to scavenge the DPPH radical can be seen in the equation (1).

$$\text{Radical scavenging activity (\%)} = \frac{A0-A1}{A0} \times 100 \qquad \text{(Eq. 1)}$$

where A_0 is the absorbance of the control and A1 is the absorbance in the presence of the sample [20].

3. Results and Discussion

Table 1 summarizes the effect of solvent polarity on the yield of extracts from *S. duplicatum* and *P. palmata*. The extract yield ranged between 0.58 to 17.22% (w/w). It was found that the methanolic extracts resulted higher yield than two other solvents. This is probably due to the polarity of methanol as a solvent. Since solvent polarity affected the extraction efficiency, methanol was effective enough to extract the compounds from the samples. As a solvent with high polarity, methanol would solubilize both polar and nonpolar compounds, so that resulted higher extraction yield. These results were in line with the previous study by Sharma and Cannoo [21]. which is working with methanol, chloroform and hexane in the extraction of *Nepeta leucophylla*. These results also go along with a study by Do [22], which is using methanol, pure ethanol and acetone in the extraction of *Limnophila aromatica*.

Figure 2 presents the TPC of all extracts. The results showed that the TPC of ethyl acetate extracts from both algae were higher than that of the other extracts. The TPC of ethyl acetate extract from *S. duplicatum* was the highest and reached up to 726.54 mg GAE/g. This result goes along with the previous research that the phenolic acid and quercetin were confirmed in ethyl acetate extract of *Sargassum plagiophyllum* [23]. Meanwhile, the TPC of methanol extracts from *S. duplicatum* was the lowest and reached up to 118.99 mg GAE/g. This indicates that phenolic constituents in the samples of *S. duplicatum* and *P. palmata* prefer to be extracted with semi-polar degree solvents such as ethyl acetate. This result was also reported in several literatures [24,25]. A study also revealed that complex formation among phenolic constituents possibly occurred and facilitated the extraction of phenolic constituents in ethyl acetate solvents [26].

Table 1. The yield of *S. duplicatum* and *P. palmata* extracts.

Macroalgae	Extracts	Yield (% w/w)
S. duplicatum	Chloroform	1.22
	Ethyl acetate	1.22
	Methanol	13.25
P. palmata	Chloroform	0.62
	Ethyl acetate	0.58
	Methanol	17.22

The antioxidant activity of *S. duplicatum* and *P. palmata* can be seen at Table 2. The results reported that the antioxidant activity of ethyl acetate extracts from *S. duplicatum* and *P. Palmata* were similar with IC_{50} values of 790.34 and 789.29 µg/mL, respectively. These results were higher than those of chloroform and methanol extracts, although the antioxidant activity of chloroform extract from *S. duplicatum* is higher than of *P. palmata*. The antioxidant activity of the ethyl acetate extracts higher than others because TPC on this extract was higher than others.

IOP Conf. Series: Materials Science and Engineering 1011 (2021) 012052 doi:10.1088/1757-899X/1011/1/012052

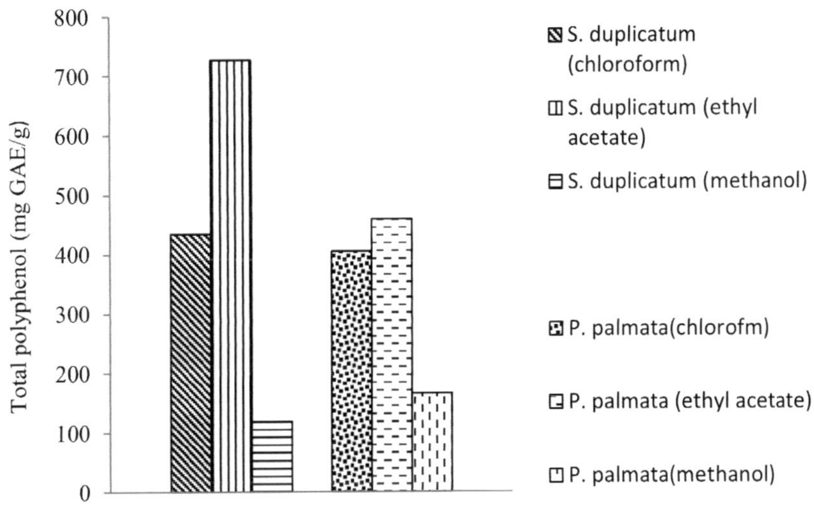

Figure 2. The total polyphenol of *S. duplicatum* and *P. palmata* extracts.

This study showed that there was a correlation between TPC and the antioxidant activity of the ethyl acetate extracts for both algae types. TPC affected antioxidant activities. TPC have ability to donate hydrogen, quench singlet oxygen and act as metal chelators, which can act as antioxidant agent [19]. The results were in line the previous research that TPC and antioxidant activity on brown algae higher than green and red algae [27-29]. In this study, all extracts were inactive against DPPH radicals (IC_{50}>250 μg/mL) [30].

Table 2. Radical scavenging activity of *S. duplicatum* and *P. palmata* extracts.

Samples	Extracts	IC_{50} (μg/mL)
S. duplicatum	Chloroform	821.301
	Ethyl acetate	790.34
	Methanol	>1000
P. palmata	Chloroform	>1000
	Ethyl acetate	789.29
	Methanol	>1000
Ascorbic acid		8.28

4. Conclusion

The yield and TPC on *S. duplicatum* and *P. palmata* extracts were affected by the polarity of solvents. Chloroform, ethyl acetate, and methanol extracts of *S. duplicatum* and *P. palmata* showed inactive antioxidant properties toward DPPH radical. It suggested that other biological activities could be conducted on these extracts.

Acknowledgements

This project was fully supported by the Indonesian Institute of Sciences, the Government of Indonesia.

References

[1] Thomas N V and Kim S K 2013 Beneficial effects of marine algal compounds in cosmeceuticals. *Mar. Drugs* **11**(1) 146-64

[2] Cikoš A M and Jokić S 2018 Overview on the Application of Modern Methods for the Extraction of Bioactive Compounds from Marine Macroalgae *Mar. Drugs* **16**(10) 348

[3] Wulanjati M P, Indrianingsih A W, Darsih C, Apriyana W and Batrisya 2020 Antioxidant and antibacterial activity of ethanolic extract from *Ulva* sp. *IOP Conf. Series: Earth and Environmental Science* **462** 012028

[4] Wells M L, Potin P, Craigie J S, Raven J A, Merchant S S, Helliwell K E, Smith A G, Camire M E and Brawley S H 2016 Algae as nutritional and functional food sources: revisiting our understanding *J. Appl. Phycol.* **29**(2) 949-82

[5] Abdel-Raouf N, Ibraheem I B M, Abdel-Hameed M S and Elyamany K M N E 2008 Evaluation of antibacterial, antifungal and antiviral activities of ten marine macroalgae from Red Sea, Egypt. *Egypt. J. Biotechnol.* **29** 157-72

[6] Barbalace M C, Malaguti M, Giusti L, Lucachinni A, Hrelia S and Angeloni C 2019 Anti-inflammatory activities of marine algae in neurodegenerative diseases. *Int. J. Mol. Sci.* **20**(12) 3061

[7] Pirian K, Moein S, Sohrabipour J, Rabiei R and Blomster J 2017 Antidiabetic and antioxidant activities of brown and red macroalgae from the Persian Gulf *J. Appl. Phycol.* **29**(6) 3151-9

[8] Ponce N M A, Pujol C A, Damonte E B, Flores M L and Stortz C A 2003 Fucoidans from the brown seaweed Adenocystis utricularis: extraction methods, antiviral activity and structural studies *Carbohydr. Res.* **338**(2) 153-65

[9] Spavieri J, Allmendinger A, Kaiser M, Itoe M A, Blunden G, Mota M M and Tasdemir D 2013 Assessment of dual life stage antiplasmodial activity of british seaweeds *Mar. Drugs* **11**(10) 4019-34

[10] Generalić Mekinić I, Skroza D, Šimat V, Hamed I, Čagalj M and Perković Z P 2019 Phenolic content of brown algae (Pheophyceae) species: extraction, identification, and quantification *Biomolecules* **9**(6) 244

[11] Firmansyah S B, Firmansyah R A and Hayati N 2016 Antioxidant activity and antibacterial seaweed methanol extract (*Sargassum duplicatum* J. Agardh) and its potential as a natural preservative alternative to salted eggs *J Nat Sci Mathematics Res* **1**(1) 133

[12] Johnson M, Asha Kanimozhi S, Renisheya Joy Jeba Malar T, Freitas S P R, Tintino S R, Menezes I R A, da Costa J G M and Coutinho H D M 2019 The antioxidative effects of bioactive products from *Sargassum polycystum* C. Agardh and *Sargassum duplicatum* J. Agardh against inflammation and other pathological issues *Complement. Ther. Med.* **46** 19-23

[13] Usoltseva R V, Anastyk S D, Shevchenko N M, Surits V V, Silchenko A S, Isakov V V, Zvyagintseva T N, Thinh P D and Ermakova S P 2017 Polysaccharides from brown algae *Sargassum duplicatum*: the structure and anticancer activity in vitro *Carbohydr. Polym.* **175** 547-56

[14] Chan Y Y, Kim K H and Cheah S H 2011 Inhibitory effects of *Sargassum polycystum* on tyrosinase activity and melanin formation in B16F10 murine melanoma cells. *J. Ethnopharmacol.* **137** 1183-8.

[15] Yende S R, Harle U N and Chaugule B B 2014 Therapeutic potential and health benefits of Sargassum species *Pharmacog. Rev.* **8**(15) 1-7

[16] Beaulieu L, Sirois M and Tamigneaux É 2016 Evaluation of the in vitro biological activity of protein hydrolysates of the edible red alga, *Palmaria palmata* (dulse) harvested from the Gaspe coast and cultivated in tanks *J. Appl. Phycol.* 28(5) 3101-15

[17] Harnedy P A, O'Keeffe M B and FitzGerald R J 2017 Fractionation and identification of antioxidant

peptides from an enzymatically hydrolysed Palmaria palmata protein isolate *Food Res. Int.* **100** 416-22
[18] Harnedy P A, Soler-Vila A, Edwards M D and FitzGerald R J 2014 The effect of time and origin of harvest on the in vitro biological activity of *Palmaria palmata* protein hydrolysates *Food Res. Int.* **62** 746-52
[19] Nisa, K., et al. 2017 Investigation of total phenolic and flavonoid contents, and evaluation of antimicrobial and antioxidant activities from *Baeckea frutescens* extracts *IOP Conf. Series: Earth and Environmental Science* **101** 012002
[20] Indrianingsih A W, Tachibana S and Itoh K 2015 *In vitro* evaluation of antioxidant and α-glucosidase inhibitory assay of several tropical and subtropical plants *Proced. Environ. Sci.* **28** 639-48
[21] Sharma A and Cannoo D S 2017 A comparative study of effects of extraction solvents/techniques on percentage yield, polyhenolic composition, and antioxidant potential of various extracts obtained from stems of Nepeta leucophylla: RP-HPLC-DAD assessment of its polyhenolic constituents. *J. Food Biochem.* **41**(2) e12337
[22] Do Q D, Angkawijaya A E, Tran-Nguyen P L, Huynh L H, Soetaredjo F E, Ismadji S and Ju Y H 2014 Effect of extraction solvent on total phenol content, total flavonoid content, and antioxidant activity of *Limnophila aromatica J. Food Drug Anal.* **22**(3) 296-302
[23] Chakraborty K, Maneesh A and Makkar F 2017 Antioxidant Activity of Brown Seaweeds. *J. Aquat. Food Prod. Technol.* **26**(4) 406-19
[24] Wijaya Y A, Widyadinata D, Irawaty W and Ayucitra A 2017 Fractinarion of phenolic compounds from Kaffir Lime *(Citrus hystrix)* peel extract and evaluation of antioxidant activity **2017** 7
[25] Anagnostopoulou M A, Kefalas P, Papageorgiou V P, Assimopoulou A N and Boskou D 2006 Radical scavenging activity of various extracts and fractions of sweet orange peel *(Citrus sinensis)* *Food Chem.* **94**(1) 19-25
[26] Zhu K X, Lian C X, Guo X N, Peng W and Zhou H M 2011 Antioxidant activities and total phenolic contents of various extracts from defatted wheat germ *Food Chem.* **126**(3) 1122-6
[27] Gupta S and Abu-Ghannam N 2011 Bioactive potential and possible health effects of edible brown seaweeds. *Trends Food Sci. Technol.* **22**(6) 315-26
[28] Balboa E M, Conde E, Moure A, Falque E and Dominguez H 2013 In vitro antioxidant properties of crude extracts and compounds from brown algae *Food Chem.* **138**(2-3) 1764-85
[29] Li Y, Fu X, Duan D, Liu X, Xu J and Gao X 2017 Extraction and identification of phlorotannins from the Brown Alga, *Sargassum fusiforme* (Harvey) Setchell *Mar. Drugs* **15**(2) 49
[30] Phongpaichit S, Nikom J, Rungjindamai N, Sakayaroj J, Hutadilok-Towotana N, Rukachaisirikul V and Kirtikara K 2007 Biological activities of extracts from endophytic fungi isolated from Garcinia plants *FEMS Immunol. Med. Microbiol.* **51**(3) 517-25

A comparative study on the degradation of 2,4-dichlorophenol and 2,5-dichlorophenol using DBD non-thermal plasma reactor

Febri Monica Diristiani, Badzlina Khairunizzahra, Samantha Suryo and Setijo Bismo*

Department of Chemical Engineering, Faculty of Engineering, Universitas Indonesia, Depok, West Java, 16424, Indonesia.

*Corresponding author: setijo.bismo@ui.ac.id

Abstract. The dichlorophenol compound group (DCP) is one of the pollutants contained in industrial wastewater. This component is difficult to degrade in the environment, subsequently leading to hazards and a severe impact on human health. Based on the location of the chlorine atom in the aromatic ring, the variety present in some isomers ought to be removed from the water, in order to reduce the impact. Therefore, the technology adopted to eliminate the DCP concentration levels in wastewater includes the application of Dielectric Barrier Discharge (DBD) plasma reactor of high efficiency. This set up implements advanced oxidation processes (AOPs) by relying on highly active species, specifically ·OH and O_3. The aim of this study, therefore, is to compare the effectiveness of DBD reactors in the degradation activity on isomer 2,4-DCP and 2,5-DCP under acidic (pH 4) and alkaline (pH 10) conditions. The results showed the highest percentage removal (89.05%) after 120 minutes process of 2.5-DCP degradation, under the following conditions: pH 10; flow rate 50 mL/minute; airflow rate of 2.5 L / min; plasmatron voltage of 19 kV. Under similar circumstances, 2,4-DCP was reduced by 86.95%. In addition, the COD content in the waste material reached 46.88% and 46.14%, respectively.

1. Introduction

Water pollution instigated by industrial waste is one of the environmental problems requiring the development of law-compliant treatments. Moreover, one of the potential driving factors is the increasing number of industries, due to the elevation in life necessities. Based on various previous research, a number of contaminant phenolic compounds have been identified in production effluents, including dichlorophenol (DCP). In addition, several industries are known to generate the liquid-based variety as waste, comprising the pharmaceutical, pesticide, pulp and paper, petrochemical, plywood, and others. However, wastewaters based on DCP compounds are highly dangerous for living things and the environment.

These dichlorophenols have very low biodegradability, hence the possibility of sustained risk to living things with direct exposure. Moreover, the presence in wastes has been identified in several isomer forms, with the most common type being 2,4-dichlorophenol and 2,5-dichlorophenol, where each confers a different adverse effect on humans. Notably, a deposit of 2,4-dichlorophenol is able to damage the liver as well as kidneys and initiate blood circulation failures, while 2,5-dichlorophenol is an endocrine-disrupting chemical with the capacity to trigger obesity and instigate type 2 diabetes [1,2]. These potential impacts are to be eliminated by phenolic compound producing industries, through the practice of proper waste treatment before disposal, with reference to waste quality standards. The maximum content is estimated at 0.5 ppm, through PERMENLH No. 5/2014, although there is no specific standard for the complete control, alongside the isomers. However, most industries in several countries pay little attention to this aspect, particularly the compounds discharged into water [3].

Furthermore, advanced oxidation processes (AOPs) are one of the waste treatment methods developed using hydroxyl radicals and ozone as active species, and dielectric barrier discharge (DBD) plasma reactor is one of the technologies involved. Moreover, a non-thermal plasma reactor is able to produce active species, including O_3 and H_2O_2 molecules, as well as hydroxyl radicals (·OH, H·, O ·), with the capacity to decompose phenolic compounds. The oxidation process with DBD has the advantage of requiring only a short time, low operating temperature, high energy efficiency, and compatibility with the environment, compared to other AOP implementations [4].

There has been extensive research on the degradation of 2,4-DCP isomers with AOP, but not with 2.5-DCP [3], especially in terms of DBD plasma reactor applications. Meanwhile, degradation depends on the chlorine atom position [5], and the entire process is determined by the solution pH, resulting from the direct influence [6]. This study, therefore, examines the degradation ability of DBD plasma against different DCP isomers, and dielectric barrier discharge (DBD) was used is the coaxial pipe configuration of spiral cylinders. The initial ph of wastewater was conditioned according to the average value stipulated in the industry.

2. Materials and Methods

2.1. Wastewater Materials

The experiments involved synthetic wastewater contaminated with DCP. Furthermore, 2,4-DCP (99%) was obtained from Sigma-Aldrich, while 2,5-DCP (99%) was obtained from Merck, and the solutions were prepared using distilled water. Also, all chemicals used in the study were of extra purity or analytical grade.

2.2. Experimental installation and procedure

The experimental setting comprised of a DBD non-thermal plasma reactor with a continuous system, alongside cylinder-spiral configuration inner and outer electrodes, respectively. This possessed the dimensions of 150 mL, 40 cm, and 22 mm for volume, height, and diameter. Figure 1 shows a diagram of the experimental arrangement scheme for this study. This research was initiated with a pump and compressor calibration to ensure precise fluid measurements. Subsequently, a reactor start-up test was conducted to ascertain the possibility of operating the plasma generating systems without significant complexities or problems. In addition, a leak test was performed by running distilled water from the reservoir through the pump, as well as air feed from the compressor to the reactor. Therefore, the duct tape is installed after a crack is detected, to prevent loss of water or injected air. The next step required a physical plasma test, involving the observation of the plasma phenomenon, characterized by the display of a purplish incandescent along the reactor. Furthermore, a hydrodynamic test was performed to ensure the fluid flows as specified.

Some preliminary tests were conducted to quantify the hydroxyl radicals, and dissolved zone. Also, hydroxyl radical analysis was performed using the iodometric titration method, based on some experimental parameters, including liquid fluid flow rates of 50, 65, and 80 mL/min, plasmatron voltage of 15, 17, and 19 kV, as well as solution pH of 4 and 10. Similarly, dissolved ozone concentration was evaluated based on the pH parameters by an ozone test kit with spectrophotometric measurements at a wavelength of 550nm. Therefore, DCP degradation was carried out with best-operating conditions implemented from the preliminary test and other prerequisites.

IOP Conf. Series: Materials Science and Engineering 1011 (2021) 012053 doi:10.1088/1757-899X/1011/1/012053

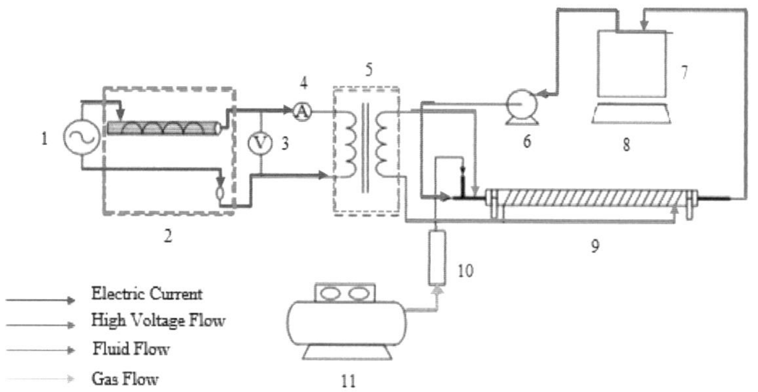

1	Electricity Source
2	Slide Regulator
3	Voltmeter
4	Amperemeter
5	NST (Neon Sign Transformer)
6	Peristaltic Pump
7	Reservoir
8	Stirrer
9	DBD Plasma Reactor
10	Flowmeter
11	Compressor

Figure 1. The schematic diagram of the experimental setup for wastewater degradation.

The initial wastewater concentration was 50 mg/L, with pH of 4 and 10, while the process duration extended for 120 minutes, with sampling at intervals of 0, 5, 10, 15, 30, 45, 60, 75, 90, 105 and 120. The amount of dichlorophenol compounds were determined through spectrometry, 2,4-DCP and 2,5-DCP were identified at a wavelength of 513.8 and 505.6 nm, respectively. Furthermore, the COD levels were evaluated using similar method and recognized at 650 nm using COD HACH.

3. Result and Discussion

3.1. Quantification of Hydroxyl Radical and Dissolved Ozone Concentration

Hydroxyl radical quantification aimed to determine the performance of DBD plasma reactors in producing ·OH as an active species, with the intrinsic capacity to degrade phenolic compounds under certain conditions. Therefore, potassium iodide solution was used as a phenolic compound substitute to obtain an accurate measurement. Based on the high reactivity and non-selective nature of OH•, the quantification process is represented by H_2O_2 concentrations, through the following reaction:

$$\cdot OH + \cdot OH \rightarrow H_2O_2 \quad (1)$$

The oxidation of H_2O_2 by KI follows a reduction-oxidation principle, as formed in the reaction below:

$$H_2O_2 + 2KI + H_2SO_4 \leftrightarrow I_2 + K_2SO_4 + 2H_2O \quad (2)$$

$$I_2 + 2Na_2S_2O_3 \leftrightarrow Na_2S_4O_6 + 2\ NaI \quad (3)$$

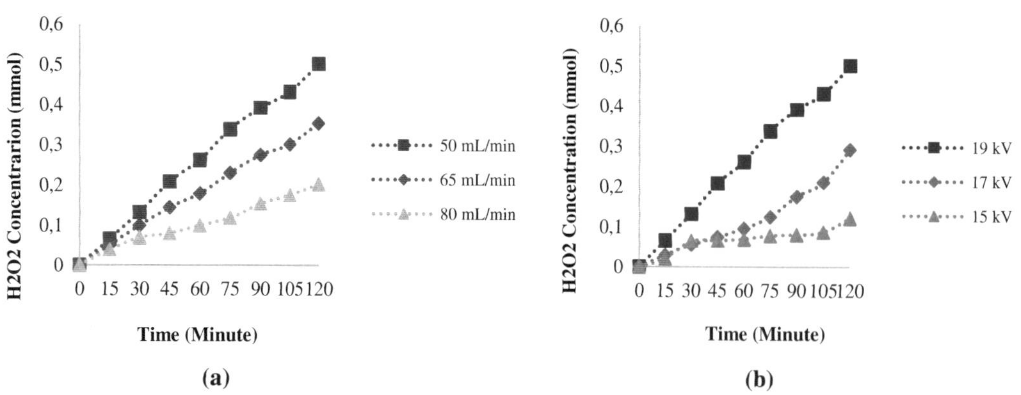

Figure 2. Quantity of (a) hydroxyl radicals at liquid flow rate variations (b) voltage variations.

IOP Conf. Series: Materials Science and Engineering 1011 (2021) 012053 doi:10.1088/1757-899X/1011/1/012053

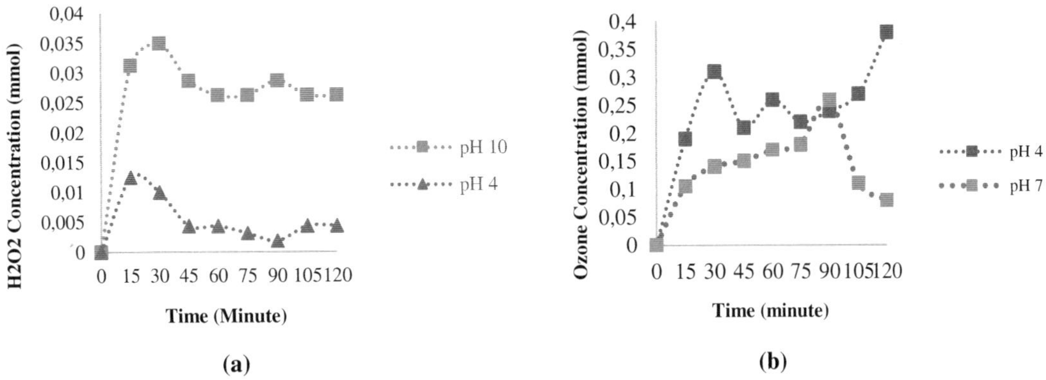

Figure 3. Quantity of (a) hydroxyl radicals (b) Dissolved ozone concentrations at pH variations.

The Hydroxyl radical quantification was evaluated by varying the flow rate of the liquid fluid feed 50, 65 and 80 mL/minute, as well as the plasmatron input voltage 15, 17, 19 kV. Figure 2(a) shows more significant production of hydrogen peroxide in samples with slower rate. This phenomenon was due to the longer fluid residence time in the reactor, and the slower circulation of entry and exit materials. In addition, the reactants formed are expected to remain in the reactor plasma for a long time to achieve a complete treatment [8]. Therefore, higher contact time perfects the hydroxyl radical formation, despite the relatively low circulation. Figure 2(b) shows a directly proportional increase in hydrogen peroxide concentrations produced along with the elevated plasmatron voltage supply. Furthermore, higher electric field intensity facilitates higher energy generation for particles and electrons. This phenomenon leads to greater collisions and more significant ionization rates [7]. Therefore, elevated voltage increases the electric field intensity, and prompts the release of an extensive number of ionized, excited, as well as dissociated molecules, and further augments the formation of active species, including ·OH.

The best conditions were selected based on the variables evaluated, including the fluid flow rate and plasmatron voltage applied to assess the hydroxyl radical quantification, as well as the ozone solubility tests with varied solution pH. Figure 3(a) shows the detection of H_2O_2 concentrations at acidic and basic pH. Furthermore, higher values were observed in the early phase of the reaction, while the consequent addition of NaOH initiated the formation of ·OH, through ozone decomposition in the plasma field [9]. This base solution influences the dissociation into hydroxide anion (OH^-) and sodium cation (Na^+). Moreover, H_2O_2 concentration required to consume species concentration is relatively high because of the reactive and non-selective nature of ·OH combined with fellow ·OH and the slight decline over time due to oxidation processes. Figure 3(b) shows a decrease in ozone solubility with increasing pH, and the value recorded tends to fluctuate with time. This phenomenon is due to the shorter ozone half-life initiated in the water sample. [10] and is also related to the presence of hydroxide ions (OH^-) at high pH, with the capacity to induce the ozone decomposition rate. Furthermore, this fluctuation in solubility is implicated in the formation of other compounds, through the interaction mechanism between molecules producing H_2O_2, O2, O3-, and ·OH [11]. Also, water temperature is another influential parameter, as lower values are strongly associated with higher levels of ozone dissolution. It occurs because the ambient temperature is erratic, therefore causing the fluctuating concentration values.

3.2. The Effect of pH of the Solution on DCP Degradation

The effect of pH on the degradation of DCP isomer waste is evaluated based on the final concentration obtained under similar operating conditions. Figure 4 shows the final output in the form of percentage degradation with varied pH treatments.

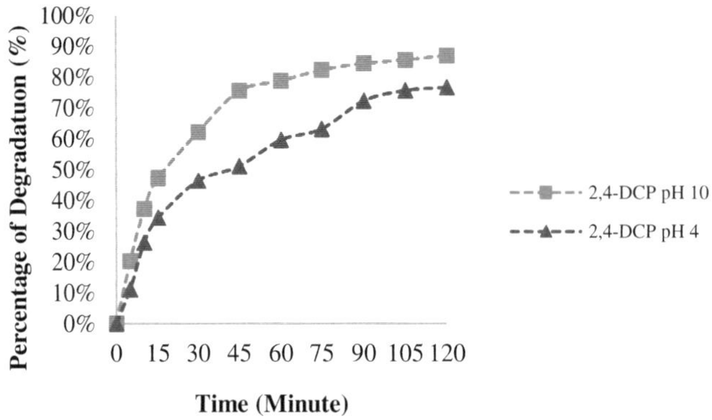

(a)

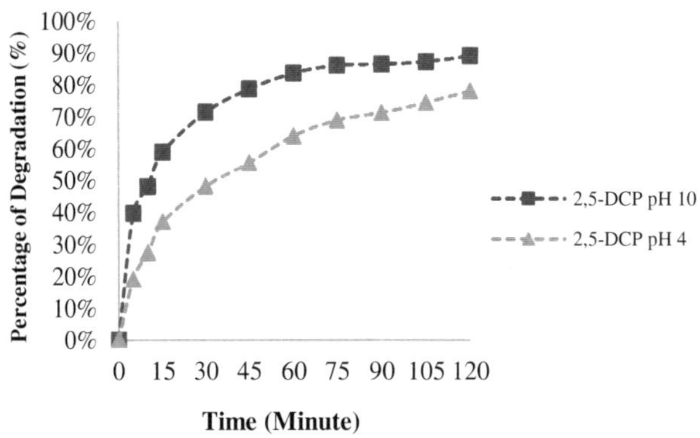

(b)

Figure 4. Degradation of (a) 2,4-DCP and (b) 2,5-DCP at pH variations.

Table 1. Comparison of the percentage degradation, initial and final concentrations of 2,4-DCP and 2,5-DCP isomers.

Dichlorophenol (DCP)	pH	Final Degradation (%)	Initial Conc. of DCP (mg/L)	Final Conc. of DCP (mg/L)
2,4-DCP	10	86.95	49.8	6,5
	4	76.63	50.5	11.8
2,5-DCP	10	89.05	53.9	5.9
	4	78.06	51.5	11.3

As seen in Figure 4 an increase in percentage degradation at the exponent of time, which shows pH influenced the entire process, as more significant extents were measured pH 10, compared to 4. This parameter is influential due to the presence of DCP in the form of either molecular or dichlorophenoxide ions [12]. The DCP compound molecule tends to dissociate into dichlorophenoxide ions at higher pH, therefore enhancing the ease to degrade. As previously explained, the presence of ·OH is higher as the pH of the solution increases while O_3, which less oxidative to attack DCP molecule, is higher at lower pH. Furthermore, the degradation yield at pH 4 somewhat smaller than at pH 10. However, DBD plasma reactors have the advantage of being able to produce ·OH under both acidic and alkaline conditions. Thus, the presence of ·OH facilitates the vulnerability of DCP compounds in molecular form to attack [13].

IOP Conf. Series: Materials Science and Engineering 1011 (2021) 012053 doi:10.1088/1757-899X/1011/1/012053

Based on Table 1, the percentage of degradation of 2,5-DCP (89.05%) was slightly greater than 2,4-DCP (86.95%) over time. 2,5-DCP, due to the relatively faster polarization and dissociation to dichlorophenoxide ions. Furthermore, chlorine atom substituent in close proximity to the hydroxyl group instigate the effect of a steric hindrance [14] and further prevents the attack tendency of ·OH against aromatic rings. The distance between each chlorine atom influences the steric hindrance strength, and the chlorination rate was determined to be more significant in 2,5-DCP, compared to 2,4 DCP [15]. Conversely, the degradation rate of 2,5-DCP was only slightly faster, although the final percentage showed no significant difference (<5%). This is because ·OH and O_3 work synergistically and complementarily under certain conditions.

3.3. Changes in pH of the Solution

The pH of samples obtained from the reservoir during the treatment was evaluated. Figure 5 shows any changes detected pertaining to the degree of acidity.

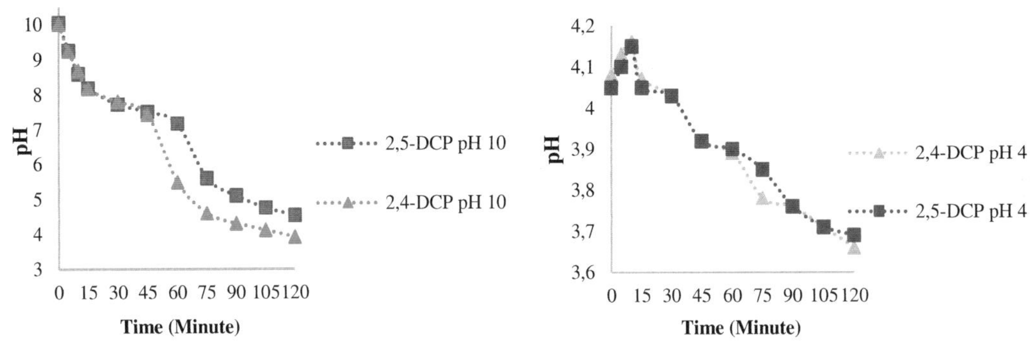

Figure 5. pH changes in degradation of 2,4-DCP and 2,5-DCP at pH variations.

Figure 5 demonstrates the ability for DCP to lower the solution pH rapidly. The oxidation process modifies some of the molecules into organic acids, including muconic acid, malonaldehyde, maleic acid and others. These are further oxidized into simpler forms, comprising oxalic and formic acid [6] to ensure the final pH is acidic. However, DCP solutions in exhibit different phenomena of pH changes under alkaline conditions, where consistently decreasing values were recorded and acidic environment characterized by an increase in pH, particularly during the initial 10 minutes. This preliminary rise in possibly occurs due to the presence of organic compounds, including carbonates and hydroxides.

3.4. COD Removal

The COD removal analysis was conducted by measuring the concentration per sample at 0, 5, 60, and 120 minutes under best conditions at pH 10, and the results are shown in Figure 6.

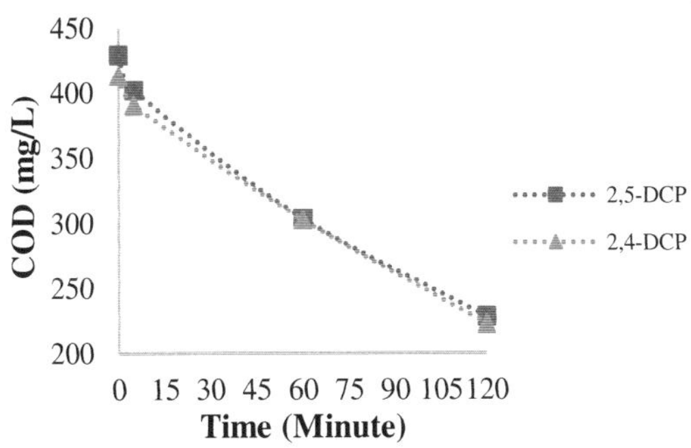

Figure 6. COD Removal at 2,4-DCP and 2,5-DCP.

Based on Figure 6, the DCP degradation process in the DBD plasma reactor was consistently able to reduce COD levels over time by 46.88% (201.14 mg / L) and 46.14% (190.46 mg / L) for 2,5-DCP and 2,4-DCP, respectively. Correspondingly, the amount removed within the initial 5 minutes was relatively low with reference to the preliminary degradation mechanisms by active species. The intermediate compound manifested in the form of an aromatic ring, as indicated by the consistently high oxygen content requirement. In addition, there was a continuous decline in the degradation levels at the exponent of time. This indicates to the phenomenon of declination or aromatic ring termination, caused by further oxidation by active species. The procedure leads to the production of simpler intermediate compounds and carboxylic acid. Based on the final percentage of COD, the removal rate was determined to be slower compared to the DCP degradation rate. The treatment result for both isomers shows high final COD levels, indicating a large number of intermediate compounds were formed [13]. These composites had not reached the end of mineralization to become CO_2 and H_2O.

4. Conclusion

DBD plasma reactors are able to degrade 2,4-DCP and 2,5-DCP isomers into simpler compounds effectively, and a final percentage of 86.95% and 89.05% were respectively recorded after a 120 minutes process. In addition, there was a simultaneous decline in the pH of the solution following the generation of acidic compounds in the final product. This elevation in the Ph of the solution increases the productivity of ·OH as the strongest oxidant. Therefore, the removal efficiency is improved to achieve lower DCP concentrations in wastewater. The COD content of the 2,5-DCP and 2,4-DCP isomers after treatment through the DBD decreased by 46.88% and 46.14%, respectively. Meanwhile, the best-operating degradation conditions include a wastewater and airflow rate of 50 mL/minute, and 2.5 L / min, correspondingly, with a plasmatron voltage of 19 kV. The results show the absence of any significant differences regarding the location of chlorine in the DCP isomer while using the DBD plasma reactor. However, the final yield was unable to attain the quality standards issued by the government PERMENLH No. 5/2014; hence there is a need for additional processes.

Acknowledgments

The authors are grateful to the PUTI Prosiding 2020 (The Grants of International Indexed Publication for the lecturers of Universitas Indonesia) No. PENG-1/UN2.RST/PPM.00.00/2020. This was organized by the Directorate of Research and Community Service (DRPM) – Universitas Indonesia. Also, there are no competing or conflicting financial interests.

References

[1] Melanie B, H Murray, and Franco S 2014 Association of urinary phenols with increased body Weight Measures and Obesity in Children and Adolescents. *J. of Ped.iatr* **165** 744-9.
[2] Wei Y and J Zhu 2015 Urinary concentrations of 2,5-dichlorophenol and diabetes in US adult *J. Expo. Sci. Environ. Epidemiol* 1-5
[3] Alexander J and Clementina R 2015 A comparative study: degradation of 2,5-dichlorophenol in wastewater and distilled water by ozone and ozone-UV *Ozone-Sci Eng,* **38** 181-93
[4] Li L, Zhang H and Huang Q 2015 New insight into the residual inactivation of Microcystis aeruginosa by dielectric barrier discharge *Sci. Rep.* **5** 13683
[5] Song-hu Y and L. Xiao-hua 2005 Comparison treatment of various chlorophenols by electro-fenton method: relationship between chlorine content and degradation *J. Hazard. Mater.* **118** 85–92
[6] Poznyak T and Vivero J 2005 Degradation of aqueous phenol and chlorinated phenols by ozone. *J. Ozone-Sci Eng*, **27** 447-58
[7] Setyaningrum D, et al 2007 Aplikasi plasma lucutan penghalang dielektrik dengan reaktor berkonfigurasi spiral-silinder menggunakan gas sumber oksigen (O_2) murni untuk enghasilkan gas ozon (O_3)
[8] Wang J and Jiang X 2009 Unusual catalytic effects of iron salts on phenol degradation by glow

discharge plasma in aqueous solution *J. of Hazard Mater* **161** 926-32

[9] Hoigne J and Bader H 1975 Ozonation of water: role of hydroxyl radicals as oxidizing intermediates *Sci* **190**(4216) 782-4

[10] Gardoni D, Vailati A and Canziani R 2012 Decay of ozone in water: a review *Ozone Sci. Eng.* **34** 233-42

[11] Ershov B and Morozov P 2009 The kinetics of ozone decomposition in water: the influence of pH and temperature *Russ. J. Phys. Chem.* **83** 1457-62

[12] Hoigne J and Bader H 1983 Rate Constant of Reactions of Ozone with Organic and Inorganic Compounds in Water: Dissociating Organic Compounds *Water Res.* **17** 185-94

[13] Zhang H, Zhang Q, Miao C and Huang Q 2018 Degradation of 2,4-dichlorophenol in aqueous solution by dielectric barrier discharge: effects of plasma-working gases, degradation pathways, and toxicity assessment *Chemosphere* **204** 351-8

[14] Tang W and C P Huang 1996 Effect of chlorine content of chlorinated phenols on their oxidation kinetics by fenton's reagent *Chemosphere* **33** (8) 1621-35

[15] Tang W 2003 *Physicochemical Treatment of Hazardous Wastes* (New York: Taylor & Francis Group) pp 182

IOP Conf. Series: Materials Science and Engineering 1011 (2021) 012054 doi:10.1088/1757-899X/1011/1/012054

Synthesis of Nano-Composite Ag/TiO_2 for Polyethylene Microplastic Degradation Applications

Didit Ardi Maulana[1], Muhammad Ibadurrohman[1] and Slamet[1,a)]

[1] Department of Chemical Engineering, Faculty of Engineering, University of Indonesia, Depok 16424, Indonesia

Corresponding author: [a)] slamet@che.ui.ac.id

Abstract. The aim of this research is to synthesize Ag/TiO_2 nano-composites to study their ability to degrade microplastics in water, followed by varying size of microplastic particles as pollutants in drinking water. The method used for synthesizing Ag/TiO_2 is Photo Assisted Deposition (PAD). Characterizations done to determine the difference between TiO_2 and Ag/TiO_2 were SEM-EDX and UV-Vis DRS. Polyethylene microplastics were selected as samples in this study. The variation of microplastic size used is 100-125, 125-150, and 150-250 micrometers with initial concentrations of 100 ppm. Magnetic stirrers used at a rotational speed of 2000 rpm during the degradation process with a UV lamp irradiation. The addition of the Ag dopant has a good effect in microplastic degradation, where the percentage of degradation reaches 100% within 120 minutes of irradiation at an initial concentration of 100 ppm. An initial concentration of 100 ppm obtained the best percent degradation at particle size 125-150 micrometers, where 100% degradation achieved at 90 minutes irradiation.

1. Intruduction

Bottled drinking water meets the standards of WHO (World Health Organization) regarding quality and eligibility for consumption. Therefore, consumers believe that consuming bottled water is safe for health. However, in 2018 a study from the State University of New York at Fredonia that tested 259 bottles of drinking water from 11 brands sold in eight countries found that 93 percent of the water in the bottles tested contain microplastics [1]. One sample taken from Indonesia on average contains 382 microplastic particles per liter with sizes ranging from 6.5 to more than 100 micrometers [1]. It is known that bottled drinking water samples taken from local products contain up to 4713 microplastic particles per liter. Whereas non-local drinking water samples contain up to 10390 particles per liter of microplastic with polypropylene, nylon, polystyrene, and polyethylene types [1]. The test is carried back to the bottled water products under the same brand in the laboratory of the faculty of Mathematics, University of Indonesia, and found a microplastic particle size of 11 to 247 micrometers [2]. Microplastic carcinogenicity of particles contain in drinking water are known to cause cancer [3]. Physically existence microplastic as pollutants can also trigger clotting in the digestive tract. It is thought that microplastics can also translocate in the human body if they are below 150 μm [4]. Therefore, it should underline that the drinking water purification technology not only separates base on the size of microplastic particle, but also can be decomposed into harmless compounds.

The most important component in a water purification unit is a technology that can be apply to disinfect pathogenic microorganisms and degrade other pollutants. Plasma ion technology has been widely apply with the main component of ion generators which can produce OH radicals for water purification. However, purification units with plasma ion technology are still relatively expensive in terms of purchase costs and operating costs. In addition, plasma technology also has a high enough risk relate to work safety when apply. Therefore, purification technology using photocatalytic nanocomposite materials has become the choice and much develope. Various studies have combine several materials into composites to obtain an effective purification unit. Previous studies have shown that TiO_2 nanoparticles compile with plastic PE particles with a composition of 1% weight, showed microplastic degradation activity of 86% with UV irradiation, but the irradiation time required was quite long at 300 hours [5]. In increasing photocatalytic ability, a photocatalyst can be combine with a metal dopant. This increase is relate to the efficiency and ability of the core dopant to capture the charge carrier (electron trapping) and decrease the rate of recombination between e^- and h^+. Researcher chose silver metal as a dopant because it has been proven to reduce band-gap energy, so it can be active under visible light and act as an electron catcher that can limit the rate of recombination so that electrons are more easily excited and increase the photocatalytic activity of TiO_2 [6]. Besides being proven to be able to support the performance of TiO_2 in degrading organic compounds, silver metal also has advantages as an antibacterial agent. [7].

In this study, Ag/TiO_2 was synthesized by PAD method in which $AgNO_3$ precursor was added to TiO_2 P25 slurry where it was put under photoreactor for 6 hours. Characterizations such as SEM-EDX (Scanning Electron Microscopy with Energy Dispersive X-ray Spectroscopy) and UV-Vis-DRS (UV-Vis Diffuse Reflectance Spectroscopy) were done to provide knowledge regarding differences between TiO_2 and Ag/TiO_2. Ag/TiO_2 was then tested for its ability to degrade polyethylene microplastics in water in the presence of a stirrer under UV irradiation. Polyethylene microplastics were selected as samples because of their wide application in a variety of needs both industrial and household [8].

2. Experimentals

The purpose of this research is to synthesize Ag/TiO_2 nano-composites to test their ability to degrade polyethylene microplastics in liquid media. This study includes several stages of synthesis Ag/TiO_2, characterization, and testing of polyethylene degradation microplastics.

2.1. Tools and Materials

Polyethylene microplastic is derived from one of the raw materials for making facial wash (scrub) obtained from an online supplier with a size of 50-300 micrometers. The silver metal (Ag) used as a dopant for TiO_2 Evonik P25 is synthesized from $AgNO_3$ salt by the PAD method using a photoreactor equipped with a UV lamp and magnetic stirrer. Supporting materials such as anhydrous methanol (96%) and nitric acid (HNO_3) 65% are needed in synthesizing nano-composites. Distilled water (aquades) was used in all experiments. During the synthesis process also used equipment such as analytical balance with an accuracy of 0.001 gram, hot plate, sonicator, centrifuge, oven, and furnace. The characterization equipment used in this study is the UV-Vis DRS Spectrophotometer to measure absorption and band-gap energy, as well as the Inspect F50 Scanning Electron Microscope (SEM-EDX device) to see the morphology on the surface of nano-composites.

2.2. Methods

This section will explain the Ag/TiO_2 synthesis method, composite characterization, and polyethylene microplastic degradation test methods.

2.2.1. Synthesis of Ag/TiO_2 nano-composite. Synthesis of Ag/TiO_2 nanocomposites was carried out using the Photo Assisted Deposition (PAD) method. The first step of this research is to make TiO_2 slurry by adding 3 grams of TiO_2 to 405 mL of distilled water and adding nitric acid to reduce the pH of the slurry to 3 before being put into the sonicator for 30 minutes. The next step is to add $AgNO_3$ to a slurry of 0.047 grams to get a 1% addition of dopant to Ag/TiO_2 based on the initial mass of TiO_2 used. Stirring was continued for 30 minutes and added 45 ml of methanol to the slurry. Next, insert the slurry into the photoreactor for irradiation for 6 hours in the presence of stirring. Then the slurry is centrifuged 2 times *to separate and increase the pH of the composite at 4000 rpm for 30 minutes. Finally, the sample was dried at 150°C then calcined at 300°C for 1 hour.*

2.2.2. Characterizations of Ag/TiO_2. The characterizations done include SEM-EDX (Scanning Electron Microscopy with Energy Dispersive X-ray Spectroscopy) to analyse the final composition and surface morphology of the photocatalyst. UV-Vis-DRS (UV-Vis Diffuse Reflectance Spectroscopy) is also performed to determine the band-gap energy and light absorption power from the photocatalyst.

2.2.3. Microplastic degradation test. Polyethylene microplastic degradation test was carried out using the mass lose method. The performance of catalysts in microplastic degradation can be measured through measured microplastic mass reduction before and after UV irradiation treatment for 2 hours. Photoreactors used to activate photocatalysts in degrading microplastic pollutants are shown in Figure 1.

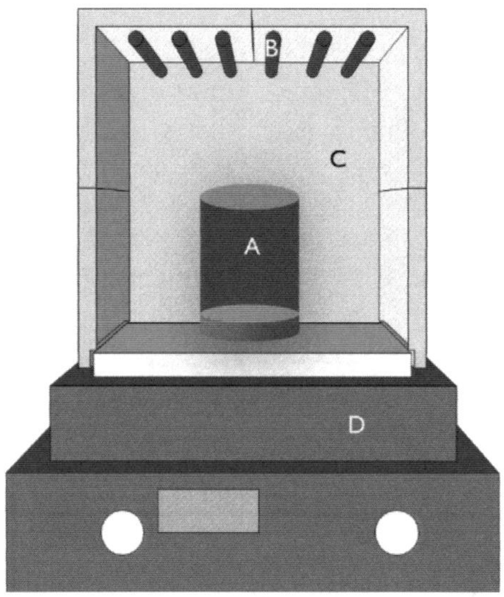

Figure 1. Front view of photoreactor with (A) beaker containing photocatalyst, microplastic and magnetic stirrer, (B) UV lamp, (C) aluminum reflector, and (D) hot plate as a base.

In this study, to determine the ability of Ag/TiO_2 in microplastic degradation testing, a variation in the size of the microplastic used are 100-125, 125-150, and 150-250 micrometers at a concentration of 100 ppm. Microplastic degradation analysis was carried out for 2 hours for each sample.

3. Results and Discussion

After synthesizing Ag/TiO_2 with the Photo Assisted Deposition method, the ability to degrade polyethylene microplastic will be tested. The researcher characterizes nano-composites to obtain relevant data that can support the results of the polyethylene degradation test. Characterization data that can be analyzed are the morphology and composition of nano-composites (SEM-EDX) and absorbance (UV-vis DRS).

3.1. Nano-composite characterizations

Data from SEM-EDX characterization and UV-vis-DRS used as a parameter of success in the manufacture of nano-composite Ag/TiO_2.

IOP Conf. Series: Materials Science and Engineering 1011 (2021) 012054 doi:10.1088/1757-899X/1011/1/012054

3.1.1. SEM-EDX characterization. The SEM-EDX characterization of the catalyst was carried out to see the surface morphology of the synthesized Ag/TiO_2 catalyst and to see the percentage loading of Ag elements in Ag/TiO_2 composites. The results of SEM characterization can be seen in Figure 2.

Figure 2. SEM-EDX characterization of (a) TiO_2 P25 and (b) 1% Ag/TiO_2.

Based on the results of the characterization in Figure 2, it can be explained that the treatment of Ag dopant loading on TiO_2 does not provide enough topological changes on the surface of the catalyst. It can be observed that samples from Ag/TiO_2 have fairly uniform morphology. The addition of dopants, in general, does not greatly affect the aggregate size of each catalyst sample. Therefore, the ability of Ag as a dopant can work optimally because the surface morphology of the catalyst has not changed significantly, where no other aggregates are found that can interfere with the photocatalytic performance of TiO_2. However, the addition of high Ag concentrations can increase the amount of Ag deposited on the surface of the TiO_2 photocatalyst [9]. This is causes the color of Ag/TiO_2 composite on SEM characterization changed gray. Besides that, the addition of large Ag loading can be a barrier on the active side of TiO_2 in absorbing photon energy for photocatalytic processes and can become a recombination center [10]. The advantage of the PAD method does not provide significant changes to the structure, size, and uniformity of the photocatalyst because Ag metal is deposited by electron photo-excitation mechanism which reduces Ag on the TiO_2 surface [11]. It is different from the hydrothermal method which substitutes metal into the crystal lattice, so it is quite influential on the structure, size, and uniformity of the crystal which has an impact on optical absorption and energy gap [12]. The sol-gel method also requires heat treatment up to 800°C, so it is quite dangerous during the synthesis process and has a significant effect on the photocatalyst crystal size [13]. Based on EDX results, it founded that the mass of silver metal propelled on TiO_2 was sufficient by the desired percentage of 1.11% Ag/TiO_2. Therefore, the PAD method used for composite synthesis is appropriate and gives quite good results that is performed by very simple procedures.

3.1.2. UV-vis-DRS characterization. UV-Vis DRS characterization was performed to determine the band-gap value of a semiconductor. In this study, characterization was carried out on a 1% Ag / TiO2 catalyst with the results shown in Figure 3.

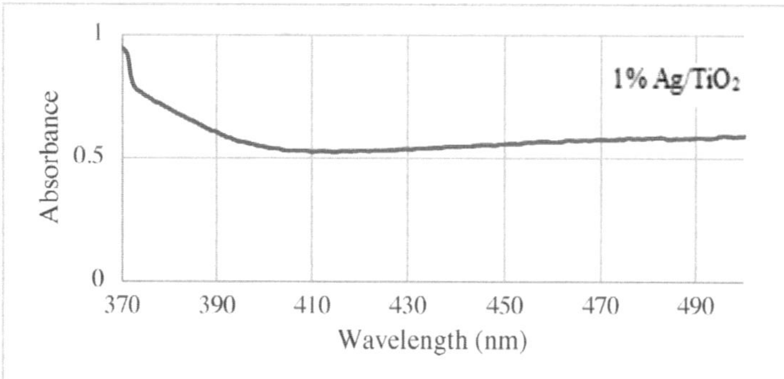

Figure 3. 1% Ag/TiO_2 absorbance.

Figure 3 shows the ability of photocatalyst absorbance to light with a wavelength of 370-490 nm which belongs to the wavelength region of visible light as shown by the x-axis. The absorbance value of the catalyst to light at wavelengths >400 nm is increasing. This shows that Ag/TiO_2 photocatalyst has responsiveness to visible light. This capability makes Ag/TiO_2 photocatalysts can be active under sunlight. In contrast to pure TiO_2 P25 which only has maximum performance under UV irradiation with a wavelength of <380 [14]. It should be noted that this absorbance measurement focuses on the intensity of light or electromagnetic radiation that can be absorbed by the composite so that an electron gets the energy to be able to move from a low energy level to a higher energy level. In the case of photocatalysts especially semiconductors, the low energy level is in the valence band position while the high energy level is in the conduction band position. This relates to the band-gap energy of a semiconductor which shows the minimum energy required by electrons to be excited from the valence band to the conduction band.

Based on the calculation of the band-gap using the Kubelka-Munk equation of 1% Ag/TiO_2 composite, it can be proven that the addition of silver metal dopants can reduce the band-gap to 3.02 eV with an optimum wavelength of 411 nm. Meanwhile, the band-gap owned by pure TiO_2 P25 is 3.19 eV with an optimum wavelength of 381 nm [15]. This decrease in band-gap energy results in the composite being able to absorb light at wavelength regions above 380 nm. Therefore, synthesized Ag/TiO_2 nanocomposites can be active under UV irradiation and also visible light irradiation. This can be seen from the increased absorbance ability of Ag/TiO_2 in the visible light region, but no peaks were found showing plasmonic resonance activity of Ag. This may be due to the relatively wide size distribution of Ag nanoparticles in Ag/TiO_2 [13]. Therefore, the role of Ag as a dopant is more dominant in its ability to capture electrons with a lower fermi level energy and a higher energy work function than TiO_2. This electron trapping process significantly reduces the electron-hole recombination rate, which results in stronger photocatalytic reactions.

IOP Conf. Series: Materials Science and Engineering 1011 (2021) 012054 doi:10.1088/1757-899X/1011/1/012054

3.2. Microplastic degradation

3.2.1. Degradation performance based on variations in the size of microplastic particles. Research with variations in microplastic particle size was carried out using an initial microplastic concentration of 100 ppm. Concentrations of 100 ppm were obtained by dispersing 10 mg of microplastic polyethylene particles into 100 ml of distilled water. Meanwhile, the nano-composite mass used was 50 mg. The degradation performance of nano-composites can be seen from the decrease in microplastic mass for 2 hours with observations every 30 minutes. The decrease in microplastic mass and the percentage of degradation can be seen in Table 1 and Table 2.

Table 1. Microplastic mass reduction based on variations in particle size.

Particle size (μm)	Mass of microplastic (mg)				
	0 min	30 min	60 min	90 min	120 min
100-125	10	6	4	2	0
125-150	10	6	2	0	0
150-250	10	8	6	2	0

Table 2. Percent degradation base on variations in particle size.

Particle size (μm)	Persent Degradation (%)				
	0 min	30 min	60 min	90 min	120 min
100-125	0	40	60	80	100
125-150	0	40	80	100	100
150-250	0	20	40	80	100

The ratio of decreased microplastic concentration due to degradation activity by Ag/TiO_2 nano-composites is presented in graphical form in Figure 4.

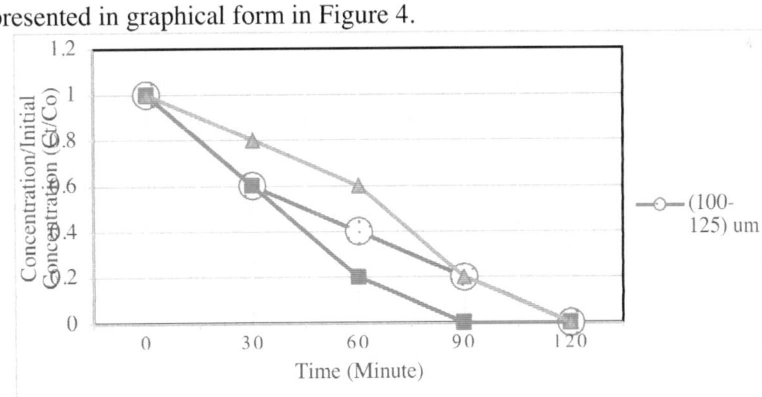

Figure 4. Decreased microplastic concentration based on variations in the initial size of microplastic particles.

The graph in Figure 4 shows that the decrease in microplastic concentration reaches 0 at the 90 minute for microplastic samples with particle sizes of 125-150 micrometers. This event is related to the ability of the microplastic to mix evenly in a solution. The mixing of solid particles is affected by the stirring carried out. Where the stirring process is carried out to suspend microplastic and nano-composite solids in aquades. The type of stirrer influences the flow pattern produced in the beaker. In this study, the magnetic stirrer used has the same concept as the paddles type stirrer, where the resulting flow pattern has a radial and tangential direction to cause the formation of a whirlpool. This whirlpool has an effect on microplastic particles in the solution.

The density of polyethylene microplastic is smaller than the density of aquades, so microplastic particles are more likely to be on the surface. Therefore, microplastic particles will experience centrifugal force. The centrifugal force is directly proportional to the mass of the particle and the speed of the particle, but inversely proportional to the radius of the path. Particle speed is influenced by the speed of stirring, where the stirring speed in this study is made constant at 2000 rpm. The radius of the track is also constant, because the diameter of the beaker used is the same. Therefore the main factor affecting the centrifugal force on microplastic particles is the mass of the particle.

The mass of a microplastic particle varies depending on the size of the particle. The particle size of 100-125 micrometers experiences a centrifugal force that is small enough that the particles cannot be mixed optimally. Meanwhile, particles with a size of 150-250 micrometers accept maximum centrifugal force, so they tend to be thrown to the outermost path which is limited by the wall of the beaker. Microplastic particle gives the optimum mass for one particle at a size of 125-150 micrometers. Therefore, the contact between the microplastic and radical species is optimal at that size. This research took 2 hours to degrade microplastics up to 100%. These results are better than previous studies by Zhao 2007, which were only able to degrade microplastics up to 86% for 300 hours of irradiation in the solid phase PE-TiO_2 (1% wt) under UV irradiations.

It should be underlined that this study uses mass reduction measurements to determine decreased microplastic concentrations after irradiation. Therefore, the accuracy of the measuring instrument will determine the quality of the data obtained. Measuring instruments that researchers use have accuracy of 0.001 grams or equivalent to 1 mg, so that the measurement error of each mass reduction data obtained is equal to 1 mg.

Some suggestions that can be given for the development of further research are conducting a microplastic degradation test with variations in stirring speed, using a measuring instrument with higher accuracy, at least four digits behind the comma so that the measurement results are more accurate, then after obtaining optimum degradation performance, it can be applied to drinking water treatment equipment to avoid health hazards due to microplastics.

4. Conclusion

The characterization of silver doped TiO_2 using SEM EDX technique revealed the dispersion of silver metal on the surface of TiO_2, while the UV-Vis DRS result show that the band-gap energy of TiO_2 decreased as a result of the presence of deposited silver metal. The silver doped TiO_2 composite catalysts show an adsorption threshold extended into the visible region. Under UV irradiation, Ag deposits exhibit the effect only as electron traps, thus leading to the enhancement in the Ag/TiO_2 photocatalytic activity. The synthesized material of Ag/TiO_2 has been tested in microplastic degradation and it and it was shown that microplastics with a size of 125-150 micrometers showed very good results. Where the degradation performance was obtained up to 100% under UV irradiation and the presence of stirring which affected the mixing of microplastic particles in the slury.

Acknowledgment

Authors would like to express gratitude to PDUPT University of Indonesia fiscal year 2020, No. Contract NKB-2864/UN2.RST/HKP.05.00/2020., that has funded this study.

References

[1] Mason S A 2018 *Synthetic Polymer Contamination in Bottled Water*. Science Complex **340** pp 6 – 12

[2] Tyree C 2018 *Mikroplastik dalam Botol Air Mineralmu*

[3] Meyer-Rochow V B 2015 *Plastic ocean and the cancer connection*. Environmental research **142** pp 575-578

[4] Lusher A 2017 *Microplastics in fisheries and aquaculture*. FAO Fisheries and Aquaculture

Technical Paper (FAO) eng no 615

[5] u Zhao 2007 *Solid-phase photocatalytic degradation of polyethylene plastic under UV and solar light irradiation*. Journal of Molecular Catalysis A: Chemical **268** pp 101-106

[6] Reddy P V L 2017 *TiO_2-based photocatalytic disinfection of microbes in aqueous media* Environmental research **154** pp 296-303

[7] Gupta K 2013 *Photocatalytic antibacterial performance of TiO_2 and Ag-doped TiO2 against S. aureus. P. aeruginosa and E. coli* Beilstein journal of nanotechnology **4** pp 345-351

[8] Tofa S T and Dutta J 2018 *Degradation of Microplastic Residuals in Water by Visible Light Photocatalysis* (Sweden: KTH Royal Institute of Technology)

[9] Sobana N, Muruganadham M & Swaminathan M 2006 *Nano-Ag particles doped TiO_2 for efficient photodegradation of direct azo dyes* Journal of Molecular Catalysis A: Chemical **258** pp 124-132

[10] Marugan J 2009 *Photocatalytic inactivation of bacteria in water using suspended and immobilized silver-TiO_2* Applied Catalysis B: Environmental **93** pp 112-118

[11] Hu H 2017 *Photo-assisted deposition of Ag nanoparticles on branched TiO_2 nanorod arrays for dye-sensitized solar cells with enhanced efficiency*. Journal of Alloys and Compounds **694** 653-661

[12] Anggelina F, Iwantono I, Umar L, & Awitdrus A 2017 *PENGARUH PENAMBAHAN MATERIAL PLASMONIK PERAK (Ag) TERHADAP SIFAT FISIS NANOROD ZnO DENGAN METODE SEED-MEDIATED HYDROTHERMAL* Indonesian Physics Communication, **14** pp 1041-1047

[13] Zaleska-Medynska A 2018 *Metal oxide-based photocatalysis: fundamentals and prospects for application*. Elsevier.

[14] Li S 2010 *Photocatalytic degradation of polyethylene plastic with polypyrrole/TiO_2 nanocomposite as photocatalyst*. Polymer-plastics technology and engineering **49** pp 400-406

[15] Suwanchawalit C 2012 *Enhancement of the photocatalytic performance of Ag-modified TiO_2 photocatalyst under visible light* Ceramics International **38** pp 5201-5207

IOP Conf. Series: Materials Science and Engineering 1011 (2021) 012055 doi:10.1088/1757-899X/1011/1/012055

Microplastic Pollutant Degradation in Water Using Modified TiO_2 Photocatalyst Under UV-Irradiation

Muhamad Haris Fadli[1c], Muhammad Ibadurrohman[1b] and Slamet Slamet[1a]

[1]Department of Chemical Engineering, Universitas Indonesia, Depok, West Java, 16424, Indonesia

[a]Corresponding Author: slamet@che.ui.ac.id
[b]ibad@che.ui.ac.id
[c]muhamad.haris@ui.ac.id

Abstract. Modified TiO_2 photocatalyst has been widely studied on its ability to degrade pollutants in water. Nowadays, one of the water pollutants that significantly polluted water is microplastic. Photocatalyst TiO_2 can create an oxidizing agent to degrade microplastic pollutants but have a very high recombination rate of electrons and holes. By modifying TiO_2 with silver dopant using Photo-assisted Deposition (PAD) and Reduced Graphene oxide (RGO) using the ultrasonic radiation method, can highly increase the ability of TiO_2 catalyst. The modified catalyst was characterized by SEM-EDX and UV-Vis DRS to analyze the properties change of the composite. The performance of composites in degrading microplastic (polyethylene) was tested under UV radiation for 4 hours. The $Ag/TiO_2/RGO$ catalyst has a significantly high degradation percentage with 76% compared to pure TiO_2 and Ag/TiO_2 with degradation percentage of 56% and 68% respectively.

1. Introduction

Microplastic has recently become a major problem for the environment because of its nature which is difficult to degrade and small in size and its impact on the environment because it is widely spread in nature [1]. New problems caused by microplastics emerged in 2018 ago. According to the results of research from the State University of New York at Fredonia which tested 259 bottles of drinking water from 11 brands sold in eight countries, the results obtained are that 93 percent of the water in the bottled bottles tested contains microplastics. Samples taken from Indonesia on average contain 382 microplastic particles per liter with sizes ranging from 6.5 to more than 100 micrometers [2]. Clarifying the results of the study, a re-testing of bottled drinking water products was conducted at the University of Indonesia's Faculty of Mathematics and Natural Sciences laboratory. With the same brand, the test results obtained that the bottled water products contain microplastics measuring 11 to 247 micrometers [3]. The danger of the presence of microplastics in water is known to trigger cancer caused by the nature of the carcinogenic it has [4]. Physically, the presence of microplastics as pollutants can generally trigger clots in the digestive tract. Microplastics are also expected to translocate in the human body if they are below 150 μm [5].

The usage of photocatalyst has an opportunity to degrade microplastic in water. Research lately have been done to prove the ability of photocatalyst to degrade microplastic. Previous study revealed that the microplastic degradation mechanism using LDPE as the model of plastic, starts with the formation of hydroxyl radicals that will react with LDPE to form polyethylene alkyl radicals and water. The alkyl radical polyethylene will then be degraded to CO_2 and H_2O after undergoing several series of reactions [6]. However, the ability to reduce microplastic mass is still very low. A study by Ariza-Tarazona and team (2019) investigated the degradation of microplastics using N-TiO_2 modified photocatalyst. From the result, the degradation of microplastic is still not efficient using the N-TiO_2 modified photocatalyst shown by low mass degradation percentage [7]. According to Ahmed and team (2010), the photocatalytic activity of titania are reliant on the surface and structural properties of the semiconductor in which including the surface area, crystal composition, particle size distribution, band gap, and others. These factors have a huge impact on the adsorption behavior during degradation of pollutants [8]. Tavakoli and team (2018) conducted a research to examine the effect of the addition of Ag metals and graphene on TiO_2 on their photocatalyst performance by assessing the degradation efficiency of the organic acid Acid 7. modeled with Acid Orange (AO7). The addition of graphene and Ag increased the efficiency of degradation from 61% to increase to 99%.

From the study above, modifying TiO_2 with Ag metals or graphene proves to be effective in increases the photocatalyst performance of TiO_2. In this study, the effect of Ag metals and graphene derivatives, which is reduced graphene oxide (RGO) to the photocatalyst activity of TiO_2 on the degradation of microplastic modeled as polyethylene scrub was investigated. The modified photocatalyst was synthesized using photo-assisted deposition (PAD) and ultrasonic radiation mixing method. The catalyst was characterized with SEM-EDX and UV-Vis DRS and the degradation efficiency was tested under UV irradiation.

2. Experimental Method

2.1. Tools and materials

The material used in this study is polyethylene scrub with a size of 100-150 μm as the model of microplastic, aquades, photocatalyst TiO_2 P25 Evonik powder, $AgNO_3$ as the precursor of Ag metal, RGO, HNO_3 and methanol as the solvent. The degradation process of the microplastic was carried out in a photoreactor with eight 8watt UV lamps, reflector from aluminium plate, and a magnetic stirrer in the reactor.

2.2. Preparation of catalyst

The Ag/TiO_2 and Ag/TiO_2/RGO photocatalyst was prepared using TiO_2 P25 Evonik powder as the titania precursor, while $AgNO_3$ and RGO used as the source of silver (Ag) and RGO respectively. HNO_3 and methanol are used as a solvent in the PAD method. The catalyst was first synthesized with PAD to get Ag/TiO_2. The amount of $AgNO_3$ was weighed based on the dopant concentration which is 3% Ag. TiO_2 P25 was dissolved in distilled water that had been added with HNO_3 solution to a pH of 3 and stirred with a magnetic stirrer and sonicated with ultrasonication. $AgNO_3$ was added to TiO_2 soles with varying concentrations and stirred with magnetic stirrers. Then, add methanol while stirring. After that, irradiated with a UV lamp for 6 hours and centrifuged for 30 minutes at 4000 rpm. Dry the sample at 150 °C using a hotplate. The sample then calcined using a furnace at 300 °C for 1 hour. The Ag/TiO_2 catalyst was then synthesized with RGO to get Ag/TiO_2/RGO. The amount of RGO is 1% wt of Ag/TiO_2. Ag/TiO_2 and RGO were mixed in 10 ml distilled water. After that, the solution was sonicated for 4 hours and irradiate with UV for 4 hours. Then the solution dried at 60°C and the solid-state calcined at 500°C for 1 hour.

2.3. Characterization of catalyst

The catalyst was characterized by using SEM-EDX (Scanning Electron Microscopy with Energy Dispersive X-ray Spectroscopy) and UV-Vis DRS (UV-Vis Diffuse Reflectance Spectroscopy). The SEM-EDX characterization of the catalyst was carried out to see the pattern or surface structure of the synthesized Ag/TiO_2 and Ag/ TiO_2/RGO catalysts and to see the percentage loading of Ag elements on Ag/TiO_2 and the percentage of carbon in Ag/TiO_2/RGO catalysts. The UV-Vis DRS characterization is intended to determine the value of the energy gap band of the TiO_2 photocatalyst composite which has been supported by silver metal.

2.4. Microplastic degradation test
Microplastic degradation test is done by adding microplastics modeled with polyethylene scrub sized 100-150 μm into distilled water. For the test 100 ml distilled water and 50 mg microplastic were added to beaker glass and then 50 mg of catalysts were added to the solution. The solution was stirred continuously during the degradation test. The tests carried out at room temperature with the help of UV irradiation Tests were conduct for 1, 2, 3 and 4 hours with the same initial amount of microplastic pollutant. The leftover of microplastic in the solution then filtered using Whatman Filter Paper 41. After that, the filter and the microplastic were dried using an oven with temperature 100 °C for 1 hour. The degradation reactor is modeled in Figure 1.

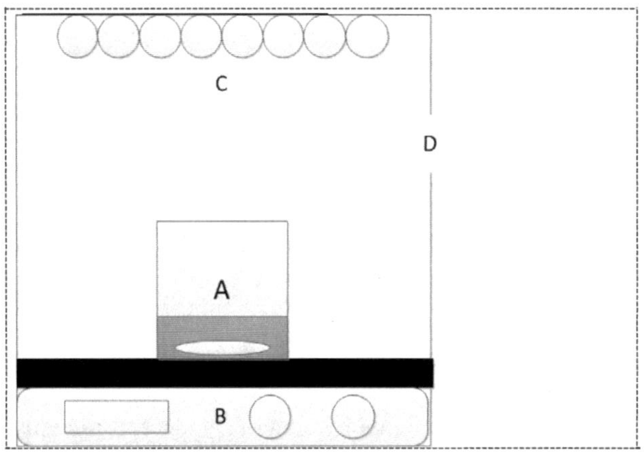

Figure 1. Design of microplastic degradation test reactor with (A) beaker containing photocatalyst, microplastic and magnetic bar, (B) magnetic stirrer, (C) UV lamp, and (D) reflector.

Determination of the performance of microplastic degradation is done by measuring the percentage of microplastic mass loss for every hour of UV irradiation. The equation for calculating the percentage using Equation (1).

$$\% = \frac{m_0 - m_t}{m_0} \times 100\% \quad (1)$$

where:
% : microplastic mass degradation percentage (%)
m_0 : microplastic initial mass (mg)
m_t : microplastic mass at time t (mg)

3. Result and Discussion

3.1. Characterization of catalyst
The SEM characterization of the catalyst is shown in Figure 2. SEM characterization aims to see the surface pattern or structure of the synthesized Ag/TiO_2 and $Ag/TiO_2/RGO$ catalysts. From the figure, it shows spherical particles, unevenly sized and a bit of agglomeration occur in certain areas. The agglomeration can occur due to the formation of a solid bridge between surface hydroxyl groups during the drying process or combination of the particle after being treat with a high temperature in the

calcination process [10,11]. It can be explained that the treatment of PAD and ultrasonic radiation does not greatly affect the aggregate size of each catalyst sample when compared to TiO_2 P25.

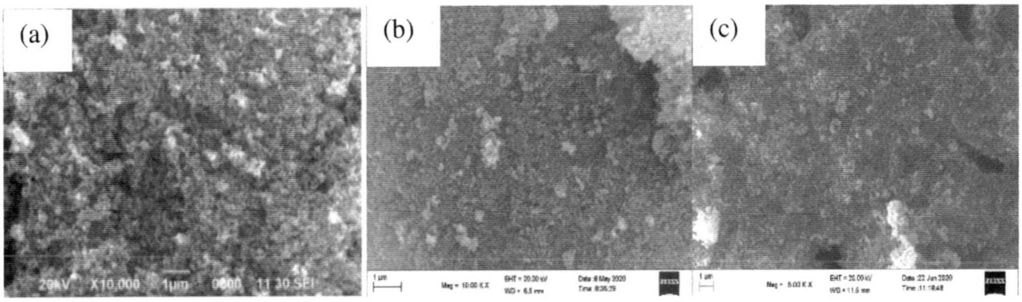

Figure 2. SEM-EDX characterization of (a) TiO_2 P25, (b) 3%Ag/TiO_2 and (c) 3% Ag/TiO_2-1%RGO

From the EDX results, it founded that the mass of silver metal-doped on TiO_2 was sufficient by the desired percentage of 2.71%wt Ag on Ag/TiO_2 and 3.94%wt on Ag/TiO_2/RGO. Therefore, the PAD method used for synthesis gives quite good results of silver dopant on photocatalyst. From the EDX, it also gives the carbon content in Ag/TiO_2/RGO with the amount of 1.56%wt which indicates that RGO was compiled with Ag/TiO_2 as a composite.

Figure 3 shown the absorbance of catalyst from UV-Vis DRS characterization result of 3%Ag/TiO_2 catalyst compared to TiO_2 P25. Based on 3, with the presence of silver dopant, the absorbance of light wavelength greater than 380 nm is increasing. Then by using the Kubelka-Munk equation the value of the band gap energy for TiO_2 and 3%Ag/TiO_2 catalyst was calculated as 3.1 eV and 2.85 eV respectively. The effect of loading silver metal dopants is quite significant in the value of the catalyst band gap energy. This can occur because the silver metal dopant acts as a good electron trapper and causes the band gap energy of the catalyst to decrease so that it can absorb light with a wavelength higher than 380 nm. Moreover, silver metal as a noble metal is highly absorbed UV and visible light and thus can be integrated into a semiconductor to improve photoactivity [12,13,14]. The higher absorbance for visible light and also has a plasmonic effect that can help to increase the electron-hole charge separation of photocatalyst [15,16].

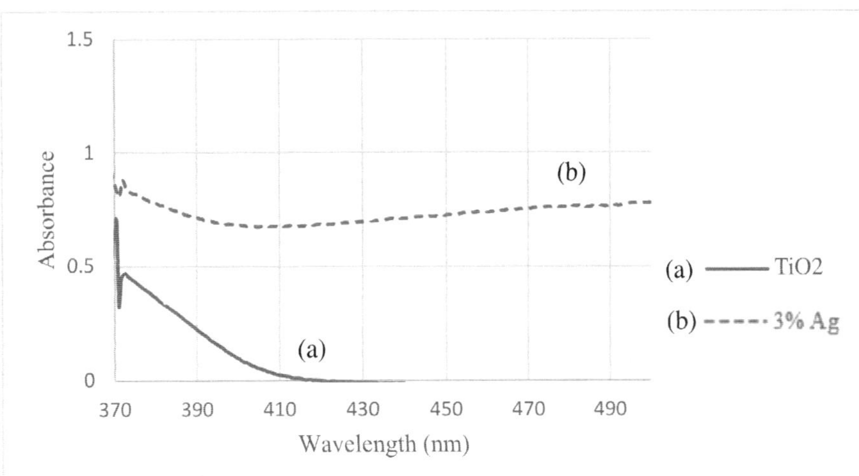

Figure 3. Absorbance of (a) TiO_2 P25 and (b) 3% Ag/TiO_2 catalyst.

3.2. Microplastic Degradation Test

In this test, there are three variations of the sample and one blank test using no catalyst to test if the degradation of microplastic can happen only with UV irradiation. The catalyst that is used for testing is pure TiO_2 P25 catalyst, 3% Ag/TiO_2, and 3%Ag/TiO_2-1%RGO. The test is carried out using an initial microplastic mass of 50 mg in 100 mL of distilled water and a catalyst of 50 mg. The experiment time is 1 hour, 2 hours, 3 hours, and 4 hours with each experiment will have the same initial amount of microplastic. The results of the experiment can be seen in Table 1.

Table 1. Microplastic mass reduction.

Catalyst	Mass of microplastic (mg) at time				
	0 h	1 h	2 h	3 h	4 h
Without catalyst	50	50	50	50	50
TiO_2 P25	50	40	32	26	22
3% Ag/TiO_2	50	30	24	19	16
3% Ag/TiO_2-1%RGO	50	28	18	14	12

Based on Table 1 the microplastic degradation performance of 3% Ag/TiO_2 and 3% Ag/TiO_2-1% RGO catalysts are greater than pure TiO_2. This is because silver metal and RGO has high electron mobility properties so that it will act as an efficient electron acceptor to increase the induced photon transfer thus inhibiting the rate of electron-hole recombination [9,12,13]. Also, silver metal is highly absorbed UV and visible light and because of the higher absorbance for visible light can help to increase the electron-hole charge separation of the photocatalyst [16]. Therefore, it will cause more active sites on the surface of the catalyst. Moreover, RGO has a large theoretical specific surface area which can help to increase the electron mobility to RGO so electron will be stored on the surface of RGO. Also because of the large surface area, it can increase the adsorption ability of catalysts so it will help the catalyst to get contact with the pollutant. Using Equation (1), the percentage of microplastic degradation of each catalyst can be calculated. The data of microplastic degradation percentage then plotted into graphic and shown in Figure 4 below.

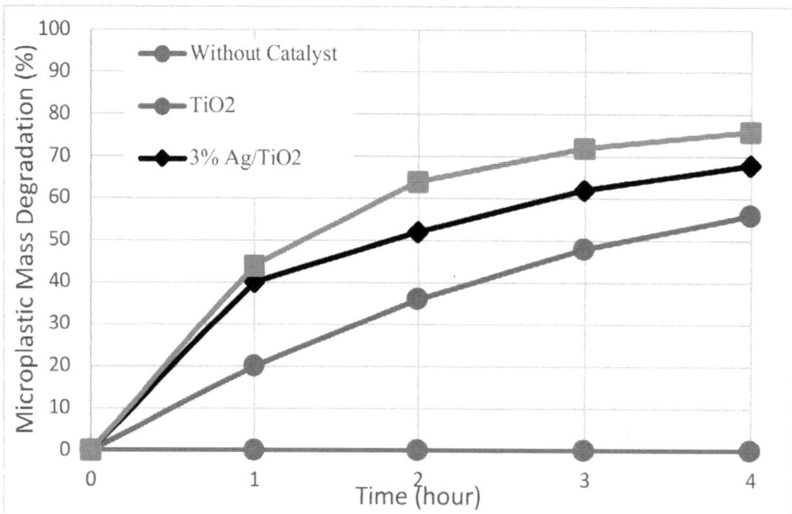

Figure 4. Microplastic mass degradation percentage of (a) TiO_2 P25 and (b) 3% Ag/TiO_2 catalyst.

4. Conclusion

From the study, it can be concluded that the addition of silver metal dopant and RGO to the TiO_2 photocatalyst significantly increases the degradation performance of microplastic in water. From the experiment that had been done, the best performance of microplastic degradation is 76% which is from the 3%Ag/TiO_2-1%RGO catalyst.

Acknowledgment

The study was funded by PUTI Proceedings of Universitas Indonesia 2020 Grant, No. Contract NKB-1107/UN2.RST/HKP.05.00/2020. The author also would thank all people and organizations that help and support the author during this research.

References

[1] Tofa T, Kunjali K, Paul S and Dutta J 2019 *Environ. Chem. Lett.* **17** 1341-6
[2] Mason S, Welch V and Neratko J 2018 *Front. Chem.* **6** 407
[3] Tyree C, Morrison D, Makitan G and Maulidar I 2018 Mikroplastik dalam Botol Air Mineralmu. Investigasi Tempo
[4] Benno Meyer-Rochow V, Valérie Gross J, Steffany F, Zeuss D and Erren T 2015 *Environ. Res.* **142** 575-8
[5] Lusher A, Hollman P and Mendoza-Hill J 2017 Microplastics in Fisheries and Aquaculture: Status of Knowledge on Their Occurrence and Implications for Aquatic Organisms and Food Safety
[6] Ali S, Qazi I, Arshad M, Khan Z, Voice T and Mehmood C 2016 *Environ. Nanotechnol. Monit. Manag.* **5** 44-53
[7] Ariza-Tarazona M, Villarreal-Chiu J, Barbieri V, Siligardi C and Cedillo-González E 2019 *Ceram. Int.* **7** 9618-24
[8] Ahmed S, Rasul M G, Martens W N, Brown R and Hashib M A 2010 *Desalination* **261** 3–18
[9] Tavakoli F, Badiei A and Ghasemi J 2019 *J. Water Environ.* **4** 31-9
[10] Sikirman A, Krishnan J, Jai J and Senusi F 2014 *Adv. Mater. Res.* **894** 245–9
[11] Krishnan J, Mohamad E N and Hadi A 2014 *Adv. Mater. Res.* **661** 63–7
[12] Rycenga M, Cobley C M, Zeng J, Li W, Moran CH, Zhang Q, Qin D and Xia Y 2011 *Chem. Rev.* **111** 3669–712
[13] Mogensen K B and Kneipp K 2014 *J. Phys. Chem. C.* **118** 28075–83
[14] Wen B, Liu C and Liu Y 2005 *Inorg. Chem.* **44** 6503–5
[15] Zhang S, Ren F, Wu W, Zhou J, Sun L, Xiao X and Jiang C 2014 *J. Colloid Interface Sci.* **427** 29–34
[16] Zaleska-Medinska A 2018 Metal oxide-based photocatalysis
[17] Tang B, Chen H, Peng H, Wang Z and Huang W 2018 *J. Nanomater.* **8** 105

IOP Conf. Series: Materials Science and Engineering 1011 (2021) 012056 doi:10.1088/1757-899X/1011/1/012056

Characterization of Fine Particulate Matters Collected in the Vicinity of Coal-fired Power Plants in Java using ED-XRF

D D Lestiani, M Santoso, I Kusmartini, D P D Atmodjo, S Kurniawati, D K Sari

Center for Applied Nuclear Science and Technology, BATAN, Jln Tamansari 71, Bandung, Indonesia

*Corresponding author's email: diahdwi@batan.go.id

Abstract. The number of coal-fired power plant in Java has been increased because of the government policy to supply the electricity in Indonesia to meet the electricity needs. This mega project is assumed to have an impact on the environmental and human health. In this study, the assessment of the air quality near the coal fired power plant has been carried out through characterization of the fine particulate matters. The particulate matters $PM_{2.5}$ were collected in four sites in the vicinity of coal-fired power plant in Banten, West Java and Central Java, Indonesia. The elemental concentrations in $PM_{2.5}$ such as Mg, Al, Si, S, Cl, K, Ca, Ti, V, Cr, Mn, Fe, Ni, Cu, Zn, Br and Pb have been identified using energy dispersive X-ray Fluorescence-EDXRF. The validation of XRF measurement was applied using standard reference material NIST SRM 2783 Air Particulate on Filter Media. The elemental concentration of $PM_{2.5}$ especially for the toxic elements as well as the mass and black carbon were compared between four sites. For some sites, the level of elemental concentrations was lower than other industrial sites in Java, due to the sampling was carried out before the commissioning/normal operation. Further and periodical monitoring will be needed to obtain a better comprehensive result to assess the impact of coal power plant to the surrounding environment.

1. Introduction

The energy consumption in Indonesia especially in Java has risen due to the rapid growth of industrial and economic activities. Perusahaan Listrik Negara (PLN), Indonesia's state-owned electricity company has a plan to build an additional power plant; the energy industry in Indonesia is currently dominated by coal-based power plants to meet this need. Indonesia government plan to develop an additional 35,000 MW of new power capacity [1,2]. The program comprises more than 40 power plant mostly coal fired in Java and Bali. Coal fired power plant covers about 53% of total electricity generation capacity in Indonesia, and it is estimated to be increased up to 66% [3]. Completion of the 35,000-Megawatt (MW) plant program is expected to decline from the original target in 2026. During 2015 to 2019, PLN had completed the construction of a 14,792.5 MW power plant from the planned 35,000 MW [4].

This huge and rapid development of coal fired power plant should be taken into consideration related to the environmental impact. The process of coal combustion from coal fired power plant is significant source of pollutant since it produces massive amounts of COx, SOx, NOx and particulate matter [3,5]. The local coal fired power plant in Indonesia use low rank coal subbituminous coal, however its sulfur content is often low, sometimes less than 1 percent by weight. These emissions increase the ambient concentration of PM less than 2.5 microns in diameter ($PM_{2.5}$) over hundreds to thousands of kilometers downwind of the plants [6,7]. Exposure to $PM_{2.5}$ has been associated with cardiopulmonary diseases, lung cancer and numerous other respiratory illnesses [8]. The concentration of elements such as mercury, lead, cadmium, arsenic, chromium, and other hazardous elements should be monitored. The high concentration of heavy metals in the fine particulate matter could be harmful to human health.

Nuclear analysis techniques such as ED-XRF are analytical techniques that are very suitable for characterizing airborne particulate samples. ED-XRF has been applied to characterization of airborne particulate samples in various pollutant source identification research. It has high capability, simultaneous measurement, multi-element, high sensitivity, short analysis time and its non-destructive which makes it ideal for this type of work [9,10]. Air particulate samples in the media filters have a mass that is very small up to 100 µg and the trace elements in it are also in trace level, making it difficult to analyze using other conventional methods. ED-XRF is an effective tool and allows the characterization of multi-element compositions and can produce large data sets for source identification [11].

There are limited numbers of publications related assessment of environmental impact from coal-fired power plant in Indonesia, with respect to the chemical composition of the fine particulate matter. Previous study has been published related the chemical composition on feed coal and fine particulate matter in Cilacap area[12]. In this study, characterization of fine particulate matter in the vicinity of a coal-fired plant in Java island covering Banten, West Java and Central Java was carried out. The objectives of this study are to provide characteristic of mass, black carbon and elemental concentrations of fine particulate matter emitted from coal-fired power plant.

2. Method

2.1. Sampling
Sampling was conducted at 4 sampling sites near the coal fired power plant Suralaya, Labuan, Cilacap and Tanjung Jati in Java island covering Banten, West Java, and Central Java provinces. The sampling locations for these sampling sites are shown in Figure 1. Airborne particulate matter was collected using Gent stacked filter unit dichotomous sampler that capable for collecting particulate matter in the $PM_{2.5-10}$ (coarse) fraction and $PM_{2.5}$ size (fine) fraction [13]. Sampling was collected on flow rate of 15-18 L/min.

2.2. PM, BC and elemental analysis
The particulate matter mass of each fine and coarse fractions was determined by gravimetric using a microbalance. The filters were stored for 24 hours prior to weighing in controlled environmental condition room. Mass concentrations of $PM_{2.5}$ and $PM_{2.5-10}$ ($\mu g.m^{-3}$) were obtained by dividing the gravimetric mass by the volume of air that passed through the filter. The concentration of black carbon (BC) in the samples was determined by reflectance measurement using an EEL Smoke Stain Reflectometer model 43D. Secondary standard of known reflectance were used to calibrate the reflectometer. Elemental analysis was performed using an energy dispersive Xray fluorescence ED-XRF Epsilon 5 (Panalytical, Ltd) by optimizing the five existing secondary targets (Al, CaF_2, Fe, Ge, and Zr) [11].

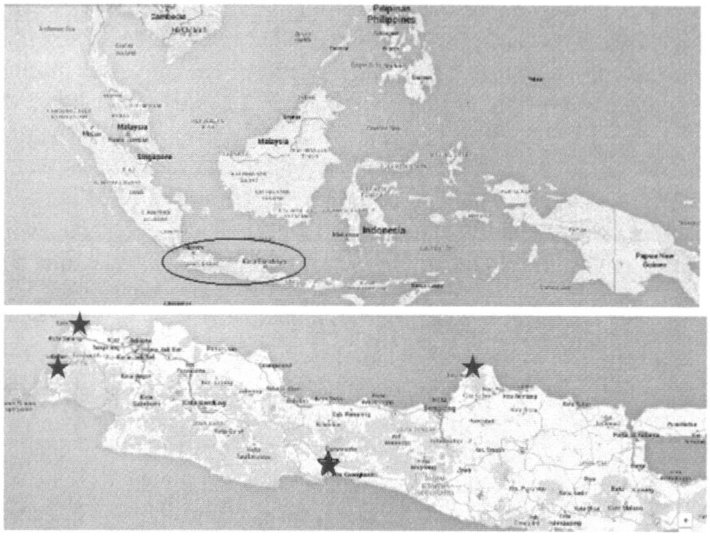

Figure 1. Geographical location of coal fired power plant in Cilacap, Central Java (source: map.google.com)

2.3. Reconstructed Mass (RCM)

The measurement of so many different elemental species allowed estimates of significant fine particle components like sea salt, ammonium sulphate, soil and smoke. We could define the RCM as a sum from sea salt=Na+Cl, ammonium sulfate=4.125*S, soil=2.20*Al+2.49*Si+1.63*Ca+1.94*Ti+2.42*Fe and smoke= K-0,6*Fe, and BC [7].

3. Results and Discussions

$PM_{2.5}$, PM_{10} and black carbon concentrations for each sampling site are summarized in Table 1 and shown in Figure 2. Twenty-four hours average concentration of $PM_{2.5}$ at four sampling sites in the vicinity of coal power plant in Cilacap, Labuan, Suralaya and Tanjung Jati were 8.24, 17.06, 14.65 and 14.46 $\mu g.m^{-3}$, respectively. The highest concentration of $PM_{2.5}$ was found in Labuan and the lowest was in Cilacap. Even though, this $PM_{2.5}$ concentration was still below the 24 hours of Indonesian national air ambient standard (65 $\mu g.m^{-3}$), but by assuming that these weeks of sampling can represent the annual mean of $PM_{2.5}$ level, the $PM_{2.5}$ concentrations in sampling site Labuan, Suralaya dan Tanjung Jati would exceed the Indonesian national ambient air quality standard of an annual mean $PM_{2.5}$ value of 15 $\mu g.m^{-3}$ (Government Regulation, 1999).

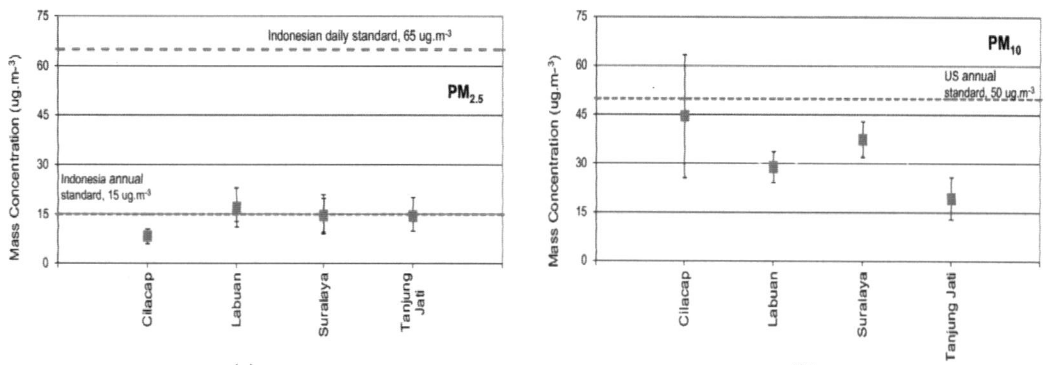

(a) (b)

Figure 2. $PM_{2.5}$ and PM_{10} mass concentration in 4 sites

Table 1. The sampling data, mean mass and BC concentrations (μg.m^{-3})

Sampling site (*n*)	$PM_{2.5}$	PM_{10}	BC
Cilacap (*2*) [12]	8.24±0.02	44.50±18.89	1.35±0.14
Labuan (*3*)	17.06±2.32	28.89±4.79	2.11±0.26
Suralaya (*3*)	14.65±5.93	37.43±5.44	2.65±0.73
Tanjung Jati (*3*)	14.46±5.11	19.37±6.51	2.52±0.71

In Cilacap sampling site, the samples were collected before the coal fired power plant was not operating and still under construction. Therefore, the $PM_{2.5}$ concentration was low due to the emission of coal power plant is not contributed significantly. On the contrary, the PM_{10} concentration in Cilacap site was higher compared to other sites since the pollutant source such as road dust or crustal element were higher during the construction activity. The ratio of $PM_{2.5}/PM_{10}$ in Cilacap was 18.5% which indicated that the coarse particulate was dominant compared to fine particulate. Compared to other $PM_{2.5}$ data in several cities in Java island, $PM_{2.5}$ concentrations in Labuan, Suralaya and Tanjung Jati were in similar level with $PM_{2.5}$ reported by Santoso et al, 2020 for seven urban cities in Java in average of 14-18 μg.m^{-3} except for Yogyakarta [14]. In this study, we are unable to identify the high concentration or increasing of concentration of $PM_{2.5}$, PM_{10} or BC from time to time, due to the short-term monitoring.

Black carbon, BC is commonly emitted to the atmosphere due to incomplete combustion, the major source of BC are transport related emissions such as diesel emission and long-range transport from fossil fuel related sources such as coal fired power plant emission, and biomass burning could be other substantial sources of BC [15]. The average BC fine contributed 12-19% of the $PM_{2.5}$ mass concentration in the four sampling sites. These values do not differ significantly with the values reported in industrial area in Serpong area in 2008 (13-26%) which used the same sampling method and analytical procedure [16]. The lowest BC was found in Cilacap, with similar reason for $PM_{2.5,}$ It is should be noted that the activity of coal combustion due to the operation of coal fired power plant in Cilacap site was not begin yet at the sampling time. This data could be used as preliminary data or background data before the emission of coal combustion will take place.

Table 2. The elemental concentrations in $PM_{2.5}$ in four sampling sites

Element (ng.m^{3})	Cilacap [9]	Labuan	Suralaya	Tanjung Jati
Na	218.33	2.63	24.41	105.35
Mg	112.15	16.27	19.15	19.21
Al	137.19	16.21	72.47	40.35
Si	160.21	17.87	128.45	50.84
S	172.35	392.24	528.74	589.16
Cl	383.18	5.11	7.27	5.69
K	174.11	168.51	188.99	174.19
Ca	87.35	14.42	57.61	42.02
Ti	8.49	3.64	5.72	4.59
Cr	1.68	1.94	2.93	5.29
Mn	4.96	4.33	11.90	5.17
Fe	269.47	187.36	302.04	259.18
Ni	1.35	1.23	2.45	1.58
Cu	0.07	1.42	1.24	1.04
Zn	8.28	21.25	52.15	12.08
Br	11.13	3.56	4.80	6.40
Pb	3.19	8.27	6.22	4.57

IOP Conf. Series: Materials Science and Engineering 1011 (2021) 012056 doi:10.1088/1757-899X/1011/1/012056

Elemental concentration of fine particulate matters collected in four sampling sites are summarized in Table 2. Elements such as Na, Mg, Al, Si, S, Cl, K, Ca, Ti, Cr, Mn, Fe, Ni, Cu, Zn, Br and Pb were identified. Elemental concentration results in fine particulate matter, $PM_{2.5}$, shown that sulfur has highest concentrations in three sites of Labuan, Suralaya and Tanjung Jati with average of 392.24, 528.74 and 589.16 $ng.m^{-3}$, respectively. Sulfur in airborne particulate matter is generally present because of the atmospheric conversion of SO_2 to sulfate particulate phase through homogeneous processes. Hanoi also reported high sulfur in $PM_{2.5}$ with average value of 3810±2700 $\mu g.m^{-3}$ due to the emission of coal power plant [7]. Compared to Hanoi study, the sulfur emission in the three sampling sites were still lower, this could be happened due to the short monitoring and limited number of samples. Even though the emission of sulfur is lower than other study, it is worth noted that the major emission of the coal power plant identified was sulfur. Long monitoring to identify the emitted pollutant from coal power plant emission should be considered to monitoring the impact to the air quality and surroundings. In Cilacap sites, the major elements were the crustal elements such as Al, Si, Ca, and Fe which indicated the road dust/soil sources are dominant. Heavy metals such as Cr, Ni, Cu, Zn and Pb were in low concentrations compared to other industrial sites in Indonesia reported by Santoso et al [16].

The calculation of RCM, the average of measured $PM_{2.5}$ and its ratio are resumed in Table 3. It can be seen that the average RCM for fine particulates reaches 25.7 to 51.0%, which means that reconstruction represents maximum only 51% of the mass of the fine particulate and the remaining 49% cannot be reconstructed due to unmeasured of organic matter such as nitrates and organic carbon. The remaining missing mass more than 49% was mainly nitrates and organic matters. Other study reported that organic matter contributes about 28±11% of mass and nitrates are between 5 to 15% of the mass [7].

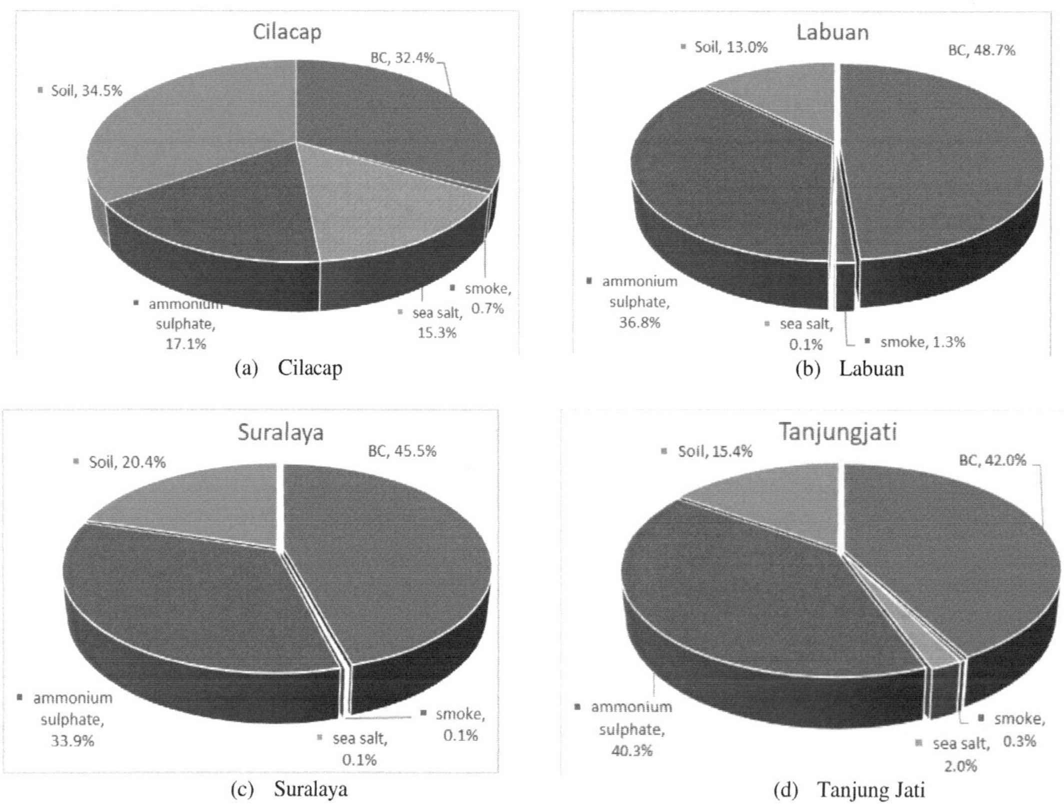

(a) Cilacap (b) Labuan (c) Suralaya (d) Tanjung Jati

Figure 3. Each component of RCM in each sampling site

Table 3. Reconstructed mass vs measured $PM_{2.5}$

Sites	RCM (ng.m^3)	Measured $PM_{2.5}$ (ng.m^3)	Ratio RCM to measured $PM_{2.5}$
Cilacap	4196.18	8236.11	51.0%
Labuan	4355.87	17056.55	25.7%
Suralaya	6472.50	16903.50	40.8%
Tanjung Jati	6009.19	14459.66	42.6%

By the calculation of RCM as a sum of sea salt, ammonium sulphate, soil, smoke and BC, the percentage of RCM components in the calculated RCM values for fine particulate matter samples are shown in Figure 3. Figure 3 showed that the average RCM of $PM_{2.5}$ mass at Labuan, Suralaya and Tanjung Jati was composed of 42-49% black carbon, 33.9-40.3% sulphate, 13-20% soil, 0.1-2.0% seasalt and 0.1-1.3% smoke. While for Cilacap, the RCM was composed of 32.4% black carbon, 17.1% sulphate, 34.5% soil, 15.3% seasalt and 0.7% smoke. Black carbon and sulphate give the largest contribution to fine particulate RCM with a contribution of a total more than 79% for three sites Labuan, Suralaya and Tanjung Jati except for Cilacap, the major contribution of BC and soil was dominant.

4. CONCLUSION

This study presented the $PM_{2.5}$, BC and elemental concentration from fine particulate matter samples collected in the vicinity of four coal power plant in Java island. The 24 hours of mass concentration of $PM_{2.5}$ were still below the threshold value of twenty-four hours of $PM_{2.5}$ concentration of Indonesia. Elemental analysis of $PM_{2.5}$ revealed that sulfur has a highest concentration for Labuan, Suralaya and Tanjung Jati sites, while in Cilacap was dominated by crustal matter. The major component of RCM in $PM_{2.5}$ in those three sites were black carbon and sulphate which dominates more than 79%, while in Cilacap were soil and with average of 34.5%. For some sites, the level of elemental concentrations was lower than other industrial sites in Java, due to the sampling was carried out before the commissioning/normal operation. Further and periodical monitoring will be needed to obtain a better comprehensive result to assess the impact of coal power plant to the surrounding environment.

ACKNOWLEDGEMENT

Authors acknowledge the financial support from the National Nuclear Energy Agency (BATAN), Indonesia and Minister of Environmental and Forestry of Indonesia. We also would like to acknowledge the help of staff at the Center of Applied Nuclear Science and Technology for their assistance throughout this study.

References

[1] Hasan MH, Mahlia TMI and Nur H 2012 A review on energy scenario and sustainable energy in Indonesia *Renew. Sustain. Energy Rev.* **16** 2316–28
[2] JICA 2010 Preparatory Survey for Indramayu Coal-fired Power Plant Project in Indonesia
[3] Córdoba P, Ochoa-Gonzalez R, Font O, Izquierdo M, Querol X, Leiva C, López-Antón MA, Díaz-Somoano M, Rosa Martinez-Tarazona M, Fernandez C and Tomás A 2012 Partitioning of trace inorganic elements in a coal-fired power plant equipped with a wet Flue Gas Desulphurisation system *Fuel* **92** 145–57
[4] Anonymous 2020 The boss of PLN opens up about the 35,000 MW project, what is the progress? *Kompas Newspaper 2020 June 18.*
[5] Destyanto AR, Hidayatno A and Amalia A 2017 Analysis of the effects of CO2 emissions from coal-fired power plants on the gross domestic regional product in Jakarta *Int. J. Technol.* **8** 1345–55

[6] Cohen DD, Crawford J, Stelcer E and Bac VT 2010 Long range transport of fine particle windblown soils and coal fired power station emissions into Hanoi between 2001 to 2008 *Atmos. Environ.* **44** 3761–9\

[7] Cohen DD, Crawford J, Stelcer E and Bac VT 2010 Characterisation and source apportionment of fine particulate sources at Hanoi from 2001 to 2008 *Atmos. Environ.* **44** 320–8

[8] WHO 2020 WHO Air quality guidelines for particulate matter, ozone, nitrogen dioxide and sulfur dioxide *Biol. Trace Elem. Res.* **195** 491–8

[9] Watson JG, Chow JC and Frazier C 1999 X-Ray Florescence Analysis of Ambient Air Samples. *Elemental Analysis of Airborne Particles, Vol. 1;* ed M Landsberger, S., Creatchman (Amsterdam: Gordon and Breach Science) pp 67–96

[10] Bamford SA, Wegrzynek D, Chinea-Cano E and Markowicz A 2004 Application of X-ray fluorescence techniques for the determination of hazardous and essential trace elements in environmental and biological materials *Nukleonika*

[11] Santoso M and Lestiani DD 2014 *IAEA XRF Newsletter: X ray Fluorescence in Member States*

[12] Lestiani DD, Santoso M, Kurniawati S, Adventini N and Prakoso DAD 2015 Characteristics of Feed Coal and Particulate Matter in the Vicinity of Coal-fired Power Plant in Cilacap, Central Java, Indonesia *Procedia Chem.* **16** 216–21

[13] Hopke PK, Xie Y, Raunemaa T, Biegalski S, Landsberger S, Maenhaut W, Artaxo P and Cohen D 1997 Characterization of the gent stacked filter unit pm10 sampler *Aerosol Sci. Technol.* **27** 726–35

[14] Santoso M, Lestiani DD, Kurniawati S, Damastuti E, Kusmartini I, Prakoso D, Sari DK, Hopke PK, Mukhtar R, Tamrin T, Tjahyadi A, Sukadi S, Nurkholik NSDA, Wahyudi D, Sitorus TD, Jasmiyati J, Riadi A, Supriyanto J, Dahyar N, Sondakh S, Hogendorp K, Wahyuni N, Bejawan IG and Suprayadi LS 2020 Assessment of Urban Air Quality in Indonesia *Aerosol Air Qual. Res.* **20**

[15] Begum BA, Biswas SK, Markwitz A and Hopke PK 2010 Identification of sources of fine and coarse particulate matter in Dhaka, Bangladesh *Aerosol Air Qual. Res.* **10** 345–53

[16] Santoso M, Lestiani DD, Mukhtar R, Hamonangan E, Syafrul H, Markwitz A and Hopke PK 2011 Preliminary study of the sources of ambient air pollution in Serpong, Indonesia *Atmos. Pollut. Res.* **2**

IOP Conf. Series: Materials Science and Engineering 1011 (2021) 012057 doi:10.1088/1757-899X/1011/1/012057

The extraction of antioxidants from *Chlorella vulgaris* for cosmetics

Siti Agustina[1]*, Novi Nur Aidha[1] and Eva Oktarina[1]

Badan Penelitian dan Pengembangan, Kementerian Perindustrian RI
Jl. Gatot Subroto kav. 52-53, Jakarta Selatan, Indonesia

*Corresponding author's email: sitiagustina.rj@gmail.com

Abstract. *Chlorella vulgaris* is microalgae that contain chlorophyll as antioxidants, which has been widely used as a functional food. Antioxidants from *Chlorella vulgaris* also have potential as active ingredients in the cosmetics industry. Nowadays, consumers prefer natural cosmetics because they aren't harmful to the skin. This study aimed to obtain chlorophyll antioxidants from *Chlorella vulgaris* and applied it to the cosmetics. The extraction process of antioxidants from *Chlorella vulgaris* was performed by maceration method, with *Chlorella vulgaris* concentration variable: 0.01; 0.05 and 0.1% in water as solvent. The antioxidant extracts powdered by spray drying method, with addition of 100 g/L maltodextrin. The observation was conducted on the characteristics of water content, ash content, carbohydrate content, protein content, antioxidant activity, and powder morphology. The extracts powder applied to cosmetics in the form of creams and lotions, then characterized for the antioxidant activity and microbial content. The best antioxidant capacity of extract powder was obtained on *Chlorella vulgaris* concentration 0.1%, which was 11.83 mg Vit C per 100 g sample. The results of antioxidant capacity in cosmetics as cream and lotion were 4.95 mg Vit. C per 100 g sample (IC_{50} 719.75 mg/ml) and 4.73 mg Vit. C per 100 g sample (IC_{50} 660 mg/ml), respectively. Both of cream and lotion weren't shown any microbial contamination. Based on this study, *Chlorella vulgaris* with its antioxidant capacity can be developed as active ingredients for various kinds of cosmetics.

1. Introduction

Many species of microalgae are valuable as cosmetics raw materials because of the diversity on biologically active compounds in them. The example of microalgae that have the potential to be developed as cosmetics compound is *Chlorella vulgaris*. *C. vulgaris* is a unicellular, eukaryotic organism and a spherical microscopic cell with 2–10 μm diameter and contains several compounds like protein, vitamins, and pigments[1,2]. This microalgae contains green pigment compounds called chlorophyll[3]. *Chlorella* has antioxidant compounds that are relatively strong, and it has an EC_{50} value of less than 50 ppm[4]. Antioxidants from *C. vulgaris* also have potential as active ingredients in the cosmetics industry. It's a rich source of protein, which reaches 42 to 58% proteins per dry weight of biomass[1] and contain essential amino acids which supplement human RNA and DNA; help in tissue growth; and repair or restructure and smooth out the surface of the skin, particularly with scared skin or aging skin in cosmetics[5].

Chlorophyll is a bioactive compound with antioxidant and anti-mutagenic properties which widely used in cosmetics, pharmaceutical, and food industries[6]. Chlorophyll has five types known as a, b, c, d, and f; whose maximum wavelengths of absorption when solved in methanol are 665, 652, 630, 696, and 707 nm, respectively[7]. Chlorophylls can absorb light in the red and blue regions, as a result, it emits a green color. It can be used as a natural coloring agent and have antioxidant, antibacterial, and deodorizing properties[3].

A compound with antioxidant activity can donate its hydrogen atom to DPPH as free radicals, which are characterized by the occurrence of purple to pale yellow[8]. Antioxidant activity of *Chlorella* compounds that rich in protein, contributed by functional groups or the residue of amino acids. The antioxidant activity as IC_{50} of *C. vulgaris* in acetone was 57.25 ppm, which can function as an scavenger of free radicals[6] and able to minimize the concentration of the ROS[9]. Antioxidants such as chlorophyll in cosmetics can work as oxidizing agents that provide glowing skin by preventing skin damage. An antioxidant also helps in skin tightening, reduction of wrinkles, and reduces inflammation[5]. Antioxidants from *Chlorella* have a weakness on its instability and sensitivity to heat, oxygen, pH, illumination, and acid degradation that can change its color. One of the technology to increase the stability of chlorophyll is encapsulation method[10].

The encapsulation process can be performed by spray drying using maltodextrin as a matrix encapsulant. Spray dryers commonly used for antioxidants encapsulation because of the fast method, low process, low temperature needed and can produce variance sized powder[11,12]. Maltodextrin used as an encapsulant because it has low solubility in water, high viscosity, and relatively low process, able to protect the bioactive compound from degradation[13]. This research aimed to obtain chlorophyll antioxidants from *C. vulgaris* and applied to cosmetics in the form of creams and lotions.

2. Materials and Methods

2.1. Materials

Chlorella vulgaris was obtained from Balai Besar Perikanan Budidaya Air Payau (BBPBAP) Jepara. The cultivating system of *C. vulgaris* was conducted on open pond. Maltodextrin was procured from Bratachem. Chemicals used in this study including acetic acid, ethanol, K_2SO_4, $CuSO_4$, H_3BO_3, NaOH, H_2SO_4, and HCl were purchased from Merck. Distilled water was obtained from Ikapharmindo Putramas, while methanol and hexan were purchased from J.T. Baker. DPPH reagent and ascorbic were obtained from Sigma-Aldrich. Material for cosmetics aquades, methyl parabens, virgin coconut oil, cethyl alcohol, propylene glycol, sodium lauryl sulfate, propylene parabens, glycerin, trienolamin, stearate acid, cethyl alcohol, and lanolin.

2.2. Extraction of C. vulgaris

Separation of *Chlorella vulgaris* from the medium was preceded by centrifugation on 17.000 rpm/min, the supernatant is *Chlorella vulgaris*. *C. vulgaris* diluted on distilled water with various concentrations 0.01; 0.05; 1 and 0.1% w/v, respectively. *C. vulgaris* was extracted with maceration for 1 hour, then separated for its filtrate and supernatant, by centrifugation on 6000 rpm for 15 minutes. The filtrate was used as chlorophyll content.

2.3. Production of chlorophyll powder using maltodextrin

The chlorophyll solution with variation concentration 0.01; 0.05 and 0.1% were blended with maltodextrin 100 mg/l in each solution. Then, chlorophyll powder was roduced using spray dyer that operated at condition inlet 120 to 130 °C, outlet 42 to 44 °C, blower 1.01m^3/min to 1.04 m^3/min and atomizing 200 to 230 kPa.

2.4. Determination of antioxidant activity

Antioxidant content was analyzed by Ascorbic acid Equivalent Antioxidant Capacity (AEAC), as the capacity of sample as antioxidant which react with DPPH as free radical. The samples were mixed with methanol DPPH then homogenized and stored for 120 minutes. The absorbance was recorded at 517 nm. Distilled water was used as blank and ascorbic acid was served as standard[13].

2.5. Analysis of water content and ash content
Water content and ash content were analyze by using gravimetric method according to SNI 01-2891-1992 point 5.1 and 6.1, respectively. Water content was determined by weighing 1 g sample and dried it at 105 °C for 3 hours. Samples were placed in the desiccator until stable, and then the weight was determined. Subtract the weight before with after drying to get water content (%). Ash content was determined by weighing 2 g sample and furnace it at 600 °C for 2 hours. The ash content (%) was calculated based on the loss of weight before and after treatment.

2.6. Analysis of total fat, protein, and carbohydrate content
Protein content was analyzed by biuret method (Kjeltec) according to SNI 01-2891-1992. Total fat was determined by using soxhlet-hydrolysis method, using hexane. Carbohydrate percentage was measure using this formula: Carbohydrate = 100 – (%water content + %ash content + %protein + %lipid).

2.7. Analysis of morphology
Chlorophyll-maltodextrin rheology was determined by using SEM, the sample was coated with gold (Au) and observed using Scanning Electron Microscope (SEM) at 500x and 4000x magnifications on 20 kV voltages.

2.8. Production of chlorophyll cream
The ingredients used for making the cream consist of the water phase and the oil phase. Ingredients as water phase were distilled water and methyl parabens, which were heated above a water bath at 65 to 70 °C. Ingredients as oil phase werere virgin coconut oil, cetyl alcohol, propylene glycol, sodium lauryl sulfate and cetyl stearil alcohol, which were melted at 65 °C – 70 °C while stirring using a hot plate magnetic stirrer at 200 rpm. Then, the oil phase was to the water phase, while stirring it by using a homogenizer at 200 rpm for 15 minutes, to prevent air bubbles formation. If both phases were already well mixed, increased to 800 rpm for 10 minutes to form a cream base. After the cream base formed, extract alga (chlorophyll powder) was added, stirring for another 5 minutes until the cream homogeneously mixed.

Table 1. Formulation of cream cosmetic.

Ingredients	Composition (%)
Extract alga	4
Distilled water	40
Methyl parabens	0.18
Propylene glycol	5
Cetyl stearil alcohol	4.5
Cetyl alcohol	5
Sodium lauryl sulfate	0.5
VCO	add until reach 100

2.9. Production of lotion
The ingredients used in making the lotion consist of the water phase and the oil phase. Ingredients served as water phase were distilled water, glycerin, triethanolamine, and methyl parabens, which were heated above a water bath at 65 to 75 °C. Ingredients employed as oil phase were stearate acid, cetyl alcohol, lanolin, and propylene parabens, which were melted at 65 to 75 °C. After all phases dissolved, the water phase was added to the phase oil while constantly stirring to form an emulsion. Then, chlorophyll powder was added in the emulsion and mixed until homogeneously.

Table 2. Formulation of lotion.

Ingredient	Composition (%)
Extract alga	3
Stearic acid	6
Methyl parabens	0.02
Lanolin	3
Trietahanolamine (TEA)	3
Cetyl alcohol	6
Prophylene parabens	30.18
Glycerin	3
Aquadest	Add until reach 150

2.10. Microbial contamination analysis

Microbial contamination analysis of bacteria refers to ISO 21149:2006, yeast and mould refers to ISO 16212:2008, *Pseudomonas aeruginosa* refers to 22717:2015, *Candida albicans* refers to ISO 18416:2015, and *Staphylococcus aureus* refers to ISO 22718:2015. Enumeration and detection for bacteria, yeast and mould were analyzed by total plate count method (TPC) which involves enumeration of microbial colonies on a non-selective agar medium on mesophilic aerobic condition. Detection of *P. aeruginosa, C. albicans* and *S. aureus* were performed by checking the absence of microbial target growth after enrichment. First step was the enrichment by using a non-selective broth medium. The second step (isolation) was performed on a selective medium followed by identification tests. *P. aeruginosa, C. albicans* and *S. aureus* can cause skin or eye infections.

3. Result and Discussion

3.1. Extraction of C. vulgaris and production of chlorophyll powder

The bioactive compound on *C. vulgaris* is chlorophyll, which is known to has antioxidant properties and has a sensitivity and unstable characteristic[10]. The extraction method was performed by maceration with distilled water as solvent, for one hour. Extraction chlorophyll by maceration method was also performed by Aryanti, Nafiunisa, and Wilis (2016). The research needs 3-hour maceration to extract chlorophyll from *Pleomele angustifolia* leaves. Maceration method can keep the bioactive compound from heating because it was done without heating, only by stirring. The maceration method also commonly used because it's fast, environmentally friendly, and effectively cost, so it easy to expanded it in the industrial scale[15]. The production of chlorophyll powder was using a spray dryer, with maltodextrin as a coating. In this research, spray dryer was commonly used for antioxidant as the process will be on low temperature condition[11,12]. Maltodextrin was commonly used as a matrix encapsulation because it can protect the bioactive compound from degradation[13].

3.2. Moisture and ash content

The moisture content of chlorophyll powder ranged between 5.34 to 5.58 g per 100 g (figure 1). The highest moisture content was on the powdered *C. vulgaris* 0.05%, and the lowest on *C. vulgaris* 0.01%. The moisture content of all *C. vulgaris* powders were below 10%. The moisture content indicates the quality of a product (powder), according to the Indonesian Pharmacopoeia or Materia Medika Indonesia, the water content in a product should be <10%. According to the previous research, encapsulation chlorophyll by spray drying produces a moisture content below 10% [14].

The ash content, ranged between 0.17 to 0.26 g per 100 g, thus increase with the addition of concentration of *C. vulgaris* (figure 2). Spray dryer reducing water content and water activity [11]. Maltodextrin has a glucose chain that effects on moisture and ash content. The ash content decrease because the addition of maltodextrin will increase of evaporation [15].

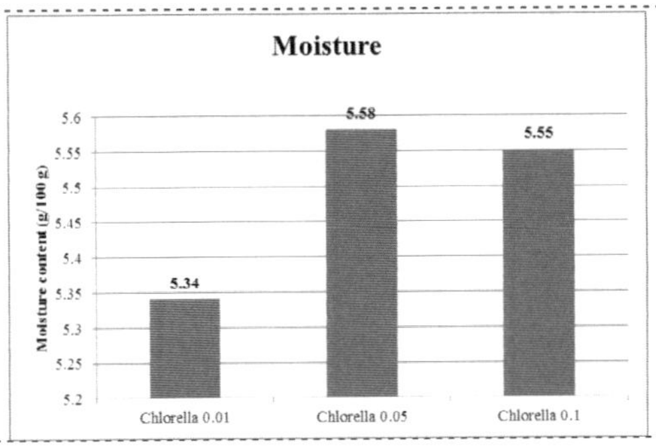

Figure 1. Moisture Content

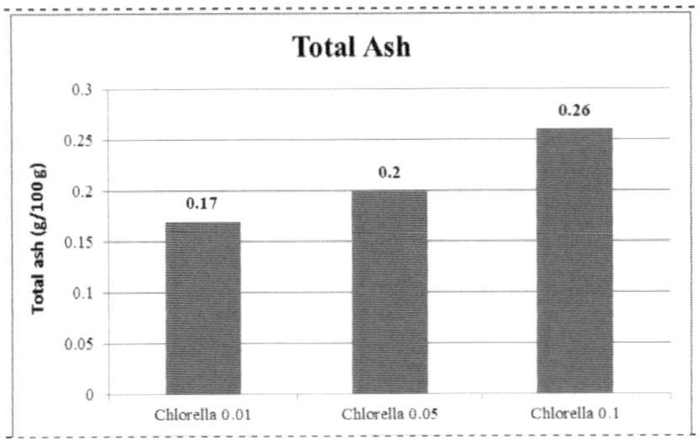

Figure 2. Ash content

3.3. Carbohydrate, protein and total fat content

The carbohydrate content of chlorophyll powder ranged between 93.5 g/100 g until 93.95 g/100 g (figure 3). The highest carbohydrate content was powdered *C. vulgaris* 0.05%. The addition of maltodextrin increased the carbohydrate content because its component is carbohydrate [13].

The highest protein content was powdered *C. vulgaris* 0.1% with protein content 0.69 g per 100 g (figure 4). The addition concentration of *C. vulgaris* increases the protein content because it's rich in protein, contributed by functional groups or the residue of amino acids. The protein content of *C. vulgaris* reaches 42–58% proteins per dry weight of biomass[1].

Total lipid content on microalgae in open pond cultivating system is 3.18%, which usually present in the form of glycerol and fatty acids. Total fatty acid content on microalgae in open pond cultivating system is 0.9 to 22.6%, which majority compound are palmitic acid (C16:0) 22.6% and stearic acid (C18:4) 21.4%[16]. *C. vulgaris* was macerated with water as polar solvent, due to hydrophobic of fatty acids that interact with non-polar or non-polar-polar solvents (biphasic system), the fat content was not detected on *C. vulgaris* various concentration. Lipid can be extracted using n-hexane-methanol which can break the bond between lipid and protein[16].

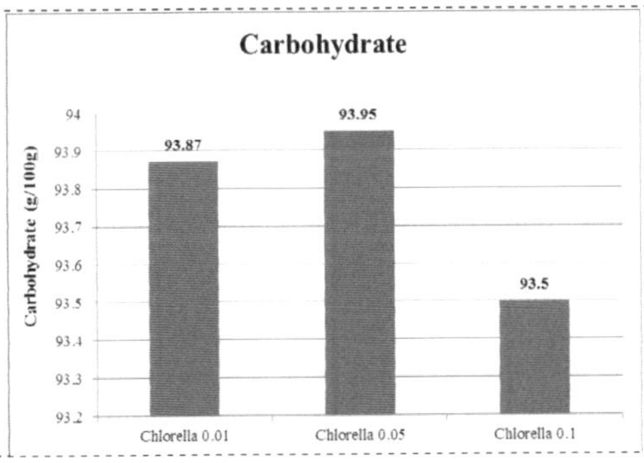

Figure 3. Carbohydrate content

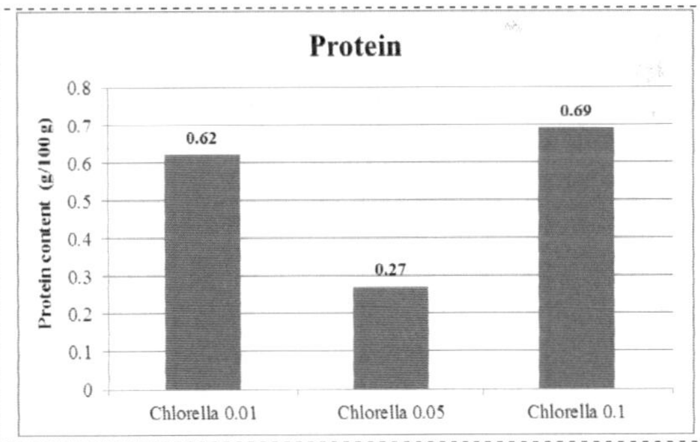

Figure 4. Protein content

3.4. Morphology by SEM

Morphology was analyzed by SEM with magnification of 500x and 4000x (figure 5). The morphology of the powder displayed a spherical structure, separate dispersedly, and has a smooth form. These results revealed a homogeneous fluorescence distribution across all-spherical particles. The homogeneous distribution of fluorescence demonstrates the mono-dispersity of chlorophyll inside each sphere. In line with the addition of concentration *Chlorella*, the form is anisotropic wavy form and tends to rupture. Based on previous studies, emulsions microencapsulation is better than powder encapsulation because the size of particles can be controlled until less than 10% variance[10]. Furthermore, the encapsulation could increase the stability of chlorophyll as an antioxidant.

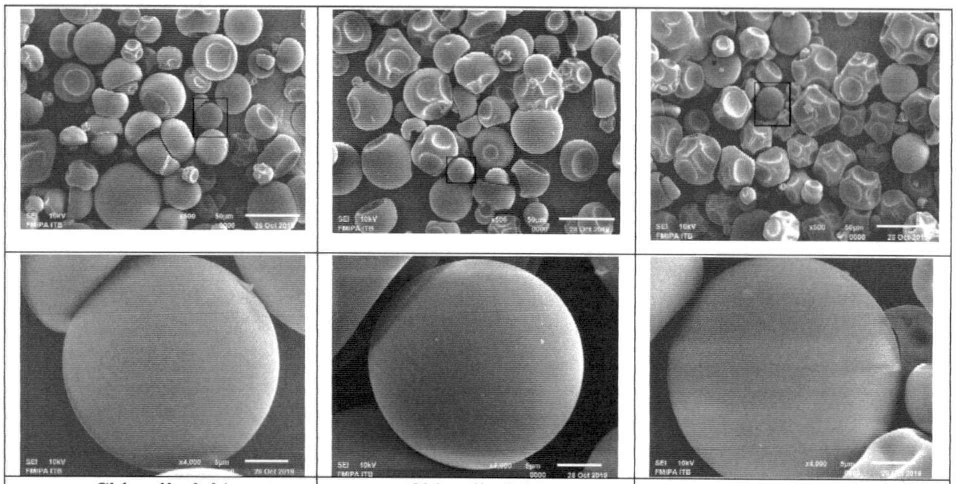

Chlorella 0.01	Chlorella 0.05	Chlorella 0.1

Figure 5. SEM analysis.

3.5. Antioxidant content

The antioxidant content of *C. vulgaris* powder was increasing with the concentration addition of *C. vulgaris*. The antioxidant content was not detected on powders with concentration *C. vulgaris* 0.01 and 0.05%. The highest antioxidant content was 11.83 mg Vit. C/100 g sample, produced by powdered *C. vulgaris* 0.1% (figure 6). The antioxidant activity test was conducted by in vitro using the free radical reduction method, by using 1,1-diphenyl-2-picryl-hydrazyl (DPPH). The powder with a concentration of *C. vulgaris* 0.1% can donate its hydrogen atom to DPPH as free radicals, which are characterized by the occurrence of purple to pale yellow[8]. *C. vulgaris* has the highest antioxidant activity compared to other Chlorella strains using DPPH method [17].

The antioxidant activity linked to bioactive compounds, such as chlorophylls. The antioxidant from *Chlorella* has a weakness, for its un-stability and sensitivity to heat, oxygen, pH, illumination, and acid degradation that can change its color. The encapsulation method using spray drying enhanced the stability of chlorophyll[10]. The encapsulant like maltodextrin able to protect antioxidant and bioactive compound release, which induced by the heat on the spray drying process. Beneficially of using spray drying process are water evaporation occurs on a very wide surface, shorten the time for drying process, and increase the thermal stability of bioactive compound because the addition of maltodextrin[15].

C. vulgaris is a species that is often used in cosmetics because it has a bioactive compound like chlorophyll, which as an antioxidant able to minimize the concentration of the ROS. Chlorophyll as antioxidant in cosmetics has a function as an oxidizing agent which providing glowing skin by preventing skin damage. Antioxidants also help in skin tightening process, reduction of wrinkles, and reduces inflammation[5,18]. In this research, chlorophyll powder with the highest antioxidant activity applied to cosmetics in form of cream and lotion. Cream cosmetics has antioxidant content was 4.95 mg Vit C / 100 g sample with antioxidant activity 719.75 mg/ml and antioxidant content on lotion was 4.73 mg Vit C / 100 g sample with antioxidant activity 660 mg/ml (figure 6 and table 3). Both cream and lotion have antioxidants content and activity which was not different descriptively. In this research, the antioxidant capacity wasn't influenced by the type of cosmetic, whether its cream or lotion.

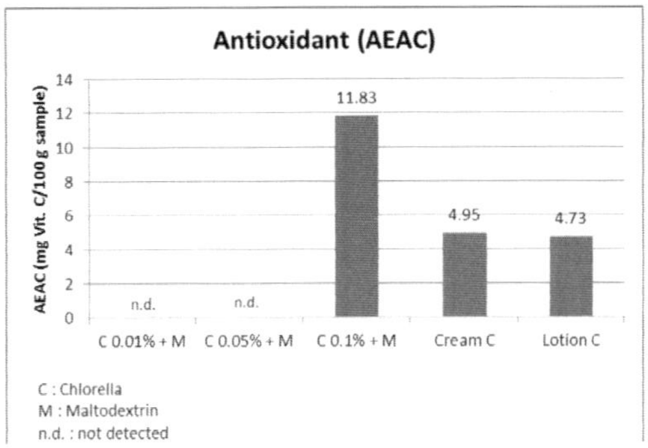

Figure 6. Antioxidant capacity of *C. vulgaris* powder and cream and lotion.

Table 3. Antioxidant Capacity of Cosmetics.

Cosmetics	IC_{50} (mg/mL)
Cream	719.75
Lotion	660

3.6. Antioxidant content

The microbial contamination in cosmetics was analyzed on Total Plate Count, yeast and mold, *P. aeruginosa, C. albicans*, and *S. aureus*, according to the Regulation of the Republic of Indonesia Drug and Food Supervisory Agency (2014) concerning Amendments to BPOM Regulation Number HK.03.1.23.07.11.6662 in 2011. Based on the analysis of microbial contamination (table 4), both of cosmetic, cream and lotion were not containing microbial contamination. It concludes that the cream with the composition above is safe to be used.

Table 4. Microbial contamination of cosmetics.

Parameter	Unit	Result	
		Cream	Lotion
Total Plate Count	CFU/g	< 1.0×10^1	< 1.0×10^1
Yeast and Mould	CFU/g	< 1.0×10^1	< 1.0×10^1
P. aeruginosa	Per g	Negative	Negative
C. albicans	Per g	Negative	Negative
S. aureus	Per g	Negative	Negative

3.7. Antioxidant content

C. vulgaris has a green color with the bioactive compound chlorophyll. The color from chlorophyll already used as a natural pigment in cosmetics. Chlorophyll also used on skin aging supplements and cosmeceuticals to support collagen repair mechanisms[19]. The chlorophyll powder product from the spray drying process with maltodextrin has a green color. From the organoleptic evaluation for cosmetics (table 5), both of cosmetics cream and lotion has a similar color with the powder. The color was stable until 8 weeks. The cosmetics (cream and lotion) were not having aroma or odor, and the texture still smooth until 8 weeks. The lotion was more liquid than cream. The homogeneity was excellently stable until 8 weeks, without bilayer formation or agglomeration. The cosmetics were relatively stable.

Table 5. Organoleptic observations of cosmetics.

Condition	Cosmetics	Weeks							
		1	2	3	4	5	6	7	8
Color	cream	green	green	green	green	green	green	green	green
	lotion	green	green	green	green	green	green	green	green
Odor	cream	no odor	no odor	no odor	no odor	no odor	no odor	no odor	no odor
	lotion	no odor	no odor	no odor	no odor	no odor	no odor	no odor	no odor
Texture	cream	smooth	smooth	smooth	smooth	smooth	smooth	smooth	smooth
	lotion	smooth	smooth	smooth	smooth	smooth	smooth	smooth	smooth
Homogenity	cream	homogen	homogen	homogen	homogen	homogen	homogen	homogen	homogen
	lotion	homogen	homogen	homogen	homogen	homogen	homogen	homogen	homogen

4. Conclusion

The bioactive compound from *C. vulgaris* was extracted by the maceration method, and then encapsulated by spray drying method. The best powder based on its antioxidant capacity, was *C. vulgaris* 0.1% which gave 11.83 mg Vit C per 100 g sample. The powders have a spherical structure, separate dispersedly, and smooth form. The best powdered *C. vulgaris* was applied to cosmetics as cream and lotion form. The antioxidant content in cosmetics type creams was 4.95 mg Vit C / 100 g sample with antioxidant activity 719.75 mg/ml and antioxidant content in the type of lotion was 4.73 mg Vit C / 100 g sample with antioxidant activity 660 mg/ml. The antioxidant capacity was not significantly different between cosmetic type of cream or lotion. The cosmetics, cream and lotion were not containing microbial contamination according to BPOM regulations, and the organoleptic analysis showed it stable for eight weeks. Based on this research, *C. vulgaris* have antioxidant activity which potential to be developed as active ingredients for various cosmetics.

Acknowledgement

This research was supported by Industrial Research and Development Agency (BPPI), Balai Besar Perikanan Budidaya Air Payau (BBPBAP) Jepara, and CV. Budhi Andhika for the cooperation.

References

[1] Stramarkou M, Papadaki S, Kyriakopoulou K and Krokida M 2017 Effect of drying and extraction conditions on the recovery of bioactive compounds from Chlorella vulgaris. *J Appl Phycol.* **29**(6) 2947-2960. doi:10.1007/s10811-017-1181-8

[2] Safi C, Zebib B, Merah O, Pontalier PY and Vaca-Garcia C 2014 Morphology, Composition, Production, Processing and Applications of Chlorella vulgaris: A review. *Renew Sustain Energy Rev* **35** 265-278. doi:10.1016/j.rser.2014.04.007

[3] Fabrowska J, Leska B, Schroeder G, Messyasz B, Pikosz M 2015 Biomass and Extracts of Algae as Material for Cosmetics. In: Se-Kwon Kim and Katarzyna Chojnacka, ed. *Marine Algae Extracts: Processes, Products, and Applications*. First Edit. (Wiley-VCH Verlag GmbH & Co. KGaA) 681-706.

[4] Fithriani D, Amini S, Melanie S and Susilawati R 2015 Uji Fitokimia , kandungan total fenol dan aktivitas antioksi dan mikroalga Spirulina sp ., Chlorella sp ., dan Nannochloropsis sp . Activity of Microalgae Spirulina sp ., Chlorella sp . *JPB Kelaut dan Perikan*. **10**(2):101-109.

[5] Joshi S, Kumari R and Upasani VN 2018 Applications of Algae in Cosmetics: An Overview. *Int J Innov Res Sci Eng Technol* **7**(2). doi:10.15680/IJIRSET.2018.0702038

[6] Rosahdi TD, Susanti Y and Suhendar D 2015 Uji Aktivitas Daya Antioksidan Biopigmen Pada Fraksi Aseton Dari Mikroalga Chlorella vulgaris. **IX**(1) 1-16.

[7] da Silva Ferreira V and Sant'Anna C 2017 Impact of culture conditions on the chlorophyll content of microalgae for biotechnological applications. *World J Microbiol Biotechnol* **33**(1). doi:10.1007/s11274-016-2181-6

[8] Ridlo A, Sedjati S and Supriyantini E 2015 Aktivitas anti oksidan fikosianin dari Spirulina Sp . Menggunakan metode transfer elektron dengan DPPH (1 , 1-difenil-2-pikrilhidrazil).**18** 58-

63.
[9] Kumar RR, Rao PH, Subramanian V V and Sivasubramanian V 2014 Enzymatic and non-enzymatic antioxidant potentials of Chlorella vulgaris grown in effluent of a confectionery industry *J Food Sci Technol* **51**(2) 322-328. doi:10.1007/s13197-011-0501-2
[10] Hsiao C, Lin J, Wen H, Lin Y and Yang C 2020 Enhancement of the stability of chlorophyll using chlorophyll-encapsulated polycaprolactone microparticles based on droplet micro fl uidics. *Food Chem.* **306** 125300. doi:10.1016/j.foodchem.2019.125300
[11] Prihapsara F, Astirin OP, Artanti AN dan, Sentot AS 2018 Pengembangan teknologi enkapsulasi fikosianin. *Pros Semin Nas Has Pengabdi Kpd Masy.*Yogykarta,:140-146.
[12] Nasyarudin M and Hadiyanto H 2016 Pembuatan Mikrokapsul Phycocyanin Menggunakan Maltodekstrin sebagai Bahan Pelapis dengan Metode Spray Drying *Pros Semin Nas Tek Kim "Kejuangan" Pengemb Teknol Kim untuk Pengolah Sumber Daya Alam Indones*.1-7.
[13] Kurniasih RA, Purnamayati L, Amalia U and Dewi EN 2018 Formulation and characterization of phycocyanin microcapsules within maltodextrin- alginate. *Agritech.* **38**(1):23-29. doi:10.22146/agritech.16752
[14] Aryanti N, Nafiunisa A and Wilis MF 2016 Ekstraksi dan karakterisasi klorofil dari daun Suji (Pleomele Angustifolia) sebagai pewarna pangan alami. *J Apl Teknol Pangan.* **5**(4):129-135. doi:10.17728/jatp.196
[15] Agustina S, Aidha NN and Oktarina E 2019 Effect of maltodextrin concentration on the characteristic of phycocyanin powder as a functional food. *AIP Conf Proc.* **2175**(November). doi:10.1063/1.5134614
[16] Jay MI, Kawaroe M and Effendi H 2018 Lipid and fatty acid composition microalgae Chlorella vulgaris using photobioreactor and open pond. *IOP Conf Ser Earth Environ Sci.* **141**(1). doi:10.1088/1755-1315/141/1/012015
[17] Chatzikonstantinou M, Kalliampakou A and Gatzogia M 2016 Comparative analyses and evaluation of the cosmeceutical potential of selected Chlorella strains *J Appl Phycol.* doi:10.1007/s10811-016-0909-1
[18] Adrade CJ De and Andrade LM De. 2017 An overview on the application of genus Chlorella in biotechnological processes. *J Adv Res Biotechnol.* doi:10.15226/2475-4714/2/1/00117
[19] Mourelle ML, Gómez CP and Legido JL 2017 The potential use of marine microalgae and cyanobacteria in cosmetics and thalassotherapy. *Cosmetics* **4**(4). doi:10.3390/cosmetics4040046

Evaluation of uncertainties measurement in EDXRF for air filter analysis

D K Sari*, D D Lestiani, S Kurniawati, D P D Atmodjo and I Kusmartini

Center for Applied Nuclear Science and Technology, National Nuclear Energy Agency (BATAN), Tamansari 71, Bandung, Indonesia

*Corresponding author's email: dykuza@gmail.com

Abstract. Energy Dispersive X-Ray Fluorescence (EDXRF) is one of the nuclear analytical techniques that has been widely used for elemental analysis, especially for air filter analysis. EDXRF Epsilon 5 has been installed in BATAN Bandung laboratory, Indonesia. EDXRF Epsilon 5 is an analytical instrument that is used to determine trace elements in many kinds of materials, especially in air filter samples. Measurement conditions have to be set up to determine several elements. Uncertainty evaluation was different according to the instrument used in each laboratory. This activity is aimed to calculate and evaluate the uncertainty measurement associated with EDXRF Epsilon 5 for air filter analysis. Evaluation of uncertainties measurement in EDXRF Epsilon 5 was carried out following the guides to express uncertainty: JCGM 100:2008 and EURACHEM. The uncertainty evaluation focused on 10 elements: Na, Si, S, K, Ca, Ti, Fe, Zn, Br, and Pb that generally existed in airborne particulate samples. The uncertainties results for 10 selected elements measured with Epsilon 5 were in a good correlation, with % relative uncertainty were ranged from 2% to 32%.

1. Introduction

Nuclear analytical techniques (NATs) have been widely used for elemental analysis in many fields. EDXRF is one of the NATs that generally used to analyze trace elements in many kinds of samples. An EDXRF named Epsilon 5 has been installed in the Center for Applied Nuclear Science and Technology-BATAN Bandung. Epsilon 5 with its 3-dimensional polarizing optics is used to analyze trace elements in samples especially dedicated to air filter analysis. EDXRF can be used for both qualitative and quantitative analysis with minimum sample preparation.

Before EDXRF Epsilon 5 being used to analyze samples, the method must be set to optimize the measurement condition according to the element to be analyzed. After the sample is measured, then it is analyzed both qualitative and quantitative. Calibration standards become an important thing for quantitative analysis. The optimized method that had been made was then used to measure samples. From the measurement then the software calculated the peak intensity of the element to be determined and converted to concentration according to the calibration curve that had been made and optimized [1].

Epsilon 5 software only showed the raw intensity and corrected intensity then they were calculated and converted to concentration automatically by the software. There were no uncertainty values showed in the final results showed in the software. Therefore, this study is aimed to calculate and estimate uncertainty from the measurement [2]. Uncertainty evaluation was calculated using the basic theory of calculating the uncertainty in XRF, especially for air filter analysis.

2. Experimental Method

2.1. Air filter analysis
Airborne particulate sampling was carried out using Gent Stacked Sampler in several locations in Indonesia. The filter used in this activity were nucleopore polycarbonate filters. The sampler can be used for collecting particulate matter for two different size fractions: fine and coarse. Sampling was performed at a flow rate of 15-18 liters/min for 24 hours.

All the exposed samples were collected and then stored in a clean room for at least 24 hours then continued to be weighed using microbalance to determine the masses of the fine and coarse fractions. Then the Black Carbon (BC) concentrations were measured using an EEL Smoke Stain Reflectometer (Diffusion System, Ltd., model 43M). The elemental determination for air filter was performed using EDXRF Epsilon 5.

2.2. Instrumentation
The instrument used in this study is 3-dimensional polarizing optics EDXRF Epsilon 5 with a maximum power of 600 Watt (100 kV and 24 mA). Epsilon 5 also has 10 secondary targets and 6 sample trays that can accommodate up to 48 samples. Its 3-dimensional polarizing optics was used to optimize the energy excitation [3, 4].

2.3. Method setup
The method should be set before Epsilon 5 is used to measure samples. There were several parameters to set the application in Epsilon 5's software such as tube voltage and current (kV and mA), analysis time, and also secondary target corresponding to the elements to be measured [5].

2.4. Calibration
The calibration was set by measuring micromatter standards. The standards used for making the calibration curves were thin-film standards from Micromatter Technologies Inc. This method used 85 micromatter standards for several elements' measurement. The concentrations were ranged from 17-64 $\mu g/cm^2$. There was no sample preparation for measuring micromatter standards. The standards were placed into the holder and measured using the method that had been set. The micromatter standards were measured as standards and then will be calculated by the software automatically and plotted as calibration curves.

There are two main calibration factors: The D and E values. The D value is the constant term in the main intensity to concentration conversion equation (intercept). The E value is the coefficient of the linear term in the main intensity to concentration conversion equation (angle of the regression line or slope). These values should be calculated by using regression from the calculation program window [6, 7].

2.5. XRF analytical uncertainties
The analytical uncertainties were calculated using international standards: JCGM 100:2008 and EURACHEM, customized according to the instrument used in the laboratory, in this case, EDXRF Epsilon 5. The process of measurement uncertainty estimation schematically is shown in Figure 1.

XRF analytical uncertainties in EDXRF Epsilon 5 calculated following an international method [2, 9, 10, 11]. The mass loading of each element for EDXRF Epsilon 5 (ng/cm^2), C_{E5}, is quantified using Equation (1):

$$C_{E5} = \frac{I_{net} - I_{blk}}{E} \qquad (1)$$

where I_{net} is the net intensity of X-rays emitted by the sample (cps/mA) and I_{blk} is the net intensity of a blank filter (cps/mA) and E is the calibration factor [(cps/mA)/(ng/cm^2)], calculated as the slope of the linear regression between elemental loadings of calibration standards and their blank subtracted (net) intensities.

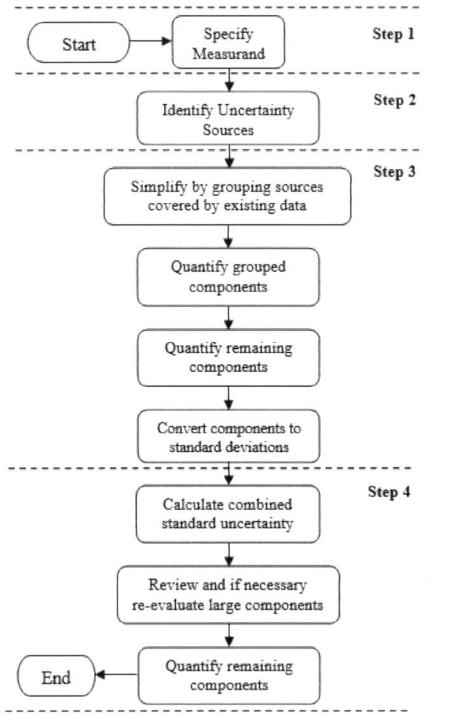

Figure 1. The uncertainty estimation process [8].

The combined analytical uncertainty of EDXRF Epsilon 5 (u_{E5}) was estimated applying GUM (Guide to Express the Uncertainty of Measurement 2008, JCGM 100:2008) methodology to Equation (1), as shown in Equation (2):

$$u_{E5} = \sqrt{\Sigma\left(\frac{\partial c_{E5}}{\partial x}\right)^2 u_x^2}$$

$$= \sqrt{\Sigma\left(E \times u_{(I_{net})}\right)^2 + \left(E \times u_{(I_{blk})}\right)^2 + \left(\left(u_{(I_{net})} - u_{(I_{blk})}\right) u(E)\right)^2} \quad (2)$$

where x is the components of Equation (1) and u_x is the uncertainty of each component. The uncertainties of I_{net} and I_{blk} were estimated as the standard deviation of the replicated sample and blank filter analysis, respectively. The uncertainty of calibration factor, u(E), was estimated from the uncertainty of Micromatter standard that was used in the method and the deviation of the linear curve based on software calculation. The combined uncertainty is converted into expanded uncertainty (U_{E5}) by using a coverage factor, *k*, as shown in Equation (3).

$$U_{E5} = k\sqrt{\Sigma\left(E \times u_{(I_{net})}\right)^2 + \left(E \times u_{(I_{blk})}\right)^2 + \left(\left(u_{(I_{net})} - u_{(I_{blk})}\right) u(E)\right)^2} \quad (3)$$

The coverage factor takes into account the distribution of uncertainties possible for a given measurement and, in this work, a coverage factor of 2 is used to give approximately the 95% confidence interval on the uncertainty value (k = 1.96 at 95% confidence level for a normal distribution) [10].

3. Results and Discussion

3.1. Measurement application

Optimum excitation conditions were set up as shown in Table 1. The method was set up for 26 elements divided in five conditions, but in this discussion only focused on the evaluation of 10 elements: Na, Si, S, K, Ca, Ti, Fe, Zn, Br, and Pb that generally existed in airborne particulate samples.

There were several parameters should be optimized in the method to generate an optimum excitation in each condition [4, 12]. The application had five conditions with different settings according to the characteristic energy for each element [13].

Table 1. Optimum excitation conditions of the method used in Epsilon 5 spectrometer.

No.	Parameter	Conditions				
		1	2	3	4	5
1.	Tube voltage (kV)	40	75	75	90	90
2.	Tube current (mA)	15	8	8	6	6
3.	Secondary target	CaF_2	Fe	Ge	Zr	CeO_2
4.	Detector setting	high resolution	standard	standard	high intensity	high instensity
5.	Energy range (keV)	0-20	0-20	0-20	0-20	0-40
6.	Analysis time (s)	600	400	400	600	400
7.	Elements analyzed	Na, Mg, Al, Si, P, S, Cl, K	Ca, Sc, Ti, V, Cr	Mn, Fe, Co, Ni, Cu, Zn	As, Se, Br, Pb	Sn, Sb, Ba
8.	Analysis line	Kα	Kα	Kα	Kα (except Pb using Lβ1)	Kα

3.2. Calibration standards

The calibration parameters used in this study are presented in Table 2. In quantitative analysis, the net intensities are converted into concentration using a calibration standard [14]. There are several parameters: D value, E value, and also correlation was used in calibration. Those values are determined by linear regression that is calculated automatically from the software. The correlations for 10 elements were in good results according to the data shown in Table 2.

There were good correlations between the expected concentrations and count rates with the correlation value close to one. The ability of an X-ray to penetrate an aerosol deposit depends on the energy of the X-ray, composition, and thickness of the deposit. As a characteristic of light elements, lower-energy X-rays are absorbed in samples greater than higher energy. However, attenuation factors for $PM_{2.5}$ samples are generally negligible even for the lightest element. $PM_{2.5}$ samples contain particles small enough so that no self-absorption corrections are needed [13].

Table 2. Calibration parameters of the method in EDXRF Epsilon 5 spectrometer.

No.	Elements	D value	E value	Correlation
1	Na	0	1.16E+04	0.9997
2	Si	0	7.58E+02	1
3	S	0	3.38E+02	1
4	K	0	9.04E+01	1
5	Ca	0	1.32E+02	1
6	Ti	0	1.04E+02	1
7	Fe	0	1.71E+02	1
8	Zn	0	7.64E+01	1
9	Br	0	2.03E+02	1
10	Pb	0	5.38E+02	1

3.3. XRF analytical uncertainties

Uncertainty evaluation is an essential part of quantitative analysis because analytical results should always be expressed with its corresponding uncertainty value [15]. In the present EDXRF uncertainty determination, the main source of uncertainty could be identified from counting statistics (gross or raw intensity, blank, and subtracted intensity) which were correlated with the calibration function. The counting statistic parameters become essential in calculating uncertainty because they were contributed to the conversion of intensity to concentration.

Generally, the element concentrations in air filter samples were very small (trace elements). Therefore, the role of the blank becomes important [16]. In this study, the blank was measured as a blank while the samples were measured as routine according to the software of Epsilon 5. The blank intensity then used to subtract the raw intensity of samples.

Table 3. Uncertainty values from air filter samples.

No.	Elements	Atomic Number (Z)	Concentration Range (ng/cm^2)	Relative Uncertainty (%)	Correlation
1	Na	11	92.6-1015	22.5	0.96
2	Si	14	49.3-1528	7.78	1.00
3	S	16	530-3739	2.44	0.91
4	K	19	70.5-756	1.58	1.00
5	Ca	20	16.4-1119	3.61	1.00
6	Ti	22	0.28-43.4	25.4	0.99
7	Fe	26	33.5-808	26.5	1.00
8	Zn	30	4.66-1031	4.49	1.00
9	Br	35	1.39-36.6	30.8	0.98
10	Pb	82	8.99-299	31.9	1.00

In this study, the blank filter was the clean filter before being used to collect the dust samples. The blank filter measured as blank in the application then used to subtract the air filter samples and it was set automatically in the software. The results from EDXRF Epsilon 5's software were the counting statistics data: net intensity (cps/mA), corrected intensity (cps/mA), and concentration (ng/cm^2). The analytical uncertainties in this paper were calculated using Equation (2) and Equation (3) for expanded uncertainty. There were three main sources of uncertainties according to the Equation (1) to get the mass loading of the air filter samples in ng/cm^2: net intensity (I_{net}), blank intensity (I_{blk}), and slope from standard calibration (E).

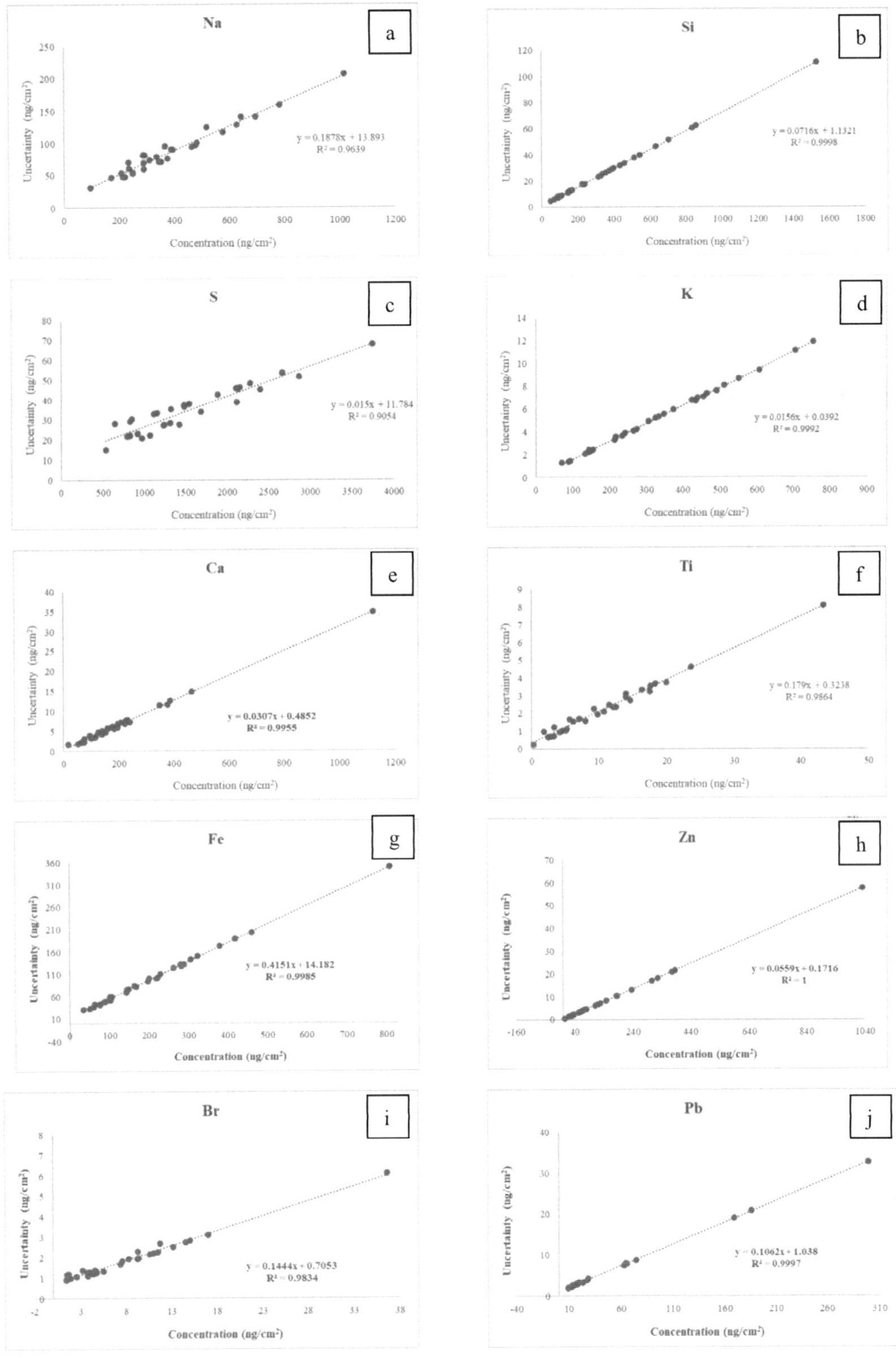

Figure 2. Total uncertainty estimation for 10 elements: (a) Na, (b) Si, (c) S, (d) K, (e) Ca, (f) Ti, (g) Fe, (h) Zn, (i) Br and (j) Pb.

The uncertainties evaluation results for 10 elements were illustrated in Figure 2. The figure shows that generally, the results of uncertainty measurements were in good correlation when compared with the concentration. The average uncertainty calculation was shown in Table 3. The results were varied from one to another elements. The averages of uncertainties from 10 elements were ranged from 2 to 32%. These results showed that uncertainties of EDXRF measurements depended on the concentration of elements that existed in the samples. The uncertainties for X-ray attenuation losses, although based on an extreme range of deposit compositions, are reasonable and conservative estimates; only exhaustive, actual measurement of particle composition and size distribution would allow calculation of more accurate values [17].

4. Conclusion

The uncertainty evaluation results demonstrate that the EDXRF Epsilon 5 spectrometer is applicable for air filter analysis both qualitatively and quantitatively and gave reliable results. Uncertainty evaluation is an essential part of quantitative analysis because analytical results should always be expressed with its corresponding uncertainty value.

Evaluation of uncertainties measurement in EDXRF Epsilon 5 was carried out following the guide to express uncertainty. The uncertainties evaluation results for 10 selected elements were in a good correlation, with % relative uncertainty were ranged from 2-32%. The results of this analysis can be utilized to determine the uncertainty value for air filter samples measured using EDXRF Epsilon 5, especially for 10 elements: Na, Si, S, K, Ca, Ti, Fe, Zn, Br, and Pb.

Acknowledgments

The authors would like to extend our thanks to all staff of the Analysis Techniques and Radiometry team for their technical assistance and also to others for their contribution to this research.

References

[1] Rousseau RM 2001 Detection Limit and Estimate of Uncertainty of Analytical Xrf *Results*. **18** (2) 33–47.

[2] Joint Committee for Guides in Metrology 2008 Evaluation of measurement data — Guide to the expression of uncertainty in measurement. *International Organization for Standardization Geneva ISBN*, *50*(September), 134. https://doi.org/10.1373/clinchem.2003.030528

[3] René EVG and Andrzej AM. 1999 *Handbook of X-ray spectrometry* second edition revised and expanded, practical spectroscopy series. Vol. 29. (New York: Marcel Dekker, Inc)

[4] PAN alytical 2009 *Analysis of solid alternative fuel materials*.

[5] RTI 2009 Standard Operating Procedure for the X-ray Fluorescence Analysis of Particulate Matter Deposits on Teflon Filters. August 19 2009 (North Carolina: Research Triangle Park).

[6] Yatkin S 2014 *IMPROVE Standard Operating Procedure for the X-Ray Fluorescence Analysis of Aerosol Deposits on PTFE Filters*. 1–17. http://airquality.crocker.ucdavis.edu/files/9014/5774/5516/SOP_301_X-Ray_Fluorescence_Analysis.pdf

[7] PANalytical 2005 *Epsilon 5 System User's Guide*. PANalytical B.V.

[8] S L R Ellison and A Williams 2012 EURACHEM/CITAC Guide CG 4: Quantifying Uncertainty in Analytical Measurement, Third Edition.

[9] Quality A 2019 *UCD CSN Standard Operating Procedure # 302 X-Ray Fluorescence Analysis of Aerosol Deposits on PTFE Filters (with PANalytical Epsilon 5)*. 1–21.

[10] Yatkin S, Amin HS, Trzepla K and Dillner AM 2016 Preparation of lead (Pb) X-ray fluorescence reference materials for the EPA Pb monitoring program and the IMPROVE network using an aerosol deposition method. *Aerosol Science and Technology*, ***50***(4) 309–320.

[11] Yatkin S, Belis CA, Gerboles M, Calzolai G, Lucarelli F, Cavalli F and Trzepla K 2016 An interlaboratory comparison study on the measurement of elements in PM10. *Atmospheric Environment* **125** 61–68.

[12] Lestiani DD and Santoso M *Analytical Methods INAA and PIXE Applied to Characterization of Airborne Particulate Matter in Bandung, Indonesia*. Atom Indonesia Vol. 37 No. 2 (2011) 52-56.
[13] Sari DK, Lestiani DD, Kurniawati S, Adventini N, Atmodjo DPD and Kusmartini I 2020 Applicability of EDXRF for elemental analysis in airborne particulate matter (APM): assessment using APM reference material. *Journal of Physics: Conference Series*, *1436*, 012137.
[14] Brouwer P 2010 Theory of XRF. In *Almelo: PANalytical BV.*
[15] Remya DPS, Trupti AC, Nicy A, Dalvi AA, Swain KK, Wagh DN and Verma R 2015 Evaluation of uncertainty in the energy dispersive X-ray fluorescence determination of platinum in alumina. *Analytical Methods*, **7**(12) 5345–5351.
[16] IAEA 2004 Quantifying uncertainty in nuclear analytical measurements. *International Atomic Energy Agency*, *July*, 247. https://doi.org/10.3850/978-981-09-0446-3_133
[17] Gutknecht W, Flanagan J, McWilliams A, Jayanty RKM, Kellogg R, Rice J, Duda P and Sarver RH 2010 Harmonization of uncertainties of x-ray fluorescence data for PM 2.5 air filter analysis. *Journal of the Air and Waste Management Association*, **60**(2) 184–194.

IOP Conf. Series: Materials Science and Engineering 1011 (2021) 012059 doi:10.1088/1757-899X/1011/1/012059

Establishment of Traceability Chain For pH Measurement in Indonesia

Ayu Hindayani, Nuryatini Hamim*, Christine Elishian, Yosi Aristiawan, Muhammad Rizky Mulyana, Eka Mardika Handayani, Eka Pratiwi, Gigin Ginanjar

Center for Research and Human Resources Development, National Standardization Agency of Indonesia, Gedung 420, Kawasan Puspiptek Serpong 15314, Tangerang Selatan - Banten, Indonesia

*Corresponding author's email: nuryatini@bsn.go.id

Abstract. As a consequence of the globalization of trade and industries, the presence of an accurate and traceable measurement has become very important for quality assurance of a product or service in the market, including pH measurement. pH measurement is a basic analysis used in a wide variety of applications, such as agriculture, wastewater treatment, industrial processes, research and development, and environmental monitoring. The reliable and comparable results of pH measurement are advantaged as one of the parameters for quality control that will affect the decision making. The Electrochemistry Laboratory under Deputy of National Measurement Standard – National Standardization Agency of Indonesia (SNSU–BSN) has a main duty to ensure the reliability, comparability, and traceability of pH measurements in Indonesia, as a commitment in building chemical metrology infrastructure. In this article, all the activities related to the establishment of traceability chain for pH measurement in Indonesia are comprehensively discussed. As of currently, the Electrochemistry Laboratory has successfully developed the secondary method for pH measurement using differential potentiometric cell, production of certified reference material (CRM) for buffer solution (oxalate, phthalate, phosphate, and borate buffer), dissemination by proficiency testing (PT) schemes for pH measurement, participation in the key comparison (KC) of APMP.QM-K91 for phthalate buffer, and claim submission for the Calibration Measurement Capabilities (CMCs) of phthalate buffer (pH 3.97-4.03) with expanded uncertainty 0.004.

1. Introduction

The viable globalization of trade and the complex manufacturing in recent years has caused a significant increase of the needs for accurate, reliable, and traceable measurements in any kind of industrial field, including pH measurement [1]. pH is defined as the negative logarithm of hydrogen ions activity [2]. pH measurement is carried out on a large scale ranging from laboratory level up to industry level, becoming pH is one of the most common quantitative measurements in many areas, such as agriculture, health care, biochemistry, wastewater treatment, industrial processes, research, and development and environmental monitoring [2-4]. For example, the pH value of food products is actively monitored to avoid the pathogenic microorganism growth that can damage food or adversely affect the taste performance and quality of the products. In the electrochemistry industry application, the plate formed will be affected by the pH of the plating solution. The incorrect pH may increase the possibility of the plate surface to peeling and produce less optimal color and final product [5].

Besides producing accurate and reliable results, the comparable result of pH measurement is necessary to prevent duplicating measurements which will consume more money and time [6]. Therefore, an accurate and reliable result of pH measurement is important as one of the parameters for quality control which will, directly and indirectly, affect every aspect of daily life and give impacts on decision-making [7]. To achieve this goal, SNSU-BSN as National Metrology Institute (NMI) for Indonesia sees the importance of the development of pH measurement infrastructure through the Electrochemistry Laboratory that is capable of providing accurate and traceable to the International System (SI) unit measurement.

NMI is a government institute appointed by national decision to hold and develop national measurement standards [8]. It holds the responsibility for establishment and maintenance of national measurement standards (primary or secondary) and methods, participation in international comparison, keep the traceability chain through dissemination of the SI unit to industry and society, ensuring the suitability of the standard for national needs, providing metrological expertise and knowledge to national users, assuring the equivalence of measurement standard, research and development in the field of metrology for government, industry, and society [1, 8-10]. The establishment of NMI in one country is very important, especially to assure the measurements used in manufacturing and international trade are accurate and traceable to the international standards [11].

Indonesia has started to establish the NMI in 2004, namely the Research Center for Calibration, Instrumentation, and Metrology (Puslit KIM–LIPI) [12], changed to Research Center for Metrology (RCM–LIPI) in 2017. However, RCM–LIPI has only capabilities in physical metrology. Therefore, the Center for Technology of Radiation Safety and Metrology/National Nuclear Energy Agency (PTKMR–BATAN) has been chosen as Designated Institute (DI) for radiation metrology and Chemical Metrology Laboratory of Research Center of Chemistry–Indonesian Institute of Science (RCChem–LIPI) as DI for chemical metrology. Then, based on Presidential Decree No. 4/2018, the government of Indonesia has moved the task of NMI from LIPI to BSN [12, 13]. So that, RCM–LIPI and Chemical Metrology Laboratory of RCChem–LIPI joint to BSN.

In BSN, NMI function is held by the Deputy for National Measurement Standards (SNSU-BSN), handling three fields of metrology; physical, biological, and chemical metrology. These three fields of metrology have a responsibility in research and development of their respective field, maintaining the national standard, and, especially for metrology in chemistry, providing certified reference materials (CRMs) for organic, inorganic, electrochemistry, and gas substances. Those CRMs are disseminated to the end-user or testing laboratory in Indonesia to keep the traceability chain to the SI unit. This article describe the establishment of traceability chain for pH measurement in Indonesia by the Electrochemistry Laboratory through development of the secondary method for pH measurement using differential potentiometric cell, production of various certified reference material (CRM) for buffer solution, dissemination by proficiency testing (PT) schemes for pH measurement, participation in the key comparison (KC) of APMP.QM-K91 for phthalate buffer, and the claim submission for the Calibration Measurement Capabilities (CMCs) of phthalate buffer.

2. Development of the secondary method for pH measurement

The traceability in metrology is led up to primary standard, whose quantity value is characterized by a primary measurement method, like as in pH measurement [4]. The primary measurement method for pH is based on the potential difference of the electrochemical cell without a liquid junction involving a selected buffer solution, a platinum hydrogen gas electrode, and a silver/silver chloride reference electrode, well-known as Harned Cell. This method is able to assign pH value for the primary buffer standard solution with an uncertainty value at 0.003. This primary buffer standard solutions are advantaged for the calibration of the cell without a liquid junction, as a secondary pH measurement. The cell works to assign the pH value of the secondary buffer standard solution with uncertainty

IOP Conf. Series: Materials Science and Engineering 1011 (2021) 012059 doi:10.1088/1757-899X/1011/1/012059

between 0.004 and 0.01. The secondary buffer standard solutions are used for calibrating commercial pH meters in common laboratories before the pH measurement in the sample [14]. This traceability guarantee that the pH measurement is reliable and valid. The metrological traceability scheme for pH is described in Figure 1.

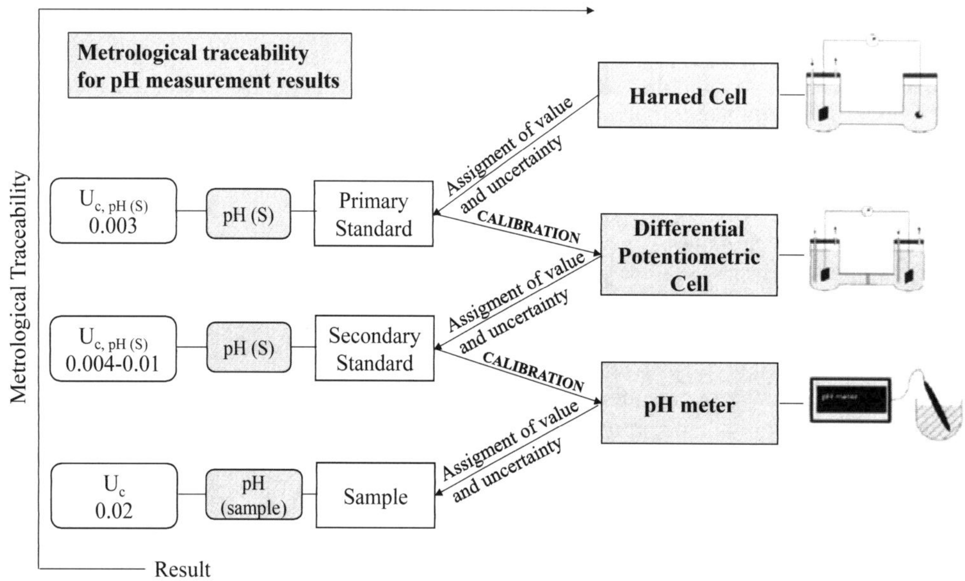

Figure 1. Metrological traceability scheme for pH (Uc is expanded uncertainty with coverage factor k=2) [4].

The primary measurement method for pH is realized by NMI as the top of the traceability chain [4]. The system requires the highest degree of accuracy and the available primary buffer standard is limited and expensive. Therefore, Indonesia under RCChem-LIPI's Electrochemistry Laboratory selected the secondary method as the highest traceability chain for pH measurement in Indonesia in 2015. In accordance with the traceability chain in Figure 1, the system is used to measure the secondary standard buffer solution with the same nominal composition as primary standard buffer solution. It used a primary buffer standard solution from primary pH measurement (Harned cell) as standard. Therefore, the secondary pH measurement is traceable to the primary buffer standard solution. The secondary standard buffer is expected to extend traceability to a greater market, offers significant economic advantages, the simpler procedure, faster, and employs cheaper instrumentation [3,14]. Then, the secondary buffer standard solutions that are produced by the Electrochemistry Laboratory can be used for calibrating the pH meter to measure the pH value of the sample and give accurate, reliable, confidence, and traceable results of pH measurement in Indonesia.

The secondary system of pH measurement developed by the Electrochemistry Laboratory was obtained from Japan. The instrument set is described in Figure 2. It consists of a U-shape cell which separated by a sintered glass disk in the middle of the cells, recognized as a Baucke cell [3]. The Baucke cell is immersed in the temperature control water bath. Each cell contains identical Platinum/ Hydrogen (Pt/H2) electrodes. One cell contains a primary buffer standard solution. Then the other cell contains secondary buffer standard solutions produced by the Electrochemistry Laboratory as a sample. As a result, the secondary pH measurement is traceable to the primary buffer standard solution. These two buffers must have the same nominal composition. Moreover, gas H_2 is firstly pre-humidified before entering the Baucke cell by passing through the gas into two chambers which contain similar nominal composition buffer solution to that of the sample and standard. The

IOP Conf. Series: Materials Science and Engineering 1011 (2021) 012059 doi:10.1088/1757-899X/1011/1/012059

Platinum/Hydrogen (Pt/H_2) electrodes then connected into the multimeter and the potential difference between two Pt/H_2 electrodes is measured. The detailed procedure of pH measurement using the secondary pH measurement system is described in the previous report [15]. The pH of the secondary buffer standard solutions is calculated using Equation (1) [3]. pH_p stands for pH value from the primary buffer standard solution, ΔE_{cell} represents the potential difference between two Pt/H_2 electrodes, F is Faraday constant, R is the universal gas constant, and T defines temperature (Kelvin).

$$pH_s = pH_p - \frac{\Delta E_{cell} F}{RT \ln 10} \quad (1)$$

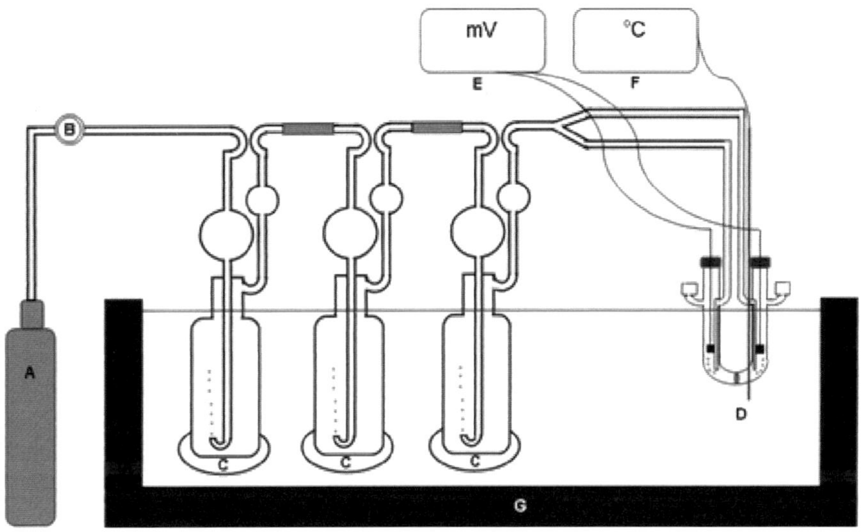

Figure 2. The secondary pH measurement system: (A) hydrogen gas, (B) gas controller, (C1&C2) chambers contain buffer, (D) Baucke cell with Pt electrodes, (E) multimeter, (F) thermometer, (G) water bath [15].

Before routine usage, this secondary pH measurement must be validated to confirm that the method is suitable for intended use as required in ISO/IEC 17025:2017. The results of the validation method for secondary pH measurement described in the previous experiment [15]. The results showed that the accuracy and precision of the method were sufficiently satisfying, with an estimated uncertainty value of 0.001 (k=2) which is a contribution of the primary buffer standard solution (56%), potential (43%), and temperature (1%) for secondary pH measurement at 25°C. Afterward, this method can be routinely used to measure the pH value of secondary buffer standard solutions.

3. Production of CRM for Buffer Solution

CRM is a reference material that has been characterized and accompanied by a certificate that provides the value of the specified property, its uncertainty, and the statement of traceability [16]. It plays a vital role in the quality of measurement, such as in pH measurement. It is used for calibration of instruments, assessment of the measurement method, and establishment of the measurement traceability results [17]. The production and certification of RM is a key activity of NMI to improve and maintenance of a worldwide coherent system of measurement [18].

IOP Conf. Series: Materials Science and Engineering 1011 (2021) 012059 doi:10.1088/1757-899X/1011/1/012059

The Electrochemistry Laboratory has started to do the research and development in produce CRM for buffer solution since 2013. The characterization, homogeneity, and stability of CRM buffer were conducted based on ISO Guide 35. In the beginning, a pH meter was used by the Electrochemistry Laboratory to measure the pH value of the CRM buffer. Various buffers have been produced, including oxalate (1.68 ± 0.03), phthalate (pH 4.00 ± 0.02), phosphate (pH 6.86 ± 0.02), and borate buffer (pH 9.18 ±0.03) at 25°C [19, 20]. This CRM buffer, however, has big uncertainty. In order to reduce the uncertainty, the Electrochemistry Laboratory started to produce CRM for buffer solution using secondary pH measurement system with a lower uncertainty in 2015. The buffers produced with this method consist of phthalate (pH 4.005 ± 0.004) and phosphate (pH 6.863± 0.004). The availability of CRM buffer in Indonesia is expected to fulfill the local needs, suppress import value, and provide the traceability of pH measurement in Indonesia to get accurate, reliable, confidence, and traceable results. The CRM buffer that has been produced by the Electrochemistry Laboratory is shown in Figure 3.

Figure 3. CRM buffer of the Electrochemistry Laboratory and its packaging.

4. Proficiency Testing Providing

Proficiency Testing (PT) is an evaluation of laboratory performance against pre-establish criteria by means of interlaboratory comparisons [21]. It is aimed to monitor laboratories work and compare their results with other laboratories and is a part of the process in assessing the capability of laboratories to perform measurements and tests for which accreditation is intended [22].

PT scheme for pH commodity in Indonesia has been conducted by the chemical metrology laboratories since the laboratory was under RCChem–LIPI. At the first, chemical metrology laboratories only provide the PT samples individually. Beginning in 2008, the PT scheme has been integrated with National Accreditation Committee (KAN) program, marking the RCChem's chemical metrology laboratory is one of the PT providers in Indonesia, which is responsible to process participant data and evaluate their performance. This activity becomes an annual agenda, serving different commodities and parameters each year, to disseminate the traceability to laboratories in Indonesia through PT sample or standard solution (packaged together with PT sample). The PT scheme for pH measurement that has been conducted by the Electrochemistry Laboratory is tabulated in Table 1.

Table 1. PT scheme for pH measurement that has been conducted by the Electrochemistry Laboratory.

Year	Commodities	Participants
2013	Phthalate (pH 4.04) and phosphate buffer (pH 6.56)	10
2014	pH in wastewater (pH 1.91)	94
2015	Citrate buffer (pH 4.92)	36
2016	Borate buffer (pH 9.186)	50

The increasing number of laboratories participating in this proficiency test shows that the laboratory realizes the importance to participate in proficiency testing. It is one of kind to assure the quality of test and calibration results as mention in ISO 17025 [23]. The PT results could be a self-evaluation for each laboratory. The satisfactory results show the performance of the laboratory is good, while for unsatisfactory results means failure to attain the minimum satisfactory score for an analyte, test, subspecialty, or specialty for a testing event. This may also be referred to as unacceptable performance and the laboratory must take the evaluation for improving their performance [24].

5. Participating in the Key Comparison (KC) and Claiming the Calibration and Measurement Capability (CMC)

KCs are international inter-laboratory studies used to establish the degree of equivalence between national measurement standards that carried out by an NMI. Participation in KCs for NMI is the best way to demonstrate the NMI confidence in its measurement results and consequently, in its measurement capabilities [25]. Therefore, the measurement results obtained will be internationally comparable and accepted, giving no need to duplicate the measurement in other countries.

The basis for working international trades, commerce, and regulatory affairs is the presence of mutual recognition in calibrations, measurements, and test certificates [26,27]. KC plays an important role to enable that mutual recognition framework. As an international inter-laboratory study, KCs establish the degree of equivalence between national measurement standards that carried out by an NMI. From the analysis of the key comparison result, the corresponding measurement capability of NMIs is validated and it will be written as calibration and measurement capabilities (CMCs) [27].

Comparability and traceability are normally established through participation in KC organized by regional or international metrology organizations like the International Bureau of Weights and Measures (BIPM) and the Asia Pacific Metrology Programme (APMP) [28]. The Electrochemistry Laboratory as part of NMI focuses on the study of comparability and traceability of pH measurements participated in the KC of APMP.QM-K91 in 2015. This was organized by TCQM of APMP to test the abilities of NMI in the APMP region to measure the pH value of phthalate buffer. This comparison was coordinated by NMIJ (National Metrology Institute of Japan). The sample was phthalate buffer with molality 0.05 mol/g prepared in a polyethylene bottle. There were 12 NMIs that participated in this KC, including the Electrochemistry Laboratory that still under the RCChem–LIPI at that time. In the KC, the Electrochemistry Laboratory employed the above-mentioned secondary pH measurement system to measure the sample of phthalate buffer. The measured value of phthalate buffer at 25°C (4.006) is close to the Key Comparison Reference Value (KCRV) (4.00765) as described in Figure 4 [29,30], showing the satisfactory of the measurement. This good results from APMP.QM-K91 can be used by the Electrochemistry Laboratory to support the CMC claim for a phthalate buffer using secondary pH measurement method (Differential Potentiometric Cell), beside the proof of the dissemination activity in national PT.

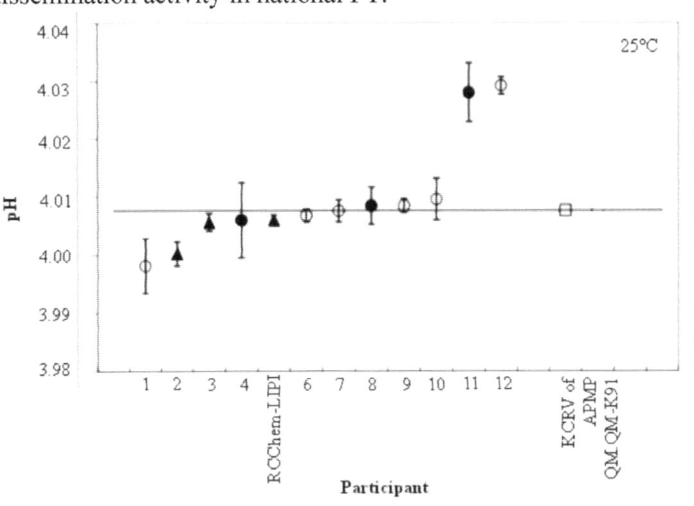

Figure 4. Results of pH measurement for phthalate buffer at 25°C compared with KCRV of APMP.QM-K91 [29, 30].

After the peer-review evaluation and document process as required, the Electrochemistry Laboratory has successfully been granted the CMC by BIPM in 2018 for pH of phthalate buffer (pH 3.97-4.03) with expanded uncertainty 0.004 [31]. The list of CMCs of all countries is openly available to customers as published in the BIPM key comparison (KCDB) of the CIPM MRA [32]. It is mainly aimed to inform customers about quantities, ranges, and achievable uncertainty of the measurement laboratory [33].

6. Conclusion

The establishment of traceability chain for pH measurement in Indonesia by Electrochemistry Laboratory is at the beginning level to accommodate the laboratory needs. The Electrochemistry Laboratory has conducted various research activities from 2015 to build the traceability chain of pH measurement in Indonesia. The activities cover the development of the secondary method for pH measurement, the production of CRM for buffer solution, the organization of the PT scheme, the participation in the KC, and finally the claim of the CMC. All activities in the efforts to develop metrology in chemistry in Indonesia, to build the pH measurement infrastructure, and furthermore, to support the quality assurance and the global trade. The metrological research will be continued in the short-term and long-term plan to another pH level and other electrochemical measurands.

Acknowledgments

The authors would like to thank to RCChem–LIPI and DIPA BSN 2020 for financially supporting this work.

References

[1] Kühne M, Krystek M and Odin A 2004 *Traceability of measurement results in industrial metrology* In: Simposio de Metrología.

[2] Buck RP, Rondinini S, Covington AK, Baucke FGK, Brett CMA, Camoes MF, Milton MJT, Mussini T, Naumann R, Pratt KW, Spitzer P and Wilson G S 2002 *Pure Appl. Chem.* 74 2169–200

[3] Gonzaga FB, Dias JC, Jehnert D, Werner B, Schrapler K and Vyskocil L 2013 *Electroanalysis* **25** 1955 – 9

[4] Spitzer P and Seitz S 2010 *MAPAN* **25** 191-6

[5] Aliouche H 2019 *Importance of Measuring pH in Industry*, available at: https://www.news-medical.net/life-sciences/Importance-of-Measuring-pH-in-Industry.aspx.

[6] Papadakis I and Wegscheider W 2000 *Accred. Qual.Assur.* **5** 388-9

[7] Samitsch C 2015 *Data Quality and its Impacts on Decision-Making* In Springer Gabler: Innsbruck, Austria.

[8] Kellermann M 2019 *Ensuring Quality to Gain Access to Global Markets* In World Bank Group: Washington, D.C.

[9] EURAMET and the Operation of NMIs 2008 *EURAMET Guide No. 1 Version 2.0 (01/2015).*

[10] Dube G 2001 *Accred. Qual. Assur.* **6** 3-7

[11] Ofori RA 2019 *The Correct Measurement*, available at https://www.quality.org/knowledge/correct-measurement.

[12] BIPM 2019 *The Republic of Indonesia/ CIPM MRA*, available at https://www.bipm.org/en/about-us/member-states/id/cipm-mra.html.

[13] President of the Republic of Indonesia, 2018, *Presidential Decree no. 4/2018 about National Standardization Body* In Jakarta.

[14] Spitzer P and Pratt K W 2011 *J. Solid State Electrochem.* **15** 69–76

[15] Hindayani A, Zuas O, Sujarwo, Krismastuti FSH, Nuryatini 2019 *Chem. Chem. Technol.* **13** 377–83

[16] ISO 2017 *ISO Guide 35:2017, in Reference materials-Guidance for characterization and assessment of homogeneity and stability* In ISO copyright office: Switzerland.

[17] Fajgelj A 2000 *Spectroscopy* **15** 6
[18] Guimarães EF, do Rego ECP, Cunha HCM, Rodrigues JM, Figueroa-Villar JD and Braz J 2014 *Chem. Soc.* **25** 2
[19] Nuryatini, Sujarwo and Hindayani A 2016 *J. Standardisasi* **18** 34-42
[20] Sujarwo and Nuryatini 2013 *JKTI* **15** 9-14
[21] ISO 2010 *ISO/IEC 17043:2010, in Conformity assessment - General requirements for proficiency testing* In Geneva.
[22] Duarte AM, Almeida C, Candeias R and Sousa C 2007 *Proficiency testing – as a quality tool in the laboratories* In The First International Proficiency Testing Conference, Sinaia, România.
[23] Sidney S 2011 *The Importance of Proficiency Testing (PT) & Why Take Part"*, In T&M 2011 Workshops, available at http://www.nla.org.za/conferences/proceedings_archive/2011/Plenary/Importance%20of%20PT.pdf
[24] CLIA 2014 *Proficiency testing and PT referral Dos and Dont's*, available at https://www.cms.gov/Regulations-and-Guidance/Legislation/CLIA/downloads/CLIAbrochure8.pdf.
[25] Oliveira JCV 2009 *Key Comparison Reference Values (Kcrv) And Their Importance To Take Final Conclusions And Decisions* In 5th Brazilian Congress of Metrology, Salvador, Bahia – Brazil.
[26] Adriana H and Guthrie WF 2005 *A Survey of Key Comparisons* In Measurement Science Conference, Anaheim, CA: NIST.
[27] Anonim 2020 *Analysis of Key Comparisons*, available at https://mathmet.org/research/kc/.
[28] Marín MM 2018 *Role of National Metrology Institutes* In Simposio metrologia, Mexico.
[29] Hamim N, Sujarwo, Hindayani A and Krismastuti FSH 2018 *MAPAN* **33** 83-90
[30] Hioki A, Asakai T, Maksimov I, *et.al.* 2016 *Report of the key comparison APMP.QM-K91 APMP comparison on pH measurement of phthalate buffer*, available at https://www.bipm.org/utils/common/pdf/final_reports/QM/K91/APMP.QM-K91.pdf.
[31] BIPM 2020 *Key Comparison Data Base (KCDB) of Indonesia in chemistry,* available at https://www.bipm.org/kcdb/cmc/quick-search?includedFilters=cmcDomain.CHEM-BIO&excludedFilters=&page=0&keywords=indonesia.
[32] BIPM 2017 *Calibration and Measurement Capabilities in the context of the CIPM MRA*, available at https://www.bipm.org/utils/common/documents/CIPM-MRA/CIPM-MRA-D-04.pdf.
[33] Horský J and Horská J 2014 *Системи обробки інформації*, p. 37-40

IOP Conf. Series: Materials Science and Engineering 1011 (2021) 012060 doi:10.1088/1757-899X/1011/1/012060

Mechanical properties of edible film from Tanduk Banana *(Musa corniculata* Rumph) peels for food packaging

Muryeti*, Rina Ningtyas, Herdian Nugroho and Aufi Sabrina

Technology Industry of Prining Packaging Department, Politeknik Negeri Jakarta, Jl. Prof. DR. G.A. Siwabessy, Kukusan Beji, Kota Depok, 16414

*Corresponding author's email: yeti.tgp@gmail.com

Abstract. The widespread use of plastic has adverse environmental impacts. Therefore, it is necessary to find environmentally friendly packaging material alternatives, derived from biopolymers. This study aimed to make biodegradable films from Tanduk Banana *(Musa corniculata* Rumph) peels and determine the characteristics of these films. The methods used were extraction of pectin from Tanduk banana peel, making of edible film and characterization of mechanical properties of edible film. The raw materials used were variation of tapioca flour (0g; 0.6g; 1.2g and 1.8g), CMC 3%, Glycerol 0.5 ml, and 6g of banana peel pectin. Various tests were carried out to determine mechanical properties of edible film, include tensile strength, elongation, modulus Young, and thickness. Test results show the optimum value of edible film thickness is 1.025 mm and obtained at the composition of 2.4g of tapioca flour. The optimum value of tensile strength is 8.22 MPa, the elongation percentage is 40.4% and obtained at the composition of 1.2g tapioca flour. The optimum value of modulus young is 0.20347 MPa and obtained at the composition of 1.2g tapioca flour.

1. Introduction

Currently the extent of plastic use in Indonesia as food packaging material in order to meet daily needs is very high, reaching 64,000,000 tons/year. Plastic is flexible, economical, durable, hard to tear, and a good barrier against oxygen, water vapour, and carbon dioxide. Plastic also has several disadvantages derived from petroleum, which is increasingly limited in number. Plastic degradation is very slow in nature. It would take years for plastic to fully degrade, and resulting in buildup of plastic waste which causes environmental pollution. Currently, the development of food packaging technology leads to environmentally friendly, biodegradable, more natural and without preservatives packaging materials [1]. Many studies have been conducted to develop biodegradable packaging that is made with natural ingredients that are safe for consumption, does not contaminate food products, can maintain the freshness and durability of the product and emerge in the form of edible thin films. Edible films have been considered an alternative as food packaging due their capability to extend the shelf life of food by reducing moisturizer, and oxidative reactions rate [2].

One of the natural ingredients used in making edible film is pectin extracted from banana peels. Previous studies have shown the yields of extracted pectins from saba banana peels range from 16.24 to 24.5% [3,4]. The formation of edible films from pectin has advantages, in that the surface smoothness in resulting film but it has low mechanical properties [5]. Adding glycerol or sorbitol as plastisizer can make edible film more flexible, smooth, increasing permeability toward water, gas and soluble substances [6]. In this study, the edible film was made from Tanduk Banana *(Musa corniculata* Rumph) peel pectins with the addition of glycerol, CMC plasticizers and tapioca which expected produce a mixture of edible film as food packaging materials with optimum mechanical properties such as tensile strength and elongation percentage. This study aimed to make biodegradable films from Tanduk Banana peels and determine the characteristic of these films.

2. Methodology

This study comprised of three stages in the implementation. The first stage was extraction of pectin from banana peels. The second stage was preparation of edible film. The final stage was characterization of mechanical properties of the edible film.

2.1. Materials

Materials used in this research were Tanduk Banana peels from traditional market in Depok, West Java, Indonesia, HCl (Merck), distillate water, tapioca flour, CMC 3%, Glycerol 0.5 ml, NaOH (Merck), Ca Propionate, K-Sorbat. The instruments used in this research include glass apparatus in the laboratory, blender, hotplate stirrer, oven, and tensile testing machines.

2.2. Extraction of pectin

One kg of banana peels was dried in the oven at 60 °C for 24 hours until dried and brown. The dried peels were then grinded into powder by using blender and then sieved using a 100 mesh sieve. The powder was dissolved in 200 mL of 0.05 N HCl and was stirred at 80 to 90 °C with magnetic stirrer for 120 minutes. The solution was filtered and then the resulting filtrate was added with 96% ethanol in a ratio of 1: 1 and stirred to form precipitated pectin. It was purified with ethanol repeatedly and dried in the oven at 50 ^{0}C [6].

2.3. The preparation of edible film

Three types of solutions were prepared in this study. The first solution was prepared by dissolving pectin powder in 150ml of distillate water. The second solution was made from tapioca flour with variation concentrations of 0 g; 0.6 g; 1.2 g; 1.8 g and 2.4 g which was dispersed in 150 ml of distillate water. The solution was then heated and stirred with a magnetic stirrer for 60 seconds. The third solution was formed by mixing CMC with 150 ml of distillate water then the solution was heated and stirred with magnetic stirrer until no bubble observed. Next, all of the prepared solutions were put into a glass beaker and stirred and then heated until homogeneous. Then 0.5 ml of glycerol was added to the mixture solution and it was further stirred and heated to 90 °C. The solution was poured into a mold and dried in oven at 90 °C for 12 hours.

2.4. Characterization of edible film

The mechanical properties of edible film including thickness, tensile strength, elongation, and modulus young were analysed.

3. Result and Discussion

Pectin was made from Tanduk Banana peel by using extraction process. The extraction started with the separation of banana peels using Hydrochloric acid (HCl). The acid was being used in pectin extraction to hydrolyze protopectin into water-soluble pectin or to prevent pectin from bonding with other compounds, such as cellulose. The use of hydrochloric acid increased yield of pectin extracted was 11.93%.

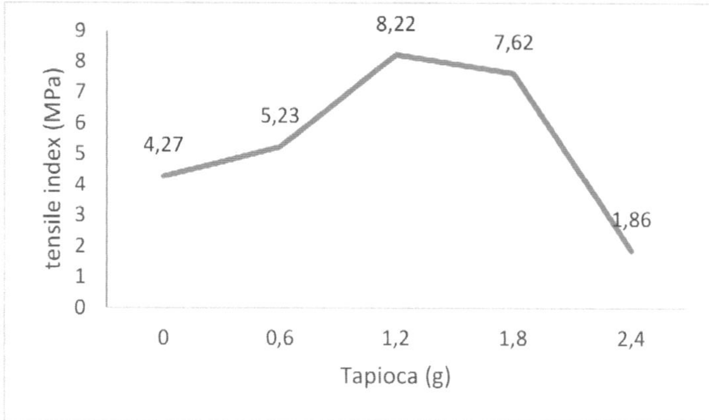

Figure 1. Tensile strength of edible film from pectin banana peels.

Test result showed that edible film made without tapioca flour had tensile index of 4.27 MPa. Edible film with tapioca flour of 0.6 g had tensile index of 5.23 MPa. Edible film with the addition of tapioca flour of 1.2g resulted in a tensile index of 8.22 Mpa. Moreover with the addition of tapioca flour of 1.8g and 2.4g, tensile index were 7.62 MPa and 1.86 Mpa (figure 1). The tensile strength test showed that tensile index of edible film increased with increasing of concentration of tapioca flour until reaching optimum tensile index at tapioca flour of 1.2 gram. It shows that tensile strength values were influenced by the composition of the tapioca flour. However, tapioca flour addition of more than 1.2g decreased the tensile index of edible film. This was due to the reduction of intermolecular interactions of protein chains so that the film matrix was less formed [7]. The formation of edible film involves gelatinization of tapioca granules by heating in excess water. It results in granules swelling and disruption as well as leaching of pectin from the granule. The gelatinization technique and drying method used to obtain edible films affected network charecteristics determining changes in physical properties. Tapioca based edible film are regarded stach systems that have been plasticized with glycerol (level of retrogradation), directly affects the mechanical properties [8].

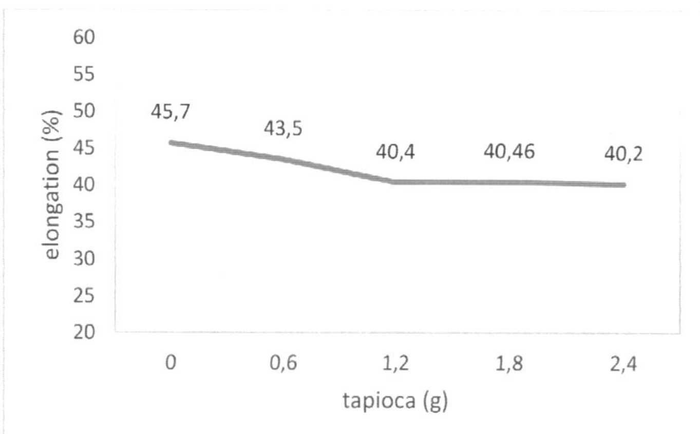

Figure 2. Elongation of edible film from pectin banana peels.

Test results showed that elongation of edible films were obtained range between 40.4 to 45.7% (figure 2). The addition of tapioca flour did not affect the elongation percentage of edible films. Percent elongation of the edible films was influenced by the plasticizer (glycerol, sorbitol, and CMC). The

IOP Conf. Series: Materials Science and Engineering 1011 (2021) 012060 doi:10.1088/1757-899X/1011/1/012060

plastisizer are intended to decreased the intermolecular forces along polymer chains, imparting increased film flexibility while decreasing the barrier properties of films [9,10].

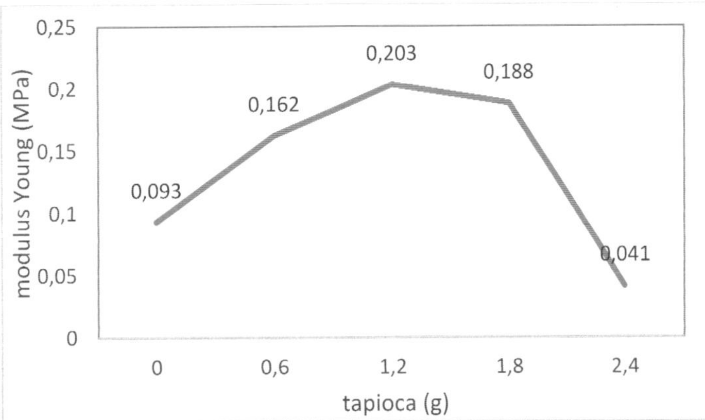

Figure 3. Modulus Young percentage of edible film.

Figure 3 demonstrated that Modulus Young percentage of edible film made without tapioca flour (0g) was 0.093 MPa. Modulus young percentage of edible film made with the addition of 0.6g tapioca flour was 0.162 MPa. Edible film with the addition of 1.2g tapioca flour resulted in modulus young percentage of 0.203 MPa. Edible film with 1.8g tapioca flour had modulus young percentage of 0.188 MPa. Moreover with the addition of 2.4g tapioca flour, modulus young percentage was 0.041 MPa. The results showed that optimum percentage of modulus young of edible film was reached at 0.203 MPa with 1.2g additional tapioca flour. Percentage of modulus Young of edible film increased with the addition of tapioca flour.

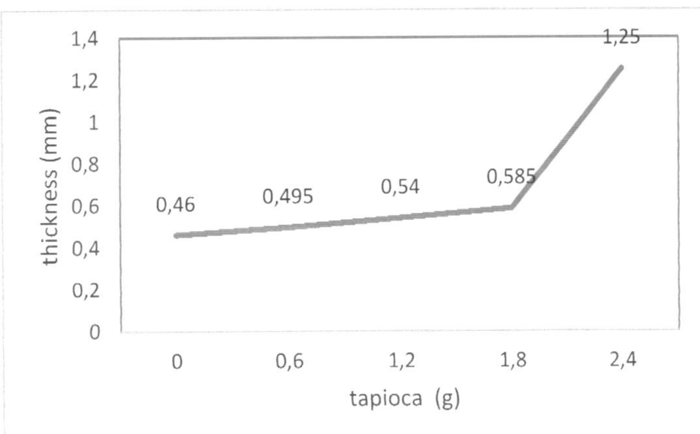

Figure 4. Thickness of edible film from pectin banana peels

Test result showed that thickness value of edible film made without tapioca flour (0 gram) was 0.46 mm. Thickness value of edible film made with the addition of 0.6g tapioca flour was 0.495 mm. Edible film with the addition of 1.2g tapioca flour resulted in thickness value of 0.54 mm. Edible film with the addition of 1.8g tapioca flour had thickness value of 0.585 mm. Moreover with the addition of 2.4g tapioca flour, thickness value was 1.025 mm (figure 4).

The addition of tapioca flour affected the thickness of edible film. The results showed that optimum thickness value of edible film was reached at 1.025 mm with addition of tapioca flour at concentration of 2.4g. The thickness value of edible film increased with increasing of concentration of tapioca flour. Thickness is an important parameter that influences the use of film in forming the product to be packaged. The thickness of film affected by retraction phenomenon that accure in film casting. It was related to composition of dry matter of film and shrinkage during drying period [9]. Thickness also affected the mechanical properties of edible film, such as tensile strength and elongation.

4. Conclusion

This study showed that edible film made from Tanduk Banana peel pectin had following mechanical properties: tensile strength, elongation, modulus young, and thickness. Mechanical properties of edible film such as tensile strength and modulus Young was affected by the addition of tapioca flour in edible film making. The optimum mechanical properties of edible film was obtained with the addition of 1.2g of tapioca flour. The value of edible film elongation and thickness were not affected by tapioca addition.

References

[1] Huang T, Qian Y, Wei J and Zhou C 2019 *Polymers* **11** 3
[2] Acevedo-Fani A, Salvia-Trujillo L, Rojas-Grau MA and Martın-Belloso O 2015 *Food Hydrocoll.* **47** 168–77
[3] Castillo-Israel K A T Baguio, et al. 2015 Extraction and characterization of pectin from Saba banana [*Musa ' saba '(Musa acuminata x Musa balbisiana*)] peel wastes : A preliminary study". *International Food Research Journal,* ***22***(1) 202–207.
[4] Khamsucharit P, Laohaphatanalert K and Gavinlertvatana P 2018 *Food Sci. and Biotechnol.* **27** 775–81
[5] Valdés A, Ramos M, Beltrán A, Jiménez A and Garrigós MC 2017 *Coatings* **7** 4
[6] Kulkarni SG and Vijayand P 2010 *LWT-Food Sci. Technol.* **43** 1026–31
[7] Bourtoom T, Chinnan MS, Jantawat P and Sanguandeekul R 2006 *Food Sci. Technol. Int.* **12** 119–26
[8] Fahem AH and Kadhim M 2018 *IJMET* **9** 1493–503
[9] Chillo S, Flores S, Mastromatteo M, Conte A, Gerschenson L and Del Nobile MA 2008 *J. Food Eng.* **88** 159–68
[10] Phan The D, Debeaufort F, Voilley A and Luu D. 2009 *J. Food Eng.* **90** 548-58

IOP Conf. Series: Materials Science and Engineering 1011 (2021) 012061 doi:10.1088/1757-899X/1011/1/012061

Phytochemical properties and antioxidant activity of wild-grown and cultivated *Ganoderma lucidum*

Septi Nur Hayati*, Cici Darsih, Vita Taufika Rosyida, Wuri Apriyana, Khoirun Nisa, Anastasia Wheni Indrianingsih and Martha Purnami Wulanjati

Research Unit for Natural Products Technology, Indonesian Institute of Sciences, Yogyakarta, Indonesia

*Corresponding author's email: hayanurrizal@gmail.com

Abstract. The most biologically active compounds of medicinal mushroom *Ganoderma lucidum* can be classified into polysaccharides and terpenoids. Most of these biological compounds are supposed to associate with its antioxidant activity. Both of wild grown and cultivated *G. lucidum* have been commercially in demand in Indonesia during the past years. Due to their different growing conditions, the wild-grown and cultivated *G. lucidum* may contain different levels of effective chemical components which affect their quality and medicinal efficacy. This present study was carried out to determine the differences between wild-grown and cultivated *G. lucidum* which might be useful in exploring the characteristic of chemical compounds of *G. lucidum* regarding its antioxidant activity. The physicochemical evaluation was determined using gravimetric method. The phytochemical evaluation includes water – soluble polysaccharides, phenolic, and terpenoids content. The antioxidant activity was evaluated by measuring the radical scavenging activity using 2,2-diphenyl-1-picrylhydrazyl (DPPH) free radical assay. Cultivated *G. lucidum* from Godean has the highest water – soluble polysaccharides (29.86±2.42 GE, mg/g dw) and phenolic content (5.07±0.39 GAE, mg/g dw) among other studied samples. Whereas, cultivated G. lucidum from Gunung Kidul has the lowest water – soluble polysaccharides (21.65±2.45 GE, mg/g dw) and phenolic content (3.21±0.87 GAE, mg/g dw). Both of wild grown *G. lucidum* have higher terpenoids content compare to all of cultivated *G.lucidum*. The cultivated – Godean revealed the highest DPPH scavenging activity (the lowest IC50, 344.15±9.57 μg/mL) among of the studied samples. Hence, the results suggested that *G. lucidum* contained high metabolites compounds and has a potential natural source of antioxidants.

1. Introduction

Ganoderma lucidum, one of polypore macro fungi growing in decomposing woods, commonly known as lingzhi, has been used as a therapy for the prevention and treatment of various diseases for centuries in Asia. The regular consumption was believed to preserve human vitality and to promote health and longevity. A number of studies on the aqueous or ethanol extracts of *G. lucidum* have discovered that the mushroom has anti-tumor, hepatoprotective, anti-inflammatory effects and shows activities in the immune, cardiovascular and central nervous systems [1,2]. Most of these therapeutic activities are believed to have association with its antioxidant activity [3]. *G. lucidum* contains bioactive components such as terpenoids, steroids, phenols, glycoproteins and polysaccharides [4]. Numerous publications have shown that the most pharmacologically active compounds of *G. lucidum* can be generally divided into triterpenes and polysaccharides [4,5]. *G. lucidum* was introduced to Indonesia in the 1990s and its large-scale cultivation was started in 1999. Both of wild grown and cultivated *G. lucidum* has been commercially in demand in Indonesia during the past several years. Due to their different growing conditions, the wild-grown and cultivated *G. lucidum* may comprise different levels of effective chemical compound which affect their quality and medicinal efficacy. This current study aimed to determine the differences between wild and cultivated *G. lucidum* which might be useful in exploring the characteristic of chemical compounds of *G. lucidum* regarding its antioxidant activity.

2. Materials and Methods

2.1. Mushroom materials
The mushrooms were dried in the sun with a sufficient amount of air flow to prevent molding. The dried wild – grown *G. lucidum* were collected from Cianjur and Gunung Kidul, Indonesia. While, the dried cultivated *G. lucidum* were collected from Kaliurang, Godean, and Gunung Kidul, Indonesia. They have been taxonomically verified at Pharmaceutical Biology Division, Faculty of Pharmacy, Universitas Gadjah Mada. The voucher specimen was retained in the Research Unit for Natural Products Technology, Indonesian Institute of Sciences. The ground dried mature fruiting bodies of *G. lucidum* were stored in air tight container for further analysis.

2.2. Preparation and extraction
Samples were macerated in ethanol 70% (1: 15) at room temperature for 72 hours. After filtration, the residue was re-extracted with the same method. The filtrates were combined and dried under reduced pressure at 60° C. The resulting extracts were further used for determination of some parameters.

2.3. Physicochemical evaluation
Determination of physicochemical characteristics include total ash content, acid insoluble ash, water soluble extractive, and ethanol soluble extractive using gravimetric method [6,7].

2.4. Determination of water-soluble polysaccharides
The water-soluble polysaccharides were determined with phenol-sulfuric acid colorimetric assay as glucose with hydrolyze polysaccharides into glucose monomer [8-11]. Samples (0.50 g) were extracted with hot water at 95°C and hydrochloric acid (HCl) 2 M in water bath (Memmert) for two hours. Filtrates were separated by filter paper and transferred on the centrifuge tube 10 mL. Then, 1 mL of filtrates were added with 5% phenol 0.5 mL (Merck) and 2.5 mL of concentrated sulfuric acid (H_2SO_4) (98% v/v) (Merck). The mixtures were shaken for 2 minutes and incubated using water bath at 100 °C for 15 minutes. The water-soluble polysaccharides were analyzed quantitatively by measuring the absorbance at 490 nm using UV/Vis spectrophotometer (Dynamica Halo RB-10). The blank solution contained 1 mL of distilled water, 0.5 mL of 5% phenol and 2.5 mL of H_2SO_4 (98% v/v). The standard glucose (Sigma, Milwaukee, WI, USA) was used as a standard solution. The results are expressed as mg glucose equivalent (GE) in gram dry weight (dw) basis.

2.5. Determination of total of phenolic content
The total of phenolic content of extracts was determined using Folin – Ciocalteu reagent [12]. The reaction mixture contained 500 μL (1 mg/mL) of samples, 500 μL of the Folin – Ciocateu reagent (Merck), and 1.5 mL of 20% sodium carbonate (Merck). The final volume was made up to 10 mL with aquadest. After two hours of incubation in a dark at room temperature, the absorbance of samples were measured by UV/Vis spectrophotometer (Dynamica Halo RB-10) at 765 nm and a gallic acid (Merck) was used for calibration purposes [13], and the results are presented as milligram gallic acid equivalent (GAE) in gram dry weight (dw) basis.

2.6. Determination of the total terpenoid content

The determination of total terpenoids was performed according to the colorimetric method of Lin *et al.* (2015) with slight modification [14]. The samples were added with 0.4 mL of vanillin – glacial acetic acid (5% w/v) (Merck) and 1.0 mL of perchloric acid solution (Merck). The tubes were then placed in a water bath (Memmert) at 60°C for 45 min. Then the mixed solution was cooled and diluted with 5 mL of acetic acid solution (Merck). The absorbance was measured at 548 nm against blank solution using UV/Vis spectrophotometer (Dynamica Halo RB-10). The standard ursolic acid (Sigma, Milwaukee, WI, USA) was used as a standard solution. The results are expressed as milligram ursolic acid equivalent (UAE) in gram dry weight (dw) basis.

2.7. DPPH Radical Scavenging Assay

The antioxidant activity of mushrooms methanolic extracts was evaluated by measuring the radical scavenging activity using 2,2-diphenyl-1-picrylhydrazyl (DPPH) free radical assay [15]. The 2 mL methanolic extracts of dried *G. lucidum* powder were added with DPPH (Merck) solution at different concentrations of 0.25, 0.5, 1.0, 1.5, and 2.0% w/v. Each of mixtures were then shaken and left to stand in the dark at room temperature for 30 minutes. Quercetin was used as a positive control. The absorbances were measured by UV/Vis spectrophotometer (Dynamica Halo RB-10) at 517 nm. The antioxidant activities presented as IC_{50} values. The IC_{50} values is the effective concentration of samples at which the DPPH radicals were inhibited by 50%. The IC_{50} value were calculated by linear regression, where the abscissa (x) represents the level concentration of tested samples and the ordinate (y) represents the average percentage of inhibitory effect.

2.8. Statistical analysis

All assays were carried out in triplicate. The data is represented as means ± standard deviation (SD). Analysis of variance and Duncan's test were used for determination of statistical significance and $p < 0.05$ were regarded as significant.

3. Results and Discussion

3.1. Physicochemical evaluation

Physicochemical evaluation is used for the preliminary identification of the natural drug to know the significance of physical and chemical properties of the substance being analyzed in terms of their observed activities and especially for the determination of their purity and quality. Physical properties are often exhibited as observables. The total ash values provide an overview of inorganic composition and other impurities in dried mushrooms. Ash content parameters related to purity and contamination. The dried mushrooms are heated at the temperature at which organic compounds and their derivatives are destructed and evaporated leaving only mineral or inorganic elements. The highest and the lowest total ash value were found to be 13.88 % of cultivated – Kaliurang and 7.63 % of cultivated – Godean, respectively. While, the highest and the lowest acid – insoluble ash value were found to be 2.11 % in cultivated – Kaliurang and 0.64 % in wild grown – Cianjur, respectively.

The acid-insoluble ash value indicates the presence of silicacious substances. The determination of extractable matter refers to the amount of constituent in material which is extracted by specific solvents. Polar compounds tend to be more attracted to and are more soluble in polar solvents. It is indicating the nature of constituents of the raw material, and also helps in detecting low grade material. Water-soluble and ethanol soluble extractive values plays an important role in evaluation of crude drugs. Less extractive value indicates addition of exhausted material, adulteration or incorrect processing. The highest water – soluble and ethanol extractive value were found in cultivated – Kaliurang (71.19 % and 59.49 %, respectively). Whereas, the lowest water – soluble and ethanol extractive value were found to be 45.25 % in wild – Gunung Kidul and 4.18 % in wild grown – Cianjur, respectively. Even though wild grown – Gunung Kidul has the lowest water – soluble extractive value, it has higher ethanol soluble extractive value compare to cultivated – Gunung Kidul and cultivated – Godean. All the results were tabulated in Table 1.

Table 1. Physicochemical properties of wild grown and cultivated *G. lucidum.*

G. lucidum	Total ash (%)	Acid insoluble ash (%)	Water soluble extractive (%)	Ethanol soluble extractive (%)
Wild grown				
Gunung Kidul	9.12 ± 2.23	0.90 ± 0.21	45.25 ± 4.45	27.2 ± 2.37
Cianjur	10.43 ± 1.45	0.64 ± 0.11	55.67 ± 3.67	4.18 ± 1.17
Cultivated				
Gunung Kidul	10.34 ± 1.40	0.83 ± 0.17	57.22 ± 2.23	12.18 ± 1.67
Kaliurang	13.88 ± 2.33	2.11 ± 0.67	71.19 ± 3.11	59.49 ± 2.15
Godean	7.63 ± 1.11	0.75 ± 0.25	54.62± 3.17	12.26 ± 1.85

3.2. Phytochemical evaluation

The results of phytochemicals evaluation of *G. lucidum* samples are presented in Tables 2. Water – soluble polysaccharides were the most abundant metabolite compound in *G. lucidum,* followed by terpenoids, and phenolic, based on the samples. Cultivated *G. lucidum* from Godean has the highest water – soluble polysaccharides (29.86 GE, mg/g dw) and phenolic content (5.07 GAE, mg/g dw) among other studied samples. Whereas, cultivated *G. lucidum* from Gunung Kidul has the lowest water – soluble polysaccharides (21.65 GE, mg/g dw) and phenolic content (3.21 GAE, mg/g dw). There was very interesting that both of wild grown *G. lucidum* have higher terpenoids content compare to all of cultivated *G. lucidum*. Wild grown *G. lucidum* are grow spontaneously in self-maintaining populations in natural or semi-natural ecosystems (usually in the forest and oil palm plantation) and can exist independently of direct human action. Cultivated *G. lucidum* have arisen through human action (composting, spawning, casing, pinning) and are grown for their produce in mushroom farm. Metabolites in mushroom could be affected by physical variations, ecological conditions, terrestrial variations, genetic factors and evolution. Another research showed that production of metabolites was influenced by environmental factors such as humidity, temperature, intensity of light, supply waters, minerals, and CO_2 [16].

Table 2. Phytochemical properties of wild grown and cultivated *G. lucidum.*

G. lucidum	Water-soluble polysaccharides (GE, mg/g dw)	Phenolic content (GAE, mg/g dw)	Terpenoids content (UAE, mg/g dw)
Wild grown			
Gunung Kidul	22.45 ± 1.08	4.18 ± 0.67	23.72 ± 2.17
Cianjur	23.90 ± 2.39	4.13 ± 0.54	23.14 ± 2.39
Cultivated			
Gunung Kidul	21.65 ± 2.45	3.21 ± 0.87	18.71 ± 2.85
Kaliurang	23.11 ± 3.23	3.90 ± 0.72	20.42 ± 2.52
Godean	29.86 ± 2.42	5.07 ± 0.39	22.80± 2.43

3.3. Antioxidant activities

Effectiveness in antioxidant activities inversely corresponded with IC_{50} value. The scavenging capacities of wild grown and cultivated *G. lucidum* in the DPPH free radical scavenging activity assay ranged between 344.15±9.57 and 686.05±12.11 μg/mL. Quercetin was used as positive controls in this experiment and exhibited IC50 of 10.55±2.23 μg/mL. The antioxidant activity of *G. lucidum* decreased in order: cultivated Godean > wild Gunung Kidul > wild Cianjur > cultivated Kaliurang > cultivated Gunung Kidul. The cultivated – Godean revealed the highest DPPH scavenging activity (the lowest IC50, 344.15±9.57 μg/mL) among of the studied samples. Both of wild grown *G. lucidum* have higher

DPPH scavenging activity than a sample from cultivated *G. lucidum* from Gunung Kidul and Kaliurang. This bioactivity was believed to be associated with the different contents of polysaccharides (b-1,3-glucans) and triterpenes (ganoderic acids and others) in each of the samples [17,18].

Table 3. Antioxidant activities of wild grown and cultivated *G. lucidum*.

G. lucidum	IC50 (μg/mL)
Wild grown	
Gunung Kidul	381.16 ± 13.52
Cianjur	387.17 ± 15.47
Cultivated	
Gunung Kidul	686.05 ± 12.11
Kaliurang	457.25 ± 10.44
Godean	344.15 ± 9.57
Quercetin	10.55 ± 2.23

4. Conclusion

Physicochemical, phytochemical includes water – soluble polysaccharides, phenolic, terpenoids content, and the antioxidant activity of 2 wild grown and 3 cultivated *G. lucidum* were evaluated. The present investigation demonstrates that the physicochemical, phytochemical and antioxidant properties of wild grown and cultivated *G. lucidum* were significantly different. Cultivated *G. lucidum* from Godean has the highest water – soluble polysaccharides and phenolic content. Both of wild grown G. lucidum have higher terpenoids content compare to all of cultivated *G. lucidum*. The cultivated – Godean revealed the highest DPPH scavenging activity among of the studied samples. Overall, the presence of primary and secondary metabolites in *G. lucidum* suggested that the mushroom has potential natural source of antioxidants.

Acknowledgements

This research was supported by National Priorities Programme 2015 - 2017 – Indonesian Institute of Sciences.

References

[1] Lei LS and Lin ZB 1993 Effect of Ganoderma polysaccharides on the activity of DNA polymerase of splenocytes and immune function in aged mice. *Acta Pharm. Sinica* **28** 577–82

[2] Lin JM, Lin CC, Chiu HF, Yang JJ and Lee SG 1993 Evaluation of the anti-inflammatory and liver protective effects of *Anoectochilus formosamus*, *Ganoderma lucidum* and *Gunostemma pentaphyllum* in rats. *Am. J. Chin. Med.* **21** 59–69

[3] Ahmad S 1995 Oxidative Stress and Antioxidant Defences in Biology (Chapman and Hall, New York) pp. 62–88.

[4] Boh B, Berovic M, Zhang J and Zhi-Bin L 2007 *Ganoderma lucidum* and its pharmaceutically active compounds. *Biotechnol. Annu. Rev.* **13** 265-301

[5] Zhou XW, Lin J, Yin YZ, Zhao JY, Sun XF and Tang KX 2007 Ganodermataceae: Natural products and their related pharmacological functions. *Am. J. Chinese Med.* **35**(4) 559-74

[6] Dep. Kes. RI 1986 Sediaan Galenik, 2-10, Departemen Kesehatan Republik Indonesia, Jakarta.

[7] Dep. Kes. RI 2000 Parameter Standar Umum Ekstrak Tumbuhan Obat, Departemen Kesehatan Republik Indonesia, Jakarta.

[8] Masuko T, Minami A, Iwasaki N, Majima T, Nishimura SI and Lee YC 2005 *Anal. Biochem.* **339** 69–72

[9] Sutherland IW 1990 Biotechnology of microbial exopolysaccharides, (Cambridge University press, New York), pp. 1-11, 20-23, 38-42, 117-151

[10] Graeme WM 1992 Yeast Physiology and piotechnology, (John Wiley and Sons, New York), pp. 122-131

[11] David HJ and Hazel P 1993 Analytical biochemistry, 2nd Ed., (John Wiley and Sons Inc., New York), pp. 338-431, 408-411
[12] Arnous A, Makris DP and Kefalas P 2002 Correlation of pigment and flavanol content with antioxidant properties in selected aged regional wines from Greece. *J Food Compos. Anal.* **15** 655
[13] Yu L, Haley S, Perret J, Harris M, Wilson J and Qian M 2002 *J. Agric. Food Chem.* **50**(6) 1619-24
[14] Lin M, Yu, Z, Wang B, Wang CH, Weng Y and Koo M 2015 *Sains Malays.* **44** 1685-91
[15] Shimada K, Fujikawa K, Yahara K and Nakamura T 1992 Antioxidative properties of xanthin on autoxidation of soybean oil in cyclodextrin emulsion *J. Agr. Food Chem.* **40** 945–8
[16] Ramakrishna A and Ravishankar G A 2011 Influence of abiotic stress signals on secondary metabolites in plants. *Plant Signal. Behav.* **6**(11) 1720-31
[17] Ferreira ICR, Vaz JA, Vasconcelos MH and Martins A 2010 Compounds from wild mushrooms with antitumor potential. Anti-cancer Ag *J. Med. Chem* **10** 424–36
[18] Martı´nez-Montemayor MM, Acevedo RR, Otero-Franqui E, Cubano LA and Dharmawardhane SF 2011 *Ganoderma lucidum* (Reishi) inhibits cancer cell growth and expression of key molecules in inflammatory breast cancer. *Nutr. Cancer* **63** 1085–94

Macro and micromineral in commercial infant formula milk in Indonesia by neutron activation analysis

Indah Kusmartini*, Syukria Kurniawati, Natalia Adventini, Endah Damastuti, Djoko Prakoso, Woro Yatu Niken, Dyah Kumala Sari, Diah Dwiana Lestiani, Muhayatun

Center for Applied Nuclear Science and Technology, National Nuclear Energy Agency of Indonesia (BATAN), Bandung, Indonesia

*Corresponding author's email: indahkus@batan.go.id

Abstract. The content of protein and minerals in infants formula milk is composed based on the composition of breast milk that provides adequate nutritional requirements for food intake in infants. Minerals as in macro and microminerals are essential for biological processes since they involve bone mineralization, enzymatic reactions, secretion of natural hormones, cells and lipid protection. The composition of macro and micromineral is one indicator in influential the quality of formula milk. Therefore, in this activity, the determination of macro and micromineral elements in infant formula milk was carried out by Neutron Activation Analysis (NAA) method and evaluated formula milk by considering the concentration of daily nutritional requirements. Formula milk samples were obtained in Indonesian market. The samples were irradiated on a lazy susan system at TRIGA 2000 Bandung Reactor with a neutron flux of ~ 10^{13} $n.cm^{-2}.s^{-1}$ for three days. The validation method was also carried out using Typical Diet standard reference material. The validation results have a good agreement with the certificate value with bias (%) ranging from 0.6 to 6.9. Determination of macro and micromineral nutrients using NAA method resulted K, Na, Fe, Se, and Zn with an average value were 0.66%; 0.25%; 82.7 μg/g; 0.14 μg/g; and 34.7 μg/g respectively. These values were still within the concentration ranges as stated in nutritional labelled by manufacturers. The daily intake of macro and micromineral K, Na, Fe, Se, and Zn were 5150, 5110, 65, 0.11, and 27 mg/day, respectively. By consuming tolerable amounts of infant formula milk, this commodity is a source of nutrition that required by the infant's nutritional adequacy rate.

1. Introduction

Formula milk is a product that required as a substitute for breast milk that can satisfy the nutritional supplies of infants from four to six months of life. Breast milk is the best nutrition for the growth and development of babies since it contains a balanced diet and plays an essential role in bioactive components such as growth factors [1-3]. These immune factors are explicitly provided in breast milk. However, for various reasons, infant formula can substitute the essential for breast milk which is desirable for the growth and development of the infant. The content of formula milk is based on the composition of modified breast milk to increase nutritional requirements for dietary intake in infants [4,5]. Commonly, formula milk comes from cow's milk; however, some formulas custom soy as a source of protein and minerals [6,7]. Minerals are essential for biological processes due to their forming bone mineralization, enzymatic reactions, secretion of natural hormones, cells and lipid protection. The mineral content of formula milk has been added and fortified as in fulfilled desires of Fe and Zn in formula milk [8-10].

On the other hand, the macro and micromineral content in milk is an indicator of the quality of formula milk for infant growth and development. Therefore, in this activity, the macro and micro mineral content in infant formula are carried out by the Neutron Activation Analysis method. NAA is a non-destructive, multi-elemental, sensitive, and widely used technique for characterizing biological, and food samples [11]. An analytical purpose for infant formula milk has been observed previously. Among of them, researchers from Poland by Chacduk *et al.* discovered the elements As, Cr, Fe and Se in formula milk using NAA and ICP methods [12]. Sorbo *et al.* studied the toxic elements (Pb, Cd, As) in infant formula powder by ICP method. They were stated that a transfer of hazardous chemicals in the food chain due to environmental pollution such as in the manufacturing process; therefore, it is necessary to monitor food safety in the Union of Europe [13]. Besides that, Arif, A et al. stated that metal concentrations such as Fe, Cu, Zn, Co, Ni, Mg, Cd, Pb, Na, K were analyzed by atomic absorption spectrometer. They also stated the metal in infant formula was calculated and estimated intake according to the recommended level and could be tolerated from these elements [14,15]. This present study aimed to determine the macro and micromineral composition using neutron activation analysis (NAA) to compare the measured values with the corresponding nutritional labelled by manufacturers on commercial milk packs and to evaluate formula milk daily intake by considering daily nutritional requirements.

2. Methodology

2.1. Sample collection

Samples of infant formula milk powder were attained from supermarkets in several regions are one of the leading infant formula milk brands in Indonesia. Standard Reference Material (SRM) Typical Diet in these activities utilized for the validation method [16].

2.2. Sample and standard preparation

About 25–40 mg infant formula milk powder were weighed and put into 0.3 mL polyethylene vials, Standard Reference Material NIST Typical Diet 1548a were used to validate the method. Mix standards for the relative process were prepared and placed in the same vials. The preparation of NAA standards was carried out by pipetting 100 μL of ICP multi-element standard (E-Merck) then dripped in the polyethylene vial. The vial dried under the infrared lamp then sealed by heating. The obtained weights of elements Fe, K, Na, Se, and Zn in mix standard were 49.7; 20.9; 22.7; 10; 49.9 μg, respectively.

2.3. Irradiation and measurement Gamma-ray

The samples, SRMs and mix standards were irradiated in the lazy susan system of Reactor TRIGA 2000 Bandung with neutron flux ~ 10^{13} n.cm^{-2}.s^{-1} for three days. The sample, SRM and standard mix after irradiation process were cooled to shelter radiation exposure. Duration time required to decay radiation exposure for medium (K, Na) and long half-time (Fe, Se, Zn) radionuclides were a day and a week, respectively. For measurement of medium and long half time radionuclides was carried out by the High Pure Germanium multichannel gamma-ray spectrometer and the GENIE 2000 software for 1000 and 50000 seconds, respectively.

2.4. Calculation of estimated daily intake

The daily intake of macro and microelements depends on the concentration of elements initiate in food and the amount of daily food consumption. Estimated daily intake of macro and microelements for 6-month-old infants were calculated based on the following formula.

$$EDI = Cmetal \text{ X } D \text{ food intake} \quad (1)$$

IOP Conf. Series: Materials Science and Engineering 1011 (2021) 012062 doi:10.1088/1757-899X/1011/1/012062

Where
EDI: Estimated Daily Intake of macro and microelements for infant six months, expressed in mg/day or μg/day as adjusted for element respectively.
C: average concentrations of the elements in samples of infant formula expressed in μg/g
D: the average daily intake of milk consumed by 6-month-old infants is expressed in L/day
An average daily milk intake equal to 780 mL/day for 6- month-old infants was assumed in this study[17]

3. Result and Discussion

In this activity, neutron activation analysis (NAA) techniques were utilized by the relative method, which is based on the comparison of the radiation activity of the samples (Asamples) to the radiation activity of the standards (Astandard) with known concentration. Standards and samples were irradiated in the similar conditions of irradiation. The method validation was carried out by testing SRM Typical Diet 1548a which has a matrix similar to the sample and using the identical method to acquire a valid analytical method and then ensuring the quality of the analytical results [18]. The measurement results of SRM and the certificate values were shown in Table 1. The certificate values and analysis results given relative bias in the range of 0.6 to 6.9%. Generally, a good agreement of the chemical element concentration in SRM Typical Diet with the certified data of reference material indicates an acceptable preference of the result obtained in the study. The ratio between the certificate value compared to the analysis results value was 0.93 to 1 (Figure 1). The result showed that there is closeness between the certificate value and the analysis results, therefore it can be concluded that the analytical method used were valid and reliable for analysis on infant formula milk samples.

Table 1. Analysis result for SRM Typical Diet 1548a.

Element	Certificate value (μg/g)	Analysis result (μg/g)	Bias %
Fe	35.3±3.7	33.4±4	0.6
K	6970±125	6600±878	5.3
Na	8132+942	8078±28.9	3.8
Se	0.245±0.028	0.235±0.064	5.5
Zn	24.6±1.79	22.9±0.35	6.9

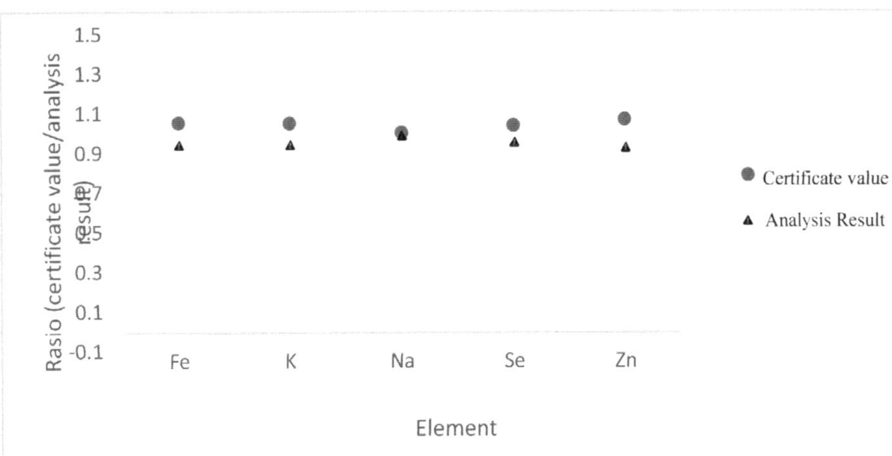

Figure 1. Ratio of certificate value and SRM Typical Diet 1548a analysis result.

The results of the analysis of macro and microelement in infant formula using the NAA method are listed in Table 2. The K, Na, Fe, Se, and Zn elements were obtained with average concentration value of 0.66%; 0.25%; 82.7 µg/g; 0.14 µg/g; 34.7 µg/g, respectively. These values were compared with the values stated on the manufacture label product with recovery values range between 96 to 118%. In comparison with pure cow's milk without fortifying was carried out by Herwig *et al.* using ICP MS method (K, Na, Fe, Se and Zn) with values were 1.39%, 0.28%, 2.4 µg/g, 26 µg/g and 0.27 µg/g, respectively, showed that the value of Potassium and Sodium almost have similar concentration value. In contrast, the concentration of iron in formula milk shows four times higher than cow's milk. Meanwhile, Zink in cow's milk has greater content than formula milk. The selenium concentration in cow's milk is smaller than formula milk [19]. The results of previous research conducted by Daniels *et al.*, 2019 analyzed the content of macro and micromineral elements in breast milk, were obtained (K, Na, Fe, Se and Zn) 43.5%, 12.2%, 220 µg/g, 990 µg/g, 10.2 µg/g, respectively. It was stated that breast milk is a source of nutrition for infant up to 6 months of life recommended by WHO. Exclusive breastfeeding can be protecting against gastrointestinal infections during the early growth of the infant [20].

Table 2. The concentration of macro and micromineral in infant milk formula by NAA Method.

Element	The present study, n=18	Label Value	%Rec	Bias
K (%)	0.66	0.66	101	0.8
Na (%)	0.25	0.26	99	1.5
Fe (µg/g)	82.77	86.21	96	4
Se (µg/g)	0.14	0.12	118	18.4
Zn (µg/g)	34.67	33.10	105	4.7

Several other studies determining the concentration of macro and microelements in formula milk have been carried out using different methods by Rossmann *et al.*, 2016 explain that elements of K, Fe, Zn with concentrations of 1.52%, 53.4 µg/g, 45.7 µg/g respectively [21]. These values are not significantly different from this study. Previously research related to the content of formula milk in Pakistan by Ahmed *et al.*, 2017 obtained results (K, Na, Fe, Se and Zn) of 0.42%, 0.12%, 94 µg/g, 101 µg/g, 36.5 µg/g, respectively [22]. It has different with Guniem, Mahmoud *et al.*, 2019 [23] researcher from Egypt attained for iron and selenium elements were 44.7 and 57.4 µg/g in formula milk. Meanwhile, other researchers in Brazil Gamela, R et al. determined (K, Na, Fe, Se) in formula milk were 0.38%. 0.11%, 82 µg/g, 100 µg/g respectively [24]. There are significant differences or similarities in determining of concentration in formula milk for several countries, due to related to the variance of methodology that have been carried out, such as the preparation of methods, qualitative and quantitative analysis, accuracy and precision of methods that influencing the measurement of results [21]. Besides that, other factors such as environmental conditions or the effect of contamination from the milking process of formula milk derived from fortified cow's milk affect the concentration of macro and micromineral contained in formula milk. However, macro and micromineral can occur several problems such as clinical respiratory system, cancer, skin disorders, anaemia, growth depression, reproductive disorders, heart failure, gastrointestinal disorders, fatigue, decreased immunity and even death, which has consequences for public health [25].

In this activity, an estimate of the daily intake of macro and micromineral in infant formula is showed in Figure 2. The estimated daily intake values for K, Na, Fe, Se, Zn were 5150, 1987, 65, 0.11, and 27 mg/day, respectively. Previous studies stated the value of daily intake of formula milk conducted by Bargellini *et al.*, 2018 iron, zinc and selenium were obtained with EDI values of 4.79, 4.99, 10.69 mg/ day [17]. This EDI value is different from the amount of this study. It was caused by the nutritional intake of formula milk required for each country is different and depent on fortification substances contained in milk formulas. The value of sodium intake in infants 6 to 12 months 170 to 370 mg/day. The main effect of increased sodium chloride is an increase in blood pressure, a reference risk for cardiovascular and kidney disease, and a blood potassium level for heart function. The cell membrane potential in the body is controlled by sodium and potassium concentrations as it provides the potential for nerve transmission of muscle contraction and blood transport for active nutrients (glucose) [26].

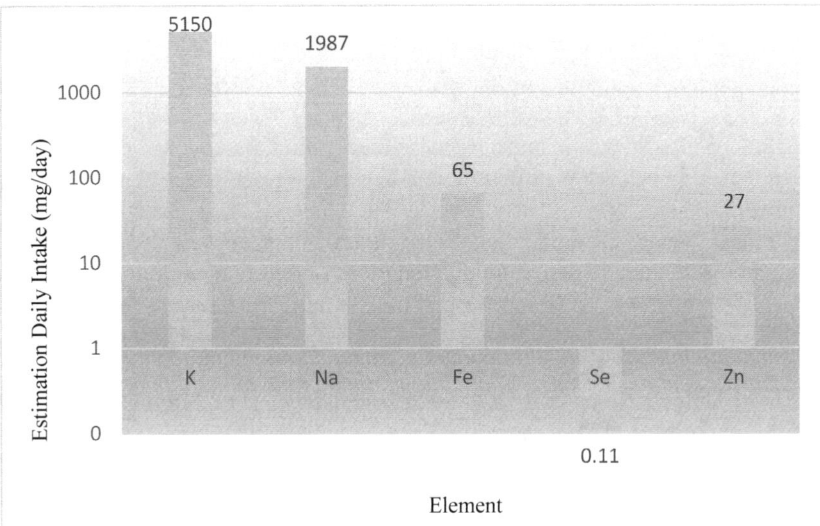

Figure 2. Daily Intakes of Macro and Micromineral from Infant Milk Formula

Iron is an essential element that has a metabolic function, including oxygen transportation. Inadequate intake causes anaemia in deficiency conditions, impaired psychomotor development and cognitive performance as well as decreased immune function. Iron absorption from breast milk is higher than absorption from infant formula and iron-fortified complementary foods. Iron intake for infants from 6 to <12 months 6.2 mg per day is assumed to be an absorption efficiency of 15% whereas 10% absorption efficiency of 9 mg per day. Significant iron deficiency in infants can cause cell-mediated immunity, while excess becomes a source of liver damage. In their study Molska *et al.*, 2014 stated that the high concentration of this element of Fe in formula milk and fruit juice could complement the deficiency of this element in the body [27].

Zinc is involved in several aspects of cell metabolism with several enzymes for catalytic activity. Regarding immune function, protein synthesis wound healing, deoxyribonucleic acid (DNA) synthesis and cell division. Excess zinc can cause microcytic anaemia or neutropenia and reduce the concentration of iron in the body. Besides that, zinc competes with magnesium in absorption rates in the intestine and parts of the bone structure. Infant formula usually fortified with zinc in a higher level than human milk (0.5-1.5 mg / 100 kcal20) to compensate for the lower absorption of zinc in the body. The results in a daily intake of around 3 to 5 mg. These levels of intake were not found to be associated with zinc deficiency in studies of Korean infants who received about 4 to 4.5 mg of zinc daily from cow's milk-based formulas or about 5.5 to 7 mg of zinc from soy-based formulas [26].

The absorption of selenium in food is not under homeostatic control and varies from 50 to 90%. Selenium deficiency, such as malabsorption syndrome, causes impaired muscle function and loss of pigment in the hair and skin. While selenium deficiency, causes Keshan disease, cardiomyopathy, degenerative osteoarthropathy [26-27]. Chronic selenium excess is characterized by hair loss and nail dystrophy, dermatitis and neurological and endocrinologic symptoms. Various concentrations of selenium that have been observed in breast milk depend on the amount of selenium consumed by the mother from natural foods. In general, the highest selenium concentration in colostrum (zero to five days, median 26 μg/L). Selenium concentrations in Europe range from 3 to 84 μg / L, with an average value of 16.3 ± 4.7 μg/L since it is assumed that the average selenium intake observed from breast milk is 12.5 μg per day (i.e. 16 μg/L × 0.8 L) is adequate for most infants in the first half-year of life from 6 to <12 months. After six months, the nonbreastfed period of children usually consumes infant formula milk, continuation formula milk and various other supplementary foods and another food intake such as vegetables, fruit. Based on research conducted by Chekri, R contributed 94% of the total macro and microelement concentration were obtained from food. Therefore, available food is an essential part of infants and toddlers in evaluating the distribution of macro and microelements in the body [26,28].

4. Conclusion

In this study, the value is still within the recommended limits stated on the nutritional labelled by manufacture. By consuming infant formula milk powder in adequate quantities, this commodity can be one source of nutrition that can fulfil the nutritional sufficiency rate of infants. The establishment of level macro and microelement concentrations in formula milk remains a challenge, particularly in infants aged 0 to 6 months. Increasing public awareness of health is essential to determine the level of macro and microelement content that is appropriate for the nutritional requirement of infants.

References

[1] Nguyen TTP, Bhandari B, Cichero J and Prakash S 2015 A comprehensive review on in vitro digestion of infant formula *Food Res. Int.* **76** 373–86

[2] Machado I, Bergmann G and Pistón M 2016 A simple and fast ultrasound-assisted extraction procedure for Fe and Zn determination in milk-based infant formulas using flame atomic absorption spectrometry (FAAS) *Food Chem.* **194** 373–6

[3] Dao DT , Anez-Bustillos L, Cho BS, Li Z, Puder M and Gura KM 2017 Assessment of micronutrient status in critically ill children: Challenges and opportunities *Nutrients* **9**(11) 1–26

[4] Lönnerdal B 2016 Bioactive proteins in human milk: health, nutrition, and implications for infant formulas *J. Pediatr.* **173** S4–S9

[5] Gómez RL 2018 Research Article Research Article *Arch. Anesthesiol. Crit. Care* **4**(4) 527–34

[6] Akhtar S, Shahzad MA, Yoo SH, Ismail A, Hameed A, Ismail T and Riaz M 2017 Determination of aflatoxin M1 and heavy metals in infant formula milk brands available in Pakistani markets *Korean J. Food Sci. Anim. Resour.* **37**(1) 79–86

[7] Kirchner U, Degenhardt K, Raffler G and Nelson M 2012 Determination of vitamin A (Retinol) in infant formula and adult nutritionals by LC 2011.15 *J. AOAC Int.* **95**(4) 933–6

[8] Chen L, Li X, Li Z and Deng L 2020 Analysis of 17 elements in cow, goat, buffalo, yak, and camel milk by inductively coupled plasma mass spectrometry (ICP-MS) *RSC Adv.* **10**(12) 6736–42

[9] Etcheverry P, Wallingford JC, Miller DD and Glahn RP 2004 Calcium, zinc, and iron bioavailabilities from a commercial human milk fortifier: A comparison study *J. Dairy Sci.* **87**(11) 3629–37

[10] Papachristodoulou C, Tsiamou MC, Sakkas H and Papadopoulou C 2018 Determination of minerals in infant milk formulae by energy dispersive X-ray fluorescence spectrometry *J. Food Compos. Anal.* **72** 39–47

[11] Birhanu WT, Chaueby AK, Teklemariam TT, Dewu BBM, Oladipo MOA, Ahmed YA and Abubakr NH 2015 Analysis of essential elements in Ethiopian finger millets *(Eeleusine coracanda)* by instrumental neutron activation analysis (INAA) *Int. J. Basic Appl. Sci.* **4**(1) 82

[12] Chajduk E and Polkowska-Motrenko H 2017 Application of ICP-MS, INAA and RNAA to the determination of some 'difficult' elements in infant formulas *J. Radioanal. Nucl. Chem.*

311(2) 1347–53
[13] Sorbo A, Turco AC, Di Gregorio M and Ciaralli L 2014 Development and validation of an analytical method for the determination of arsenic, cadmium and lead content in powdered infant formula by means of quadrupole Inductively Coupled Plasma Mass Spectrometry *Food Control* **44** 159–65
[14] Arif A, Khan B, Majeed K, Shahzad MS, Nadeem MS and Ahmed R 2018 Spectrophotometric estimation of metal ions and nutrient value of formulae milk and baby foods available in Pakistan under different brand names *South Asian J. Life Sci.* **6** 1
[15] Yang L, Zhang M, Lin S, Chen D and Zheng M 2003 Minor and trace element analysis in breast milk using inductively coupled plasma mass spectrometry (ICP-MS) *Mikrochim. Acta* **142**(1–2) 85–8
[16] Method T *et al.* 2018 Certificate of Analysis Certificate of Analysis," no. November, pp. 2–4
[17] Bargellini A, Venturelli F, Casali E, Ferrari A, Marchesi I and Borella P 2018 Trace elements in starter infant formula: dietary intake and safety assessment *Environ. Sci. Pollut. Res.* **25**(3) 2035–44
[18] Lestiani DD, Santoso M and Adventini N 2010 Asesmen kandungan unsur pada cuplikan batubara *Pros. Semin. Nas. AAN 2010* **2085**
[19] Herwig N, Stephan K, Panne U, Pritzkow W and Vogl J 2011 Multi-element screening in milk and feed by SF-ICP-MS *Food Chem.* **124**(3) 1223–30
[20] Daniels L, Gibson RS, Diana A, Haszard JJ, Rahmannia S, Luftimas DE, *et al.* 2019 Micronutrient intakes of lactating mothers and their association with breast milk concentrations and micronutrient adequacy of exclusively breastfed Indonesian infants *Am. J. Clin. Nutr.* **110** (2) 391–400
[21] Rossman M, Zaichick S and Zaichick V 2016 Determination of key chemical elements by energy dispersive X-Ray fluorescence analysis in commercially available infant and toddler formulas consumed in UK," *Nutr. Food Technol. Open Access* **2** 4
[22] Ahmed M, Khaleeq A, Munir R, Huma R, Muddassar M and Naseem S 2017 Determination of macronutrients, micronutrients, non-essential elements in milk based-infant formulas by atomic absorption and inductively coupled plasma-optical emission spectroscopy after microwave and ultrasound assisted digestion *Spectrosc. Lett.* **50**(2) 88–95
[23] Ghuniem MM, Khorshed MA and Khalil MMH 2020 Determination of some essential and toxic elements composition of commercial infant formula in the Egyptian market and their contribution to dietary intake of infants *Int. J. Environ. Anal. Chem.* **100**(5) 525–48
[24] Gamela RR, Barrera EG, Duarte AT, Boschetti W, da Silva MM, Vale MGR, *et al.* 2019 Fast sequential determination of Zn, Fe, Mg, Ca, Na, and K in infant formulas by high-resolution continuum source flame atomic absorption spectrometry using ultrasound-assisted xxtraction *Food Anal. Methods* **12**(6) 1420–8
[25] Pereira JSF, Pereira LSF, Schmidt L, Moreira CM, Barin JS and Flores EMM 2013 Metals determination in milk powder samples for adult and infant nutrition after focused-microwave induced combustion *Microchem. J.* **109** 29–35
[26] Panel E and Nda A 2013 Scientific Opinion on nutrient requirements and dietary intakes of infants and young children in the European Union *EFSA J.* **11**(10) 1–103
[27] Molska A, Gutowska I, Baranowska-Bosiacka I, Noceń I and Chlubek D 2014 The content of elements in infant formulas and drinks against mineral requirements of children," *Biol. Trace Elem. Res.* **158**(3) 422–7
[28] Chekri R, Le calves E, Zinck J, Leblanc JC, Sirot V, Hulin M, *et al.* 2019 Trace element contents in foods from the first French total diet study on infants and toddlers *J. Food Compos. Anal.* **78** 108–20

IOP Conf. Series: Materials Science and Engineering 1011 (2021) 012063 doi:10.1088/1757-899X/1011/1/012063

Drug Development from Peptide-derived Marine Natural Products

L Sukmarini

Research Center for Biotechnology, Indonesian Institute of Sciences (LIPI), Jl. Raya Bogor Km. 46 Cibinong 16911, Bogor, West Java, Indonesia

E-mail: linda.sukmarini@lipi.go.id

Abstract. Marine natural products (MNPs) provide an ample source of pharmacologically bioactive compounds with great chemical diversity and complexity. Peptide-derived MNPs are a class of compounds with diverse and interesting biological properties, such as antiinfective, anticancer, cytotoxic, and antiallodynic. Their biosynthetic production machinery from the ribosomal origin or non-ribosomal origin and the hybrid with other compounds confer these compounds remarkable diversity in terms of structural complexity as well as biological activity. The interest in these peptide compounds is their potential to produce valuable therapeutics and pharmaceuticals. Some of these peptides or their derivatives have reached the pharmaceutical market, while some are currently under clinical trials for developing new drugs. This review focuses on marine peptide-based compounds in advanced drug development for clinical use or pharmaceutical applications. The drug development process and clinical status of ribosomal peptides, non-ribosomal peptides and their hybrids, or the derivatives, are hereby described.

1. Introduction

Natural products, especially those from marine origin, have been an important source of structurally diverse and biologically active array of secondary metabolites. Marine life covers a significant part of Earth's biodiversity (an estimated 25% of the total number of species). The extreme oceanic ecosystems lead marine species to develop their self-protection system by producing unique compounds. These abundant and unprecedented chemodiversity compounds of marine natural products (MNPs) offer an extensive resource of novel compounds in the field of drug discovery and development [1- 2].

Many peptide natural products have received great attention in the drug discovery and development. Peptide-derived MNPs are a class of compounds that present diverse and interesting biological properties of mainly pharmaceutical interests, such as antimicrobial, antiviral, anticancer, cytotoxic, and antiallodynic or analgesic [3]. Peptides are structurally defined as amino acid-based compounds containing at least one amide (peptide) bond. Generally, (oligo)peptides refer to oligomers formed by 10-50 amino acid residues. The structural features of their particular interest are the configuration and side-chain variability of the constituent amino acids, the formation of conformationally stabilized cyclic structures, and the structural variation or modification of peptide bonds to avoid hydrolytic cleavage and prolong bioactivities [4]. Marine peptides exhibit remarkable diversity both in terms of structural complexity and biological activity due to the versatility of their biosynthetic production machinery. They are mainly synthesized either by ribosomal machinery or the overly majority of medically used peptides are by non-ribosomal pathways, including the hybrids of non-ribosomal peptides (NRPs) with other compounds [5].

Extensive literatures on MNPs-based drugs, peptide drugs, and biological activities of marine peptides have been reviewed so far [2-3, 5-14]. However, this present short review focuses on peptide-derived MNPs that have been approved, developed, or are currently being evaluated in clinical trials. The development of peptide-derived MNPs from some selected examples of ribosomal or non-ribosomal origin as viable pharmaceuticals or therapeutic drugs is herein featured.

2. Marine peptide compounds: The biosynthetic machinery

The peptide class of metabolites occurs in almost all domains including plants, bacteria, fungi, or marine organisms. As mentioned above, their assembly can be effectively explained by the two main biosynthetic machineries (Figure 1). Firstly, ribosomal peptides termed as the ribosomally synthesized and post-translationally modified peptides (RiPPs) and secondly, those synthesized by the non-ribosomal peptide synthetases (NRPSs) termed as non-ribosomal peptides (NRPs) [15-16].

Peptides synthesized from the ribosome (Figure 1a) depend on a set of only 20 standard proteinogenic amino acids or canonical amino acids in which diversity is generated by post-translational modifications (PTMs). Their precursor sequence consists of an N-terminal leader and a C-terminal core sequence. For eukaryotic peptide such as conopeptides, a signal sequence is often found N-terminal to the leader peptide that directs the peptide to the specific cellular compartments where PTMs will undergo. Cyanobactins and conopeptides are among the RiPPs families found in marine organisms. The latter discussed in this review has been already commercialized in the market [16-17].

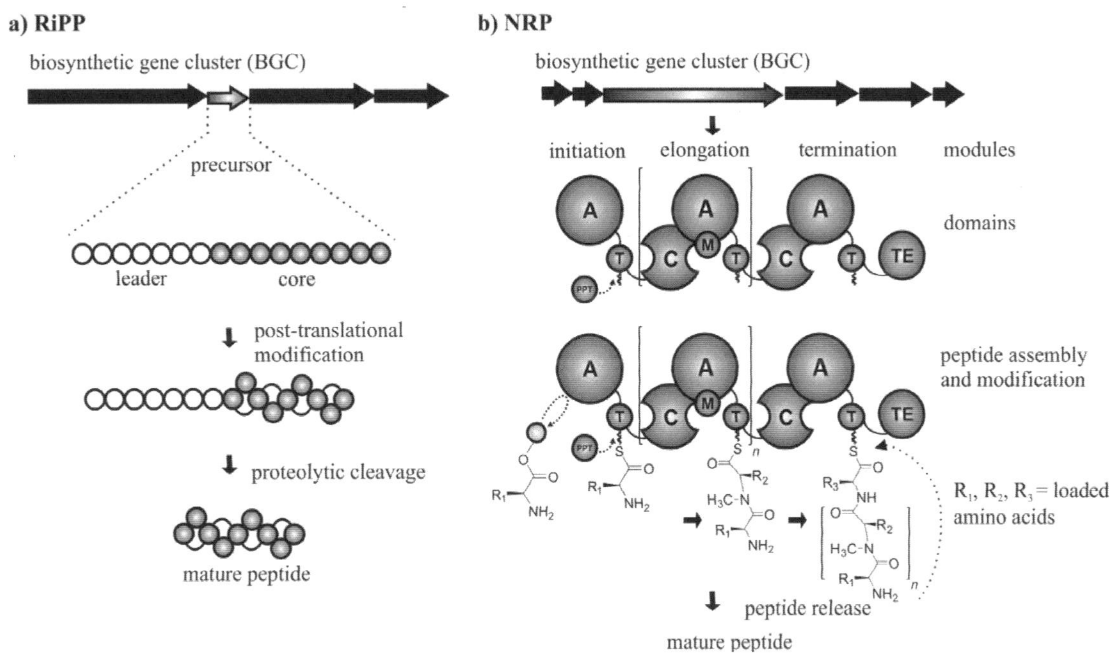

Figure 1. General biosynthetic architecture of RiPPs and NRPs. a) Peptides synthesized by the RiPP biosynthesis undergo post-translational modifications, while b) NRP biosynthesis employs modular NRPS assembly machinery. Adapted from Dang & Süssmuth (2017) [16].

Unlike RiPPs, the non-ribosomal peptides are decoupled off the ribosome and are synthesized from a large pool building blocks of both proteinogenic and non-proteinogenic amino acids. These non-proteinogenic building blocks include *N*-methylated amino acids, D-amino acids, fatty acids, and hydroxy acids, which are synthesized by a large multifunctional modular enzyme complex of the NRPS machinery [18]. Basically, an NRPS assembly line (Figure 1b) consists of the initiation, elongation, and termination module. Each module includes adenylation (A), thiolation (T), and condensation (C). The A domain is responsible for the selection of a specific amino acid building block at the expense of adenosine triphosphate (ATP), then activates its carbonyl group to the corresponding aminoacyl adenylate form that is tethered to the thiolation domain via a thioester linkage. Further, the C domain catalyzes amide of the peptide bond formation coupled to adjacent modules leading to peptide elongation. In addition, optional domains such as methylation, formylation, epimerization, and thioesterase domains among others can be present during chain elongation. The termination or product release from NRPS assembly can be achieved in different ways yielding linear, cyclic, or cyclodepsipeptides. Moreover, the presence of a thioesterase domain in an assembly line in many cases is known to catalyze the cleavage of the mature peptide and connect the macrocyclization process. After the termination of NRPS assembly, post-modification, and decoration may also occur and recruits tailoring enzymes that carry out acylations, glycosylations, or halogenations. Moreover, some of the marine NRPs also appear as hybrids of NRPs and polyketides (PKs), termed as PKS-NRPS hybrids [19-20].

Most of the marine RiPPs and NRPs drugs have been the subject of thorough structure-activity relationship studies and, in consequence, have reached the market as semisynthetic or synthetic compounds in multiple variations unless due to a shortage supply as described in this review.

3. Ribosomally formed conopeptides

The venoms of predatory marine snails (e.g. the fish-hunting cone snails) are a complex mixture of RiPPs. These marine cone snail peptides are generally known as conopeptides or conotoxins. Conopeptides consist of a more or less continuous spectrum of PTM, with several of those widely distributed in all animal venom peptides, such as proteolytic cleavage, disulphide bond formation, C-terminal amidation, and N-terminal glutamine residues to pyroglutamate. The defining structural feature of conotoxins are mostly smaller peptide (10-30 amino acid residues) with a significantly higher density of disulphide crosslinks than most animal venom peptides [17].

An already approved commercial RiPP drug and the first US Food and Drug Administration (FDA) approved marine-derived drug is ziconotide, which is now known under the trade name Prialt® for the specific indication of chronic pain. Ziconotide (Figure 2) is a synthetic analog of the ω-conotoxin MVIIA. The naturally occurring conopeptide ω-conotoxin MVIIA is originally isolated from the Pacific piscivorous cone snails *Conus magus*. This conopeptide is a linear 25 amino acid-polycationic peptide containing six cysteine residues linked by three disulphide bridges that stabilize its well-defined three-dimensional structure. The analgesic properties of ziconotide are attributed to their ability to block specific voltage-activated calcium channels in neurons. The nerve signal conduction is inhibited, thereby, resulting in pain relief [21-22].

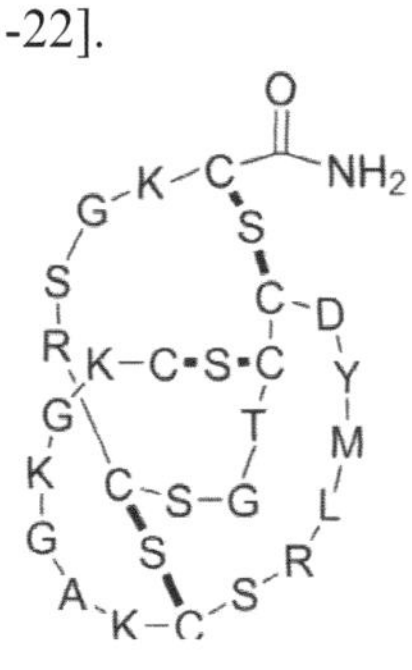

Figure 2. Chemical structure of ziconotide (Prialt®) [5].

The approval by the FDA and the European Medicines Evaluation Agency (EMEA) of ziconotide prompted investigations on other modified conotoxins such as contulakin G and χ-conotoxin-MrIA. The first peptide reached Phase I clinical trial, while the latter, which inhibits the norepinephrine transporter, was in clinical development and. However, unfortunately, these peptide toxins have been discontinued due to side-effect toxicity issues [17, 23].

4. 'Head-to-side-chain' cyclodepsipeptides of non-ribosomal origin

Cyclic depsipeptides or cyclodepsipeptides (CDPs) are non-ribosomal cyclic peptides in which at least one amide group is replaced by a corresponding ester bond due to the presence of a hydroxylated carboxylic acid in the peptide structure. CDPs are widely distributed in bacteria, fungi, plants, algae, sponges, and other marine organisms. Most of the CDPs of marine belong to a subclass 'Head-to-side-chain' CDPs with a distinctive macrocyclic region closed through an ester bond between the C terminus and a β-hydroxyl group. Interestingly, they have either a terminal PK moiety or a more simple branched aliphatic acid at the exocyclic arm. Moreover, both *N*-methyl and unnatural and complex amino acids are also found in the sequence of these peptides [24, 25].

The distinctive 'head-to-side-chain' has been strongly pointed out as a critical feature for the remarkable biological functions of these CDPs, which are mostly as cytotoxic and anticancer. Such compounds include didemnins and the kahalalide family. These CDPs have been extensively investigated and developed to enter clinical trials and market commercialization [13, 25].

Plitidepsin (Aplidin®; formally known as dehydrodidemnin B) was originally isolated from the Mediterranean tunicate *Aplidium albicans*, but it is currently obtained by total synthesis (Figure 3). The chemical structure of plitidepsin closely resembles that of didemnin B isolated from the Caribbean tunicate *Trididemmum solidum*. However, due to severe fatigue and anaphylaxis experienced by patients for various cancer treatments, the clinical trials of didemnin B were then put on hold and discontinued [3, 6, 13, 25]. Hence, pharmacologically similar plitidepsin replaced didemnin B in multiple Phase I, II, and III clinical trials as an anticancer agent. Recently, this NRP drug developed by a Spanish big pharmaceutical company PharmaMar, S.A. has been approved by the Australian Regulatory Agency (TGA) for the treatment of multiple myeloma. Moreover, the recent *in vitro studies* on the human coronavirus HCoV-229E revealing antiviral properties have led PharmaMar, S.A. to initiate the Phase II clinical trials to treat patients with COVID-19 [26].

Another marine CDP is elisidepsin (Irvalec®, PM02734) which belongs to the kahalalide family. It was reported that this compound induced oncolytic rather than apoptotic cell death. However, similar to didemnin B that failed in the Phase II clinical trial, elisidepsin was also put on hold for the development of cancer treatment due to safety concerns [13, 25].

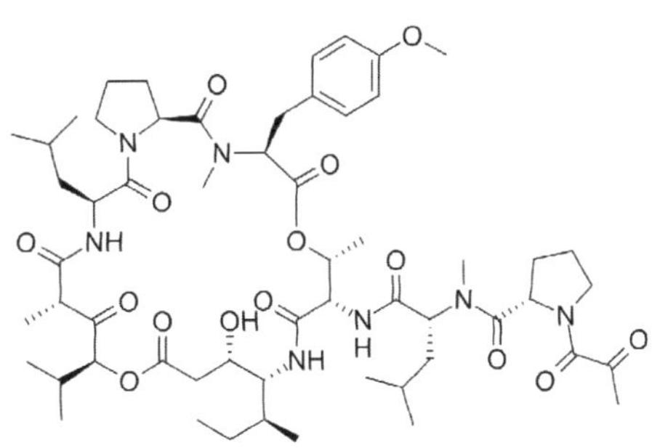

Figure 3. Chemical structure of plitidepsin (Aplidin®) [14].

IOP Conf. Series: Materials Science and Engineering 1011 (2021) 012063 doi:10.1088/1757-899X/1011/1/012063

5. Non-ribosomal peptides hybrids

In addition to CDP plitidepsin that has been mentioned above, trabectedin (ET-743), lurbinectedin (PM01183), and brentuximab vedotin (SGN-35) (Figure 4) are some successful examples of the NRP hybrid-derived drugs that obtained clinical approval in the US and/or Europe. Some molecules such as marizomib and glembatumubab vedotin are currently being evaluated in Phase II and III clinical trials, respectively. Despite their success story, there were also some unsatisfactory evaluation results in different phases of clinical trials that have been discontinued for antibody-drug conjugate-NRPS hybrids such as SGN-75 and ASG-5ME, which are not discussed further in this review [17, 23].

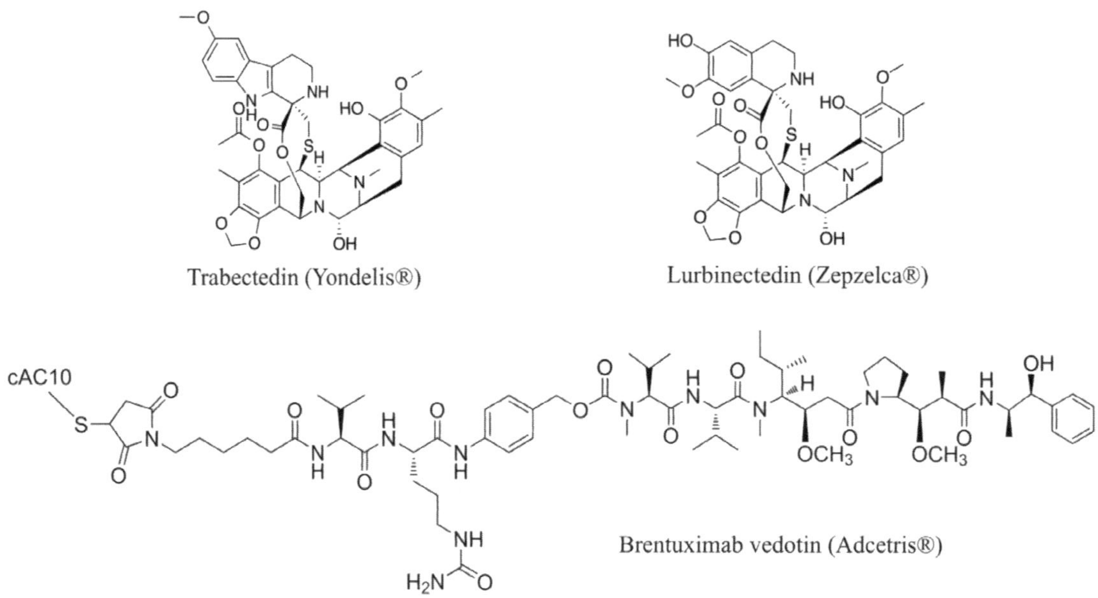

Figure 4. Chemical structure of selected marketed NRP hybrid-derived drugs: trabectedin (Yondelis®), lurbinectedin (Zepzelca®), and brentuximab vedotin (Adcetris®) [5].

Trabectedin demonstrated an impressive milestone in the development of MNPs-derived drugs. It would be nearly 30 years until the structure of the anticancer active compound extracted from the Caribbean tunicate *Ecteinascidia turbinata* trabectedin (ecteinascidin 743: ET-743), was finally elucidated [27, 28]. Trabectedin was identified as a tetrahydroisoquinoline alkaloid. Commercialized by PharmaMar under the trade name Yondelis®, its development was severely hindered by a lack of supply. Eventually, this need was supplied by semisynthesis from the microbial-derived natural product of cyanosafracin B, a fermentation product of bacterium *Pseudomonas fluorescens* [29]. The advancement of metagenomic sequencing, however, revealed that trabectedin is basically made by microbial symbionts using NRPS machinery based on the biosynthetic gene cluster profiles of the symbionts. A recent patent on that biosynthetic pathway finding has been taken out for the commercial production of NRPS-derived trabectedin [30-32]. Trabectedin was the first MNPs-based drug approved for clinical use in cancer chemotherapy (the treatment of refractory soft-tissue sarcomas) and has been commercialized in Europe. The FDA, however, approved the use of Yondelis® only in 2015 [13, 25].

Continuing the success of Yondelis®, an analog of trabectedin namely lurbinectedin (PM01183) has also been developed. This compound is structurally similar to trabectedin, although the tetrahydroisoquinoline present in trabectedin is replaced with a tetrahydro β-carboline, which enables lurbinectedin to exhibit an increased antitumor activity compared with trabectedin. Recently, lurbinectedin known under the trade name Zepzelca® has been approved by the FDA for the treatment of adult patients with metastatic small cell lung cancer [32-33] and is ready for market in the US by July 2020.

An other recent addition to the arsenal of approved and marketed anticancer agents is brentuximab vedotin (SGN-35), an antibody-drug conjugate-NRPS hybrid that targets cell membrane protein CD30. This chimeric antibody is attached through a protease-cleavage linker to a synthetic derivative of the potent tubulin agent dolastatin 10, monomethyl auristatin E (MMAE). Dolastatin 10 belongs to a linear pentapeptide group generated by NRPS-PKS machinery. Brentuximab vedotin, commercially known as Adcetris®, was developed by the US-based pharmaceutical company Seattle Genetics, Inc. and approved in 2012 for the treatment of Hodgkin's lymphoma (a type of cancer that is distinguished by the high expression of CD30) [34]. Clinical trial Phase II is undergoing to evaluate similar drug candidate glembatumumab vedotin that has shown yet modest activity and an acceptable safety profile in patients with advanced melanoma [35].

Another promising NRPS-derived anticancer drug candidate under human trial evaluation is a potent proteasome inhibitor, marizomib (salinosporamide A). Salinosporamide A is assembled by NRPS-PKS machinery and possesses a complex, densely functionalized γ-lactam-β-lactone. Following its successful assessment in phase I trials in patients with newly diagnosed as well as recurrent glioblastoma, marizomib is now being investigated in a phase III trial [36-38].

6. Concluding remarks and future perspectives

Peptide-derived MNPs show fascinating chemical structures and remarkable biological properties and as such many of them are successfully commercialized in the market. However, there are still challenges to develop those peptide drugs, such as supply shortage and synthetic difficulties. Thereby, progress in the advancement of biosynthetic pathway engineering can disclose new doors for harnessing newly designed marine peptide drugs with improved biological and pharmaceutical properties.

References

[1] Newman DJ and Cragg GM 2012 *J. Nat. Prod.* **75** 311-335
[2] Jaspars M, De Pascale D, Andersen JH, Reyes F, Crawford AD and Ianora A 2015 *J. Mar. Biol. Assoc. U. K.* **96** 151-158
[3] Cheung RCF, Ng TB and Wong JH 2015 *Mar. Drugs.* **13** 4006-4043
[4] Von Döhren H 2010 Peptides In *Encyclopedia of Industrial Biotechnology,* ed MC Flickinger (New Jersey: John Wiley & Sons, Inc.) pp 1-20
[5] Gerwick WH and Moore BS 2012 *Chem. Biol.* **19** 85-98
[6] Molinski TF, Dalisay DS, Lievens SL and Saludes JP 2009 *Nat. Rev. Drug Discov.* **8** 69-85
[7] Martins A, Vieira H, Gaspar H and Santos S 2014 *Mar. Drugs.* **12** 1066-1101
[8] Lindequist U 2016 *Biomol. Ther.* **24** 561-571
[9] Carroll AR, Copp BR, Davis RA, Keyzers RA and Prinsep MR 2020 *Nat. Prod. Rep.* **37** 175-223
[10] Villa FA and Gerwick L 2010 *Immunopharm. Immunot.* **32** 228-237
[11] Henniot A, Collins JC and Nuss JM 2018 *J. Med. Chem.* **61** 1382-1414
[12] Blunt JW, Copp BR, Keyzers RA, Munro MH and Prinsep MR 2015 *Nat. Prod. Rep.* **31** 160-258
[13] Mayer AMS, Glaser KB, Cuevas C, Jacobs RS, Kem W, Little RD, McIntosh JM, Newman DJ, Potts BC and Shuster DE 2010 *Trends Pharmacol. Sci.* **31** 255-265
[14] Mayer AMS, Rodríguez AD, Taglialatela-Scafati O and Fusetani N 2013 *Mar. Drugs.* **11**, 2510-2573
[15] McIntosh JA, Donia MS and Schmidt EW 2009 *Nat. Prod. Rep.* **26** 537-559
[16] Dang T and Süssmuth RD 2017 *Acc. Chem. Res.* **50** 1566-1576

[17] Arnison PG, Bibb MJ, Bierbaum G, Bowers AA, Bugni TS, Bulaj G, Camarero JA, Campopiano DJ, Challis GL, Clardy J, *et al*. 2013 *Nat. Prod. Rep*. **30** 108-160
[18] Walsh CT, Brien RVO and Khosla C 2013 *Angew. Chemie Int. Ed*. **52** 7098-7124
[19] Süssmuth RD and Mainz A 2017 *Angew. Chemie Int. Ed*. **56** 3770-3821
[20] Marahiel MA, Stachelhaus T and Mootz HD. 1997. *Chem. Rev*. **97**, 2651-2674
[21] Scheidler CM, Kick LM and Schneider S 2019 *ChemBioChem* **20** 1479-1486
[22] Akondi KB, Muttenthaler M, Duterte S, Kaas Q, Craik DJ, Lewis RJ and Alewood PF 2014 *Chem. Rev*. **114** 5817-5847
[23] Pennington MW, Czerwinski A and Norton RS 2018 *Bioorg. Med. Chem*. **26** 2738-2758
[24] Andavan GSB and Lemmens-Gruber R 2010 *Mar. Drugs*. **8** 810-834
[25] Pelay-Gimeno M, Tulla-Puche J and Albericio F 2013 *Mar. Drugs*. **11** 1693-1717
[26] Fernández R, Bayu A, Hadi TA, Bueno S, Pérez M, Cuevas C and Putra MY 2020 *Mar. Drugs*. **18** 396
[27] Rinehart K, Holt TG, Fregeau NL, Stroh JG, Keifer PA, Sun F, Li LH and Martin DG 1990 *J. Org. Chem*. **55** 4512-4515
[28] Wright AE, Gunawardana GP, Gunasekera SP, Koehn FE and McConnell OJ 1990 *J. Org. Chem*. **55** 4508-4512
[29] Cuevas C, Pérez M, Martin MJ, Chicarro JL, Fernández-Rivas C, Flores M, Francesch A, Gallego P, Zuerzelo M, de La Calle F, García J, Polanco C, Rodríguez and Manzanares 2000 *Org. Lett*. **2** 2545-2548
[30] Rath CM, Janto B, Earl J, Ahmed A, Hu FZ, Hiller L, Dahlgren M, Kreft R, Yu F, Wolf JJ, *et al*. 2011 *ACS Chem. Biol*. **6** 1244-1256
[31] Sherman DH, Ehrlich GD, Janto B, Williams RM and Rath CM 2017 9,611,491.*U.S.Patent* 2017 Apr 4
[32] Watters DJ 2018 *Mar. Drugs*. **16** 162
[33] Trigo J, Subbiah V, Besse B, Moreno V, López R, Sala MA, Peters S, Ponce S, Fernández C, Alfaro V, *et al*. 2020 *Lancet Oncol*. https://doi.org/10.1016/S1470-2045(20)30097-8
[34] Katz J, Janik JE and Younes A 2011 *Clin. Cancer Res*. **17** 6428-6436
[35] Ott PA, Pavlick AC, Johnson DB, Hart LL, Infante JR, Luke JJ, Lutzky J, Rothschild NE, Spitler LE, Cowey CL, *et al*. 2019 *Cancer* **125** 1113-1123
[36] Gulder TAM and Moore BS 2010 *Angew. Chem. Int. Ed*. **49** 9346-9367
[37] Potts BC, Albitar MX, Anderson KC, Baritaki S, Berkers C, Bonavida B, Chandra J, Chauhan D, Cusack JC, Fenical W 2011 *Curr. Cancer Drug Targets* **11** 254-284
[38] Roth P, Reijneveld JC, Gorlia T, Dhermain F, et al. 2019 *J. Clin. Oncol*. **37** 15

IOP Conf. Series: Materials Science and Engineering 1011 (2021) 012064 doi:10.1088/1757-899X/1011/1/012064

Determination of Ca, Na, Mg Impurities in Demineralized Water Bandung Triga 2000 Reactor using Flame Atomic Absorption Spectrophotometry

Natalia Adventini*, Endah Damastuti, Neni Ratnawati, Fuji Octa Indah Suciati, Woro Yatu Niken Syahfitri

Center for Applied Nuclear Science and Technology-BATAN, Tamansari 71, Bandung, West Java, Indonesia

*nadventinigm@gmail.com

Abstract. Demineralized water is commonly used as cooling water, neutron moderator and as a radiation shield in a nuclear reactor such as Bandung Triga 2000 Reactor. According to its function, demineralized water has to fulfilled some requirements as stated in Safety Analysis Report (SAR) of Bandung Triga 2000 Reactor. The requirements were impurities of Ca, Na and Mg, pH and conductivity electricity in demineralized water reactor. Determination of Ca, Na and Mg as impurities were carried out using Flame Atomic Absorption Spectrophotometry (FAAS). In this activity, instrument and method verification were carried out as quality control. Instrument verification determined were sensitivity, precision and detection limit of FAAS. Method verification was carried out by using spike demineralized water with Na, Ca, and Mg standard. Accuracy and precision of the method were in the acceptance criteria according to AOAC then implemented to determined impurities of Ca, NA, Mg in demineralized water reactor samples. Na and Mg in demineralized water using FAAS gave values less than 1 mg/L, pH values were in the range 5,5- 6,5 meanwhile conductivity electricity gave values were less than 3 μmho fulfilled the requirements in SAR. These results were indispensable to support the performance of Bandung Triga 2000 Reactor which needed by public and university for the development of science and its application in the industry, health, environment and other fields.

1. Introduction

Low power research reactors such as Bandung Triga 2000 reactor commonly used in many fields. Bandung Triga 2000 reactor could be implemented in various fields such as health, environtment, industry and other applications that provide many benefit to mankind [1]. Bandung Triga 2000 Reactor and other type of reactor require water which is its function as water coolant, neutron moderator and radiation shield [2]. Water sources of Bandung Triga 2000 Reactor were Perusahaaan Daerah Air Minum (PDAM) and groundwater that were processed using a water softening system and purified using reserve osmosis process [3]. Both water sources caused demineralized water reactor may contain elements of calcium and chlorine derived from PDAM and elements of ground water such as magnesium, iron and many compounds such as carbonate-bicarbonate and sulphate meanwhile Na originated from the ion exchange process. Demineralized water used by Bandung Triga 2000 Reactor must meet requirements such as pH, conductivity and Si, Ca, Mg, Na content which values were stated in the Safety Analysis Report (SAR).

Determination of mineral impurities in demineralized water reactor could be done using Flame Atomic Absorption Spectrophotometry (FAAS) and other methods. Samin *et all* and Supriyanto *et all* studied the demineralized water reactor quality usng FAAS meanwhile Neutron Activatin Analysis was used by Elisabeth R *et all* [4, 5, 6, 7]. AAS method had been common used for determination elements in various sample matrices [8, 9] such in environment field. In this activity, FAAS method was used for determination impurities of Ca, Na, Mg in demineralized water Bandung Triga 2000 Reactor. FAAS instrument and the used method must be checked before sample measurement were carried out. Instrument verification includes were sensitivity, precision, and limit of detection (LOD). Method verification of the specified method includes the accuracy and precision parameters of Ca, Na and Mg in spike demineralized water based on the acceptance criteria of accuracy and precision from AOAC [9].

In this activity, determination of Ca, Na, Mg impurities, pH values and conductivity parameter in demineralized water Bandung Triga 2000 reactor was carried out periodically since 2017 after having a new operating permit from national nuclear energy regulatory body (Bapeten) for 2017-2027 period.

The purpose of this activity is to observe the characteristics and dissolved chemical concentration in reactor demineralized water for the study of reactor safety assesment and the study of Bandung Triga 2000 reactor aging [10,11] in suppoting performance of Bandung Triga 2000 Reactor to develope science and its applications.

2. Materials and methods

2.1. Instrumentation/procedur

Flame atomic absorption spectrofotometer (FAAS) from GBC Avanta P, Australia, equipped with Deuterium (D_2) background corrector was used as main device. To reach optimum absorbance signal of each elements, the device was set as follows: setting of flow rate (mL/minutes), the height of burner and adjusting of gas flow rates (oxidant and fuel). Calcium, Sodium and Magnesium hollow cathode lamps were used as light sources. Instrumental parameters used was listed in Table 1.

Table 1. FAAS instrument parameter of Ca, Na and Mg

Analyte	Wavelength (nm)	Slit width (nm)	Lamp current (mA)	Flame	Optimum working range of FAAS (mg/L)
Ca	422.7	0.5	5.0	C_2H_2-air	0.01 – 4.0
Na	589.0	0.5	5.0	C_2H_2-air	0.01 – 0.7
Mg	285.2	0.5	3.0	C_2H_2-air	0.01 – 0.4

2.2. Reagents

All reagents such as HNO_3 65%, stock standard solutions Ca, Na, Mg Titrisol (1000 ± 0,002mg) each were obtained from E. Merck pro analysi (pa) grade. Deionised water with ± 18,1 MΩ resistivity was used throughout the experimental work produced by AriumPro deionised water machine from Sartorius, Germany. Laboratory glasswares/plasticwares were soaked overnight in 5 N HNO_3 solution. All of the apparatus then were washed with deionised water and dried.

2.3. Demineralized water reactor sampling

Demineralized water reactor samples were collected from the surface of reactor tank and bulk shielding using plastic measuring cup then placed into the plastic bottles 100 mL volume. Before the samples keep in the refrigerator, a drop of HNO_3 65% was added.

2.4. Preparation of working standard solution series
FAAS working standard solution series was made according to the optimum working range of the instrument. Stock standard of Ca prepared to be a mix working standard solution into 0.1; 0.2; 0.3; 0.4; 1; 2; 3 mg/L meanwhile Na, Mg were prepared into 0.1; 0.2; 0.3; 0.4mg/L with HNO_3 0.1 N used in adjusting standard solution in 100 mL volumetric flask. Strontium 50,000 mg/L solution then added to have 2000 mg/L for each concentration of mix working standard solution. Strontium (Sr) 2000 mg/L in the mix working standard solution was meant as a releasing agent in Ca analysis.

2.5. Preparation of blank solution for instrument verification
Reagent blank was made from 0.1 N HNO_3 by adding strontium solution until the concentration was 2000 mg/L.

2.6. Preparation of spike demineralized water as Ca, Na, Mg method verification
Ten mL of Ca standard, 1 mL of Na and 2 mL Mg standards with concentration of 10 mg/L each were added into a 100 mL volumetric flask, added by releasing agent 2000 mg/L Strotium and adjusted using demineralized water. Concentration of spike standards in demineralized water were 1; 0,1; 0,2 mg/L for Ca, Na and Mg respectively. Demineralized water as sample also included in the measurement process.

2.7. FAAS instrument verification
Instrument verification was carried out by constracting standard curve of Ca, Na, Mg using working standard solution series with 10 replicates each to determine instrument sensitivity and precision of each element while measuring. Blank solution then aspirated to determine LOD of the instrument. Sensitivity was determined as follow:

$$S=0.0044 \text{ x } \frac{C}{A} \qquad (1)$$

where S=sensitivity; C = analyte concentration in standard solution; A= mean analyte absorbance of standard. The sensitivity values (mg/L) of the instrument were stated in GBC Avanta P Series AAS method manual book [10] as reference sensitivity. Ratio sensitivity observed to sensitivity reference must be ≤ 125 as our laboratory criteria acceptance.

Precision which is stated as Relative Standard Deviation (RSD) was calculated using this formula:

$$RSD = \frac{SD}{mean} \text{ x } 100\%. \text{ Precision value must meet to} \leq 1\% \qquad (2)$$

Limit of detection is the smallest amount of analyte in the blank solution that can still be detected by the instrument which calculated with formula:

$$LOD = \text{mean blank concentration} + (3 \text{ SD blank concentration}) \qquad (3)$$

In addition to LOD, the limit of quantification (LOQ) is also determined according to the LOQ formula:

$$LOQ = 10 \text{ LOD} \qquad (4)$$

LOQ is intended to determine the smallest analyte signal in a blank solution that can be quantified. Both unit are in mg/L.

2.8. Method verification
Method verification was carried out by 2 parameters those were accuracy and precision using spike standards in demineralized water. Accuracy is expressed as percent recovery was calculated according to the formula as follow:

$$\%Rec = \frac{a-b}{c} \text{ x } 100\% \qquad (5)$$

where: a = concentration of demineralized water + concentration of standard addition (mg/L); b = demineralized water sample concentration (ppm); c = theoritic concentration of standard addition (mg/L).

Precision of method verification is stated as Horwitz precision (RSDr) calculated from the equation:

$$RSDr = 2^{(1-0.5 \log C)} \qquad (6)$$

where C is concentration fraction. For example: concentration fraction of ppm is 10^{-6}. Precision acceptance is set by our laboratory as HorRat value as much as 0,3–0,67 as fully acceptable recommended range according to Association of Official Analytical Chemists (AOAC) [11].

HorRat values is calculated as follow:

$$\text{HorRat} = \frac{\%\text{RSD}}{\text{RSDr}} \quad (7)$$

2.9. Quantitative calculation

Analyte concentration in demineralized water reactor samples were calculated using this formula:

$$C_{sample} = a \times F \quad (8)$$

Where a= concentration (mg/L) of the element in the demineralized water reactor sample was obtained from calibration curve, F=dilution factor of the sample; concentration sample unit in mg/L.

3. Result and discussion

Verification parameters for sensitivity, precision and LOD of Ca, Na and Mg can be seen in Table 2 below. From Table 2 it showed that Ca was in good agreement with the sensitivity and precision acceptance criteria of instrument verification. Unfortunately, Natrium and Magnesium sensitivity values were out of criteria acceptance but they had good results in precision.

Table 2. Instrument verification of Ca, Na and Mg

Parameters	Ca	Na	Mg	References Values	Unit
Sens.Obs	0.020	0.016	0.0038	-	mg/L
Sens. Ref	0.020	0.004	0.0030	-	mg/L
Sens. (Obs/Ref)	100	288	127	≤ 125	-
Precision	0.700	0.600	0.640	≤ 1	%
LOD	0.020	0.007	0.0007	-	mg/L
LOQ	0.20	0.07	0.007	-	mg/L

Sensitivity problem may caused not only by instrument flow rate which was not suitable for Na and Mg but also corrosive nebulizer. These conditions, will decrease instrument sensitivity and resulted in LOD values were increased. Chemical interferences in air-acetylene flame can be reduced by additon of a releasing agent 2000 mg/L Strontium to remove the interference [10]. This step can minimized sensitivity problem.

Method verification using spike demineralized water with standards presented in Table 3. Accuracy refers to how close a measurement is to the true value which is stated in percent recovery as a function of analyte concentration [11].

Table 3. Method verification using spike demineralized water

Element	Spike conc. (mg/L)	Demin conc. (mg/L)	Std conc. (mg/L)	Theoretical Conc. (mg/L)	%Rec	%RSD	$\%RSD_r$	Horrat
Ca	2.90	0.80	1.05	1.85	112	4.88	13.63	0.36
Na	0.20	0.07	0.12	0.19	71	8.86	20.40	0.43
Mg	0.19	0.02	0.19	0.21	78	8.32	20.58	0.40

Accuracy Ca at 2,90 mg/L concentration obtained 112% Recovery meanwhile Na and Mg at concentration 0.2 ppm and 0.19 mg/L had 71 and 78%Recovery respectively. It meant that accuracy of Ca, Na, Mg in spike demineralized water were suitable with the percent recovery range values determined by AOAC. HorRat values from Table 3 were in the range 0.36 up to 0.43. The results showed that accuracy and precision of the method fullfilled the value as AOAC reccomendation so it could be implemented to determined impurities of Ca, Na, Mg in demineralized water of Bandung Triga 2000 reactor.

Tables 4 presented impurities of all elements in spike demineralized water reactor samples which collected from reactor tank and bulk shielding. The results showed that concentration of the impurities throughout 2017-2019 were in the range 0.01-0.96 mg/L. Based on the Table 2 it can be seen that the LOQ concentration of blank solution for the three elements had values were 0.2; 0.07; 0.007 mg/L for Ca, Na, Mg which indicated that these concentrations of blank solution signal could be quantified properly as analyte by the instrument.

Table 4. Minerals impurities in demineralized water Bandung Triga 2000 reactor 2017-2020

Date	Samples	Ca (mg/L)	Mg (mg/L)	Na (mg/L)
Apr-17	Water tank	0.51	0.02	0.26
Apr-17	Bulk shielding	0.50	0.03	0.21
Mei-17	Water tank	0.26	0.03	0.37
Mei-17	Bulk shielding	0.96	0.08	0.91
Sep-17	Water tank	0.31	0.03	0.50
Sep-17	Bulk shielding	0.37	0.02	0.11
Feb-19	Water tank	0.28	0.01	0.23
Feb-19	Bulk shielding	0.28	0.01	0.27
Jan-20	Water tank	LOD	LOD	LOD
Jan-20	Bulk shielding	LOD	LOD	LOD
Jun-20	Water tank	LOD	LOD	LOD
Jun-20	Bulk shielding	LOD	LOD	LOD
Jul-20	Water tank	LOD	LOD	LOD
Jul-20	Bulk shielding	LOD	LOD	LOD

However, since Since 2020 concentration of three elements were less than LOD presented in Table 2. It means, the smallest blank concentrations of Ca, Na, Mg which could be detected by the instrument were 0.02; 0.007 and 0.0007 mg/L respectively. This could be stated that the blank solution signal was a noise signal. Measurement results which provided values around the lower limit of the working range will be better if method detection limit (MDL) is also determined as a study by Rona M. Napitupulu *etal,* 2019 [12]. Table 4 showed that minerals impurities in demineralized water Bandung Triga 2000 reactor was less then 1 mg/L. Low mineral content will reduce corrosion rate up to maximum and moves the primary cooling system. This condition not only to minimize reaction between source radiation against dissolved chemical particles but also to prevent radioactive contamination on all components of the primary cooling system from excessive radiation exposure.

Other parameters that supported reactor demineralized water quality presented in Figures 1 and 2 in time series.

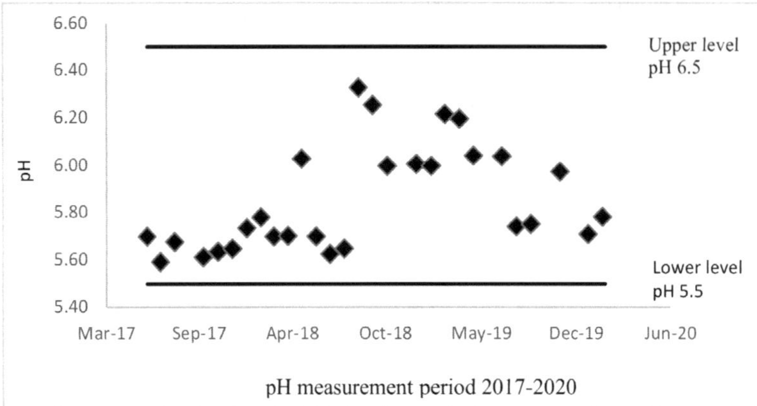

Figure 1. pH values in demineralized water Bandung Triga 2000 reactor 2017-2020

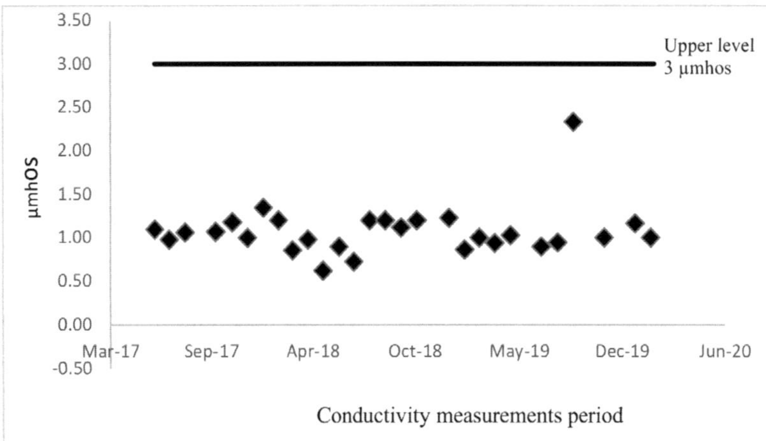

Figure 2. Conductivity in demineralized water Bandung Triga 2000 reactor 2017-2020

Based on the SAR of the Bandung TRIGA 2000 reactor, document No. R 093/KN 01 01 /SNT 4, Revi 0, Issue 2, 2017, water (H_2O) as one of the components in the primary cooling system must have several requirements. Figure 1 presented pH values through 2017-2020 were in the range 5,5-6,5 while conductivity were less than 3 showed in Figure 2. To maintain the quality of primary cooling water, inside the reactor during operation, demineralized water is cleaned continuously using a water purification system designed separately from the primary cooling system. This system consists of filters and resins. Filter system must be able to filter particles up to 5 µm while mixed-bed resin system purify water until demineralized water's conductivity is less than 3.5 µmhos. The values of pH and electrical conductivity need to be determined because they are demineralized water impurities indicators which are influenced by dissolved chemical species in the water. While the reactor is not operating, purifier system is operated periodically and at least once a month the cooling water parameters must be measured to ensure demineralized water reactor quality is always maintained. From the obtained data, these activities were needed for safety reactor assessment and study of the Bandung TRIGA 2000 reactor Aging [13,14].

4. Conclusion

Determination impurities of Ca, Na and Mg in demineralised water Bandung Triga 2000 reactor using FAAS gave a good results meanwhile pH and conductivity gave values were in the range between lower and upper level. These data could be used for Assessment Periodic Safety and Study of Bandung TRIGA 2000 reactor aging in supporting performance of Bandung Triga 2000 Reactor.

References

[1] Sayed A. El-Mongy, Overview of reactor research worldwide and their applications, Nuclear regulatory authority of Egypt, Article May, 2018.

[2] J Tominaga, Advances in Steam Turbines for Modern Power Plants, Science Direct, (2017) : 41-56

[3] F.N. Kemmer; The Nalco water handbook; 2. Edition; 1988

[4] Samin, Pemutakhiran Metode SSA, UV-VIS dan Potensiometri-ESI untuk Kontrol Kualitas Air Tangki Reaktor, Prosiding Pertemuan dan Presentasi Ilmiah – Penelitian Dasar Ilmu Pengetahuan dan Teknologi Nuklir 2014 Pusat Sains dan Teknologi Akselerator – BATAN, Yogyakarta, 10-11 Juni 2014

[5] Supriyanto C, Iswani G, Kualifikasi Air Tangki Reaktor (ATR) Kartinini Berdaarkan Data Dukung Metoda Nyala Spektrometri Serapan Aatom (SSA_ dan Ion Selective ElecTrode (ISE), Prodising PPI-PDIPTN (2007).

[6] Elisabeth R, M. Gading Permadi, Dicky Tri Jatmiko, Analisis Unsur Pengotor di dalam Air Pendingin Primer Reaktor RSG-GAS, Jurnal Forum Nuklir 9JFN) Vol. 12, Nomor 2 (2018).

[7] Elisabeth Ratnawati, Diyah Erlina Lestari, T. Rina Mulyaningsih, Studi Pengotor Pada Pendingin Primer Reaktor RSG GAS Setelah 30 Tahun Beroperasi, Jurnal Iptek Nuklir Ganendra, Vol. 22 No. 1 (2018): 17-24

[8] Alfan Hidayatulloha, Taslimaha, Abdul Harisa, Penentuan Kandungan Logam Magnetik Komponen Penyusun Abu Layang Batubara, Jurnal Kimia Sains dan Aplikasi 14 (1) (2011) 1 – 3

[9] Idha Yulia Ikhsani, Eki Naidania Dida, Sri Yudawati Cahyarini, Evaluasi Penggunaan Spektrofotometri Serapan Atom Nyala (FAAS) untuk Analisis Konsentrasi Sr/Ca dalam Karang *Porites* dari Teluk Ambon dan Pulau Jukung, Jurnal Ilmu dan Teknologi Kelautan Tropis, Vol 9, No. 1 (2017) 247-253

[10] GBC Avanta P Series AAS method manual book, May 2009.

[11] Standard Format and Guidance for AOAC Standard Method Performance Requirement Documenets, Version 12.1, 31 Januari 2011

[12] Rona M. Napitupulu, Dirgarini Julia, Aman Sentosa Panggabean, Validasi metode penentuan Mn dalam oli lubrikan dengan metode pengenceran langsung menggunakan spektrofotomeer serapan atom, Indo. J. Chem. Res., 2019 , 6(20), 94-100.

[13] Peraturan Kepala Badan Pengawas Tenaga Nuklir Nomor 2 Tahun 2015 tentang Verifikasi dan Penilaian Keselamatan Reaktor NonDaya

[14] Peraturan Kepala Badan Pengawas Tenaga Nuklir Nomor 8 TAHUN 2008 tentang Ketentuan Keselamatan Manajemen Penuaan Reaktor NonDaya

IOP Conf. Series: Materials Science and Engineering 1011 (2021) 012065 doi:10.1088/1757-899X/1011/1/012065

Nuclear analytical techniques for identification of elemental composition of fine and coarse airborne particulate matter Collected in Bandung, Indonesia

M Santoso[1a], D D Lestiani[1], I Kusmartini[1], D K Sari[1], S Kurniawati[1], E Damastuti[1], D P D Atmodjo[1], A Tjahyadi[2], W D Windiandi[2]

[1]Center for Applied Nuclear Science and Technology, National Nuclear Energy Agency of Indonesia (BATAN)
[2]The Environmental Protection Agency of West Java Province, Bandung, Indonesia

[a)] Corresponding author: hayat@batan.go.id

Abstract. Air pollution has received serious attention from both the government and the public in Bandung, Indonesia. Nuclear analytical techniques (NAT) such as X-ray fluorescence (XRF) have been implemented for the quantification of environmental pollutants, especially airborne particulate matter (APM). Sampling for fine APM ($PM_{2.5}$) and coarse APM ($PM_{2.5\text{-}10}$) have been conducted in Bandung using a GENT sampler, once a week for 24 hrs from 2015 to 2017. The samples were then analyzed for their mass, black carbon (BC), and elemental composition using gravimetric, smoke stain reflectometer, and ED-XRF, respectively. The results obtained indicate that the annual average of $PM_{2.5}$ is in the range of 17.85-20.90 $\mu g/m^3$ which has exceeded the Indonesian annual standard (15 $\mu g/m^3$). While for PM_{10} it is still below the standard. Compared to $PM_{2.5}$ and $PM_{2.5\text{-}10}$ in 2002-2004, the concentration has increased by around 35% and 29%, respectively. This situation needs serious attention especially because of the adverse effects of $PM_{2.5}$ on human health. The average BC fraction at $PM_{2.5}$ is around 19%. In this paper, chemical composition, annual concentration trends, and correlations between elements will be discussed in detail including the advantages and disadvantages of the NATs method. For the next stage, this information can be used to identify sources of pollutants.

1. Introduction

Increased urbanization and various economic activities such as transportation and industry have resulted in air pollution. Urban air quality in Indonesia, including Bandung, has shown a declining trend in the past two decades. This condition occurs because the source of pollutants has exceeded the carrying capacity of the environment so it cannot be neutralized naturally. This air pollution has a significant impact on human health, ecosystems, climate change and global warming [1]. The health risks associated with air pollution in urban areas have generally attracted a lot of attention in recent decades [2]. The main parameter of air pollution that has a significant impact on health is particulate matter (PM). Particulates present in the atmosphere are generally up to 50 μm in size, which varies in existence depending on size. Air particulates that are less than 2.5 μm ($PM_{2.5}$), called fine particulates, are very dangerous because they can penetrate into the deepest parts of the lungs and heart, causing health problems including acute respiratory infections, lung cancer, cardiovascular disease and even death [3]. Fine particulates matter ($PM_{2.5}$) are estimated to have a large contribution to mortality due to health problems related to air pollution. $PM_{2.5}$ generally consist of micro-sized particles originating from anthropogenic sources such as motor vehicles, biomass combustion, and fuel combustion. Besides $PM_{2.5}$, also known as PM_{10} which is a particulate matter that is less than 10 μm in size, whereas total suspended particulate (TSP) is all suspended substances that are generally less than 50 μm in size.

One of the critical aspects in air quality management and air pollution control is the implementation of continuous air quality monitoring. In Indonesia, including Bandung, the identification of $PM_{2.5}$ and PM_{10} are still very rare, because it requires advanced analytical techniques that are able to detect the concentrations of elements at the nanograms per cubic meter level. Nuclear analytical techniques (NATs) such as neutron activation analysis (AAN), X-ray fluorescence (XRF), and proton X-ray emission (PIXE) are suitable techniques because they have high sensitivity, simultaneous, selective, and the detection limit reaches the part per billion (ppb) level. NATs have been widely applied in the characterization of APM [4]. The advantages of NATs are expected to be a new breakthrough in the environmental field.

Air quality in Bandung in 2002-2004 was reported by Santoso (2008) which discussed in detail the chemical composition of the fine and coarse APM, including the identification of the source of the pollution. Motor vehicles, biomass burning, secondary sulfate, soil, and road dust are sources of pollutants. The results obtained also indicate that biomass burning contributes around 20% [5]. Kim Oanh (2006) reports the results of particulate monitoring conducted in 6 cities in Asia, including in Bandung. $PM_{2.5}$ and PM_{10} concentrations in Bandung in the dry season reached 53 and 83 $\mu g/m^3$ respectively while in the rainy season they reached 38 and 62 $\mu g/m^3$, respectively [6]. Lestari 2009 reported that different source factors were obtained in the dry and wet season in Bandung. Secondary aerosols, vehicle emissions, electroplating industry, and biomass burning are sources for fine particulates. While the sources of coarse particulates during the dry season are the electroplating industry, aged sea salt, volcanic dust, soil dust, and lime dust. In the wet season, vehicle emissions and secondary aerosols are the main sources of fine particulates [7]. The objective of the current study is to provide an overview of the current status of Bandung's air quality, including the level of mass concentration and elemental composition of the APM and its comparison with similar data from 2002-2004.

2. Methods

2.1. Sampling

Sampling of ambient $PM_{2.5}$ and PM_{10} were carried out on the roof of the office of West Java Provincial Environmental Protection Agency (S 06°55'10.9", E 107°36'39.1", Figure 1), using the GENT dichotomous sampler. In each sampling, 2 types of Nucleopore polycarbonate filters were used, namely fine and coarse filters, each of which had porous of 0.4 and 8 μm, respectively. Sampling was carried out at an altitude of about 20 m above the ground, for 24 hours and once a week during 2015-2017. This sampling produces APM samples with particulate sizes up to 2.5 μm in fine filters and between 2.5-10 μm in coarse filters.

2.2. Particulate matter analysis

Determination of mass concentration ($PM_{2.5}$ and PM_{10}) was done by gravimetric method using Mettler Toledo micro balance. Fine APM ($PM_{2.5}$) and coarse APM ($PM_{2.5\text{-}10}$) are obtained from the weighing of the sample weights on the fine and coarse filter, respectively. While, PM_{10} is obtained from the total mass of particulates from fine and coarse filters. Before weighing, the filter must be conditioned in a room with humidity between 45-55% and a temperature of 18-25°C.

2.3. Black carbon analysis

Determination of carbon in the filter is based on the process of light reflection. The absorption and reflection of visible light by airborne particulate matter in the filter depends on particle concentration, density, refractive index and size. Because in atmospheric particles, BC is a component that has the highest absorption ability, so it is assumed that the absorption of light in the filter is caused by BC. The light reflection method is used in BC measurements, where the light coming from a light source is transmitted through the annular photocell to the surface of the sample filter, then the light is reflected back to the photocell, the length of the light path is twice the length of the transmission path. The amount of light absorbed will be proportional to the amount of BC in the filter. In this study BC was determined using a Digital Smoke Stain Reflectometer. Measurement of BC in the filter was done using the assumption that the average coefficient of particle mass absorption is 5.7 m^2/g [8].

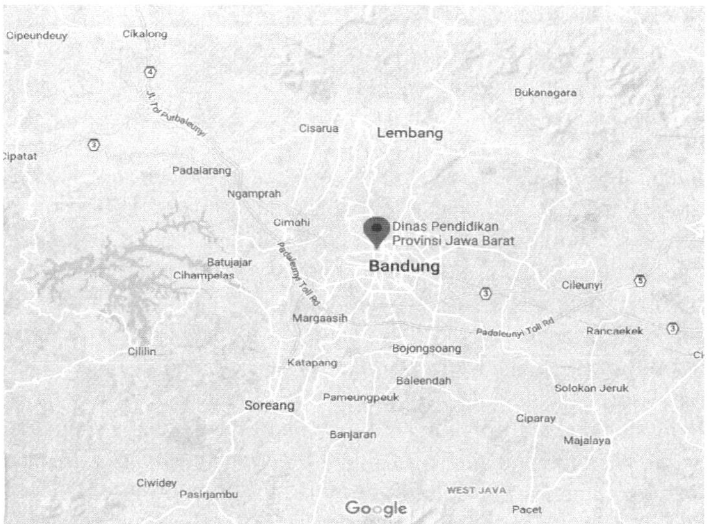

Figure 1. Sampling site of airborne particulate matter in Bandung, Indonesia (source: Google Maps [9]).

2.4. Multi elemental analysis

Determination of the elements contained in airborne particulate matter is an important step in identifying sources of air pollutants. APM with a mass of less than 1 mg are very small mass to be analyzed using conventional methods. Therefore, nuclear analytical techniques which have high sensitivity, simultaneous, selective, and a very low detection limits are needed. This is the main reason for selecting the use of nuclear-based analytical methods for elemental characterization in APM. This method is suitable for analysis of relatively large sample sizes, sometimes reaching hundreds of filters and a small sample weight per filter; 100 - 600 μg. The method used for elemental analysis has been validated using SRM (Standard Reference Material) Air Particulate on Filter Media. The samples were analyzed using ED-XRF with 9 secondary targets (Fe, CaF_2, Ge, Zr, CeO_2, Mo, Ag, Al and one Barkla polarizing target (Al_2O_3)). Calibration was carried out using a standard single element MicroMatter, whereas for method validation, NIST SRM 2783 Air Particulate Matter was used in the filter media. Except for light elements, the ratio between measurement results and certificate values is in the range of 0.92-1.09% [4]. XRF analysis methods including the way of calibration and validation are explained in detail by [4,10].

3. Results and Discussions

3.1. Particulate Matter Concentration

The results of 24-hour average mass fraction of the $PM_{2.5}$ and $PM_{2.5\text{-}10}$ in Bandung from 2015 to 2017 are shown in Figure 2, while the annual averages are presented in Table 1. The results obtained showed that the annual average concentration of 2015 (20.9 $\mu g/m^3$) was higher than in 2016 (17.85 $\mu g/m^3$) and 2017 (18.94 $\mu g/m^3$), due to the El Nino effect on 2015, this effect occurs periodically every 5 to 7 years which causes a longer dry season. In Figure 2 and Table 1, it can be seen that in general the concentration of coarse particulate matter is higher than that of fine particulate, this is probably because natural pollutant sources have a relatively higher contribution than anthropogenic sources.

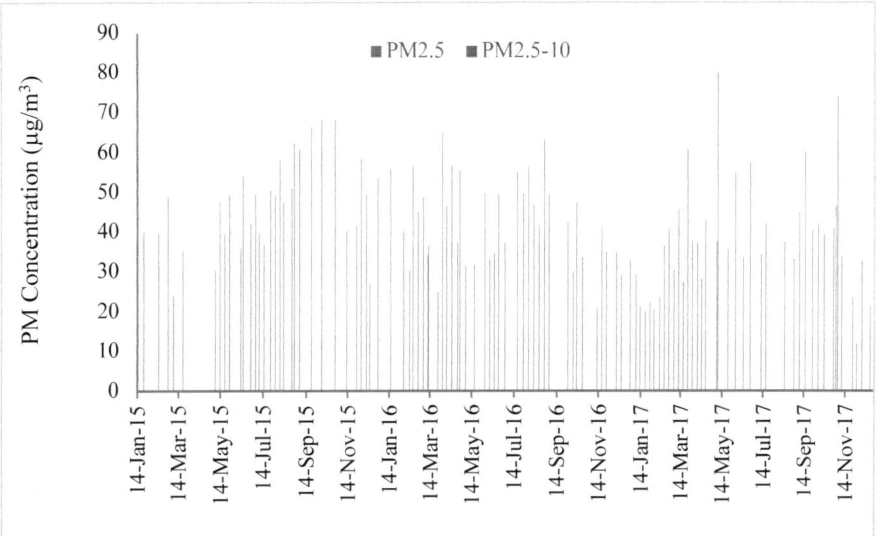

Figure 2. 24-hour averages for $PM_{2.5}$, $PM_{2.5\text{-}10}$ and PM_{10} ($\mu g/m^3$) in Bandung, Indonesia.

The average level of mass and the major components of fine and coarse APM in Bandung, 2015-2017 are shown in Table 2, while the comparison with 2002-2004 data is presented in Table 3. Comparison with 2002-2004 data was chosen because the sampling and analysis methods in both periods (2002-2004 and 2015-2017) were carried out in the same procedure. The concentrations of $PM_{2.5}$, $PM_{2.5\text{-}10}$ and PM_{10} for 2002-2004 were 14.03, 17.64 and 31.67 $\mu g/m^3$, respectively. Compared to 2002-2004 data, after 13-15 years, mass concentrations of fine and coarse APM increased by 49%, 27%, 35%, and 49%, 38%, 6% for 2015, 2016, 2017, respectively. An increase in $PM_{2.5}$ level is expected due to population growth, number of motorized vehicles and industrial activities. Based on data from the Central Bureau of Statistics, the total population of the city of Bandung in 2010 was 2,394,873; while in 2017 there were 2,412,458. In Indonesia, the increase in vehicles for 10 years reached 3 to 4 times. The number of motorcycle vehicles in 2002 was 17,002,130 and in 2012 it became 76,381,183 (increase by more than 400%), while the car vehicles from 3,403,433 to 10,432,259 (increase by more than 300%). In Bandung, the increase in the number of vehicles also occurs with the same tendency as the national level, this is one of the main reasons for the increase of $PM_{2.5}$ in addition to the increase of population and industrial activity in the city of Bandung. The increase of coarse APM in 2015 and 2016 which is reached more than 35% is most likely due to the large number of road infrastructure developments including the fly over in the city of Bandung. Increasing the concentration of particulates needs serious attention, especially for fine particulates which are very dangerous to human health because they are able to penetrate deeply into the lung. In addition, the $PM_{2.5}$ generated from the GENT sampler tends to underestimate because the sample collection point is closer to 2.2 μm and not 2.5 μm [11]. Therefore, the measured $PM_{2.5}$ level should be greater than what is observed. $PM_{2.5}$ concentrations that has exceeded the annual standard of either the Indonesian or WHO standards will pose a major health risk to the surrounding community. For this reason, an appropriate strategy is needed to reduce $PM_{2.5}$ concentrations in Bandung.

Table 1. Annual averages for $PM_{2.5}$, $PM_{2.5\text{-}10}$ and PM_{10} ($\mu g/m^3$) in Bandung, Indonesia

	2015				2016				2017			
	Average	SD	Min	Max	Average	SD	Min	Max	Average	SD	Min	Max
PM2.5	20.90	5.31	8.52	30.01	17.85	6.57	6.21	30.16	18.94	10.01	4.51	55.71
PM2.5-10	26.29	7.84	15.36	43.10	24.42	6.90	11.71	37.72	18.72	7.41	5.40	38.99
PM10	47.19	11.50	23.87	68.17	42.27	10.94	20.74	64.97	37.66	14.39	11.87	80.06

3.2. Black carbon concentrations

Black carbon concentrations in $PM_{2.5}$ for the 2015-2017 period are presented in Figure 2 and the average BC is shown in Table 3. BC is generally emitted into the atmosphere due to incomplete combustion. To avoid inaccuracies in reflectance and absorbance for large particle sizes, the BC measurements are only applied to fine APM. Black carbon is a form of impurity from carbon that results from incomplete combustion of fossil fuels or biomass burning. Black Carbon has a complex effect on climate change, which is warming of the atmosphere as well as cooling the earth's surface. Reducing BC pollution is believed to be a good strategy in reducing or minimizing the effects of global warming.

In Figure 2, BC concentrations are in the range of 1.1 to 5.7 $\mu g/m^3$ with an average of 3.6 $\mu g/m^3$ which contributes 19% to $PM_{2.5}$. The main sources of BC are anthropogenic combustion sources including biomass combustion, motor vehicles and industrial emissions such as coal combustion. Compared to 2002-2004 data, the average BC level has decreased by 15%. This indicates the reduction in BC emission sources originating from both motor vehicles and biomass burning. This is in line with various regulations that have been issued by the local government related to the prohibition of open burning (biomass burning), as well as improving the quality of motor vehicles in order to reduce emissions, one of which is through the implementation of vehicle emission tests. The average BC in Bandung (3.6 $\mu g/m^3$) is lower compared to Kuala Lumpur, Malaysia (4.5 $\mu g/m^3$) [12], but higher than the results found in Mumbai India (3.58 $\mu g/m^3$) and Ulaanbaatar Mongolia (3.29 $\mu g/m^3$) [13].

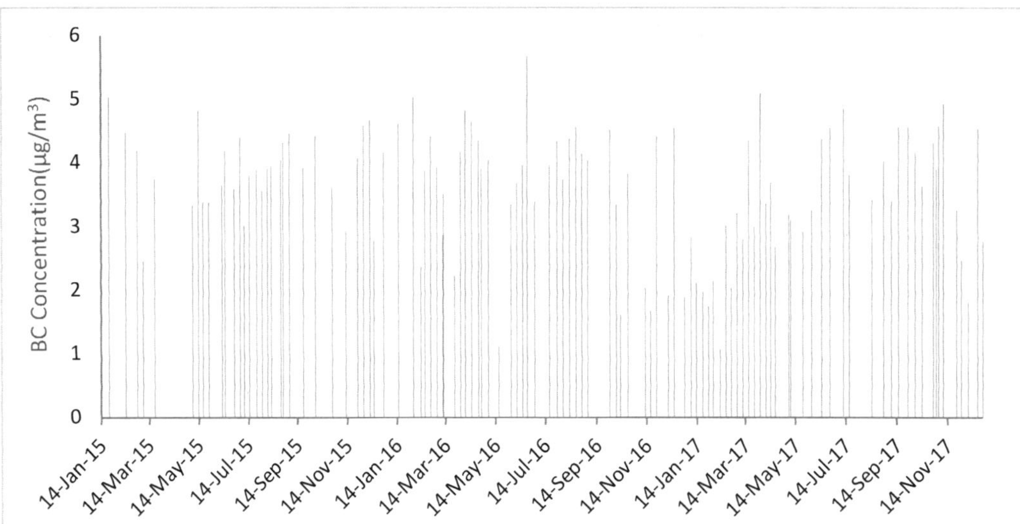

Figure 3. The 24-hour concentration of BC ($\mu g/m^3$) in Bandung, Indonesia

3.3. Multi elemental concentrations

Multi-element concentrations measured by ED-XRF with 9 secondary standards are presented in Table 2. The results obtained indicate that the chemical composition of fine particulates is dominated by S and BC which are anthropogenic sources, while coarse particulates are dominated by crustal elements such as Si, Al, Ca, and Fe. A good correlation and the same trend between these elements shown in Figure 3 and Figure 4, can then be used to estimate the source of these particulate pollutants. A good correlation between S and BC can indicate coming from vehicle sources or industrial activities, while the Al, Si, and Ca correlations originate from soil dust or road dust.

Table 2. The mean and standard deviations of mass and major components of fine and coarse APM (ng/m^3) in Bandung, 2015-2017

	Fine fraction ($PM_{2.5}$)				Coarse fraction ($PM_{2.5-10}$)			
	Average	SD	Min	Max	Average	SD	Min	Max
Mass	19019	7798	4507	55712	22795	7981	5403	43105
BC	3609	962	1069	5681				
Na	262.9	229.0	5.7	933.3	220.9	172.8	12.2	858.3
Mg	26.2	21.4	1.4	119.4	141.0	55.7	20.8	300.7
Al	70.7	39.8	1.7	168.9	606.4	215.9	48.5	1215.2
Si	129.4	49.6	32.0	266.9	1044.6	356.9	198.8	1935.9
S	1226.1	591.8	166.5	2868.4	623.1	255.7	133.0	1326.3
K	197.3	86.0	48.1	571.9	155.2	55.8	32.5	351.3
Ca	97.6	33.8	25.5	230.7	888.3	294.4	221.9	1648.8
Ti	7.7	2.8	2.0	18.6	50.8	17.5	8.4	87.4
V	0.5	0.3	0.01	1.5	1.27	0.74	0.01	3.89
Cr	0.9	0.6	0.01	2.3	3.3	1.6	0.4	10.4
Mn	4.9	5.7	0.4	51.7	17.1	6.2	3.9	42.2
Fe	112.2	54.4	27.8	516.5	735.1	249.0	144.8	1279.9
Ni	1.09	0.55	0.05	2.52	1.28	0.71	0.01	2.94
Cu	2.87	2.40	0.02	14.52	5.41	4.17	0.29	30.55
Zn	34.3	15.9	8.7	97.5	35.7	18.5	7.7	115.8
Pb	19.0	16.0	0.3	82.8	12.7	12.8	0.6	75.1

Compared with 2002-2004 data (Table 3), it can be shown that in 2015-2017, BC concentrations have decreased, but sulfur (S) and potassium (K) concentrations have increased. In general, S is a key element of vehicle and industrial emissions, while K is from biomass burning. The increase in S concentration can be due to the increasing use of coal both for power plants and industries as well as the high concentration of Sulfur in vehicle fuels. Meanwhile, the increase in K is probably due to open burning activities that are still being carried out in Bandung and its surroundings, especially to reduce the volume of household waste. The increase in Zn and Mn elements also needs attention because these elements are indicators of metal industry pollutants. Meanwhile, reducing the Pb concentration can show the success of the unleaded gasoline program implemented in Indonesia since July 2006. Even though unleaded gasoline has been applied, the problem of heavy metal lead pollution in ambient air still occurs as in Tangerang Indonesia [14] and Dhaka Bangladesh [15]. In the coarse fraction APM, Si is a major chemical species followed by other elements namely Ca, Al and Fe. These elements (Si, Ca, Al, Fe) are crustal elements which generally have a significant contribution to the coarse APM. These APM data can then be used to determine multi-element correlation which is very useful for estimating pollutant sources. For example, if there is a good correlation between S against BC, K, Al, and Si, it can be used as an indicator of emissions from brick kiln sources around Bandung. These data can then be used to estimate pollutant sources using receptor model. The results obtained from this study will then be used to determine the type of pollutant source both qualitatively and quantitatively.

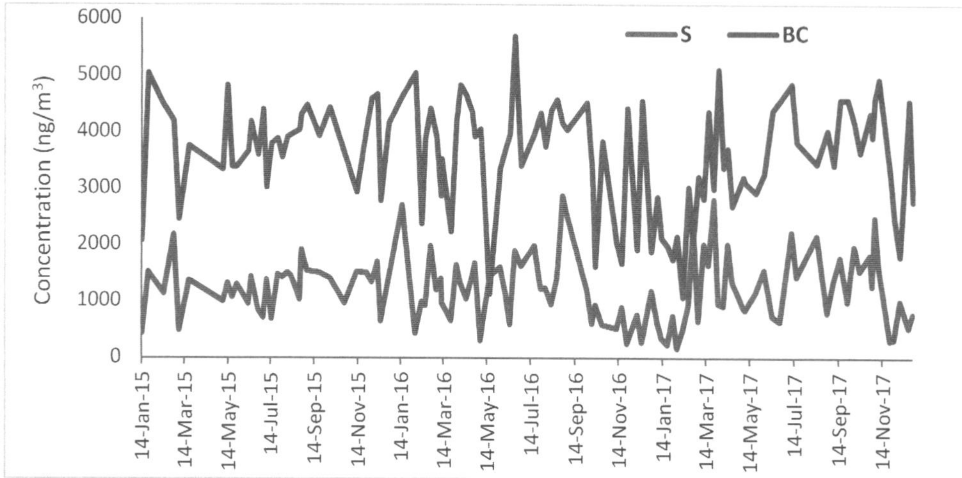

(a)

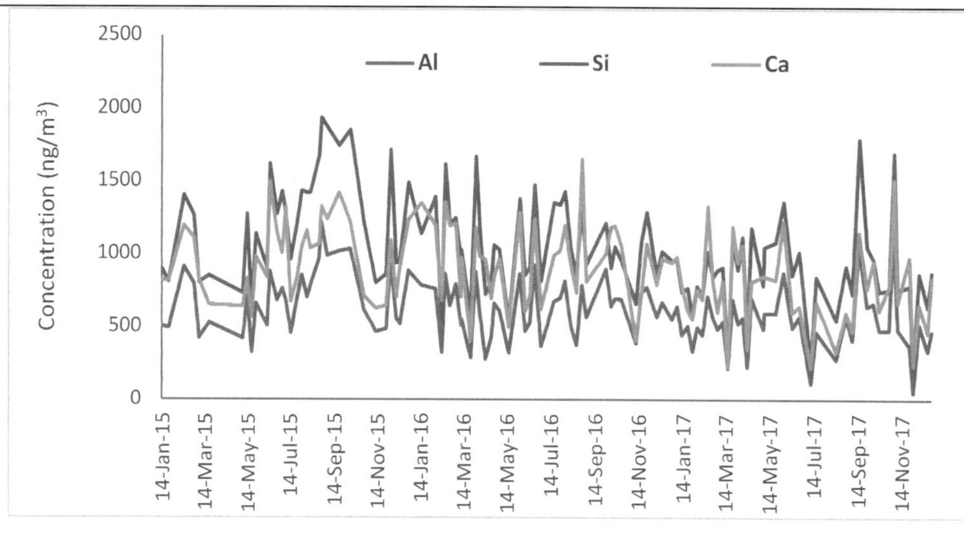

(b)

Figure 4. (a) BC and S concentrations of $PM_{2.5}$ (b) Al, Si and Ca concentrations of $PM_{2.5\text{-}10}$.

Table 3. The mean and standard deviations of mass and major components of fine and coarse particulate (ng/m^3) in Bandung during 2015-2017 and 2002-2004

	2015-2017				**2002-2004[5]**			
	$PM_{2.5}$		$PM_{2.5-10}$		$PM_{2.5}$		$PM_{2.5-10}$	
	Average	SD	Average	SD	Average	SD	Average	SD
Mass	19019	7798	22795	7981	14030	6860	17640	9420
BC	3609	962			4244	1777		
Na	262.87	229.03	220.93	172.76	76.49	48.79	260.36	209.75
Mg	26.22	21.39	140.97	55.71				
Al	70.73	39.85	606.45	215.89	89.69	88.61	727.75	459.82
Si	129.37	49.57	1044.56	356.86	150.03	73.06		
S	1226	592	623	256	1049	684		
K	197.28	86.01	155.22	55.81	123.78	95.37	122.25	54.29
Ca	97.62	33.79	888.35	294.36	52.72	25.14	583.58	240.31
Ti	7.72	2.77	50.84	17.47	9.84	14.60	43.96	21.40
V	0.47	0.33	1.27	0.74	0.89	0.99	1.96	1.57
Cr	0.86	0.57	3.34	1.61	6.92	14.49		
Mn	4.95	5.65	17.13	6.22	2.09	1.89	13.22	9.96
Fe	112.17	54.43	735.13	249.04	168.25	319.45	448.72	457.49
Zn	34.27	15.91	35.69	18.53	15.49	13.73	14.05	14.14
Pb	19.00	15.98	12.65	12.75	59.25	45.39		

4. Conclusions

The annual average of $PM_{2.5}$ in Bandung during 2015-2017 is in the range of 17.85-20.9 $\mu g/m^3$ and has exceeded Indonesian standards (15 $\mu g/m^3$). Compared to 2002-2004 data, $PM_{2.5}$ level have increased by 35%. Although the annual average of PM_{10} is still below the standard, but $PM_{2.5}$ levels that have exceeded the standard and tend to increase from time to time, serious attention and appropriate strategies are needed in order to reduce health risks for the community. The characteristics of chemical composition both in fine and or coarse APM can be used as a preliminary study in identifying the source of pollutants as well as to assess the effectiveness of an air quality program.

Acknowledgement

This study is financially supported by the National Nuclear Energy Agency of Indonesia (BATAN) and International Atomic Energy Agency (IAEA) through the RCA RAS 7029, TC INS 7007, and TC INS 1029 projects. The authors gratefully acknowledge the help of staff at the Center of Applied Nuclear Science and Technology and Environmental Protection Agency of West Java Province for their assistance throughout this study. Muhayatun Santoso, Diah Dwiana Lestiani, Endah Damastuti and Syukria Kurniawati are the main and have same contribution authorship of this article.

References

[1] Thunis P, Clappier A, Tarrason L, Cuvelier C, Monteiro A, Pisoni E, Wesseling J, Belis A, Pirovano G, Janssen S, Guerreiro C, Peduzzi E 2019 *Environ. Int.* **130** 104825
[2] Rönkkö TJ, Jalava PI, MS Happo, Kasurinen S, Sippula O, Leskinen A, Koponen H, Kuuspalo K, Ruusunen J, Väisänen O, Hao L, Ruuskanen A, Orasche J, Fang D, Zhang L, Lehtinen KEJ, Zhao Y, Gu C, Wang Q, Jokiniemi J, Komppula M and Hirvonen MR 2018 *China. Sci. Total Environ.* **639** 1290–310
[3] Janssen NAH, Fischer P, Marra M, Ameling C and Cassee FR 2013 *Sci. Total Environ.* **463-464**

IOP Conf. Series: Materials Science and Engineering 1011 (2021) 012065 doi:10.1088/1757-899X/1011/1/012065

20-6
[4] Santoso M and Lestiani DD 2014 *X ray Fluorescence in Member States Application of ED XRF in supporting national program of air quality*. IAEA Newsletter (Vienna: IAEA)
[5] Santoso M, Hopke PK, Hidayat A and Lestiani DD 2008 *Sci. Total Environ.* **397** 229-37
[6] Kim Oanh NT, Upadhyay N, Zhuang YH, Hao ZP, Murthy DVS, Lestari P, Villarin JT, Chengchua K, Co HX, Dung NT and Lindgren ES 2006 *Atmos. Environ.* **40** 3367–80
[7] Lestari P and Mauliadi 2009 *Atmos. Environ.* **43**(10) 1760–70
[8] Seneviratne MCS, Waduge VA, Hadagiripathira L, Sanjeewani S, Attanayake T, Jayaratne N and Hopke PK 2011 *Atmos. Pollut. Res.* **2** 207–12
[9] Google map 2020 Sampling sites. Retrieved from https://www.google.com/maps/place/Dinas+Lingkungan+Hidup+Jawa+Barat/@-6.9271484,107.601458,13z/data=!4m8!1m2!2m1!1smap+dinas+lingkungan+hidup+Jawa+Barat!3m4!1s0x2e68e62f971b3425:0xf99a7f4a115129e8!8m2!3d-6.9196703!4d107.6109418
[10] Landsberger S, M. (Ed). Creatchman, 1999. *Elemental Analysis of Airborne Particles (1st ed.)*. Amsterdam, Netherland: Gordon and Breach Science Publisher.
[11] Hopke PK, Xie Y, Raunemaa T, Biegalski S, Landsberger S, Maenhaut W, Artaxo P and Cohen DD 1997 *Aerosol Sci. Technol.* **27** 726–35
[12] Rahman SA, Hamzah MS, Elias MS, Salim NAA, Hashim A, Shukor S, Siong WB and Wood AK 2015 *Aerosol Air Qual. Res.* **15** 2291-304
[13] Salako GO, Hopke PK, Cohen DD, Begum BA, Biswas SK, Pandit GG, Chung YS, Rahman SA, Hamzah MS, Davy P, Markwitz, Shagjjamba D, Lodoysamba S, Wimolwattanapun W and Bunprapob S 2012 *Aerosol Air Qual. Res.* **12** 1-7
[14] Santoso M, Lestiani DD, Mukhtar R, Hamonangan E, Syafrul H, Markwitz A and Hopke PK 2011 *Atmos. Pollut. Res.* **2** 190-6
[15] Begum BA, Biswas SK 2008 *J. Bangladesh Acad. Sci.* **32**(2) 209-17

IOP Conf. Series: Materials Science and Engineering 1011 (2021) 012066 doi:10.1088/1757-899X/1011/1/012066

The Screening of Antidiabetic Activity and The Cultivation Study of Local Marine Microalgae

S Priatni[*], D Ratnaningrum and W Kosasih

Research Unit for Clean Technology, Indonesian Institute of Sciences. Jl. Sangkuriang, Bandung 40135, West Java, Indonesia.

*sripriatni@gmail.com

Abstract. Microalgae are rich source of primary and secondary metabolites which have potential bioactive compounds for application in the pharmaceutical industry. The study of antidiabetic activity from microalgae have been carried out because it can be cultivated in a small area and efficient in light capturing. This study aimed to screen the local marine microalgae which potential as the antidiabetic source. Five microalgae strains (*Chlorella* sp A and B, *Nanochlropsis* sp, *Porphyridium* sp, and *Skeletonema* sp were screened according to their growth rates, antidiabetic activity (α-glucosidase inhibition) and biomass yield. Results showed that *Porphyridium* sp cells were grown very active compared to other strains. After 8 days of cultivation, the cells' number of *Porphyridium* sp was $345x10^3$ cells/ml. The highest activity in α-glucosidase inhibition was detected in *Porphyridium* sp biomass (12.63%). The biomass yield of Porphyridium sp (6.71 g) was obtained after cultivated at 5 L scale production. The exponential growth rate of *Porphyridium* sp which cultivated with aeration pump system (Rexp= 0.177, TD= 3.8 days) was higher than cultivated with shaker incubator system (Rexp= 0.045, TD= 15.4 days). We concluded that *Porphyridium* sp is the best microalgae compared to other microalgae both its productivity and antidiabetic activity. The cultivation of *Porphyridium* sp by using aeration pump system is much recommended.

1. Introduction

Diabetes mellitus is a metabolic disease due to the glucose content in the blood exceeds the allowed standard. This is caused by the insufficient production of insulin which causes increase glucose content and induces mismatches in the metabolism of fat and protein [1, 2]. According to the data in 2011, about eight million people in Indonesia were affected by diabetes and it is predicted to increase to 21.3 million sufferers in 2030 [3]. Lifestyle, obesity, food, and other medical conditions implicated in the increase of diabetes prevalence [2]. The disease is characterized also by hyperglycemia that highly contributes to type 2 diabetes development. Therapy for type 2 diabetes by various oral anti-diabetic drugs are available in the market. However, these drugs have proven side effects significantly. Alternative therapy has been explored by using natural products [4].

There are several ways for obtaining the anti-diabetic compounds from natural products, including by extraction of the metabolites from animals, plants, and microbes. Tropical marine microalgae are interesting subjects to study due to its potential as a new source of bioactive compounds. Microalgae is a rich source of primary and secondary metabolites that has potential bioactive compounds for application in the pharmaceutical industry. Microalgae are important sources of proteins and bioactive compounds that benefit pharmaceutical and nutritional purposes [6]. The study of antidiabetic activity from microalgae has been carried out by some researchers. Biomass of *Rhodella* showed was potential in lowering the level of insulin and glucose of rodents. The other study presented the capacity of sulfated polysaccharides from *Porphyridium* was the potential in lowering the glucose level in the blood of diabetic mice [7].

Marine microalgae have the advantage because it can be cultured and cultivated in a small area, the cultivation time is between 7-10 days and very efficient in light capturing [8,9]. The growth of microalgae is influenced by cultivation conditions which have four types: photoautotrophic, heterotrophic, mixotrophic, and photoheterotrophic cultivation [10]. The culture growth of *Porphyridium purpureum* was optimized at specific conditions to increase the phycobiliproteins content, yield, and the quality of the product. The growth process of *P. purpureum* is following the exponential kinetic model which used light and sodium bicarbonate concentration as a variable experimental design [11]. In our previous study, antidiabetic activity screening has been carried out to extracellular and intracellular polysaccharide and the total biomass of marine cyanobacteria. *Pseudanabaena* sp showed the highest activity in α -glucosidase inhibition. This study aimed to screen the local marine microalgae which potential as antidiabetic sources, so that can be applied in the pharmaceutical industry. In this study was used the microalgae strain: *Chlorella* sp, *Nanochlropsis* sp, *Porphyridium* sp, and *Skeletonema*. The culture growth condition of selected microalgae was then evaluated in increasing the biomass production.

2. Experimental

2.1. Materials

Five marine microalgae isolates (*Chlorella* sp A, *Chlorella* sp B, *Nanochlropsis* sp, *Porphyridium* sp, and *Skeletonema*) were obtained from Microbiology Laboratory – Research Center for Oceanography, Indonesian Institute of Sciences. The modification of F2 medium was the processed seawater which enriched with the following additives (per liter): $NaNO_3$ 0.075 g, Na_2HPO_4 0.05 g, Na-EDTA 0.00436 g, $ZnSO_4.7H_2O$ 0.022 g, $Na_2MoO_4.H_2O$ 0.0063 g, $CuSO_4.5H_2O$ 0.0098 g, $MnCl_2.4H_2O$ 0.18 g, $FeCl_3$ 6 H_2O 0.00315 g, $CoSO_4$ 0.00100 g. The medium was sterilized (120°C, 20 min) before use. Cyanocobalamin 1 mg, thiamin HCl 0.2 mg and biotin 1 mg (per liter) was added to the sterile medium.

2.2. Methods

2.2.1. Microalgae cultivation. The cultivation of marine microalgae was carried by a modified method [12], as follows: (i) 1 ml stock culture was cultivated in 5 ml of F2 medium, continued to 50 ml of F2 medium. The cultivation was carried out in an incubator at 25°C for 8 days. (ii) The cultivation of marine microalgae was increased to 500 ml of medium. Microalgae in this experiment were cultivated in bottles that connected to an aeration pump (pump output: 70 L per minute) exposed to 2x10 W white lamp equal to 500-2000 lux. The growth of microalgae cells was monitored and counted the cell number for every 2 days.

2.2.2. Cell number counting. The microalgae cells in the cultures were counted with a cell counter on a light microscope at a magnificent 400x. the cell number (N) was calculated as follow:

$$N = \frac{(\sum N1 + \sum N2)}{2} \times \frac{1}{1\,mm \times 0.2\,mm \times 0.1\,mm} \times \frac{1\,mm^3}{10^{-3}\,mL}$$

2.2.3. Biomass extraction. The biomass of microalgae was extracted by using methanol with a ratio of 1:20 and the extraction was carried out at 60°C, 100 rpm for 2 hours. The methanol extract was then centrifuged at 4°C, 10.000 rpm for 10 minutes and evaporated to have the concentrate of extract.

2.2.4. Intracellular polysaccharides (IPS) extraction. The extraction of IPS was carried out by the modified method [13]. IPS was extracted by homogenizing the biomass in distilled water (50 mL) and heated in a water bath at 95°C for 6 hours. The extracts were filtrated through the Whatman No.2 filter paper to separate the rest of the biomass. The filtrate was then precipitated with 5% ethanol (ratio 1:4), stirred vigorously and keep for overnight at 4°C. The precipitated IPS was recovered by centrifugation at 8.000 rpm for 15 min and the supernatant was discarded.

2.2.5. Determination of inhibition of α-glucosidase activity [1]. The inhibition of α- glucosidase activity was determined by using p-nitrophenyl -D-glucopyranose (pNPG) as a substrate. The sample assay contained 0.25 mL of 0.1 M phosphate buffer (pH 7.0), 0.25 mL of substrate solution (2.5 mM pNPG in 0.1 M phosphate buffer) and 0.1 mL of sample solution in DMSO. The solution was then incubated at 37°C for 5 minutes and added with 0.25 mL of enzyme solution (0.2 U/mL α-glucosidase in 0.01 M phosphate buffer containing 0.2% BSA). The solution was continued the incubation for 15 min at 37°C. The reaction was stopped by dropped 0.1 mL of 0.2 M sodium carbonate. The amount of p-nitrophenol released was measured by the absorbances at 400 nm. The % inhibition of α-glucosidase was calculated as follow:

$$\% \text{ inhibition} = \frac{\text{Abs (Control)} - \text{Abs (Sample)}}{\text{Abs (Control)}} \text{ x } 100$$

2.2.6. The kinetic calculation of microalgae growth. The growth of selected microalgae was evaluated the process kinetic by the modified method [11]. Table 1 presented the equations for kinetic calculation of microalgae.

Table 1. The equations for kinetic calculation

Parameters	Equation
The rate of cell growth at a certain time	$\frac{dN}{dt} = RexpN$
The total cell number at the exponential growth	$N = N_0 \exp(\mathrm{Re}\, xpT)$
The exponential growth rate of microalgae	$\ln \frac{N}{N_0} = Re\ xpt$
The doubling time (TD)	$TD = \frac{0.6931}{\mathrm{Re}\, xp}$
The doubling number (ND)	$ND = \frac{\mathrm{Re}\, xp}{0.6931}$

3. Results and Discussion

The screening of local marine microalgae for antidiabetic have been carried out by monitoring the growth of microalgae culture samples by using F2 medium in the processed seawater. The important parameters in microalgae growth are nutrient, light, pH, aeration, salinity, and temperature [11]. In this study, the cultivation was carried out in a laboratory-scale and monitored every two days. The results analysis of cell counting during microalgae cultivation is presented in Figure 1. The result trends of microalgae growth displayed that *Porphyridium* sp cells were grown very active compared to other

IOP Conf. Series: Materials Science and Engineering 1011 (2021) 012066 doi:10.1088/1757-899X/1011/1/012066

strains. After 8 days of cultivation, cell number data of *Porphyridium* sp was $345x10^3$ cells/ml. The intensity of light is important for algaculture, it will depend on the cell concentrations and the culture depth that the light will penetrate through the culture, 1000 lux is suitable for Erlenmeyer scale [14]. The light absorption and the number of photosynthetic is the key factor in the light reaction [15].

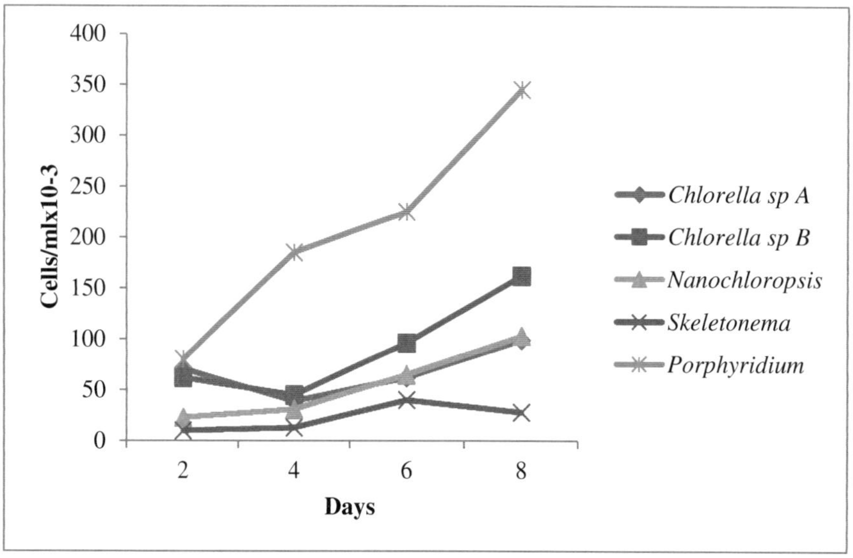

Figure 1. Growth curve of five microalgae strains (Chlorella sp A, Chlorella sp B, Nanochlropsis sp, Porphyridium sp and Skeletonema)

The screening of antidiabetic activity has been carried out to five strains of local marine microalgae. The screening was carried out by extraction of the total biomass with methanol and extraction IPS of microalgae (Table 1). The results demonstrated that all methanol extract samples can inhibit the activity of α-glucosidase and *Porphyridium* sp showed the highest activity (12.63%). However, for IPS samples were not found positive data in α-glucosidase inhibition. Some metabolites from natural resources have been evaluated the inhibition of glucose production and absorption in the intestine or gut. α-glucosidase is located around the surface membranes of intestinal which has a role in carbohydrate digestion. This enzyme catalyzes the cleaved of α-1,4-glucosidic linkages of carbohydrates into simpler sugars such as glucose. Phlorotannins from *Ecklonia cava* have been extracted and analyzed their α-Glucosidase inhibitory effects which showed % of inhibitions were between 10.8 to 49.5% [4]. Pigments from *Spirulina* sp, such as anthocyanin and phycocyanin, have potential as antihyperglycemic agents which can inhibit the increase of glucose level in rat experiments [16]. In our study, the pigments of *Porphyridium* sp (12.63%) and *Chlorella* sp A (12.55%) which contained in methanol extract potentially showed α-glucosidase inhibition. Our previous study reported the growth of *Porphyridium* sp in media containing Fe^{3+} and Co^{2+} promotes the production of phycoerythrins, phycocyanins and allophycocyanins pigments.

Table 2. Antidiabetic activity of an IPS and methanol extract of local marine microalgae

Microalgae	Antidiabetic activity (% of inhibition)	
	IPS	Methanol extract
Control	91.05	91.05
Chlorella sp *A*	-4.73	12.55
Chlorella sp *B*	-4.10	9.38
Nanochlropsis sp	-5.36	7.29
Porphyridium sp	-6.64	12.63
Skeletonema sp	-9.31	11.26

Based on the data in Table 2, the study was continued to evaluate the biomass yield of *Porphyridium* sp and *Chlorella* sp A, both in 1 L scale and 5 L scale for ten days cultivation. Figure 2 showed that at 5 L scale cultivation the biomass yield of *Porphyridium* sp (6.71 g) was higher than *Chlorella* sp A (4.40 g). However, these two strains show a similar yield in 1 L scale cultivation.

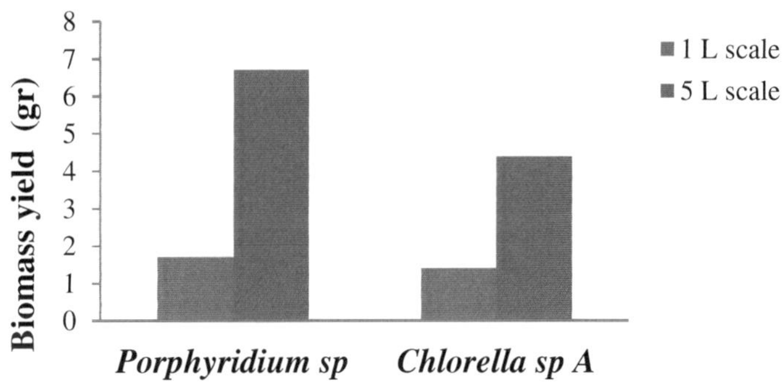

Figure 2. Biomass yield of *Porphyridium* sp and *Chlorella* sp A at 1 L and 5 L scale cultivation

The results showed that *Porphyridium* sp is the best microalgae compared to other microalgae both its productivity and antidiabetic activity. *P. purpureum* produces macromolecular products such as polysaccharides, phycobiliproteins, polyunsaturated and fatty acids. The biomass composition is highly variable and that product formation is regulated in a complex way [15]. Sulfated polysaccharides (sPS) from *Porphyridium* sp significantly decreased the glucose levels in the blood of diabetic mice, causing no modifications in the pancreatic island cells and no fibrosis, or hemorrhagic necrosis in cells, either. The experiments suggested the potential of sulfated polysaccharides for hypolipidaemic and hypoglycaemic agents [7]. According to our study, the antidiabetic activity of intracellular polysaccharide (IPS) of microalgae samples were shown the negative result. We assumed that the polysaccharides were not in the sPS molecule structure. According to these data, *Porphyridium sp* was chosen for further study. To optimize the growth of *Porphyridium* sp, the effect of aeration systems on growth rates was evaluated. The cultivation was carried out with two aeration systems by using the shaker incubator and aeration pump. The rate of cell number growth was monitored every day until day 10 (Figure 3).

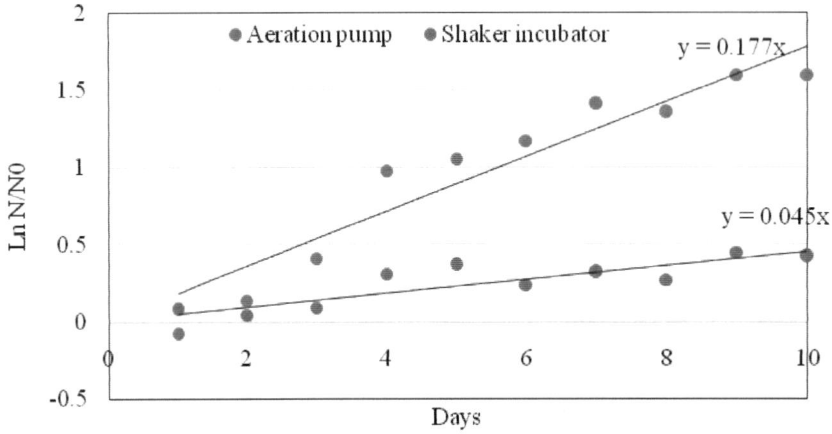

Figure 3. The exponential growth rate of *Porphyridium* sp (Rexp) cultivated by using aeration pump and shaker incubator system

Data on Figure 3 shows the exponential growth rate of *Porphyridium* sp which cultivated with aeration pump system (Rexp= 0.177, TD= 3.8 days) was higher than cultivated with shaker incubator system (Rexp= 0.045, TD= 15.4 days). Cultivation of microalgae with aeration pump system can accelerate the mixing process that is necessary for preventing the sedimentation of the algae. The mixing process will ensure all cells are equally exposed to the light and nutrients, avoid thermal stratification, and support gas exchange between the medium and the air [14]. Aeration is a technique that involves introducing dissolved oxygen into a lake. In some lakes, aeration can help increase dissolved oxygen content, control algae growth, decrease internal recycling of phosphorus and convert ammonium to nitrate [17].

4. Conclusion

The antidiabetic activity from *Chlorella* sp A, *Chlorella* sp B, *Nanochlropsis* sp, *Porphyridium* sp, and *Skeletonema* has been studied and evaluated by cultivated of these microalgae in F2 medium. The results showed that *Porphyridium* sp is the best microalgae compared to other microalgae both its productivity and antidiabetic activity. The cultivation of *Porphyridium* sp by using aeration pump system is much recommended.

Acknowledgments

This research was supported by Research Unit for Clean Technology and Research Centre for Oceanography - Indonesian Institute of Sciences (LIPI).

References

[1] Yang JP, Hsu T, Lin F, Hsu W and Chen Y 2012 *Carbohydr. Polym.* **90** 174-80
[2] Islam MS, Ali S, Rahman M, Islam R, Ali A, Azad AA and Islam MR 2011 *J. Med. Plant Res* **5** 3745–50
[3] Mulyanti S, Musthapa I and Aisyah S 2010 *Jurnal Sain dan Teknologi Kimia* **1** 1–9
[4] Lee SH and Jeon YJ 2013 *Fitoterapia* **86** 129–36
[5] Morada NJ, Metillo EB, Uy MM and Oclarit JM 2011 *International Conference on Asia Agriculture and Animal IPCBEE* **13** 197–200
[6] Priyadarshani I and Rath B. 2012 *J. Algal Biomass Utln.* **3** 89-100

[7] de Jesus Raposo MF, de Morais RMSC and de Morais AMMB 2013 *Mar. Drugs* **11** 233-52
[8] Kabinawa INK 2014 *Jurnal Aplikasi Teknologi Pangan* **3** 103–9
[9] Huang G, Chen F, Wei D, Zhang X and Chen G 2010 *Appl. Energy* **87** 38–46
[10] Chen CY, Yeh KL, Aisyah R, Lee DJ and Chang JS 2011 *Bioresour. Technol.* **102** 71–81
[11] Velea S, Ilie L and Filipescu L 2011 *U.P.B. Sci. Bull. Series B* **73** 4
[12] Priatni S, Budiwati TA, Ratnaningrum D, Kosasih W, Andryani R, Susanti H and Susilaningsih D 2016 *Biodiversitas* **17** 642-6
[13] El-sheekh MM, Khairy HM and El-Shenody R 2012 *Iranian J. Environ. Health Sci. Eng.* **9** 10
[14] Coutteau P 1996 *FAO Fisheries Technical Paper* **7** 41
[15] Csőgör Z, Kiessling B, Perner I, Fleck P, Posten C 2001 *J. Appl. Phycol.* **13** 317–24
[16] Setyaningsih I, Bintang M and Madina N 2015 *Procedia Chem.* **14** 211–15
[17] Priyadarshani I, Thajuddin N and Rath B 2014 *Int. J. Curr. Microbiol. App. Sci.* **3** 173–82

IOP Conf. Series: Materials Science and Engineering 1011 (2021) 012067 doi:10.1088/1757-899X/1011/1/012067

Antibacterial activities of Solo garlic

F Setiyoningrum[1*], N Herlina[1], F Afiati[1] and G Priadi[1]

Research Center for Biotechnology, Indonesian Institute of Sciences, Cibinong Science Center, Jalan Raya Bogor Km 46 16911

*email: fitri.setiyoningrum@gmail.com

Abstract. Solo garlic, either in fresh form or its black garlic has a lot of bioactive compounds, such as flavonoids and polyphenols. The aim of this research was to evaluate the antibacterial activity of organic solvents extract's solo garlic inside the fresh and black garlic form against pathogenic bacteria such as *Escherichia coli* and *Staphylococcus aureus*, by using agar diffusion and dilution methods. The diffusion test shown that BG4 had the highest inhibitor activity comparing the other treatments. Concentration of BG4 1 mg/ml, either in isopropanol or ethyl acetate was able to inhibit the growth of *E. coli* and *S. aureus*. Based on that results, solo black garlic especially 3 or 4 weeks-fermented solo black garlic could be considered as a therapeutic agent.

1. Introduction

Indonesia is well known around the world for its biodiversity. It is also mention as one of the world's richest nations in terms of its *biodiversity. For centuries,* local people have been exploring and utilizing the benefit of this, especially parts of plants as herbal medicine. According to Agarwal and Chauhan [1], herbal medicines are the people's preference because they are cheap and relatively safe. Moreover, WHO reports due to the occurrence of prolonged illness and increased mortality caused by antibacterial resistance affecting people to live with natural products [2]. Secondary metabolites from plants are believed to have antibacterial properties. Some studies mention groups of compounds that are having antibacterial effects, include tannins [3], flavonoids, phenolics [4], alkaloids and saponins [5].

Antibacterial mechanisms are divides into bacteriostatic and bactericidal. Bacteriostatic can inhibit the multiplication of bacterial populations while bactericidal is an antibacterial that can kill bacteria. Bacteriostatics can play a role as bacteriocides in high concentrations [6]. Furthermore, Pelczar and Chan [7] mentioned that the mechanism of antibacterial from compounds in general is done by damaging cell walls, changing membrane permeability, disrupting protein synthesis, and inhibiting the work of enzymes. Antibacterial activity can be tested by two methods, by agar diffusion and dilution. The diffusion method is divided into the disk, wells and trenches. The dilution method is divided into broth dilution and solid dilution. The difference between the methods occurs based on the media used.

In this study, an antibacterial activity test of black garlic was done to *Staphylococcus aureus* and *Escherichia coli.* Solo black garlic is single garlic that is fermented spontaneously at a certain temperature and humidity. Processing with incubation at high temperature and humidity causes garlic to changed its color from white to brown and eventually became black due to the Maillard and Browning reaction process [8,9]. Solo black garlic is known to contain high amounts of flavonoid compounds, antioxidants and polyphenols [10,11].

According to Wang et al. [12], *E. coli* is a pathogenic bacterium in humans which causes digestive disorders and disrupts the working system of the stomach. This bacterium is also a major cause of

IOP Conf. Series: Materials Science and Engineering 1011 (2021) 012067 doi:10.1088/1757-899X/1011/1/012067

morbidity and mortality worldwide. *E. coli* and *S. aureus* are the most common pathogenic bacterial which cause of infection in the world, with varying severity of infections, ranging from minor infections of the skin (furunculosis and impetigo), urinary tract infections, respiratory tract infections, respiratory infections and the Central Nervous System (CNS) [13]. The purpose of this study was to investigate antibacterial activity of solo garlic, either in fresh form or black garlic in organic solvent by diffusion and dilution methods.

2. Materials and Methods

2.1. Materials

Garlic samples used in this study were fresh solo garlic (FG), one week fermented solo garlic (BG1), two weeks fermented solo garlic (BG2), three weeks fermented solo garlic (BG3) and four weeks fermented solo garlic (BG4). Solo garlic was fermented spontaneously on an incubator at 72°C and humidity close to 80%. The tested bacteria used were *S. aureus* ATCC 25923 and *E. coli* ATCC 25922. Chloramphenicol was used as a positive control. The media used are Nutrient Agar (NA) and Nutrient Broth (NB).

2.2. Extraction of active component of solo garlic

Solo garlic samples both FG and BG were extracted by maceration with isopropanol or ethyl acetate. Forty grams was dissolved in 200 ml of solvent and macerated at room temperature for 24 hours. The maceration solution was evaporated by using vacuum until the volume becomes 40 ml. So, the initial concentration of the extract was 1 g/ml.

2.3. Tested Bacteria Preparation

Tested bacteria were rejuvenated with NB media. One inoculant of tested bacteria was subjected in 5 ml of sterile NB and incubated for 24 hours at 37°C. Rejuvenation is repeated up to 3 times. The fresh culture of the tested bacteria viability was determined by the method of Total Plate Count (TPC).

2.4. Determination of Flavonoid

A mixture of 30 µl ethanol, 50 µl aluminum chloride, 50 µl potassium acetate and 600 ml water were added to a 50 µl of sample. The sample mixture was incubated for 30 minutes. Next, sample absorbance is measured on wavelength of 415 nm. Flavonoids identified as quercetin equivalent (QE) is derived by comparing sample absorbance to dilution absorbance of standard quercetin [14].

2.5. Antibacterial activity of black garlic using diffusion method

One inoculant of tested bacteria was dissolved into 10 ml NB and incubated at 37°C for 24 hours. Bacterial concentration was determined by optical density (OD) and TPC measurements. The concentration of bacteria used was 10^6 CFU / m. A hundred µl of bacterial suspension was spread on NA media. Six mm diameter disc paper was prepared and 10 µL fractions with a concentration of 1 mg/ml were dropped. The solvent used to extract solo black garlic was used as negative control, while 1 mg/ml of chloramphenicol (CMP) was served as positive control. Each disc was placed on the surface of the agar medium and incubated at 37°C for 24 hours. Antibacterial activity was determined based on the clear zone produced. Each treatment was carried out three times [15].

2.6. Antibacterial activity of black garlic using dilution method

Solo black garlic samples which showed the highest results in the diffusion method were made in the following dilution series 0.25 mg/ml; 0.5 mg/ml; 0.75 mg/ml and 1 mg/ml, both with isopropanol and ethyl acetate solvents. Antibacterial activity test was done by determining Minimum Inhibitor Concentration (MIC) against *S. aureus* and *E. coli*. The results of the antibacterial activity test of each test solution level were compared with controls. Determination of MIC from each series of levels can be seen from the turbidity of the test tube, or by measuring the absorbance at 500 nm absorption [16].

3. Result and Discussion

In this study, a solo garlic, either in the form of fresh garlic or black garlic is extracted using maceration method. The benefit of this method is its feasibility and trait not to utilize heat to perform the process, as a result its natural elements is less likely to deteriorate and break down. Dilution is preferred based on its dilution and polarity level which favorable to separate bio-active compound in the sample. Provided maceration process runs fairly lengthy and static enabling many compounds can be extracted. During extraction, solvents diffuse into the solid plant material and solubilize compounds with similar polarity [17]. In this study, an organic solvent i.e. isopropanol and ethyl acetate were chosen considering plant extracts in the organic solvent have been known to have higher bactericidal activities than that of wate [18].

Total flavonoids content obtained from different solvents was shown in Table 1. Both fresh garlic and black garlic showed higher number in isopropanol than ethyl acetate solvents. Black garlic fermented in 4 weeks presented 19 times higher than fresh garlic. Heat treatment in black garlic processing has a large influence on flavonoid availability. Ioannou et al. [19] stated that heat treatment possibly increases the total flavonoids contents. This result in accordance with Choi et al. [20] which reported that the total flavonoid content of BG was not only significantly higher than that of fresh garlic, but also increased significantly up to the 21st day of aging, after which the continued to increase slightly for the remainder of the aging period.

Table 1. Flavonoid concentration resulted from solo garlics (fresh and black garlic).

Sample	**The total flavonoid content in solo garlic (mg QE/g)**	
	isopropanol	**ethyl acetate**
FG	20.90	22.84
BG1	53.56	49.85
BG2	64.56	62.12
BG3	375.11	308.49
BG4	408.62	391.12

Flavonoids are semi-polar compounds and can be extracted in semi-polar solvents. Therefore, in this study, isopropanol and ethyl acetate were used to compare between polar and non-polar solvents in garlic activities. The highest number of total flavonoids was obtained in BG4 extracted in isopropanol which contain 408.62 mg QE/g. The result was contrary to Bozin et al. [21] and Jang et al. [22], which stated that total flavonoids content in polar solvents were found better than non-polar. According to Bozin et al. [21], total phenol and flavonoids contents of 80% methanol extract from black garlic higher (0.98 mg GAE/g and 6.99 lg QE/g, respectively) than that from fresh garlic. This result was corresponded with Jang et al. [22] with the findings that water extract from black garlic showed the highest total phenols and flavonoid contents. The significant increase in total phenol and flavonoids in black garlic might be due to the conversion of some components in garlic into these highly hydrophilic compounds. Another study revealed that flavonoid compounds in plants extracted in a less amount in another organic solvent such as methanol and petroleum ether [17]. This evidence may be due to another possibility i.e. dielectric constant in ethyl acetate is lower than that in isopropanol solvent [23].

Table 2. demonstrated antibacterial activities of solo garlics in its fresh and black-garlic form. Generally, it can be seen that the sample resistance-zone was relatively bigger in *S. aureus* than *E. coli*. The result was possibly due to the structure of cell wall of Gram-positive bacteria which is simpler and enables anti-bacterial compounds to penetrate easily in to the cells and finding a suitable place to operate. On the contrary, the structure of cell wall of Gram-negative bacteria is more complex, comprised of three layers of lipoprotein, peptidoglycan and lipopolysaccharide. Likewise, a similar result reported by [24]. A study done by Nguyen et al. [25] mentioned fermented black garlic to have antibacterial activities towards *Listeria monocytogenes* and *Salmonella* sp.

Table 2. Antibacterial activity of fresh solo garlic and its black garlic by diffusion method

Bacteria	Solvent	Sample (1 mg/ml)	Resistance zone (cm)
S. aureus	Ethyl acetate	FG 1	0.50
		BG1	0.61
		BG2	0.85
		BG3	1.79
		BG4	2.07
	Isopropanol	FG1	0.58
		BG1	0.63
		BG2	2.73
		BG3	2.71
		BG4	2.82
E. coli	Ethyl acetate	FG 1	0.74
		BG1	0.75
		BG2	0.86
		BG3	1.52
		BG4	1.89
	Isopropanol	FG1	0.70
		BG1	1.19
		BG2	2.27
		BG3	2.20
		BG4	2.40
Control	Positive: CMP (1 mg/ml)		1.47
	Negative 1: Isopropanol		0.10
	Negative 2: Ethyl acetate		0.15

Among approaches of the sample, BG4 possessed the biggest resistance zone relative to the other samples. This was possibly due to formed flavonoid compounds in BG4 was higher than other samples which was clearly shown in Table 1. A study conducted by Setiyoningrum et al. [10] stated that a longer fermentation period of solo garlic would result in a higher amount of flavonoid compounds. Taken together, antibacterial compounds produced through maceration using isopropanol create relatively wider resistance zone compared to ethyl acetate, both in *S. aureus* and *E. coli*. This may be evident because flavonoid or other anti-bacterial compounds are extracted more in isopropanol than in ethyl acetate. Average resistance zone of BG3 in all solvents was higher than antibiotic chloramphenicol used as positive control. Hence, these finding indicate that fresh or black form of solo garlics could be potentially used as an antibacterial agent.

MIC is defined as the highest dilution or least concentration of extracts that inhibits organism growth. MIC determination is essential in a lab diagnostic since it is pivotal in assisting confirmation process of microorganism-resistance towards anti-microbe genes and monitoring the activities of new anti-microbe agents [26]. The MIC test was performed to BG4 sample which has the largest resistance zone during diffusion method. The MIC BG4 value in isopropanol or ethyl acetate solvent is provided in Table 3. Isopropanol extract of BG4 showed the MIC value of 0.75 mg/ml on *S. aureus* while ethyl acetate extract showed a value of 1 mg/ml. Preethi et al. [27] reported that alcohol extract possesses a trait as a better anti-bacterial solvent than other organic solvents. Isopropanol extract and ethyl acetate extract have the same MIC level with *E. coli* i.e. 1 mg/ml. This was possible because *E. coli* cell wall is more complex than *S. aureus* and the antibacterial mechanism of plant extract involves the simultaneous disruption of the cell membrane [28].

Table 3. Minimum Inhibitory Concentration (MIC) of BG4

Bacteria	Solvent	Concentration of extract (mg/ml)	Turbidity*
S. aureus	Ethyl acetate	1	-
		0.75	+
		0.5	+
		0.25	+
	Isopropanol	1	-
		0.75	-
		0.5	+
		0.25	+
E. coli	Ethyl acetate	1	-
		0.75	+
		0.5	+
		0.25	+
	Isopropanol	1	-
		0.75	+
		0.5	+
		0.25	+

4. Conclusion

The BG4 had the biggest resistance zone relative to the other samples, in isopropanol extract or ethyl acetate. Both of extract solvents of BG4 had the MIC level against *E. coli* i.e. 1 g/ml. Otherwise, isopropanol extract of BG4 showed lower MIC against *S. aureus* than ethyl acetate.

Acknowledgments

This research was supported by Ministry of Research, Technology and Higher Education of the Republic of Indonesia through INSINAS 2018 programme.

References

[1] Agarwal M and Chauhan S 2015 *Int. J. Advan. Res. Bot.* **1** 1-9
[2] WHO 2014 https://www.who.int/gho/publications/world_health_statistics/2014/en/.
[3] Zhang X, Li N, Lu X, Liu P and Qiao X 2016 *J. Sci. Food Agric.* **96** 2366-2372
[4] Parvin MS, Das N, Jahan N, Akhter MA, Nahar L and Ekramul Islam M 2012 *BMC Res. Notes* **8** 412.
[5] Ejelonu BC, Lasisi AA, Olaremu AG and Ejelonu OJ 2011 *African Journal of Biotechnology*. **10** 19631-6
[6] Schunack W, Mayer K and Haake M 1990 *Senyawa Obat, diterjemahkan oleh bagian farmakologi FK UNAIR Edisi II* (Yogyakarta: Gadjah Mada University Press) p 187
[7] Pelczar MJ and Chan ESC 2008 Dasar-*dasar Mikrobiologi Jilid 1* (Jakarta: UI Press)
[8] Kimura S, Tung YC, Pan MH, Sun NW, Lai YJ and Cheng KC 2017 *J. Food Drug Anal.* **25** 62-70.
[9] Ngan N, Giang M and Tu N 2017 *Indian J. Pharm. Edu. Res.* **51** 672-8
[10] Setiyoningrum F, Priadi G, Afiati F, Herlina N, Solikhin A and Lisani N 2018 *Asian J. Agric.* **2** 48–51

[11] Afiati F, Setiyoningrum F, Priadi G and Handoyo CH 2020 *Conf Series: Earth and Environmental Science* **439** 012053
[12] Wang B, Bachaiyappan B, Bruber J, Schmidt MG, Zhang Y and Woster PM 2016 *J. Med. Chem.* **59** 3140–51
[13] DeLeo FR, Otto M, Kreiswirth BN and Chambers NF 2010 *Laboratory of Human Bacterial Pathogenesis. Rocky Mountain Laboratories. -National Institute of Allergy and Infectious Diseases* 1.1
[14] Mathur RR and Vijayvergia R 2017 *Int. J. Pharm. Sci. Res.* **8** 5282-5
[15] Natta L, Orapin K, Krittika N and Pantip B 2008 Int. *Food Res.* **15** 337–46
[16] He F, Li K, Zhang X, Yang Y, Fang Y and Xiang F 2019 *LWT-Food Sci. Technol.* **101** 389-94
[17] Sruthi DR and Indira G 2016 *J. Phar Phyto.* **5** 386-9
[18] Cowan M 1999 *Clin. Microbiol. Rev.* **12** 564-58
[19] Ioannou, I, Hafsa I, Hamdi S, Charbonnel C and Ghoul M 2012 *J. Food Eng.* **111** 208–17
[20] Choi IS, Cha HS and Lee YS 2014 *Molecules* **19** 16811-23
[21] Bozin B, Mimica-Dukic N, Samojlik I, Goran A and Igic R 2008 *Food Chem.* **111** 925–9
[22] Jang H, Lee H, Yoon D, Ji D, Kim J and Lee C 2018 *Food Sci. Biotechnol.* **27** 219–25
[23] Adhikari P, Pandey A, Agnihotri V and Pande V 2018 *Research in Pharmacy* **8** 1-9
[24] Kusmiyati and Agustini NWS 2007 Biodiversitas 8 48-53
[25] Nguyen N, Giang M and Nguyen T 2017 *Indian J. Pharma. Edu. Res.* **51** 672 – 8
[26] Sen A and Batra A 2012 *Intl. J. Curt. Pharm. Res.* **4** 67 – 73
[27] Preethi R, Devanathan VV and Loganathan M 2010 *Adv. Bio. Res.* **4** 122-5
[28] He F, Wang W, Wu M, Fang Y, Wang, S Yang Y, Ye C and Yiang F 2020 *Ind. Crop. Prod.* **153** 112552

IOP Conf. Series: Materials Science and Engineering 1011 (2021) 012068 doi:10.1088/1757-899X/1011/1/012068

Toxicity and Total Phenolic Content of *Saurauia vulcani* Extracts from Cultivation

Gunawan Pasaribu[1*], Emil Budianto[2], Antonius Herry Cahyana[2]

[1]Postgraduated student at Department of Chemistry, Faculty of Mathematics and Natural Sciences (FMIPA), Universitas Indonesia, Depok 16424, Indonesia
[2]Department of Chemistry, Faculty of Mathematics and Natural Sciences (FMIPA), Universitas Indonesia, Depok 16424, Indonesia

Corresponding author: *gunawan.tri91@ui.ac.id; gun_pa1000@yahoo.com

Abstract. Saurauia vulcanii plant has been widely used to treat a variety of diseases suffered by villagers as antidiabetic and digestive problems' remedy. The Forestry Research and Development Agency has cultivated this plant at Forest Area for Specific Purpose in Northern Sumatra. The aims of this study were to determine the toxicity and total phenolic content from n-hexane, ethyl acetate and methanol Saurauia vulcani extracts. The toxicity was carried out using the Brine Shrimp Lethality Test (BSLT) method and Folin Ciocalteu method is used for total phenolic content. The result showed that the toxicity of n-hexane, ethyl acetate and methanol extracts i.e. LC50 values of 365.19 ppm, 715.28 and 225.77 ppm, respectively. The total phenolic content of hexane, ethyl acetate and methanol extracts were 7,155 mg GAE/g, 13,702 mg GAE/g and 16,560 mg GAE/g, respectively.

1. Introduction

Plants of the genus Saurauia are widely distributed in Indonesia. *Saurauia vulcani* is one of species that is quite widely distributed. This species grows widely in the Lake Toba's Catchment Area, located in North Sumatra, Indonesia. The spread of this species is found in several areas in North Sumatra such as in Toba, North Tapanuli, Simalungun and Tana Karo districts. This simplicia has been widely trafficked on the roads across Sumatra with claims to be anti-diabetic herbs. The Tapanuli villagers said this plant as *pirdot* which is believed able to treat various diseases related to the digestive system.

Saurauia vulcani leave is one of woody plant species have been used to cure diabetes at Karo's local wisdom. Situmorang *et al.* said that the in practice of diabetes treatment is done by boiling 5-8 dry leaves, then boiling with a liter of water to leave half of the water. This herb is consumed 2-3 times a day.[1] Traditionally, the people of the Batak Simalungun sub-ethnic group, North Sumatra Province, have used parts of *Saurauia vulcanii* leaves for treatment such as diarrhea, gastrointestinal disorders, and wounds.[2]

Research conducted by Sitorus showed that the characteristics of the ethanol extract of *Saurauia vulcani* showed 6.65% and 7.25% in water content. Meanwhile the water soluble extract values were 23.55% and 64.25%.. [3]

Research on antioxidants has been carried out on the aqueous extract of the *Saurauia vulcani* plant, where the IC50 value is 22.9182 ± 1.32 ppm while the IC50 quercetin is 4.96 ± 0.02 ppm. The results of this study indicate that *Saurauia vulcani* has a strong antioxidant. This is based on the strong category of antioxidants with a concentration value of 50-100 μg / mL.[4]

There are several chemical compounds that have been isolated from the genus Saurauia, namely 3β-hydroxy-Olean-12-en-28-oat acid; 3,19-dihydroxyurs-12-en-28-oic acid; some monoterpene lactones; triterpenoids, fatty acids; and two steroids, stigmasterol; and β-sitosterol [5][6][7]. There are several biological activities afforded by *Saurauia vulcani* i.e. antioxidant activity, anti-cholesterol, antidiabetic activity, antihyperlipidemic, analgetic activity, antimicrobial activity, and wound-healing activity. [8][9][10][11]

Research on Saurauia plants usually uses wild plants as research objects. In this study, the material plant used is the result of cultivation at the age of 4 years. This research will investigate how the differences in secondary metabolites produced from wild plants and cultivated plants.

The brine shrimp lethality test value has corelated with cytotoxicity activity in plant extracts. Simultaneous bioassay testing can be used simultaneously to test various natural marine products for pharmacological activity.[12]. Furthermore, the thotal phenolic content can predict the antioxidant activity of the extract. Polyphenols have properties in preventing oxidative stress associated with cancer. Further, polyphenols also have other biological activities in prevention and treatment.[13]

The aims of this study were to determine the toxicity and total phenolic content from n-hexane, ethyl acetate and methanol *Saurauia vulcani* extracts.

2. Experimental details

2.1. Plant material

Saurauia vulcani was collected from Sipiso-Piso, Tana Karo Regency, at Forest Area for Specific Purpose in Northern Sumatra. The collected plant material was air-dried and powdered.

2.2. Extraction

Dried ground leaves (1 kg) of Pirdot plant (*Saurauia* vulcani) leaf and flowers were extracted with multilevel extraction method by used three solvents i.e. n-hexane, ethyl acetate and methanol. The filtrated obtained is concentrated using a rotary vacuum evaporator.

2.3. Brime Shrimp Letality Test

The fractions of n-hexane, ethyl acetate and metanol were evaluated in a test for lethality to brine shrimp larvae. Toxicities of compounds were tested at 1, 10, 100, and 1000 ppm in 10 mL sea-water solutions with 1% DMSO (*v/v*). In this test ten larvae were used in each test and counted how many were alive after 24 hours. At each concentration 3 replications were set. Distilled water is used as a control blank. The lethality concentration (LC50) was calculated using probit analysis. The LC50 value is calculated using a regression line obtained by plotting the concentration against the percentage of mortality on a probit scale.[14]

2.4. Total phenolic contents in Saurauis vulcanii extracts

The spectrophotometric method can be used to measure phenol content in plant extracts.[15]. Methanolic solution of the extract in the concentration of 1 mg/ml was used in the analysis. The reaction mixture was prepared by mixing 0.5 ml of solution of extract, 2.5 ml of 10% Folin-Ciocalteu's reagent dissolved in water and 2.5 ml 7.5% $NaHCO_3$. Blank was concomitantly prepared, containing 0.5 ml methanol, 2.5 ml 10% Folin-Ciocalteu's reagent dissolved in water and 2.5 ml of 7.5% of $NaHCO_3$. The samples were there after incubated in a thermostat at 45 oC for 45 min. The absorbance was determined using spectrophotometer at λmax = 765 nm. The samples were prepared in triplicate for each analysis and the mean value of absorbance was obtained. The same procedure is performed on the standard solution of gallic acid and the calibration line. The measured absorbance value, the phenolic concentration can be read from the calibration line. The phenolic content in the extract is expressed as mg of gallic acid equivalent (mg GA /g of extract).

3. Result and Discussion

3.1. Toxicity Activity

Brine Shrimp Lethality Assay is a fairly practical method for initial screening to determine the toxicity of plant extracts. The advantages of this method are fast and simple. However, it is necessary to control the research conditions such as temperature, pH, salinity, aeration and lighting. The concentrations used in this study were in the range of 10, 100, and 1000 μg / ml. In general, the determination of toxicity is determined after 24 hours.[16]

The results of the toxicity test on n-hexane, ethyl acetate and methanol extracts i.e. LC_{50} values of 365.19 ppm, 715.28 and 225.77 ppm, respectively. Toxicity value is clasified as non toxic if the LC_{50} more than 1000 ppm. The low toxic is classified if the LC_{50} from 500 to1000 ppm. The medium toxic is classified if the LC_{50} from 100 to 500 ppm. The high toxic is classified if the LC50 from 0 to 100 ppm.[17]

Hamidi et al., reported the LC_{50} values from ethanol extract of suren seeds, mahogany, mimba, and saga were 75, 84, 323 and 449 ppm, respectively. The four extracts are classified from medium toxic to high toxic.[18]. Based on Albuntana et.al., stated that Bohadcshia argus extract has the active species indicated by LC_{50} value 69,254 ppm from their research. Furthermore, water fraction of Holothuria leucospilota's crude extract is the most active fraction indicated by LC_{50} at 50,968 ppm.[19]

The toxicity activity of some plant in Meliaceae famil have been reported i.e Azedirachta indica, Azedirachta indica var. siamensis, Melia azedarach, Sandoricum indicum and Swietenia macrophylla species. The toxicity of stem bark extracts showed the significant toxicity from 8.63 to 234.06 ppm, whereas part of leaves were inactive. Researcher reported that the leaves and seed part have the higher potency than the stem bark part. [14]

The LC_{50} of Annona reticulata with Allium fistolisum and Brassica oleraceae alkoholic extract were 24.162, 13.433 and 10.818 ppm, respectively. The toxicity of those plant is classified as high toxic and potential as anticancer agent.[20]

Based on Puspitasari that the leaves extract Avicennia marina, Rhizopora mucronata and Sonneratia alba have the LC_{50} i.e 403.44 ppm, 709.7 and 801.75 ppm, respectively. The Avicennia marina leaf extract had a potential to be the strongest toxic among other extracts.[21]

Innocent et al., Have reported the results of a toxicity study on 30 medicinal plants in the Bukoba district, northwestern Tanzania. The solvents used are dichloromethane and/or ethanol. The toxicity of the ethanol extract of Lantana trifolia (LC_{50} 32.3 ppm), Vernonia bradycalyx (LC_{50} 33.9 ppm), Antiaris toxicaria (LC_{50} 38.2 ppm) and Rubus rigidus (LC_{50} 41.7 ppm) and dichloromethane extract from Gynura skandens (LC_{50} 36.5 ppm) and Bridelia micrantha (LC_{50} 32.0 ppm) and categorized as mild toxic. Meanwhile, the dichloromethane extracts of Picralima nitide (LC_{50} 18.3 ppm) and Rubus rigidus (LC_{50} 19.8 ppm) were classified as quite toxic. Only 1.1 and 1.2 Picralima nitida and Rubus rigidus extracts were classified as less toxic when compared to the standard drug, cyclophosphamide (LC_{50} 16.3 ppm) [22].

3.2. Total Phenolic Content of Saurauia vulcani

The total phenolic content of *Saurauia vulcani* are 7.155 mg GAE/g, 13.702 mg GAE/g and 16.560 mg GAE/g of dry weight of extract, respectively for hexane, ethyl acetate, and methanol extract. This value indicates that in every gram of extracts of n-hexane, ethyl acetate and methanol, the equivalent of 7.155, 13.702 and 16.560 mg of gallic acid.

Lovena *et.al.,* reports that the antioxidant activity of *Saurauia vulcani* from wild plant is strong activity [4]. The value of total phenolic content can be used as approach to predict the antioxidant activity. The relationship between total phenolic content of noni leaf extract and antioxidant activity has a coefficient of determination of 0.9820 [23]. Trisharyanti *et.al.* said that phenolic content and antioxidant activity of ethanol extract Cashew leaves showed a positive correlation of 58.8% [24].

When compared with the results of total phenolic content in wild and cultivated *Saurauia vulanii,* it shows different results. Total phenolic content of plant from cultivation is tendency smaller than from wild. This is probably influenced by the age of the plant which is still quite young (4 years). In general, the Saurauia vulcanii plant taken from nature is a tree that is mature and bigger. Total phenols, flavonoids, saponins, alkaloids, triterpenes and steroids increased with age of the *Neonotonia wightii* tree. [25]

Almost the same results were reported by Hossain and Shah with this study. They conducted research on total phenolic content in the *Merremia borneensis* plant. The total phenol content of hexane, ethyl acetate, chloroform, butanol and ethanol extracts were 5.90, 6.68, 20.25, 6.84 and 30.35 GAE/g, respectively [26].

Different study results are reported by Herawati on mangrove plants used hexane, chloroform, ethyl acetate and methanol as solvents. The results showed that the total phenolic content of the three fractions (hexane, chloroform, ethyl acetate) and the methanol extract of *Sonneratia alba's* bark were 17.84 ± 0.974, 156.310 ± 0.703, 226.89 ± 0.605, 249.56 ± 0942 respectively. GAE µg/g [27]. A corresponding study was carried out by Sembiring et al that reported variations in the total phenolic content of surian (*Toona sinensis*) leaves. The results showed that the total phenolic content varied from 276.62 to 444.68 (mg GAE/g). [28]. While the four crude extracts of *Caesalpinia bonduc* (L.) Roxb, the leaves contained the highest amount (31.05 ± 0.35 mgQE / g).[29]

Research on total phenolic content is not carried out only on medicinal plants. Total phenolic content is also carried out in vegetable and fruit crops. The vegetables and fruit that were tested were the result of planting in the Tower Garden aeroponic system and in the ground. The results showed that the vegetables that had the highest phenol content were chaletard 57.73 mgGAE/g in aeroponic growing system and 53.45 GAE/g in the field followed by basil, red cabbage, and parsley. This study showed that vegetables grown in soil had a slightly higher phenolic content in basil, turnips, and parsley than those grown in an aeroponic system. Meanwhile, the total phenolic content, red kale is slightly higher in aeroponic [30].

Furthermore, Aryal et. al reported the results of a study on phenolic content in eight selected wild vegetables from Nepal (*Alternanthera sessilis, Basella alba, Cassia tora, Digera muricata, Ipomoea aquatica, Leucas cephalotes, Portulaca oleracea* and *Solanum nigrum*). This research uses methanol as a solvent in its extraction. The phenolic compound content ranges from 292.65 to 72.66 mg GAE/g. Vegetables *Alternanthera sessilis, Cassia tora* and *Portulaca oleracea* had the largest phenolic content (292.65 ± 0.42, 287.73 ± 0.16 and 216.96 ± 0.87 mg GAE/g, respectively). Meanwhile, the smallest phenolic content was found in *Basella alba, Ipomoea aquatica,* and *Solanum nigrum* (72.66 ± 0.46, 77.06 ± 0.70 and 97.96 ± 0.62 mg GAE /g respectively).

Stankovi reported there was a significant correlation between the total phenolic content and the antioxidant activity of the plant extracts studied. Researcher reported that five different extracts of the entire herb *Marrubium peregrinum* L. (Lamiaceae). The results of the spectrometric study showed that the total phenolic content of the extract ranged from 27.26 to 89.78 GAE/g. Researchers also proved that the higher the phenolic compound content, the higher the antioxidant activity. [22]

There are several things that affect the total phenol content. The total phenol content was studied in terms of susceptibility to degradation of aqueous solutions in plant extracts, the effect of light, during storage at room temperature for fortification of food products. The total phenol content in the model solution (olive leaf, green tea, red wine, red wine, 5: 1 PE pine bark, 95% PE pine bark, resveratrol), ranged from 11.10 mg GAE/100 mL to 92.19 mg GAE/100 mL [31].

Noreen et al., Reported variations in the total phenolic content in different solvents. The highest total phenolic content was found in the ethanol extract of *Coronopus didymus* (47.8 mM GAE) and the lowest in the dichloromethane extract (3.13 mM GAE). Furthermore, using a calibration curve, ethanol extract showed the highest phenolic content (47.8 mM GAE) followed by acetone (38.6 mM GAE) and n-hexane (27.2 mM GAE) while chloroform and dichloromethane extracts showed the lowest. The total phenolic content variation reaches 15 times [32]. The polarity of solvent used in extraction can affect the total phenolic contents in plant extracts. In polar solvents have a high solubility of phenols will provide a high concentration.

4. Summary

The results of the toxicity study showed that the hexane and methanol *Saurauia vulcani* extracts were classified as moderate toxic, while the ethyl acetate extract was classified as low toxic. Meanwhile, the total phenolic content of hexane, ethyl acetate and methanol extracts were 7,155 mg GAE/g, 13,702 mg GAE/g and 16,560 mg GAE/g, respectively.

Acknowledments

The authors gratefully acknowledge financial support from PUTI Dr (Q3) Universitas Indonesia.

References

[1] R. O. P. Situmorang, A. H. Harianja, J. Silalahi, and P. City, "Karo' S Local Wisdom : The Use Of Woody Plants for Traditional Diabetic Medicines," *Indones. Journa; For. Res.*, vol. 2, no. 2, pp. 121–130, 2015.

[2] M. Silalahi, J. Supriatna, and E. K. O. B. Walujo, "Local knowledge of medicinal plants in sub-ethnic Batak Simalungun of North Sumatra , Indonesia," *BIODIVERSITAS*, vol. 16, no. 1, pp. 44–54, 2015.

[3] P. Sitorus, "Characterization Simplisia and Ethanolic Extract of Pirdot (Saurauia Vulcani , Korth) Leaves and Study of Antidiabetic Effect in Alloxan Induced Diabetic Mice," *Int. J. ChemTech Res.*, vol. 8, no. 6, pp. 789–794, 2015.

[4] T. N. Lovena, U. Harahap, P. Sitorus, and D. Satria, "Antioxidant activity and α-glucosidase inhibition effect of water extract Saurauia vulcani Korth leaves," *J. Innov. Pharm. Biol. Sci.*, vol. 5, no. 3, pp. 47–51, 2018.

[5] W. J. A. Musa, B. Situmeang, and J. Sianturi, "Anti-cholesterol Triterpenoid Acids from Saurauia vulcani Korth. (Actinidiaceae)," *Int. J. Food Prop.*, vol. 22, no. 1, pp. 1439–1444, 2019.

[6] Y. Ahmed, S. Rahman, P. Akhtar, F. Islam, M. Rahman, and Z. Yaakob, "Isolation of Steroids from n-hexane Extract of the Leaves of Saurauia roxburghii," *Int. Food Res. J.*, vol. 20, no. 5, pp. 2939–2943, 2013.

[7] B. Situmeang, A. R. Suparman, M. Kadarusman, S. Parumbak, and T. Herlina, "Isolasi Senyawa Triterpenoid dari Ekstrak Etil Asetat Pirdot (Saurauia vulcani . Kurth) Isolation of Triterpenoid Compound from Ethyl Acetate Extract of Pirdot," *J. Kim. Val.*, vol. 4, no. 2, pp. 93–97, 2018.

[8] L. Zhu, S. Zhu, and Y. Tian, "Antioxidant and Antimicrobial Activities of Essential Oil and Extracts of Saurauia lantsangensis Hu Root," *A J. Biosci.*, vol. 67, no. 5, pp. 282–290, 2012.

[9] S. Hutahaean, M. Tanjung, D. P. Sari, and V. E. Ningsih, "Antihyperglycemic and antihyperlipidemic effects of pirdot (saurauia vulcani korth .) leaves extract in mice Antihyperglycemic and antihyperlipidemic effects of pirdot (saurauia vulcani korth .) leaves extract in mice," in *Earth and Environmental Science*, 2018, pp. 1–8.

[10] E. Salim, C. Fatimah, and D. Y. Fanny, "ANALGETIC ACTIVITY OF CEP-CEPAN (Saurauia cauliflora DC .) LEAVES EXTRACT," *J. Nat.*, vol. 17, no. 1, pp. 31–38, 2017.

[11] G. Anastasia, P. Sitorus, and D. Satria, "Wound Healing Activity of Saurauia vulcani Korth Aqueous Leaves Extract Evaluation on Excision Wound in Hyperglycemia Rats," *J. Innov. Pharm. Biol. Sci.*, vol. 5, no. 3, pp. 52–57, 2018.
[12] J. L. Carballo, Z. L. Hernández-Inda, P. Pérez, and M. D. García-Grávalos, "A comparison between two brine shrimp assays to detect in vitro cytotoxicity in marine natural products," *BMC Biotechnol.*, vol. 2, no. 17, pp. 1–5, 2014.
[13] J. Dai and R. J. Mumper, "Plant phenolics: extraction, analysis and their antioxidant and anticancer properties," *Molecules*, vol. 15, no. 10, pp. 7313–7352, 2010.
[14] S. Pisutthanan, P. Plianbangchang, and N. Pisutthanan, "Brine Shrimp Lethality Activity of Thai Medicinal Plants in the Family Meliaceae," *Naresuan Univ. J.*, vol. 12, no. 2, pp. 13–18, 2004.
[15] V. L. Singleton, R. Orthofer, and R. M. Lamuela-Raventos, "Analysis of total phenols and other oxidation substrates and antioxidants by means of Folin-Ciocalteu reagent.," *Method Enzym.*, vol. 299, no. 1999, pp. 152–178, 1999.
[16] M. R. Hamidi, B. Jovanova, and T. K. Panovska, "Toxic ological evaluation of the plant products using Brine Shrimp (Artemia salina L .) model," *Maced. Pharm. Bull.*, vol. 60, no. 1, pp. 9–18, 2014.
[17] M. Hamidi, B. Jovanova, and T. Panovska, "Toxicological evaluation of the plant products using Brine Shrimp (Artemia salina L .) model.," *Maced. Pharm. Bull.*, vol. 60, no. 1, pp. 9–18, 2014.
[18] Zuraida, "Toxicity Analysis of Forestry Plants Using Brine Shrimp Lethality Test (BSLT) Method," *J. Penelit. Has. Hutan*, vol. 36, no. 3, pp. 239–246, 2018.
[19] A. Albuntana, Yasman, and W. Wardhana, "TOXICITY TEST OF EXTRACTS OF THE FOUR SEA CUCUMBER (FAMILY HOLOTHURIIDAE) FROM EAST PENJALIRAN ISLAND, SERIBU ISLANDS, JAKARTA BASED ON THE BRINE SHRIMP LETHALITY TEST (BSLT)," *J. Ilmu dan Teknol. Kelaut. Trop.*, vol. 3, no. 1, pp. 65–72, 2011.
[20] S. Waghulde, M. K. Kale, and V. Patil, "Brine Shrimp Lethality Assay of the Aqueous and Ethanolic Extracts of the Selected Species of Medicinal Plants," in *International Electronic Conference on Synthetic Organic Chemistry*, 2019, vol. 2, pp. 1–12.
[21] E. Puspitasari, U. Sriwijaya, and R. Rozirwan, "Uji Toksisitas dengan Menggunakan Metode Brine Shrimp Lethality Test (Bslt) Pada Ekstrak Mangrove (Avicennia Marina , Rhizophora Mucronata , Sonneratia Alba dan Xylocarpus Granatum)yang Berasal dari Banyuasin, Sumatera Selatan," *J. Biol. Trop.*, vol. 18, no. 1, pp. 91–103, 2018.
[22] M. S. Stankovi, "Total phenolic content, flavonoid concentration and antioxidant activity of Marrubium peregrinum L . Extracts," *Kragujev. J.Sci*, vol. 33, no. 2011, pp. 63–72, 2011.
[23] A. Rohman, S. Riyanto, and N. K. Hidayati, "Antioxidant Activity, Total Phenolics and Total Flavonoid Contents of Mengkudu (Morinda citrifolia L) Leaves," *Agritech*, vol. 27, no. 4, pp. 147–151, 2007.
[24] I. Trisharyanti, D. Kusumowati, R. Melannisa, and K. Ratri, "Korelasi kandungan fenolik dan aktivitas antioksidan daun jambu mete," *Biomedika*, vol. 3, no. 2, pp. 25–30, 2011.
[25] D. M. Verdecia *et al.*, "Effect of re-growth age in the content of secondary metabolites from Effect of re-growth age in the content of secondary metabolites from Neonotonia wightii in the Valle del Cauto , Cuba," *Cuba. J. Agric. Sci.*, vol. 48, no. 2, 2014.

[26] M. A. Hossain and M. D. Shah, "A study on the total phenols content and antioxidant activity of essential oil and different solvent extracts of endemic plant Merremia borneensis," *Arab. J. Chem.*, vol. 8, no. 2015, pp. 66–71, 2015.
[27] N. Herawati, "Total Phenolic and Antioxidant Activity of Sonneratia alba Bark," *J. Sainsmat*, vol. I, no. 1, pp. 93–99, 2012.
[28] T. I. Maulana, S. Falah, and D. Andrianto, "Total phenolic content , total flavonoid content , and antioxidant activity of water and ethanol extract from Surian (Toona sinensis) leaves Total phenolic content , total flavonoid content , and antioxidant activity of water and ethanol extract from Suri," in *Earth and Environmental Science 299*, 2019, pp. 1–10.
[29] E. N. Sembiring, B. Elya, R. Sauriasari, E. N. Sembiring, B. Elya, and R. Sauriasari, "Phytochemical Screening , Total Flavonoid and Total Phenolic Content and Antioxidant Activity of Different Parts of Caesalpinia bonduc (L .) Roxb," *Pharmacogn. J.*, vol. 10, no. 1, pp. 123–127, 2018.
[30] S. Chandra *et al.*, "Assessment of Total Phenolic and Flavonoid Content , Antioxidant Properties , and Yield of Aeroponically and Conventionally Grown Leafy Vegetables and Fruit Crops : A Comparative Study," *Evidence-Based Complement. Altern. Med.*, vol. 2014, no. ID 253875, pp. 1–9, 2014.
[31] M. Kopjar, V. Piližota, J. Hribar, and M. Sim, "Total phenol content and antioxidant activity of water solutions of plant extracts," *Croat. J. Food Sci. Technol.*, vol. 1, no. 1, pp. 1–7, 2009.
[32] H. Noreen, N. Semmar, M. Farman, and J. S. O. Mccullagh, "Measurement of total phenolic content and antioxidant activity of aerial parts of medicinal plant Coronopus didymus," *Asian Pac. J. Trop. Med.*, vol. 10, no. 8, pp. 792–801, 2017.

IOP Conf. Series: Materials Science and Engineering 1011 (2021) 012069 doi:10.1088/1757-899X/1011/1/012069

Preliminary assessment of heavy metals in street vended foods in the surrounding of elementary school area in Bandung city, Indonesia.

Endah Damastuti[1*], Natalia Adventini[1], Katerina Oginawati[2], Woro Yatu Niken Syahfitri[1], Syukria Kurniawati[1] and Muhayatun Santoso[1].

[1]Center for Applied Nuclear Science and Technology, National Nuclear Energy Agency (BATAN), Tamansari 71, Bandung 40132, Indonesia.
[2]Environmental Engineering- Faculty of Civil and Environmental Engineering, ITB, Ganesha 10, Bandung, Indonesia.

*Email: endah_dt@batan.go.id

Abstract. Street vended foods are ready to eat foods sold by the vendors in the street and easily found in the school surrounding area. The availability and affordability of street vended foods made it widely consumed by large of population include children, especially when it is packaged attractively. Unfortunately, the safety of street vended foods especially related to the heavy metal contaminant is still unknown. Therefore, assessing the heavy metal level of street vended foods that widely consumed by children need to carried out. A number of 24 street vended food samples were collected in four elementary school surrounding area in Bandung city. The determination of heavy metal content in this food samples was conducted by neutron activation analysis and atomic absorption spectrometry as well. The quality of analytical results was evaluated using Standard Reference Material (SRM) NIST 1548a Typical Diet and 1567a Wheat Flour. Compare to the maximum permissible limit established by Indonesian Food and Drug Control Agency (BPOM), it was found that Cu, Hg and Zn concentration of street vended foods were below their maximum permissible limit. However, health risk assessment through Hazard Index (HI) formulation indicated that children in Bandung city may possess non-carcinogenic health risk due to heavy metals from the consumption of several street vended foods since the HI value are in the range of 0.06 – 2.32 with more than 10% of the foods observed were have HI > 1. These may be harmful situation to the children health. Further supervision of the preparation and handling of the vended foods are needed and guidance about the importance of hygiene and quality of vended foods to the vendors need to be improved.

1. Introduction

Street vended foods are ready-to-eat foods and beverages prepared and/or sold by vendors and hawkers especially in streets and other similar public places for immediate consumption or consumption at a later time without further processing or preparation [1,2]. This definition emphasizes the retail location 'on the street', with foods sold from pushcarts, bicycles, baskets or balance poles, stalls, kiosks or shops. Street vended foods are categorized as follows: food prepared in small factories and sold by mobile vendors; food prepared at mobile vendors home and offered for sale to the public; and food prepared and sold on the street [2]. Street vended food is present even in less developed countries, and has occasionally to be the hallmark of the early development of fast food [3]. These foods vending is very popular to a large number of people, among commuters and urban dwellers, in many cities and towns in developing countries because they are usually perceived to be inexpensive, convenient and attractive. It is also easily accessible, and generally affordable to large parts of the population to meet up their needs of food [1,2,4,5]. Street foods vending plays an important role in assuring food security for low-income urban populations [6]. It is a large source of employment in many cities of developing countries [7] including in Indonesia and it can be easily found in surrounding area of the school. The dependence of many people on street vended foods, likely or not, has made the enforcement of food safety has been transferred from individuals and families to the food vendors who rarely conduct of such practices [7]. Most handlers of street food in the developing countries are poor, not adequately educated, and ignorant of basic food safety. Consequently, street foods are commonly exposed to various contaminants at different stages of handling [1,2,6,8].

Food safety is a major public concern worldwide. The risks of food safety are contributing to the burden of illness [9]. During the last decades, the increasing demand of food safety has stimulated research related to the risk associated with consumption of food stuffs contaminated by pesticides, heavy metals and/or toxins [10]. The implication associated with heavy metal contamination is of great concern. Heavy metals, in general are not biodegradable, having long biological half-lives and having the potential for accumulation in the different body organs leading to unwanted side effects [11]. Heavy metals have positive and negative effects on the human health, it can be classified as essential for biochemical reactions in the body (copper, iron, zinc, etc.) and potentially toxic (lead, cadmium, arsenic, mercury etc.) [10,12]. Long-term exposures to toxic elements can pose a significant health risk even at low concentration but the intake of too much of essential metals can also cause produce toxic effects [10,11]. Street vended foods are perceived to be a major health risk and vulnerable to contaminated of pathogens, toxin and heavy metals as well. Quality assessment of street foods shown that these foods were positive vectors of food borne illnesses and every person that eating these foods is at the risk of food borne illness [13].

The contamination of heavy metal can seriously affect the children health as well since children are the most sensitive and vulnerable age group to any kind of contamination in the food chain [14]. It was stated by Asiegbu that up to 2 million people per year has died, most of are children, because of diseases caused by consumption of contaminated food and water [15]. Children is human resources asset and the future generation of a nation. Efforts to improve the quality of human resources begin with the fulfillment of basic human needs with primary focus on the process of growth and development of the child. The nutritional adequacy and safety greatly affect intelligence and work productivity of human. School-age children, especially in elementary school, represent an important and diverse target group for health and nutrition interventions. At this age, children do more activities that requires more nutrition as well. Unfortunately, this requirement does not always fill perfectly. High rushing often makes the parents do not have time to provide adequate lunches and snacks for their children. This condition, and also the savory of tastes and attractive packaging make children more attracted to buy street vended foods.

Several assessments related to quality, hygienic and sanitation of street vended foods in Indonesia have been carried out [16]. The chemical contaminants such as coloring agent, additives and preservatives were also found in abnormal level of some street vended foods. But the assessment related to the heavy metal contaminant in street vended foods that commonly consumed by children is still scarce. Therefore, this research of assessing the level and the risk of heavy metals in street vended foods consumed by school-age children was conducted. The determination of heavy metal was carried out using Neutron Activation Analysis (NAA) and Atomic Absorption Spectrometry (AAS). In assuring the quality and reliability of analytical results, several reference materials were analyzed along with the food samples. The results from this study will provide information that can verify the nature of the nutrition problem within a society and the effectiveness of the specific solution or intervention as it can be used as the science-based reference for in charge institutions to develop a better quality of street vended foods.

IOP Conf. Series: Materials Science and Engineering 1011 (2021) 012069 doi:10.1088/1757-899X/1011/1/012069

2. Methodology

2.1. Food Sampling and Preparation

Food samples were collected from 4 elementary schools' area in Bandung city which are Setiabudi, Pelesiran, Sindanglaya and Soka, as described in Figure 1. The four elementary schools were chosen as they are located in the populated area near the arterial road of Bandung city and supposed to be represent the north, south, east and west of Bandung city. Five to six kinds of food were taken from street vendors in surrounding areas of each school that often consumed by its students. The selection of the foods was based on questionnaire results that previously distributed. As much as 20 types of foods from 24 foods samples were collected as listed in Table 1 and the description of each street vended foods were given in Table 2.

Each part of one type of street fended food were weighing individually and then mashed all parts together with titanium blade blender. Weighing each of a street vended food was used to calculate the total calories of every street vended foods. Food samples that have been smooth and homogeneous were weighted and placed into a small tube, then stored in the freezer until frozen. Tube which contains frozen food sample attached on freeze dryer. Freeze drying was done by the help of vacuum pump and the temperature was set at -55°C. Drying process took 3x24 hours until the samples were dry and had constant weight. Dried samples then refined into fine powder using a Teflon mortar and pestle, and then placed into polyethylene container.

Table 1. List of collected street vended foods

Sample ID	School Area	Street vended foods
1	Setiabudi	Bakwan
2		Martabak
3		Potato
4		Creepes
5		Noodle1
6		Cakwe1
7	Sindanglaya	Egg1
8		Cireng
9		Cilok
10		Cakwe2
11		Bakso Tahu
12	Pelesiran	Batagor
13		Gelatin
14		Banana
15		Cakwe3
16		Sausage
17		Egg2
18	Soka	Bacil
19		Lumpia Basah
20		Noodle2
21		Cimol
22		'Cubit' cake
23		Cracker ball
24		Biting

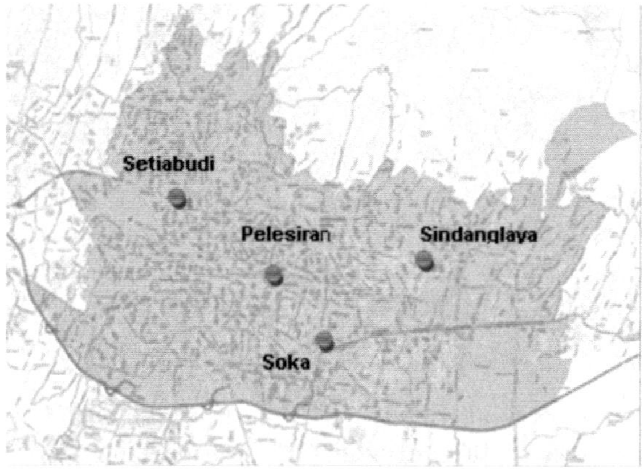

Figure 1. Sampling location of street vended foods in Bandung city, Indonesia.

Table 2. List of collected street vended foods

Street vended foods	Description
Bakwan	Deep fried mixture of wheat flour, carrot, cabbage and onion.
Martabak	Deep fried thick folded creep filled with lot of leek, egg, and some minced meat.
Potato	Fried potato covered by flour and served with spicy powder.
Creepes	Filled with many variations such as tuna fish, smoked beef, burger etc.
Noodle	Fried noodle served with meatball, and some vegetables.
Cakwe	Fried bread stick served with chili and sour sauce.
Egg	Fried egg served with chili sauce.
Cireng	Fried cornstarch filled with oncom.
Cilok	Boiled cornstarch served with soy sauce, chili sauce.
Bakso Tahu	Steam tofu filled with dough of fish and cornstarch and served with peanut dressing.
Batagor	Fried tofu filled with fish and cornstrach and served with peanut dressing.
Gelatin	Served with variation of fruit flavours.
Banana	Deep fried of folded creep filled with banana and chocolate.
Sausage	Fried sausage served with chili sauce and mayo.
Bacil	Boiled small cornstarch served with soy sauce, chili sauce.
Lumpia Basah	Folded creep filled with saute bean sprouts, egg and bamboo shoot with sweet and hot sauce.
Cimol	Fried cornstrach ball served with peanut sauce or chili powder.
'Cubit' cake	Cake made from dough of wheat flour, egg and sugar.
Cracker ball	Savory crackers ball with variations of flavour.
Biting	Dried oven noodle served with spicy and savory powder.

2.2. Neutron Activation Analysis (NAA) Procedure.

Neutron activation analysis is a method for qualitative and quantitative determination of elements based on the measurement of characteristic radiation from radionuclide formed by neutron irradiation of the material. The elemental concentrations were calculated by the comparative method. In this experiment, as much as 0.1 g powder dried samples were placed into 0.274 mL polyethylene vial respectively and each vial then sealed by heating [17]. An amount of 0.1 g samples was used to analyze long half-life elements such as chromium, iron, zinc and mercury. For long half-life elements, the samples irradiated along with SRM NIST 1548a Typical Diet and mix standard, consist of 0.2, 10 and 20 μg of Cr, Zn and Fe respectively. Irradiation went for an hour at 15MW at rabbit system facilities of multipurpose reactor G.A. Siwabessy, Serpong. Samples then allowed to cooling for a month then counted for 10000s using gamma spectrometer with low resolution of HPGe detector. The resolution (FWHM) of the system was

1.90 keV for 1332 keV gamma-ray of 60-Co. The radionuclides measured were identified according to their half-lives and gamma-ray energies. The spectrum observed using Genie 2000 software at 320.08 keV, 1099.25 keV and 1115.5 keV for Cr, Fe and Zn respectively.

2.3. Atomic Absorption Spectrometry (AAS) Procedure
An amount of 0.5g of food sample powder was dissolved by adding 2.5 mL sterile water; 6.5 mL concentrated HNO_3 solution and 1 mL of $HClO_4$ using microwave digestion (ETHOS 1, Microwave Digestion, Milestone) for 20 minutes at 150°C and power 1000 W [18]. The same steps were also applied to standard reference material (SRM) NIST 1548a Typical Diet. Measurement of samples were done by flame-AAS methods using GBC Avanta P. The wavelength observed was 324.7 nm and 279.5 nm for Cu and Mn respectively. The dissolved sample then diluted by aquadest into 25 mL flask and transferred to polyethylene container. The optimum working range for Cu and Mn were 1-5 mg/kg and 1-3.6 mg/kg, respectively.

2.4. Estimated Daily Intake
The estimated daily intake (EDI) depends on the metal concentration, ingestion/consumption rate and body weight. The EDI was determined using the following equation [19]:

$$EDI = (MC \times FCR) \times BW^{-1} \quad (1)$$

where MC is metal concentration in food (mg/kg on wet basis), FCR represent the average of food consumption rate per day and BW is the average of the body weight. It was assumed that the children were consuming the street vended food one portion per day which means that the consumption rate will be depend on the kind of the vended foods. And the body weight of the children was assumed to be 30 kg.

2.5. Determination of Target Hazard Quotient (THQ) and Hazard Index (HI)
The health risk from consumption of street vended foods by elementary school children were assessed based on THQ. The THQ is a ratio of the exposure dose to the reference dose (RfD). If the ratio is less than 1, the exposed population is unlikely to experience obvious adverse effects. The THQ value was determined by the following equation [19,20]:

$$THQ = ((EF.ED.CR.MC)(RfD.BW.ATn)^{-1}) \times 10^{-3} \quad (2)$$

where EF is exposure frequency (365 days/year), ED is exposure duration (assuming 60 years as life time people still consume street vended foods), CR is consumption rate (g/day), MC is metal concentration (mg/kg wet basis), RfD is reference dose (mg/kg/day), BW is body weight (30 kg, the average weight of elementary school age children), ATn is the average exposure time for noncarcinogens (365 days/year x ED). To evaluate the possible health effects that may cause by the combination of all heavy metals observed, the total Hazard Index were applied. The HI was expressed as the arithmetic sum of the individual metal THQ values [18]. The RfD values used in this study were 0.003; 0.04; 0.7; 0.0005; 0.14 and 0.3 mg/kg/day for Cr, Cu, Fe, Hg, Mn and Zn, respectively [19-21].

2.6. Quality control and data analysis
The standard reference material (SRM) was used as quality control assesment of data validity. The results of analysis SRM were compared with its certificate value and evaluated its accuracy and precission by %Recovery and %CV calculation. Net areas from Genie 2000 were used to calculate the concentrations of elements in the samples using a comparative method as well as absorbance from AAS GBC Avanta P.

IOP Conf. Series: Materials Science and Engineering 1011 (2021) 012069 doi:10.1088/1757-899X/1011/1/012069

3. Results and Discussions

3.1. Quality control and data analysis

To assess the reliability of the result obtained and the accuracy of analitycal methods used in this study, the standard reference material (SRM) which has similar to the matrix of sample, was also analyzed in the same conditions as the samples. The results of SRM NIST 1548a Typical Diet and SRM NIST 1567a Wheat Flour for some elements were listed in Table 3. It is shown that the results obtained were in good agreement with its certified values except Cr and Hg since they were not exist in both reference materials. Analytical accuracy, described as %Recovery, were in the range 94 to 103%; while analytical precision, described as %CV were in the range 1-5%. Both analytical accuracy and precision were acceptable according to AOAC guidelines [22].

Table 3. Quality control assesment using SRM NIST 1548a Typical Diet (TD) and SRM NIST 1567a Wheat Flour (WF)

Element	SRM	Result	Certified value	%Rec	%CV
Cu	WF	1.98 ± 0.10	2.1 ± 0.2	94	3
Fe	TD	35.8 ± 1.81	35.3 ± 3.77	101	5
Mn	WF	5.87 ± 0.17	5.75 ± 0.17	102	3
Zn	WF	11.9 ± 0.50	11.6 0.4	103	4

3.2. Heavy Metals in Street Vended Foods

The heavy metal concentrations obtained from measurement were in dry mass basis. The heavy metal concentrations of the samples must be converted to wet mass basis to compare with the value of maximum limit of heavy metals content in food by determining the water content of each samples. Figure 2 – 7 presents the concentration, in wet basis, of Copper, Chromium, Iron, Manganese, Mercury and Zinc respectively in street vended foods.

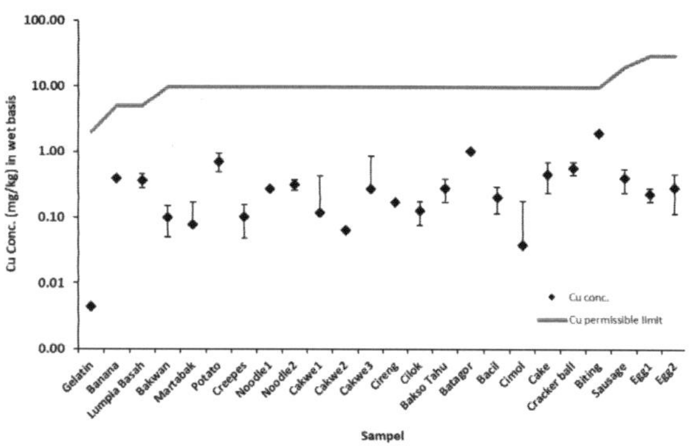

Figure 2. Copper concentration of street vended foods.

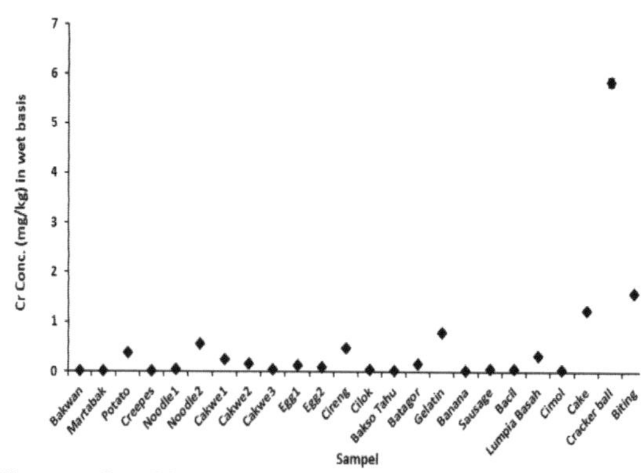

Figure 3. Chromium concentration of street vended foods.

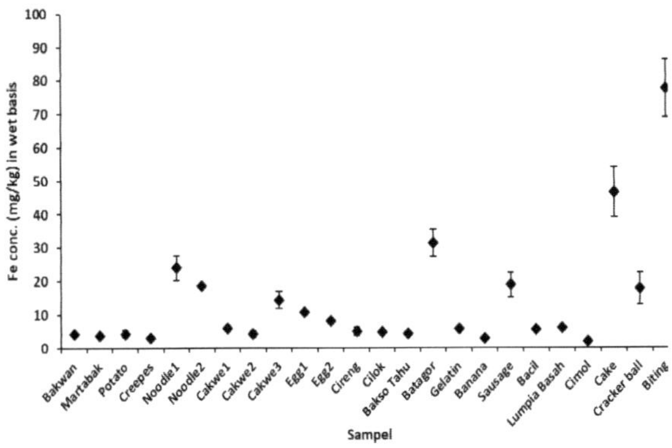

Figure 4. Iron concentration of street vended foods.

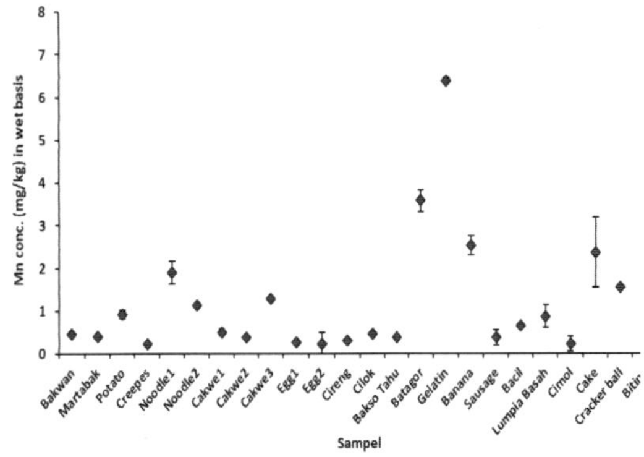

Figure 5. Manganese concentration of street vended foods.

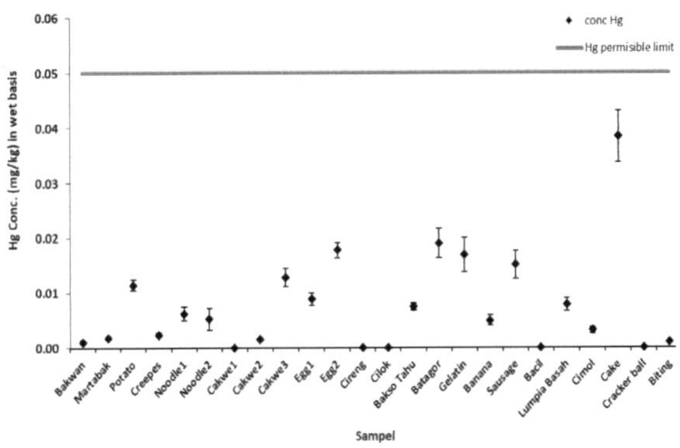

Figure 6. Mercury concentration in street vended foods.

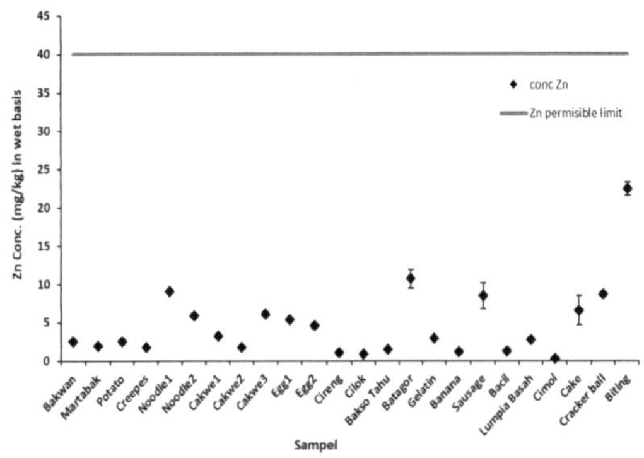

Figure 7. Zinc concentration of street vended foods.

Figure 2 shows the concentration of Cu and its comparison by maximum permissible limit of Cu in foods issued by National Agency of Drugs and Food Control (BPOM) No 03725/B/SK/VII/89 about the Maximum Limitation of Metal Contamination in Food [23]. The maximum value of Cu contained in the regulation was vary based on the main ingredient of the food. It can be seen that the concentration of Cu in this present study, ranged from 0.004 mg/kg to 1.93 mg/kg, were found to be in lower concentration than the maximum permissible limit value of BPOM, 2-30 mg/kg. This indicated that the foods are still safe to be consumed. Copper has essential function in haemoglobin synthesis as a constituent of some metallo enzymes and metabolic growth as catalysis [6]. In excessive intake, Cu may cause adverse health problems such as liver and kidney damage [21]. The possible contamination of Cu could came from the water of the copper pipe, that used to prepare the food or to wash the cooking utensils [24]. The low quality of cooking utensils itself are also have most probability to contaminate the food, since the material of low quality of cooking utensils is easier to peeled off [24,25].

Meanwhile Figure 3 shows the concentration of chromium (Cr) in street vended food samples. The concentration of chromium in street vended foods was vary from 0.01 – 5.86 mg/kg (wet weight basis). These values can not be compared since Indonesia doesn't have an established maximum permissible limit value of chromium in food. However, we were trying to compare with other countries maximum permissible limit of chromium. The maximum permissible value of Cr issued by Food Adulteration Regulations of Hongkong government is 1 mg/kg [26]. Compared to this, as much as 12.5% of the food samples were above the value. Precautions should be taken since excessive chromium intake has pose a risk of the health, such as decreased body weight, heart liver damages, skin diseases, cancer, headache and respiratory illness [24].

The iron concentration in street vended foods are described in Figure 4. Similar with chromium, Iron was also did not have established maximum permissible value in food of Indonesia. The concentration of Iron in street vended foods were in the range 1.87 – 77.3 mg/kg wet weight basis. Compare to other metals, Fe was found in higher concentration among others since it was the most abundant element [21]. Iron is needed by human body as component of haemoglobin and numerous enzymes and prevent microcytic hypochromic anaemia but at the excessive consumption, it may cause adverse effect to human body such as gastrointestinal distress, heart diseases and others [24,27,28].

Manganese concentrations in street vended foods are describe in Figure 5. The concentration of Manganese was varied 0.2 – 6.6 mg/kg. Similar with Fe, the maximum permissible limit of Manganese is difficult to be find elsewhere. Manganese is also classified as an essential element, it has involvement in bone formation and in metabolism of amino acid, cholesterol and carbohydrate. Though, similar to the other elements, excessive Mn consumption may cause neuropsychiatric disorders [17].

Mercury concentration in street vended foods are presented in Figure 6. Mercury in some of the food samples were below the detection limit of NAA methods due to the inference spectrum with Se-75. The concentration of mercury in street vended foods were in the range n.d (not detectable) – 0.04 mg/kg. The maximum permissible limit of mercury was derived from National Agency of Food and Drugs Control (0.05 mg/kg) [23]. Figure 6 shows that all of street vended foods observed have Hg concentration below the maximum permissible limits, only one of them were close to this value, which is 'cubit' cake. Mercury is classified as toxic elements. Due to its high toxicity, excessive intake of it could pose serious health risk. The health risk due to mercury intake through consumption rate will be assessed through THQ value.

Figure 8 presents the zinc concentration of street vended foods. The maximum permissible limit of Zn in foods was derived from National Agency of Food and Drugs control (40 mg/kg) [23]. It was found that Zn concentration were vary from 0.31 to 22.4 mg/kg. All of the samples were still below the limit value. Almost all of 24 vended food samples showed Zn concentration < 15 mg/kg, only one samples, biting, has significant higher Zn concentration among other. The Zn in foods is mostly derived from cooking utensils where this material is commonly used for preparation of the foods. The aging of the cooking utensils could made the leaching out of the material occurred. The zinc contamination in street vended foods could also be caused by road dust pollution since the foods vending in the street have opening stall. Zinc was known to be used as an additives to gasoline and also auto-lubricants [29]. The excessive consumption of Zn leads to reducing of copper status in the body, imbalance of electrolyter, nausea and lethargy as well [30].

3.3. Total Hazard Quotient and Hazard Index

The health risk assessment to children in Bandung city from consumption of street vended foods were assessed based on THQ. As mentioned in the previous section of experimental methods, the THQ value less than 1 means the exposed population is unlikely to experience obvious adverse effects. The overall values of THQ and HI for elementary school children population consumer of street vended foods in their surrounding school area were shown in Table 4. The THQ values of Cu, Fe, Hg, Mn and Zn were all found < 1, while for Cr the range of THQ were 0.01 – 1.41. Based on this THQ result for Cu, Fe, Hg, Mn and Zn, the consumer of street vended foods indicated no potential health risk. However, biggest concern should be given for Cr which some of the foods have THQ > 1. Hence, there is a potential of health risk presence due to Cr and exposure of children that consume the foods.

The results of HI, as can be seen in Table 4, were in the range of 0.06 – 2.32 and gives the mean HI value of 0.64. This result gives a clearly view that some of the street vended foods may have potential noncarcinogenic risks of heavy metals for children through the consumption of the street vended foods. From 24 kind of street vended foods, as much as 21% have HI value exceed 1. From the Table 4, it can be seen that the major contribution of the high HI values was contributed by Cr THQ which were already > 1, and followed by Hg THQ. The average of Cr THQ (0.33) and Hg THQ (0.20) gives contribution to the average HI as much as 51% and 32% respectively. The possible sources of chromium are cigarette smoke, utensils, fertilizers that contain in raw material of the food and the packaging of the food [24,31].

However, this Cr results need further analysis for its speciation, since Cr that analyzed in this study was Cr total, while the highest toxicity were in the form of Cr (VI) and the one that needed by human body was Cr (III). While the high level of Hg can be due to contamination from soil or water [19]. According to Proietti, the sources of metal contaminant can be various such as from leaching of utensils, the depositions of urban air particulate and unsafe water [4]. These results should be scientific based reason to government and authorized party to make appropriate strategic plan to solve these problems. Further supervision and guidance to the vendors about the importance of hygiene and quality of vended foods should be improved. Children and the family are also should be given understanding of the safety of the foods they eat and accustomed to prefer eating the foods or snacks brought from their home.

Table 4. Target Hazard Quotient (THQ) and total Hazard Index (HI) of heavy metals for street vended foods consumption.

Street vended food	THQ values						HI
	Cr	Cu	Fe	Hg	Mn	Zn	
Bakwan	0.02	0.01	0.01	0.01	0.01	0.02	0.08
Martabak	0.01	0.01	0.01	0.03	0.01	0.02	0.08
Potato	0.21	0.03	0.01	0.12	0.01	0.01	0.40
Creepes	0.01	0.004	0.01	0.02	0.003	0.01	0.06
Noodle1	0.04	0.02	0.08	0.09	0.03	0.07	0.33
Noodle2	0.76	0.03	0.11	0.13	0.03	0.08	1.14
Cakwe1	0.31	0.01	0.03	*n.d*	0.01	0.04	0.40
Cakwe2	0.28	0.01	0.03	0.05	0.01	0.03	0.40
Cakwe3	0.03	0.01	0.04	0.16	0.02	0.04	0.31
Egg1	0.06	0.01	0.02	0.07	0.002	0.02	0.18
Egg2	0.04	0.01	0.02	0.15	0.002	0.02	0.24
Cireng	0.80	0.02	0.03	*n.d*	0.01	0.02	0.88
Cilok	0.10	0.02	0.04	*n.d*	0.02	0.02	0.19
Bakso Tahu	0.11	0.06	0.05	0.37	0.02	0.04	0.64
Batagor	0.27	0.13	0.22	0.59	0.13	0.18	1.52
Gelatin	1.41	0.001	0.04	0.56	0.24	0.05	2.32
Banana	0.07	0.06	0.02	0.17	0.10	0.02	0.44
Sausage	0.02	0.02	0.03	0.09	0.003	0.03	0.19
Bacil	0.23	0.05	0.08	*n.d*	0.05	0.04	0.44
Lumpia Basah	0.76	0.06	0.06	0.34	0.04	0.06	1.33
Cimol	0.06	0.003	0.01	0.07	0.01	0.003	0.15
'Cubit' Cake	1.11	0.11	0.18	0.64	0.05	0.06	2.14
Cracker ball	0.65	0.005	0.01	*n.d*	0.004	0.01	0.68
Biting	0.53	0.05	0.11	*n.d*	0.05	0.07	0.81

n.d : not detectable

4. Conclusion

Assessment of heavy metals content in street vended foods that widely consumed by elementary school children in Bandung city has been carried out. Six heavy metals; Cu, Cr, Fe, Hg, Mn and Zn, were determined in 24 samples of street vended fooods. It was found that Cu, Hg and Zn concentration of street vended foods were below their maximum permissible limit. The health risk assessment based on THQ and HI showed that high frequent of consumption of some street vended foods may poses non-carcinogenic health risk due to its heavy metals content since the HI values are in the range of 0.06 – 2.32 with 21% of the foods observed were have HI > 1. These may put children in the harmful situation. Preventive and settlement action should be taken and further supervision and guidance to the vendors about the importance of hygiene and quality of vended foods are still needed and have to be improved.

Acknowledgement
The authors would like to give acknowledgement to BATAN for the financial support, FTSL- ITB for the cooperative research and the colleagues of PSTNT for their contributions to this research.

References
[1] Abdulmajid N, Bautista MK, Bautista S, Chavez E, Dimaano W and Barcelon E 2014 *Int. Food. Res. J.* **21** 2127-31
[2] Imathiu S 2017 *Curr. Res. Nutr. Food. Sci.* **05** 55-65
[3] Buscemi S, Barile A, Maniaci V, Batsis JA, Mattina A and Verga S 2011. *Nutr. J.* **10** 119
[4] Proietti I, Frazzoli C, Mantovani A 2014 *Food Chem. Toxicol.* **63** 143-52
[5] Elsisy SF, Ibrahim HM, Hassan MA and Ali AM 2017 *Benha Veterinary Medical Journal*, **32(1)** 54-8
[6] Atter A, Ofori H, Anyebuno GA. Amoo-Gyasi M, Amoa-Awua WK 2015 *Food Control* **55** 200-5
[7] Oyolela OT, AfolabiI MI, Ajiboshin IO and Banjoko IO 2013 *Int. J. Res. Med. Health Sci.* **3** 28-32.
[8] Azanu D, Kyei SK and Oppong A 2015 *Int. J. Sci.Technol.* **4** 99-104
[9] Unnevehr L 2013 *Global Food Security* **4** 24-9
[10] Davarynejad G, Zarel M, Nagy PT 2013 *Not. Sci. Biol.* **5** 438-44
[11] Hussain RTH, Ebraheem MKH, Moker HM 2012 *Bas. J. Vet. Res.* **11** 43-51
[12] Folake IA, Adegunwa Mojisola O, Adebowale Abdulrasaq A, Olawale O, Sanni Lateef O 2015 *Food Sci.Qual. Manag.* **39** 20-5
[13] Idowui MA, Atolagbe YM, Henshaw FO, Akpan I and Oduse K 2012 *Afr. J. Food Sci.* **6** 474-82
[14] Prakash P, Rajan G, Elavarasi V, Priyadharsari TKP, Thamaraiselvi K. Contamination (Nickel and Arsenic) using GF-AAS in Local Brand Chocolates and Candies from Tiruchirappalli, India. International Journal of Computer Applications® (IJCA) (0975 – 8887) National Conference cum Workshop on Bioinformatics and Computational Biology, NCWBCB (2014) 25-8.
[15] Asiegbu CV, Lebelo SL, Tabit FT 2016 *Food Control* **60** 422-9
[16] Patriasih R, Widiaty I, Dewi M, Sukandar D 2010 *J. Nutr. Food.* **5** 177–83
[17] Basuki K, Muhayati N, Santoso M 2013 *Proc. 2nd. Int. Conf. of the Indonesian Chemical Society* p 82-90
[18] Gray PJ, Mindak WR, Cheng J 2013 *FDA Elemental Analysis Manual for Food and Related Products*. Ver 1 (US Food and Drug Administration) Section 4.7 p 1-22
[19] Bortey-Sam N, Nakayama SMM, Ikenaka Y, Akoto O, Baidoo E, Yohannes YB, Mizukawa H, Ishizuka M 2015 *Ecotoxicol. Environ. Saf.* **111** 160-7
[20] Yi Y, Tang C, Yi T, Yang Z, Zhang S 2017 *Ecotoxicol. Environ. Saf.* **145** 295-302
[21] Yap CK, Cheng WH, Karami A, Ismail A 2016 *Sci. Total Environ.* **553** 285-96
[22] AOAC Official Methods of Analysis 2013 *Guidelines for Dietary Supplements and Botanical* (AOAC International) Appendix K p 1-32
[23] BPOM, The Decree of General Directur of National Agency of Food and Drug Control (BPOM) No 03725/B/SK/VII/89 about the Maximum Limitation of Metal Contamination in Food.
[24] Thirulogachandar A, Rajeswari M, Ramya S 2014 *International Journal of Scientific and Research Publications* **4** 1-8.
[25] Handayani EH, Oginawati K, Santoso M. Analysis of Zn and Cu in Primary School Children's Street Foods with Atomic Absorption Spectrophotomery (AAS) Method. Thesis of Bachelor Degree, ITB, 2010.
[26] CENTER OF FOOD SAFETY, Food Legislation CAP 132V Food Adulteration (Metallic Contamination) Regulations. Hongkong, 1997.
[27] Goldhaber SB 2003 *Regul. Toxicol. Pharmacol.* **38** 232-42
[28] Damastuti E, Syahfitri WYN, Santoso M and Lestiani DD 2012 *Atom Indones.* **38** 29-34

[29] Lanre-Iyanda TY and Lanre-Iyanda IMA 2012 *Afr. J. Food, Agric. Nutr. Dev.* **12** 6156-69
[30] Onianwa PC, Adeyemo AO, Idowu OE and Ogabiela EE 2001 *Food Chem.* **72** 89-95
[31] Kim KC, Park YB, Lee MJ, Kim JB, Huh JW, Kim DH, Lee JB, Kim JC 2008 *Food Res. Int.* **41** 411-8

IOP Conf. Series: Materials Science and Engineering 1011 (2021) 012070 doi:10.1088/1757-899X/1011/1/012070

The analysis of several types of catalysts for ozone decomposition

Veny Luvita[1,2)], Setijo Bismo[1,*] and Anto Tri Sugiarto[2]

[1]Laboratorium Intensifikasi Proses, Chemical Engineering, Faculty of Engineering, Universitas Indonesia, Kampus UI Depok 16424, Indonesia
[2]Technical Implementation for Instrumentation Development Unit – LIPI, Bandung, Indonesia

*Email: setijo.bismo@ui.ac.id

Abstract. There has been rapid technological development in the aspect of water and wastewater treatment, including the application of ozone technology, commonly known as Advanced Oxidation Processes (AOPs). This involves the use of ozone gas converted to oxygen and $OH^{\bullet}$ radicals, which possess the capacity to degrade the unwanted compounds present, although there is need for performance optimization to enhance the ease of formation and decomposition. This process is faced with serious problems that require immediate attention, particularly the presence of excess ozone, which is harmful to the health of living things, especially humans. Ozone decomposition uses an MnOx active core catalyst with a nominal loading of 0–2%. As a comparison used catalyst supports such as Zeolite Alam Lampung (ZAL), Granular Activated Carbon (GAC) and Green Sand (GS) with diameter 18–100 mesh. However, the challenge is comprehensively attenuated by using catalysts possessing the capacity to degrade the excesses, and promote procedure optimization, subsequently enhancing conversion and decomposition. Furthermore, previous studies have demonstrated the potential to elevate the production of OH radicals, and it was also established that catalysts from natural origin are less expensive, easy to operate, highly selective and difficult to damage under various conditions. The result of this study is GAC is more effective and efficient compared to others for decomposition.

1. Introduction

Catalysts are substances with the capacity to increase the reaction rate to reach equilibrium, without reacting with the reactants. Furthermore, they are generally characterized by the ability to convert raw materials into desired products, with activeness obtained from a combination of chemical and mineralogical materials. Conversely, selectivity is defined as the ability of a catalyst to produce the desired products as one of the substances involved serves as a barrier in other processes, hence the need to examine each potential catalytic material (Nasikin, 2010).

About 3 factors are known to affect function during performance, including fluid flow, catalyst activity, and stability. These are possibly improved through the proper selection of components, materials and preparation methods needed to provide optimum surface area, due to the responsibility of catalyst active core towards the reaction principle. Furthermore, the first step to be taken in the design requires the selection of active components, which is performed on the basis of conductivity.

IOP Conf. Series: Materials Science and Engineering 1011 (2021) 012070 doi:10.1088/1757-899X/1011/1/012070

The use of Advanced Oxidation Processes (AOPs) involves adopting the right type of reactor, optimization, and the need for additional ozone decomposition, which particularly involves the use of catalysts. This causes an upsurge in demand, due to the excessive use of ozone in numerous industries. Furthermore, ozone is a powerful oxidant and disinfectant, used to eliminate unwanted chemical compounds, bacteria, and viruses (Koltunski E., Plumridge J., 2005).

The process of ozone decomposition involves the oxidation of complex organic compounds into simple units, leading to an improvement in biodegradability, and a decline in toxicity. This occurs directly and indirectly, through the activation of free radical hydroxide (OH •), hydroperoxide radicals (HO_2•), superoxide radical ions (O_2^-) and ozonide radical ions (O_3^-) formed in the process (Chu and Ching., 2003).

This research will compare the quality and performance of the catalyst with three support used Zeolite Alam Lampung (ZAL), Granular Activated Carbon (GAC) and Green Sand (GS). Performance of the catalyst is shown based on the conversion of ozone decomposition. Catalyst will be applied in a continuous fixed bed reactor in which ozone enter to the reactor and react with the catalyst.

2. Methodology

In this research are using 3 variables, namely independent variables, variables bound, and control variable. The independent variable in this study is the type of catalyst support namely Zeolite Alam Lampung (ZAL), Granular Activated Carbon (GAC) and Green Sand (GS), catalyst support diameter, namely 18-35 mesh, 35-60 mesh, and 60-100 mesh, and the nominal % loading of MnOx is 0% -w, 1% -w, and 2% -w. Dependent variable in this study the conversion produced by the catalyst and ozone concentration after bypassed the catalyst. The control variables of this study are atmospheric pressure, temperature space, the volume of catalyst used is 5 mL, and the ozone flow rate is 3 Lpm.

Scheme of the experiment can see in Fig 1. The experiment is using ozone generator system. Free air enters to the fixed bed reactor then the air will be converted into ozone. Ozone which out will go into the fixed bed reactor containing the catalyst. Inside the fixed bed reactor will occur ozone decomposition process, then the concentration of ozone will be tested using iodometric analysis method.

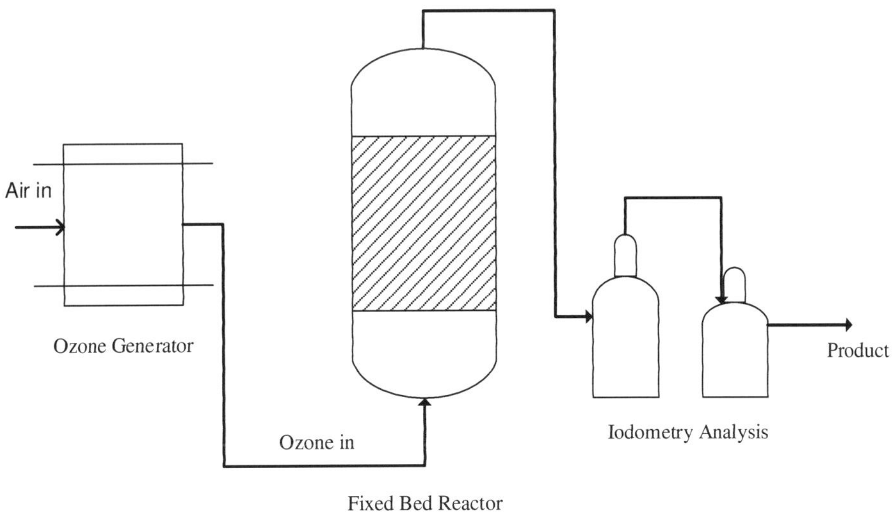

Figure 1. Research scheme.

3. Result and Discussion

3.1. Catalytic ozonation
Catalytic ozonation is aimed at activating the ozone, characterized by an increase in selectivity and change in reaction rate, as well as reduced ozone consumption. This phase is classified as (1) homogeneous, typically involving the use of transition metals, expressed as the rate of $OH^{\bullet}$ radical formation or contaminant degradation, and (2) heterogeneous, which generally requires metal oxides, activated carbon and zeolites, which involves adsorption, reaction and desorption.

Activated Carbon is an oxidative or adsorptive material that assists in the decomposition of hydrophobic non-polar compounds present in water with ozone. Furthermore, the surface area possesses acid, represented by carbonyl, carboxyl, hydroxyl phenoleate, lactone and quinon, and also base groups, including ether and carbonyl (Laszlo *et al.*, 2001). These collectively serve as a buffer in heterogeneous catalysts with several advantages, including high stability in acidic or basic media, and mechanical resistance resulting from the ease of recovery from the dispersed active phase through combustion, high surface area and porosity. This is also an inexpensive process, characterized by the ready market availability, and the stability of carbon in numerous conditions. This includes the procedure of methane or oxidation catalysis reactions, which is dependent on the nature of calcination/reduction.

Granular Carbon Active (GAC) is another catalyst widely used as a support, due to the intrinsic favorable adsorbent feature and the large surface area. In addition, they also possess an oxygen functional group that is possibly used as a catalyst for ozone decomposition, with the purpose of increasing the rate of conversion (Wang M. *et al.*, 2014).

The process of ozone decomposition conducted by A. Pradyasti *et al.*, (2017) involves the effective use of ZnO catalysts and GAC support, characterized by high conversion values using varied sizes and loading percentages. Meanwhile, the combination prepared with a size 60-100 mesh and a loading percentage of 2% -w, reached the highest conversion of 100% in 30 minutes. Hence, a relationship was established between both parameter and ozone conversion, where smaller ZnO catalyst exhibited easier dispersion potential, and a higher percentage comprised of more oxygen holes that enhance the modification to oxygen molecules.

3.2. Ozone decomposition reaction using a catalyst
Based on the problems above, there is need for adopting simple and inexpensive technological solution, including the use of catalysts as an alternative in the ozonation process, especially in cases where the contaminants are difficult to completely decompose. This approach accelerates the conversion of ozone to hydroxide radicals ($OH^{\bullet}$), which subsequently plays a role in the degradation reaction. Moreover, some examples of catalysts used include activated carbon, iron oxide, zeolite and manganese oxide.

The use of activated carbon in the ozonation process accelerates the formation of $OH^{\bullet}$ and oxygen radicals responsible for the degradation of contaminants on the surface of the catalyst, subsequently leading to a decline in ozone demand. Furthermore, the effect of utilizing solid materials at an average rate varies greatly, including 0.25 mg/min. grams of catalyst for activated carbon and 0.02-0.25 mg/min. gram of catalyst for zeolites, while Al_2O_3 and SiO_2 have the similar value of 0.01- 0.25 mg/min. gram of catalyst.

From Table 1 and 2, the ZAL and GS buffered catalysts have the highest performance on size 60-100 mesh with a nominal loading concentration of 1%. This is because the smaller size of catalyst will give the greater surface area of catalyst reacting. Buffer catalysts with a nominal loading concentration of 2% have lower performance than catalysts with a nominal loading of 1% because the catalyst has been saturated with the active core so that the buffer catalysts play less role in the ozone decomposition process (Rohimmahtunnissa Azhar, 2012).

Table 1. Ozone decomposition with Zeolite Alam Lampung (ZAL) catalyst.

	Decomposition (%)							
Time (min)/MnOx Loading/Size (mesh)	*5*	*10*	*15*	*20*	*30*	*40*	*50*	*60*
0%/60-100	24.89	23.68	26.1	26.1	29.74	34.58	37	37.61
1%/60-100	73.35	70.92	70.92	72.14	77.59	78.19	75.77	76.98
2%/60-100	39.43	35.79	35.19	36.4	38.22	36.4	40.64	44.27
0%/35-60	22.77	26.73	26.73	24.75	21.78	21.78	24.2	26.24
1%/35-60	45.83	44.79	44.79	42.71	36.98	38.02	38.02	34.9
2%/35-60	22.51	22.51	23.62	25.28	25.83	30.26	31.37	38.01
0%/18-35	9.65	8.83	11.29	10.47	13.76	14.58	17.04	17.04
1%/18-35	26.13	26.13	21.62	16.22	10.81	10.81	10.81	11.26
2%/18-35	17.59	15.74	11.11	11.11	12.04	13.89	14.81	15.28

Table 2. Ozone decomposition with Green Sand (GS) catalyst.

	Time (min)	*5*	*10*	*15*	*20*	*30*	*40*	*50*	*60*
Decomposition (%)	0%/60-100	93.09	79.28	75.33	75.33	75.33	74.34	73.36	72.86
	1%/60-100	100	100	100	100	98.03	94.08	89.14	86.18
	2%/60-100	100	100	100	95.07	84.7	80.26	77.8	77.3
	0%/35-60	48.87	33.22	33.22	34.26	35.83	37.91	39.48	38.96
	1%/35-60	92.7	65.57	59.3	60.35	62.43	65.04	67.13	68.17
	2%/35-60	79.52	52.4	54.61	57.38	62.92	62.36	65.68	63.47
	0%/18-35	14.04	14.89	14.89	16.57	11.52	14.04	8.15	8.99
	1%/18-35	32.65	17.78	16.91	17.78	15.6	19.1	20.85	21.72
	2%/18-35	24.07	20.37	13.89	8.33	12.04	11.11	10.19	12.5

The performance of the GAC-supported catalyst is very high, see from Table 3, the catalyst has a percentage of 100% up to the 24th hour for the size of the catalyst support 35-60 mesh and 60-100 mesh with nominal loading concentrations from 0% to 2%.

Table 3. Ozone decomposition with Granular Active Carbon (GAC) catalyst.

	Time (min)	*5*	*10*	*15*	*20*	*30*	*40*	*50*	*60*
Decomposition (%)	0%/60-100	100	100	100	100	100	100	100	100
	1%/60-100	100	100	100	100	100	100	100	100
	2%/60-100	100	100	100	100	100	100	100	100
	0%/35-60	100	100	100	100	100	100	100	100
	1%/35-60	100	100	100	100	100	100	100	100
	2%/35-60	100	100	100	100	100	100	100	100
	0%/18-35	100	100	100	100	100	100	100	100
	1%/18-35	100	100	100	100	100	100	100	100
	2%/18-35	100	100	100	100	100	100	100	100

Ozone used for industrial purposes is synthesized from pure oxygen through thermal, photochemical, chemical and electrochemical processes. Furthermore, they are created in the form of electrical discharges and particle flow, which is attained using the following reactions:

$O_2 + (e^-, hv, T) \rightarrow 2O\,(O_2^*)$

$O_2 + O + M \rightarrow O_3 + M$

$O_2^* + O_2 \rightarrow O_3 + O$

In which: M is the third particle (Batakliev *et al.*, 2014)

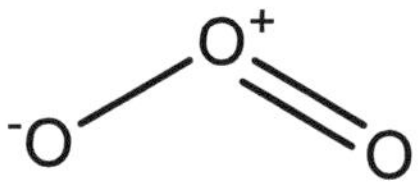

Figure 2. Structure of O_3 molecule.

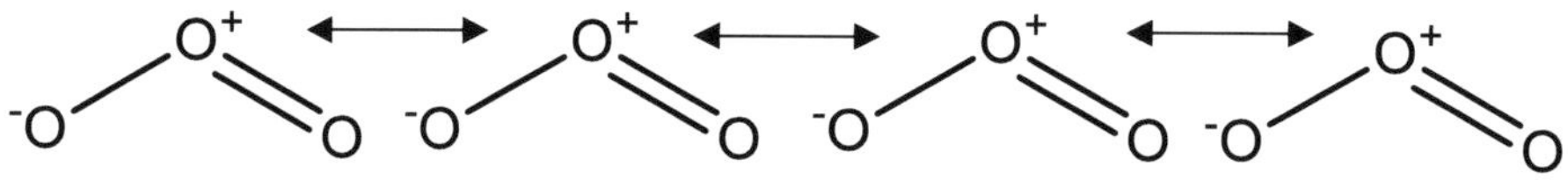

Figure 3. Resonance structures of ozone.

The ozone structure is resonance stabilized, which is one of the reasons for its resistance against decomposition at low temperatures (Figure 2 and 3). At high temperatures, the decomposition reaction occurs spontaneously and very quickly, although the process is difficult and expensive as a result of the high energy requirement. The kinetics study of ozone thermal decomposition is also complicated by the fact that above a certain critical temperature the steady kinetic decomposition is transformed into explosion, subsequently passing to detonation (Rakovsky et al., 2007). Ozone detonation is observed above 105 °C. The gaseous ozone is characterized by different times of half-life, depending on the temperature The life time *(half-life)* of ozone at certain temperatures entails decomposition at 1.5 sec using a temperature of 250°C, and up to 3 days at room temperature, due to reduced resonance (Batakliev, *et al.*, 2014).

Ozone decomposition reactions with activated carbon as a catalyst by Beltran *et al.*, (2002) are stated as follows:

Homogeneous decomposition:

$O_3 + OH^- \rightarrow HO_2^- + O_2$ (Eq. 1)

$O_3 + HO_2^- \rightarrow HO_2^- + O_3^{-\bullet}$ (Eq. 2)

$O_3 + In \rightarrow O_3^{-\bullet} + In^+$ (Eq. 3)

$HO_2^\bullet \leftrightarrow O_2^{-\bullet} + H^+$ (Eq. 4)

Whereas heterogeneous decomposition on the reaction surface for pH 2 to 6 include:

$O_3 + S \leftrightarrow O_3\text{-}S$ (Eq. 5)

$O_3\text{-}S \leftrightarrow O\text{-}S + O_2$ (Eq. 6)

$O_3 + O\text{-}S \leftrightarrow 2O_2 + S$ (Eq. 7)

At the pH ≥ 6, the reaction formed are :

$OH\text{-} + S \leftrightarrow OH\text{-}S$ (Eq. 8)

$O_3 + OH\text{-}S \leftrightarrow {}^\bullet O_3\text{-}S + HO^\bullet$ (Eq. 9)

${}^\bullet O_3\text{-}S + HO^\bullet \leftrightarrow {}^\bullet O\text{-}S + O_2$ (Eq. 10)

$O_3 + {}^\bullet O\text{-}S \leftrightarrow O_2^{-\bullet}S + S + O_2$ (Eq. 11)

Homogeneous propagation and termination reactions:

$O_3 + O_2^{-\bullet} \rightarrow O_3^{-\bullet} + O_2$ (Eq. 12)

$O_3^{-\bullet} + H^+ \rightarrow HO_3^\bullet$ (Eq. 13)

$HO_3^\bullet + P \rightarrow HO^\bullet + O_2$ (Eq. 14)

$HO^\bullet + P \rightarrow$ *End product* (Eq. 15)

In which: S is the catalyst surface and in is the initiator.

In this mechanism, reactions [1] to [4] and [12] to [15] constitute the homogeneous decomposition contribution to the total reaction rate, while reactions [5] to [11] are the steps of the surface chemical reaction involving adsorption, surface reactions and desorption processes. These reactions would be followed in the water phase by reactions [12] to [15] which support conclusions about the direct formation of free radicals in the water phase. In reactions [3] and [15], In and P account for the presence of initiating agents of ozone decomposition and scavengers of hydroxyl radicals, respectively (Beltran *et al.*, 2002).

4. Conclusion

The processing of contamination/waste using advanced oxidation process with ozone decomposition catalyst is identified as an alternative. The best quality of the catalyst is seen from the greater surface area, pore volume, and the optimum pore diameter and actual loading composition will be even higher catalyst performance. This can be seen from the quality of GAC-supported catalysts. The performance of GAC supported MnOx catalysts is based on ozone decomposition conversion has the highest performance compared to ZAL and GS magnitude of conversion of 100% for 24 hours for the diameter of the catalyst buffer 35-100 mesh and nominal loading concentrations of 0-2%.

Acknowledgement

The authors are grateful to the financial support provided for this research by the University of Indonesia and by-research program of Indonesian Institute of Science.

References

[1] Pradyasti A, Azhariyah A S, Karamah E F and Bismo S 2018 Preparation of zink oxides catalyst with activated carbon support for ozone decomposition. *IOP Conf. Series: Earth and Environmental Science* **105** 012013

[2] Azhariyah A S, Pradyasti A, Dianty A G and Bismo S 2018 Somparative study of activated carbon, natural zeolite, and green sand supports for CuOx and ZnO sites as ozone decomposition catalyst *IOP Conf. Series: Materials Science and Engineering* **334** 012075

[3] Beltran F J, Rivas J, Alvarez Pand Montero-de-Espinosa R 2002 Kinetics of hetereneous catalytic ozone decomposition in water on an activated carbon *Ozone Sci. Eng.* **24** 227-37

[4] Gao G, Kang J, Shen J, Chen Z and Chu W 2016 Investigation on kinetics of heterogeneous catalytic ozone decomposition in aqueous solution over composite iron-manganese silicate oxides *Ozone Sci Eng.* **38** 6

[5] Kasprzyk-Hordern B, Ziolek M and Nawrocki J 2003 Catalytic ozonation and methods on enhancing molecular ozone reaction in water treatment *Appl. Catal.* **46** 639-69

[6] Koltunski E and Plumridge J 2005 *Ozone as a Disinfecting Agent in the Reuse of Wastewater* (Duebendorf, Switzerland: Ozonia Ltd)

[7] Yuan L, Chen Z, Liu Y and Liu Y 2011 Catalytic ozonation for the degradation of p-Chloronitrobenzene in aqueous solution by pumice. *Adv. Mater. Res.* **239-242** 1123-8

[8] Nasikin M and Susanto B H 2010 Katalisis Heterogen. Universitas Indonesia.

[9] Chokshi N P, Bhutia H, Chotalia A, Kadiwala A and Ruparelia J P 2017 Heterogeneous catalytic ozonation of Reactive Black 5 with cobalt oxide. *Int. J. Chem. Tech Res.* **10**(9) 402-9

[10] Zhu Q, Chen F, Guo S, Chen X and Chen J 2017 Variation of catalyst structure and catalytic activity during catalyst preparation for catalytic ozonation of heavy oil produced water *Water Air Soil Pollut.* **228** 168.

[11] Rakovsky S K and Zaikov G E 2007 Kinetic and Mechanism of Ozone Reactions with Organic and Polymeric Compounds in Liquid Phase – 2nd edition (New Yoek: Nova Sci. Publ., Inc.)

[12] Rohimmahtunnissa A and Setijo Bismo 2012 Studi Komparatif Zeolit Alam, Karbon Aktif dan Pasir Hijau sebagai Penyangga Katalis dengan Inti Aktif MnOx untuk Dekomposisi Ozon, Thesis submitted to the University of Indonesia.
[13] Batakliev T, Georgiev V, Anachkov M, Rakovsky S and Zaikov G E 2014 Ozone decomposition *Interdiscip. Toxicol.* 7(2) 47-59
[14] Wang M, Zhang P, Li J and Jiang C 2014 The effects of Mn loading on the structure and ozone decomposition activity of MnOx supported on activated carbon *Chinese J. Catal.* **35** 335-41
[15] Li W 1998 Kinetics and Mechanism of Ozone Decomposition and Oxidation of Ethanol on Manganese Oxide Catalysts. Dissertation submitted to the Faculty of the Virginia Polytechnic Institute and State University

AUTHOR INDEX